SCHAUM'S OUTLINE OF

THEORY AND PROBLEMS

of

ORGANIC CHEMISTRY

•

by

HERBERT MEISLICH, Ph.D.
Professor of Chemistry
City College of CUNY

HOWARD NECHAMKIN, Ph.D.
Chairman, Department of Chemistry
Trenton State College

and

JACOB SHAREFKIN, Ph.D.
Professor of Chemistry
Brooklyn College of CUNY

SCHAUM'S OUTLINE SERIES
McGRAW-HILL BOOK COMPANY
New York   St. Louis   San Francisco   Auckland   Bogotá   Düsseldorf   Johannesburg
London   Madrid   Mexico   Montreal   New Delhi   Panama   Paris
São Paulo   Singapore   Sydney   Tokyo   Toronto

0-07-041457-2

5 6 7 8 9 10 11 12 13 14 15 SH SH 8 7 6 5 4 3 2 1

Library of Congress Cataloging in Publication Data

Meislich, Herbert.
  Schaum's outline of theory and problems of organic chemistry.

  (Schaum's outline series)
  Includes index.
  1. Chemistry, Organic—Problems, exercises, etc.
2. Chemistry organic. I. Nechamkin, Howard, 1918–
joint author. II. Sharefkin, Jacob George, 1909–
joint author. III. Title.
QD257.M44   547   77-9318
ISBN 0-07-041457-2

To Amy Nechamkin

Preface

The beginning student in Organic Chemistry is often overwhelmed by facts, concepts, and new language. Each year, textbooks of Organic Chemistry grow in quantity of subject matter and in level of sophistication. This Schaum's Outline was undertaken (initially by the senior author, J. S.) to give a clear view of first-year Organic Chemistry through the solution of illustrative problems. Completely worked-out problems make up over 80% of the book, the remainder being a concise presentation of theory. The reader learns by thinking and doing rather than by being told.

This outline can be used in support of a standard text, as a text to supplement a good set of lecture notes, as a review book for taking professional examinations, and as a vehicle for self-instruction.

Our thanks and appreciation to Mr. Larry Alemany for his expert criticism from a student's viewpoint and for his careful proofreading, to Mr. David Beckwith for his editorial assistance, and to Mrs. Joyce Gaiser for her meticulous typing.

HERBERT MEISLICH
HOWARD NECHAMKIN
JACOB SHAREFKIN

Contents

CONTENTS

CONTENTS

CONTENTS

Chapter 1

Structure and Properties

1.1 ORGANIC COMPOUNDS

Organic chemistry is the study of carbon (C) compounds. Most organic compounds consist of individual molecules whose atoms are held together by covalent bonds but some also have ionic bonds. Carbon atoms can bond to each other to form chains as in **open-chain (acyclic)** compounds or to form rings as in **cyclic** compounds. Both types can also have branches of C atoms. Cyclic compounds having at least one atom other than C (a **heteroatom**) are called **heterocyclic**. The heteroatoms are usually oxygen (O), nitrogen (N) or sulfur (S). C's can be bonded to each other by:

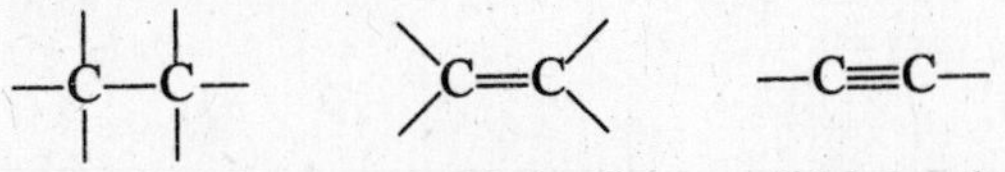

Single bond   Double bond   Triple bond

Hydrocarbons contain only C and hydrogen (H). H's in hydrocarbons can be replaced by other atoms or groups of atoms. These replacements, called **functional groups**, are the reactive sites in molecules. The C to C double and triple bonds are considered to be functional groups. Some common functional groups are the halogens, —OH, —NH$_2$,

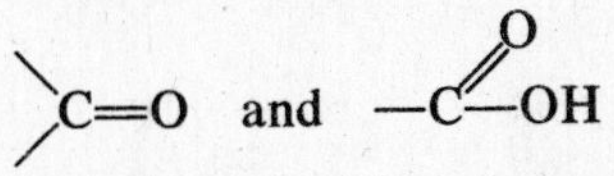

Compounds with the same functional group form a **homologous series** having similar chemical properties and often exhibiting a regular gradation in physical properties with increasing molecular weight.

Organic compounds show a widespread occurrence of **isomers**, which are compounds having the same *molecular* formula but different *structural* formulas and therefore different properties. **Structural formulas** show the arrangement of atoms in a molecule. These formulas are written by assigning the following numbers of bonds: 1 for H and halogen, 2 for O, 3 for N, and 4 for C. These numbers are the **covalences** of the atoms.

Most carbon-containing molecules have three-dimensional shapes. In methane, the bonds of C make equal angles of 109.5° with each other, and each of the four H's is at a corner of a regular tetrahedron whose center is occupied by the C atom. The spatial relationship is indicated as in Fig. 1-1(*a*) (Newman projection) or in Fig. 1-1(*b*) ("wedge" projection).

H^f's project toward viewer
H^b's project away from viewer

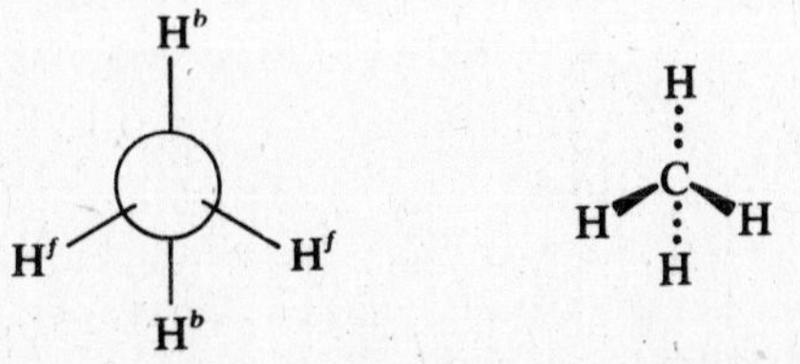

... projects in back of plane of paper
— projects out of plane of paper

(*a*)   (*b*)

Fig. 1-1

1

Structural formulas for some hydrocarbons are:

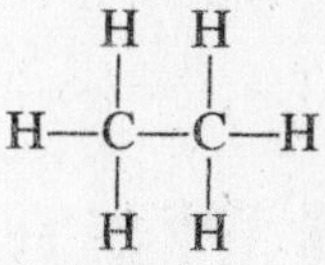

$$
\begin{array}{cc}
\overset{\displaystyle H \quad H}{H-\underset{\displaystyle H \quad H}{C-C}-H} &
\overset{\displaystyle H \quad H \quad H}{H-\underset{\displaystyle H \quad H \quad H}{C-C-C}-H}
\end{array}
$$

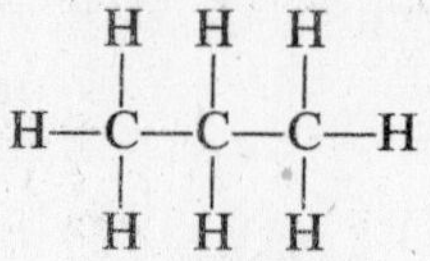

CH_3CH_3 $CH_3CH_2CH_3$

Ethane   Propane

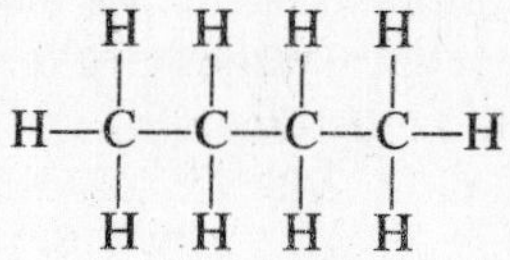

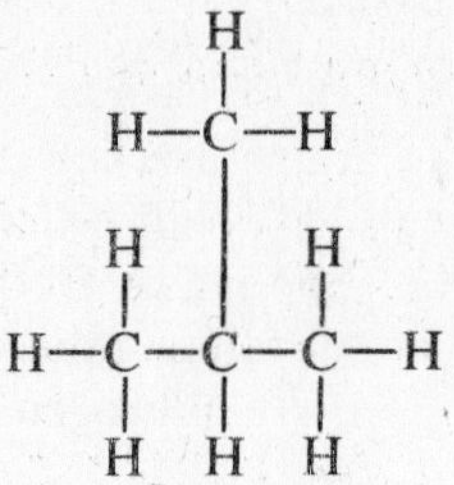

$CH_3CH_2CH_2CH_3$ $CH_3CH(CH_3)CH_3$

n-Butane   Isobutane

These are flat projections of the three-dimensional structures. **Condensed** structural formulas are shown beneath the molecular formulas. **Lewis (electron-dot)** structural formulas show all shared and unshared pairs of electrons, e.g.

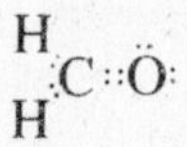

$$
\underset{\displaystyle H}{\overset{\displaystyle H}{}} :C::\ddot{O}:
$$

Problem 1.1 Why are there so many compounds that contain carbon? ◀

Bonds between C's are covalent and strong, so that C's can form long chains and rings, both of which may have branches. C's can bond to almost every element in the periodic table. Also, the number of isomers increases as the organic molecules become more complex.

Problem 1.2 How do the boiling points, melting points and solubilities of covalent organic compounds differ from those of inorganic salts? Account for the differences. ◀

Covalent organic compounds have much lower boiling and melting points because the forces attracting molecules to each other are weak. The electrostatic forces attracting the oppositely charged ions in inorganic salts are very strong. The forces of attraction between the atoms within a covalent molecule are, however, very strong. Inorganic salts are generally soluble in water because water aids in the separation of the ions. Inorganic salts are not soluble in organic solvents such as ether and benzene. Most organic compounds are insoluble in water but dissolve in organic solvents.

Problem 1.3 Give four differences in chemical reactivity between an organic compound, such as a hydrocarbon, and an inorganic salt. ◀

1. Ionic reactions often occur *instantaneously*. Reactions between covalent organic molecules proceed *more slowly*; they often need higher temperatures and/or catalysts.
2. Many reactions of organic compounds give *mixtures of products*.
3. Organic compounds are *less stable* to heat. They usually decompose at temperatures above 700 °C, which break many of their covalent bonds.
4. Organic compounds are more *susceptible to oxidation*. Hydrocarbons burn in O_2 to give CO_2 and H_2O. Inorganic compounds generally remain unchanged even after intense heating.

Problem 1.4 Write structural and condensed formulas for the three isomers of pentane, C_5H_{12}.　◀

Carbon forms four covalent bonds; hydrogen forms one. The carbons can bond to each other in a chain:

$$
\begin{array}{ccccccccc}
 & H & & H & & H & & H & & H \\
 & | & & | & & | & & | & & | \\
H- & C & - & C & - & C & - & C & - & C & -H \\
 & | & & | & & | & & | & & | \\
 & H & & H & & H & & H & & H
\end{array}
$$

$$CH_3(CH_2)_3CH_3$$
n-Pentane

or they can be "branches" (shown circled in Fig. 1-2) on the linear backbone (shown in a rectangle).

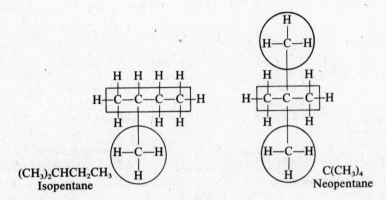

$(CH_3)_2CHCH_2CH_3$　　　　　　$C(CH_3)_4$
Isopentane　　　　　　　　　　Neopentane

Fig. 1-2

Problem 1.5 Write structural formulas for: (*a*) hydrazine (N_2H_4), (*b*) carbonyl chloride ($COCl_2$), (*c*) nitrous acid (HNO_2).　◀

(*a*) N needs three covalent bonds and H needs one. Each N is bonded to the other N and to two H's:

$$
\begin{array}{ccc}
H & & H \\
| & & | \\
H-N & - & N-H
\end{array}
$$

(*b*) The tetravalent C is bonded to divalent O by a double bond and to each Cl by a single bond:

$$
\begin{array}{c}
:\ddot{O}: \\
\| \\
:\ddot{C}l-C-\ddot{C}l:
\end{array}
$$

(*c*) The valences of one, two and three for H, O and N, respectively, are satisfied by one bond of O to H and another of the same O to N. The second O is double-bonded to N: $H-\ddot{O}-\ddot{N}=\ddot{O}$.

Problem 1.6 (*a*) Write possible structural formulas for (1) CH_4O, (2) CH_2O, (3) CH_2O_2, (4) CH_5N, (5) CH_2BrI. (*b*) Indicate the functional group in each case.　◀

The atom with the higher valence is usually the one to which most of the other atoms are bonded.

(*a*) (1) $H-\overset{\displaystyle H}{\underset{\displaystyle H}{\overset{|}{\underset{|}{C}}}}-\ddot{O}-H$　　(2) $\overset{\displaystyle H}{\underset{\displaystyle H}{>}}C=\ddot{O}:$　　(3) $H-C\overset{\displaystyle \ddot{O}:}{\underset{\displaystyle \ddot{O}-H}{<}}$　　(4) $H-\overset{\displaystyle H}{\underset{\displaystyle H}{\overset{|}{\underset{|}{C}}}}-\overset{}{\underset{\displaystyle H}{\ddot{N}}}-H$　　(5) $H-\overset{\displaystyle H}{\underset{\displaystyle :\ddot{I}:}{\overset{|}{\underset{|}{C}}}}-\ddot{B}r:$

(*b*) (1) $-\ddot{O}H$　　(2) $>C=\ddot{O}:$　　(3) $-C\overset{\displaystyle \ddot{O}:}{\underset{\displaystyle \ddot{O}-H}{<}}$　　(4) $-\ddot{N}H_2$　　(5) $-\ddot{B}r:$ and $-\ddot{I}:$

　　ALCOHOL　　　　KETONE　　　CARBOXYLIC　　AMINE
　　　　　　　　　　　　　　　　　　ACID

Problem 1.7 Write structural formulas for (a) C_2H_4, (b) C_3H_6, (c) C_2H_2. ◄

In each case there is at least one bond between the two C atoms. There are not enough H's to satisfy the tetravalence of C. Multiple bonds must be used.

(a)

$$\underset{\displaystyle}{H-\overset{\displaystyle H}{\underset{\displaystyle}{C}}=\overset{\displaystyle H}{\underset{\displaystyle}{C}}-H}$$ Ethylene

(b)

$$H-\overset{\displaystyle H}{\underset{\displaystyle H}{C}}-\overset{\displaystyle}{\underset{\displaystyle H}{C}}=\overset{\displaystyle H}{\underset{\displaystyle}{C}}-H$$ Propylene

(c) $H-C\equiv C-H$ Acetylene

Problem 1.8 Is propylene the only compound with the formula C_3H_6? ◄

The valence of four for C can be met if the bonding C's form a ring instead of using a double bond. Cyclopropane,

is an isomer of propylene.

Problem 1.9 What physical properties are used to determine the purity of liquids and solids? ◄

For liquids, the physical properties are boiling point, refractive index, density, and various spectra such as ultraviolet, infrared, nuclear magnetic resonance and mass. Melting points and spectra are most frequently used for solids.

Problem 1.10 Use the Lewis-Langmuir octet rule to write Lewis electron-dot structures for: (a) HCN, (b) CO_2, (c) CCl_4 and (d) CH_3CH_2OH. ◄

According to the octet rule, elements tend to combine so as to acquire the electron configuration of the nearest noble gas. This configuration is 2 for H and Li, which are close to He, and is 8 (octet) for elements of the second and third periods of the periodic table.

(a) The covalencies of H, N and C are one, three and four, respectively, and the structural formula is $H-C\equiv N$. H and C both have completed their outer shells with two and eight electrons, respectively, but N has only six electrons. This formula shows only eight outer electrons (two in each bond). It should have ten, since H, C and N are in families I, IV and V of the periodic table, and their outer electron shells have one, four and five electrons, respectively. The missing pair of electrons is placed on N to give $H:C:::N:$, thereby also completing the octet of N.

(b) The valences 2 for O and 4 for C are used to deduce the structural formula $O::C::O$. There are 4 bonding electrons in each of the two double bonds between O and C. C has an octet, but each O has only 4 electrons. The octet of each O may be completed by adding two unshared electron pairs to each. There are 16 electrons in the complete electron-dot formula, $:\ddot{O}=C=\ddot{O}:$, in agreement with the total number of outer electrons (4 from one C and 6 from each of the two O's).

(c) The structural formula

$$\underset{\displaystyle \overset{\textstyle |}{Cl}}{\overset{\displaystyle \overset{\textstyle Cl}{|}}{Cl-C-Cl}}$$

is deduced from the covalencies of 1 for Cl and 4 for C. C has an octet but each Cl shares only 2 electrons. Three unshared electron pairs are placed on each of the four Cl's to provide the complete electron-dot structure

$$\overset{\displaystyle :\ddot{Cl}:}{\underset{\displaystyle :\ddot{Cl}:}{:\ddot{Cl}:\ddot{C}:\ddot{Cl}:}}$$

The 32 electrons in this structure agrees with the sum of 4 electrons from C and 28 from 4 Cl's, each of which has 7 outer electrons.

(d) The procedure outlined in Problem 1.5 gives the formula

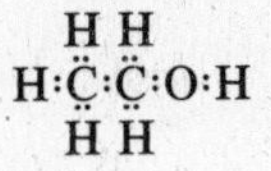

$$
\begin{array}{c}
\text{H H} \\
\text{H:C:C:O:H} \\
\text{H H}
\end{array}
$$

Only O with 4 electrons lacks an octet. Placing two unshared electron pairs on O gives the complete formula

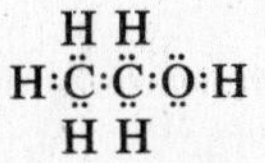

$$
\begin{array}{c}
\text{H H} \\
\text{H:C:C:O:H} \\
\text{H H}
\end{array}
$$

This shows 20 outer electrons, which agrees with the sum of 8 electrons from C (2×4), 6 from H (6×1) and 6 from O.

Problem 1.11 Determine the positive or negative charge, if any, on the following species:

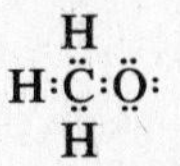

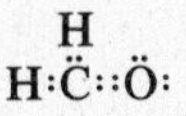

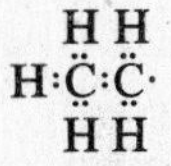

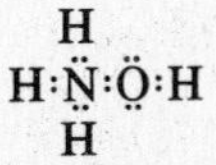

$$
(a)\quad \begin{array}{c} \text{H} \\ \text{H:C:O:} \\ \text{H} \end{array} \qquad
(b)\quad \text{H:C::O:} \qquad
(c)\quad \begin{array}{c} \text{H H} \\ \text{H:C:C:} \\ \text{H H} \end{array}
$$

$$
(d)\quad \begin{array}{c} \text{H} \\ \text{H:N:O:H} \\ \text{H} \end{array} \qquad
(e)\quad \begin{array}{c} \text{:Cl:} \\ \text{:C:Cl:} \\ \text{:Cl:} \end{array} \qquad\qquad \blacktriangleleft
$$

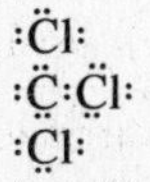

The charge on a species equals the sum of the outer electrons minus the total number of electrons shown.

(a) The sum of the outer electrons (6 for O, 4 for C, and 3 for three H's) is 13. The electron-dot formula shows 14 electrons. The net charge is -1 $(13 - 14)$ and the species is the methoxide anion, $CH_3O^{:-}$.

(b) There is no charge on the formaldehyde molecule because the 12 electrons in the structure equals the number of outer electrons, i.e. 6 for O, 4 for C, and 2 for two H's.

(c) This species is neutral because there are 13 electrons shown in the formula and 13 outer electrons: 8 from two C's and 5 from five H's.

(d) There are 15 outer electrons: 6 from O, 5 from N, and 4 from four H's. The Lewis dot structure shows 14 electrons. It has a charge of $+1$ $(15 - 14)$ and is the hydroxylammonium cation, $[H_3NOH]^+$.

(e) There are 25 outer electrons, 21 from three Cl's and 4 from C. The Lewis dot formula shows 26 electrons. It has a charge of -1 $(25 - 26)$ and is the trichloromethide anion, CCl_3^-.

Problem 1.12 Determine the **formal charge** on each atom in the following species: (a) H_3NBF_3, (b) $CH_3NH_3^+$ and (c) SO_4^{2-}. $\blacktriangleleft$

The formal charge on an atom equals the number of outer electrons minus the number of electrons assigned to the atom in its bonding state. The assigned number is one-half the sum of all shared electrons, plus all unshared electrons.

(a)

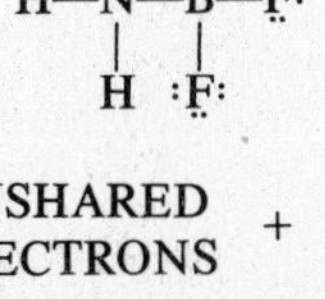

$$
\begin{array}{c}
\text{H} \quad \text{:F:} \\
\text{|} \quad\quad \text{|} \\
\text{H—N—B—F:} \\
\text{|} \quad\quad \text{|} \\
\text{H} \quad \text{:F:}
\end{array}
$$

	OUTER ELECTRONS	$-$	UNSHARED ELECTRONS	$+$ 1/2	SHARED ELECTRONS	$=$	FORMAL CHARGE
H atoms	1	$-$	0	$+$	1	$=$	0
F atoms	7	$-$	6	$+$	1	$=$	0
N atom	5	$-$	0	$+$	4	$=$	$+1$
B atom	3	$-$	0	$+$	4	$=$	-1

The sum of all formal charges equals the charge on the species.   In this case, the $+1$ on N and the -1 on B cancel and the species is an uncharged molecule.

(b)

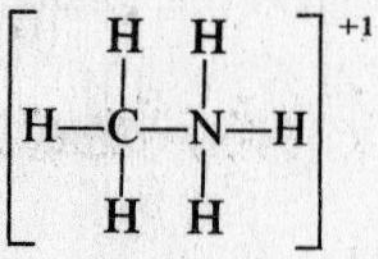

$$
\begin{bmatrix}
& \text{H} & \text{H} & \\
& | & | & \\
\text{H}-&\text{C}-&\text{N}-&\text{H} \\
& | & | & \\
& \text{H} & \text{H} &
\end{bmatrix}^{+1}
$$

	OUTER ELECTRONS	−	UNSHARED ELECTRONS	+ 1/2	SHARED ELECTRONS	=	FORMAL CHARGE
C atom	4	−	0	+	4	=	0
N atom	5	−	0	+	4	=	+1
H atoms	1	−	0	+	1	=	0

Net charge on species = +1

(c)

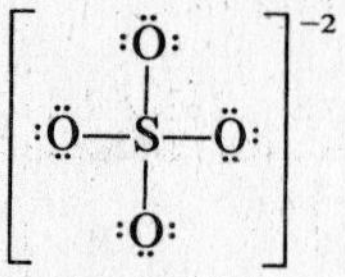

$$
\begin{bmatrix}
& & :\!\ddot{\text{O}}\!: & \\
& & | & \\
:\!\ddot{\text{O}}-&\!\!\text{S}-&\!\!\ddot{\text{O}}\!: & \\
& & | & \\
& & :\!\ddot{\text{O}}\!: &
\end{bmatrix}^{-2}
$$

	OUTER ELECTRONS	−	UNSHARED ELECTRONS	+ 1/2	SHARED ELECTRONS	=	FORMAL CHARGE
S atom	6	−	0	+	4	=	+2
each O atom	6	−	6	+	1	=	−1

Net charge is $+2 + 4(-1)$ =   −2

In this book the signs + and − will be used to indicate both formal charge and actual net charge; some texts employ ⊕ and ⊖ for formal charge.

Chapter 2

Bonding and Molecular Structure

2.1 ATOMIC ORBITALS

An **atomic orbital** (AO) is a region of space about the nucleus in which there is a high probability of finding an electron. An electron has a given energy as designated by: (*a*) the principal energy level (quantum number), *n*, related to the size of the orbital; (*b*) the sublevel, *s*, *p*, *d*, *f* or *g*, related to the shape of the orbital; (*c*) except for the *s*, each sublevel has some number of equal-energy (**degenerate**) orbitals differing in their spatial orientation; (*d*) the electron spin, designated ↑ or ↓. Table 2-1 shows the distribution and designation of orbitals.

Table 2-1

Principal Energy Level, n	1 K	2 L		3 M			4 N			
Maximum No. of Electrons, $2n^2$	2	8		18			32			
Sublevels	$1s$	$2s$,	$2p$	$3s$,	$3p$,	$3d$	$4s$,	$4p$,	$4d$,	$4f$
Designation of Filled Orbitals	$1s^2$	$2s^2$,	$2p^6$	$3s^2$,	$3p^6$,	$3d^{10}$	$4s^2$,	$4p^6$,	$4d^{10}$,	$4f^{14}$
Maximum Electrons per Sublevel	2	2,	6	2,	6,	10	2,	6,	10,	14
Orbitals per Sublevel	1	1,	3	1,	3,	5	1,	3,	5,	7

The *s* orbital is a sphere around the nucleus, as shown in cross section in Fig. 2-1(*a*). A *p* orbital is two spherical lobes touching on opposite sides of the nucleus. The three *p* orbitals are labeled p_x, p_y and p_z because they are oriented along the *x*-, *y*- and *z*-axes, respectively (Fig. 2-1(*b*)). In a *p* orbital there is no chance of finding an electron at the nucleus—the nucleus is called a **node point**. Regions of an orbital separated by a node are assigned + and − signs. These signs are *not associated with electrical or ionic charges*. The *s* orbital has no node and is usually assigned a +.

Three principles are used to distribute electrons in orbitals.

1. **"Aufbau" or Building-up Principle.** Orbitals are filled in order of increasing energy: $1s$, $2s$, $2p$, $3s$, $3p$, $4s$, $3d$, $4p$, $5s$, $4d$, $5p$, $6s$, $4f$, $5d$, $6p$, etc.
2. **Pauli Exclusion Principle.** No more than two electrons can occupy an orbital and then only if they have opposite spins.
3. **Hund's Rule.** One electron is placed in each equal-energy orbital so that the electrons have parallel spins, before pairing occurs. (Substances with unpaired electrons are **paramagnetic**—they are attracted to a magnetic field.)

Problem 2.1 Show the distribution of electrons in the atomic orbitals of (*a*) carbon and (*b*) oxygen. ◄

A dash represents an orbital; a horizontal space between dashes indicates an energy difference. Energy increases from left to right.

(*a*) Atomic number of C is 6.

$$\underset{1s}{\uparrow\downarrow} \quad \underset{2s}{\uparrow\downarrow} \quad \underset{2p_x}{\uparrow} \; \underset{2p_y}{\uparrow} \; \underset{2p_z}{\quad}$$

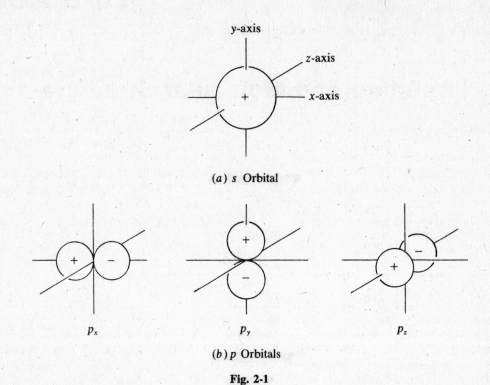

(a) s Orbital

p_x p_y p_z

(b) p Orbitals

Fig. 2-1

The two $2p$ electrons are unpaired in each of two p orbitals (Hund's rule).

(b)   Atomic number of O is 8.

$$\underset{1s}{\uparrow\downarrow} \quad \underset{2s}{\uparrow\downarrow} \quad \underset{2p_x}{\uparrow\downarrow} \underset{2p_y}{\uparrow} \underset{2p_z}{\uparrow}$$

Problem 2.2 Show how the ionic compound Li^+F^- forms from atoms of Li and F.   ◄

These elements react to achieve a stable noble-gas electron configuration (NGEC).   Li(3) has 1 electron more than He and loses it.   F(9) has 1 electron less than Ne and therefore accepts the electron from Li.   In this *transfer* of an electron, oppositely charged ions are formed which attract each other to create an **ionic bond**.

$$Li\cdot \; + \; \cdot\ddot{\underset{..}{F}}: \longrightarrow Li^+:\ddot{\underset{..}{F}}:^- \qquad \text{(or simply LiF)}$$

2.2 COVALENT BOND FORMATION–MOLECULAR ORBITAL (MO) METHOD

A covalent bond forms by overlap (fusion) of two AO's—one from each atom.   This overlap produces a new orbital, called a **molecular orbital** (MO), which embraces both atoms.   The interaction of two AO's actually creates two MO's.   If orbitals with like signs overlap, a **bonding MO** results which has a high electron density between the atoms and therefore has a lower energy (greater stability) than the individual AO's.   If AO's of unlike signs overlap, an **antibonding MO*** results which has a node (no electron density) between the atoms and therefore has a higher energy than the individual AO's.

Head-to-head overlap of AO's gives a **sigma** (σ) **MO**—the bonds are called σ bonds, Fig. 2-2(a).   The corresponding antibonding MO* is designated σ^*, Fig. 2-2(b).   The imaginary line joining the nuclei of the bonding atoms is the **bond axis**, whose length is the **bond length**.

Two parallel p orbitals overlap side-to-side to form a **pi** (π) bond, Fig. 2-3(a), or a π^* bond, Fig. 2-3(b).   The bond axis lies in a nodal plane (plane of no electronic density) perpendicular to the cross-sectional plane of the π bond.

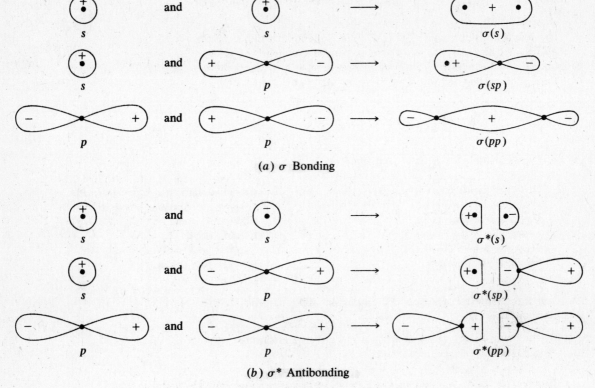

(a) σ Bonding

(b) σ^* Antibonding

Fig. 2-2

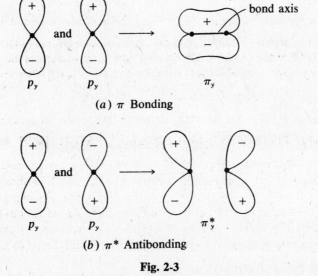

(a) π Bonding

(b) π^* Antibonding

Fig. 2-3

Single bonds are σ bonds.   A double bond is one σ and one π bond.   A triple bond is one σ and two π bonds (a π_z and a π_y, if the triple bond is along the x-axis).

Although MO's encompass the entire molecule, it is best to visualize most of them as being localized between pairs of bonding atoms.

Problem 2.3 What type of MO results from side-to-side overlap of an s and a p orbital? ◄

Fig. 2-4

The overlap is depicted in Fig. 2-4. The bonding strength generated from the overlap between the $+ s$ AO and the $+$ portion of the p orbital is canceled by the antibonding effect generated from overlap between the $+ s$ and the $-$ portion of the p. The MO is **nonbonding** (n); it is no better than two isolated AO's.

Problem 2.4 List the differences between a σ bond and a π bond. ◄

σ Bond	π Bond
1. Formed by head-to-head overlap of AO's.	1. Formed by lateral overlap of p orbitals (or p and d orbitals).
2. Has cylindrical charge symmetry about bond axis.	2. Has maximum charge density in the cross-sectional plane of the orbitals.
3. Has free rotation.	3. No free rotation.
4. Lower energy.	4. Higher energy.
5. Only one bond can exist between two atoms.	5. One or two bonds can exist between two atoms.

Problem 2.5 Show the electron distribution in MO's of (a) H_2, (b) H_2^+, (c) H_2^-, (d) He_2. Predict which are unstable. ◄

Fill the lower-energy MO first with no more than 2 electrons.

(a) H_2 has a total of 2 electrons, therefore

$$\frac{\uparrow \downarrow}{\sigma} \quad \frac{}{\sigma^*}$$

Stable (excess of 2 bonding electrons).

(b) H_2^+, formed from H^+ and $H\cdot$, has one electron:

$$\frac{\uparrow}{\sigma} \quad \frac{}{\sigma^*}$$

Stable (excess of one bonding electron). Has less bonding strength than H_2.

(c) H_2^-, formed theoretically from H^- and $H\cdot$, has 3 electrons:

$$\frac{\uparrow \downarrow}{\sigma} \quad \frac{\uparrow}{\sigma^*}$$

Stable (has net bond strength of one bonding electron). The antibonding electron cancels the bonding strength of one of the bonding electrons.

(d) He_2 has 4 electrons, two from each He atom. The electron distribution is

$$\frac{\uparrow \downarrow}{\sigma} \quad \frac{\uparrow \downarrow}{\sigma^*}$$

Not stable (antibonding and bonding electrons cancel and there is no net bonding). Two He atoms are more stable than a He_2 molecule.

2.3 HYBRIDIZATION OF ATOMIC ORBITALS

A carbon atom must provide 4 equal-energy AO's in order to form 4 equivalent sigma bonds as in methane, CH_4. It is assumed that the 4 equivalent AO's are formed by blending the $2s$ and the three $2p$ orbitals. This blending is called **hybridization**, Fig. 2-5. These 4 **hybrid** orbitals are called sp^3 hybrid AO's. The shape of the sp^3 hybrid orbital is shown in Fig. 2-6. The small tail is often omitted when depicting hybrid orbitals (see Fig. 2-10).

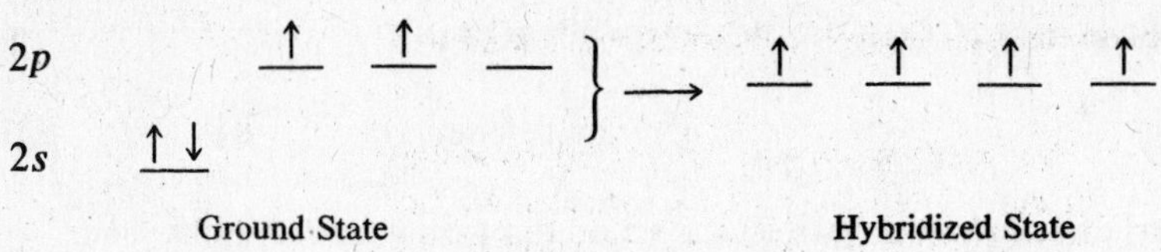

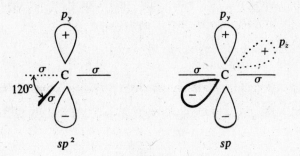

$$2p \quad \underline{\uparrow} \; \underline{\uparrow} \; \underline{\quad} \Bigg\} \rightarrow \underline{\uparrow} \; \underline{\uparrow} \; \underline{\uparrow} \; \underline{\uparrow}$$

$$2s \quad \underline{\uparrow\downarrow}$$

Ground State   Hybridized State

Fig. 2-5 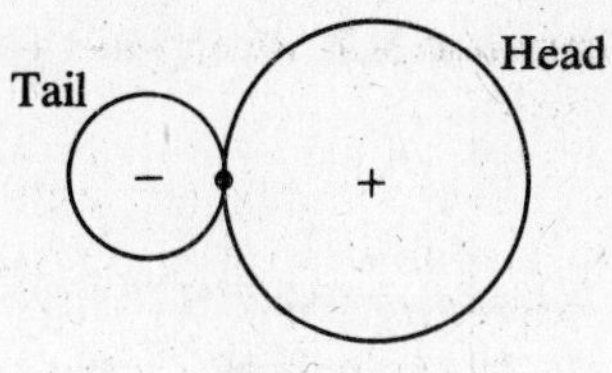   **Fig. 2-6**

The AO's of carbon can hybridize in ways other than sp^3, as shown in Fig. 2-7.

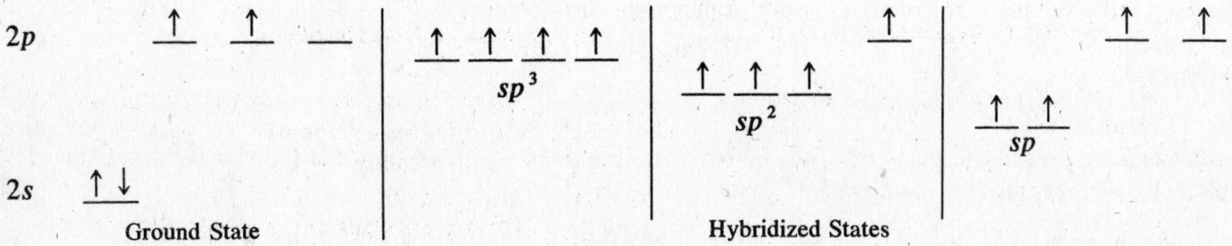

$$2p \quad \underline{\uparrow} \; \underline{\uparrow} \; \underline{\quad} \Bigg| \underbrace{\underline{\uparrow} \; \underline{\uparrow} \; \underline{\uparrow} \; \underline{\uparrow}}_{sp^3} \Bigg| \underbrace{\underline{\uparrow} \; \underline{\uparrow} \; \underline{\uparrow}}_{sp^2} \; \underline{\uparrow} \Bigg| \underbrace{\underline{\uparrow} \; \underline{\uparrow}}_{sp} \; \underline{\uparrow} \; \underline{\uparrow}$$

$$2s \quad \underline{\uparrow\downarrow}$$

Ground State   Hybridized States

Fig. 2-7

Repulsion between pairs of electrons causes these hybrid orbitals to have the maximum bond angles and geometries summarized in Table 2-2. The sp and sp^2 hybrid AO's have the shapes shown in Fig. 2-8.

Table 2-2

Type	Bond Angle	Geometry	Number of Remaining p's	Type of Bond Formed
sp^3	109.5°	tetrahedral	0	σ
sp^2	120°	trigonal planar	1	σ
sp	180°	linear	2	σ

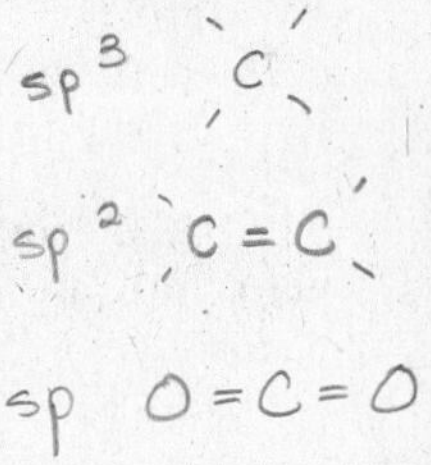

$$sp^3 \quad \diagdown C \diagup$$
$$sp^2 \quad C = C$$
$$sp \quad O = C = O$$

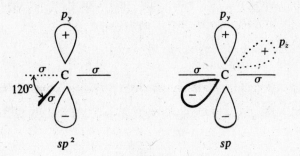

Fig. 2-8

Problem 2.6 The H_2O molecule has a bond angle of 105°. What type of AO's does O use to form the two equivalent σ bonds with H? ◄

$$_8O = \underset{1s}{\underline{\uparrow\downarrow}} \; \underset{2s}{\underline{\uparrow\downarrow}} \; \underset{2p_x}{\underline{\uparrow\downarrow}} \; \underset{2p_y}{\underline{\uparrow}} \; \underset{2p_z}{\underline{\uparrow}} \quad \text{(ground state)}$$

O has two degenerate orbitals, the p_y and p_z, with which to form two equivalent bonds to H. However, if O used these AO's, the bond angle would be 90°, which is the angle between the y- and z-axes. Since the angle is actually 105°, which is close to 109.5°, O is presumed to use sp^3 hybrid AO's.

$$_8O = \underset{1s}{\underline{\uparrow\downarrow}} \; \underset{2sp^3}{\underline{\uparrow\downarrow} \; \underline{\uparrow\downarrow} \; \underline{\uparrow} \; \underline{\uparrow}} \quad (sp^3 \text{ hybrid AO's})$$

Problem 2.7 Each H—N—H bond angle in $:NH_3$ is 107°. What type of AO's does N use? ◄

$$_7N = \frac{\uparrow\downarrow}{1s}\quad\frac{\uparrow\downarrow}{2s}\quad\frac{\uparrow}{2p_x}\quad\frac{\uparrow}{2p_y}\quad\frac{\uparrow}{2p_z}\qquad\text{(ground state)}$$

If the ground-state N atom were to use its three equal-energy p AO's to form three equivalent N—H bonds, each H—N—H bond angle would be 90°. Since the actual bond angle is 107° rather than 90°, N, like O, uses sp^3 hybrid AO's

$$_7N = \frac{\uparrow\downarrow}{1s}\qquad\frac{\uparrow\downarrow\quad\uparrow\quad\uparrow\quad\uparrow}{2sp^3}\qquad(sp^3\text{ hybrid AO's})$$

Apparently, for atoms in the second period forming more than one covalent bond (Be, B, C, N and O), a hybrid AO must be provided for each σ bond and each unshared pair of electrons. Atoms in higher periods also often use hybrid AO's.

Problem 2.8 Predict the shape of (a) the boron trifluoride molecule (BF_3) and (b) the boron tetrafluoride anion (BF_4^-). All bonds are equivalent. ◄

(a) The AO's used by the central atom, in this case B, determine the shape of the molecule.

$$_5B = \frac{\uparrow\downarrow}{1s}\quad\frac{\uparrow\downarrow}{2s}\quad\frac{\uparrow}{2p_x}\quad\frac{}{2p_y}\quad\frac{}{2p_z}\qquad\text{(ground state)}$$

There are three sigma bonds in BF_3 and no unshared pairs; therefore, three hybrid AO's are needed. Hence, B uses sp^2 hybrid AO's, and the shape is trigonal planar. Each F—B—F bond angle is 120°.

$$_5B = \frac{\uparrow\downarrow}{1s}\qquad\frac{\uparrow\quad\uparrow\quad\uparrow}{2sp^2}\qquad\frac{}{2p_z}\qquad(sp^2\text{ hybrid state})$$

The empty p_z orbital is at right angles to the plane of the molecule.

(b) B in BF_4^- has four σ bonds and needs four hybrid AO's. B is now in an sp^3 hybrid state.

$$_5B = \frac{\uparrow\downarrow}{1s}\qquad\frac{\uparrow\quad\uparrow\quad\uparrow}{2sp^3}\quad\frac{}{}\qquad(sp^3\text{ hybrid state})$$

The empty sp^3 hybrid orbital overlaps with a filled orbital of F^- which holds two electrons,

$$F^- + BF_3 \longrightarrow BF_4^-$$

The shape is tetrahedral; the bond angles are 109.5°.

Problem 2.9 Arrange the s, p and the three sp-type hybrid AO's in order of decreasing energy. ◄

The more s character in the AO, the lower the energy. Therefore, the order of decreasing energy is

$$p > sp^3 > sp^2 > sp > s$$

Problem 2.10 What effect does hybridization have on the stability of bonds? ◄

Hybrid orbitals can (a) overlap better and (b) provide greater bond angles, thereby minimizing the repulsion between pairs of electrons and making for greater stability.

2.4 ELECTRONEGATIVITY AND POLARITY

The relative tendency of a bonded atom in a molecule to attract electrons is expressed by the term **electronegativity**. The higher the electronegativity, the more effectively does the atom attract and hold electrons. A bond formed by atoms of dissimilar electronegativities is called **polar**. A **nonpolar** covalent bond exists between atoms having a very small or zero difference in electronegativity.　A few relative electronegativities are

$$F\,(4.0) > O\,(3.5) > Cl, N\,(3.0) > Br\,(2.8) > S, C, I\,(2.5) > H\,(2.1)$$

The more electronegative element of a covalent bond is relatively negative in charge, while the less electronegative element is relatively positive.　The symbols $\delta+$ and $\delta-$ represent partial charges (**bond polarity**).　These partial charges should not be confused with ionic charges.　Polar bonds are indicated by $\leftrightarrow$; the head points towards the more electronegative atom.

The vector sum of all individual bond moments gives the net **dipole moment** of the molecule.

Problem 2.11　What do the molecular dipole moments $\mu = 0$ for CO_2 and $\mu = 1.84\,D$ for H_2O tell you about the shapes of these molecules?　◄

In CO_2

$$\overset{\delta-}{:\ddot{O}}=\overset{\delta+}{C}=\overset{\delta-}{\ddot{O}:}$$

O is more electronegative than C, and each C—O bond is polar as shown.　A zero dipole moment indicates a symmetrical distribution of $\delta-$ charges about the $\delta+$ carbon.　The geometry must be linear; in this way, individual bond moments cancel:

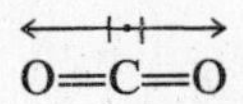

$$\overset{\longleftarrow \;\;\vdash\!\cdot\!\dashv\;\; \longrightarrow}{O=C=O}$$

H_2O has polar bonds

$$\overset{\delta+}{H}-\overset{\delta-}{O}-\overset{\delta+}{H}$$

Since there is a net dipole moment, the individual bond moments do not cancel, and the molecule must have a *bent* shape:

2.5　OXIDATION NUMBER

The oxidation number (ON) is a value assigned to an atom based on relative electronegativities.　It equals the number of outer electrons minus the number of assigned electrons, when the bonding electrons are assigned to the more electronegative atom.　The sum of all (ON)'s equals the charge on the species.

Problem 2.12　Determine the oxidation number of each C, $(ON)_C$, in:　(*a*) CH_4,　(*b*) CH_3OH,　(*c*) CH_3NH_2,　(*d*) $H_2C{=}CH_2$.　Use the data　$(ON)_N = -3$; $(ON)_H = 1$; $(ON)_O = -2$.　◄

All examples are molecules; therefore the sum of all (ON) values is 0.

(*a*)　$(ON)_C + 4(ON)_H = 0$;　　　$(ON)_C + (4 \times 1) = 0$;　$(ON)_C = -4$

(*b*)　$(ON)_C + (ON)_O + 4(ON)_H = 0$;　　　$(ON)_C + (-2) + 4 = 0$;　$(ON)_C = -2$

(*c*)　$(ON)_C + (ON)_N + 5(ON)_H = 0$;　　　$(ON)_C + (-3) + 5 = 0$;　$(ON)_C = -2$

(*d*)　Since both C atoms are equivalent,

$$2(ON)_C + 4(ON)_H = 0;\quad\ 2(ON)_C + 4 = 0;\quad (ON)_C = -2$$

2.6　INTERMOLECULAR (VAN DER WAALS) FORCES

(**a**)　**Dipole-dipole** interaction results from the attraction of the $\delta+$ end of one polar molecule for the $\delta-$ end of another polar molecule.

(b) **Hydrogen-bond.** X—H and :Y may be bridged X—H---:Y if X and Y are small, highly electronegative atoms, usually F, O and N. H-bonds also occur intramolecularly.

(c) **London forces.** Electrons of a nonpolar molecule may momentarily cause an imbalance of charge distribution in neighboring molecules, thereby inducing a temporary dipole moment. Although constantly changing, these induced dipoles result in a weak net attractive force. The greater the molecular weight of the molecule, the greater the number of electrons and the greater these forces.

The order of attraction is

$$\text{H-bond} \geqslant \text{dipole-dipole} > \text{London forces}$$

Problem 2.13 Account for the following progression in boiling point: $CH_4 = -161.5\ °C$, $Cl_2 = -34\ °C$, $CH_3Cl = -24\ °C$. ◀

The greater the intermolecular force, the higher the boiling point. Polarity and molecular weight must be considered. Only CH_3Cl is polar, and it has the highest boiling point. CH_4 has a lower molecular weight (16 g/mole) than has Cl_2 (71 g/mole) and therefore has the lower boiling point.

Problem 2.14 Account for the following progression in boiling point: $CH_3Cl = -24\ °C$, $CH_3Br = 5\ °C$, $CH_3I = 43\ °C$. ◀

The order of polarity is $CH_3Cl > CH_3Br > CH_3I$. The order of molecular weight is $CH_3I > CH_3Br > CH_3Cl$. The two trends oppose each other in affecting the boiling point. The order of molecular weight predominates here.

Problem 2.15 The boiling points of *n*-pentane and its isomer neopentane are 36.2 °C and 9.5 °C, respectively. Account for this difference. (See Problem 1.4 for the structural formulas.) ◀

These isomers are both nonpolar. Therefore, another factor, the shape of the molecule, influences the boiling point. The shape of *n*-pentane is rodlike, whereas that of neopentane is spherelike. Rods can touch along their entire length; spheres touch only at a point. The more contact between molecules, the greater the London forces. Thus, the b.p. of *n*-pentane is higher.

Problem 2.16 NaCl dissolves in water but not in *n*-hexane, $CH_3(CH_2)_4CH_3$. Explain. ◀

Solution of a salt, such as NaCl, necessitates separation of the attracting ions. In the strongly polar solvent H_2O, each positive ion is surrounded by water molecules due to ion-dipole attraction:

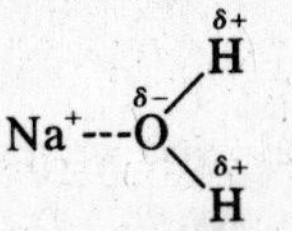

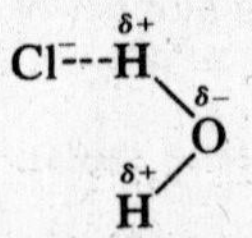

The negative ion is H-bonded to H_2O. These interactions, called **solvation** (more specifically, **hydration**, since the solvent is water), cause the ions to separate and disperse in the solvent. Nonpolar solvents such as *n*-hexane cannot solvate and therefore cannot dissolve ionic compounds. Few organic compounds are polar enough to dissolve salts.

Problem 2.17 Mineral oil, a mixture of high-molecular-weight hydrocarbons, dissolves in *n*-hexane but not in water or ethyl alcohol, CH_3CH_2OH. Explain. ◀

Attractive forces between nonpolar molecules such as mineral oil and *n*-hexane are very weak. Therefore, such molecules can mutually mix and solution is easy. The attractive forces between polar H_2O or C_2H_5OH molecules are strong H-bonds. Most nonpolar molecules cannot overcome these H-bonds and therefore do not dissolve is such **polar protic** solvents.

Problem 2.18 Which of the following substances would resemble H_2O as a solvent: CCl_4, CH_3OH (methyl alcohol), liquefied NH_3? ◀

NH_3 and CH_3OH, like H_2O, are polar molecules capable of forming H-bonds and so resemble H_2O as a solvent.

Problem 2.19 Suggest a good solvent for removing a butter stain from a tablecloth. ◀

Butter is composed largely of organic compounds with low polarity. Water, being polar, is not good for dissolving a butter stain. Far better are the common nonpolar organic solvents such as carbon tetrachloride (CCl_4) or benzene (C_6H_6).

Problem 2.20 Ethyl alcohol, C_2H_5OH, boils at 78.3 °C. Its isomer, dimethyl ether, CH_3OCH_3, boils at −24 °C. Explain. ◀

Both are polar molecules and have dipole-dipole attraction. However, in ethyl alcohol, H-bonding can exist:

$$\underset{\displaystyle H_3CCH_2O}{} \overset{\displaystyle H}{\underset{\displaystyle |}{}} \cdots H-O-CH_2CH_3$$

In CH_3OCH_3 the H's are on C and so cannot H-bond.

2.7 RESONANCE AND DELOCALIZED p ELECTRONS

Resonance theory describes species for which a single Lewis electron structure cannot be written, as exemplified by dinitrogen oxide, N_2O:

$$:\!\overset{\frown}{N}\!\!=\!\!\overset{+}{N}\!\!=\!\!\overset{\frown}{O}: \longleftrightarrow :N\!\!=\!\!\overset{+}{N}\!\!-\!\!\overset{\frown}{\ddot{O}}:^{-} \quad \text{(with formal charges)}$$

Calculated Bond Length, Å	1.20	1.15	1.10	1.47
Observed Bond Length, Å	1.12	1.19	1.12	1.19

A comparison of the calculated and observed bond lengths shows that neither structure is correct. However, these structures contribute to the description of the actual structure, called a **resonance hybrid**, for which a Lewis structure cannot be written. In other words, these structures tell us that the hybrid has some double-bond character between N and O and some triple-bond character between N and N. We can replace the above **contributing (resonance) structures** with a non-Lewis structure,

$$:\!\overset{\delta-}{N}\!\!=\!\!\overset{+}{N}\!\!=\!\!\overset{\delta-}{\ddot{O}}:$$

The broken lines stand for the partial bonds in which there are delocalized p electrons in an extended π bond created from overlap of a p orbital on each atom. The symbol ↔ denotes resonance, *not equilibrium*.

The energy of the hybrid, E_h, is always less than the calculated energy of any hypothetical contributing structure, E_c. The difference between these energies is the **resonance (delocalization) energy**, E_r:

$$E_r = E_c - E_h$$

The more nearly equal in energy the contributing structures, the greater the resonance energy and the less the hybrid looks like the contributing structures. When contributing structures have dissimilar energies, the hybrid looks most like the lowest-energy structure.

Contributing structures (*a*) differ only in positions of electrons (atomic nuclei must have the same positions) and (*b*) must have the same number of paired electrons. Relative energies of contributing structures are assessed by the following rules.

1. Structures with the greatest number of covalent bonds are most stable. However, for second-period elements (C, O, N) the octet rule must be observed.
2. With few exceptions, structures with the least number or amount of formal charges are most stable.
3. If all structures have formal charge, the most stable (lowest energy) one has − on the more electronegative atom and + on the more electropositive atom.
4. Structures with like formal charges on adjacent atoms have very high energy.
5. Resonance structures with electron-deficient, positively charged atoms have very high energy.

Problem 2.21 Write contributing structures, showing formal charges when necessary, for (a) ozone, O_3; (b) CO_2; (c) hydrazoic acid, HN_3; (d) isocyanic acid, HNCO. Indicate the most and least stable structures and give reasons for your choices.   Give the structure of the hybrid.   ◀

(a) $:\ddot{O}{=}\ddot{O}{-}\ddot{O}: \longleftrightarrow :\ddot{O}{-}\ddot{O}{=}\ddot{O}:$ (equal-energy structures).   The hybrid is $:\overset{\delta-}{O}{=}\overset{+}{O}{=}\overset{\delta-}{O}:$.

(b) $:\ddot{O}{=}C{=}\ddot{O}: \longleftrightarrow {}^-:\ddot{O}{-}C{\equiv}O:^+ \longleftrightarrow {}^+:O{\equiv}C{-}\ddot{O}:^-$

                    (1)   (2)   (3)

(1) is most stable; it has no formal charge.   (2) and (3) have equal energy and are least stable because they have formal charges.   In addition, in both (2) and (3), one O, an electronegative element, bears a + formal charge.   Since (1) is so much more stable than (2) and (3), the hybrid is $:\ddot{O}{=}C{=}\ddot{O}:$, which is just (1).

(c) $H{-}\ddot{N}{=}\overset{+}{N}{=}\ddot{N}: \longleftrightarrow H{-}\overset{+}{N}{\equiv}\overset{+}{N}{\equiv}\ddot{N}: \longleftrightarrow H{-}\overset{+}{N}{=}N{-}\overset{2-}{\ddot{N}}: \longleftrightarrow H{-}\overset{2+}{N}{=}\ddot{N}{-}\overset{2-}{\ddot{N}}:$

        (1)   (2)   (3)   (4)

(1) and (2) have about the same energy and are the most stable, since they have the least amount of formal charge.   (3) has very high energy since it has + charge on adjacent atoms and, in terms of absolute value, a total formal charge of 4.   (4) has a very high energy because the N bonded to H only has 6 electrons.   The hybrid, composed of (1) and (2), is

$$H{-}\overset{\delta-}{\ddot{N}}{=}\overset{+}{N}{=}\overset{\delta-}{N}:$$

(d) $H{-}\ddot{N}{=}C{=}\ddot{O}: \longleftrightarrow H{-}\overset{-}{\ddot{N}}{=}C{=}\overset{+}{O}: \longleftrightarrow H{-}\overset{+}{N}{\equiv}C{-}\ddot{O}:^-$

        (1)   (2)   (3)

(1) has no formal charge and is most stable.   (2) is least stable since the − charge is on N rather than on the more electronegative O as in (3).   The hybrid is $H{-}\ddot{N}{=}C{=}\ddot{O}:$ (the same as (1)), the most stable contributing structure.

Problem 2.22 (a) Write contributing structures and the delocalized structure for (i) NO_2^- and (ii) NO_3^-. (b) Compare the stability of the hybrids of each.   ◀

(a)   (i) $:\ddot{O}{-}\ddot{N}{=}\ddot{O}: \longleftrightarrow :\ddot{O}{=}N{-}\ddot{O}:$ or $\left[:\overset{-1/2}{\ddot{O}}{=}N{=}\overset{-1/2}{\ddot{O}}: \right]^-$

The − is delocalized over both O's so that each can be assumed to have a $-\frac{1}{2}$ charge.   Each N—O bond has the same bond length.

(ii) $\begin{matrix} & \overset{\ddot{O}:}{\|} & \\ {}^-:\ddot{O}{-}\overset{+}{N} & & \\ & \overset{\diagdown}{\ddot{O}:^-} & \end{matrix} \longleftrightarrow \begin{matrix} & \overset{\ddot{O}:^-}{\|} & \\ :\ddot{O}{=}\overset{+}{N} & & \\ & \overset{\diagdown}{\ddot{O}:^-} & \end{matrix} \longleftrightarrow \begin{matrix} & \overset{\ddot{O}:^-}{} & \\ {}^-:\ddot{O}{-}\overset{+}{N} & & \\ & \overset{\diagdown\!\|}{\ddot{O}:} & \end{matrix}$ or $\left[\begin{matrix} & \overset{-1/3}{\overset{\ddot{O}:}{}} & \\ \overset{-1/3}{:\ddot{O}}{=}N & & \\ & \overset{\diagdown}{\underset{-1/3}{\ddot{O}:}} & \end{matrix} \right]^-$

The − on the ion is delocalized over three O's so that each has a $-\frac{1}{3}$ charge.

(b) We can use resonance theory to compare the stability of these two ions because they differ in only one feature—the number of O's on each N, which is related to the oxidation numbers of the N's. We could not, for example, compare NO_3^- and HSO_3^-, since they differ in more than one way; N and S are in different groups and periods of the periodic table. NO_3^- is more stable than NO_2^- since the charge on NO_3^- is delocalized (dispersed) over a greater number of O's and since NO_3^- has a more extended π bond system.

Problem 2.23 Indicate which one of each of the following pairs of resonance structures is the less stable and is an unlikely contributing structure. Give reasons in each case.

(a)   (b) $H_2\ddot{C}-\ddot{O}: \longleftrightarrow H_2\ddot{C}-\overset{+}{\ddot{O}}:$

I   II   III   IV

(c) $H_2C\overset{\frown}{=}CH-\bar{\ddot{C}}H_2 \longleftrightarrow H_2\overset{+}{C}-\bar{\ddot{C}}H-\bar{\ddot{C}}H_2$ (d) $H-\bar{\ddot{C}}::N: \longleftrightarrow H-\overset{\frown}{\ddot{C}}::\overset{+}{N}:$

V   VI   VII   VIII

(e) $H_3C-\ddot{\ddot{C}}l: \longleftrightarrow H_3\bar{\ddot{C}}=\overset{+}{\ddot{C}}l:$

IX   X

(a) I has fewer covalent bonds, more formal charge and an electron-deficient N.
(b) IV has + on the more electronegative O.
(c) VI has similar − charges on adjacent C's, fewer covalent bonds, more formal charge and an electron-deficient C.
(d) VII has fewer covalent bonds and a + on the more electronegative N, which is also electron-deficient.
(e) C in X has 10 electrons; this is not possible with the elements of the second period.

Supplementary Problems

Problem 2.24 State the difference between an AO, a hybrid AO, an MO and a localized MO.

An AO is a region of space in an *atom* in which an electron may exist. A hybrid AO is mathematically fabricated from some number of AO's to explain equivalency of bonds. An MO is a region of space about the *entire molecule* capable of accommodating electrons. A localized MO is a region of space between a pair of bonded atoms in which the bonding electrons are assumed to be present.

Problem 2.25 Show the orbital population of electrons for N in (a) ground state, (b) sp^3, (c) sp^2, and (d) sp hybrid states.

(a)   (c)

(b)   (d)

Note that since the energy difference between hybrid and *p* orbitals is so small, Hund's rule prevails over the Aufbau principle.

Problem 2.26 Roughly, what kind of hybridization occurs if the interorbital bond angle is (a) 107°, (b) 118°, (c) 165°?

(a) sp^3, (b) sp^2, (c) sp.

Problem 2.27 State the geometric shapes of molecules with the interorbital bond angles given in Problem 2.26.

(a) tetrahedral, (b) trigonal planar, (c) linear.

Problem 2.28 Since the σ MO formed from $2s$ AO's has a higher energy than the σ^* MO formed from $1s$ AO's, predict whether (a) Li_2, (b) Be_2 can exist. ◄

The MO levels are: $\sigma_{1s}\sigma_{1s}^*\sigma_{2s}\sigma_{2s}^*$.

(a) Li_2 has 6 electrons which fill the MO levels to give

$$\frac{\uparrow\downarrow}{\sigma_{1s}} \quad \frac{\uparrow\downarrow}{\sigma_{1s}^*} \quad \frac{\uparrow\downarrow}{\sigma_{2s}} \quad \frac{}{\sigma_{2s}^*}$$

designated $(\sigma_{1s})^2(\sigma_{1s}^*)^2(\sigma_{2s})^2$. Li_2 has an excess of 2 electrons in bonding MO's and therefore can exist; it is by no means the most stable form of lithium.

(b) Be_2 would have 8 electrons:

$$\frac{\uparrow\downarrow}{\sigma_{1s}} \quad \frac{\uparrow\downarrow}{\sigma_{1s}^*} \quad \frac{\uparrow\downarrow}{\sigma_{2s}} \quad \frac{\uparrow\downarrow}{\sigma_{2s}^*}$$

There are no net bonding electrons, and Be_2 does not exist.

Problem 2.29 The MO's formed when the two sets of the three $2p$ orbitals overlap are

$$\pi_{2p_y}\pi_{2p_z}\sigma_{2p_x}\pi_{2p_y}^*\pi_{2p_z}^*\sigma_{2p_x}^*$$

(the π and π^* pairs are degenerate). (a) Show how MO theory predicts the paramagnetism of O_2. (b) What is the net bonding (**bond order**) in O_2? ◄

The complete sequence of MO's formed from overlap of the $n = 1$ and $n = 2$ AO's of diatomic molecules is:

$$\sigma_{1s}\sigma_{1s}^*\sigma_{2s}\sigma_{2s}^*\pi_{2p_y}\pi_{2p_z}\sigma_{2p_x}\pi_{2p_y}^*\pi_{2p_z}^*\sigma_{2p_x}^*$$

O_2 has 16 electrons to be placed in these MO's, giving

$$(\sigma_{1s})^2(\sigma_{1s}^*)^2(\sigma_{2s})^2(\sigma_{2s}^*)^2(\pi_{2p_y})^2(\pi_{2p_z})^2(\sigma_{2p_x})^2(\pi_{2p_y}^*)^1(\pi_{2p_z}^*)^1$$

(a) The electrons in the two π^* MO's are unpaired; therefore, O_2 is paramagnetic.

(b) Electrons in the first 4 MO's cancel each other's effect. There are 6 electrons in the next 3 bonding orbitals and 2 electrons in the next 2 antibonding orbitals. There is a net bonding effect due to 4 electrons. The bond order is 1/2 of 4, or 2; the two O's are joined by a net double bond.

Problem 2.30 (a) NO_2^+ is linear, (b) NO_2^- is bent. Explain in terms of the hybrid orbitals used by N. ◄

(a) NO_2^+, $:\ddot{O}{=}\overset{+}{N}{=}\ddot{O}:$. N has two σ bonds, no unshared pairs of electrons and therefore needs two hybrid orbitals. N uses sp hybrid orbitals and the σ bonds are linear. The geometry is controlled by the arrangement of the sigma bonds.

(b) NO_2^-, $:\ddot{O}{=}\ddot{N}:\ddot{O}:^-$. N has two σ bonds, one unshared pair of electrons and, therefore, needs three hybrid orbitals. N uses sp^2 hybrid AO's, and the bond angle is about $120°$.

Problem 2.31 Draw an orbital representation for the cyanide ion, $:C{\equiv}N:^-$. ◄

See Fig. 2-9. The C and N each have one σ bond and one unshared pair of electrons, and therefore each needs two sp hybrid AO's. On each atom one sp hybrid orbital forms a σ bond while the other has the unshared pair. Each atom has a p_y AO and a p_z AO. The two p_y orbitals overlap to form a π_y bond in the xy-plane; the two p_z orbitals overlap to form a π_z bond in the xz-plane. Thus, two π bonds at right angles to each other and a σ bond exist between the C and N atoms.

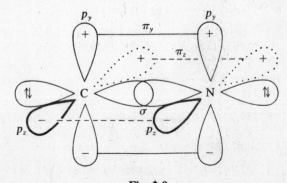

Fig. 2-9

Problem 2.32 (a) Which of the following molecules possess polar bonds: F_2, HF, BrCl, CH_4, $CHCl_3$, CH_3OH? (b) Which are polar molecules? ◄

(a) HF, BrCl, CH_4, $CHCl_3$, CH_3OH.
(b) HF, BrCl, $CHCl_3$, CH_3OH. The symmetrical individual bond moments in CH_4 cancel.

Problem 2.33 Considering the difference in electronegativity between O and S, would H_2O or H_2S exhibit greater (a) dipole-dipole attraction, (b) H-bonding? ◄

(a) H_2O, (b) H_2O.

Problem 2.34 Nitrogen trifluoride (NF_3) and ammonia (NH_3) have an electron pair at the fourth corner of a tetrahedron and have similar electronegativity *differences* between the elements (1.0 for N and F and 0.9 for N and H). Explain the larger dipole moment of ammonia (1.46 D) as compared with that of NF_3 (0.24 D). ◄

The dipoles in the three N—F bonds are toward F, see Fig. 2-10(a), and oppose and tend to cancel the effect of the unshared electron pair on N. In NH_3, the moments for the three N—H bonds are toward N, see Fig. 2-10(b), and add to the effect of the electron pair.

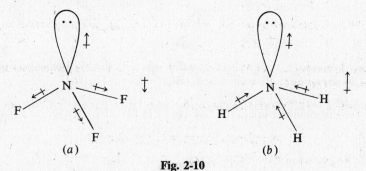

Fig. 2-10

Problem 2.35 Explain why the dipole moment of CH_2Cl_2 is greater than that of $CHCl_3$. ◄

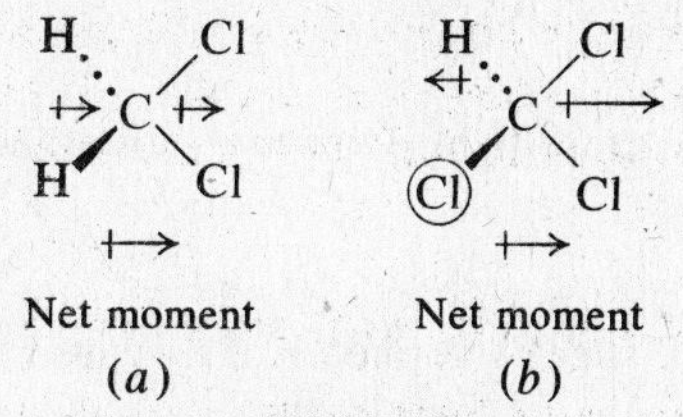

Net moment   Net moment
(a)   (b)

Fig. 2-11

The resultant moment of the H and the encircled Cl in $CHCl_3$ opposes the resultant moment of the other two Cl's, as shown in Fig. 2-11(b). The actual values for CH_2Cl_2 and $CHCl_3$ are 1.6 D and 1.0 D, respectively.

Problem 2.36 Assume there is a compound AB_2C_2 with B and C each bonded to A. Assume the electronegativities of these atoms are A = 4, B = 3, C = 2. Which of the following molecular structures would give rise to dipoles? (a) Tetrahedron. (b) The atoms B and C are arranged at the corners of a square with A in the center and with (i) the like atoms adjacent to each other and (ii) the like atoms on diagonally opposite corners. ◄

See Fig. 2-12. In (a) and (b)(i) dipoles are present. In (b)(ii) the individual bond moments cancel, and there is no dipole.

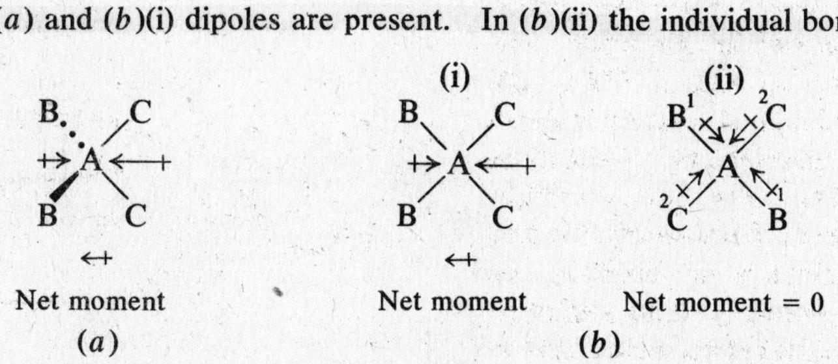

(i)   (ii)

Net moment   Net moment   Net moment = 0
(a)   (b)

Fig. 2-12

Problem 2.37 When 50 ml of C_2H_5OH and 50 ml of H_2O are mixed, the volume of the solution is less than 100 ml.   Explain. ◄

The H-bond between H_2O and C_2H_5OH molecules is stronger than the H-bond between the like molecules of the individual compounds.   The H_2O and C_2H_5OH molecules are closer when mixed, and the volume shrinks.

Problem 2.38 NH_4^+ salts are much more soluble in water than are the corresponding Na^+ salts.   Explain. ◄

Na^+ is solvated merely by an ion-dipole interaction. NH_4^+ is solvated by H-bonding,

$$H_3N^+\!\!-\!\!H\text{---}\overset{\delta-}{\underset{\underset{H}{|}}{:\!\ddot{O}}}\!\!-\!\!\overset{\delta+}{H}$$

which is a stronger attractive force.

Problem 2.39 The F^- of dissolved NaF is more reactive in dimethyl sulfoxide,

$$\overset{\displaystyle O}{\overset{\displaystyle \|}{CH_3SCH_3}}$$

and in acetonitrile, $CH_3C\!\equiv\!N$, than in CH_3OH.   Explain. ◄

H-bonding prevails in CH_3OH (a protic solvent), $CH_3OH\text{---}F^-$, thereby decreasing the reactivity of F^-. CH_3SOCH_3 and CH_3CN are aprotic solvents; their C—H H's do not H-bond.

Problem 2.40 Find the oxidation number of the C in   (a) CH_3Cl,   (b) CH_2Cl_2,   (c) H_2CO,   (d) HCOOH, and (e) CO_2, if $(ON)_{Cl} = -1$. ◄

From Section 2.5:

(a) $(ON)_C + (3 \times 1) + (-1) = 0;\ (ON)_C = -2$ (d) $(ON)_C + 2 + (-4) = 0;\ (ON)_C = 2$

(b) $(ON)_C + (2 \times 1) + [2 \times (-1)] = 0;\ (ON)_C = 0$ (e) $(ON)_C + (-4) = 0;\ (ON)_C = 4$

(c) $(ON)_C + (2 \times 1) + [1 \times (-2)] = 0;\ (ON)_C = 0$

Problem 2.41 Irradiation with ultraviolet (uv) light permits rotation about a π bond.   Explain in terms of bonding and antibonding MO's. ◄

Two p AO's overlap to form two pi MO's, π (bonding) and π^* (antibonding).   The two electrons in the original p AO's fill only the π MO (ground state).   A photon of uv causes excitation of one electron from π to π^* (excited state).

$$\underset{\pi}{\underline{\uparrow\downarrow}}\quad\underset{\pi^*}{\underline{}}\quad\text{(ground state)}\ \overset{uv}{\longrightarrow}\ \underset{\pi}{\underline{\uparrow}}\quad\underset{\pi^*}{\underline{\downarrow}}\quad\text{(excited state)}$$

(Initially the excited electron does not change its spin.)   The bonding effects of the two electrons cancel. There is now only a sigma bond between the bonded atoms, and rotation about the bond can occur.

Problem 2.42 Write the contributing resonance structures and the delocalized hybrid for   (a) BCl_3, (b) H_2CN_2 (diazomethane). ◄

(a) Boron has 6 electrons in its outer shell in BCl_3 and can accommodate 8 electrons by having a B—Cl bond assume some double-bond character.

all equivalent

(b) $H_2C\!=\!\overset{+}{N}\!=\!\overset{..}{\underset{..}{N}}\!:\ \longleftrightarrow\ H_2\overset{..}{\underset{..}{C}}\!-\!\overset{+}{N}\!\equiv\!N\!:\ \equiv\ H_2\overset{\delta-}{C}\text{---}\overset{+}{N}\!\equiv\!\overset{\delta-}{N}\!:$

Problem 2.43 Arrange the contributing structures for (*a*) vinyl chloride, $H_2C=CHCl$, and (*b*) formic acid, HCOOH, in order of increasing importance (increasing stability) by assigning numbers starting with 1 for most important and stable.   ◄

(*a*)

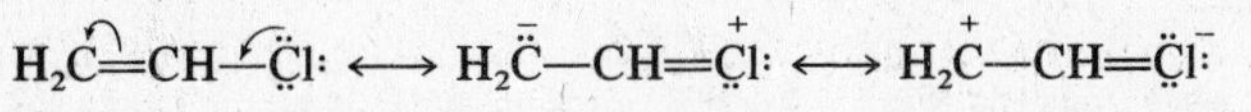

$$H_2\overset{\frown}{C}=CH\overset{\frown}{}\ddot{C}l: \longleftrightarrow H_2\bar{\ddot{C}}-CH=\overset{+}{\ddot{C}}l: \longleftrightarrow H_2\overset{+}{C}-CH=\ddot{C}l:^-$$

$$\text{I} \qquad\qquad\qquad \text{II} \qquad\qquad\qquad \text{III}$$

I is most stable because it has no formal charge.   III is least stable since it has an electron-deficient C.   In III, Cl uses an empty 3*d* orbital to accommodate a fifth pair of electrons.   Fluorine could not do this.   The order of stability is

$$\text{I}(1) > \text{II}(2) > \text{III}(3)$$

(*b*)

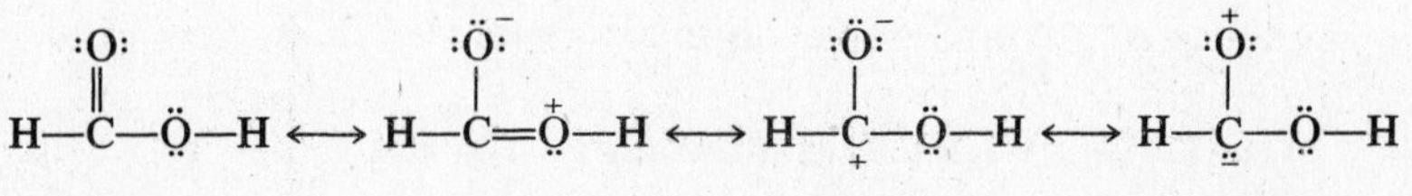

$$H-\overset{\displaystyle :\overset{..}{O}:}{\underset{\displaystyle \|}{C}}-\overset{..}{O}-H \longleftrightarrow H-\overset{\displaystyle :\overset{..}{O}:^-}{\underset{\displaystyle \|}{C}}=\overset{+}{\underset{..}{O}}-H \longleftrightarrow H-\overset{\displaystyle :\overset{..}{O}:^-}{\underset{\displaystyle |}{\underset{+}{C}}}-\overset{..}{O}-H \longleftrightarrow H-\overset{\displaystyle :\overset{+}{\overset{..}{O}}:}{\underset{\displaystyle |}{C}}-\overset{..}{O}-H$$

$$\text{V} \qquad\qquad \text{VI} \qquad\qquad \text{VII} \qquad\qquad \text{VIII}$$

V and VI have the greater number of covalent bonds and are more stable than either VII or VIII.   V has no formal charge and is more stable than VI.   VIII is less stable than VII since VIII's electron deficiency is on O, which is a more electronegative atom than the electron-deficient C of VII.   The order of stability is

$$\text{V}(1) > \text{VI}(2) > \text{VII}(3) > \text{VIII}(4)$$

Chapter 3

Chemical Reactivity and Organic Reactions

3.1 REACTION MECHANISM

The way in which a reaction occurs is called a **mechanism**. A reaction may occur in one step or, more often, by a sequence of several steps. For example, $A + B \rightarrow X + Y$ may proceed in two steps:

$$(1) \quad A \longrightarrow I + X \quad \text{followed by} \quad (2) \quad B + I \longrightarrow Y$$

Substances such as I, formed in intermediate steps and consumed in later steps, are called **intermediates**. Sometimes the same reactants can give two sets of products by two different mechanisms.

3.2 CARBON-CONTAINING INTERMEDIATES

Carbon-containing intermediates often arise from two types of bond cleavage:

Heterolytic (polar) reactions. Both electrons go with one group, e.g.

$$A:B \longrightarrow A^+ + :B^- \quad \text{or} \quad A:^- + B^+$$

Homolytic (radical) reactions. Each separating group takes one electron, e.g.

$$A:B \longrightarrow A\cdot + \cdot B$$

1. **Carbonium ions** or **carbocations** are positively charged species containing a carbon atom having only six electrons in three bonds:

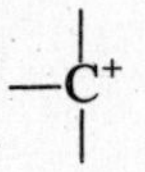

$$-\overset{|}{\underset{|}{C}}{}^+$$

2. **Carbanions** are negatively charged species containing a carbon atom with three bonds and an unshared pair of electrons:

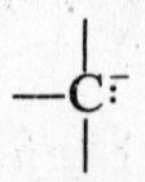

$$-\overset{|}{\underset{|}{C}}{}^{:-}$$

3. **Radicals** are species with at least one unpaired electron. This is a broad category in which carbon radicals,

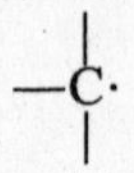

$$-\overset{|}{\underset{|}{C}}\cdot$$

are just one example.

4. **Carbenes** are neutral species having a carbon atom with two bonds and two electrons. There are two kinds: **singlet**

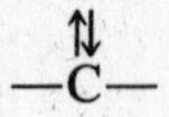

$$-\overset{\uparrow\downarrow}{C}-$$

in which the two electrons have opposite spins and are paired in one orbital, and **triplet**

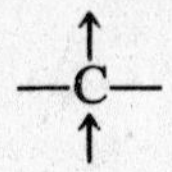

$$-\overset{\uparrow}{\underset{\uparrow}{C}}-$$

in which the two electrons have the same spin and are in different orbitals.

3.3   TYPES OF ORGANIC REACTIONS

1. **Displacement (substitution).**   An atom or group of atoms is replaced by another atom or group.

2. **Addition.**   Two molecules combine to yield a single molecule.   Addition frequently occurs at a double or triple bond and sometimes at small-size rings.

3. **Elimination.**   This is the reverse of addition.   Two atoms or groups are removed from a molecule.   Removal of the atoms or groups from different atoms produces another bond between these atoms.   If the atoms or groups are taken from adjacent atoms (β-**elimination**), a multiple bond is formed; if they are taken from other than adjacent atoms, a ring results.   Removal of atoms or groups from the same atom (α-**elimination**) produces a carbene.

4. **Rearrangements.**   Bonds in the reactant are scrambled as in conversion of a compound to an isomer.

5. **Oxidation-reduction (redox).**   These reactions involve transfer of electrons or change in oxidation number.   A decrease in the number of H atoms bonded to C and an increase in the number of bonds to other atoms such as C, O, N, Cl, Br, F and S signals oxidation.

Problem 3.1   The following represents the steps in the mechanism for chlorination of methane:

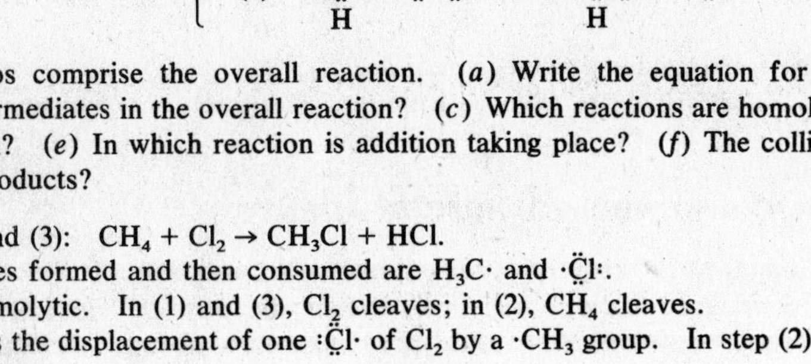

Initiation Step   (1) $:\ddot{C}l:\ddot{C}l: + \text{energy} \longrightarrow :\ddot{C}l\cdot + \cdot\ddot{C}l:$

Chlorine radicals

Propagation Steps

(2) $H:\overset{H}{\underset{H}{C}}:H + \cdot\ddot{C}l: \longrightarrow H:\ddot{C}l: + H:\overset{H}{\underset{H}{C}}\cdot$

Methyl radical

(3) $H:\overset{H}{\underset{H}{C}}\cdot + :\ddot{C}l:\ddot{C}l: \longrightarrow H:\overset{H}{\underset{H}{C}}:\ddot{C}l: + \cdot\ddot{C}l:$

The propagation steps comprise the overall reaction.   (a) Write the equation for the overall reaction. (b) What are the intermediates in the overall reaction?   (c) Which reactions are homolytic?   (d) Which is a displacement reaction?   (e) In which reaction is addition taking place?   (f) The collision of which species would lead to side products?   ◄

(a)   Add steps (2) and (3): $CH_4 + Cl_2 \rightarrow CH_3Cl + HCl$.
(b)   The intermediates formed and then consumed are $H_3C\cdot$ and $\cdot\ddot{C}l:$.
(c)   Each step is homolytic.   In (1) and (3), Cl_2 cleaves; in (2), CH_4 cleaves.
(d)   Step (3) involves the displacement of one $:\ddot{C}l\cdot$ of Cl_2 by a $\cdot CH_3$ group.   In step (2), $:\ddot{C}l\cdot$ displaces a $\cdot CH_3$ group from an H.
(e)   None.
(f) $H_3C\cdot + \cdot CH_3 \rightarrow H_3CCH_3$ (ethane)

Problem 3.2   Identify each of the following as (1) carbonium ions, (2) carbanions, (3) radicals or (4) carbenes:

(a) $(CH_3)_2C:$ (d) $(CH_3)_3C:^-$ (g) $C_6H_5\dot{C}HCH_3$
(b) $(CH_3)_3C\cdot$ (e) $CH_3CH_2\dot{C}H_2$ (h) $CH_3\ddot{C}H$
(c) $(CH_3)_3C^+$ (f) $CH_3CH{=}\underset{+}{CH}$ ◄

(1) (c), (f).   (2) (d).   (3) (b), (e), (g).   (4) (a), (h).

Problem 3.3 Write formulas for the species resulting from the (a) homolytic cleavage, (b) heterolytic cleavage of the C—C bond in ethane, C_2H_6, and classify these species. ◀

(a)
$$H_3C:CH_3 \longrightarrow H_3C\cdot + \cdot CH_3$$
Ethane   Methyl radicals

(b)
$$H_3C:CH_3 \longrightarrow H_3C^+ + {}^-CH_3$$
Ethane   Carbonium   Carbanion
ion

Problem 3.4 Classify the following as substitution, addition, elimination, rearrangement or redox reactions. (A reaction may have more than one designation.)

(a) $\quad CH_2{=}CH_2 + Br_2 \longrightarrow CH_2BrCH_2Br$

(b) $\quad C_2H_5OH + HCl \longrightarrow C_2H_5Cl + H_2O$

(c) $\quad CH_3CHClCHClCH_3 + Zn \longrightarrow CH_3CH{=}CHCH_3 + ZnCl_2$

(d) $\quad NH_4^+(CNO)^- \longrightarrow H_2NCNH_2$
$$\overset{\displaystyle \|}{O}$$

(e) $\quad CH_3CH_2CH_2CH_3 \longrightarrow (CH_3)_3CH$

(f)
$$\overset{\displaystyle CH_2}{\underset{}{H_2C{-\!\!-\!}CH_2}} + Br_2 \rightarrow BrCH_2CH_2CH_2Br$$

(g) $\quad 3CH_3CHO + 2MnO_4^- + OH^- \overset{\Delta}{\longrightarrow} 3CH_3COO^- + 2MnO_2 + 2H_2O \quad$ (Δ means heat.)

(h) $\quad HCCl_3 + OH^- \longrightarrow :CCl_2 + H_2O + Cl^-$ ◀

(a) Addition and redox. In this reaction the two Br's add to the two double bonded C atoms (1,2-addition). The oxidation number (ON) for C has changed from $4 - 2(2) - 2 = -2$ to $4 - 2(2) - 1 = -1$; (ON) for Br has changed from $7 - 7 = 0$ to $7 - 8 = -1$.

(b) Substitution of a Cl for an OH.

(c) Elimination and redox. Zn removes two Cl atoms from adjoining C atoms to form a double bond and $ZnCl_2$ (a β-elimination). The organic compound is reduced and Zn is oxidized.

(d) Rearrangement.

(e) Rearrangement (isomerization).

(f) Addition and redox. The Br's add to two C atoms of the ring. These C's are oxidized and the Br's are reduced.

(g) Redox. CH_3CHO is oxidized and MnO_4^- is reduced.

(h) Elimination. An H^+ and Cl^- were removed from the same carbon (α-elimination).

3.4 ELECTROPHILIC AND NUCLEOPHILIC REAGENTS

Reactions generally occur at the reactive sites of molecules and ions. These sites fall mainly into two categories. One category has a high electron density because the site (a) has an unshared pair of electrons or (b) is the $\delta-$ end of a polar bond or (c) has π electrons. Such electron-rich sites are **nucleophilic** and the species possessing such sites are called **nucleophiles** or **electron-donors**. The second category (a) is capable of acquiring more electrons or (b) is the $\delta+$ end of a polar bond. These electron-deficient sites are **electrophilic** and the species possessing such sites are called **electrophiles** or **electron-acceptors**. Many reactions occur by bond formation between a nucleophilic and an electrophilic site.

Problem 3.5 Classify the following species as being (1) nucleophiles or (2) electrophiles and give the reason for your classification: (a) $H\ddot{O}^-$, (b) $:C{\equiv}N^-$, (c) $:\ddot{B}r^+$, (d) BF_3, (e) $H_2\ddot{O}:$, (f) $AlCl_3$, (g) $:NH_3$, (h) H_3C^- (a carbanion), (i) SiF_4, (j) Ag^+, (k) H_3C^+ (a carbonium ion), (l) $H_2C:$ (a carbene), (m) $:\ddot{I}^-$. ◀

(1) (a), (b), (e), (g), (h), and (m). They all have unshared pairs of electrons. All anions are potential nucleophiles.

(2) (d) and (f) are molecules whose central atoms (B and Al) have only 6 electrons rather than the more desirable octet; they are electron-deficient. (c), (j) and (k) have positive charges and therefore are electron-deficient. Most cations are potential electrophiles. The Si in (i) can acquire more than 8 electrons by utilizing its d orbitals, e.g.

$$SiF_4 + 2 \colon\!\ddot{F}^{-} \longrightarrow SiF_6^{2-}$$

Although the C in (l) has an unshared pair of electrons, (l) is electrophilic because the C has only 6 electrons.

Problem 3.6 Why is the reaction $CH_3Br + OH^- \to CH_3OH + Br^-$ a nucleophilic displacement? ◄

The $\colon\!\ddot{O}H^-$ has unshared electrons and is a nucleophile. Because of the polar nature of the bond

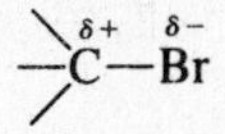

$$\overset{\delta+}{\underset{}{C}}\!-\!\overset{\delta-}{Br}$$

C acts as an electrophilic site. The displacement of Br^- by OH^- is initiated by the nucleophile $H\ddot{O}\colon^{-}$.

3.5 THERMODYNAMICS

The thermodynamics and the rate of a reaction determine whether the reaction proceeds. The thermodynamics of a system is described in terms of several important functions:

(1) ΔE, the change in **energy**, equals q_v, the heat transferred to or from a system at constant volume: $\Delta E = q_v$.

(2) ΔH, the change in **enthalpy**, equals q_p, the heat transferred to or from a system at constant pressure: $\Delta H = q_p$. Since most organic reactions are performed at atmospheric pressure in open vessels, ΔH is used more often than is ΔE. For reactions involving only liquids or solids: $\Delta E = \Delta H$. ΔH of a chemical reaction is the difference in the enthalpies of the products, H_P, and the reactants, H_R:

$$\Delta H = H_P - H_R$$

If the bonds in the products are more stable than the bonds in the reactants, ΔH is negative (reaction is **exothermic**).

(3) ΔS is the change in **entropy**. Entropy is a measure of randomness. The more the randomness, the greater is S; the greater the order, the smaller is S. For a reaction,

$$\Delta S = S_P - S_R$$

(4) ΔG is the change in **free energy**. At constant temperature,

$$\Delta G = \Delta H - T\,\Delta S \quad (T = \text{absolute temperature})$$

Problem 3.7 State whether the following reactions have a positive or negative ΔS and give a reason for your choice.

(a) $H_2 + H_2C\!\!=\!\!CH_2 \longrightarrow H_3CCH_3$

(b) $H_2C\overset{\displaystyle CH_2}{-\!\!\!-\!\!\!-}CH_2 \overset{\Delta}{\longrightarrow} H_3C\!-\!CH\!\!=\!\!CH_2$

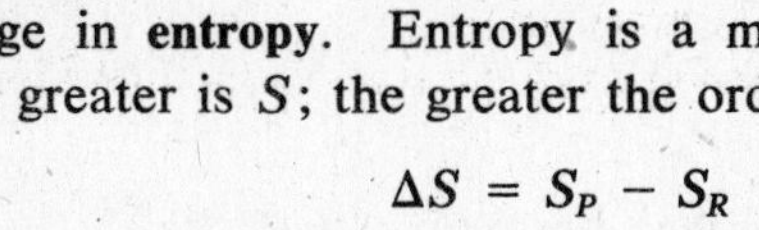

(c) $CH_3COO^-(aq) + H_3O^+(aq) \longrightarrow CH_3COOH + H_2O$ ◄

(a) Negative. Two molecules are changing into one molecule and there is more order (less randomness) in the product ($S_P < S_R$).

(b) Positive. The rigid ring opens to give an acyclic compound that now has free rotation about the C—C single bond ($S_R < S_P$).

(c) Positive. The ions are solvated by more H_2O molecules than is CH_3COOH. When ions form molecules, many of these H_2O molecules are set free and therefore have more randomness ($S_P > S_R$).

Problem 3.8 Predict the most stable state of H_2O (steam, liquid or ice) in terms of   (a) enthalpy, (b) entropy, and (c) free energy.   ◀

(a) Gas → Liquid → Solid   are exothermic processes and, therefore, ice has the least enthalpy. For this reason, ice should be most stable.

(b) Solid → Liquid → Gas   shows increasing randomness and therefore increasing entropy. For this reason, steam should be most stable.

(c) Here the trends to lowest enthalpy and highest entropy are in opposition and neither can be used independently to predict the favored state. Only G, which gives the balance between H and S, can be used. The state with lowest G or the reaction with the most negative ΔG is favored. For H_2O, this is the liquid state, a fact which cannot be predicted until a calculation is made using the equation $G = H - TS$.

3.6 BOND DISSOCIATION ENERGIES

The **bond dissociation energy** (given as ΔH in kcal/mole or kJ/mol) is the energy needed for the endothermic homolysis of a covalent bond $A{:}B \to A{\cdot} + {\cdot}B$; ΔH is positive. Bond *formation*, the reverse of this reaction, is exothermic and the ΔH values are negative. The ΔH of reaction is the sum of all the (positive) ΔH values for bond cleavages *plus* the sum of all the (negative) ΔH values for bond formations.

Problem 3.9 Calculate ΔH for the reaction $CH_4 + Cl_2 \to CH_3Cl + HCl$. The bond dissociation energies in kcal/mole are 102 for C—H, 58 for Cl—Cl, 81 for C—Cl, and 103 for H—Cl.   ◀

The values are shown under the bonds involved:

$$H_3C{-}H + Cl{-}Cl \longrightarrow H_3C{-}Cl + H{-}Cl$$

$$\underbrace{102 \ + \ 58}_{\text{cleavages (endothermic)}} \ + \ \underbrace{(-81) \ + (-103)}_{\text{formations (exothermic)}} = -24$$

The reaction is exothermic, with $\Delta H = -24$ kcal/mole.

3.7 CHEMICAL EQUILIBRIUM

Every chemical reaction can proceed in either direction, $dA + eB \rightleftarrows fX + gY$, even if it goes in one direction to a microscopic extent. A **state of equilibrium** is reached when *the concentrations of A, B, X and Y no longer change* even though the reverse and forward reactions are taking place.

Every reversible reaction has an equilibrium expression in which K_e, the **equilibrium constant**, is defined in terms of molar concentrations (mole/liter) as indicated by the square brackets:

$$K_e = \frac{[X]^f[Y]^g}{[A]^d[B]^e} \qquad \begin{array}{l} \text{Products favored; } K_e \text{ is large} \\ \text{Reactants favored; } K_e \text{ is small} \end{array}$$

K_e varies only with temperature.

The ΔG of a reaction is related to K_e by

$$\Delta G = -2.303\, RT \log K_e$$

where R is the molar gas constant (1.987 cal $\text{mole}^{-1}\,°K^{-1}$ = 8.314 J $\text{mol}^{-1}\,K^{-1}$) and T is the absolute temperature.

Problem 3.10   Given the reversible reaction

$$C_2H_5OH + CH_3COOH \rightleftharpoons CH_3COOC_2H_5 + H_2O$$

what changes could you make to increase the yield of $CH_3COOC_2H_5$?   ◄

The equilibrium must be shifted to the right, the side of the equilibrium where $CH_3COOC_2H_5$ exists.   This is achieved by any combination of the following: adding C_2H_5OH, adding CH_3COOH, removing H_2O, removing $CH_3COOC_2H_5$.

Problem 3.11   Summarize the relationships between the signs of ΔH, $T\Delta S$ and ΔG, and the magnitude of K_e, and state whether a reaction proceeds to the right or to the left for the reaction equation as written.   ◄

See Table 3-1.

Table 3-1

ΔH	$-$ $T\Delta S$	$=$ ΔG	Reaction Direction	K_e
$-$	$+$	$-$	Forward → right	>1
$+$	$-$	$+$	Reverse → left	<1
$-$	$-$	Usually $-$ if $\Delta H < -15$ kcal/mole	Depends on conditions	?
$+$	$+$	Usually $+$ if $\Delta H > +15$ kcal/mole	Depends on conditions	?

Problem 3.12   Given

$$CH_3CH_2CH_2CH_3 \underset{}{\overset{AlCl_3}{\rightleftharpoons}} CH_3\underset{\underset{CH_3}{|}}{C}HCH_3 \qquad \Delta H = -2000 \text{ cal/mole}$$
$$\qquad \Delta S = -3.69 \text{ cal/mole-}°K$$

n-Butane   Isobutane

Account for the observation that at 25 °C isobutane is the more stable isomer, but at 269 °C both isomers are present.   Assume one starts with 1 mole/liter of n-butane.   ◄

The change in

$$K_e = \frac{[\text{isobutane}]}{[n\text{-butane}]}$$

with temperature reflects a change in the free energy of the system.   Calculate ΔG at 25 °C and 269 °C from $\Delta G = \Delta H - T\Delta S$.

$$\Delta G_{25} = -2000\frac{\text{cal}}{\text{mole}} - (298 \text{ °K})\left(-3.69 \frac{\text{cal}}{\text{mole-°K}}\right) = -900 \frac{\text{cal}}{\text{mole}}$$

$$\Delta G_{269} = -2000\frac{\text{cal}}{\text{mole}} - (542 \text{ °K})\left(-3.69 \frac{\text{cal}}{\text{mole-°K}}\right) = \quad 0 \frac{\text{cal}}{\text{mole}}$$

At 25 °C the relation $\Delta G = -2.303 \, RT \log K_e$ gives

$$-900 = -2.303(1.987)(298) \log K_e$$

whence $K_e = 4.57$, and

$$\frac{x}{1-x} = 4.57$$

where $x = $ [isobutane], $1 - x = $ [n-butane].   Solving, one finds $x = 0.82$, i.e. 82% of the mixture is isobutane.   At 269 °C, $\Delta G = 0$ and therefore $\log K_e = 0$ and $K_e = 1$.   There is a 1-to-1 mixture of isomers.   At temperatures higher than 269 °C n-butane is more stable because ΔG is positive.

Problem 3.13 At 25 °C when 1 mole each of ethanol and acetic acid are reacted, 0.667 mole of ethyl acetate is present at equilibrium.

$$CH_3CH_2OH + CH_3COOH \rightleftharpoons CH_3COOCH_2CH_3 + H_2O$$

Ethanol   Acetic   Ethyl   Water
acid   acetate

Calculate K_e. ◀

$$K_e = \frac{[CH_3COOC_2H_5][H_2O]}{[CH_3CH_2OH][CH_3COOH]} = \frac{(0.667)(0.667)}{(1 - 0.667)(1 - 0.667)} = 4.0$$

Problem 3.14 Calculate ΔG at 25 °C for the reaction in Problem 3.13. ◀

$$\Delta G = -2.303 \, RT \log K_e$$

$$= -2.303 \left(1.987 \, \frac{cal}{mole\text{-}°K}\right) (298 \, °K) \log 4.0 = -821 \, \frac{cal}{mole}$$

The negative sign of ΔG indicates that the products ethyl acetate and water are favored.

Problem 3.15 At 25 °C the formation of a cyclic ester (lactone),

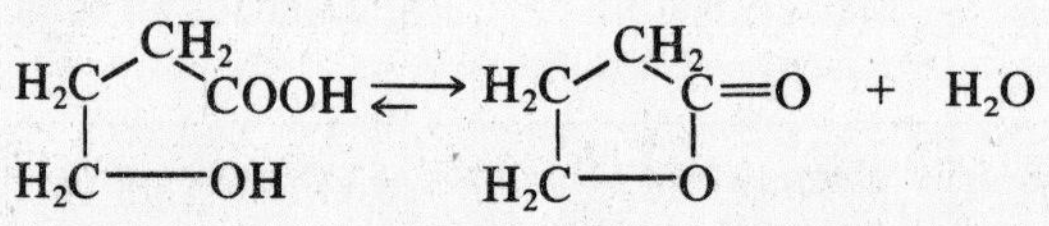

$$\begin{array}{c} H_2C{-}CH_2 \\ | \quad \backslash COOH \\ H_2C{-}OH \end{array} \rightleftharpoons \begin{array}{c} H_2C{-}CH_2 \\ | \quad \backslash C{=}O \\ H_2C{-}O \end{array} + H_2O$$

has a K_e of about 1000. Since the changes in chemical bonding in this reaction and those in Problem 3.13 are similar, both reactions have about the same ΔH. Use thermodynamic functions to explain why this reaction has a much greater K_e value than the one in Problem 3.13. ◀

A larger K_e means a more negative ΔG. Since ΔH is about the same for both reactions, a more negative ΔG means that ΔS for this reaction is more positive. A more positive ΔS (greater randomness) is expected in this reaction because one molecule is converted into two molecules, whereas in the reaction in Problem 3.13 two molecules are changed into two other molecules. When two reacting sites, such as OH and COOH, are in the same molecule, the reaction is **intramolecular**. When reaction sites are in different molecules, as in Problem 3.13, the reaction is **intermolecular**. Intramolecular reactions often have a more positive ΔS than similar intermolecular reactions.

3.8 RATES OF REACTIONS

The rate of the reaction

$$dA + eB \longrightarrow fC + gD$$

is given by

$$\text{rate} = k[A]^x[B]^y$$

where k is the **rate constant** at temperature T. The numerical values of the exponents x and y are determined experimentally; they need not be the same as d and e, the coefficients of the chemical reaction. The sum of the values of the exponents is defined as the **order** of the reaction.

Experimental conditions, other than concentrations, that affect rates of reactions are:

Temperature. A rough rule is that the value of k doubles for every rise in temperature of 10 °C.
Particle size. Increasing the surface area of solids by pulverization increases the reaction rate.
Catalysts and inhibitors. A catalyst is a substance that increases the rate of a reaction but is recovered unchanged at the end of the reaction. Inhibitors decrease the rate.

At a given set of conditions, the factors that determine the rate of a given reaction are:

1. **Number of collisions per unit time.**   The greater the chances for molecular collision, the faster the reaction.   Probability of collision is related to the number of molecules of each reactant and is proportional to the molar concentrations.

2. **Enthalpy of activation (activation energy, E_{act}) ($\Delta H^{\ddagger}$).**   Reaction may take place only when colliding molecules have some enthalpy content, $\Delta H^{\ddagger}$, in excess of the average.   *The smaller the value of $\Delta H^{\ddagger}$, the more successful will be the collisions and the faster the reaction.* ($\Delta H^{\ddagger} = E_{act}$ at constant pressure.)

3. **Entropy of activation ($\Delta S^{\ddagger}$).**   Not all collisions between molecules possessing the requisite $\Delta H^{\ddagger}$ result in reaction.   Often collisions between molecules must also occur *in a certain orientation*, reflected by the value of $\Delta S^{\ddagger}$.   *The more organized or less random the required orientation of the colliding molecules, the lower the entropy of activation and the slower the reaction.*

Problem 3.16   Predict the effect on the rate of a reaction if a change in the solvent causes: (*a*) an increase in $\Delta H^{\ddagger}$ and a decrease in $\Delta S^{\ddagger}$, (*b*) a decrease in $\Delta H^{\ddagger}$ and an increase in $\Delta S^{\ddagger}$, (*c*) an increase in $\Delta H^{\ddagger}$ and in $\Delta S^{\ddagger}$, (*d*) a decrease in $\Delta H^{\ddagger}$ and in $\Delta S^{\ddagger}$.   ◄

(*a*) Decrease in rate.   (*b*) Increase in rate.   (*c*) The change in $\Delta H^{\ddagger}$ tends to decrease the rate but the change in $\Delta S^{\ddagger}$ tends to increase the rate.   The combined effect is unpredictable.   (*d*) The trends here are opposite to those in part (*c*); the effect is also unpredictable.   In many cases the change in $\Delta H^{\ddagger}$ is more important than the change in $\Delta S^{\ddagger}$ in affecting the rate of reaction.

3.9   TRANSITION-STATE THEORY AND ENTHALPY DIAGRAMS

When reactants have collided with sufficient enthalpy of activation and with the proper orientation, they pass through a hypothetical **transition state** in which some bonds are breaking while others may also be forming.

The relationship of the transition state (TS) to the reactants (R) and products (P) is shown by the enthalpy (energy) diagram, Fig. 3-1, for a one-step exothermic reaction $A + B \rightarrow C + D$.   At equilibrium, formation of molecules with lower enthalpy is favored, i.e., $C + D$.   However, this applies only if ΔH of reaction predominates over $T \Delta S$ of reaction in determining the equilibrium state.

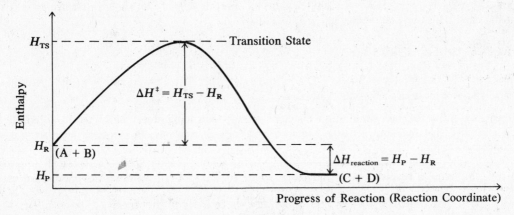

Fig. 3-1

In multistep reactions, each step has its own transition state.   The step with the highest-enthalpy transition state is the slowest step and determines the overall reaction rate.

The number of species colliding in the rate-determining step is called the **molecularity**.   If only one species breaks down, the reaction is **unimolecular**.   If two species collide and react, the reaction is **bimolecular**.   Rarely do three species collide (**termolecular**) at the same instant.

The rate equation gives the species and number of the reactant molecules involved in the slow step and in any preceding fast steps.

Problem 3.17 Draw an enthalpy diagram for a one-step endothermic reaction. Indicate ΔH of reaction and $\Delta H^\ddagger$.   ◀

See Fig. 3-2.

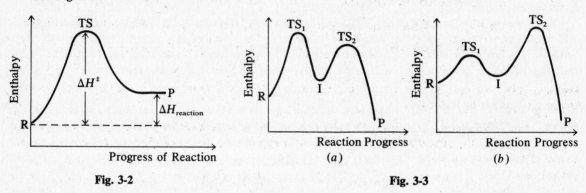

Fig. 3-2   Fig. 3-3

Problem 3.18 Draw a reaction-enthalpy diagram for an exothermic two-step reaction in which (a) the first step is slow, (b) the second step is slow.   ◀

See Fig. 3-3, in which R = reactants, I = intermediates, P = products, TS_1 = transition state of first step, TS_2 = transition state of second step. Because the reactions are exothermic, $H_P < H_R$.

Problem 3.19 In Problem 3.18(b), the first step is not only fast but is also reversible. Explain.   ◀

The $\Delta H^\ddagger$ for I to revert to reactants R is less than that for I to form the products P. Therefore, most I's re-form R, so that the first step is fast and reversible. A few I's have enough enthalpy to go through the higher-enthalpy TS_2 and form the products. The $\Delta H^\ddagger$ for P to revert to I is prohibitively high; hence the products accumulate, and the second step is at best insignificantly reversible.

Problem 3.20 Catalysts generally speed up reactions by lowering $\Delta H^\ddagger$. Explain how this occurs in terms of ground-state and transition-state enthalpies (H_R and H_{TS}).   ◀

$\Delta H^\ddagger$ can be decreased by (a) lowering H_{TS}, (b) raising H_R or (c) both of these.

Problem 3.21 The reaction A + B → C + D has (a) rate = $k[A][B]$, or (b) rate = $k[A]$. Offer possible mechanisms consistent with these rate expressions.   ◀

(a) Molecules A and B must collide in a bimolecular rate-controlling step. Since the balanced chemical equation calls for reaction of one A molecule with one B molecule, the reaction must have a single (concerted) step.

(b) The rate-determining step is unimolecular and involves only an A molecule. There can be no prior fast steps. Molecule B reacts in the second step, which is fast. A possible two-step mechanism is:

$$\text{Step 1:}\quad \text{A} \longrightarrow \text{C} + \text{I}\quad (\text{I = intermediate})$$
$$\text{Step 2:}\quad \text{B} + \text{I} \longrightarrow \text{D}$$

Adding the two steps gives the balanced chemical equation: A + B → C + D.

Problem 3.22 For the reaction 2A + 2B → C + D, rate = $k[A]^2[B]$. Give a mechanism using only unimolecular or bimolecular steps.   ◀

One B molecule and two A molecules are needed to give the species for the slow step. The three molecules do not collide simultaneously since we are disregarding the very rare termolecular steps. There must then be some number of prior fast steps to furnish at least one intermediate needed for the slow step. The second B molecule which appears in the reaction equation must be consumed in a fast step following the slow step.

Mechanism 1   **Mechanism 2**

$$A + B \xrightarrow{\text{fast}} AB \quad (\text{intermediate}) \qquad A + A \xrightarrow{\text{fast}} A_2 \quad (\text{intermediate})$$

$$AB + A \xrightarrow{\text{slow}} ABA \quad (\text{intermediate}) \qquad A_2 + B \xrightarrow{\text{slow}} A_2B \quad (\text{intermediate})$$

$$ABA + B \xrightarrow{\text{fast}} C + D \qquad\qquad A_2B + B \xrightarrow{\text{fast}} C + D$$

Problem 3.23 For the reaction $A + 2B \rightarrow C + D$, rate = $k[A][B]^2$. Offer a mechanism in which the rate-determining step is unimolecular. ◄

The slow step needs an intermediate formed from one A molecule and two B molecules. Since the rate expression involves the same kinds and numbers of molecules as does the chemical equation, there are no fast steps following the slow step.

Mechanism 1   **Mechanism 2**

$$A + B \xrightarrow{\text{fast}} AB \qquad\qquad B + B \xrightarrow{\text{fast}} B_2$$

$$AB + B \xrightarrow{\text{fast}} AB_2 \qquad\qquad B_2 + A \xrightarrow{\text{fast}} B_2A$$

$$AB_2 \xrightarrow{\text{slow}} C + D \qquad\qquad B_2A \xrightarrow{\text{slow}} C + D$$

Notice that often the rate expression is insufficient to allow the suggestion of an unequivocal mechanism. More experimental information is often needed.

3.10 ACIDS AND BASES

BRÖNSTED DEFINITION

An *acid donates a proton* and a *base accepts a proton*. The strengths of acids and bases are measured by the extent to which they lose or gain protons, respectively. In these reactions acids are converted to their **conjugate bases** and bases to their **conjugate acids**. Acid-base reactions go in the direction of *forming the weaker acid and the weaker base*.

Problem 3.24 Show the conjugate acids and bases in the reaction of H_2O with gaseous (a) HCl, (b) :NH$_3$. ◄

(a) H_2O, a Brönsted base, accepts a proton from HCl, the Brönsted acid. They are converted to the conjugate acid H_3O^+ and the conjugate base Cl^-, respectively.

$$HCl + H_2O \rightleftharpoons H_3O^+ + Cl^-$$

Acid$_1$ Base$_2$ Acid$_2$ Base$_1$
(stronger) (stronger) (weaker) (weaker)

The conjugate acid-base pairs have the same subscript and are bracketed together. This reaction goes almost to completion because HCl is a good proton donor and hence a strong acid.

(b) H_2O is **amphoteric** and can also act as an acid by donating a proton to :NH$_3$. H_2O is converted to its conjugate base, OH^-, and :NH$_3$ to its conjugate acid, NH_4^+.

$$H{:}OH + NH_3 \rightleftharpoons NH_4^+ + {:}OH^-$$

Acid$_1$ Base$_2$ Acid$_2$ Base$_1$

:NH$_3$ is a poor proton acceptor (a weak base); the arrows are written to show that the equilibrium lies mainly to the left.

The basicity of a species depends on the reactivity of the unshared pair of electrons in accepting the proton. The more spread out (dispersed, delocalized) is the electron density arising from the presence of the unshared pair of electrons, the less basic is the species. Charge and electron density can be delocalized by extended π bonding (resonance). NO_3^- is a weaker base than NO_2^- because the $-$ charge of NO_3^- is delocalized to three O's rather than two O's (see Problem 2.22). Charge can also be dispersed by the **inductive effect**, whereby an electronegative atom transmits its electron-withdrawing effect through a chain of σ bonds. The effect decreases with increasing chain length.

Problem 3.25 Compare and account for the acidity of the underlined H in:

(a)
$$R-O-\underline{H} \quad \text{and} \quad R-\overset{\|}{\underset{O}{C}}-O-\underline{H}$$

(b)

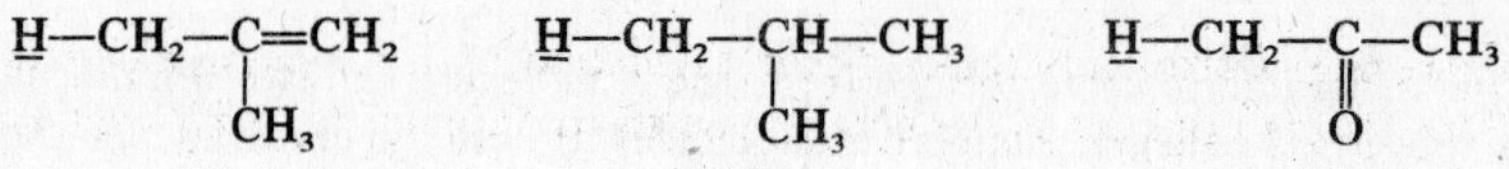

$$\underset{(I)}{\underline{H}-CH_2-\overset{CH_3}{\underset{|}{C}}=CH_2} \qquad \underset{(II)}{\underline{H}-CH_2-\overset{CH_3}{\underset{|}{CH}}-CH_3} \qquad \underset{(III)}{\underline{H}-CH_2-\overset{\|}{\underset{O}{C}}-CH_3} \qquad \blacktriangleleft$$

Compare the stability of the conjugate bases in each case.

(a) In

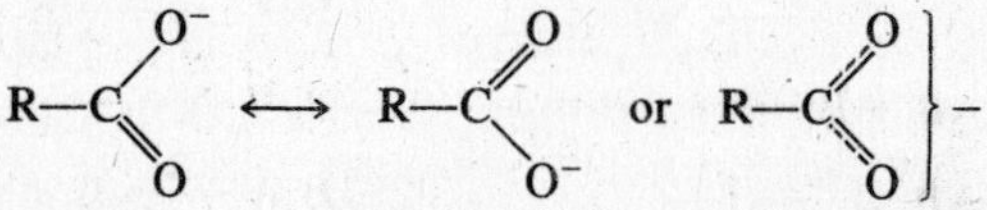

$$R-C{\overset{O^-}{\underset{O}{}}} \longleftrightarrow R-C{\overset{O}{\underset{O^-}{}}} \quad \text{or} \quad \left. R-C{\overset{O}{\underset{O}{}}} \right\}^-$$

the C and O participate in extended π bonding so that the $-$ is distributed to each O. In RO^- the $-$ is localized on the O. Hence $RCOO^-$ is a weaker base than RO^-, and $RCOOH$ is a stronger acid than ROH.

(b) The stability of the carbanions and relative acidity of these compounds is

$$(III) > (I) > (II)$$

Both (I) and (III) have a double bond not present in (II) that permits delocalization by extended π bonding. Delocalization is more effective in (III) because charge is delocalized to the electronegative O.

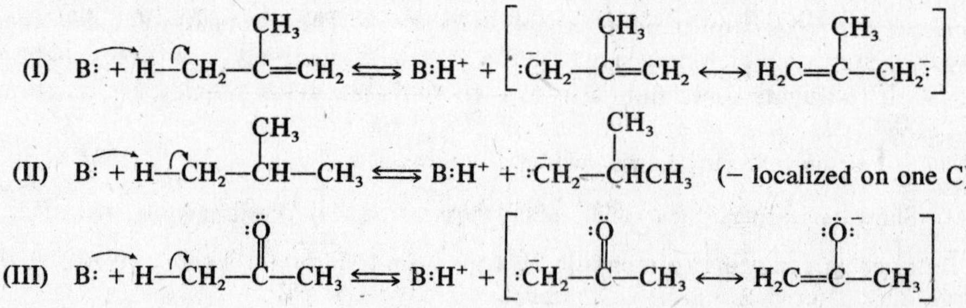

$$(I) \quad B:^- + H-CH_2-\overset{CH_3}{\underset{|}{C}}=CH_2 \rightleftharpoons B:H^+ + \left[:\bar{C}H_2-\overset{CH_3}{\underset{|}{C}}=CH_2 \longleftrightarrow H_2C=\overset{CH_3}{\underset{|}{C}}-CH_2:^- \right]$$

$$(II) \quad B:^- + H-CH_2-\overset{CH_3}{\underset{|}{CH}}-CH_3 \rightleftharpoons B:H^+ + :\bar{C}H_2-\overset{CH_3}{\underset{|}{CHCH_3}} \quad (- \text{ localized on one C})$$

$$(III) \quad B:^- + H-CH_2-\overset{:\ddot{O}:}{\overset{\|}{C}}-CH_3 \rightleftharpoons B:H^+ + \left[:\bar{C}H_2-\overset{:\ddot{O}:}{\overset{\|}{C}}-CH_3 \longleftrightarrow H_2C=\overset{:\ddot{O}:^-}{\underset{|}{C}}-CH_3 \right]$$

Problem 3.26 In terms of delocalization explain why $HCCl_3$ is more acidic than HCF_3. $\blacktriangleleft$

$Cl_3C:^-$ is less basic than $F_3C:^-$. F can disperse charge only by an inductive effect [see Fig. 3-4(a)]. In addition to an inductive effect, Cl uses its empty $3d$ orbitals to disperse charge by p-d π bonding, see Fig. 3-4(b). F is a second-period element with no $2d$ orbitals.

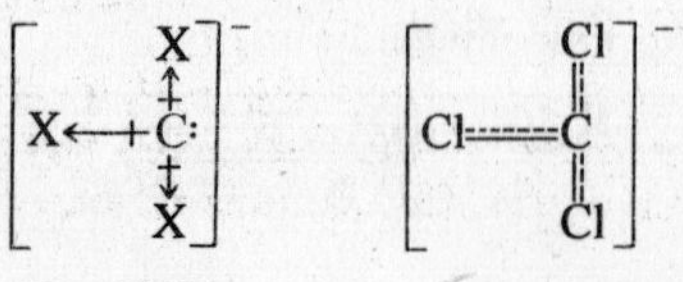

$$\left[\overset{X}{\underset{X}{X \leftarrow +C:}} \right]^- \qquad \left[\overset{Cl}{\underset{Cl}{Cl====C}} \right]^-$$

X = Cl, F

(a)　　　　　(b)

Fig. 3-4

Relative quantitative strengths of acids and bases are given either by their ionization constants, K_a and K_b, or by their pK_a and pK_b values as defined by:

$$pK_a = -\log K_a \qquad pK_b = -\log K_b$$

The stronger an acid or base, the larger its ionization constant and the smaller its pK value. The strengths of bases can be evaluated from those of their conjugate acids; the strengths of acids can be evaluated from those of their conjugate bases. The *strongest acids* have the *weakest conjugate bases*, and the *strongest bases* have the *weakest conjugate acids*. This follows from the relationships

$$K_w = (K_a)(K_b) = 10^{-14} \qquad pK_a + pK_b = pK_w = 14$$

in which K_w = ion-product of water = $[H_3O^+][OH^-]$.

LEWIS ACIDS AND BASES

A **Lewis acid (electrophile)** shares an electron pair furnished by a **Lewis base (nucleophile)** to form a covalent bond. The Lewis concept is especially useful to explain the acidity of an aprotic acid (no available proton) such as BF_3:

The three types of nucleophiles are listed in Section 3.4.

Problem 3.27 Explain the observation that neither pure H_2SO_4 nor pure $HClO_4$ conducts an electric current but a mixture of the two does. ◄

Neither pure acid is ionized. In the mixture the stronger acid, $HClO_4$, donates a proton to H_2SO_4, which acts as a base because of the unshared electron pairs on the O's.

$$H{:}\ddot{O}{:}ClO_3 + H{:}\ddot{O}{:}SO_3H \rightleftharpoons \overset{H^+}{H{:}\ddot{O}{:}SO_3H} + {:}\ddot{O}{:}ClO_3^-$$

| Acid$_1$ | Base$_2$ | Acid$_2$ | Base$_1$ |

Problem 3.28 Given the following Lewis acid-base reactions:

Lewis base	+	Lewis acid	⟶	Product
$4H_3N{:}$	+	Cu^{2+}	⟶	$Cu(NH_3)_4^{2+}$
$2{:}\ddot{F}{:}^-$	+	SiF_4	⟶	SiF_6^{2-}
$H{:}\ddot{O}{:}^-$	+	${:}O{=}C{=}O{:}$	⟶	${}^-\ddot{O}{-}C{=}\ddot{O}{:}$ with OH
$H_2C{=}CH_2$	+	H^+ (from a Brönsted acid)	⟶	$H_2\overset{+}{C}{-}CH_2$ with H
$H_2C{=}\ddot{O}{:}$	+	BF_3	⟶	$H_2C{=}\overset{+}{O}{:}\bar{B}F_3$

(*a*) Group the bases as follows: (1) anions, (2) molecules with an unshared pair of electrons, (3) negative end of a π bond dipole, and (4) available π electrons. (*b*) Group the acids as follows: (1) cations, (2) species with electron-deficient atoms, (3) species with an atom capable of expanding an octet, and (4) positive end of a π bond dipole. ◄

(*a*) (1) OH^-, F^- (2) ${:}NH_3$, $H_2C{=}\ddot{O}$ (3) $\overset{\delta+}{H_2C}{=}\overset{\delta-}{\ddot{O}{:}}$ (4) $H_2C{=}CH_2$

(*b*) (1) Cu^{2+}, H^+ (2) BF_3 (3) SiF_4 (4) $\overset{\delta-}{:}\ddot{O}{=}\overset{\delta+}{C}{=}\overset{\delta-}{\ddot{O}{:}}$

Problem 3.29   Methyl mercaptan, CH_3SH, is a stronger acid than methanol, CH_3OH.   Explain by comparing the "dispersal of charge" of their conjugate bases.   ◄

Since CH_3SH is the stronger acid, its conjugate base, CH_3S^-, is weaker than CH_3O^-, the conjugate base of CH_3OH.   S and O are both in Group VI of the periodic table.   As S is larger, the − charge in CH_3S^- is dispersed over a larger space.   Hence CH_3S^- is the weaker base.   The same reasoning applies in general to bases of the type H_mX^- or $H_nY:$ with central atoms in the same group of the periodic table.

Problem 3.30 H_3C^- is a much stronger base than $:\ddot{F}:^-$.   Explain.   ◄

The electron density of $:\ddot{F}:^-$ is dispersed over its entire surface.   The electron density of H_3C^- is dispersed over only one-fourth of its surface; the remaining three-fourths is occupied by 3 H's.   Since $:\ddot{F}:^-$ has more dispersal of its electron density, it is the weaker base.

Supplementary Problems

Problem 3.31   Describe the hybrid orbitals of the underlined C in each of the following species.   Give the approximate bond angles.   Assume that σ bonds and unshared pairs of electrons require hybrid orbitals and the π bonds and single electrons require p orbitals.

(a) $H_2\underset{\text{↿⇂}}{\underline{C}}$ (singlet carbene)   (c) $H_2C{=}\overset{+}{\underline{C}}{-}H$

(b) $H_2\underset{\text{↿↿}}{\underline{C}}$ (triplet carbene)   (d) $H_2C{=}\overset{\cdot\cdot}{\underline{C}}{-}H$ ◄

(a)   Three sp^2 hybrid orbitals for two σ bonds and an unshared pair of electrons; an empty p orbital at right angles to the plane of σ bonds.   120° bond angles.

(b)   Two sp hybrid orbitals for the two σ bonds.   Each of the two remaining p orbitals has one electron.   180°.

(c)   Two sp hybrid orbitals, one for each σ bond.   One p orbital is needed for the π bond and the second p orbital is empty.   180°.

(d)   Three sp^2 hybrid orbitals, one for each σ bond and one for the unshared electron pair.   The remaining p orbital is used for forming the π bond.   120°.

Problem 3.32   Write the formula for the carbon intermediate indicated by ?, and label as to type.

(a) $H_3C{-}\ddot{N}{=}\ddot{N}{-}CH_3 \longrightarrow ? + :N{\equiv}N:$

(b) $(CH_3)_2Hg \longrightarrow ? + \cdot Hg\cdot$

(c) $H_2\ddot{C}{-}\overset{+}{N}{\equiv}N: \longrightarrow ? + N_2$

(d) $(CH_3)_3C\ddot{O}H + H^+ \longrightarrow ? + H_2\ddot{O}:$

(e) $H{-}C{\equiv}C{-}H + Na\cdot \longrightarrow ? + Na^+ + \tfrac{1}{2}H_2$

(f) $H_2C{=}CH_2 + D{-}Br \longrightarrow ? + Br^-$

(g) $H_2CI_2 + Zn: \longrightarrow ? + Zn^{2+} + 2I^-$

(h) $(CH_3)_3C{-}Cl + AlCl_3 \longrightarrow ? + AlCl_4^-$ ◄

(a) and (b): $H_3C\cdot$, a radical.   (c) and (g): $H_2C:$, a carbene.   (d) and (h): $(CH_3)_3C^+$, a carbonium ion.   (e): $H{-}C{\equiv}C^-$, a carbanion.   (f): $H_2\overset{+}{C}{-}CH_2{-}D$, a carbonium ion.

Problem 3.33 Classify the following reactions by type.

(a)
$$H_2C-CH_2-Br + OH^- \longrightarrow H_2C-CH_2 + H_2O + Br^-$$
with OH on first carbon of reactant and O bridging the two carbons in product

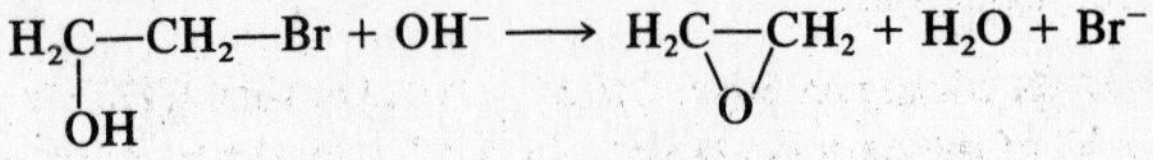

(b)
$$(CH_3)_2CHOH \xrightarrow{Cu,\ heat} (CH_3)_2C{=}O + H_2$$

(c)
$$H_2C{-}CH_2 \xrightarrow{heat} H_2C{=}CHCH_3$$
with CH$_2$ bridge

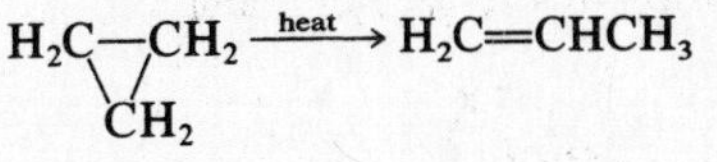

(d)
$$H_3C-CH_2Br + :H^- \longrightarrow H_3C-CH_3 + :Br^-$$

(e)
$$H_2C{=}CH_2 + H_2 \xrightarrow{Pt} H_3C-CH_3$$

(f)
$$C_6H_6 + HNO_3 \xrightarrow{H_2SO_4} C_6H_5NO_2 + H_2O$$

(g)
$$HCOOH \xrightarrow{heat} H_2O + CO$$

(h)
$$CH_2{=}C{=}O + H_2O \longrightarrow CH_3COOH$$

(i)
$$H_2C{=}O + 2Ag(NH_3)_2^+ + 3OH^- \longrightarrow HCOO^- + 2Ag + 4NH_3 + 2H_2O$$

(j)
$$H_2C{=}O + HCN \longrightarrow H_2C(OH)CN \qquad \blacktriangleleft$$

(a) Elimination and an intramolecular displacement; a C—O bond is formed in place of a C—Br bond. (b) Elimination and redox; the alcohol is oxidized to a ketone. (c) Rearrangement. (d) Displacement and redox; H_3CCH_2Br is reduced. (e) Addition and redox; $H_2C{=}CH_2$ is reduced. (f) Substitution. (g) Elimination. (h) Addition. (i) Redox. (j) Addition.

Problem 3.34 Which of the following species behave as (1) a nucleophile, (2) an electrophile, (3) both, or (4) neither?

(a) $:\ddot{C}l:^-$ (d) $AlBr_3$ (g) Br^+ (j) NO_2^+ (m) H_2

(b) $H_2\ddot{O}:$ (e) $CH_3\ddot{O}H$ (h) Cr^{3+} (k) $H_2C{=}\ddot{O}$ (n) CH_4

(c) H^+ (f) $BeCl_2$ (i) $SnCl_4$ (l) $CH_3C{\equiv}\ddot{N}$ (o) $H_2C{=}CHCH_3$ ◀

(1): (a), (b), (e), (o).   (2): (c), (d), (f), (g), (h), (i), (j).   (3): (k) and (l) (because carbon is electrophilic; oxygen and nitrogen are nucleophilic).   (4): (m), (n).

Problem 3.35 Formulate the following as a two-step reaction and label nucleophiles and electrophiles.

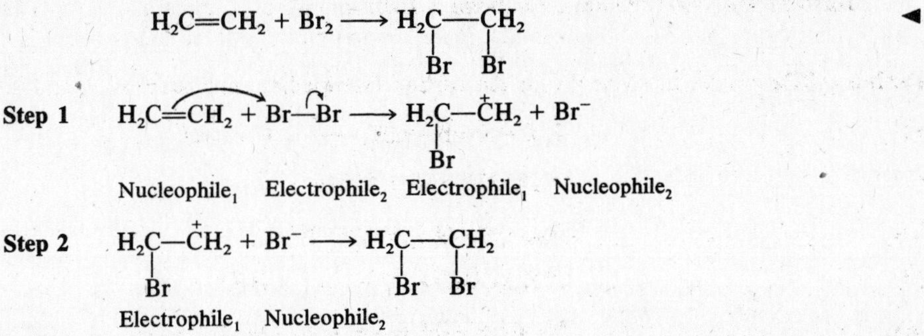

$$H_2C{=}CH_2 + Br_2 \longrightarrow H_2C-CH_2 \qquad \blacktriangleleft$$
with Br Br substituents

Step 1 $H_2C{=}CH_2 + Br{-}Br \longrightarrow H_2C-\overset{+}{C}H_2 + Br^-$
with Br substituent

Nucleophile$_1$ Electrophile$_2$ Electrophile$_1$ Nucleophile$_2$

Step 2 $H_2C-\overset{+}{C}H_2 + Br^- \longrightarrow H_2C-CH_2$
with Br substituent on left, Br Br on right

Electrophile$_1$ Nucleophile$_2$

Problem 3.36 Will the following reaction proceed at 300 °C if $\Delta S = 9.6$ cal/mole-°K and $\Delta H = +11.7$ kcal/mole? Why is ΔS positive?

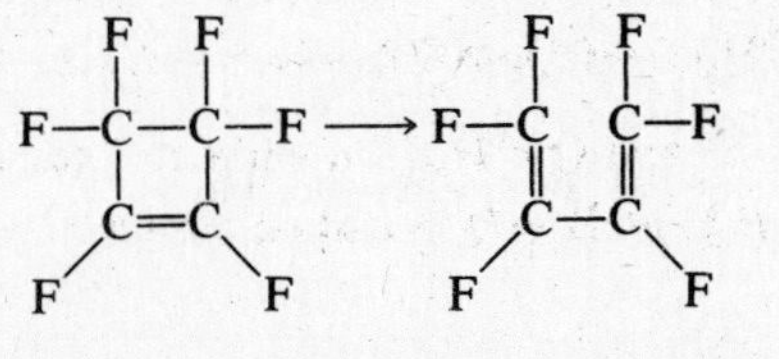

◀

No. Substituting into $\Delta G = \Delta H - T\,\Delta S$ gives

$$\Delta G = 11.7\ \frac{\text{kcal}}{\text{mole}} - (573\ ^\circ\text{K})\left(\frac{9.6\ \text{cal}}{\text{mole-}^\circ\text{K}}\ \frac{1\ \text{kcal}}{1000\ \text{cal}}\right) = 6.2\ \frac{\text{kcal}}{\text{mole}}$$

ΔS is positive because the rigid ring opens to form a more flexible chain.

Problem 3.37 The addition of 3 moles of H_2 to benzene, C_6H_6,

$$3H_2 + C_6H_6 \underset{300\ ^\circ\text{C}}{\overset{\text{Pd(rt)}}{\rightleftharpoons}} C_6H_{12}$$

occurs at room temperature (rt). The reverse elimination reaction proceeds at 300 °C. For the addition reaction, ΔH and ΔS are both negative. Explain in terms of thermodynamic functions: (a) why ΔS is negative and (b) why the addition doesn't proceed at room temperature without a catalyst. ◄

(a) A negative ΔH tends to make ΔG negative, but a negative ΔS tends to make ΔG positive. At room temperature, ΔH exceeds $T\,\Delta S$ and therefore ΔG is negative. At the higher temperature (300 °C), $T\,\Delta S$ exceeds ΔH, and ΔG is then positive. ΔS is negative because four molecules become one molecule, thereby reducing the randomness of the system.

(b) The addition has a very high $\Delta H^{\ddagger}$, and the rate without catalyst is extremely slow.

Problem 3.38 The reaction $CH_4 + I_2 \rightarrow CH_3I + HI$ does not occur as written because the equilibrium lies to the left. Explain in terms of the bond dissociation energies of 102, 36, 53, and 74 kcal/mole for C—H, I—I, C—I, and HI, respectively. ◄

The (endothermic) bond-breaking energies are $+102$ (C—H) and $+36$ (I—I), for a total of $+138$ kcal/mole. The (exothermic) bond-forming energies are -53 (C—I) and -71 (H—I), for a total of -124 kcal/mole. The net ΔH is

$$+138 - 124 = +14\ \text{kcal/mole}$$

and the reaction is endothermic. The reactants and products have similar structures and the ΔS term is insignificant. Reaction does not occur because ΔH and ΔG are positive.

Problem 3.39 Which of the isomers ethyl alcohol, H_3CCH_2OH, or dimethyl ether, H_3COCH_3, has the lower enthalpy? The bond dissociation energies (in kcal/mole) are 85, 99, 86 and 111 for C—C, C—H, C—O and O—H, respectively.

C_2H_5OH has 1 C—C bond (85 kcal/mole), 5 C—H bonds (5×99 kcal/mole), one C—O bond (86 kcal/mole) and one O—H bond (111 kcal/mole), giving a total energy of 777 kcal/mole. CH_3OCH_3 has 6 C—H bonds (6×99 kcal/mole) and 2 C—O bonds (2×86 kcal/mole), and the total bond energy is 766 kcal/mole. C_2H_5OH has the higher total bond energy and therefore the lower enthalpy of formation. More energy is needed to decompose C_2H_5OH into its elements.

Problem 3.40 Consider the following sequence of steps:

$$(1)\quad A \longrightarrow B \qquad (2)\quad B + C \xrightarrow{\text{slow}} D + E \qquad (3)\quad E + A \longrightarrow 2F$$

(a) Which species may be described as (i) reactant, (ii) product and (iii) intermediate? (b) Write the net chemical equation. (c) Indicate the molecularity of each step. (d) If the second step is rate-determining, write the rate expression. (e) Draw a plausible reaction-enthalpy diagram. ◄

(a) (i) A, C; (ii) D, F; (iii) B, E.

(b) $2A + C \longrightarrow D + 2F$ (add steps 1, 2 and 3).

(c) (1) unimolecular, (2) bimolecular, (3) bimolecular.

(d) Rate $= k[C][A]$, since A is needed to make the intermediate, B.

(e) See Fig. 3-5.

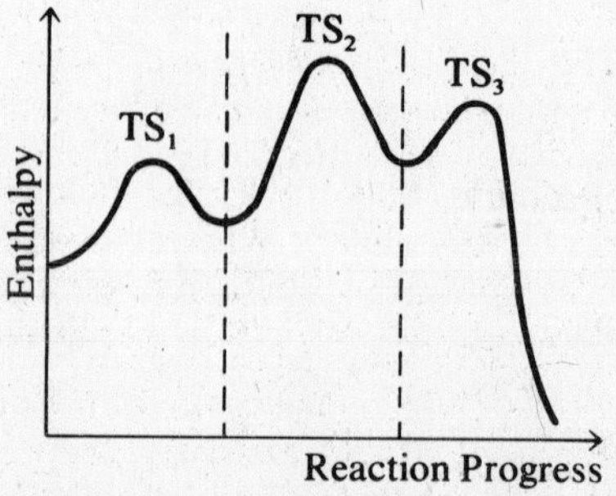

Fig. 3-5

Problem 3.41 A minor step in Problem 3.40 is $2E \rightarrow G$. What is G?   ◄

A side product.

Problem 3.42 The rate expression for the reaction

$$(CH_3)_3C—Br + CH_3COO^- + Ag^+ \longrightarrow CH_3COOC(CH_3)_3 + AgBr$$

is $$\text{rate} = k[(CH_3)_3C—Br][Ag^+]$$

Suggest a plausible two-step mechanism showing the reacting electrophiles and nucleophiles.   ◄

The rate-determining step involves only $(CH_3)_3C—Br$ and Ag^+. The acetate ion CH_3COO^- must participate in an ensuing fast step.

Step 1 $(CH_3)_3\overset{\delta+}{C}—\overset{\delta-}{Br}\!: + Ag^+ \xoverset{\text{slow}}{\longrightarrow} (CH_3)_3C^+ + AgBr$
                    $\underbrace{\qquad\qquad\qquad}_{\text{Nucleophilic}}$ Electrophile
                    site

Step 2 $CH_3COO^- + \overset{+}{C}(CH_3)_3 \xoverset{\text{fast}}{\longrightarrow} CH_3COOC(CH_3)_3$
                Nucleophile   Electrophile

Problem 3.43 Give the conjugate acid of (a) H_2O, (b) Cl^-, (c) CH_3NH_2, (d) CH_3O^-, (e) HNO_3,
(f) CH_3OH, (g) $:H^-$, (h) $:CH_3^-$, (i) $H_2C{=}CH_2$.   ◄

(a) H_3O^+, (b) HCl, (c) $CH_3NH_3^+$, (d) CH_3OH, (e) $H_2NO_3^+$, (f) $CH_3OH_2^+$, (g) H_2, (h) CH_4,
(i) $H_3CCH_2^+$.

Problem 3.44 What are the conjugate bases, if any, for the substances in Problem 3.43?   ◄

(a) OH^-, (b) none, (c) $CH_3\overset{..}{N}H^-$, (d) $:CH_2O^{2-}$, (e) NO_3^-, (f) CH_3O^-, (g) none, (h) $H_2\overset{..}{C}:^{2-}$,
(i) $H_2C{=}\overset{..}{C}H^-$. The bases in (d) and (h) are extremely difficult to form; from a *practical* point of view CH_3O^- and H_3C^- have no conjugate bases.

Problem 3.45 Are any of the following substances amphoteric? (a) H_2O, (b) NH_3, (c) NH_4^+, (d) Cl^-,
(e) HCO_3^-, (f) HF.   ◄

(a) Yes, gives H_3O^+ and OH^-. (b) Yes, gives NH_4^+ and H_2N^-. (c) No, cannot accept H^+. (d) No, cannot donate H^+. (e) Yes, gives H_2CO_3 $(CO_2 + H_2O)$ and CO_3^{2-}. (f) Yes, gives H_2F^+ and F^-.

Problem 3.46 Account for the fact that acetic acid, CH_3COOH, is a stronger acid in water than in methanol, CH_3OH.   ◄

The equilibrium

$$CH_3COOH + H_2O \rightleftharpoons CH_3COO^- + H_3O^+$$

lies more to the right than does

$$CH_3COOH + CH_3OH \rightleftharpoons CH_3COO^- + CH_3OH_2^+$$

This difference could result if CH_3OH were a weaker base than H_2O. However, this might not be the case. The significant difference arises from solvation of the ions. Water solvates ions better than does methanol; thus the equilibrium is shifted more toward the right to form ions that are solvated by water.

Problem 3.47 Which is the stronger base in each of the following pairs? Explain your choice. (a) NH_3 and PH_3, (b) Cl^- and Br^-, (c) NH_2^- and OH^-, (d) HS^- and F^-.   ◄

(a) NH_3; N is smaller than P and therefore has less dispersal of electron density. (b) Cl^- for the same reason as in (a). (c) NH_2^-; the electron density is dispersed over half of the surface, whereas in OH^- the electron density is dispersed over three-fourths of the surface. (d) Cannot be compared because S and F are in different groups and different periods of the periodic table.

Chapter 4

Alkanes

4.1 DEFINITION

Alkanes are open-chain (acyclic) hydrocarbons comprising the homologous series with the general formula C_nH_{2n+2} where n is an integer. They have only single bonds and therefore are said to be **saturated**.

Problem 4.1 (a) Use the superscripts $1, 2, 3$, etc., to indicate the different kinds of equivalent H atoms in propane, $CH_3CH_2CH_3$. (b) Replace one of each kind of H by a CH_3 group. (c) How many isomers of butane, C_4H_{10}, exist? ◄

(a)
$$CH_3^1CH_2^2CH_3^1 \equiv \begin{array}{ccc} H^1 & H^2 & H^1 \\ H^1C\!-\!C\!-\!C\,H^1 \\ H^1 & H^2 & H^1 \end{array}$$

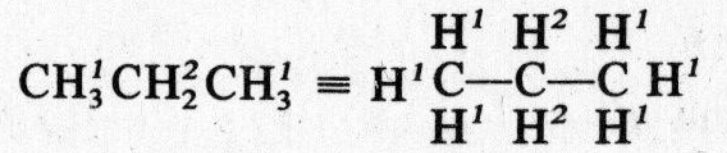

(b)
$$CH_3^1CH_2^2CH_3^1 \underset{+\boxed{CH_3}}{\overset{-H^1}{=\!=}} CH_3CH_2CH_2\,\boxed{CH_3}$$

$$CH_3^1CH_2^2CH_3^1 \underset{+\boxed{CH_3}}{\overset{-H^2}{=\!=}} CH_3\overset{\boxed{CH_3}}{\underset{}{C}}HCH_3$$

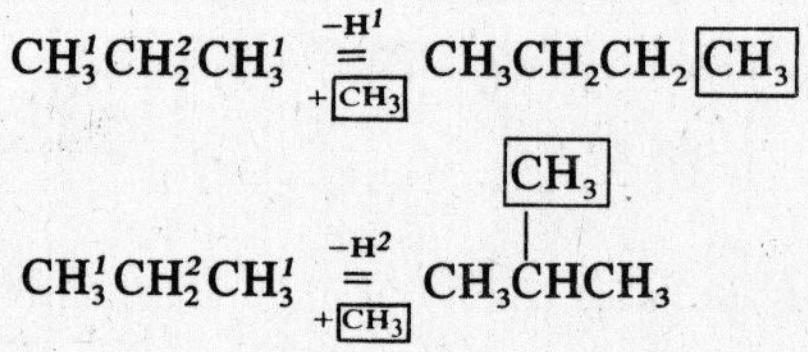

n-butane and isobutane.

Problem 4.2 (a) Use the superscripts $1, 2, 3, 4$, etc., to indicate the different kinds of equivalent H's in (1) n-butane and (2) isobutane. (b) Replace one of each kind of H in the two butanes by a CH_3. (c) Give the number of isomers of pentane, C_5H_{12}. ◄

(a) (1) $CH_3^1CH_2^2CH_2^2CH_3^1$ or $\begin{array}{cccc} {}_1H^1 & H^2 & H^2 & H^1 \\ HC\!-\!C\!-\!C\!-\!CH^1 \\ H^1 & H^2 & H^2 & H^1 \end{array}$ (2) $CH_3^3\overset{4}{C}HCH_3^3$ or $CH(CH_3)_3$ or $\begin{array}{ccc} & {}^3CH_3{}^3 & \\ {}_3H & | & H_3 \\ HC\!-\!C\!-\!CH \\ H^3 & H^4 & H^3 \end{array}$
$$\qquad\qquad\qquad \underset{CH_3^3}{\overset{}{|}}$$

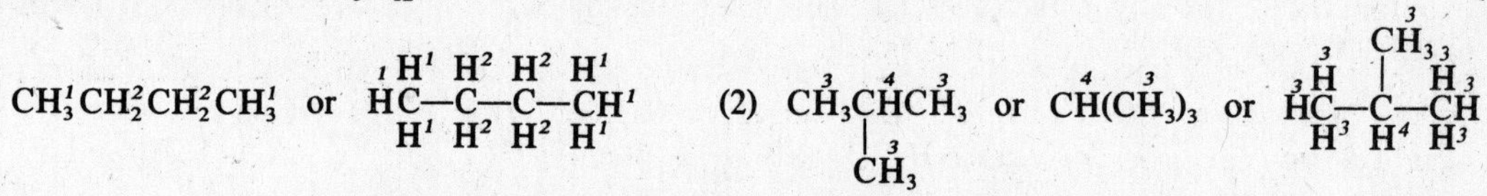

(b)
$$CH_3^1CH_2^2CH_2^2CH_3^1 \underset{+\boxed{CH_3}}{\overset{-H^1}{=\!=}} CH_3CH_2CH_2CH_2\,\boxed{CH_3}$$
n-Pentane

$$CH_3^1CH_2^2CH_2^2CH_3^1 \underset{+\boxed{CH_3}}{\overset{-H^2}{=\!=}} CH_3\overset{\boxed{CH_3}}{\underset{}{C}}HCH_2CH_3$$
Isopentane

$$CH_3^3\overset{CH_3^3}{\underset{}{C}}H^4CH_3^3 \underset{+\boxed{CH_3}}{\overset{-H^3}{=\!=}} CH_3CH_2CH_2\overset{CH_3}{\underset{}{\boxed{CH_3}}}$$
Isopentane

$$CH_3^3\overset{{}^3CH_3}{\underset{H^4}{C}}CH_3^3 \underset{+\boxed{CH_3}}{\overset{-H^4}{=\!=}} CH_3\overset{CH_3}{\underset{CH_3}{C}}CH_3$$
Neopentane

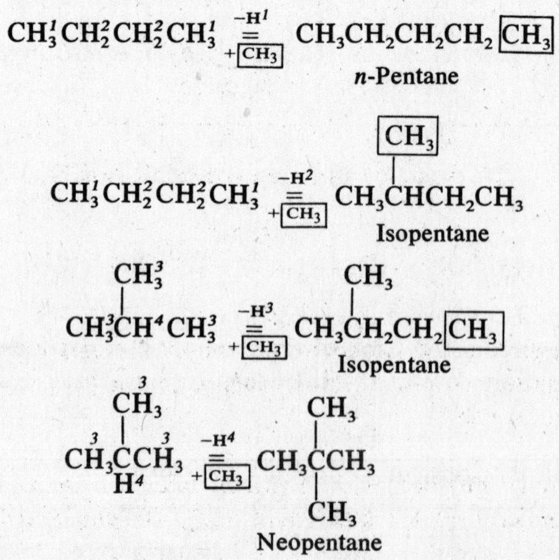

(c) Three: n-pentane, isopentane, and neopentane.

Sigma-bonded C's can rotate about the C—C bond and hence a chain of singly-bonded C's can be arranged in any zigzag shape (**conformation**). Two such arrangements, for four consecutive C's, are shown in Fig. 4-1. *Since conformations cannot be isolated, they are not isomers.*

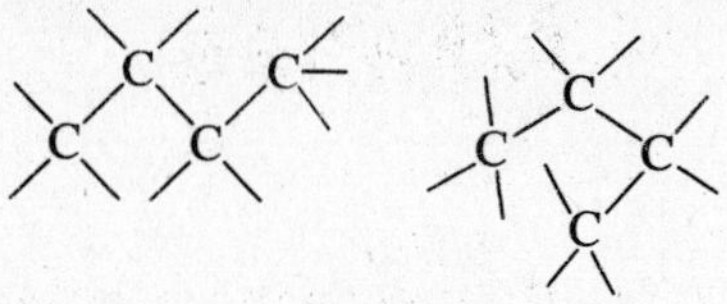

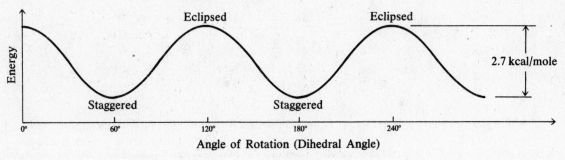

Fig. 4-1

The two extreme conformations of ethane are called **eclipsed** [Fig. 4-2(a)] and **staggered** [Fig. 4-2(b)].

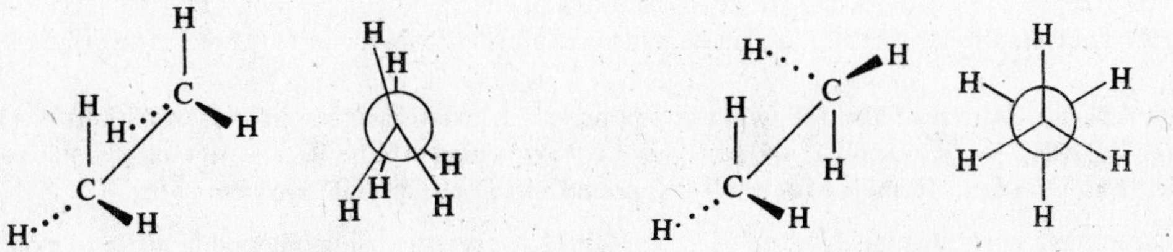

(a) Eclipsed   (b) Staggered

Fig. 4-2

Figure 4-3 traces the energies of the conformations when one CH_3 of ethane is rotated 360°.

Fig. 4-3

Problem 4.3 (a) Are the staggered and eclipsed conformations the only ones possible for ethane? (b) Indicate the preferential conformation of ethane molecules at room temperature. (c) What conformational changes occur as the temperature rises? ◄

(a) No. There are an infinite number with energies between those of the staggered and eclipsed conformations. For simplicity we are concerned only with conformations at minimum and maximum energies.

(b) The staggered form has the minimal energy and hence is the preferred conformation.

(c) The eclipsed-like conformations become more prevalent.

Problem 4.4 How many different compounds do the following structural formulas represent?

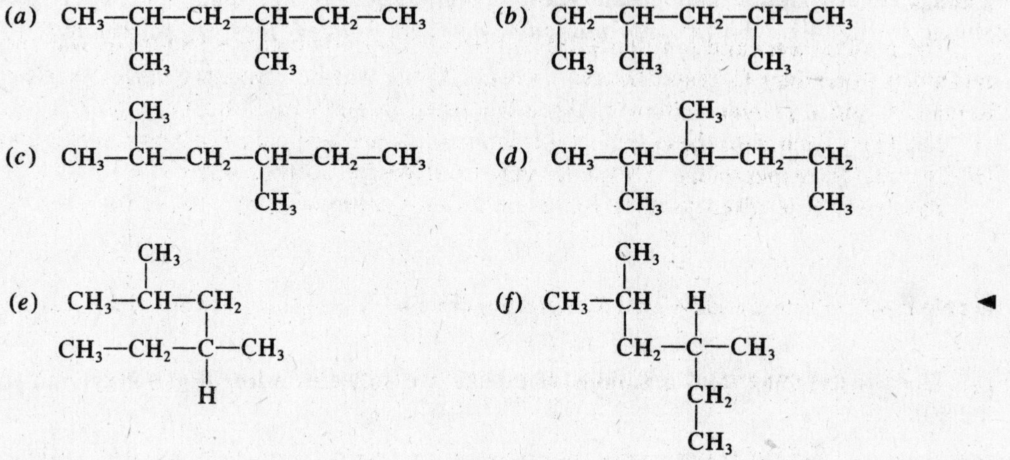

(a) CH₃—CH—CH₂—CH—CH₂—CH₃
      | |
      CH₃   CH₃

(b) CH₂—CH—CH₂—CH—CH₃
    | | |
    CH₃ CH₃   CH₃

(c)
          CH₃
          |
CH₃—CH—CH₂—CH—CH₂—CH₃
               |
              CH₃

(d)
          CH₃
          |
CH₃—CH—CH—CH₂—CH₂
      | |
      CH₃   CH₃

(e)
          CH₃
          |
    CH₃—CH—CH₂
  CH₃—CH₂—C—CH₃
           |
           H

(f)
          CH₃
          |
  CH₃—CH H
         CH₂—C—CH₃
              |
             CH₂
              |
             CH₃

◄

Two. (a), (b), (c), (e) and (f) are conformations of the same compound. This becomes obvious when the longest chain of carbons, in this case six, is written in a linear fashion. (d) represents a different compound.

Problem 4.5 (a) Which of the following compounds can exist in different conformations? (1) hydrogen peroxide, HOOH; (2) ammonia, NH_3; (3) hydroxylamine, H_2NOH; (4) methyl alcohol, H_3COH. (b) Draw two structural formulas for each compound in (a) possessing conformations. ◄

(a) A compound must have a sequence of at least three consecutive single bonds in order to exist in different conformations. (1), (3), and (4) have such a sequence. In (2),

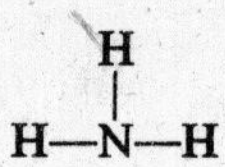

         H
         |
    H—N—H

the three single bonds are not consecutive.

(b)

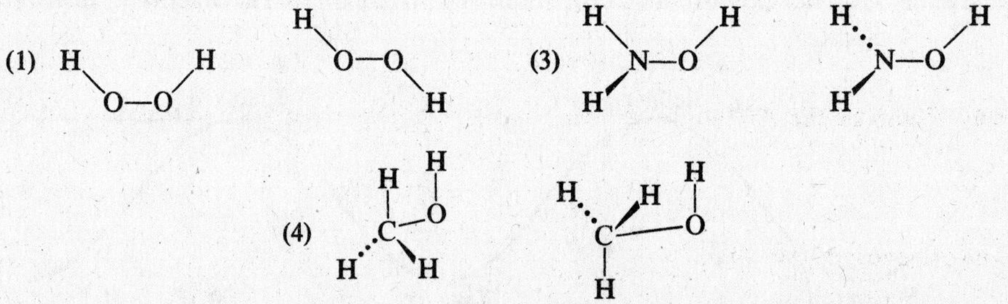

(1) H   H   O—O
     O—O   H

(3) H   H   H   H
   N—O   N—O
 H   H

(4)
      H   H   H   H
       C—O   C—O
   H   H   H

The first-drawn structure in each case is the eclipsed conformation; the second one is staggered.

Problem 4.6 Explain the fact that the calculated entropy for ethane is much greater than the experimentally determined value. ◄

The calculated value incorrectly assumes unrestricted free rotation so that all conformations are equally probable. Since most molecules of ethane have the staggered conformation, the structural randomness is less than calculated, and the actual observed entropy is less. This discrepancy led to the concept of conformations with different energies.

4.2 NOMENCLATURE OF ALKANES

The letter n (for *normal*), as in n-butane, denotes an unbranched chain of C atoms. The prefix *iso*-(*i*-) indicates a CH₃ branch on the second C from the end; e.g. isopentane is

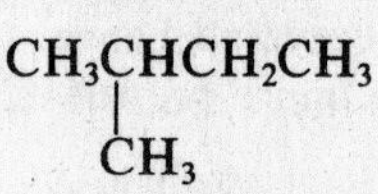

    CH₃CHCH₂CH₃
        |
       CH₃

Alkyl groups, such as methyl (H_3C) and ethyl (CH_3CH_2), are derived by removing one H from alkanes.

The prefixes *sec-* and *tert-* before the name of the group indicate that the H was removed from a **secondary** or **tertiary** C, respectively. A secondary C has bonds to two other C's, a tertiary to three other C's, and a **primary** either to three H's or to two H's and *one* C.

The H's attached to these types of carbon atoms are also called *primary*, *secondary* and *tertiary* ($1°$, $2°$ and $3°$), respectively. A **quaternary** C is bonded to four other C's.

The letter R is often used to represent any alkyl group.

Problem 4.7 Name the alkyl groups originating from (*a*) propane, (*b*) *n*-butane, (*c*) isobutane. ◄

(*a*) $CH_3CH_2CH_2$— is *n*-propyl (*n*-Pr); $CH_3\overset{|}{C}HCH_3$ is isopropyl (*i*-Pr).

(*b*) $CH_3CH_2CH_2CH_2$— is *n*-butyl (*n*-Bu); $CH_3\overset{|}{C}HCH_2CH_3$ is *sec*-butyl (*s*-Bu).

(*c*) CH_3—$\overset{|}{C}HCH_2$— is isobutyl (*i*-Bu); $CH_3\overset{|}{C}CH_3$ is *tert*-butyl (*t*-Bu).
                 | |
                 CH_3 CH_3

Problem 4.8 Use numbers *1, 2, 3* and *4* to designate the $1°$, $2°$, $3°$ and $4°$ C's, respectively, in $CH_3CH_2C(CH_3)_2CH_2CH(CH_3)_2$. Use letters *a, b, c,* etc., to indicate the different kinds of $1°$ and $2°$ C's. ◄

$$\underset{1a}{CH_3}\underset{2a}{CH_2}-\overset{\overset{1b}{CH_3}}{\underset{\underset{1b}{CH_3}}{\overset{|}{\underset{|}{C}}}}{}^{4}-\underset{2b}{CH_2}-\overset{\overset{H}{}}{\underset{\underset{1c}{CH_3}}{\overset{|}{\underset{|}{C}}}}{}^{3}\underset{1c}{CH_3}$$

Problem 4.9 Name by the IUPAC system the isomers of pentane derived in Problem 4.2. ◄

(*a*) $CH_3CH_2CH_2CH_2CH_3$ Pentane

(*b*) The longest consecutive chain in

$$\overset{\overset{CH_3}{|}}{\underset{1}{CH_3}\underset{2}{CH}\underset{3}{CH_2}\underset{4}{CH_3}}$$

has 4 C's and therefore the IUPAC name is a substituted butane. Number the C's as shown so that the branch CH_3 is on the C with the lower number, in this case C^2. The name is 2-methylbutane and not 3-methylbutane. Note that numbers are separated from letters by a hyphen.

(*c*) The longest consecutive chain in

$$CH_3-\overset{\overset{CH_3}{|}}{\underset{\underset{CH_3}{|}}{C}}-CH_3$$

has three C's; the parent is propane. The IUPAC name is 2,2-dimethylpropane. Note the use of the prefix *di-* to show two CH_3 branches, and the repetition of the number 2 to show that both CH_3's are on C^2. *Commas separate numbers and hyphens separate numbers and words.*

Problem 4.10 Name the compound in Fig. 4-4(a) by the IUPAC system.   ◄

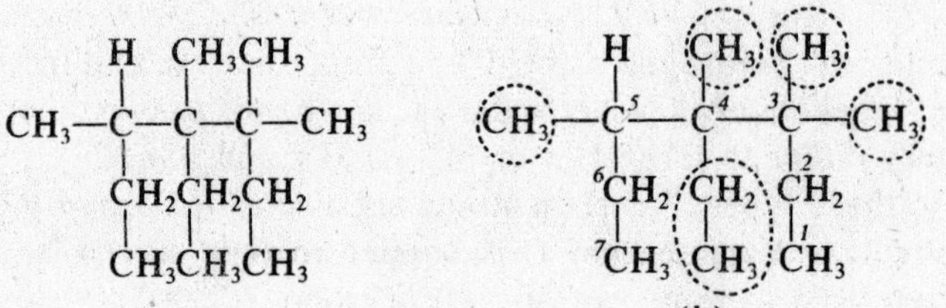

Fig. 4-4

The longest chain of consecutive C's has 7 C's [see Fig. 4-4(b)], and the compound is named as a heptane. Note that, as written, this longest chain is bent and not linear. Circle the branch alkyl groups and consecutively number the C's in the chain so that the lower-numbered C's hold the most branch groups. The name is 3,3,4,5-tetramethyl-4-ethylheptane.

4.3 PREPARATION OF ALKANES

REACTIONS WITH NO CHANGE IN CARBON SKELETON

1. Reduction of Alkyl Halides (RX, X = F, Cl, Br or I) **(Substitution of Halogen by Hydrogen)**

(a) $RX + Zn\!: + H^+ \longrightarrow RH + Zn^{2+} + :X^-$

(b) $4RX + LiAlH_4 \longrightarrow 4RH + LiX + AlX_3 \quad (X \neq F)$

or $RX + H\!:^- \longrightarrow RH + :X^- \quad (H\!:^-$ comes from $LiAlH_4)$

(c) $RX + (n\text{-}C_4H_9)_3SnH \longrightarrow RH + (n\text{-}C_4H_9)_3SnX$

(d)   Via organometallic compounds (Grignard reagent). Alkyl halides react with either Mg or Li in dry ether to give **organometallics** having a basic carbanionic site.

$$RX + 2Li \xrightarrow{\text{dry ether}} \bar{R}\overset{+}{Li} + LiX \quad \text{then} \quad \bar{R}\overset{+}{Li} + H_2O \longrightarrow RH + \overset{+}{Li}\bar{O}H$$

Alkyllithium

$$RX + Mg \xrightarrow{\text{dry ether}} \bar{R}(\overset{+}{MgX}) \quad \text{then} \quad \bar{R}(\overset{+}{MgX}) + H_2O \longrightarrow RH + \overset{2+}{Mg}\begin{smallmatrix} X^- \\ \\ OH^- \end{smallmatrix}$$

Grignard reagent

The net effect is replacement of X by H.

2. Hydrogenation of $\overset{\diagup}{\underset{\diagdown}{C}}=\overset{\diagup}{\underset{\diagdown}{C}}$ **(alkenes)**

$$CH_3-\overset{\overset{\displaystyle CH_3}{|}}{C}=CH_2 + H_2 \xrightarrow[\text{or Ni}]{\text{Pt}} CH_3-\overset{\overset{\displaystyle CH_3}{|}}{C}H-CH_3$$

Isobutylene   Isobutane

PREPARATION OF ALKANES WITH MORE C's THAN THE STARTING COMPOUNDS

Two R groups can be **coupled** by reacting RBr, RCl or RI with Na or K. Yields of product are best for 1° (60%) and poorest for 3° (10%) alkyl halides (**Wurtz reaction**).

$$2RX + 2Na \longrightarrow R{-}R + 2NaX$$

$$2Na + 2CH_3CH_2CH_2Cl \longrightarrow CH_3CH_2CH_2CH_2CH_2CH_3 + 2NaCl$$

n-Propyl chloride n-Hexane

A superior method for coupling is the **Corey-House synthesis**:

$$RMgX \text{ or } RLi \xrightarrow[\text{2. R'X}]{\text{1. CuX}} R{-}R' \quad (R = 1°, 2°, \text{ or } 3°; R' = 1°)$$

Problem 4.11 Write equations to show the products obtained from the reactions:

        (a) 2-Bromo-2-methylpropane + magnesium in dry ether
        (b) Product of (a) + H_2O
        (c) Product of (a) + D_2O

(a) 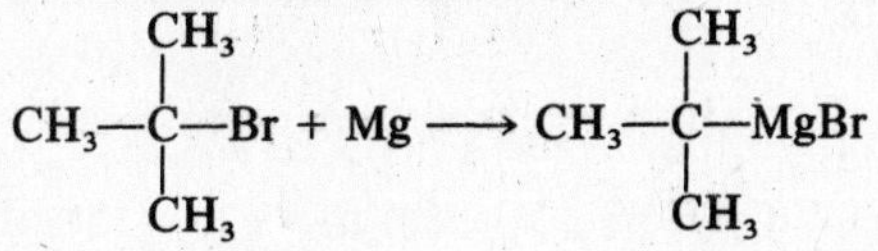
$$CH_3-\underset{\underset{CH_3}{|}}{\overset{\overset{CH_3}{|}}{C}}-Br + Mg \longrightarrow CH_3-\underset{\underset{CH_3}{|}}{\overset{\overset{CH_3}{|}}{C}}-MgBr \quad \text{t-Butylmagnesium bromide}$$

(b)
$$CH-\underset{\underset{CH_3}{|}}{\overset{\overset{CH_3}{|}}{C}}{}^-(MgBr)^+ + HOH \longrightarrow CH_3-\underset{\underset{CH_3}{|}}{\overset{\overset{CH_3}{|}}{C}}-H + (MgBr^+)(OH^-)$$

    Base 1   Acid 2   Acid 1   Base 2

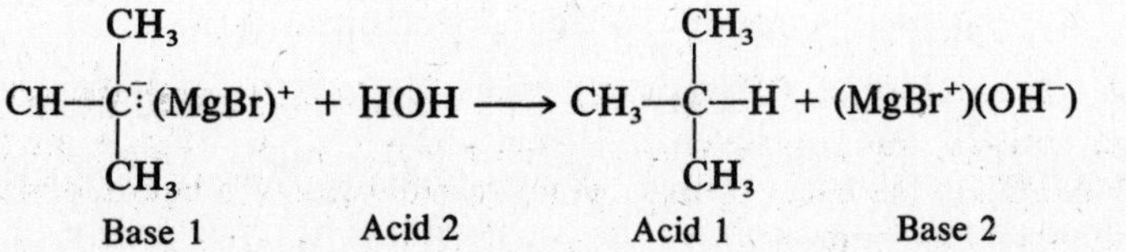

(c) The t-butyl carbanion accepts a deuterium cation to form 2-methyl-2-deuteropropane, $(CH_3)_3CD$.

Problem 4.12 Write structural formulas and give IUPAC names for the alkanes formed when a mixture of 1 mole each of 1-bromopropane and 2-bromopropane is reacted with 2 moles of sodium.

A mixture of three alkanes results:

$$CH_3CH_2CH_2Br + 2Na + BrCH_2CH_2CH_3 \longrightarrow CH_3CH_2CH_2-CH_2CH_2CH_3 \quad \text{n-Hexane}$$

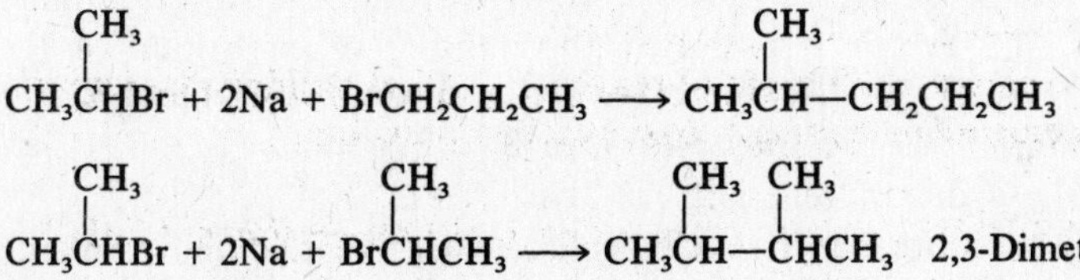

$$CH_3\overset{\overset{CH_3}{|}}{C}HBr + 2Na + BrCH_2CH_2CH_3 \longrightarrow CH_3\overset{\overset{CH_3}{|}}{C}H-CH_2CH_2CH_3 \quad \text{2-Methylpentane}$$

$$CH_3\overset{\overset{CH_3}{|}}{C}HBr + 2Na + Br\overset{\overset{CH_3}{|}}{C}HCH_3 \longrightarrow CH_3\overset{\overset{CH_3}{|}}{C}H-\overset{\overset{CH_3}{|}}{C}HCH_3 \quad \text{2,3-Dimethylbutane}$$

4.4 CHEMICAL PROPERTIES OF ALKANES

Alkanes are unreactive except under vigorous conditions.

1. Pyrolytic Cracking (heat (Δ) in absence of O_2)

$$\text{Alkane} \overset{\Delta}{\longrightarrow} \text{mixture of smaller hydrocarbons}$$

2. Combustion

$$CH_4 + O_2 \xrightarrow{600\,°C} CO_2 + 2H_2O \quad \Delta H \text{ of combustion} = -193.4 \text{ kcal/mole}$$

Problem 4.13 (a) Why are alkanes inert? (b) Why do the C—C rather than the C—H bonds break when alkanes are pyrolyzed? (c) Although combustion of alkanes is a strongly exothermic process, it does not occur at moderate temperatures. Explain.

(a) A reactive site in a molecule usually has one or more unshared pairs of electrons, a polar bond, an electron-deficient atom or an atom with an expandable octet. Alkanes have none of these.
(b) The C—C bond has a lower bond energy ($\Delta H = +83$ kcal/mole) than does the C—H bond ($\Delta H = +99$ kcal/mole).
(c) The reaction is very slow at room temperature because of a very high $\Delta H^\ddagger$.

3. Halogenation

$$RH + X_2 \xrightarrow[\text{or }\Delta]{uv} RX + HX$$

(Reactivity of X_2: $F_2 > Cl_2 > Br_2$. I_2 does not react.)

The mechanism of methane chlorination is:

Initiation Step $\qquad$ Cl:Cl $\xrightarrow[\text{or }\Delta]{\text{uv}}$ 2Cl· $\quad$ $\Delta H = +58$ kcal/mole

The required enthalpy comes from ultraviolet (uv) light or heat.

Propagation Steps

(i) $\quad$ H$_3$C:H + Cl· $\longrightarrow$ H$_3$C· + H:Cl $\quad$ $\Delta H = -1$ kcal/mole $\quad$ (rate-determining)

(ii) $\quad$ H$_3$C· + Cl:Cl $\longrightarrow$ H$_3$C:Cl + Cl· $\quad$ $\Delta H = -23$ kcal/mole

The sum of the two propagation steps is the overall reaction,

$$CH_4 + Cl_2 \longrightarrow CH_3Cl + HCl \quad \Delta H = -24 \text{ kcal/mole}$$

In propagation steps, the same free-radical intermediates, here Cl· and H$_3$C·, are being formed and consumed. Chains terminate on those rare occasions when two free-radical intermediates form a covalent bond:

$$Cl· + Cl· \longrightarrow Cl_2, \quad H_3C· + Cl· \longrightarrow H_3C:Cl, \quad H_3C· + ·CH_3 \longrightarrow H_3C:CH_3$$

Inhibitors stop chain propagation by reacting with free-radical intermediates, e.g.

$$H_3C· + ·\ddot{O}—\ddot{O}· \longrightarrow H_3C\ddot{O}—\ddot{O}·$$

The inhibitor must be consumed before chlorination can occur.

In more complex alkanes, the abstraction of each different kind of H atom gives a different isomeric product. Three factors determine the relative yields of the isomeric product.

(1) **Probability factor**. This factor is based on the number of each kind of H atom in the molecule. For example, in CH$_3$CH$_2$CH$_2$CH$_3$ there are *six* equivalent 1° H's and *four* equivalent 2° H's. The odds on abstracting a 1° H are thus 6 to 4, or 3 to 2.

(2) **Reactivity of H**. The order of reactivity of H is 3° > 2° > 1°.

(3) **Reactivity of X·**. The more reactive Cl· is less selective and more influenced by the probability factor. The less reactive Br· is more selective and less influenced by the probability factor. As summarized by the **Reactivity-Selectivity Principle**: If the attacking species is more reactive, it will be less selective, and the yields will be closer to those expected from the probability factor.

Problem 4.14 (*a*) List the monobromo derivatives of (i) CH$_3$CH$_2$CH$_2$CH$_3$ and (ii) (CH$_3$)$_2$CHCH$_3$. (*b*) Predict the predominant isomer in each case. The order of reactivity of H for bromination is

$$3°(1600) > 2°(82) > 1°(1) \qquad \blacktriangleleft$$

(*a*) There are two kinds of H's, and there are two possible isomers for each compound: (i) CH$_3$CH$_2$CH$_2$CH$_2$Br and CH$_3$CHBrCH$_2$CH$_3$; (ii) (CH$_3$)$_2$CHCH$_2$Br and (CH$_3$)$_2$CBrCH$_3$.

(*b*) In bromination, in general, the difference in reactivity completely overshadows the probability effect in determining product yields. (i) CH$_3$CHBrCH$_2$CH$_3$ is formed by replacing a 2° H; (ii) (CH$_3$)$_2$CBrCH$_3$ is formed by replacing the 3° H.

Problem 4.15 Using the bond dissociation energies for X$_2$

X$_2$	F$_2$	Cl$_2$	Br$_2$	I$_2$
ΔH, kcal/mole	+37	+58	+46	+36

show that the initiation step for halogenation of alkanes,

$$X_2 \xrightarrow[\text{or }\Delta]{\text{uv}} 2X·$$

is not rate-determining. $\qquad \blacktriangleleft$

$\Delta H^{\ddagger}$ is usually not related to ΔH of the reaction. In this reaction, however, $\Delta H^{\ddagger}$ and ΔH are identical. In simple homolytic dissociations of this type, the formed free radicals have the same enthalpy as does the transition state. On this basis alone, iodine, having the smallest ΔH and $\Delta H^{\ddagger}$, should react fastest. Similarly, chlorine, with the largest ΔH and $\Delta H^{\ddagger}$, should react slowest. But the actual order of reaction rates is

$$F_2 > Cl_2 > Br_2 > I_2$$

Therefore, the initiation step is not rate-determining.

Problem 4.16 Draw the reactants, transition state, and products for

$$Br\cdot + CH_4 \longrightarrow HBr + \cdot CH_3$$ ◀

In the transition state, Br is losing radical character while C is becoming a radical; both atoms have partial radical character as indicated by $\delta\cdot$. The C atom undergoes a change in hybridization as indicated:

REACTANTS	TRANSITION STATE	PRODUCTS

C is sp^3 (tetrahedral)	C is becoming sp^2 (trigonal planar)	C is sp^2 (trigonal planar)

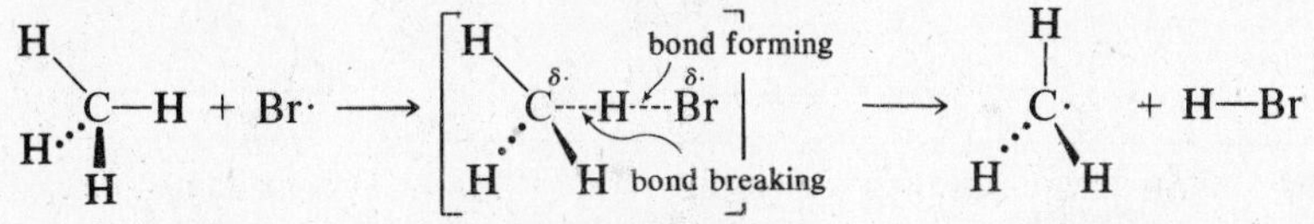

Problem 4.17 Bromination of methane, like chlorination, is exothermic, but it proceeds at a slower rate under the same conditions. Explain in terms of the factors that affect the rate, assuming that the rate-controlling step is

$$X\cdot + CH_4 \longrightarrow HX + \cdot CH_3$$ ◀

Given the same concentration of CH_4 and Cl· or Br·, the frequency of collisions should be the same. Because of the similarity of the two reactions, $\Delta S^{\ddagger}$ for each is about the same. The difference must be due to the $\Delta H^{\ddagger}$, which is less (4 kcal/mole) for Cl· than for Br· (18 kcal/mole).

Problem 4.18 2-Methylbutane has 1°, 2° and 3° H's as indicated:

$$\overset{1°}{(CH_3)_2}\overset{3°}{CH}\overset{2°}{CH_2}\overset{1°}{CH_3}$$

(*a*) Use enthalpy–reaction progress diagrams for the abstraction of each kind of hydrogen by ·X. (*b*) Summarize the relationships of relative (i) stabilities of transition states, (ii) $\Delta H^{\ddagger}$ values, (iii) stabilities of alkyl radicals and (iv) rates of H-abstraction. ◀

(*a*) See Fig. 4-5.
(*b*) (i) 3° > 2° > 1° since the enthalpy of the $TS_{1°}$ is the greatest and the enthalpy of the $TS_{3°}$ is the smallest. (ii) $\Delta H^{\ddagger}_{3°} < \Delta H^{\ddagger}_{2°} < \Delta H^{\ddagger}_{1°}$. (iii) 3° > 2° > 1°. (iv) 3° > 2° > 1°.

Problem 4.19 List and compare the differences in the properties of the transition states during chlorination and bromination that account for the different reactivities for 1°, 2°, and 3° H's. ◀

The differences may be summarized as follows:

	Chlorination	Bromination
1. Time of formation of transition state	Earlier in reaction	Later in reaction
2. Amount of breaking of C—H bond	Less, H_3C---H-----Cl	More, H_3C-----H---Br
3. Free-radical character ($\delta\cdot$) of carbon	Less	More
4. Transition state more closely resembles	Reactants	Products

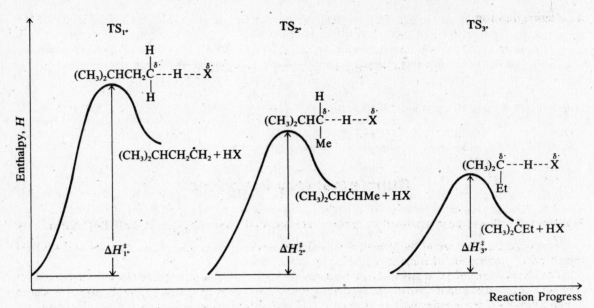

Fig. 4-5

These show that the greater selectivity in bromination is attributable to the greater free-radical character of carbon. With greater radical character, the differences in stability between 1°, 2° and 3° radicals become more important, and reactivity of H (3° > 2° > 1°) also becomes more significant.

Problem 4.20 Calculate the percentages of isomers expected during the monochlorination of $CH_3CH_2CH_3$ at room temperature. The relative reactivities for 2° and 1° H atoms are 3.8 and 1.0, respectively. ◄

The products are $CH_2ClCH_2CH_3$ from substitution of any one of the *six* 1° terminal H's and $CH_3CHClCH_3$ from substitution of either of the *two* 2° central H's. Therefore the probability factor for the 1° over the 2° reaction product is 6 to 2. We also take into account the 1.0 to 3.8 ratio of reactivities. The number of each kind of H whose removal gives a different product is multiplied by the relative rate for that kind of H to give a weighted reactivity. These weighted reactivities are totaled, and the proportion and percentage of each isomer is obtained as shown.

	H Atoms			Rel. Reactivity		Weighted				Percent
	Kind	Number	×	of H Atom	=	Reactivity	Proportion × 100%	=		Yield
$CH_3CH_2CH_2Cl$	1°	6	×	1.0	=	6.0	6/13.6		=	44
$CH_3CHClCH_3$	2°	2	×	3.8	=	7.6	7.6/13.6		=	56
					TOTAL	13.6				

4. Methylene (Carbene) Insertion

Singlet methylene (page 22), $H_2C:$, may be generated as shown:

$$H_2C{=}\overset{+}{N}{=}\overset{-}{N} \xrightarrow{uv} H_2C: + N_2 \quad \text{or} \quad H_2C{=}C{=}O \xrightarrow{uv} H_2C: + CO$$

Diazomethane   Ketene

It can insert between a C—H bond as shown for pentane.

$$CH_3CH_2CH_2CH_2CH_3 + \boxed{H_2C:} \longrightarrow$$

$$H{-}\boxed{\overset{H}{\underset{H}{C}}}{-}CH_2CH_2CH_2CH_2CH_3 \overset{48\%}{} + CH_3CHCH_2CH_2CH_3 \overset{35\%}{} + CH_3CH_2CHCH_2CH_3 \overset{17\%}{}$$

$$\boxed{HCH} \qquad \boxed{HCH}$$
$$H \qquad\qquad H$$

5. Isomerization

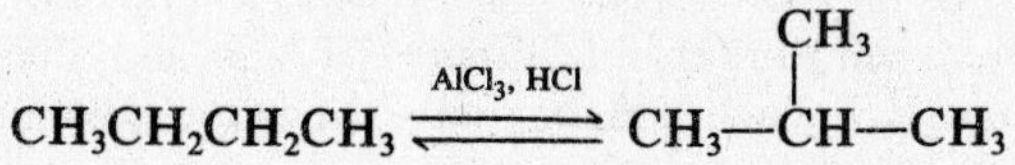

$$CH_3CH_2CH_2CH_3 \underset{}{\overset{AlCl_3,\ HCl}{\rightleftharpoons}} CH_3-\underset{\underset{CH_3}{|}}{CH}-CH_3$$

Supplementary Problems

Problem 4.21   Derive the structures of all isomers of pentane, C_5H_{12}, and name them by the IUPAC system.   ◄

First write the structure of the longest chain and then successively smaller chains with one or more alkyl substituents.   Duplication of isomers is avoided by correct naming.

The longest chain, of 5 C's, is that of pentane itself, $CH_3CH_2CH_2CH_2CH_3$.   The next-longest chain has 4 C's (butane).   A CH_3 is placed somewhere on the chain to form a five-carbon compound.   The CH_3 cannot be placed on either terminal C because this again gives pentane.   In looking for isomers, never substitute an alkyl group on a terminal C because this will merely lengthen the chain.   Substitution of CH_3 on either of the two central 2° C's yields 2-methylbutane:

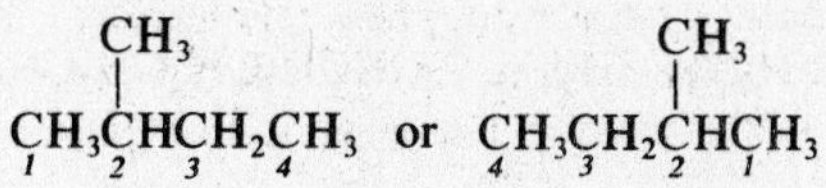

$$\underset{\underset{1\ \ 2\ \ \ 3\ \ \ 4}{}}{CH_3\underset{\underset{CH_3}{|}}{C}HCH_2CH_3} \quad \text{or} \quad \underset{\underset{4\ \ 3\ \ \ 2\ \ \ 1}{}}{CH_3CH_2\underset{\underset{CH_3}{|}}{C}HCH_3}$$

Note that the chain is numbered consecutively so that the C with the lowest possible number holds the methyl group.   Since both structural formulas shown above have the same name, they are identical.

A three-carbon chain is considered next.   On this chain, two C's must be added, neither of which can be placed on a terminal C.   Placing an ethyl group (CH_3CH_2-) on the central C also gives 2-methylbutane, which is the same compound noted above.

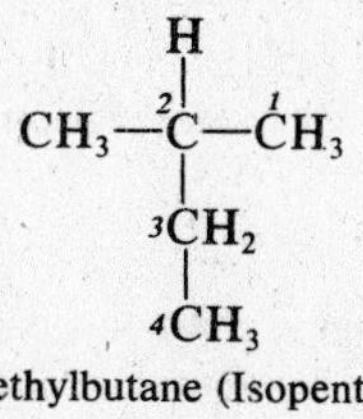

$$CH_3-\overset{\overset{H}{|}}{\underset{\underset{^3CH_2}{|}}{C}}-{}^1CH_3$$
$$\underset{^4CH_3}{|}$$

2-Methylbutane (Isopentane)

Therefore, the two remaining C's must be placed as two CH_3's on the central C to form 2,2-dimethylpropane.

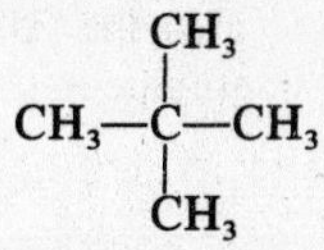

$$CH_3-\overset{\overset{CH_3}{|}}{\underset{\underset{CH_3}{|}}{C}}-CH_3$$

2,2-Dimethylpropane (Neopentane)

Problem 4.22   Write structural formulas for the five isomeric hexanes and name them by the IUPAC system.   ◄

The longest chain is hexane itself, $CH_3CH_2CH_2CH_2CH_2CH_3$.   If we use a five-carbon chain, a CH_3 may be placed either on C^2 or C^4 to produce 2-methylpentane, or on C^3 to give another isomer, 3-methylpentane.

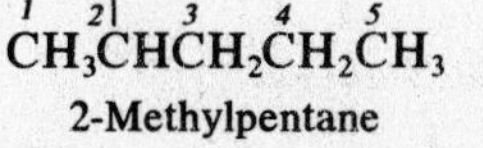

$$\underset{\underset{1\ \ \ 2\ \ \ 3\ \ \ 4\ \ \ 5}{}}{CH_3\overset{\overset{CH_3}{|}}{C}HCH_2CH_2CH_3} \qquad\qquad \underset{\underset{1\ \ \ 2\ \ \ 3\ \ \ 4\ \ \ 5}{}}{CH_3CH_2\overset{\overset{CH_3}{|}}{C}HCH_2CH_3}$$

2-Methylpentane 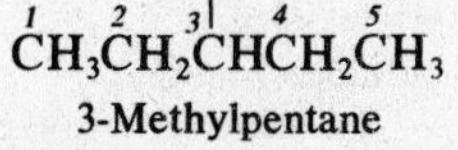   3-Methylpentane

With a four-carbon chain either a CH_3CH_2 or two CH_3's must be added as branches to give a total of 6 C's. Placing CH_3CH_2 anywhere on the chain is ruled out because it lengthens the chain. Two CH_3's are added, but only to central C's to avoid extending the chain. If both CH_3's are introduced on the same C, the isomer is 2,2-dimethylbutane. Placing one CH_3 on each of the central C's gives the remaining isomer, 2,3-dimethylbutane.

$$
\begin{array}{cc}
& CH_3 \\
CH_3-\!\!\!\overset{|}{\underset{|}{C}}\!\!\!-CH_2CH_3 \\
& CH_3
\end{array}
\qquad
\begin{array}{cc}
CH_3 \quad CH_3 \\
CH_3-\!\!\!\overset{|}{\underset{|}{C}}\!\!\!-\!\!\!\overset{|}{\underset{|}{C}}\!\!\!-\!CH_3 \\
H \quad\;\; H
\end{array}
$$

2,2-Dimethylbutane   2,3-Dimethylbutane

Problem 4.23 Write the structural formulas for (a) 3,4-dichloro-2,5-dimethylhexane; (b) 5-(1,2-dimethylpropyl)-6-methyldodecane. (Complex branch groups are usually enclosed in parentheses.) ◄

(a)
$$
\begin{array}{c}
CH_3 \;\; Cl \quad\; Cl \;\; CH_3 \\
CH_3-\!\!CH-\!\!CH-\!\!CH-\!\!CH-\!\!CH_3
\end{array}
$$

(b) The group in parentheses is bonded to the fifth C. It is a propyl group with CH_3's on its first and second C's (denoted $1'$ and $2'$) counting from the attached C.

$$
\begin{array}{c}
\qquad\qquad H \quad CH_3 \\
CH_3CH_2CH_2CH_2-\!\!\overset{5}{C}-\!\!CHCH_2CH_2CH_2CH_2CH_2CH_3 \\
\qquad\qquad H\overset{1'}{C}-\!\!CH_3 \\
\qquad\qquad H\overset{2'}{C}-\!\!CH_3 \\
\qquad\qquad\overset{3'}{C}H_3
\end{array}
$$

Problem 4.24 Name by the IUPAC system

(a) $(CH_3)_3CCH(CH_3)C_2H_5$ (b) $\overset{10}{C}H_3(CH_2)_4\overset{5}{C}H(CH_2)_3\overset{1}{C}H_3$ ◄

$$
\qquad\qquad\qquad\qquad\qquad\qquad\qquad\qquad\qquad H_2CCH(CH_3)_2
$$

(a) 2,2,3-Trimethylpentane. (b) The longest chain has ten C's, and there is an isobutyl group on the fifth C. The name is 5-(2-methylpropyl)decane or 5-isobutyldecane.

Problem 4.25 Write the structural formulas and give the IUPAC names for all monochloro derivatives of (a) isopentane, $(CH_3)_2CHCH_2CH_3$; (b) 2,2,4-trimethylpentane, $(CH_3)_3CCH_2CH(CH_3)_2$. ◄

(a) Since there are 4 kinds of equivalent H's,

$$
\overset{1}{(CH_3)_2}\overset{2}{C}H\overset{3}{C}H_2\overset{4}{C}H_3
$$

there are 4 isomers:

$$
\begin{array}{cccc}
CH_3 & CH_3 & CH_3 & CH_3 \\
ClCH_2\overset{|}{\underset{|}{C}}CH_2CH_3 & CH_3-\overset{|}{\underset{|}{C}}-CH_2CH_2 & CH_3-\overset{|}{\underset{|}{C}}-CHCH_3 & CH_3-\overset{|}{\underset{|}{C}}-CH_2CH_2Cl \\
H & Cl & H \;\; Cl & H
\end{array}
$$

1-Chloro-2-methyl-butane   2-Chloro-2-methyl-butane   2-Chloro-3-methyl-butane   1-Chloro-3-methyl-butane

(b) There are 4 isomers because there are 4 kinds of H's: $(CH_3)_3CCH_2CH(CH_3)_2$.

$$ClCH_2-\underset{\underset{CH_3}{|}}{\overset{\overset{CH_3}{|}}{C}}-CH_2-\underset{\underset{H}{|}}{\overset{\overset{CH_3}{|}}{C}}-CH_3$$

1-Chloro-2,2,4-
trimethylpentane

$$CH_3-\underset{\underset{CH_3}{|}}{\overset{\overset{CH_3}{|}}{C}}-\underset{\underset{Cl}{|}}{\overset{\overset{H}{|}}{C}}-\underset{\underset{H}{|}}{\overset{\overset{CH_3}{|}}{C}}-CH_3$$

3-Chloro-2,2,4-
trimethylpentane

$$CH_3-\underset{\underset{CH_3}{|}}{\overset{\overset{CH_3}{|}}{C}}-CH_2-\underset{\underset{Cl}{|}}{\overset{\overset{CH_3}{|}}{C}}-CH_3$$

2-Chloro-2,4,4-
trimethylpentane

$$CH_3-\underset{\underset{CH_3}{|}}{\overset{\overset{CH_3}{|}}{C}}-CH_2-\underset{\underset{H}{|}}{\overset{\overset{CH_3}{|}}{C}}-CH_2Cl$$

1-Chloro-2,4,4-
trimethylpentane

Problem 4.26 Derive the structural formulas and give the IUPAC names for all dibromo derivatives of propane. ◄

The two Br's are placed first on the same C and then on different C's.

$$Br-\underset{\underset{Br}{|}}{CH}CH_2CH_3 \qquad CH_3\underset{\underset{Br}{|}}{\overset{\overset{Br}{|}}{C}}CH_3 \qquad BrCH_2CHBrCH_3 \qquad BrCH_2CH_2CH_2Br$$

1,1-Dibromopropane   2,2-Dibromopropane   1,2-Dibromopropane   1,3-Dibromopropane

Problem 4.27 Write topological formulations for the structural formulas for (a) propane, (b) butane, (c) isobutane, (d) 2,2-dimethylpropane, (e) 2,3-dimethylbutane, (f) 3-ethylpentane and (g) 1-chloro-3-methylbutane. ◄

In this method one writes only the C—C bonds and all functional groups bonded to C. The approximate bond angles are used.

(a) (b) (c) (d) (e) (f)

(g)

Problem 4.28 Assign the numbers from (1) for the LOWEST to (3) for the HIGHEST to the relative boiling points of the following isomers without using a handbook or table: 2,2-dimethylbutane, 3-methylpentane, and n-hexane. Point out the basis for your order. ◄

(Review Problem 2.15.) First choose the extreme cases. n-Hexane (3) has the longest chain and has the highest boiling point. 2,2-Dimethylbutane (1) is the most spherical, has the smallest surface area, and so has the lowest boiling point. 3-Methylpentane is (2).

Problem 4.29 Synthesize (a) 2-methylpentane from $CH_3CH=CH-CH(CH_3)_2$, (b) isobutane from isobutyl chloride, (c) 2-methyl-2-deuterobutane from 2-chloro-2-methylbutane. Show all steps. ◄

(a) The alkane and the starting compound have the same carbon skeleton.

$$CH_3-CH=CH-\underset{\underset{CH_3}{|}}{\overset{\overset{CH_3}{|}}{CH}}-CH_3 \xrightarrow{H_2/Pt} CH_3-CH_2-CH_2-\underset{\underset{CH_3}{|}}{\overset{\overset{CH_3}{|}}{CH}}-CH_3$$

(b) The alkyl chloride and alkane have the same carbon skeleton.

$$\underset{\substack{|\\CH_3}}{CH_3CHCH_2}{-}Cl \xrightarrow[\text{or LiAlH}_4]{\text{Zn, HCl}} \underset{\substack{|\\CH_3}}{CH_3CHCH_2}{-}H$$

(c) Deuterium can be bonded to C by the reaction of D_2O with a Grignard reagent.

$$\underset{\substack{|\\CH_3}}{\overset{\substack{CH_3\\|}}{CH_3CH_2C}}{-}Cl \xrightarrow[\text{dry ether}]{Mg} \underset{\substack{|\\CH_3}}{\overset{\substack{CH_3\\|}}{CH_3CH_2C}}{-}MgCl \xrightarrow{D_2O} \underset{\substack{|\\CH_3}}{\overset{\substack{CH_3\\|}}{CH_3CH_2C}}{-}D$$

| 2-Chloro-2-methyl-butane | Grignard reagent | 2-Deutero-2-methyl-butane |

Problem 4.30 RCl is treated with Li in ether solution to form RLi. RLi reacts with H_2O to form isopentane. RCl also reacts with Na to form 2,7-dimethyloctane. What is the structure of RCl? ◄

To determine the structure of a compound from its reactions, the structures of the products are first considered and their formation is then deduced from the reactions. The product of the reaction with Na (Wurtz) must be a symmetrical molecule whose carbon-to-carbon bond was formed between C^4 and C^5 of 2,7-dimethyloctane. The only RCl which will give this product is isopentyl chloride:

$$\underset{\substack{|\\CH_3}}{CH_3CHCH_2CH_2Cl} + 2Na + \underset{\substack{|\\CH_3}}{ClCH_2CH_2CHCH_3} \longrightarrow \underset{\substack{|\\CH_3}}{CH_3CHCH_2CH_2}{-}\overset{\text{formed bond}}{\underset{\substack{|\\CH_3}}{CH_2CH_2CHCH_3}}$$

Isopentyl chloride   2,7-Dimethyloctane

This alkyl halide will also yield isopentane.

$$\underset{\substack{|\\CH_3}}{CH_3CHCH_2CH_2Cl} \xrightarrow[\text{ether}]{Li} \underset{\substack{|\\CH_3}}{CH_3CHCH_2CH_2Li} \xrightarrow{H_2O} \underset{\substack{|\\CH_3}}{CH_3CHCH_2CH_3}$$

Problem 4.31 Give steps for the following syntheses: (a) propane to $(CH_3)_2CHCH(CH_3)_2$, (b) propane to 2-methylpentane, (c) $^{14}CH_3Cl$ to $^{14}CH_3\,^{14}CH_2\,^{14}CH_2\,^{14}CH_3$. ◄

(a) The symmetrical molecule is prepared by coupling an isopropyl halide. Bromination of propane is preferred over chlorination because the ratio of isopropyl to n-propyl halide is 96%/4% in bromination and only 56%/44% in chlorination (see Problem 4.20).

$$CH_3CH_2CH_3 \xrightarrow{Br_2\ (127\ °C)} (CH_3)_2CHBr \xrightarrow[\substack{2.\ \ \text{CuI}\\3.\ \ CH_3CHBrCH_3}]{1.\ \ \text{Li}} \underset{\substack{|\\CH_3}}{\overset{\substack{CH_3\\|}}{CH_3CH}}{-}\underset{\substack{|\\CH_3}}{CHCH_3}$$

(b)

$$CH_3CH_2CH_3 \xrightarrow[\text{uv}]{Cl_2} CH_3CH_2CH_2Cl + (CH_3)_2CHCl \quad (\textit{separate the mixture})$$

$$(CH_3)_2CHCl \xrightarrow[\text{ether}]{Li} (CH_3)_2CHLi$$

$$(CH_3)_2CHLi \xrightarrow[\substack{2.\ \ CH_3CH_2CH_2Cl}]{1.\ \ \text{CuI}} (CH_3)_2CHCH_2CH_2CH_3$$

(c)

$$^{14}CH_3Cl \xrightarrow[\substack{2.\ \ \text{CuI}\\3.\ \ ^{14}CH_3Cl}]{1.\ \ \text{Li}} {}^{14}CH_3\,^{14}CH_3 \xrightarrow[\text{uv}]{Cl_2} {}^{14}CH_3\,^{14}CH_2Cl \xrightarrow[\substack{2.\ \ \text{CuI}\\3.\ \ ^{14}CH_3\,^{14}CH_2Cl}]{1.\ \ \text{Li}} {}^{14}CH_3(^{14}CH_2)_2\,^{14}CH_3$$

Problem 4.32 Synthesize the following deuterated compounds: (a) CH_3CH_2D, (b) CH_2DCH_2D, (c) CH_3CD_2H. ◄

(a)

$$CH_3CH_2Br \xrightarrow[\text{ether}]{Mg} CH_3CH_2MgBr \xrightarrow{D_2O} CH_3CH_2D$$

(b)

$$H_2C{=}CH_2 + D_2 \xrightarrow{Pt} H_2CDCH_2D$$

(c)

$$CH_4 + CD_2N_2 \xrightarrow{uv} CH_3CD_2H + N_2$$

Problem 4.33 In the dark at 150 °C, tetraethyl lead, $Pb(C_2H_5)_4$, catalyzes the chlorination of CH_4. Explain in terms of the mechanism. ◄

$Pb(C_2H_5)_4$ readily undergoes thermal homolysis of the Pb—C bond.

$$Pb(C_2H_5)_4 \longrightarrow \cdot\dot{P}b\cdot + 4CH_3CH_2\cdot$$

The $CH_3CH_2\cdot$ then generates the $Cl\cdot$ that initiates the propagation steps.

$$CH_3CH_2\cdot + Cl{:}Cl \longrightarrow CH_3CH_2Cl + Cl\cdot$$

Problem 4.34 Hydrocarbons are monochlorinated with *tert*-butyl hypochlorite, *t*-BuOCl.

$$\textit{t}\text{-BuOCl} + RH \longrightarrow RCl + \textit{t}\text{-BuOH}$$

Write the propagating steps for this reaction if the initiating step is

$$\textit{t}\text{-BuOCl} \longrightarrow \textit{t}\text{-BuO}\cdot + Cl\cdot \qquad ◄$$

The propagating steps must give the products and also form chain-carrying free radicals. The formation of *t*-BuOH suggests H-abstraction from RH by *t*-BuO·, not by Cl·. The steps are:

$$RH + \textit{t}\text{-BuO}\cdot \longrightarrow R\cdot + \textit{t}\text{-BuOH}$$
$$R\cdot + \textit{t}\text{-BuOCl} \longrightarrow RCl + \textit{t}\text{-BuO}\cdot$$

R· and *t*-BuO· are the chain-carrying radicals.

Problem 4.35 Calculate the heat of combustion of methane at 25 °C. The bond energies for C—H, O=O, C=O and O—H are respectively 98.7, 119.1, 192.0 and 110.6 kcal/mole. ◄

First, write the balanced equation for the reaction.

$$CH_4 + 2O_2 \longrightarrow CO_2 + 2H_2O$$

The energies for the bonds broken are calculated. These are endothermic processes and ΔH is positive.

$$CH_4 \longrightarrow C + 4H \quad \Delta H = 4 \times (+98.7) = +394.8 \text{ kcal/mole}$$
$$2O_2 \longrightarrow 4O \qquad \Delta H = 2 \times (+119.1) = +238.2 \text{ kcal/mole}$$

Next, the energies are calculated for the bonds formed. Bond formation is exothermic, so the ΔH values are made negative.

$$C + 2O \longrightarrow O{=}C{=}O \quad \Delta H = 2 \times (-192.0) = -384.0 \text{ kcal/mole}$$
$$4H + 2O \longrightarrow 2\,H{-}O{-}H \quad \Delta H = 4 \times (-110.6) = -442.4 \text{ kcal/mole}$$

The enthalpy for the reaction is the sum of these values:

$$+394.8 + 238.2 - 384.0 - 442.4 = -193.4 \text{ kcal/mole}$$

Problem 4.36 Why is the following mechanism for the chlorination of methane not reasonable? The energy of light needed to initiate the chlorination of methane is equivalent to 70 kcal/mole.

$$CH_4 \longrightarrow H_3C\cdot + \cdot H \qquad \Delta H = +102 \text{ kcal/mole}$$
$$H_3C\cdot + Cl_2 \longrightarrow H_3CCl + Cl\cdot \quad \Delta H = -23 \text{ kcal/mole}$$
$$H\cdot + Cl\cdot \longrightarrow HCl \qquad \Delta H = -103 \text{ kcal/mole} \qquad ◄$$

An energy of 70 kcal/mole is insufficient to dissociate the C—H bond, which requires 102 kcal/mole. Therefore the dissociation of methane cannot be the initiation step. The last step is unreasonable since it is chain-terminating and is not consistent with the fact that several thousand molecules of methane are chlorinated by the stimulus of one photon of energy.

Problem 4.37 Predict the percentages of isomers formed during monochlorination at room temperature of (a) 2,3-dimethylbutane, (b) 2-methylbutane. Relative order of reactivity is 3° (5.0), 2° (3.8), 1° (1.0). ◄

(See Problem 4.20.)

(a) $\overset{1°}{(CH_3)_2}\overset{3°}{CH}\overset{3°}{CH}\overset{1°}{(CH_3)_2}$ has two types of equivalent H's. Substitution of one of the 12 equivalent 1° H's yields 1-chloro-2,3-dimethylbutane, $ClCH_2CHCH_3CH(CH_3)_2$. Replacing one of the two 3° H's gives 2-chloro-2,3-dimethylbutane, $(CH_3)_2CClCH(CH_3)_2$. 3° H's are five times more reactive than are 1° H's.

	No. of H's	×	Relative Reactivity	=	Weighted Reactivity	Proportion × 100%	=	Percent Yield
1-chloro	12	×	1	=	12	12/22	=	54.5
2-chloro	2	×	5	=	10	10/22	=	45.5
				TOTAL	22			

(b) $\overset{1°}{(CH_3)_2}\overset{3°}{CH}\overset{2°}{CH_2}\overset{1'°}{CH_3}$ has 4 kinds of H's as indicated by superscripts 1°, 2°, 3°, and 1'°.

	H Atoms Kind	H Atoms Number	×	Relative Reactivity	=	Weighted Reactivity	Proportion × 100%	=	Percent Yield
Me|ClCH₂CHEt	1°	6	×	1.0	=	6.0	6.0/21.6	=	27.8
Me₂CHCH₂CH₂Cl	1'°	3	×	1.0	=	3.0	3.0/21.6	=	13.9
Me₂CClCH₂CH₃	3°	1	×	5.0	=	5.0	5.0/21.6	=	23.1
Me₂CHCHClMe	2°	2	×	3.8	=	7.6	7.6/21.6	=	35.2
					TOTAL	21.6			

Problem 4.38 Make a statement about the selectivity and reactivity of H_2C: from the yields of methylene insertion products formed from

$$\overset{1}{CH_3}\overset{2}{CH_2}\overset{3}{CH_2}\overset{2}{CH_2}\overset{1}{CH_3}$$

(page 46.) ◄

Calculate the yield assuming that only the probability factor is important; then compare the calculated and observed yields.

	H Atoms Kind	H Atoms Number	Proportion × 100%	=	Calc. % Yield	Observed % Yield
Hexane	1	6	6/12	=	50	48
Me₂CH-n-Pr	2	4	4/12	=	33.3	35
Et₂CHMe	3	2	2/12	=	16.7	17
	TOTAL	12				

The almost identical agreement validates the assumption that methylene is one of the most reactive and least selective species in organic chemistry.

Problem 4.39 List the alkanes having 1 to 8 C's that can react with CH_2N_2 to undergo methylene insertion to give a single product. ◄

Since methylene insertion is random, only those compounds in which all H's are equivalent can give a single product. These are CH_4, CH_3CH_3, $(CH_3)_4C$ and $(CH_3)_3C-C(CH_3)_3$.

Problem 4.40 (a) Give the formulas of all possible compounds formed on the chlorination of CH_4. (b) What experimental conditions would assure a good yield of the monochloro derivative? ◄

(a) CH_3Cl, CH_2Cl_2, $CHCl_3$, CCl_4.
(b) It is difficult to stop this halogenation at CH_3Cl. When equimolar amounts of CH_4 and Cl_2 react, a mixture of all 4 of these products is formed. If, however, an excess of CH_4 is used, the yield of CH_3Cl is greatly improved since the chances of a Cl· colliding with a CH_4 molecule rather than a chlorinated derivative are enhanced.

Chapter 5

Stereochemistry

5.1 STEREOISOMERISM

Stereoisomers have the same bonding order of atoms but differ in the way these atoms are arranged in space. They are classified by their symmetry properties in terms of certain symmetry elements. The two most important are:

1. A **symmetry plane** divides a molecule into equivalent halves. It is like a mirror placed so that half the molecule is a mirror image of the other half.
2. A **center (point) of symmetry** is a point in the *center* of a molecule to which a line can be drawn from any atom such that, when extended an equal distance past the center, the line meets another atom of the same kind.

A **chiral** stereoisomer is *not* superimposable on its mirror image. It does *not* possess a plane or center of symmetry. The nonsuperimposable mirror images are called **enantiomers**. A mixture of equal numbers of molecules of each enantiomer is a **racemic form**. The conversion of an enantiomer into a racemic form is called **racemization**. **Resolution** is the separation of a racemic form into individual enantiomers. Stereoisomers which are not mirror images are called **diastereomers**.

Molecules with a plane or center of symmetry have superimposable mirror images; they are **achiral**.

5.2 OPTICAL ISOMERISM

Plane-polarized light (light vibrating in only one plane) passed through a chiral substance emerges vibrating in a different plane. The enantiomer that rotates the plane of polarized light clockwise (to the right) as seen by observer is **dextrorotatory**; the enantiomer rotating to the left is **levorotatory**. The symbols (+) and (−) designate rotation to the right and left respectively. Because of this optical activity, enantiomers are called **optical isomers**. The racemic form (±) is optically inactive since it does not rotate the plane of polarized light.

The **specific rotation** $[\alpha]_\lambda^T$ is an inherent physical property of an enantiomer which, however, varies with the solvent used, temperature (T in °C), and wavelength of light used (λ). It is calculated from the observed rotation α as follows:

$$[\alpha]_\lambda^T = \frac{\alpha}{\ell c}$$

where ℓ = length of tube in decimeters (dm)

$c = \begin{cases} \text{concentration in g/ml, for a solution} \\ \text{density in g/ml, for a pure liquid} \end{cases}$

Problem 5.1 One and one-half grams of an enantiomer is dissolved in ethanol to make 50 ml of solution. Find the specific rotation at 20 °C for sodium light (λ = 5893 Å, the D line), if the solution has an observed rotation of +2.79° in a 10-cm polarimeter tube. ◄

53

First change the data to the appropriate units; 10 cm is 1 decimeter (dm) and the concentration is $1.5 \text{ g}/50 \text{ ml} = 0.03 \text{ g/ml}$. The specific rotation is then

$$[\alpha]_D^{20} = \frac{\alpha}{\ell c} = \frac{+2.79°}{(1)(0.03)} = +93° \text{ (in ethanol)}$$

Problem 5.2 Calculate the observed rotation in Problem 5.1 if: (a) the measurement is made in a 5-cm polarimeter tube, (b) the solution is diluted from 50 ml to 150 ml and the determination is carried out in a 10-cm tube. ◀

(a) The observed rotation is

$$\alpha = [\alpha]_D^{20}\ell c = (+93°)(0.5)(0.03) = +1.39°$$

Halving the length of the polarimeter tube allows only $\frac{1}{2}$ as many molecules to act on the plane-polarized light and the rotation is $\frac{1}{2}$ the previously observed value ($+2.79°$).

(b) Dilution from 50 ml to 150 ml causes a decrease of 1/3 in concentration and the observed rotation is decreased proportionately; doubling tube length doubles rotation and observed rotation is 2/3 that in (a).

$$\alpha = (+93°)(1.0)(0.01) = +0.93°$$

Problem 5.3 How can it be established that an observed dextrorotation of $+60°$ is not in fact a levorotation of $-300°$? ◀

Halving the concentration or the length of the tube would halve the number of molecules acting on the plane-polarized light and the rotation would then be $+30°$ if the substance is dextrorotatory or $-150°$ if levorotatory.

Many chiral organic molecules have at least one C (indicated by an asterisk) *bonded to four different atoms or groups*. Such a C is called a **chiral center**. Chiral enantiomers can be represented by planar projection formulas, as shown for lactic acid, $H_3CCH(OH)COOH$, in Fig. 5-1. In the **Fischer projection**, Fig. 5-1(c), the chiral C is at the intersection; horizontal groups project toward the viewer; vertical groups project away from the viewer.

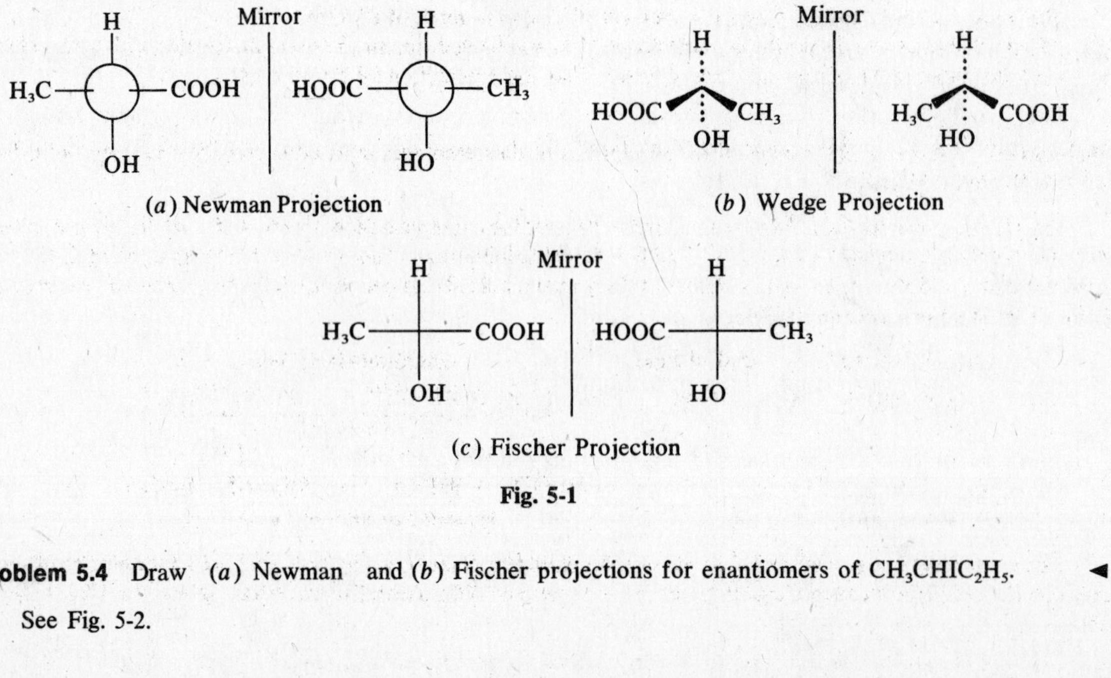

(a) Newman Projection   (b) Wedge Projection

(c) Fischer Projection

Fig. 5-1

Problem 5.4 Draw (a) Newman and (b) Fischer projections for enantiomers of $CH_3CHIC_2H_5$. ◀

See Fig. 5-2.

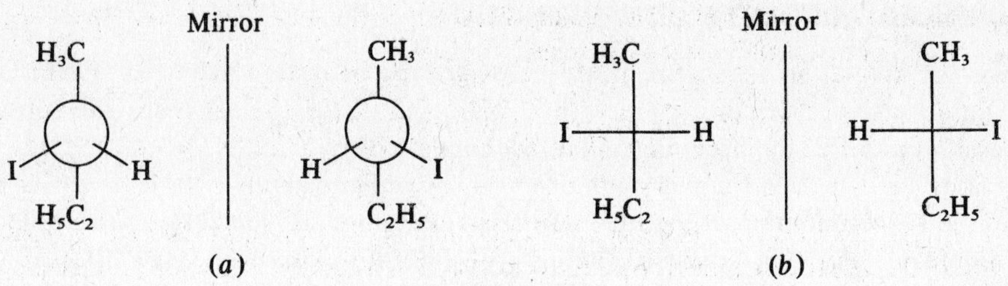

Fig. 5-2

Problem 5.5 Write structural formulas for the monochloroisopentanes.   Place an asterisk on any chiral C and indicate the 4 different groups about the C*. ◀

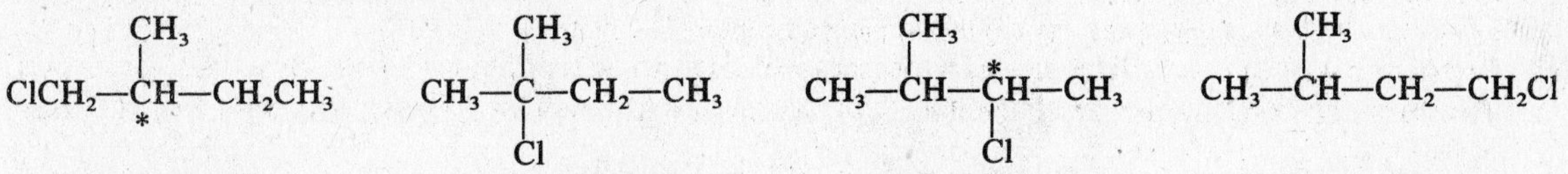

1-Chloro-2-methyl-   2-Chloro-2-methyl-   2-Chloro-3-methyl-   1-Chloro-3-methyl-
    butane (I)   butane (II)   butane (III)   butane (IV)

(H, CH₃, CH₂CH₃, ClCH₂)   (H, Cl, CH₃, (CH₃)₂CH)

In looking for chirality one considers the entire group, e.g. CH_2CH_3, attached to the C* and not just the adjacent atom.

Problem 5.6 Compare physical and chemical properties of (a) enantiomers, (b) an enantiomer and its racemic form, and (c) diastereomers. ◀

(a) With the exception of rotation of plane-polarized light, enantiomers have identical physical properties, e.g. boiling point, melting point, solubility.   Their chemical properties are the *same* toward *achiral* reagents, solvents and conditions.   Towards *chiral* reagents, solvents and catalysts, enantiomers react at *different* rates.   The transition states produced from the chiral reactant and the individual enantiomers are diastereomeric and hence have different energies; the $\Delta H^{\ddagger}$ values are different, as are the rates of reaction.

(b) Enantiomers are optically active; the racemic form is optically inactive.   Other physical properties of an enantiomer and its racemic form may differ depending on the racemic form.   The chemical properties are the *same* towards *achiral* reagents, but *chiral* reagents react at *different* rates.

(c) Diastereomers have different physical properties, and have different chemical properties with both achiral and chiral reagents.   The rates are different and the products may be different.

Problem 5.7 How can differences in the solubilities of diastereomers be used to resolve a racemic form into individual enantiomers? ◀

The reaction of a racemic form with a chiral reagent, for example a racemic (±) acid with a (−) base, yields two diastereomeric salts (+)(−) and (−)(−) with different solubilities.   These salts can be separated by fractional crystallization, and then each salt is treated with a strong acid (HCl) which liberates the enantiomeric organic acid.   This is shown schematically:

Racemic Form		Chiral Base		Diastereomeric Salts

$$(\pm)RCOOH \quad + \quad (-)B \quad \longrightarrow \quad \underbrace{(+)RCOO^-(-)BH^+ + (-)RCOO^-(-)BH^+}_{separated}$$

$$\qquad\qquad\qquad\qquad\qquad {}_{HCl}\downarrow \qquad\qquad\qquad \downarrow {}_{HCl}$$

$$\qquad\qquad\qquad (+)RCOOH + BH^+Cl^- \qquad (-)RCOOH + BH^+Cl^-$$
$$\qquad\qquad\qquad\qquad\qquad separated\ enantiomeric\ acids$$

The most frequently used chiral bases are the naturally occurring alkaloids such as strychnine, morphine and quinine.   Similarly, racemic organic bases are resolved with naturally occurring, optically active, organic acids.

5.3 RELATIVE AND ABSOLUTE CONFIGURATION

Configuration is the spatial arrangement of atoms or groups in a stereoisomer. Enantiomers have opposite configurations. For enantiomers with a single chiral site, to pass from one configuration to the other (**inversion**) requires the breaking and interchanging of two bonds. A second interchange of bonds causes a return to the original configuration. Configurations may change as a result of chemical reactions. To understand the mechanism of reactions, it is necessary to assign configurations to enantiomers. For this purpose, the sign of rotation cannot be used because *there is no relationship between configuration and sign of rotation.*

Problem 5.8 Esterification of (+)-lactic acid with methyl alcohol gives (−)-methyl lactate. Has the configuration changed?

$$
\underset{\substack{| \\ \text{OH}}}{\overset{\substack{\text{H} \\ |}}{CH_3-C^*-COOH}} \xrightarrow[\text{HCl}]{\text{CH}_3\text{OH}} \underset{\substack{| \\ \text{OH}}}{\overset{\substack{\text{H} \\ |}}{CH_3-C^*-COOCH_3}}
$$

$$(+)\text{-Lactic acid} = +3.3° \qquad (-)\text{-Methyl lactate} = -8.2°$$ ◄

No; even though the sign of rotation changes, there is no breaking of bonds to the chiral C*.

The Cahn-Ingold-Prelog rules are used to assign configuration to chiral compounds.

Rule 1.

Groups and atoms bonded to the chiral C are assigned *decreasing priorities* based on *decreasing atomic number* of the atom bonded directly to the C. (For isotopes, the one with higher mass has the higher priority, e.g. deuterium over hydrogen.) The priorities of the groups will be given by numbers in parentheses, using (4) for the highest and (1) for the lowest. The lowest-priority group (1) must project away from the viewer and be in the back of the paper, leaving the other three groups projecting forward. In the Fischer projection the priority (1) group must be in a vertical position (if necessary, make two interchanges of groups to achieve this configuration). Then, for the remaining three groups or atoms, if the sequence of decreasing priority, (4) to (3) to (2), is *counterclockwise*, the configuration is designated S; if it is *clockwise*, the configuration is designated R. The rule is illustrated for 1-chloro-1-bromoethane in Fig. 5-3. Both configuration and sign of optical rotation are included in the complete name of a species, e.g. (S)-(+)-1-chloro-1-bromoethane.

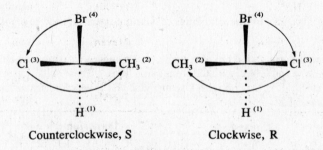

Counterclockwise, S   Clockwise, R

Fig. 5-3

Rule 2.

If the first bonded atom is the same in at least 2 groups, the priority is determined by comparing the *next* atoms in each of these groups. Thus, ethyl (H_3CCH_2-), with one C and two H's on the first bonded C, has priority over methyl ($-CH_3$), with 3 H's on the C. For butyl groups, the order of decreasing priority is

$$(CH_3)_3C- \; > \; CH_3CH_2\overset{\underset{\displaystyle CH_3}{|}}{C}H- \; > \; CH_3\overset{\underset{\displaystyle CH_3}{|}}{C}HCH_2- \; > \; CH_3CH_2CH_2CH_2-$$

Rule 3.

For purposes of assigning priorities, replace:

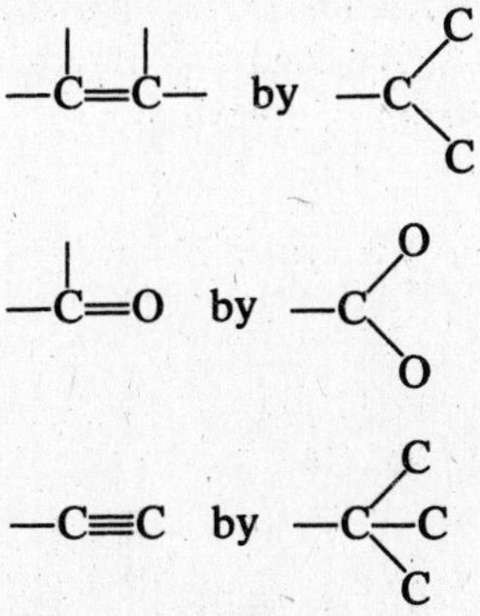

$$-\overset{|}{C}=\overset{|}{C}- \quad by \quad -C\hspace{-2pt}<\hspace{-4pt}\begin{array}{c} C \\ C \end{array}$$

$$-\overset{|}{C}=O \quad by \quad -C\hspace{-2pt}<\hspace{-4pt}\begin{array}{c} O \\ O \end{array}$$

$$-C\equiv C \quad by \quad -\overset{\underset{\displaystyle C}{|}}{C}-C$$

Problem 5.9 Structures of CHClBrF are written below in seven Fischer projection formulas. Relate (b) through (g) to structure (a).

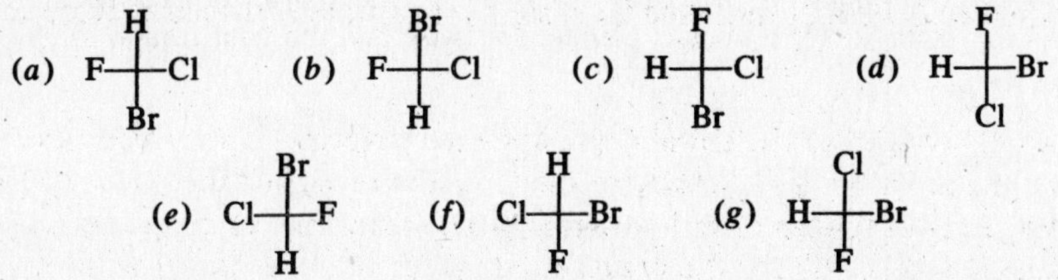

(a) $\overset{\displaystyle H}{\underset{\displaystyle Br}{F-\!\!\!\!\;\!|\!\!\!\;\!-Cl}}$　　(b) $\overset{\displaystyle Br}{\underset{\displaystyle H}{F-\!\!\!\!\;\!|\!\!\!\;\!-Cl}}$　　(c) $\overset{\displaystyle F}{\underset{\displaystyle Br}{H-\!\!\!\!\;\!|\!\!\!\;\!-Cl}}$　　(d) $\overset{\displaystyle F}{\underset{\displaystyle Cl}{H-\!\!\!\!\;\!|\!\!\!\;\!-Br}}$

(e) $\overset{\displaystyle Br}{\underset{\displaystyle H}{Cl-\!\!\!\!\;\!|\!\!\!\;\!-F}}$　　(f) $\overset{\displaystyle H}{\underset{\displaystyle F}{Cl-\!\!\!\!\;\!|\!\!\!\;\!-Br}}$　　(g) $\overset{\displaystyle Cl}{\underset{\displaystyle F}{H-\!\!\!\!\;\!|\!\!\!\;\!-Br}}$　　◄

If two structural formulas differ by an odd number of interchanges, they are enantiomers; if by an even number, they are identical. See Table 5-1.

Table 5-1

	Sequence of Group Interchanges	Number of Interchanges	Relationship to (a)
(b)	H, Br	1 (odd)	enantiomer
(c)	H, F	1 (odd)	enantiomer
(d)	H, F; Br, Cl	2 (even)	same
(e)	H, Br; Cl, F	2 (even)	same
(f)	F, Br; F, Cl	2 (even)	same
(g)	F, Br; Br, Cl; H, Cl	3 (odd)	enantiomer

Problem 5.10 Put the following groups in decreasing order of priority.

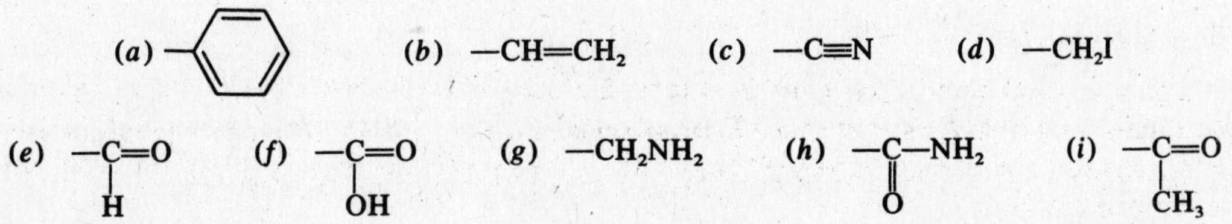

(a) —⬡　　　(b) —CH=CH$_2$　　(c) —C≡N　　(d) —CH$_2$I

(e) $-\overset{}{\underset{\displaystyle H}{C}}=O$　　(f) $-\overset{}{\underset{\displaystyle OH}{C}}=O$　　(g) —CH$_2$NH$_2$　　(h) $-\overset{}{\underset{\displaystyle O}{C}}-NH_2$　　(i) $-\overset{}{\underset{\displaystyle CH_3}{C}}=O$　　◄

In each case, the first bonded atom is a C. Therefore, the second bonded atom determines the priority. In decreasing order of priority, these are I > O > N > C. The equivalencies are

(a) $-\overset{\text{C}}{\underset{\text{C}}{\text{C}}}-\text{C}$ (b) $-\overset{\text{H}}{\underset{\text{C}}{\text{C}}}-\text{C}$ (c) $-\overset{\text{N}}{\underset{\text{N}}{\text{C}}}-\text{N}$ (d) $-\overset{\text{H}}{\underset{\text{I}}{\text{C}}}-\text{H}$ (e) $-\overset{\text{H}}{\underset{\text{O}}{\text{C}}}-\text{O}$

(f) $-\overset{\text{O}}{\underset{\text{O}}{\text{C}}}-\text{O}$ (g) $-\overset{\text{H}}{\underset{\text{N}}{\text{C}}}-\text{H}$ (h) $-\overset{\text{O}}{\underset{\text{N}}{\text{C}}}-\text{O}$ (i) $-\overset{\text{C}}{\underset{\text{O}}{\text{C}}}-\text{O}$

The order of decreasing priority is: (d) > (f) > (h) > (i) > (e) > (c) > (g) > (a) > (b). In (d), one I has a greater priority than 3 O's in (f).

Problem 5.11 Designate as R or S the configuration of

(a) $\text{ClCH}_2-\overset{\text{Cl}}{\underset{\text{CH}_3}{\text{C}}}-\text{CH(CH}_3)_2$ (b) $\text{H}_2\text{C}=\text{CH}-\overset{\text{H}}{\underset{\text{Br}}{\text{C}}}-\text{CH}_2\text{CH}_3$ (c) $\text{H}-\overset{\text{NH}_2}{\underset{\text{COOC}_2\text{H}_5}{\text{C}}}-\text{COOH}$ ◄

(a) The order of priorities is Cl (4), CH_2Cl (3), $\text{CH(CH}_3)_2$ (2) and CH_3 (1). CH_3, with the lowest priority, is projected in back of the plane of the paper and is not considered in the sequence. The sequence of decreasing priority of the other groups is counterclockwise and the configuration is S.

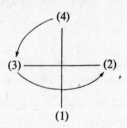

The compound is (S)-1,2-dichloro-2,3-dimethylbutane.

(b) The sequence of priorities is Br (4), $\text{H}_2\text{C}=\text{CH}-$ (3), CH_3CH_2- (2) and H (1).

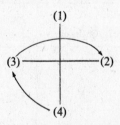

The name is (R)-3-bromo-1-pentene.

(c) Two exchanges are made to get H(1) in a vertical position without changing the configuration. Now the other three groups can be projected forward with no change in sequence. A possible identical structure is

$$\overset{(1)}{\text{H}}$$
$$\overset{(4)}{\text{H}_2\text{N}}-\overset{(3)}{\text{C}}\text{OOC}_2\text{H}_5$$
$$\underset{(2)}{\text{C}}\text{OOH}$$

The sequence is clockwise or R.

Problem 5.12 Draw and specify as R and S the enantiomers, if any, of all the monochloropentanes. ◄

n-Pentane has 3 monochloro-substituted products, 1-chloro, 2-chloro, and 3-chloropentane

$\text{ClCH}_2\text{CH}_2\text{CH}_2\text{CH}_2\text{CH}_3$ $\text{CH}_3\overset{*}{\text{C}}\text{HClCH}_2\text{CH}_2\text{CH}_3$ $\text{CH}_3\text{CH}_2\text{CHClCH}_2\text{CH}_3$

Only 2-chloropentane has a chiral C, whose substituent groups in order of decreasing priority are Cl (4), CH_3CH_2 (3), CH_3 (2) and H (1).   The configurations are:

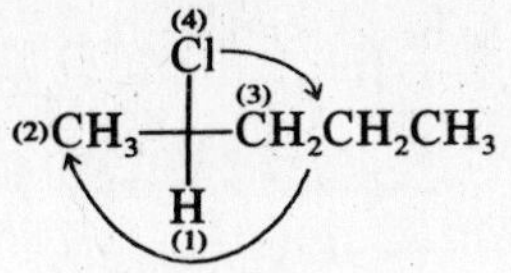

(R)-2-Chloropentane 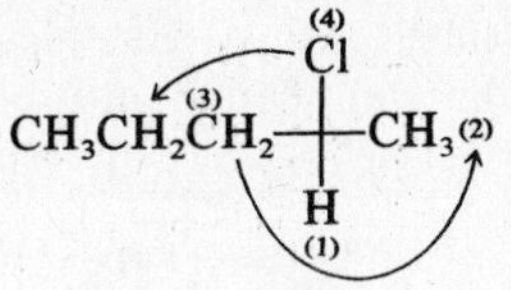   (S)-2-Chloropentane

The structures of the monochloroisopentanes are given in Problem 5.5. The sequence of decreasing priority of substituents for structure I is $ClCH_2 > CH_2CH_3 > CH_3 > H$, while that for III is $Cl > (CH_3)_2CH > CH_3 > H$.   The configurations are:

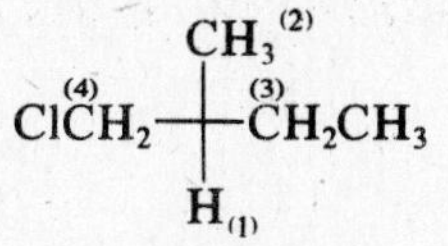

(S)-(I) 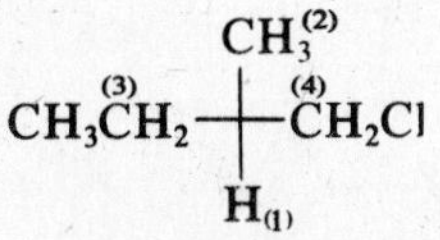   (R)-(I) 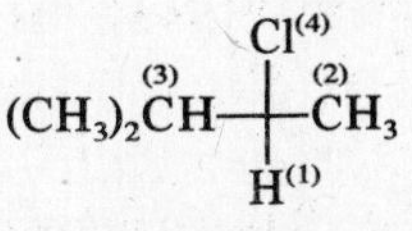   (S)-(III) 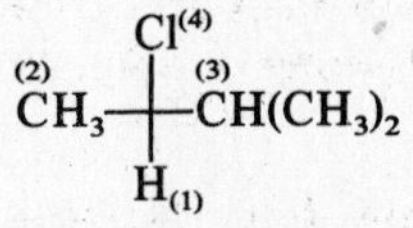   (R)-(III)

Neopentyl chloride, $(CH_3)_3CCH_2Cl$, has no chiral C and therefore no enantiomers.

Problem 5.13   Write structural formulas for the following, indicating enantiomers, if any, and their configurations: (a) 3-methyl-3-pentanol,   (b) 2,2-dimethyl-3-bromohexane,   (c) 3-phenyl-3-chloro-1-propene.   ◄

(a)   The alcohol

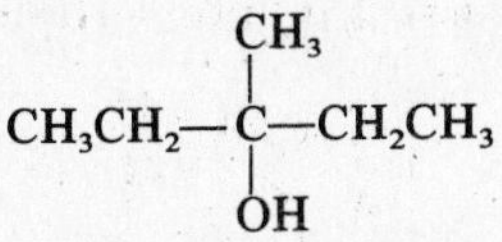

has no atom bonded to four different groups and no enantiomers.

(b) $(H_3C)_3C\overset{*}{C}HBr(CH_2)_2CH_3$ has one chiral C (shown with an asterisk).

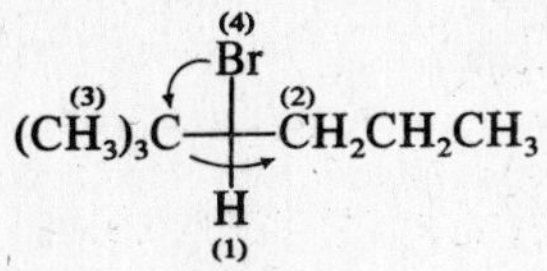

(S) 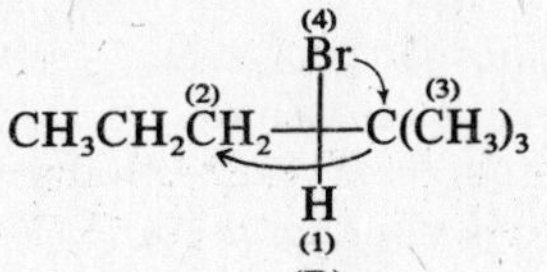   (R)

(c)

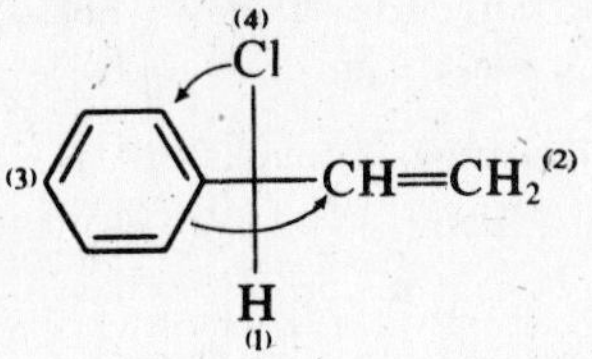

(S) 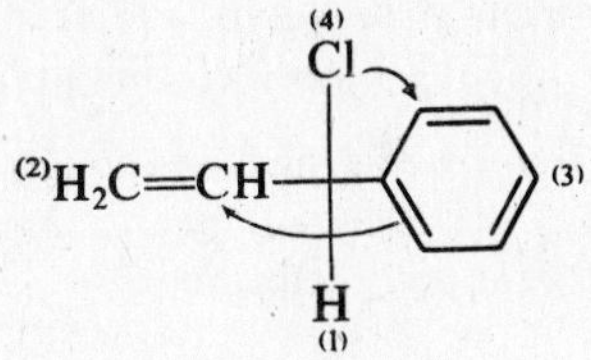   (R)

The D,L method for assigning **relative configurations** uses glyceraldehyde, $HOCH_2\overset{*}{C}H(OH)CHO$, as the reference molecule.   The (+) and (−) enantiomers were arbitrarily assigned the configurations shown below in Fischer projections and were designated as D and L respectively.

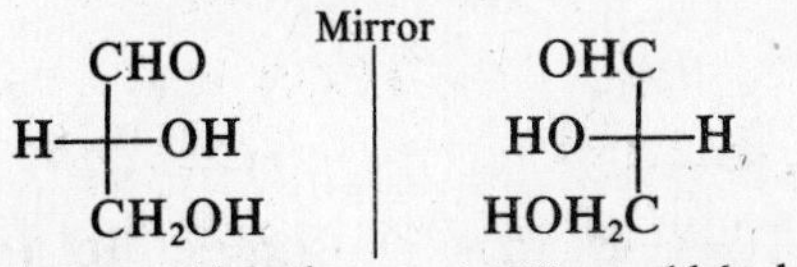

D-(+)-Glyceraldehyde   L-(−)-Glyceraldehyde

The arbitrary assignment has recently been shown to be the correct (absolute) configuration.   Relative configurations of chiral C's in other compounds are established by synthesis of these compounds from glyceraldehyde.

Problem 5.14 D-(+)-Glyceraldehyde is oxidized to (−)-glyceric acid, $HOCH_2CH(OH)COOH$. Give the D,L designation of the acid. ◄

Oxidation of the D-(+)-aldehyde does not affect any of the bonds to the chiral C. The acid has the same D configuration even though the sign of rotation is changed.

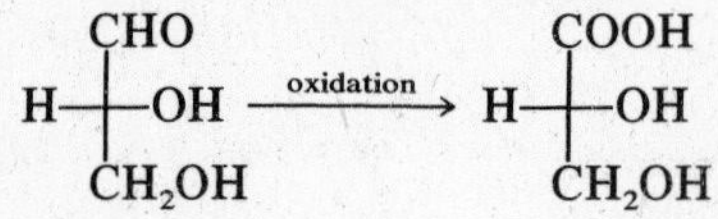

$$
\begin{array}{ccc}
\text{CHO} & & \text{COOH} \\
\text{H}-\!\!\!-\text{OH} & \xrightarrow{\text{oxidation}} & \text{H}-\!\!\!-\text{OH} \\
\text{CH}_2\text{OH} & & \text{CH}_2\text{OH}
\end{array}
$$

D-(+)-Glyceraldehyde   D-(−)-Glyceric acid

Problem 5.15 Does configuration change in the following reactions? Designate products (D,L) and (R,S).

(a)

$$
\underset{\text{(D/S)}}{\overset{(4)}{\underset{\text{H}}{\overset{\text{Cl}}{\underset{(3)}{\text{CH}_3\text{CH}_2}}}}-\overset{(2)}{\text{CH}_3}}
$$

$\xrightarrow{\text{Cl}_2,\ \text{light}}$ $\overset{(2)}{\text{CH}_3\text{CH}_2}-\underset{\text{H}}{\overset{(4)\ \text{Cl}}{\text{C}}}-\overset{(3)}{\text{CH}_2\text{Cl}}$

(b)

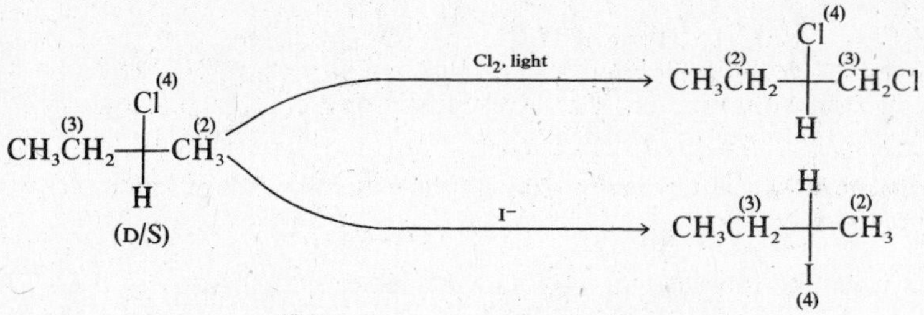

$\xrightarrow{\text{I}^-}$ $\overset{(3)}{\text{CH}_3\text{CH}_2}-\underset{(4)\ \text{I}}{\overset{\text{H}}{\text{C}}}-\overset{(2)}{\text{CH}_3}$

(c)

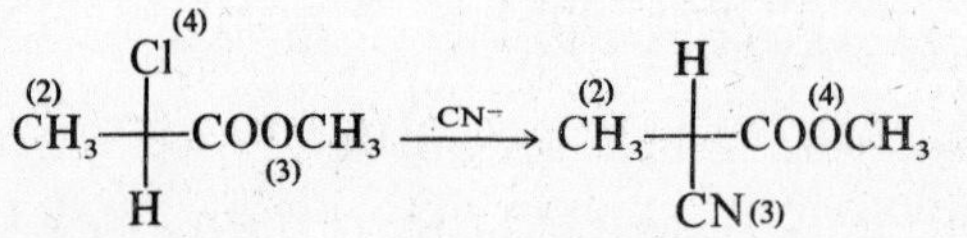

$$
\underset{\text{(D/R)}}{\overset{(2)}{\text{CH}_3}-\underset{\text{H}}{\overset{(4)\ \text{Cl}}{\text{C}}}-\underset{(3)}{\text{COOCH}_3}}
$$

$\xrightarrow[(3)]{\text{CN}^-}$ $\overset{(2)}{\text{CH}_3}-\underset{\text{CN}(3)}{\overset{\text{H}}{\text{C}}}-\overset{(4)}{\text{COOCH}_3}$ ◄

(a) No bond to the chiral C is broken and the configuration is unchanged. Therefore, the configuration of both reactant and product is D. The product is R; S converts to R because there is a priority change. Thus, a change from R to S does not necessarily signal an inversion of configuration; it does only if the order of priority is unchanged.

(b) A bond to the chiral C is broken when I⁻ displaces Cl⁻. An inversion of configuration has taken place and there is a change from D to L. The product is R. This change from S to R also shows inversion, because there is no change in priorities.

(c) Inversion has occurred and there is a change from D to L. The product is R. Even though an inversion of configuration occurred, the reactant and product are both R. This is so because there is a change in the order of priority. The displaced Cl has priority (4) but the incoming CN⁻ has priority (3).

In general, the D,L convention signals a configuration change, if any, as follows: D → D or L → L means no change (retention), and D → L or L → D means change (inversion). The R,S convention can also be used, but it requires working out the priority sequences.

5.4 MOLECULES WITH MORE THAN ONE CHIRAL CENTER

With n dissimilar chiral atoms the number of stereoisomers is 2^n and the number of racemic forms is 2^{n-1}, as illustrated below for 2-chloro-3-bromobutane ($n = 2$). The R,S configuration is shown next to the C's.

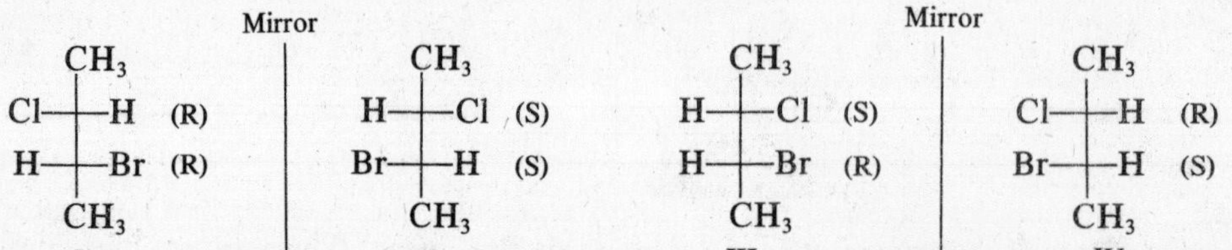

$$
\begin{array}{cccc}
\text{CH}_3 & \text{CH}_3 & \text{CH}_3 & \text{CH}_3 \\
\text{Cl}-\!\!\!-\text{H}\ (R) & \text{H}-\!\!\!-\text{Cl}\ (S) & \text{H}-\!\!\!-\text{Cl}\ (S) & \text{Cl}-\!\!\!-\text{H}\ (R) \\
\text{H}-\!\!\!-\text{Br}\ (R) & \text{Br}-\!\!\!-\text{H}\ (S) & \text{H}-\!\!\!-\text{Br}\ (R) & \text{Br}-\!\!\!-\text{H}\ (S) \\
\text{CH}_3 & \text{CH}_3 & \text{CH}_3 & \text{CH}_3 \\
\text{I} & \text{II} & \text{III} & \text{IV}
\end{array}
$$

Mirror	Mirror
Enantiomers,	*Enantiomers,*
racemic form 1	*racemic form 2*

If $n = 2$ and the two chiral atoms are identical in that each holds the same four different groups, there are only 3 stereoisomers, as illustrated for 2,3-dichlorobutane.

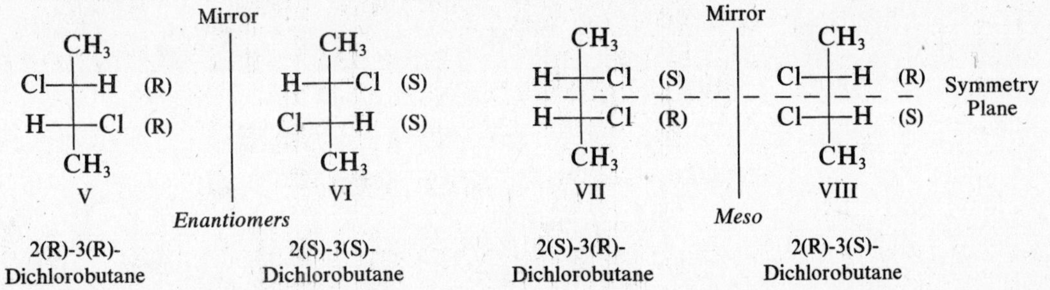

$$
\begin{array}{cccc}
\text{CH}_3 & \text{CH}_3 & \text{CH}_3 & \text{CH}_3 \\
\text{Cl}\!-\!\!-\!\text{H} \ (R) & \text{H}\!-\!\!-\!\text{Cl} \ (S) & \text{H}\!-\!\!-\!\text{Cl} \ (S) & \text{Cl}\!-\!\!-\!\text{H} \ (R) \\
\text{H}\!-\!\!-\!\text{Cl} \ (R) & \text{Cl}\!-\!\!-\!\text{H} \ (S) & \text{H}\!-\!\!-\!\text{Cl} \ (R) & \text{Cl}\!-\!\!-\!\text{H} \ (S) \\
\text{CH}_3 & \text{CH}_3 & \text{CH}_3 & \text{CH}_3 \\
\text{V} & \text{VI} & \text{VII} & \text{VIII}
\end{array}
$$

Mirror (between V and VI) — *Enantiomers*
Mirror (between VII and VIII) — *Meso* — Symmetry Plane

2(R)-3(R)-Dichlorobutane 2(S)-3(S)-Dichlorobutane 2(S)-3(R)-Dichlorobutane 2(R)-3(S)-Dichlorobutane

Structures VII and VIII are identical because rotating either one 180° in the plane of the paper makes it superposable with the other one. VII possesses a symmetry plane and is achiral. Achiral stereoisomers which have chiral centers are called *meso*. The *meso* structure is a diastereomer of either of the enantiomers. The *meso* structure with two chiral sites always has the (R, S) configuration.

5.5 SYNTHESIS AND OPTICAL ACTIVITY

1. Optically inactive reactants with achiral catalysts or solvents yield optically inactive products; with a chiral catalyst, such as an enzyme, optically active products are formed.

2. A second chiral center generated in a chiral compound may not have an equal chance for R and S configurations; a 50:50 mixture of diastereomers is *not* usually obtained.

3. Replacement of a group or atom on a chiral center can occur with retention or inversion of configuration or with a mixture of the two (complete or partial racemization), depending on the mechanism of the reaction.

Problem 5.16 (a) What is the stereochemistry of $CH_3CHClCH_2CH_3$ isolated from the chlorination of butane? (b) How does the mechanism of chlorination of RH account for the formation of the product in (a)? ◄

(a) Since all reactants are achiral and optically inactive, the product is optically inactive. Since the product has a chiral center, C^2, it must be a racemic form.

(b) See Fig. 5-4. The first propagation step produces $CH_3\dot{C}HCH_2CH_3$. The C of this radical uses sp^2 hybridized orbitals and its three bonds are coplanar. The p orbital with the odd electron is at right angles to this plane. In the second propagation step, the p orbital of the *sec*-butyl radical can add a Cl atom from Cl_2 on either side of the planar skeleton to give one product with R and one with S configuration. Since the probability of attack is identical on both sides of the plane, chlorination gives equal amounts of enantiomers and hence a racemic form.

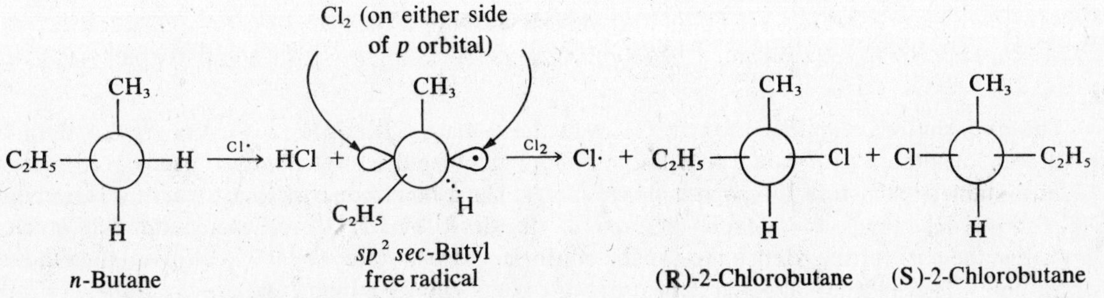

Fig. 5-4

Problem 5.17 (*a*) What two products are obtained when C^3 of (R)-2-chlorobutane is chlorinated? (*b*) Are these diastereomers formed in equal amounts? (*c*) In terms of mechanism account for the fact that

$$rac\,(\pm)\text{-ClCH}_2\text{C(Cl)CH}_2\text{CH}_3$$
$$|$$
$$\text{CH}_3$$

is obtained when (R)-ClCH$_2$-$\overset{*}{\text{C}}$H(CH$_3$)CH$_2$CH$_3$ is chlorinated. ◄

(*a*)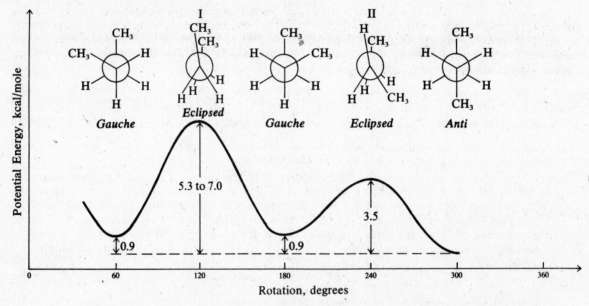

In the products C^2 retains the R configuration since none of its bonds were broken and there was no change in priority. The configuration at C^3, the newly created chiral center, can be either R or S.

(*b*) No. The numbers of molecules with S and R configurations at C^3 are not equal.

(*c*) Removal of the H from the chiral C leaves the achiral free radical ClCH$_2\dot{\text{C}}$(CH$_3$)CH$_2$CH$_3$. Like the radical in Problem 5.16, it reacts with Cl$_2$ to give a racemic form.

5.6 CONFORMATION AND STEREOISOMERISM

Figure 5-5 shows the conformations of *n*-butane.

Fig. 5-5

The *anti* conformation has the two CH$_3$'s farthest apart. It has the lowest energy, is the most stable and constitutes the greatest population of butane molecules. The two eclipsed conformations are least stable. Structure I, having eclipsed CH$_3$'s, has a higher energy than structure II, in which CH$_3$ eclipses H. In the two *gauche* or *skew* conformers, the CH$_3$'s are not separated as much as they are in the *anti* form. Hence the gauche conformer is less stable (by 0.9 kcal/mole) than the *anti* conformer. These three staggered conformations are at energy minima and are the stable conformers of butane. They are stereoisomers because they have the same structural formulas but different arrangements in space. However, since they can be readily interconverted by twisting a single bond, they are **conformational stereoisomers**.

The two *gauche* forms are nonsuperimposable mirror images; they are **conformational enantiomers**. They are both diastereomeric with the *anti* conformer because they are stereo-isomers, but not mirror images, of the *anti* form.

The energy barriers between the conformers are sufficiently low so that they are interconvertible at room temperature and therefore cannot be separated. There is a higher population of the more stable *anti* conformer and lower populations of the two less stable *gauche* conformers (which have equal stabilities).

Conformational isomers differ from configurational isomers in that configurational isomers may be interconverted by breaking and making chemical bonds. The energy needed for such changes is of the order of 50–150 kcal/mole, which is large enough to permit isolation of the isomers and is much larger than the energy required for interconversion of conformers.

Supplementary Problems

Problem 5.18 (*a*) What is the necessary and sufficient condition for the existence of enantiomers? (*b*) What is the necessary and sufficient condition for measurement of optical activity? (*c*) Are all substances with chiral atoms optically active and resolvable? (*d*) Are enantiomers possible in molecules that do not have chiral carbon atoms? ◄

(*a*) Chirality in molecules having nonsuperimposable mirror images. (*b*) An excess of one enantiomer and a specific rotation large enough to be measured. (*c*) No. Racemic forms are not optically active but are resolvable. *Meso* compounds are inactive and not resolvable. (*d*) Yes. The presence of a chiral atom is a sufficient but not necessary condition for enantiomerism. For example, properly disubstituted allenes have no plane or center of symmetry but are chiral.

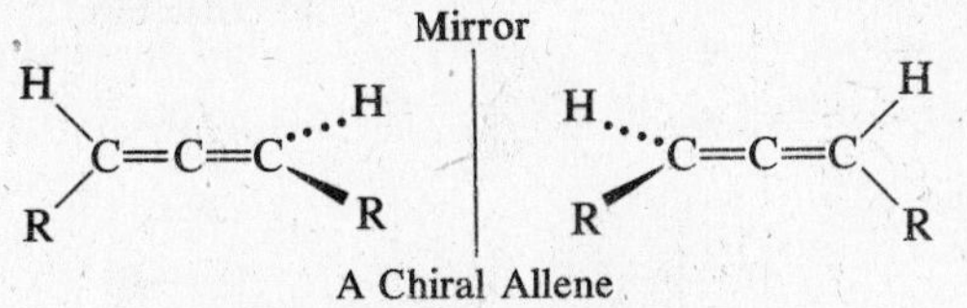

A Chiral Allene
(nonsuperimposable enantiomers)

Problem 5.19 Point out which of the following are chiral.

(*a*) Screwdriver	(*e*) Tree	(*i*) Nail	(*m*) Shoe
(*b*) Screw	(*f*) Your ear	(*j*) T-shirt	(*n*) Coat with buttons
(*c*) Automobile	(*g*) Your nose	(*k*) Helix	
(*d*) Pullover sweater	(*h*) Your foot	(*l*) Spool of thread	◄

(*b*), (*c*), (*e*), (*f*), (*h*), (*k*), (*l*), (*m*), (*n*).

Problem 5.20 Relative configurations of chiral atoms are sometimes established by using reactions in which there is no change in configuration because no bonds to the chiral atom are broken. Which of the following reactions can be used to establish relative configurations?

(*a*) $\quad$ (S)-$CH_3CHClCH_2CH_3$ + $Na^+OCH_3^-$ $\longrightarrow$ $CH_3CH(OCH_3)CH_2CH_3$ + Na^+Cl^-

(*b*) $\quad$ (S)-$CH_3CH_2\overset{\underset{\displaystyle |}{CH_3}}{C}HO^-Na^+$ + CH_3Br $\longrightarrow$ $CH_3CH_2\overset{\underset{\displaystyle |}{CH_3}}{C}HOCH_3$ + Na^+Br^-

$$\text{CH}_3 \qquad\qquad\qquad\qquad \text{CH}_3$$

(c)   (R)-CH₃CH₂ĊOHCH₂Cl + PCl₅ ⟶ CH₃CH₂ĊClCH₂Cl + POCl₃ + HCl

(d)   (S)-(CH₃)₂C(OH)CHBrCH₃ + Na⁺CN⁻ ⟶ (CH₃)₂C(OH)CH(CN)CH₃ + Na⁺Br⁻

(e)   (R)-CH₃CH₂CHCH₃ + Na ⟶ CH₃CH₂CHCH₃ + ½H₂   ◀
                 | |
                OH   O⁻Na⁺

   (b) and (e).   The others involve breaking bonds to the chiral C.

Problem 5.21   Account for the disappearance of optical activity observed when (R)-2-butanol is allowed to stand in aqueous H₂SO₄ and when (S)-2-iodooctane is treated with aqueous KI solution.   ◀

   Optically active compounds become inactive if they lose their chirality because the chiral center no longer has 4 different groups, or if they undergo racemization.   In the two reactions cited, C remains chiral and it must be concluded that in both reactions racemization occurs.

Problem 5.22   For the following compounds, draw projection formulas for all stereoisomers and point out their R,S specifications, optical activity (where present), and *meso* compounds: (a) 1,2,3,4-tetrahydroxybutane, (b) 1-chloro-2,3-dibromobutane,   (c) 2,4-diiodopentane,   (d) 2,3-tribromohexane,   (e) 2,3,4-tribromopentane.   ◀

(a)   HOCH₂ĊHOHĊHOHCH₂OH, with two similar chiral C's, has one *meso* form and two optically active enantiomers.

```
   ¹CH₂OH              ¹CH₂OH              ¹CH₂OH
 H─²─OH              H─²─OH             HO─²─H
 H─³─OH             HO─³─H              H─³─OH
   ⁴CH₂OH              ⁴CH₂OH              ⁴CH₂OH
   2S,3R               2S,3S               2R,3R
   meso      └──────────────────────────────┘
                         Racemic form
```

(b)   ClCH₂ĊHBrĊHBrCH₃ has 2 different chiral C's.   There are four (2²) optically active enantiomers.

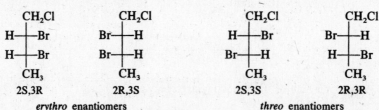

```
   CH₂Cl        CH₂Cl           CH₂Cl         CH₂Cl
 H─┼─Br      Br─┼─H          H─┼─Br       Br─┼─H
 H─┼─Br      Br─┼─H         Br─┼─H         H─┼─Br
   CH₃          CH₃             CH₃           CH₃
   2S,3R        2R,3S           2S,3S         2R,3R
     erythro enantiomers         threo enantiomers
```

   The two sets of diastereomers are differentiated by the prefix *erythro* for the set in which at least two identical or similar substituents on chiral C's are eclipsed.   The other set is called *threo*.

(c)   CH₃ĊHICH₂ĊHICH₃ has two similar chiral C's, C² and C⁴, separated by a CH₂ group.   There are two enantiomers comprising a (±) pair and one *meso* compound.

```
   ¹CH₃            CH₃            CH₃
 H─²─I          H─┼─I          I─┼─H
   ³CH₂            CH₂            CH₂
 H─⁴─I          I─┼─H          H─┼─I
   ⁵CH₃            CH₃            CH₃
   2S,4R          2S,4S          2R,4R
   meso
```

(d) With 3 different chiral C's in CH₃C̊HBrC̊HBrC̊HBrCH₂CH₃, there are 8 (2³) enantiomers and 4 racemic forms.

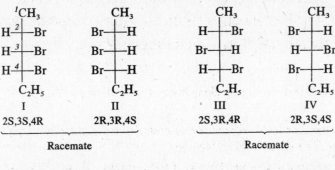

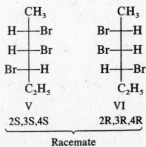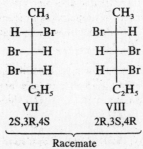

(e) $\overset{1}{C}H_3\overset{2}{\overset{*}{C}}HBr\overset{3}{C}HBr\overset{4}{\overset{*}{C}}HBr\overset{5}{C}H_3$ has two similar chiral atoms (C² and C⁴). There are 2 enantiomers in which the configurations of C² and C⁴ are the same, RR or SS. When C² and C⁴ have different configurations, one R and one S, C³ acquires two different configurations and therefore there are two *meso* forms.

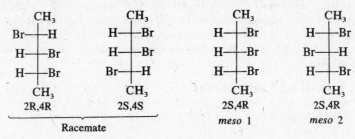

Problem 5.23 The specific rotation of (R)-(−)-2-bromooctane is −36°. What is the percentage composition of a mixture of enantiomers of 2-bromooctane whose rotation is +18°?   ◄

Let x = mole fraction of R, 1 − x = mole fraction of S.

$$x(-36°) + (1 - x)(36°) = 18° \quad or \quad x = 1/4$$

The mixture has 25% R and 75% S; it is 50% racemic and 50% S.

Problem 5.24 (S)-2-Chlorobutane yields a mixture of 71% *meso* and 29% 2(S)-3(S)-2,3-dichlorobutane, while (R)-2-chlorobutane similarly gives a mixture of 71% *meso* and 29% 2(R)-3(R)-2,3-dichlorobutane. Account for these products by conformational analysis.   ◄

See Fig. 5-6. In the first step the free radical CH₃C̊HCl—C̊H—CH₃ (III through VI) is formed on an atom adjacent to a chiral C with either an S or R configuration. The formation of the free radical does not change the configuration of the chiral center, which is S in one case and R in the other. A second chiral center, which may be R or S, is generated after step two.

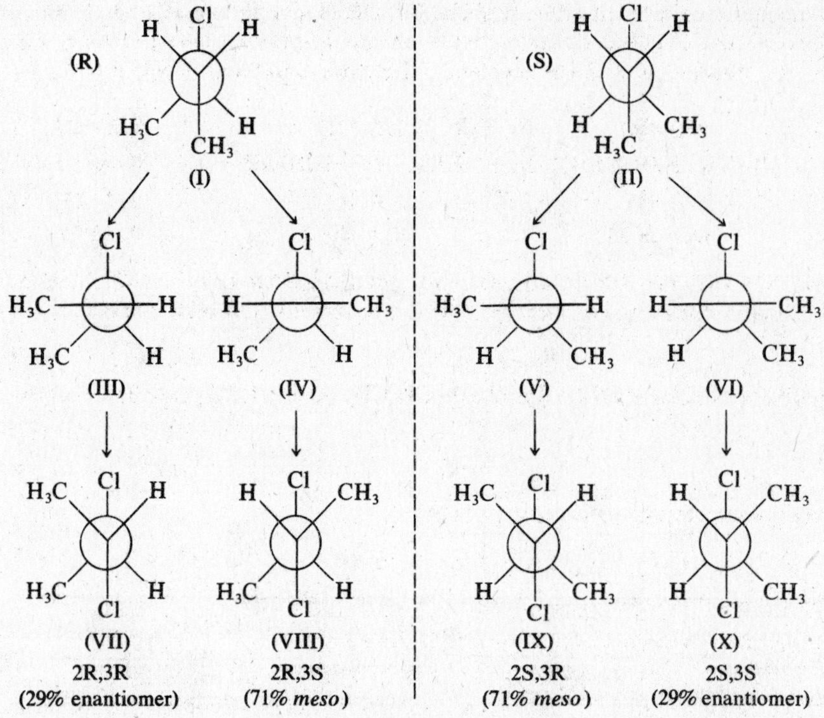

Fig. 5-6

(R)-2-Chlorobutane (I) forms free radicals (III and IV) which are conformational diastereomers with different stabilities and populations. This is also true of I's enantiomer II, (S)-2-chlorobutane, which gives free radicals V and VI. The more stable free-radical conformers are IV and V because their CH$_3$'s are *anti*-like. The transition states for Cl abstractions arising from conformations IV and V have lower $\Delta H^\ddagger$ values than the diastereomeric transition states from the more crowded *gauche*-like conformations, III and VI. The major product is therefore the *meso* compound, VIII $\equiv$ IX.

Problem 5.25 Predict the yields of stereoisomeric products, and the optical activity of the mixture of products, formed from chlorination of a racemic mixture of 2-chlorobutane to give 2,3-dichlorobutane (see Problem 5.24).   ◄

The (S)-2-chlorobutane comprising 50% of the racemic mixture gives 35.5% of the *meso* (RS) product and 14.5% of the SS enantiomer. The R enantiomer gives 35.5% *meso* and 14.5% SS products. The total yield of *meso* product is 71% and the combination of 14.5% RR and 14.5% SS gives 29% racemic product. The total reaction mixture is optically inactive. This result confirms the generalization that optically inactive starting materials, reagents and solvents always lead to optically inactive products.

Problem 5.26 For the following reactions give the number of stereoisomers that are isolated, their R,S configurations and their optical activities. Use Fischer projections.

(a)   *meso*-HOCH$_2$CHOHCHOHCH$_2$OH $\xrightarrow{\text{oxidation}}$ HOCH$_2$CHOHCHOHCOOH

(b)   (R)-ClCH$_2$CH(CH$_3$)CH$_2$CH$_3$ $\xrightarrow{\text{Na}}$ CH$_3$CH$_2$CH(CH$_3$)CH$_2$CH$_2$CH(CH$_3$)CH$_2$CH$_3$

(c) $rac(\pm)$-CH$_3$—$\overset{\overset{\text{O}}{\|}}{\text{C}}$—CHOH—CH$_3$ $\xrightarrow{\text{H}_2/\text{Ni}}$ CH$_3$CH(OH)CH(OH)CH$_3$

(d)   (S)-CH$_3$CH$_2$—$\overset{\overset{\text{CH}_2}{\|}}{\text{C}}$—CH(CH$_3$)CH$_2CH_3$ $\xrightarrow{\text{H}_2/\text{Ni}}$ CH$_3$CH$_2$CH(CH$_3$)CH(CH$_3$)CH$_2$CH$_3$ ◄

(a) This *meso* alcohol is oxidized at either terminal CH_2OH to give an optically inactive racemic form. The chiral C next to the oxidized C undergoes a change in priority order; CH_2OH (2) goes to COOH (3). Therefore, if this C is R in the reactant, it becomes S in the product; if S it goes to R.

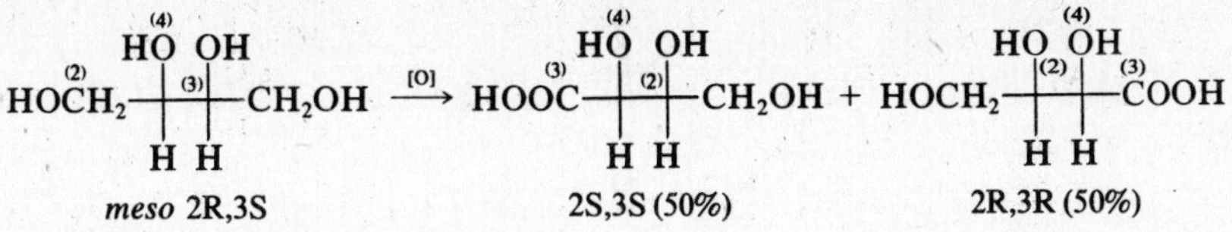

meso 2R,3S   2S,3S (50%)   2R,3R (50%)

(b) Replacement of Cl by the isopentyl group does not change the priorities of the groups on the chiral C. There is one optically active product, whose two chiral C's have R configurations.

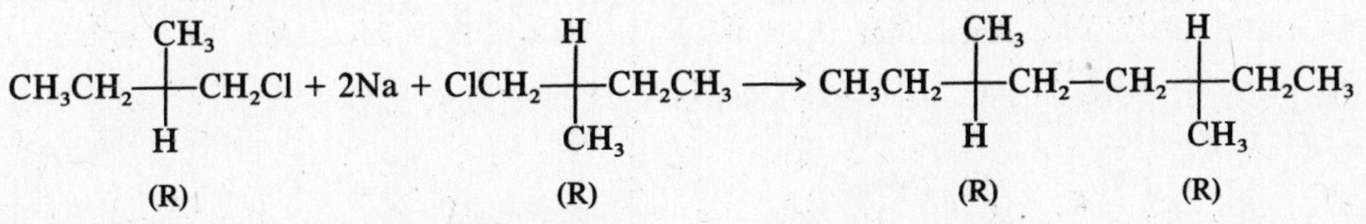

          (R)   (R)   (R)   (R)

(c) This reduction generates a second chiral center.

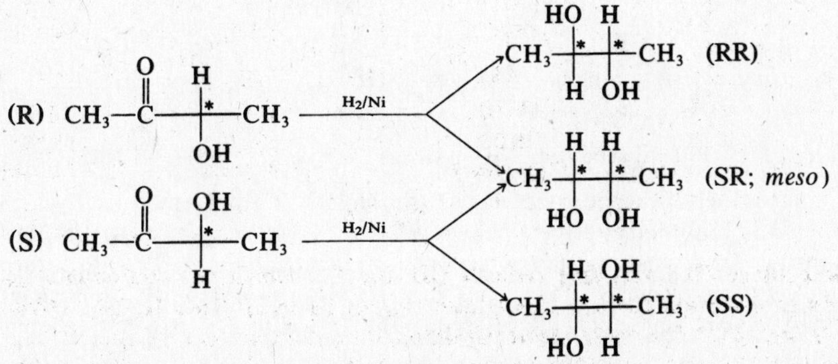

RR and SS enantiomers are formed in equal amounts to give a racemic form. The *meso* and racemic forms are in unequal amounts.

(d) Reduction of the double bond makes C^3 chiral. Reduction occurs on either side of the planar π bond to form molecules with R and molecules with S configurations at C^3. These are in unequal amounts because of the adjacent chiral C that has an S configuration. Since both chiral atoms in the product are structurally identical, the products are a *meso* structure (RS) and an optically active diastereomer (SS).

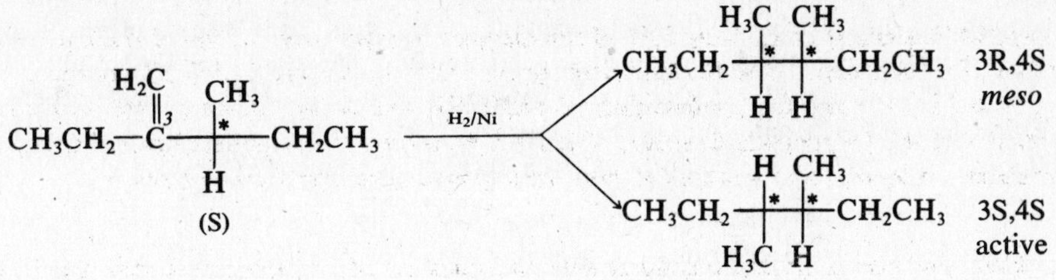

Problem 5.27 Designate the following compounds as *erythro* or *threo* structures.

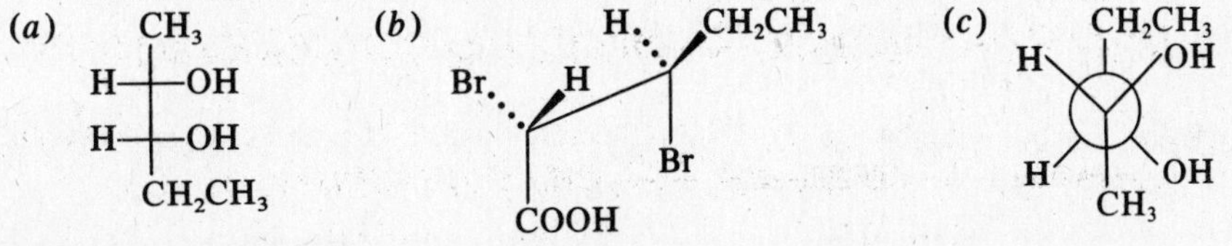

(a) *Erythro* (see Problem 5.22).
(b) *Erythro*; it is best to examine eclipsed conformations. If either of the chiral C's is rotated 120° to an

eclipsed conformation for the two Br's, the H's are also eclipsed.

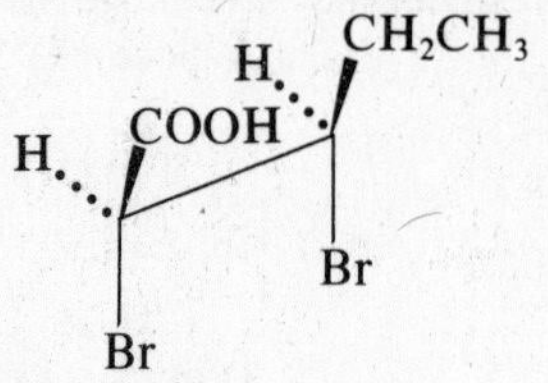

$$\begin{array}{c} \text{H}_{\cdots} \quad \text{CH}_2\text{CH}_3 \\ \text{H}_{\cdots}\diagdown\text{COOH}_{\cdots}\diagdown \\ \qquad\qquad\qquad\qquad \text{Br} \\ \qquad\quad \text{Br} \end{array}$$

(c) *Threo*; a 60° rotation of one of the chiral C's eclipses the H's but not the OH's.

Problem 5.28 CH_3CH_2OH is oxidized in the presence of an enzyme to CH_3CHO. When racemic CH_3CHDOH is similarly oxidized, optically active CH_3CHDOH is isolated after the reaction. Explain. ◄

The enzyme is chiral and causes oxidation of only one enantiomer. The unreacted, optically active enantiomer is isolated.

Problem 5.29 Glyceraldehyde can be converted to lactic acid by the two routes shown below. These results reveal an ambiguity in the assignment of relative D,L configuration. Explain.

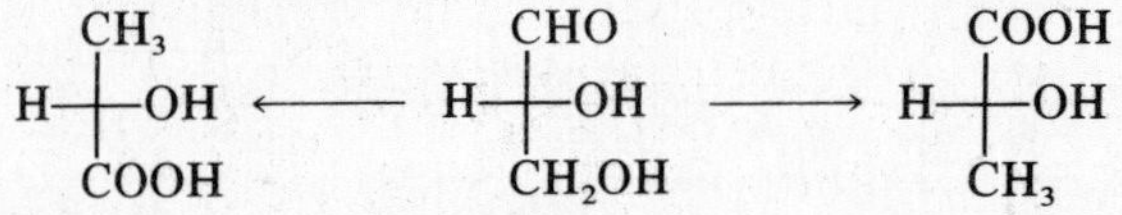

$$\begin{array}{ccccc} \text{CH}_3 & & \text{CHO} & & \text{COOH} \\ \text{H}\!-\!\!\!\!-\!\!\!\!-\!\text{OH} & \longleftarrow & \text{H}\!-\!\!\!\!-\!\!\!\!-\!\text{OH} & \longrightarrow & \text{H}\!-\!\!\!\!-\!\!\!\!-\!\text{OH} \\ \text{COOH} & & \text{CH}_2\text{OH} & & \text{CH}_3 \\ \text{(R)-(+)-Lactic acid} & \text{D-(+)-Glyceraldehyde} & & \text{(S)-(−)-Lactic acid} \end{array}$$ ◄

In neither route is there a change in the bonds to the chiral C. Apparently, both lactic acids should have the D configuration, since the original glyceraldehyde was D. However, since the CH_3 and COOH groups are interchanged, the two lactic acids must be enantiomers. Indeed, one is (+) and the other is (−). This shows that for unambiguous assignment of D or L it is necessary to specify the reactions in the chemical change. Because of such ambiguity, R,S is used. The (+) lactic acid is S, the (−) enantiomer is R.

Problem 5.30 Draw a graph of potential energy plotted against angle of rotation for conformers of (a) 2,3-dimethylbutane, (b) 2-methylbutane. Point out the factors responsible for energy differences and indicate the conformational enantiomers. ◄

Start with the conformer having a pair of CH_3's *anti*. Write the conformations resulting from successive rotations about the central bond of 60°.

(a) As shown in Fig. 5-7(a), structure IV has each pair of CH_3's eclipsed and has the highest energy. Structures II and VI have the next highest energy and they are conformational enantiomers. The stable conformers at energy minima are I, III, and V. Structure I has both pairs of CH_3 groups *anti* and has the lowest energy. It is achiral. Structures III and V are conformational enantiomers which have one pair of CH_3's *anti* and one pair *gauche*.

(b) As shown in Fig. 5-7(b), the conformations in decreasing order of energy are:

1. IX and XI; conformational enantiomers; have eclipsing CH_3's.
2. XIII; CH_3 and H eclipsing; has a plane of symmetry and is achiral.
3. X; CH_3's are all *gauche*; has a plane of symmetry and is achiral.
4. VIII and XII; conformational enantiomers; have a pair of *anti* CH_3's.

Problem 5.31 Deduce the structural formula for an optically active alkene, C_6H_{12}, which reacts with H_2 to form an optically inactive alkane, C_6H_{14}. ◄

The alkene has a group attached to the chiral C which must react with H_2 to give a group identical to one already attached, resulting in loss of chirality.

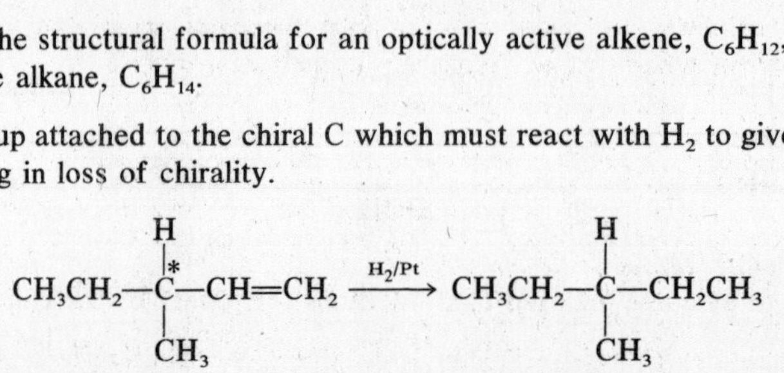

$$\begin{array}{ccc} \overset{\text{H}}{\underset{\overset{|}{\text{CH}_3}}{\overset{|*}{\text{CH}_3\text{CH}_2\!-\!\text{C}\!-\!\text{CH}\!=\!\text{CH}_2}}} & \xrightarrow{\text{H}_2/\text{Pt}} & \overset{\text{H}}{\underset{\overset{|}{\text{CH}_3}}{\overset{|}{\text{CH}_3\text{CH}_2\!-\!\text{C}\!-\!\text{CH}_2\text{CH}_3}}} \end{array}$$

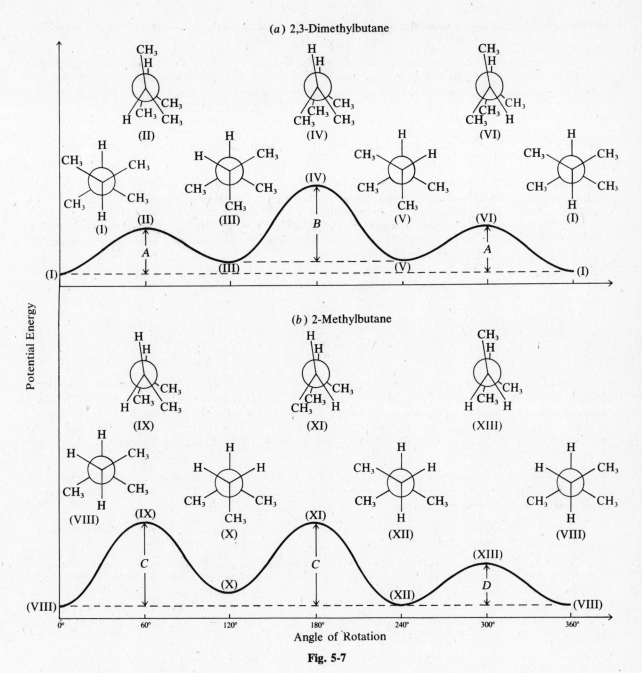

Fig. 5-7

Problem 5.32 1,2-Dibromoethane has a zero dipole moment, whereas ethylene glycol, CH_2OHCH_2OH, has a measurable dipole moment.   Explain.   ◀

1,2-Dibromoethane exists in the *anti* form, so that the C ↔ Br dipoles cancel and the net dipole moment is zero.   When the glycol exists in the *gauche* form, intramolecular H-bonding occurs.   Intramolecular H-bonding is a stabilizing effect which cannot occur in the *anti* conformer.

anti CH₂BrCH₂Br   *gauche* CH₂OHCH₂OH

Problem 5.33 Use conformational theory to evaluate the validity of the following statement: "*meso*-2,3-Dibromobutane is achiral because it possesses a plane of symmetry." ◄

The statement is true only for the high-energy eclipsed conformation, which has an extremely low population. The more stable *anti* conformers have a center rather than a plane of symmetry. Including the *gauche*, there are an infinite number of conformations between the *anti* and eclipsed forms. These all exist as conformational enantiomers comprising an infinite number of conformational racemic forms.

Problem 5.34 Tetra-*sec*-butylmethane has four optically active and one optically inactive stereoisomer. List the isomers in terms of R,S designation. ◄

Each *sec*-butyl group has a chiral C that can be R or S. Since all four groups are equivalent, the order of writing the designation is immaterial, RRRS is the same as RRSR. The possibilities are: RRRR, RRRS, RRSS, SSRR, SSSR and SSSS. RRRR and SSSS, and RRRS and SSSR, are enantiomeric pairs. These are the four optically active isomers. The mirror image of RRSS is SSRR. These are identical and therefore the isomer is *meso*. This isomer is a rare example of a compound which is achiral because it has only an improper axis of symmetry—it has no plane or center of symmetry.

Chapter 6

Alkenes

6.1 NOMENCLATURE AND STRUCTURE

Alkenes (olefins) contain the structural unit

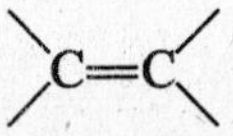

$$\text{C}=\text{C}$$

and have the general formula C_nH_{2n}. These **unsaturated** hydrocarbons are isomeric with the saturated **cycloalkanes**.

$$CH_3CH{=}CH_2 \qquad H_2C{-}{-}{-}CH_2$$
$$\boxed{C_3H_6} \qquad\qquad CH_2$$
Propylene $\qquad$ Cyclopropane

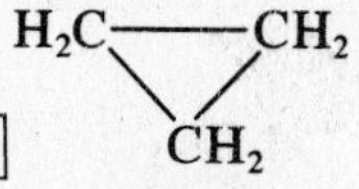

In the IUPAC system the longest continuous chain of C's *containing the double bond* is assigned the name of the corresponding alkane, with the suffix changed from *-ane* to *-ene*. The chain is numbered so that the position of the double bond is designated by assigning the lower possible number to the first doubly bonded C:

$$CH_3CH{=}CHCH_2CH_2CH_3$$
2-Hexene

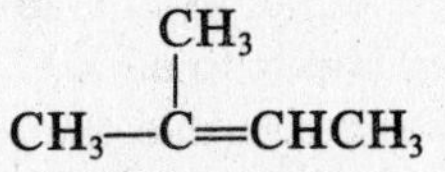

$$CH_3{-}\overset{\displaystyle CH_3}{\underset{}{C}}{=}CHCH_3$$
3-Methyl-2-butene

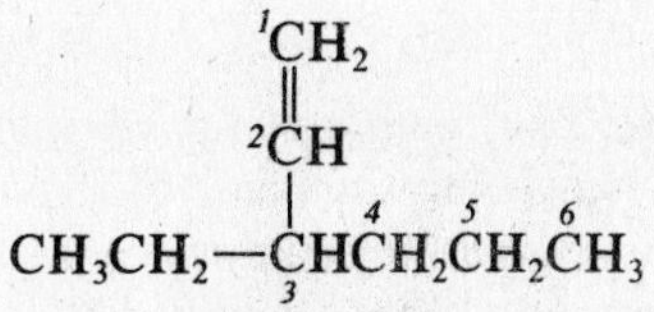

$$\overset{1}{CH_2}$$
$$\overset{2}{CH}$$
$$CH_3CH_2{-}\overset{}{\underset{3}{CH}}\overset{4}{CH_2}\overset{5}{CH_2}\overset{6}{CH_3}$$
3-Ethyl-1-hexene

A few important groups that have trivial names are: $H_2C{=}CH{-}$ (Vinyl), $H_2C{=}CH{-}CH_2{-}$ (Allyl), and $CH_3CH{=}CH{-}$ (Propenyl).

Problem 6.1 Write structural formulas for (*a*) 3-bromo-2-pentene, (*b*) 2,4-dimethyl-3-hexene, (*c*) 2,4,4-trimethyl-2-pentene, (*d*) 3-ethylcyclohexene. ◄

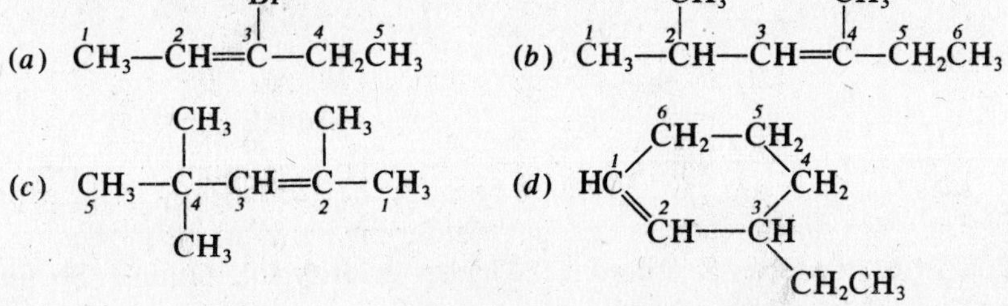

(*a*) $\overset{1}{CH_3}{-}\overset{2}{CH}{=}\overset{3}{\underset{}{C}}{-}\overset{4}{CH_2}\overset{5}{CH_3}$ with Br on C-3

(*b*) $\overset{1}{CH_3}{-}\overset{2}{\underset{}{CH}}{-}\overset{3}{CH}{=}\overset{4}{\underset{}{C}}{-}\overset{5}{CH_2}\overset{6}{CH_3}$ with CH_3 on C-2 and C-4

(*c*) $\underset{5}{CH_3}{-}\underset{4}{\overset{CH_3}{\underset{CH_3}{C}}}{-}\underset{3}{CH}{=}\underset{2}{\overset{CH_3}{C}}{-}\underset{1}{CH_3}$

(*d*) cyclohexene ring: $\overset{1}{HC}{=}\underset{2}{CH}{-}\underset{3}{CH}{-}\underset{4}{CH_2}{-}\underset{5}{CH_2}{-}\underset{6}{CH_2}$ with CH_2CH_3 on C-3

Problem 6.2 Supply the structural formula and IUPAC name for (*a*) trichloroethylene, (*b*) *sec*-butylethylene, (*c*) *sym*-divinylethylene. ◄

Alkenes are also named as derivatives of ethylene. The ethylene unit is shown in a box.

(a) 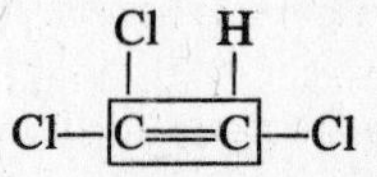
$$\text{Cl}-\boxed{\text{C}=\text{C}}-\text{Cl} \qquad \text{1,1,2-Trichloroethene}$$
with Cl and H above the boxed carbons

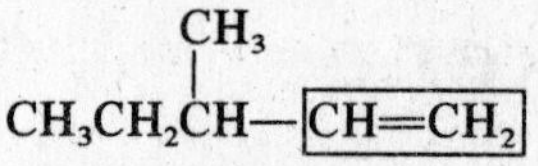

(b)
$$\text{CH}_3\text{CH}_2\text{CH}-\boxed{\text{CH}=\text{CH}_2} \qquad \text{3-Methyl-1-pentene}$$
with CH₃ above the CH

(c)
$$\text{H}_2\text{C}=\text{CH}-\boxed{\text{CH}=\text{CH}}-\text{CH}=\text{CH}_2 \qquad \text{1,3,5-Hexatriene}$$

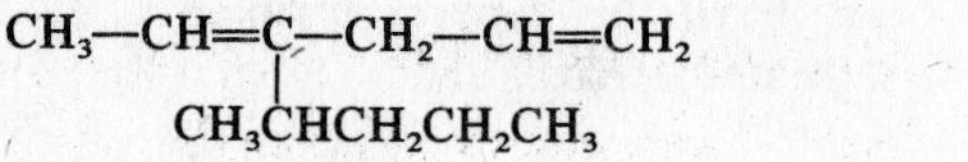

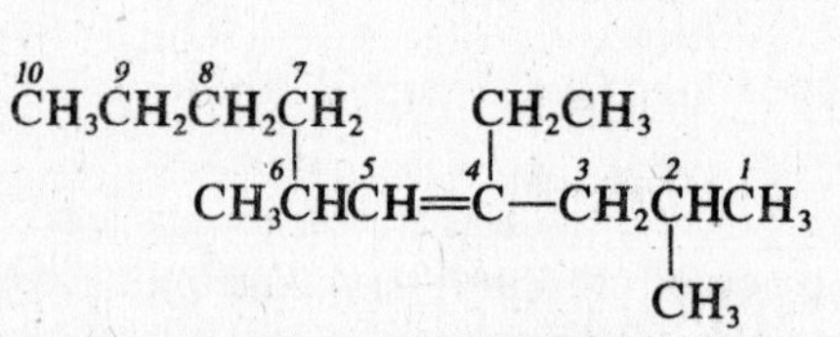

Problem 6.3 Give the IUPAC name for:

(a) cyclopropyl ring $\text{CH}-\text{CH}_2-\text{CH}=\text{CH}_2$

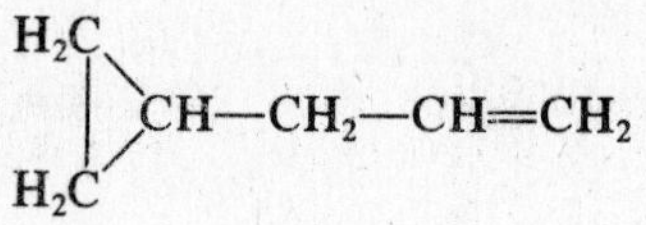

(b) $\text{CH}_3-\text{CH}=\text{C}-\text{CH}_2-\text{CH}=\text{CH}_2$
with $\text{CH}_3\text{CHCH}_2\text{CH}_2\text{CH}_3$ below

(c) $\overset{10}{\text{C}}\text{H}_3\overset{9}{\text{C}}\text{H}_2\overset{8}{\text{C}}\text{H}_2\overset{7}{\text{C}}\text{H}_2 \quad \text{CH}_2\text{CH}_3$
$\overset{6}{\text{CH}_3}\overset{5}{\text{CH}}\overset{4}{\text{CH}}=\overset{}{\text{C}}-\overset{3}{\text{CH}_2}\overset{2}{\text{CH}}\overset{1}{\text{CHCH}_3}$
with CH₃ below

(d) seven-membered ring with three double bonds (CH=CH, HC=CH, CH=CH, and CH₂)

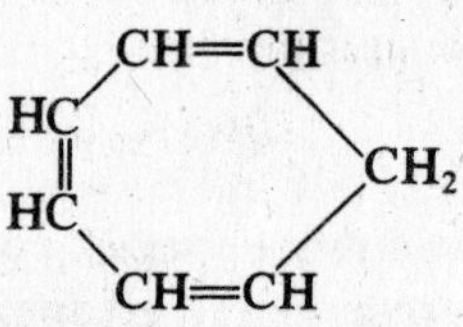

(a) 3-Cyclopropyl-1-propene. (b) 4-(1-Methylbutyl)-1,4-hexadiene (pick longest chain with both double bonds). (c) 2,6-Dimethyl-4-ethyl-4-decene. (d) 1,3,5-Cycloheptatriene.

The C=C consists of a σ bond and a π bond, in a plane at right angles to the plane of the single σ bonds to each C (Fig. 6-1). The π bond is weaker and more reactive than the σ bond. The reactivity of the π bond imparts the property of unsaturation to alkenes; alkenes therefore readily undergo addition reactions. The π bond prevents free rotation about the C=C and therefore an alkene having two different substituents on each doubly bonded C has geometric isomers. For example, there are two 2-butenes:

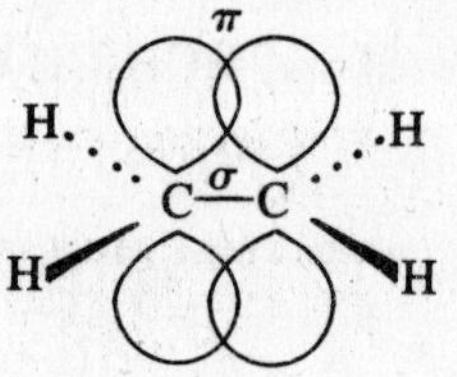

Fig. 6-1

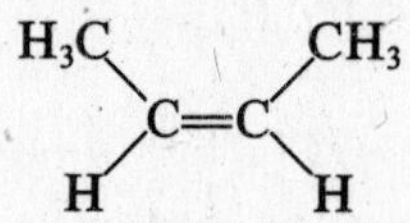

H₃C and CH₃ on same side; called *cis-*

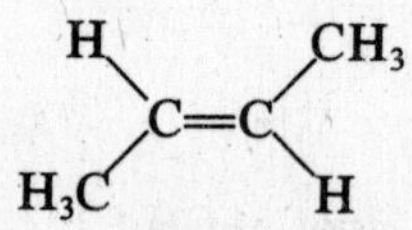

H and CH₃ on opposite sides; called *trans-*

Geometric (*cis-trans*) isomers are stereoisomers because they differ only in the spatial arrangement of the groups. They are diastereomers and have different physical properties (m.p., b.p., etc.).

In place of *cis-trans*, the letter Z is used if the higher-priority substituents (Section 5.3) on each C are on the same side of the double bond. The letter E is used if they are on opposite sides.

Problem 6.4 Predict (a) the geometry of ethylene, $H_2C=CH_2$; (b) the relative C-to-C bond lengths in ethylene and ethane; (c) the relative C—H bond lengths and bond strengths in ethylene and ethane.

(a) The C's in ethylene use sp^2 hybrid orbitals to form three trigonal σ bonds. All five σ bonds must lie in the same plane; ethylene is a planar molecule. All bond angles are approximately 120°.

(b) The $C=C$ atoms have 4 electrons between them and they can get closer to each other than can the $C-C$ atoms, which are separated by only 2 electrons. Hence, the $C=C$ length (1.34 Å) is less than the $C-C$ length (1.54 Å).

(c) The more s character in the hybrid orbital used by C to form a σ bond, the closer the electrons are to the nucleus and the shorter is the σ bond. Thus, the $C-H$ bond length in ethylene (1.08 Å) is less than the length in ethane (1.10 Å). The shorter bond is also the stronger bond.

Problem 6.5 Which of the following alkenes exhibit geometric isomerism? Supply structural formulas and names for the isomers.

(a)
$$CH_3CH_2\underset{\underset{C_2H_5}{|}}{\overset{\overset{CH_3}{|}}{C}}=CCH_2CH_3$$

(b) $H_2C=C(Cl)CH_3$

(c)
$$C_2H_5\overset{\overset{H\ \ H}{|\ \ |}}{C}=C-CH_2I$$

(d) $CH_3CH=CH-CH=CH_2$ (e) $CH_3CH=CH-CH=CHCH_2CH_3$ (f) $CH_3CH=CH-CH=CHCH_3$ ◄

(a) No geometric isomers because one double-bonded C has 2 C_2H_5's.

(b) No geometric isomers; one double-bonded C has 2 H's.

(c) Has geometric isomers because each double-bonded C has 2 different substituents:

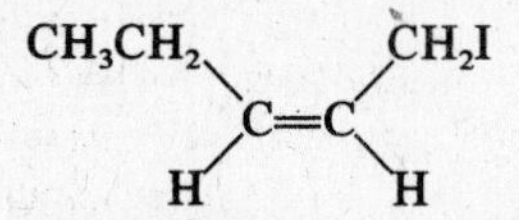

cis- or (Z)-1-Iodo-2-pentene 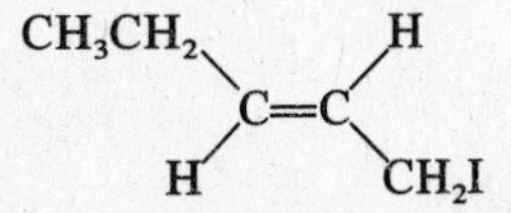　　　　trans- or (E)-1-Iodo-2-pentene

(d) There are 2 geometric isomers because one of the double bonds has two different substituents.

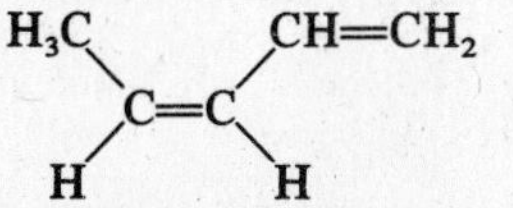

(Z)-1,3-Pentadiene (cis) 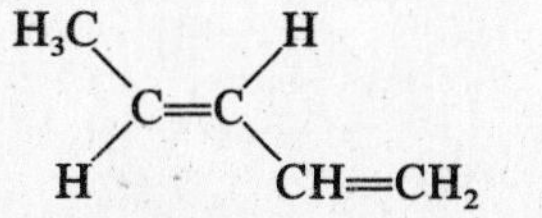　　　(E)-1,3-Pentadiene (trans)

(e) Both double bonds meet the conditions for geometric isomers and there are 4 diastereomers of 2,4-heptadiene.

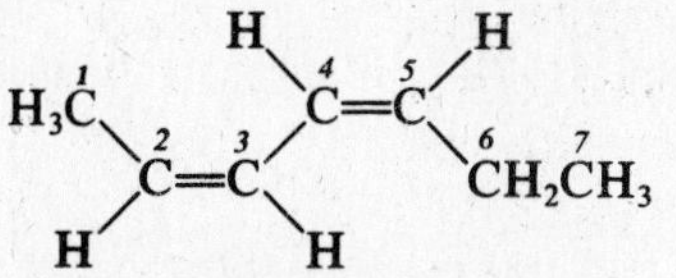

cis,cis or (Z, Z) 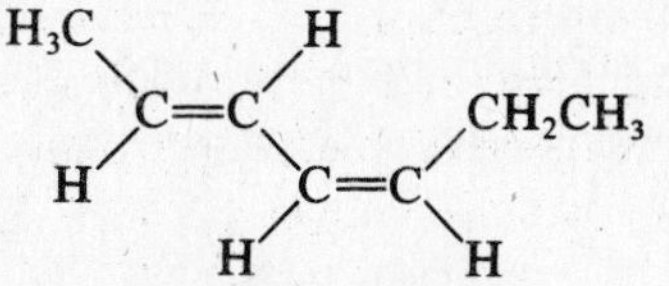　　　　　　trans, cis or (E, Z)

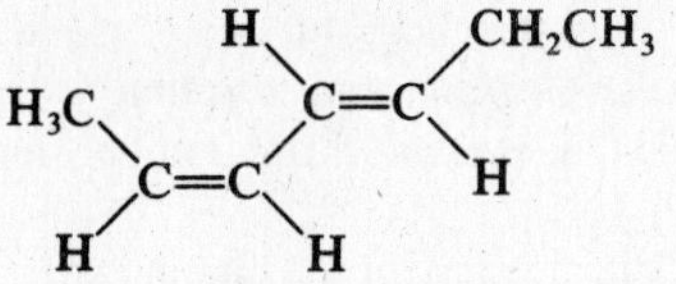

cis, trans or (Z, E) 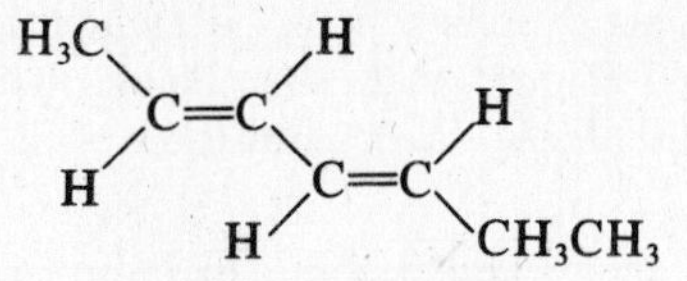　　　　　trans, trans or (E, E)

Note that cis and trans and E and Z are listed in the *same order* as the bonds are *numbered*.

(f) There are now only 3 isomers because cis-trans and trans-cis geometries are identical.

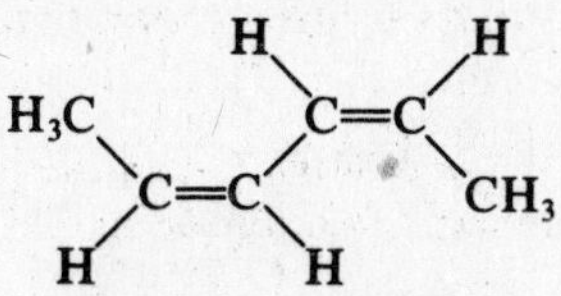

cis, cis-2,4-Hexadiene or　　　cis, trans-2,4-Hexadiene or　　　trans, trans-2,4-Hexadiene or
(Z, Z)-2,4-Hexadiene　　　　　(Z, E)-2,4-Hexadiene　　　　　　(E, E)-2,4-Hexadiene

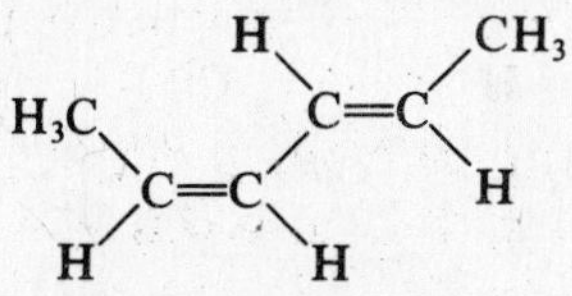

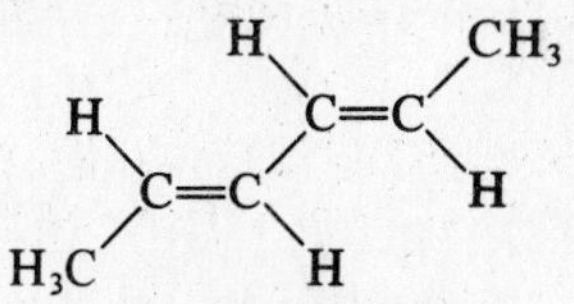

Problem 6.6   Write structural formulas for   (a) (E)-2-methyl-3-hexene (trans),   (b) (S)-3-chloro-1-pentene, (c) (R),(Z)-2-chloro-3-heptene (cis).   ◄

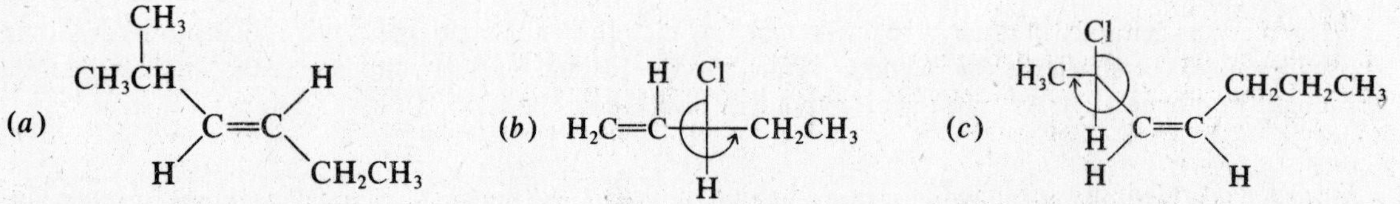

Problem 6.7 Supply structural formulas and systematic names for all pentene isomers including stereoisomers.
◄

These are derived by first writing the carbon skeletons of all pentane isomers and then introducing one double bond.   A double bond can be placed in n-pentane in 2 ways to give 2 isomers,

$$H_2C=CHCH_2CH_2CH_3 \qquad CH_3CH=CHCH_2CH_3$$
$$\text{1-Pentene} \qquad\qquad \text{2-Pentene}$$

2-Pentene has geometric isomers:

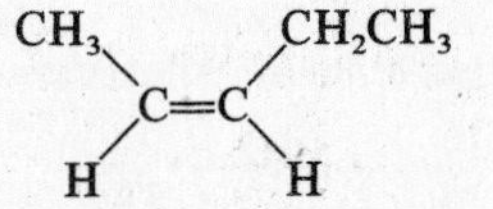

cis- or (Z)-2-Pentene 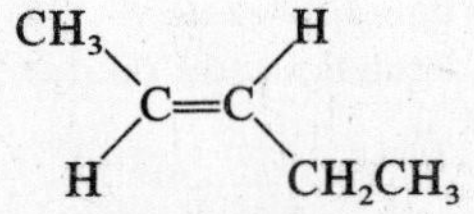   trans- or (E)-2-Pentene

From isopentane we get 3 isomeric alkenes but none has geometric isomers.

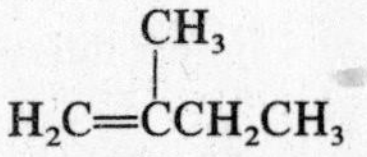

No alkene can be formed from neopentane

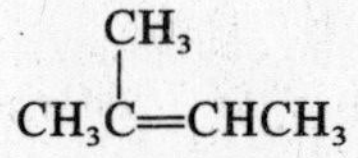

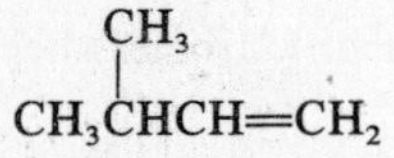

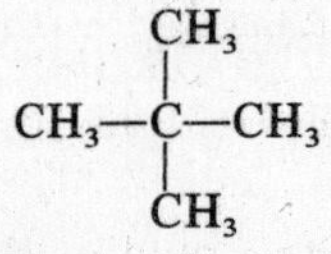

because a double bond would give the middle C five bonds.   None of the isomeric pentenes are chiral.   Atoms with multiple bonds cannot be chiral centers.

Problem 6.8   How do boiling points and solubilities of alkenes compare with those of corresponding alkanes?◄

Alkanes and alkenes are nonpolar compounds whose corresponding structures have almost identical molecular weights.   Boiling points of alkenes are close to those of alkanes and have corresponding 20° increments per C atom.   Both are soluble in nonpolar solvents and insoluble in water, except that lower-molecular-weight alkenes are slightly more water-soluble because of attraction between the pi bond and H_2O.

Problem 6.9   Show the directions of individual bond dipoles and net dipole of the molecule for   (a) 1,1-dichloroethylene,   (b) cis- and trans-1,2-dichloroethylene.   ◄

The individual dipoles are shown by the arrows on the bonds between C and Cl.   The net dipole for the molecule is represented by an arrow that bisects the angle between the two Cl's.   C—H dipoles are insignificant and are disregarded.

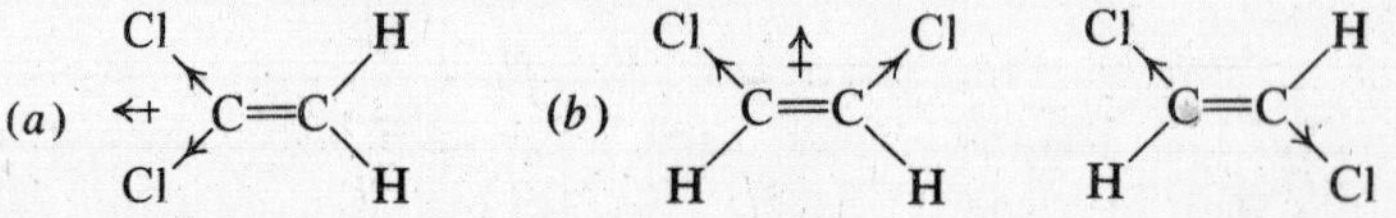

cis   trans

In the trans isomer the C—Cl moments are equal but in opposite directions; they cancel and the trans isomer has a zero dipole moment.

Problem 6.10   How can heats of combustion be used to compare the differences in stability of the geometric isomers of alkenes?   ◄

The thermodynamic stability of isomeric hydrocarbons is determined by burning them to CO_2 and H_2O and comparing the heat evolved per mole ($-\Delta H$ combustion).   The more stable isomer has the smaller ($-\Delta H$) value.   *Trans* alkenes have the smaller values and hence are more stable than the *cis* isomers.   This is supported by the exothermic (ΔH negative) conversion of *cis* to *trans* isomers by ultraviolet light and some chemical reagents.

The *cis* isomer has higher energy because there is greater repulsion between its alkyl groups on the same side of the double bond than between an alkyl group and an H in the *trans* isomer.   These repulsions are greater with larger alkyl groups, which produce larger energy differences between geometric isomers.

Problem 6.11   Suggest a mechanism to account for interconversion of *cis* and *trans* isomers by electromagnetic radiation of wavelength less than 300 nm (nanometers).   ◄

See Problem 2.41.   The excited molecule has the indicated major conformation.   Return of the excited electron from this conformation to the ground-state π bond results in formation of a mixture of geometric isomers.

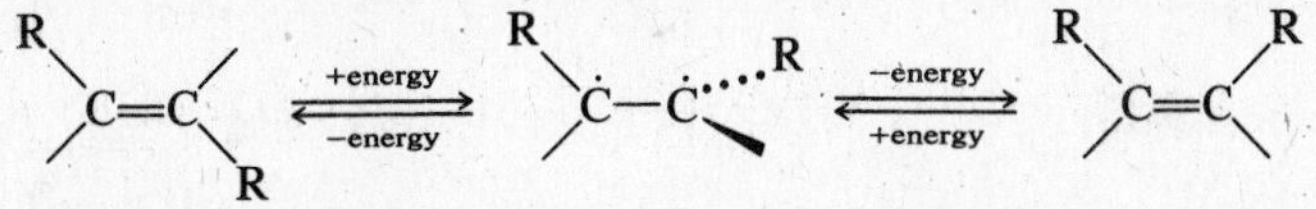

6.2   PREPARATION OF ALKENES

Cracking of petroleum hydrocarbons is the source of commercial alkenes.

(a)   (Mainly a special industrial process)

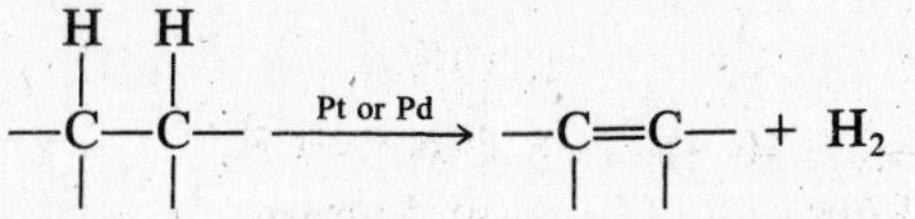

Most alkenes are made in the laboratory by β-elimination reactions.

(b)   (Dehydrogenation)   OF ALKYL HALIDE

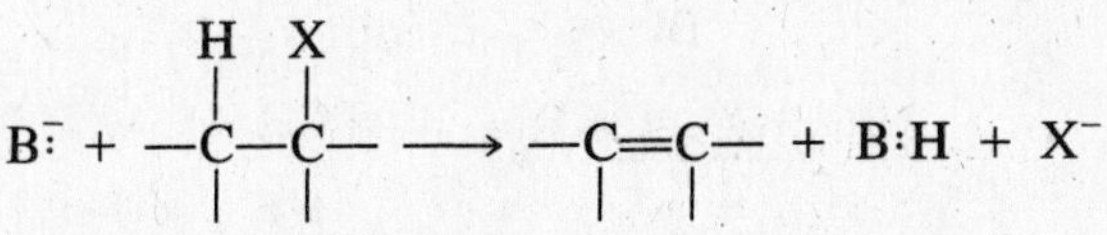

KOH in ethanol is most often used as the source of the base, $B^{\overline{\cdot}}$, which then is $C_2H_5O^-$.

(c)   (Dehydration)   OF ALCOHOL

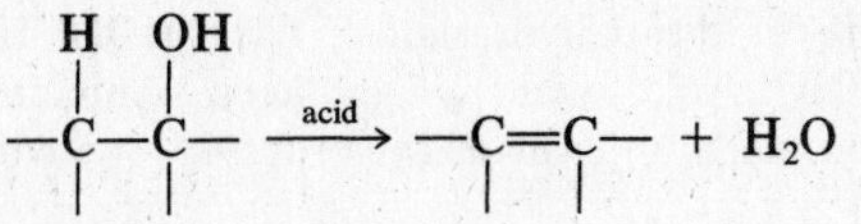

(d) Mg (or Zn) +   (Dehalogenation)

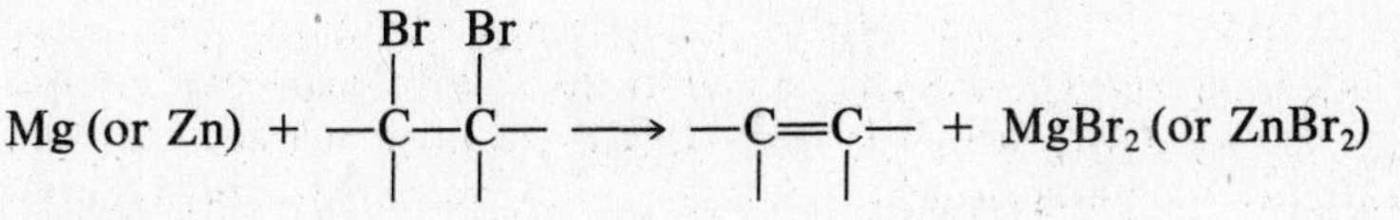

In dehydration and dehydrohalogenation the preferential order for removal of an H is $3° > 2° > 1°$ (**Saytzeff rule**).   We can say "the poor get poorer."   That is because the more R's on the C=C group, the more stable is the alkene.   The stability of alkenes in decreasing order of substitution by R is

$$R_2C{=}CR_2 > R_2C{=}CRH > R_2C{=}CH_2, \quad RCH{=}CHR > RCH{=}CH_2 > CH_2{=}CH_2$$

Alkynes can also be partially reduced to alkenes.

(e)

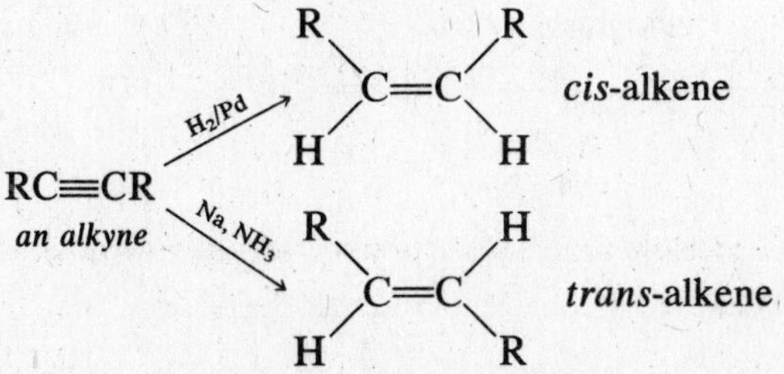

Problem 6.12 Give the structural formulas for the alkenes formed on dehydrobromination of the following alkyl bromides and underline the principal product in each reaction: (*a*) 1-bromobutane, (*b*) 2-bromobutane, (*c*) 3-bromopentane, (*d*) 2-bromo-2-methylpentane, (*e*) 3-bromo-2-methylpentane, (*f*) 3-bromo-2,3-dimethylpentane. ◄

The Br is removed with an H from an adjacent C.

(*a*) $\quad$ H$_2$C—CHCH$_2$CH$_3$ $\longrightarrow$ H$_2$C=CHCH$_2$CH$_3$ (only 1 adjacent H·; only 1 product)
$\qquad\qquad$ | |
$\qquad\qquad$ Br H^I

(*b*) $\quad$ H$_2$C—CHCHCH$_3$ $\longrightarrow$ H$_2$C=CHCH$_2$CH$_3$ + *cis*- and *trans*-CH$_3$CH=CHCH$_3$
$\qquad\qquad$ | | |
$\qquad\qquad$ H^I Br H^2 $\qquad\qquad\qquad$ (−H^I) $\qquad\qquad\qquad$ (−H^2) di-R-substituted

(*c*) $\quad$ CH$_3$CHCHCHCH$_3$ $\longrightarrow$ *cis*- and *trans*-CH$_3$CH=CHCH$_2$CH$_3$ (adjacent H's are equivalent)
$\qquad\qquad$ | | |
$\qquad\qquad$ H^I Br H^I

(*d*)
$\qquad\qquad$ H$_2$CH^I $\qquad\qquad\qquad$ CH$_3$ $\qquad\qquad\qquad$ CH$_3$
$\qquad\qquad$ | $\qquad\qquad\qquad\qquad$ | $\qquad\qquad\qquad\qquad$ |
$\quad$ H$_2$C—C—CHCH$_2$CH$_3$ $\longrightarrow$ H$_2$C=C—CH$_2$CH$_2$CH$_3$ and $\underline{CH_3C=CHCH_2CH_3}$
$\qquad\qquad$ | | |
$\qquad\qquad$ H^I Br H^2 $\qquad\qquad\qquad$ (−H^I) $\qquad\qquad\qquad$ (−H^2)

(*e*) $\quad$ (CH$_3$)$_2$C—CHCHCH$_3$ $\longrightarrow$ $\underline{(CH_3)_2C=CHCH_2CH_3}$ + *cis*- and *trans*-(CH$_3$)$_2$CHCH=CHCH$_3$
$\qquad\qquad$ | | |
$\qquad\qquad$ H^I Br H^2 $\qquad\qquad$ (−H^I) tri-R-substituted $\qquad\qquad$ (−H^2) di-R-substituted

(*f*)
$\qquad\qquad$ H$_2$CH^2 $\qquad\qquad\qquad\qquad$ CH$_3$
$\qquad\qquad$ | $\qquad\qquad\qquad\qquad\qquad$ |
$\quad$ CH$_3$CH—C—C(CH$_3$)$_2$ $\longrightarrow$ *cis*- and *trans*-CH$_3$CH=CCH(CH$_3$)$_2$
$\qquad\qquad$ | | |
$\qquad\qquad$ H^I Br H^3 $\qquad\qquad\qquad$ (−H^I) tri-R-alkene

$\qquad\qquad\qquad\qquad\qquad\qquad$ CH$_3$ $\qquad\qquad\qquad\qquad$ CH$_2$
$\qquad\qquad\qquad\qquad\qquad\qquad$ | $\qquad\qquad\qquad\qquad\qquad$ ||
$\quad$ + $\underline{CH_3CH_2C=C(CH_3)_2}$ + CH$_3$CH$_2$—C—CH(CH$_3$)$_2$
$\qquad\qquad\qquad$ (−H^3) tetra-R-alkene $\qquad\qquad$ (−H^2) di-R-alkene

Problem 6.13 (*a*) Suggest a mechanism for the dehydration of CH$_3$CHOHCH$_3$ that proceeds through a carbonium ion intermediate. Assign a catalytic role to the acid and keep in mind that the O in ROH is a basic site like the O in H$_2$O. (*b*) Use transition states to explain the order of reactivity of ROH: 3° > 2° > 1°. ◄

(*a*) **Step 1**
$\qquad\qquad\qquad$ H $\qquad\qquad\qquad\qquad\qquad\qquad$ H
$\qquad\qquad\qquad$ | $\qquad\qquad\qquad\qquad\qquad\qquad$ |
$\quad$ CH$_3$CHCH$_2$ + H$_2$SO$_4$ $\underset{}{\overset{fast}{\rightleftharpoons}}$ CH$_3$CHCH$_2$ + HSO$_4^-$
$\qquad\quad$ | $\qquad\qquad\qquad\qquad\qquad\qquad$ |
$\qquad\quad$:ÖH $\qquad\qquad\qquad\qquad\qquad\qquad$ H:ÖH
$\qquad\qquad\qquad\qquad\qquad\qquad\qquad\qquad\qquad$ +

$\qquad\quad$ Base$_1$ $\qquad\quad$ Acid$_2$ $\qquad\qquad$ Acid$_1$ $\qquad\quad$ Base$_2$
$\qquad\qquad\qquad\qquad\qquad\qquad\qquad$ *an onium ion*

Step 2 $\quad$ CH$_3$C—CH$_2$ $\overset{slow}{\longrightarrow}$ CH$_3$$\overset{+}{C}HCH_2$ + H$_2$O
$\qquad\qquad\qquad$ | |
$\qquad\qquad\qquad$ H$_2$O$^+$ H $\qquad\qquad\qquad\qquad$ H
$\qquad\qquad\qquad\qquad$ Dimethylcarbonium ion
$\qquad\qquad\qquad\qquad\quad$ (Isopropyl cation)

Step 3 $\quad$ CH$_3$$\overset{+}{C}$H—CH$_2$ + HSO$_4^-$ $\overset{fast}{\longrightarrow}$ CH$_3$CH=CH$_2$ + H$_2$SO$_4$
$\qquad\qquad\qquad\qquad$ |
$\qquad\qquad\qquad\qquad$ H
$\qquad\qquad$ *very strong*
$\qquad\qquad\quad$ Acid$_1$ $\qquad\quad$ Base$_2$ $\qquad\qquad$ Base$_1$ $\qquad\quad$ Acid$_2$

Instead of HSO$_4^-$, a molecule of alcohol could act as the base in Step 3 to give ROH$_2^+$.

The term *carbonium ion* is used generically, but specific carbonium ions are named according to two conventions. In **carbonium ion nomenclature** the positive carbon is called the carbonium ion and the names of the bonded groups are adjoined as a prefix; in **cation nomenclature** the name of the corresponding parent group is used. Thus $CH_3CH_2^+$ and $(CH_3)_2CH^+$ are called either methyl and dimethylcarbonium ions or ethyl and isopropyl cations respectively. The cation system is used more frequently to avoid ambiguity.

(b) The order of reactivity of the alcohols reflects the order of stability of the incipient carbonium ($3° > 2° > 1°$) in the TS of Step 2, the rate-determining step.

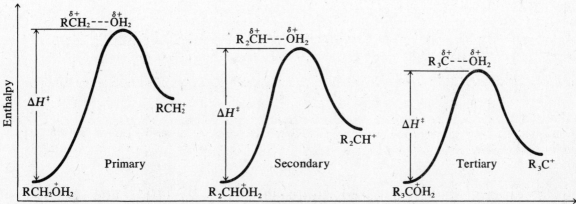

Fig. 6-2

Problem 6.14 Account for the fact that dehydration of: (a) $CH_3CH_2CH_2CH_2OH$ yields mainly $CH_3CH=CHCH_3$ rather than $CH_3CH_2CH=CH_2$, (b) $(CH_3)_3CCHOHCH_3$ yields mainly $(CH_3)_2C=C(CH_3)_2$. ◄

(a) The carbonium ion (R^+) formed in a reaction like Step 2 of Problem 6.13(a) is 1° and rearranges to a more stable 2° R_2CH^+ by a **hydride shift** (indicated as $\sim H:$; the H migrates with its bonding pair of electrons).

$$CH_3CH_2\overset{H}{\underset{\underset{\textstyle \overset{\frown}{H}}{|}}{C}}-\overset{+}{C}H_2 \xrightarrow{\sim H:} CH_3CH_2\overset{+}{\underset{\overset{|}{\ddot{H}}}{C}}CH_2 \xrightarrow[-H^+]{+ROH} CH_3CH=CHCH_3 + \overset{+}{R}\overset{..}{O}H_2$$

1° RCH_2^+ 2° R_2CH^+
(*n*-Butyl cation)   (*sec*-Butyl cation)

(b) The 2° R_2CH^+ formed undergoes a methyl shift ($\sim :CH_3$) to the more stable 3° R_3C^+.

$$(CH_3)_2\overset{..}{\underset{\underset{\ddot{C}H_3}{|}}{C}}-\overset{+}{C}HCH_3 \xrightarrow{\sim :CH_3} (CH_3)_2\overset{+}{\underset{\overset{|}{\ddot{C}H_3}}{C}}CHCH_3 \xrightarrow[-H^+]{+ROH} (CH_3)_2C=C(CH_3)_2 + \overset{+}{R}\overset{..}{O}H_2$$

2° R_2CH^+ 3° R_3C^+
(3,3-Dimethyl-   (2,3-Dimethyl-
2-butyl cation)   2-butyl cation)

Reactions that proceed through carbonium ions are always prone to undergo rearrangement.

Problem 6.15 Supply structural formulas for the compounds formed by dehydrating the following alcohols and underline the major product.

$$(a) \quad CH_3-\overset{\overset{\textstyle H_3C}{|}}{\underset{\underset{\textstyle HO}{|}}{C}}-\overset{\overset{\textstyle H}{|}}{\underset{\underset{\textstyle CH_3}{|}}{C}}-CH_3 \qquad (b) \quad CH_3-\overset{\overset{\textstyle CH_3}{|}}{\underset{\underset{\textstyle CH_3}{|}}{C}}-CH_2OH \qquad\qquad ◄$$

(a)

$$CH_3-\underset{\underset{CH_3}{|}}{\overset{\overset{CH_3}{|}}{C}}=C-CH_3 \;+\; H_2C=\underset{\underset{CH_3}{|}}{\overset{\overset{CH_3}{|}}{C}}-CH-CH_3$$

The principal product has more alkyl groups on the unsaturated C's.

(b) Neopentyl alcohol cannot be dehydrated to an alkene without rearrangement because there is no H on the adjacent (β) C. The 1° neopentyl cation RCH_2^+ rearranges to a more stable 3° R_3C^+ by a $:CH_3$ shift; this is followed by loss of an H^+.

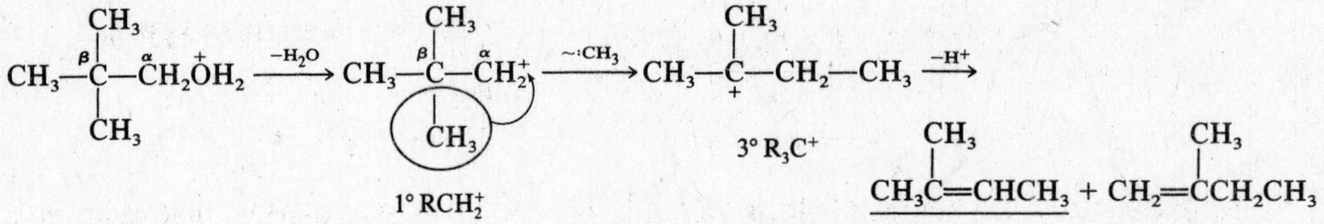

$$CH_3-\underset{\underset{CH_3}{|}}{\overset{\overset{CH_3}{|}}{\underset{\beta}{C}}}-\overset{\alpha}{CH_2}\overset{+}{O}H_2 \xrightarrow{-H_2O} CH_3-\underset{\underset{CH_3}{|}}{\overset{\overset{CH_3}{|}}{\underset{\beta}{C}}}-\overset{\alpha}{CH_2^+} \xrightarrow{\sim:CH_3} CH_3-\underset{+}{C}-CH_2-CH_3 \xrightarrow{-H^+}$$

1° RCH_2^+ 3° R_3C^+

$$\underset{\underset{CH_3}{|}}{CH_3C}=CHCH_3 \;+\; CH_2=\underset{\underset{CH_3}{|}}{C}CH_2CH_3$$

Problem 6.16 Assign numbers from 1 for LEAST to 3 for MOST to indicate the relative ease of dehydration and justify your choices.

(a) $CH_3\underset{\underset{CH_3}{|}}{CH}CH_2CH_2OH$ (b) $CH_3-\underset{\underset{OH}{|}}{\overset{\overset{CH_3}{|}}{C}}-CH_2CH_3$ (c) $CH_3\underset{\underset{CH_3}{|}}{CH}-\underset{\underset{OH}{|}}{CH}-CH_3$ ◀

(a) 1, (b) 3, (c) 2. The ease of dehydration depends on the relative ease of forming an R^+, which depends on the stability of R^+. This is greatest for the 3° alcohol (b) and least for the 1° alcohol (a).

Problem 6.17 Give structural formulas for the reactants that form 2-butene when treated with the following reagents: (a) heating with conc. H_2SO_4, (b) alcoholic KOH, (c) zinc dust and alcohol, (d) hydrogen and a catalyst. ◀

(a) $CH_3CHOHCH_2CH_3$, (b) $CH_3CHBrCH_2CH_3$, (c) $CH_3CHBrCHBrCH_3$, (d) $CH_3C\equiv CCH_3$.

Problem 6.18 Write the structural formula and name of the principal organic compound formed in the following reactions:

(a) $CH_3\underset{\underset{}{\overset{\overset{CH_3}{|}}{C}}}ClCH_2CH_3$ + alc. KOH $\longrightarrow$

(b) $HOCH_2CH_2CH_2CH_2OH + BF_3$, heat $\longrightarrow$

(c) 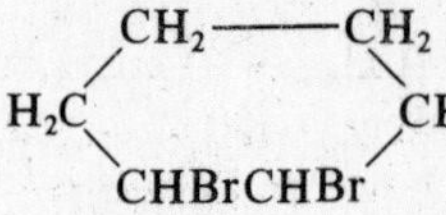 $H_2C\underset{CHBrCHBr}{\overset{CH_2-CH_2}{<\qquad>}}CH_2$ + Zn in alcohol $\longrightarrow$

(d) $CH_3-\underset{\underset{Br}{|}}{\overset{\overset{CH_3}{|}}{C}}-CH_3 + CH_3COO^- \longrightarrow$ ◀

(a) $CH_3\underset{\underset{}{\overset{\overset{CH_3}{|}}{C}}}=CHCH_3$ 2-Methyl-2-butene   (b) $H_2C=CH-CH=CH_2$ 1,3-Butadiene

(c) 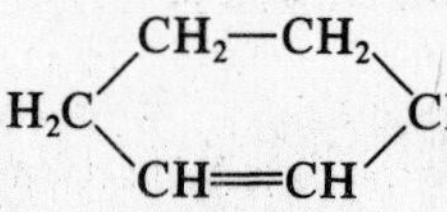 $H_2C\underset{CH=CH}{\overset{CH_2-CH_2}{<\qquad>}}CH_2$ Cyclohexene   (d) $CH_3-\underset{}{\overset{\overset{CH_3}{|}}{C}}=CH_2$ Isobutylene

6.3 CHEMICAL PROPERTIES OF ALKENES

Alkenes undergo addition reactions at the double bond. The π electrons of alkenes are a nucleophilic site and they react with electrophiles by three mechanisms (see Problem 3.35).

$$\underset{\text{Intermediate } R^+}{\diagdown C \overset{\cdots}{\cdots} C \diagup + E\overset{\cdots}{\colon}Nu \longrightarrow \diagdown \overset{+}{C} - \overset{|}{\underset{E}{C}} \diagup + (\colon Nu^-) \longrightarrow \diagdown \overset{|}{\underset{E}{C}} - \overset{|}{\underset{Nu}{C}} \diagup} \quad \text{(Ionic)}$$

$$\underset{\text{Intermediate } R\cdot}{E\cdot + \diagdown C \overset{\cdots}{\cdots} C \diagup \longrightarrow \diagdown \overset{|}{\underset{E}{C}} - \overset{\cdot}{C} \diagup + E\overset{\cdot}{\colon}E \longrightarrow \diagdown \overset{|}{\underset{E}{C}} - \overset{|}{\underset{E}{C}} \diagup + E\cdot} \quad \text{(Free radical)}$$

$$\text{RCH}{=}\text{CHR} + \text{E}{-}\text{Nu} \longrightarrow \left[\begin{array}{c} \text{RCH}{=\!=\!=}\text{CHR} \\ | \quad\quad | \\ \text{E}{\text{-}\text{-}\text{-}\text{-}\text{-}}\text{Nu} \end{array} \right]^{\ddagger} \longrightarrow \underset{\text{E} \quad\quad \text{Nu}}{\text{RCH}{-}\text{CHR}} \quad \text{(Cyclic, one-step, rare)}$$

Transition State

CONVERSION TO ALKANES

$$\text{RCH}{=}\text{CHR} + \text{H}_2 \xrightarrow{\text{Pt, Pd or Ni}} \text{RCH}_2\text{CH}_2\text{R} \quad \text{(Heterogeneous catalysis)}$$

The relative rates of hydrogenation

$$\text{H}_2\text{C}{=}\text{CH}_2 > \text{RCH}{=}\text{CH}_2 > \text{R}_2\text{C}{=}\text{CH}_2, \quad \text{RCH}{=}\text{CHR} > \text{R}_2\text{C}{=}\text{CHR} > \text{R}_2\text{C}{=}\text{CR}_2$$

indicate that the rate is decreased by steric hindrance.

$$\text{RCH}{=}\text{CHR} \xrightarrow[\text{from B}_2\text{H}_6]{[\text{BH}_3]} \underset{\text{H} \quad\quad \text{BH}_2}{\text{RCH}{-}\text{CHR}} \xrightarrow{\text{CH}_3\text{COOH}} \text{RCH}_2\text{CH}_2\text{R} \quad \text{(Homogeneous)}$$

Alkylborane

$$\underset{\text{Hydrazine}}{\text{H}_2\ddot{\text{N}}\ddot{\text{N}}\text{H}_2} \xrightarrow[\text{Cu}^{2+}]{\text{H}_2\text{O}_2} \underset{\text{Diimide}}{[\text{H}\ddot{\text{N}}{=}\ddot{\text{N}}\text{H}]} \xrightarrow{+\text{RCH}{=}\text{CHR}} \left[\begin{array}{c} \text{H} \quad \text{H} \\ | \quad\;\; | \\ \text{R}{-}\text{C}{=\!=}\text{C}{-}\text{R} \\ | \quad\;\; | \\ \text{H} \quad\;\; \text{H} \\ \vdots \quad\;\; \vdots \\ \ddot{\text{N}}{=\!=}\ddot{\text{N}} \end{array} \right]^{\ddagger} \longrightarrow \text{RCH}_2\text{CH}_2\text{R} + \colon\text{N}{\equiv}\text{N}\colon$$

Transition State

Problem 6.19 Given the following heats of hydrogenation, $-\Delta H_h$, in kcal/mole: 1-pentene, 30.1; *cis*-2-pentene, 28.6; *trans*-2-pentene, 27.6. (*a*) Use an enthalpy diagram to derive two generalizations about the relative stabilities of alkenes. (*b*) Would the ΔH_h of 2-methyl-2-butene be helpful in making your generalizations? (*c*) The corresponding heats of combustion, $-\Delta H_c$, are: 806.9, 805.3, and 804.3 kcal/mole. Are these values consistent with your generalizations in part (*a*)? (*d*) Would the ΔH_c of 2-methyl-2-butene be helpful in your comparison? (*e*) Suggest a relative value for the ΔH_c of 2-methyl-2-butene. ◄

(*a*) See Fig. 6-3. The lower ΔH_h, the more stable the alkene. (1) The alkene with more alkyl groups on the double bond is more stable; 2-pentene > 1-pentene. (2) The *trans* isomer is usually more stable than the *cis*. Bulky alkyl groups are *anti*-like in the *trans* isomer and eclipsed-like in the *cis* isomer.

(*b*) No. The alkenes being compared *must* give the same product on hydrogenation.

(*c*) Yes. Again the highest value indicates the least stable isomer.

(*d*) Yes. On combustion all four isomers give the same products, H_2O and CO_2.

(*e*) Less than 804.3 kcal/mole, since this isomer is a trisubstituted alkene and the 2-pentenes are disubstituted.

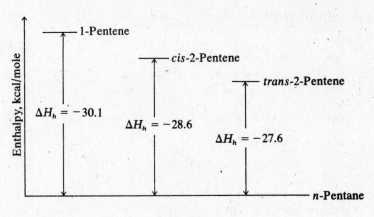

Fig. 6-3

6.4 ELECTROPHILIC POLAR ADDITION REACTIONS

Table 6-1 shows the results of electrophilic addition of polar reagents to ethylene.

Table 6-1

REAGENT		PRODUCT	
Name	Structure	Name	Structure
Halogens (Cl_2, Br_2 only)	$X\!:\!X$	Ethylene dihalide	CH_2XCH_2X
Hydrohalic acids	$\overset{\delta+\ \delta-}{H\!:\!X}$	Ethyl halide	CH_3CH_2X
Hypohalous acids	$\overset{\delta+\ \delta-}{X\!:\!OH}$	Ethylene halohydrin	CH_2XCH_2OH
Sulfuric acid (cold)	$\overset{\delta+\ \delta-}{H\!:\!OSO_3OH}$	Ethyl bisulfate	$CH_3CH_2OSO_3H$
Water (dil. H_3O^+)	$\overset{\delta+\ \delta-}{H\!:\!OH}$	Ethyl alcohol	CH_3CH_2OH
Borane	$H_2\overset{\delta+\ \delta-}{B\!:\!H}$	Ethyl borane	$[CH_3CH_2BH_2] \to (CH_3CH_2)_3B$
Peroxyformic acid	$H\!:\!\overset{\delta+}{O}\!-\!\overset{\delta-}{O}CH$ $\underset{O}{\parallel}$	Ethylene glycol	$CH_2OHCH_2OCH \xrightarrow{H^+} CH_2OHCH_2OH$ $\underset{O}{\parallel}$

Problem 6.20 What is the stereochemistry of the heterogeneous catalytic addition of H_2 if *trans*-$CH_3CBr{=}CBrCH_3$ gives *rac*-$CH_3CHBrCHBrCH_3$ and its *cis* isomer gives the *meso* product?   ◄

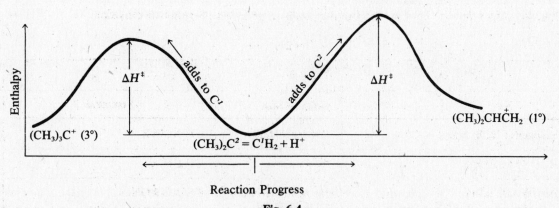

The H's as atoms are adsorbed on the surface of the solid catalyst and add *syn* (*cis*) to the π bond of the adsorbed alkene.

Problem 6.21   Unsymmetrical polar reagents like HX add to unsymmetrical alkenes such as propene according to **Markovnikov's rule**: the positive portion, e.g. H of HX, adds to the C that has more H's ("the rich get richer").   Explain by stability of the intermediate cation.   ◄

The more stable cation ($3° > 2° > 1°$) has a lower $\Delta H^{\ddagger}$ for the transition state and forms more rapidly (Fig. 6-4).

Reaction Progress

Fig. 6-4

Problem 6.22   Give the structural formula of the major organic product formed from the reaction of $CH_3CH{=}CH_2$ with: (*a*) Br_2, (*b*) HI, (*c*) BrOH, (*d*) H_2O in acid, (*e*) cold H_2SO_4, (*f*) BH_3 from B_2H_6, (*g*) peroxyformic acid H_2O_2 and HCOOH (formic acid).   ◄

The positive ($\delta+$) moiety of the addendum is an electrophile (E^+) which forms $CH_3\overset{+}{C}HCH_2E$ rather than $CH_3CHE\overset{+}{C}H_2$.

The E^+ is in a box; the Nu^- is encircled.

(*a*) $CH_3{-}CH{-}CH_2$ (*b*) $CH_3{-}CH{-}CH_2$ (*c*) $CH_3{-}CH{-}CH_2$

ⓑⓡ ⎡Br⎤   Ⓘ ⎡H⎤   ⒪Ⓗ ⎡Br⎤

(*d*) $CH_3{-}CH{-}CH_2$ (*e*) $CH{-}CH{-}CH_2$

⒪Ⓗ ⎡H⎤   ⒣ⓞ₃ⓢⓞ ⎡H⎤

(*f*) ⎡$CH_3{-}CH{-}CH_2$⎤ $\xrightarrow{2CH_3CH=CH_2}$ $(CH_3CH_2CH_2)_3B$ (*g*) $CH_3{-}CH{-}CH_2$

   Ⓗ ⎡BH₂⎤   ⒪Ⓗ ⎡OH⎤

(Anti-Markovnikov orientation; with nonbulky alkyl
groups all H's of BH_3 add to form a trialkylborane)

Problem 6.23 (*a*) What principle is used to account for the identical mechanisms for dehydration of alcohols and hydration of alkenes? (*b*) What conditions favor dehydration rather than hydration reactions? ◀

(*a*) The Principle of Microscopic Reversibility, which prescribes identical mechanisms for forward and reverse reactions when a reaction is reversible.

$$RCH_2CH_2OH \underset{}{\overset{H^+}{\rightleftharpoons}} RCH{=}CH_2 + H_2O$$

(*b*) Low H_2O concentration and high temperature favor alkene formation because the volatile alkene distills out of the reaction mixture and shifts the equilibrium. Hydration of alkenes occurs at low temperature and with dilute acid which provides a high concentration of H_2O as reactant.

Problem 6.24 Why are dry gaseous hydrohalogen acids and not their aqueous solutions used to prepare alkyl halides from alkenes? ◀

Dry hydrogen halides are stronger acids and better electrophiles than the H_3O^+ formed in their water solutions. Furthermore, H_2O is a nucleophile that can react with R^+ to give an alcohol.

Problem 6.25 Isobutylene gas dissolves in 63% H_2SO_4 to form a deliquescent white solid. If the H_2SO_4 solution is diluted with H_2O and heated, the organic compound obtained is a liquid boiling at 83 °C. Explain. ◀

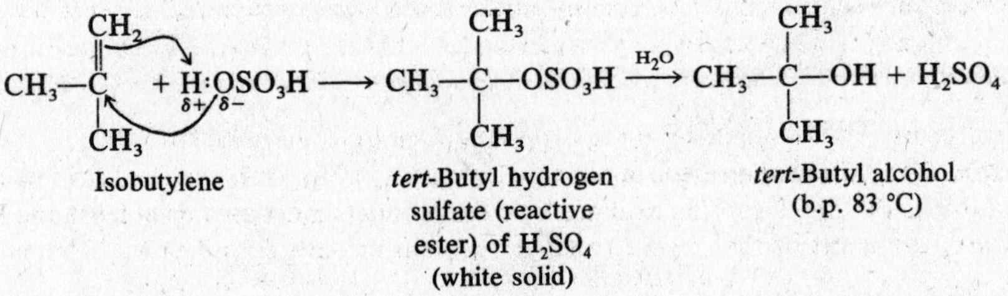

Isobutylene

tert-Butyl hydrogen
sulfate (reactive
ester) of H_2SO_4
(white solid)

tert-Butyl alcohol
(b.p. 83 °C)

Problem 6.26 Arrange the following alkenes in order of increasing reactivity on addition of hydrohalogen acids: (*a*) $H_2C{=}CH_2$, (*b*) $(CH_3)_2C{=}CH_2$, (*c*) $CH_3CH{=}CHCH_3$. ◀

The relative reactivities are directly related to the stabilities of the intermediate R^+'s. Isobutylene, (*b*), is most reactive because it forms the 3° $(CH_3)_2\overset{+}{C}CH_3$. The next most reactive compound is 2-butene, (*c*), which forms the 2° $CH_3\overset{+}{C}HCH_2CH_3$. Ethylene forms the 1° $CH_3\overset{+}{C}H_2$ and is least reactive. The order of increasing reactivity is: (*a*) < (*c*) < (*b*).

Problem 6.27 The addition of HBr to some alkenes gives a mixture of the expected alkyl bromide and an isomer formed by rearrangement. Outline the mechanism of formation and structures of products from the reaction of HBr with (*a*) 3-methyl-1-butene, (*b*) 3,3-dimethyl-1-butene. ◀

No matter how formed, an R^+ can undergo $H{:}$ or ${:}CH_3$ (or other alkyl) shifts to form a more stable R'^+.

(*a*)

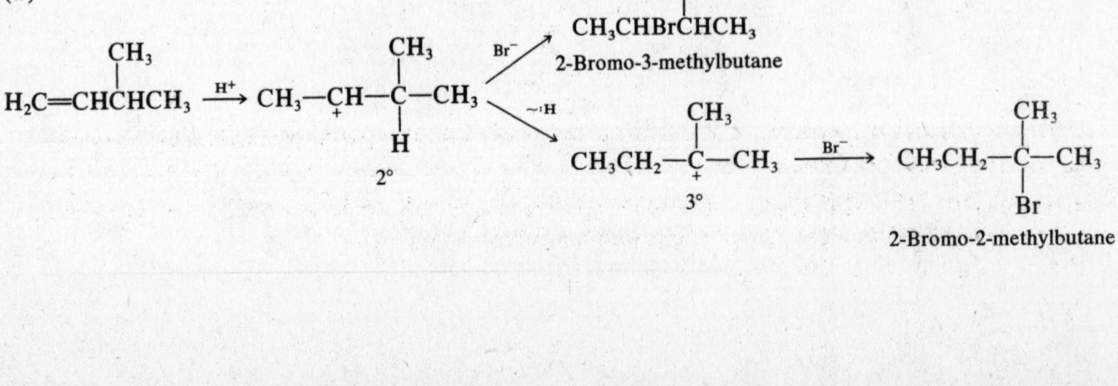

(b)

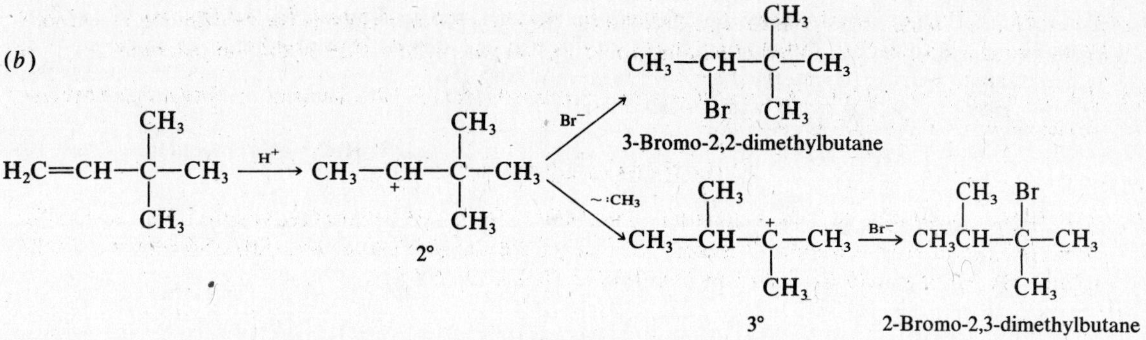

Problem 6.28 Compare and explain the relative rates of addition to alkenes (reactivities) of HCl, HBr and HI. ◄

The relative reactivity depends on the ability of HX to donate an H^+ (acidity) to form an R^+ in the rate-controlling first step. The acidity and reactivity order is HI > HBr > HCl.

Problem 6.29 (a) What does each of the following observations tell you about the mechanism of the addition of Br_2 to an alkene? (i) In the presence of a Cl^- salt, in addition to the *vic*-dibromide, some *vic*-bromochloroalkane is isolated but no dichloride is obtained. (ii) With *cis*-2-butene only *rac*-2,3-dibromobutane is formed. (iii) With *trans*-2-butene only *meso*-2,3-dibromobutane is produced. (b) Give a mechanism compatible with these observations. ◄

(a) (i) Br_2 adds in 2 steps. If Br_2 added in one step no bromochloroalkane would be formed. Furthermore, the first step must be the addition of an electrophile (the Br^+ moiety of Br_2) followed by addition of a nucleophile, which could now be Br^- or Cl^-. This explains why the products must contain at least one Br. (ii) One Br adds from above the plane of the double bond, the second Br adds from below. This is an *anti* (*trans*) addition. Since a Br^+ can add from above to either C, a racemic form results.

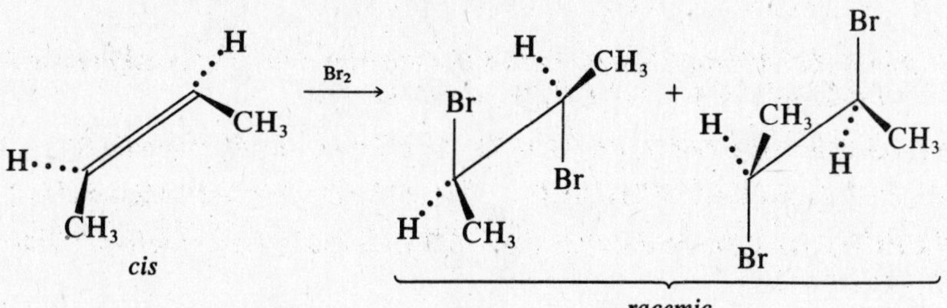

(iii) This substantiates the *anti* addition.

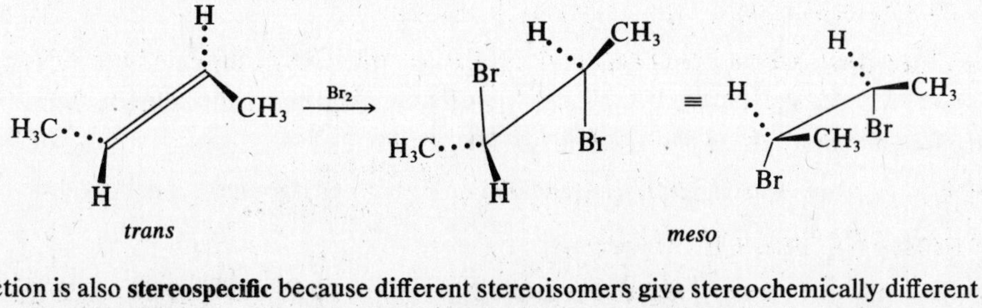

The reaction is also **stereospecific** because different stereoisomers give stereochemically different products, e.g. *cis* → racemic and *trans* → *meso*. Because of this stereospecificity, the intermediate *cannot* be the free carbonium $CH_3CHBrCHCH_3$. The same carbonium ion would arise from either *cis*- or *trans*-2-butene, and the product distribution from both reactants would be identical.

(*b*) To account for the stereospecificity the open carbonium ion is replaced by a cyclic bridged ion having Br^+ partially bonded to each C (**bromonium ion**). In this way the stereochemical differences of the starting materials are retained in the intermediate. In the second step, the nucleophile attacks the side *opposite* the bridging group to yield the *anti* addition product.

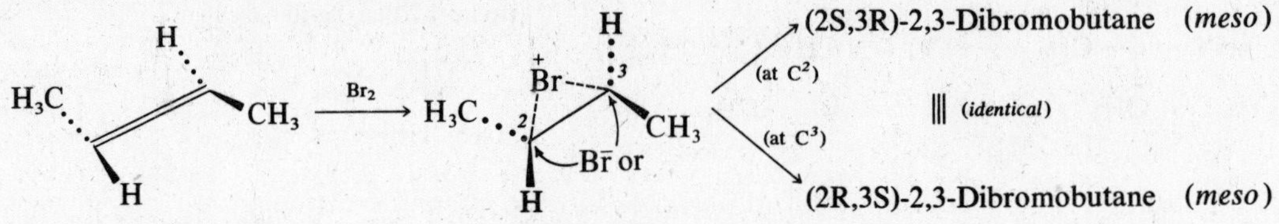

(2S,3R)-2,3-Dibromobutane (*meso*)

(identical)

(2R,3S)-2,3-Dibromobutane (*meso*)

trans   *a bromonium ion*

Problem 6.30 (*a*) Describe the stereochemistry of glycol formation with peroxyformic acid (HCO_3H) if *cis*-2-butene gives a racemic glycol and *trans*-2-butene gives the *meso* form. (*b*) Suggest a mechanism. ◄

(*a*) The reaction is a stereospecific *anti* addition similar to that of addition of Br_2.

a protonated epoxide

C^1 attack C^2 attack

Racemate

Racemate

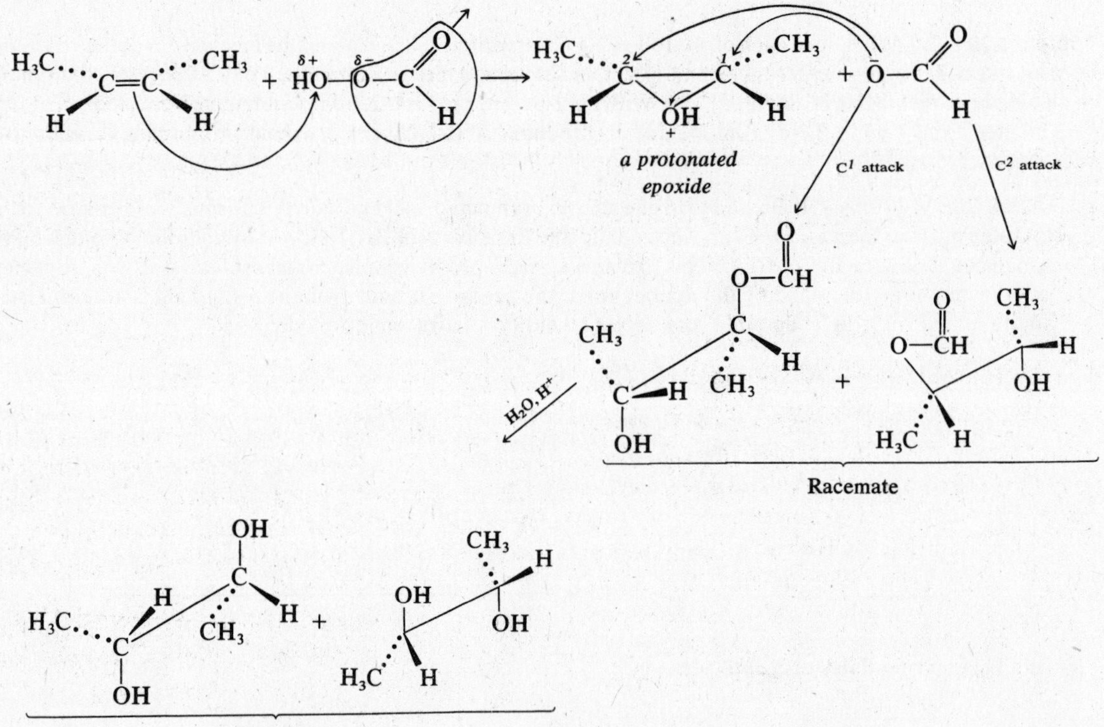

Problem 6.31 Describe the stereochemistry of glycol formation with cold alkaline aqueous $KMnO_4$ if *cis*-2-butene gives the *meso* glycol and trans-2-butene gives the racemate. ◄

The reaction is a stereospecific *syn* (*cis*) addition since both OH groups bond from the same side.

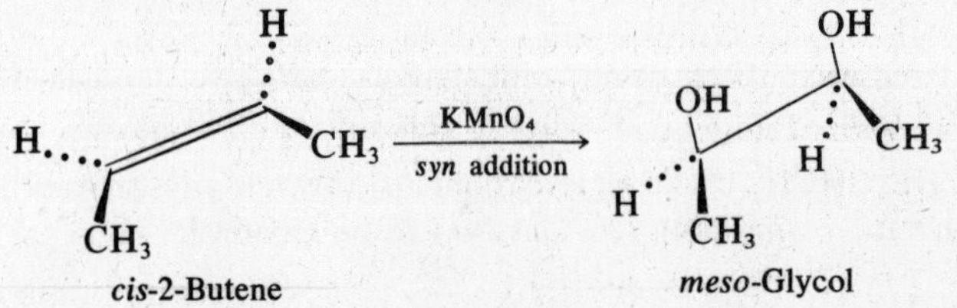

cis-2-Butene   *meso*-Glycol

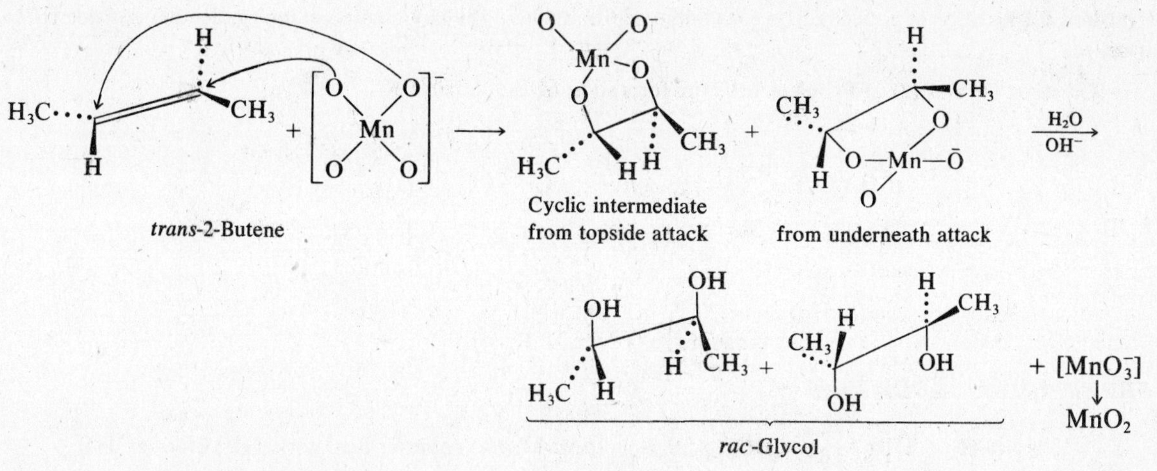

trans-2-Butene + [MnO$_4$] → Cyclic intermediate from topside attack + from underneath attack $\xrightarrow[\text{OH}^-]{\text{H}_2\text{O}}$

rac-Glycol + [MnO$_3^-$] → MnO$_2$

DIMERIZATION OF ALKENES

$$(CH_3)_2C{=}CH_2 \xrightarrow{\text{H}_2\text{SO}_4} (CH_3)_3C{-}CH_2{-}\overset{\overset{\displaystyle CH_3}{|}}{C}{=}CH_2 + (CH_3)_3C{-}CH{=}C(CH_3)_2$$

Major product

ADDITION OF ALKANES

$$(CH_3)_2C{=}CH_2 + HC(CH_3)_3 \xrightarrow[0°]{\text{HF}} (CH_3)_2CHCH_2C(CH_3)_3$$

2,2,4-Trimethylpentane

Problem 6.32 Suggest a mechanism for the dimerization of $(CH_3)_2C{=}CH_2$ that involves an intermediate R$^+$. ◄

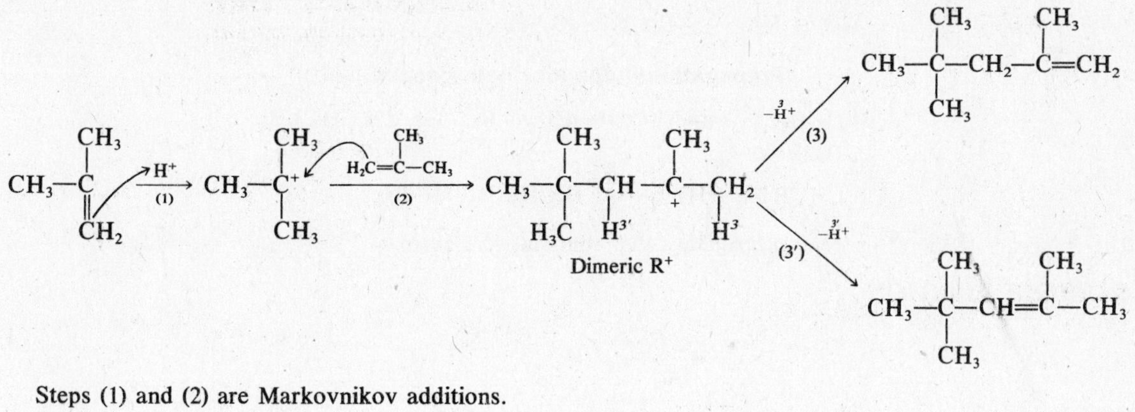

Dimeric R$^+$

Steps (1) and (2) are Markovnikov additions.

Problem 6.33 Write the structural formula for (a) the major trimeric alkene formed from $(CH_3)_2C{=}CH_2$, (b) the dimeric alkene from $CH_3CH{=}CH_2$. [Indicate the dimeric R$^+$.] ◄

(a)   (CH$_3$)$_3$C— | CH$_2$—C— | CH=C(CH$_3$)$_2$
                        with CH$_3$ above and CH$_3$ below the central C

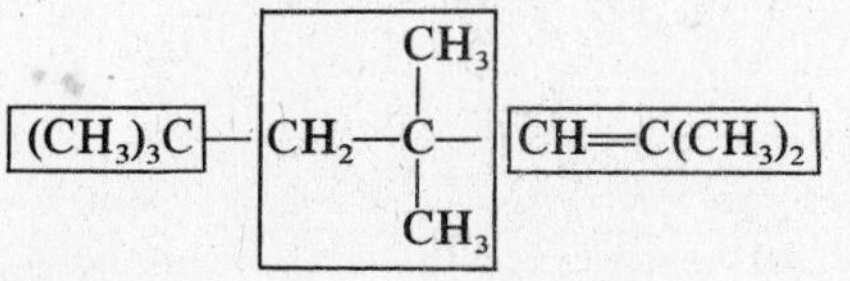

The individual combining units are boxed.

(b) $(CH_3)_2CHCH{=}CHCH_3$; $[(CH_3)_2CHCH_2\overset{+}{C}HCH_3]$

Problem 6.34 Suggest a mechanism for alkane addition where the key step is an intermolecular **hydride** (H:) transfer. ◄

See steps (1) and (2) in Problem 6.32 for formation of the dimeric R^+.

$$\underset{\text{Dimeric } R^+}{(CH_3)_3CCH_2\overset{CH_3}{\underset{CH_3}{C^+}}(H)C(CH_3)_3} \longrightarrow (CH_3)_3CCH_2\overset{CH_3}{\underset{CH_3}{C}}(H) + \overset{+}{C}(CH_3)_3$$

This intermolecular H: transfer forms the $(CH_3)_3C^+$ ion which adds to another molecule of $(CH_3)_2C=CH_2$ to continue the chain. A 3° H usually transfers to leave a 3° R^+.

FREE-RADICAL ADDITIONS

$$RCH=CH_2 + HBr \xrightarrow[\text{ROOR}]{O_2 \text{ or}} RCH_2CH_2Br \quad \text{(anti-Markovnikov; not with HF, HCl or HI)}$$

$$RCH=CH_2 + HSH \xrightarrow{\text{ROOR}} RCH_2CH_2SH \quad \text{(anti-Markovnikov)}$$

$$RCH=CH_2 + HCCl_3 \xrightarrow{\text{ROOR}} RCHCH_2\boxed{CCl_3}$$
$$\boxed{H}$$

$$RCH=CH_2 + BrCCl_3 \xrightarrow{\text{ROOR}} RCH\boxed{Br}CH_2\boxed{CCl_3}$$

Problem 6.35 Suggest a chain-propagating free-radical mechanism for addition of HBr in which Br· attacks the alkene to form the more stable carbon radical. ◄

Initiation Steps

$$R—O—O—R \xrightarrow{\text{heat}} 2R—O· \quad (—O—O— \text{ bond is weak})$$

$$RO· + HBr \longrightarrow Br· + R—O—H$$

Propagation Steps for Chain Reaction

$$\underset{\text{(1° radical)}}{CH_3CHBr\dot{C}H_2} \overset{\times}{\longleftarrow} CH_3CH=CH_2 + Br· \longrightarrow \underset{\text{(2° radical)}}{CH_3\dot{C}HCH_2Br}$$

$$CH_3\dot{C}HCH_2Br + HBr \longrightarrow CH_3CH_2CH_2Br + Br·$$

The Br· generated in the second propagation step continues the chain.

CARBENE ADDITION

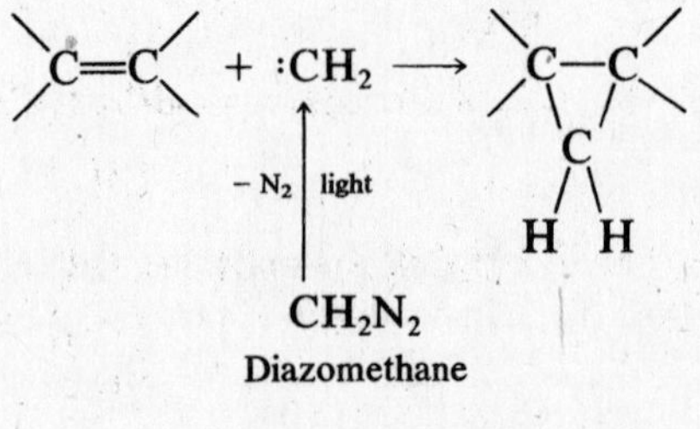

$$\underset{}{\overset{}{C}=\overset{}{C}} + :CH_2 \longrightarrow \overset{}{\underset{}{C—C}}\underset{\underset{H \quad H}{C}}{}$$

$$\overset{-N_2 \Big| \text{light}}{}$$

$$\underset{\text{Diazomethane}}{CH_2N_2}$$

CLEAVAGE REACTIONS

Ozonolysis

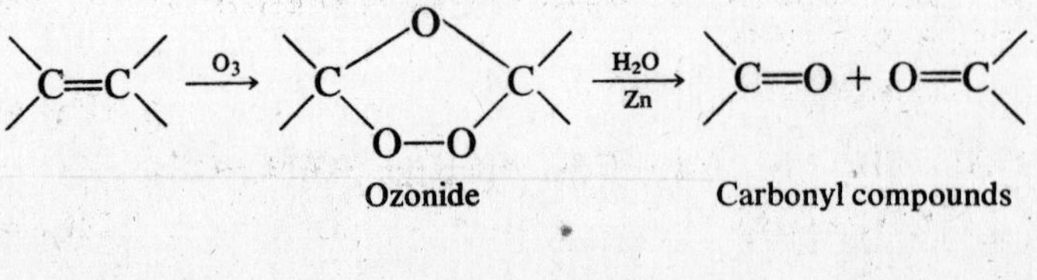

$$\overset{}{C}=\overset{}{C} \xrightarrow{O_3} \underset{\underset{\text{Ozonide}}{O—O}}{\overset{O}{C \quad C}} \xrightarrow[\text{Zn}]{H_2O} \underset{\text{Carbonyl compounds}}{C=O + O=C}$$

Vigorous Oxidation with $KMnO_4$

$$H_2C=CHCH_2CH_3 \xrightarrow{KMnO_4} CO_2 + HOOCCH_2CH_3 \quad (H_2C= \text{ gives } CO_2)$$

$$CH_3CH=CHCH_3 \xrightarrow{KMnO_4} CH_3COOH + HOOCCH_3 \quad (RCH= \text{ gives } RCOOH)$$

$$\underset{\displaystyle CH_3-\overset{\displaystyle \overset{CH_3}{|}}{C}=CHCH_2CH_2CH_3}{} \xrightarrow{KMnO_4} CH_3-\overset{\overset{CH_3}{|}}{C}=O + HOOCCH_2CH_2CH_3 \quad (R_2C= \text{ gives } R_2C=O)$$

Problem 6.36 Give the products formed on ozonolysis of (a) $H_2C=CHCH_2CH_3$, (b) $CH_3CH=CHCH_3$, (c) $(CH_3)_2C=CHCH_2CH_3$, (d) cyclobutene, (e) $H_2C=CHCH_2CH=CHCH_3$. ◄

To get the correct answers, erase the double bond and attach a $=O$ to each of the formerly double-bonded C's. The total numbers of C's in the carbonyl products and in the alkene reactant must be equal.

(a) $H_2C=O + O=CHCH_2CH_3$.
(b) $CH_3CH=O$; the alkene is symmetrical and only one carbonyl compound is formed.
(c) $(CH_3)_2C=O + O=CHCH_2CH_3$.
(d) $O=CHCH_2CH_2CH=O$; a cycloalkene gives only a dicarbonyl compound.
(e) $H_2C=O + O=CHCH_2CH=O + O=CHCH_3$. Polyenes give mixtures of mono and dicarbonyl compounds.

Problem 6.37 Deduce the structures of the following alkenes.

(a) An alkene $C_{10}H_{20}$ on ozonolysis yields only $CH_3-\overset{\overset{\displaystyle O}{\|}}{C}-CH_2CH_2CH_3$.

(b) An alkene C_9H_{18} on ozonolysis gives $(CH_3)_3C\overset{\overset{\displaystyle H}{|}}{C}=O$ and $CH_3-\overset{\overset{\displaystyle O}{\|}}{C}-CH_2CH_3$.

(c) A compound C_8H_{14} adds one mole of H_2 and forms on ozonolysis the dialdehyde

$$O=CH\overset{\overset{\displaystyle CH_3}{|}}{C}HCH_2CH_2\overset{\overset{\displaystyle CH_3}{|}}{C}HCH=O$$

(d) A compound C_8H_{12} adds two moles of H_2 and undergoes ozonolysis to give two moles of the dialdehyde $O=CHCH_2CH_2CH=O$. ◄

(a) The formation of only one carbonyl compound indicates that the alkene is symmetrical about the double bond. Write the structure of the ketone twice so that the $C=O$ groups face each other. Replacement of the two O's by a double bond gives the alkene structure.

$$CH_3CH_2CH_2\overset{\overset{\displaystyle CH_3}{|}}{C}=O + O=\overset{\overset{\displaystyle CH_3}{|}}{C}CH_2CH_2CH_3 \longleftarrow CH_3CH_2CH_2-\overset{\overset{\displaystyle H_3C}{|}}{C}=\overset{\overset{\displaystyle CH_3}{|}}{C}CH_2CH_2CH_3$$

(b)
$$CH_3-\overset{\overset{\displaystyle CH_3}{|}}{\underset{\underset{\displaystyle CH_3}{|}}{C}}-CH=O + O=\overset{\overset{\displaystyle CH_3}{|}}{C}-CH_2CH_3 \longleftarrow CH_3-\overset{\overset{\displaystyle CH_3}{|}}{\underset{\underset{\displaystyle CH_3}{|}}{C}}-\overset{\overset{\displaystyle CH_3}{|}}{C}=\overset{\underset{\displaystyle H}{|}}{C}-CH_2CH_3 \quad \textit{cis or trans}$$

(c) C_8H_{14} has 4 fewer H's than the corresponding alkane, C_8H_{18}. There are 2 degrees of unsaturation. One of these is accounted for by a $C=C$ because the alkene adds 1 mole of H_2. The second degree of unsaturation is a ring structure. The compound is a cycloalkene whose structure is found by writing the two terminal carbonyl groups facing each other.

$$CH_3-\overset{\displaystyle CH=O}{\underset{\displaystyle CH_2-CH_2}{CH}} \quad \overset{\displaystyle O=CH}{\underset{}{CH-CH_3}} \quad \underset{\overset{\longleftarrow}{\substack{1.\ O_3 \\ 2.\ H_2O\ (Zn)}}}{} \quad CH_3-\overset{\displaystyle CH=CH}{\underset{\displaystyle CH_2CH_2}{CH}}CH-CH_3$$

(d) The difference of 6 H's between C_8H_{12} and the alkane C_8H_{18} shows 3 degrees of unsaturation. The 2 moles of H_2 absorbed indicates 2 double bonds. The third degree of unsaturation is a ring structure. When 2 molecules of product are written with the pairs of C=O groups facing each other, the compound is seen to be a cyclic diene.

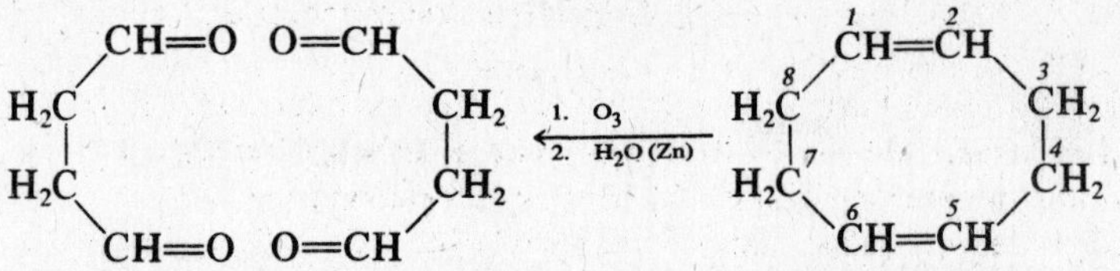

1,5-Cyclooctadiene

Problem 6.38 Give the products formed on hot permanganate cleavage of the following compounds:

(a) $H_2C{=}CH_2$ (b) $CH_3CH{=}CHCH(CH_3)_2$ (c) $(CH_3)_2C{=}C(C_2H_5)_2$

(d)   (e)

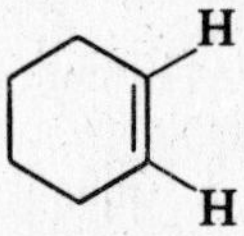

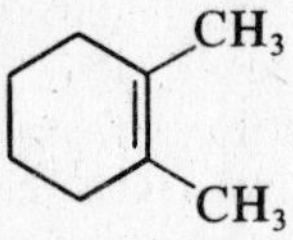

A double-bonded C with 2 H's (a terminal double bond) forms CO_2; a C with 1 H gives a carboxylic acid, RCOOH; a C with no H's gives a ketone, $R_2C{=}O$.

(a) CO_2 (only one product because alkene is symmetrical).
(b) $CH_3COOH + HOOCCH(CH_3)_2$.
(c) $(CH_3)_2C{=}O + O{=}C(C_2H_5)_2$.
(d) $HOOC(CH_2)_4COOH$.

(e) $CH_3{-}\overset{O}{\overset{\|}{C}}{-}(CH_2)_4{-}\overset{O}{\overset{\|}{C}}{-}CH_3$.

SUBSTITUTION REACTIONS AT ALLYLIC POSITION

$$Cl_2 + H_2C{=}CHCH_3 \xrightarrow{\text{high temperature}} H_2C{=}CHCH_2Cl + HCl$$

$$Br_2 + H_2C{=}CHCH_3 \xrightarrow[\text{of } Br_2]{\text{low concentration}} H_2C{=}CHCH_2Br + HBr$$

The low concentration of Br_2 comes from N-bromosuccinimide (NBS).

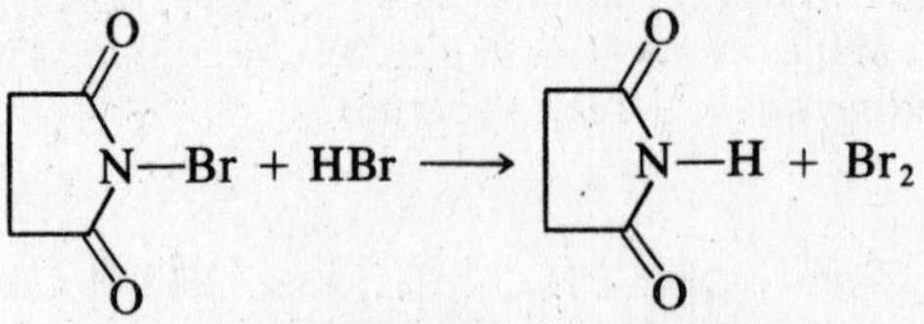

NBS   product of   Succinimide
         bromination

$$SO_2Cl_2 + H_2C{=}CHCH_3 \xrightarrow[\text{peroxide}]{\text{uv or}} H_2C{=}CHCH_2Cl + HCl + SO_2$$

Sulfuryl
chloride

These halogenations are like free-radical substitutions of alkanes (see page 44). The order of reactivity of H-abstraction is

$$\text{allyl} > 3° > 2° > 1° > \text{vinyl}$$

Problem 6.39 Use the concepts of (a) resonance and (b) extended π orbital overlap (delocalization) to account for the extraordinary stability of the allyl-type radical

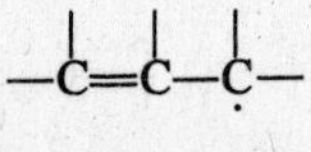

(a) Two equivalent resonance structures can be written:

$$-\overset{|}{C}=\overset{|}{C}-\overset{|}{\underset{.}{C}}- \longleftrightarrow -\overset{|}{\underset{.}{C}}-\overset{|}{C}=\overset{|}{C}-$$

therefore the allyl-type radical has considerable resonance energy
(Section 2.7) and is relatively stable.

(b) The 3 C's in the allyl unit are sp^2 hybridized and they each have a p
orbital lying in a common plane (Fig. 6-5). These three p orbitals
overlap forming an extended π system, thereby delocalizing the odd
electron. Such delocalization stabilizes the allyl-type free radical.

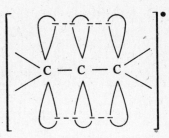

Fig. 6-5

Problem 6.40 Designate the type of each set of H's in $CH_3CH{=}CHCH_2CH_2{-}CH(CH_3)_2$ (e.g. 3°, allylic, etc.)
and show their relative reactivity toward a Br· atom, using (1) for the most reactive, (2) for the next, etc. ◄

Labeling the H's as

$$\overset{(a)\ \ (b)}{CH_3CH}{=}\overset{(b)\ (c)\ (d)\ (e)\ \ (f)}{CHCH_2CH_2CH(CH_3)_2}$$

we have: (a) 1°, allylic (2); (b) vinylic (6); (c) 2°, allylic (1); (d) 2° (4); (e) 3° (3); (f) 1° (5).

6.5 SUMMARY OF ALKENE CHEMISTRY

PREPARATION

1. Dehydrohalogenation

  $RCHXCH_3$, RCH_2CH_2X + alc. KOH

2. Dehydration

  $RCHOHCH_3$, RCH_2CH_2OH + H_2SO_4

  heat

3. Dehalogenation

  $RCHXCH_2X$ + Zn

4. Dehydrogenation

  RCH_2CH_3, heat, Pt-Pd

5. Addition

  $R{-}C{\equiv}CH$ + H_2

$$\rightarrow RCH{=}CH_2$$

PROPERTIES

1. Addition Reactions

(a) Hydrogenation

Heterogeneous: H_2/Pt

Chemical: (BH_3), CH_3CO_2H
  or H_2NNH_2, Cu^{2+}

$$\rightarrow RCH_2CH_3$$

(b) Polar Mechanism

+ $X_2 \rightarrow RCHXCH_2X$ (X = Cl, Br)

+ HX $\rightarrow RCHXCH_3$

+ HOX $\rightarrow RCH(OH)CH_2X$

+ $H_2O \rightarrow RCH(OH)CH_3$

+ $H_2SO_4 \rightarrow RCH(OSO_3H)CH_3$

+ BH_3, H_2O_2 + NaOH $\rightarrow RCH_2CH_2OH$

+ dil. cold $KMnO_4 \rightarrow RCH(OH)CH_2OH$

+ hot $KMnO_4 \rightarrow RCOOH + CO_2$

+ RCO_3H, $H_3O^+ \rightarrow RCH(OH)CH_2OH$

+ H^+, $RCH{=}CH_2 \rightarrow RCH(CH_3)CH{=}CHR$

+ O_3, Zn, $H_2O \rightarrow RCH{=}O + CH_2{=}O$

(c) Free-Radical Mechanism

+ HBr $\rightarrow RCH_2CH_2Br$

+ $CHCl_3 \rightarrow RCH_2CH_2CCl_3$

2. Substitution Reactions

$$R{-}CH_2{-}CH{=}CH_2 + X_2 \xrightarrow[\text{or uv}]{\Delta}$$

$$R{-}CHX{-}CH{=}CH_2$$

Supplementary Problems

Problem 6.41 Write structural formulas for the organic compounds designated by a ? and show the stereochemistry where requested.

(a)

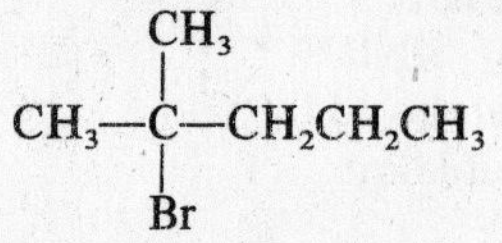

$$CH_3-\underset{\underset{Br}{|}}{\overset{\overset{CH_3}{|}}{C}}-CH_2CH_2CH_3 + \text{alcoholic KOH} \longrightarrow ? \text{(principal)} + ? \text{(minor)}$$

(b)

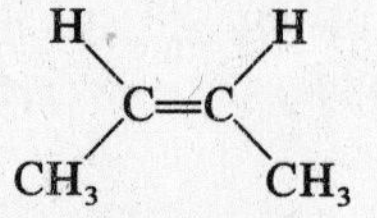

$$\underset{CH_3}{\overset{H}{}}C=C\underset{CH_3}{\overset{H}{}} + Br_2 \longrightarrow ? + ? \text{ (Stereochemistry)}$$

(c)

$$? \text{(alkene)} + ? \text{(reagent)} \longrightarrow CH_3-\underset{\underset{H}{|}}{\overset{\overset{CH_3}{|}}{C}}-CH_2-\underset{\underset{OH}{|}}{\overset{\overset{CH_3}{|}}{C}}-CH_3$$

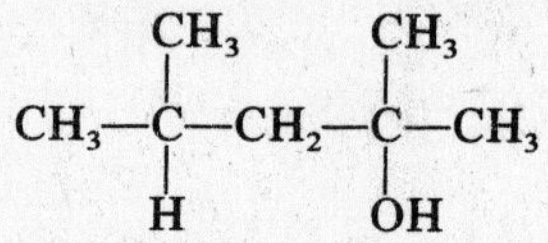

(d)

$$CH_3CH_2CH_2CH_2CH=CH_2 + CHBr_3 + \text{peroxide} \longrightarrow ?$$

(e)

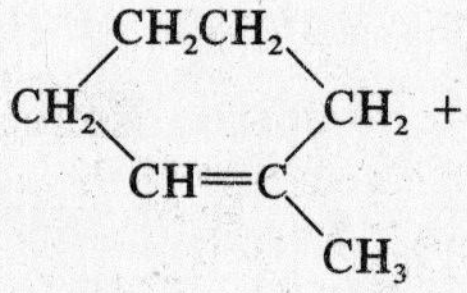

$$\begin{array}{c} CH_2CH_2 \\ CH_2 \qquad\qquad CH_2 \\ CH=C \\ \qquad CH_3 \end{array} + HOBr \longrightarrow ? \text{ (Stereochemistry)}$$

(f)

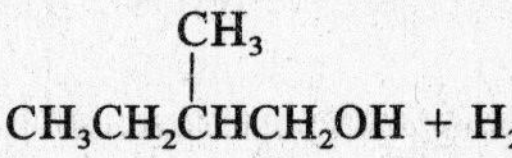

$$CH_3CH_2\underset{\underset{CH_3}{|}}{\overset{}{C}}HCH_2OH + H_2SO_4, \text{heat} \longrightarrow ? \text{(principal)} + ? \text{(minor)}$$

◄

(a)

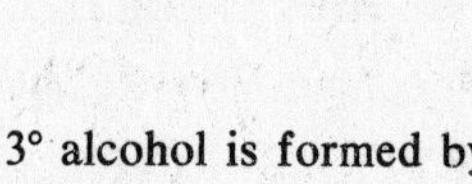

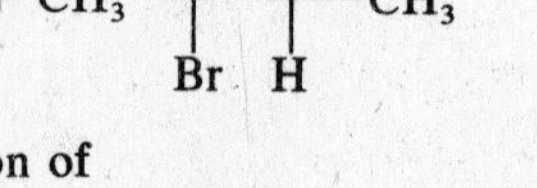

$$\underset{CH_3}{\overset{\boxed{CH_3}}{C}}=CH\boxed{CH_2CH_3} \qquad\qquad H_2C=\underset{\overset{\boxed{CH_3}}{}}{C}-CH_2\boxed{CH_2CH_3}$$

Principal: has 3 R's on the C=C.   Minor: has only 2 R's.

(b) *Anti* addition to a *cis* diastereomer gives a racemic mixture,

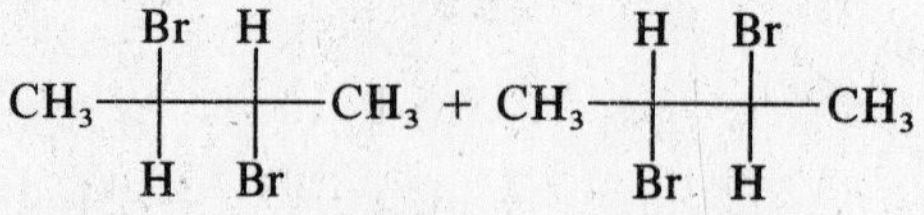

$$CH_3-\underset{\underset{H\quad Br}{}}{\overset{\overset{Br\quad H}{}}{|\quad|}}-CH_3 + CH_3-\underset{\underset{Br\quad H}{}}{\overset{\overset{H\quad Br}{}}{|\quad|}}-CH_3$$

(c) The 3° alcohol is formed by acid-catalyzed hydration of

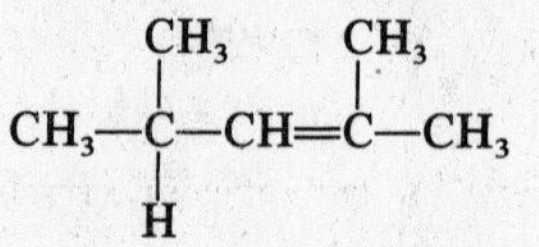

$$CH_3-\underset{\underset{H}{|}}{\overset{\overset{CH_3}{|}}{C}}-CH=\overset{\overset{CH_3}{|}}{C}-CH_3 \quad \text{or} \quad CH_3-\underset{\underset{H}{|}}{\overset{\overset{CH_3}{|}}{C}}-CH_2-\overset{\overset{CH_3}{|}}{C}=CH_2$$

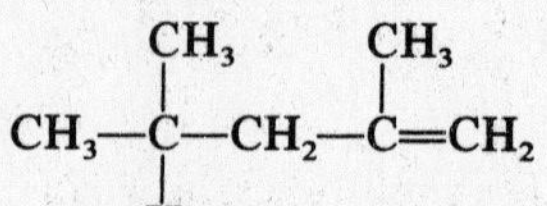

The reagent is dilute aq. H_2SO_4.

(d) $CH_3CH_2CH_2CH_2CH_2-CH_2CBr_3$.

(e) In this polar Markovnikov addition, the positive Br adds to the C having the H. The addition is *anti*, so the Br will be *trans* to OH but *cis* to CH_3.

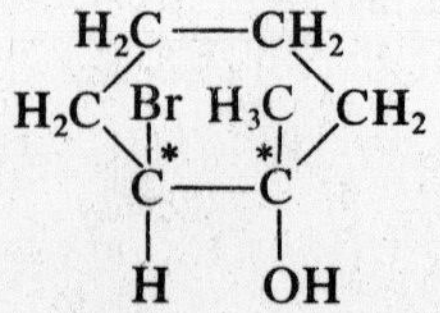

$$\begin{array}{c} H_2C-\!\!\!-CH_2 \\ H_2C \quad Br \;\; H_3C \quad CH_2 \\ \overset{*}{C}\qquad\overset{*}{C} \\ H \qquad OH \end{array}$$

rac

(*f*)   The formation of products is shown:

$$CH_3CH_2CHCH_2OH \xrightarrow{H^+} CH_3CH_2CHCH_2\overset{+}{O}H \xrightarrow{-H_2O} CH_3CH_2CH\overset{+}{C}H_2 \xrightarrow{-H^+} CH_3CH_2C{=}CH_2$$

with CH₃ groups and (1°), ~H: shift to

$$CH_3CH{=}C(CH_3)_2 \xleftarrow{-H^+} CH_3CH_2{-}\overset{+}{C}(CH_3)_2$$

(major)   (3°)

(minor)

Problem 6.42 Draw an enthalpy–reaction progress diagram for addition of Br_2 to an alkene.   ◄

See Fig. 6-6.

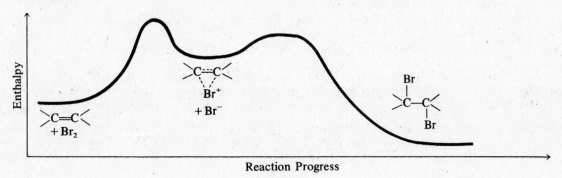

Reaction Progress

Fig. 6-6

Problem 6.43 Write the initiation and the propagation steps for a free-radical-catalyzed (RO·) addition of $CH_3CH{=}O$ to 1-hexene to form methyl *n*-hexyl ketone,

$$CH_3CH_2CH_2CH_2CH_2CH_2{-}\overset{O}{\overset{\|}{C}}{-}CH_3$$

◄

The initiation step is

$$CH_3{-}\overset{O}{\overset{\|}{C}}{:}H + RO\cdot \longrightarrow CH_3{-}\overset{O}{\overset{\|}{C}}\cdot + RO{\cdot}H$$

and the propagation steps are

$$n\text{-}C_4H_9{-}CH{=}CH_2 + \cdot\overset{O}{\overset{\|}{C}}{-}CH_3 \longrightarrow n\text{-}C_4H_9{-}\overset{\cdot}{C}H{-}CH_2{-}\overset{O}{\overset{\|}{C}}{-}CH_3 \quad \text{(adds to give } 2° \text{ R·)}$$

$$n\text{-}C_4H_9{-}\overset{\cdot}{C}H{-}CH_2{-}\overset{O}{\overset{\|}{C}}{-}CH_3 + H{-}\overset{O}{\overset{\|}{C}}{-}CH_3 \longrightarrow n\text{-}C_4H_9{-}CH_2CH_2{-}\overset{O}{\overset{\|}{C}}{-}CH_3 + \cdot\overset{O}{\overset{\|}{C}}{-}CH_3 \quad \text{(regenerates)}$$

Problem 6.44 Suggest a radical mechanism to account for the interconversion of *cis* and *trans* isomers by heating with I_2.   ◄

I_2 has a low bond dissociation energy (36 kcal/mole) and forms 2I· on heating.   I· adds to the C=C to form a carbon radical which rotates about its sigma bond and assumes a different conformation.   However, the C—I bond is also weak (56 kcal/mole) and the radical loses I· under these conditions.   The double bond is reformed and the 2 conformations produce a mixture of *cis* and *trans* isomers.

$$\underset{trans}{\overset{R}{\underset{R}{>}}C{=}C{<}} \underset{-I\cdot}{\overset{+I\cdot}{\rightleftarrows}} \overset{R}{>}C{-}\overset{R}{C}{<} \rightleftarrows \overset{R}{>}C{-}\overset{R}{C}{<} \underset{+I\cdot}{\overset{-I\cdot}{\rightleftarrows}} \underset{cis}{\overset{R}{>}C{=}C\overset{R}{<}}$$

Problem 6.45 Write structures for the products of the following polar addition reactions:

(a) $(CH_3)_2C=CHCH_3 + I—Cl \longrightarrow$?   (b) $(CH_3)_2—C=CH_2 + HSCH_3 \longrightarrow$?

(c) $(CH_3)_3\overset{+}{N}—CH=CH_2 + HI \longrightarrow$?   (d) $H_2C=CHCF_3 + HCl \longrightarrow$? ◄

(a) $(CH_3)_2C\boxed{Cl}CH\boxed{I}CH_3$.   I is less electronegative than Cl in $\overset{\delta+}{I}—\overset{\delta-}{Cl}$ and adds to the C with more H's.

(b)
$$(CH_3)_2—\underset{\underset{CH_3S}{|}}{C}—\underset{\underset{H}{|}}{CH_2}$$

H is less electronegative than S; $\overset{\delta+}{H}—\overset{\delta-}{S}CH_3$.

(c)
$$(CH_3)_3\overset{+}{N}—CH_2—CH_2—I$$

The + charge on N destabilizes an adjacent + charge. $(CH_3)_3\overset{+}{N}CH_2\overset{+}{C}H_2$ is more stable than $(CH_3)_3\overset{+}{N}\overset{+}{C}HCH_3$.   Addition is anti-Markovnikov.

(d) $ClCH_2CH_2CF_3$.   The strong electron-attracting CF_3 group destabilizes an adjacent + charge so that $\overset{+}{C}H_2CH_2CF_3$ is the intermediate rather than $CH_3\overset{+}{C}HCF_3$.

Problem 6.46 Explain the following observations: (a) Br_2 and propene in C_2H_5OH gives not only $BrCH_2CHBrCH_3$ but also $BrCH_2CH(OC_2H_5)CH_3$. (b) Isobutylene is more reactive than 1-butene towards peroxide-catalyzed addition of CCl_4. (c) The presence of Ag^+ salts enhances the solubility of alkenes in H_2O. ◄

(a) The intermediate bromonium ion reacts with both Br^- and $C_2H_5\ddot{O}H$ as nucleophiles to give the 2 products. (b) The more stable the intermediate free radical, the more reactive the alkene. $H_2C=CHCH_2CH_3$ adds $\cdot CCl_3$ to give the less stable 2° radical $Cl_3CCH_2\dot{C}HCH_2CH_3$, whereas $H_2C=C(CH_3)_2$ reacts to give the more stable 3° radical $Cl_3CCH_2\dot{C}(CH_3)_2$. (c) Ag^+ coordinates with the alkene by p-d π bonding to give an ion similar to a bromonium ion, but more stable:

$$\overset{\overset{+}{Ag}}{\underset{}{C\!\!=\!\!C}}$$

Problem 6.47 Supply the structural formulas of the alkenes and the reagents which react to form:
(a) $(CH_3)_3CI$, (b) CH_3CHBr_2, (c) $BrCH_2CHClCH_3$, (d) $BrCH_2CHOHCH_2Cl$. ◄

(a) $CH_3—\underset{\underset{}{|}}{\overset{\overset{CH_3}{|}}{C}}=CH_2 + HI$ (b) $H_2C=CHBr + HBr$

(c) $H_2C=CCl—CH_3 + HBr + $ peroxide   (d) $BrCH_2—CH=CH_2 + HOCl$

or $H_2C=CHCH_3 + BrCl$ or $H_2C=CH—CH_2Cl + HOBr$

Problem 6.48 Outline the steps needed for the following syntheses in reasonable yield. Inorganic reagents and solvents may also be used. (a) 1-Chloropentane to 1,2-dichloropentane. (b) 1-Chloropentane to 2-chloropentane. (c) 1-Chloropentane to 1-bromopentane. (d) 1-Bromobutane to 1,2-dihydroxybutane. (e) Isobutyl chloride to

$$CH_3—\underset{\underset{CH_3}{|}}{\overset{\overset{CH_3}{|}}{C}}—CH_2—\underset{\underset{I}{|}}{\overset{\overset{CH_3}{|}}{C}}—CH_3$$ ◄

Syntheses are best done by working backwards, keeping in mind your starting material.

(a) The desired product is a *vic*-dichloride made by adding Cl_2 to the appropriate alkene, which in turn is made by dehydrochlorinating the starting material.

$$ClCH_2CH_2CH_2CH_2CH_3 \xrightarrow[\text{KOH}]{\text{alc.}} H_2C=CHCH_2CH_2CH_3 \xrightarrow{Cl_2} ClCH_2CHClCH_2CH_2CH_3$$

(b) To get a pure product add HCl to 1-pentene as made in part (a).

$$H_2C=CHCH_2CH_2CH_3 + HCl \longrightarrow H_3CCHClCH_2CH_2CH_3$$

(c) An anti-Markovnikov addition of HBr to 1-pentene [part (a)].

$$H_2C=CHCH_2CH_2CH_3 + HBr \xrightarrow{peroxide} BrCH_2CH_2CH_2CH_2CH_3$$

(d) Glycols are made by mild oxidation of alkenes.

$$BrCH_2CH_2CH_2CH_3 \xrightarrow[KOH]{alc.} H_2C=CHCH_2CH_3 \xrightarrow[RT]{KMnO_4} HOCH_2CHOHCH_2CH_3$$

(e) The product has twice as many C's as does the starting material. The skeleton of C's in the product corresponds to that of the dimer of $(CH_3)_2C=CH_2$.

$$(CH_3)_2CHCH_2Cl \xrightarrow[KOH]{alc.} (CH_3)_2C=CH_2 \xrightarrow{H_2SO_4} (CH_3)_3CCH=C(CH_3)_2 + (CH_3)_3CCH_2\overset{\overset{\displaystyle CH_3}{|}}{C}=CH_2$$

$$\downarrow HI$$

$$(CH_3)_3CCH_2Cl(CH_3)_2$$

Problem 6.49 Show how propene can be converted to (a) 1,5-hexadiene, (b) 1-bromopropene, (c) 4-methyl-1-pentene. ◄

(a) $$CH_3CH=CH_2 \xrightarrow{Cl_2, 500\ °C} ClCH_2CH=CH_2 \xrightarrow{Na} H_2C=CHCH_2CH_2CH=CH_2$$

(b) $$CH_3CH=CH_2 \xrightarrow{Br_2(CCl_4)} CH_3CHBrCH_2Br \xrightarrow[KOH]{alc.} CH_3CH=CHBr$$

(Little $CH_3CBr=CH_2$ is formed because the 2° H of $—CH_2Br$ is more acidic than the 3° H of $—\overset{|}{C}HBr$.)

(c)

$$CH_3CH=CH_2 \xrightarrow{HBr} CH_3—\underset{\underset{\displaystyle Br}{|}}{C}H—CH_3 \xrightarrow{Mg} CH_3—\underset{\underset{\displaystyle MgBr}{|}}{C}H—CH_3$$

$$+$$

$$CH_3CH=CH_2 \xrightarrow{Cl_2, 500\ °C} ClCH_2CH=CH_2$$

$$\left.\vphantom{\begin{array}{c}a\\b\\c\end{array}}\right\} \xrightarrow{CuBr} CH_3—\overset{\overset{\displaystyle CH_3}{|}}{C}H—CH_2CH=CH_2$$

Problem 6.50 Dehydration of 3,3-dimethyl-2-butanol, $(CH_3)_3CCHOHCH_3$, yields two alkenes, neither of which is $(CH_3)_3CCH=CH_2$. What are their structures? ◄

The initially formed $(CH_3)_3C\overset{+}{C}HCH_3$ undergoes a $:CH_3$ shift to give

$$(CH_3)_2\overset{+}{C}—\overset{\overset{\displaystyle H}{|}}{C}(CH_3)_2$$

which forms

$$CH_3—\underset{\underset{\displaystyle CH_3}{|}}{\overset{\overset{\displaystyle CH_3}{|}}{C}}=C—CH_3 \quad \text{and} \quad H_2C=\overset{\overset{\displaystyle CH_3}{|}}{C}—\underset{\underset{\displaystyle CH_3}{|}}{C}H—CH_3$$

Problem 6.51 (a) Br_2 is added to (S)-$\overset{1}{H_2}\overset{2}{C}=\overset{3}{C}H\overset{}{C}HBr\overset{4}{C}H_3$. Give Fischer projections and R,S designations for the products. Are the products optically active? (b) Repeat (a) for HBr. ◄

(a) C^2 becomes chiral and the configuration of C^3 is unchanged. There are 2 optically active diastereomers of 1,2,3-tribromobutane. It is best to draw formulas with H's on vertical lines.

$$H_2\overset{1}{C}=\overset{2}{C}H—\overset{\overset{\displaystyle Br}{|}}{\underset{\underset{\displaystyle H}{|}}{C}}{}^3—\overset{4}{C}H_3 \longrightarrow H_2\overset{1}{C}(Br)—\overset{\overset{\displaystyle Br}{|}}{\underset{\underset{\displaystyle H}{|}}{C}}{}^2—\overset{\overset{\displaystyle Br}{|}}{\underset{\underset{\displaystyle H}{|}}{C}}{}^3—\overset{4}{C}H_3 + H_2\overset{1}{C}(Br)—\overset{\overset{\displaystyle H}{|}}{\underset{\underset{\displaystyle Br}{|}}{C}}{}^2—\overset{\overset{\displaystyle Br}{|}}{\underset{\underset{\displaystyle H}{|}}{C}}{}^3—\overset{4}{C}H_3$$

S   2R,3S   2S,3S

(optically active)   (optically active)

(b) There are two diastereomers of 2,3-dibromobutane:

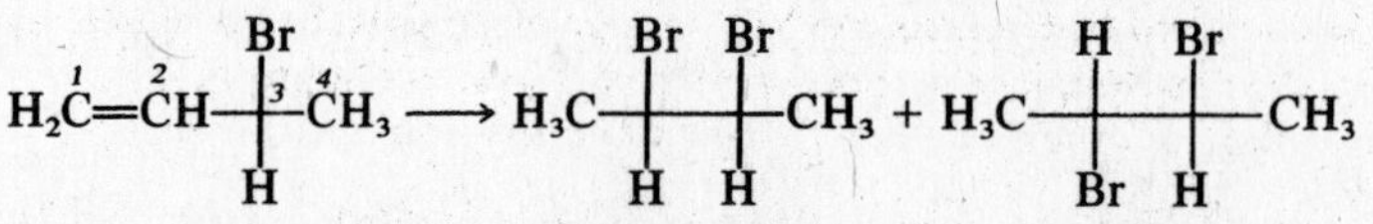

Problem 6.52 Polypropylene can be synthesized by the acid-catalyzed polymerization of propylene. (a) Show the first three steps. (b) Indicate the repeating unit (mer). ◄

(a) CH_3CH=CH_2 $\xrightarrow{H^+}$ $(CH_3)_2C^+$ ...

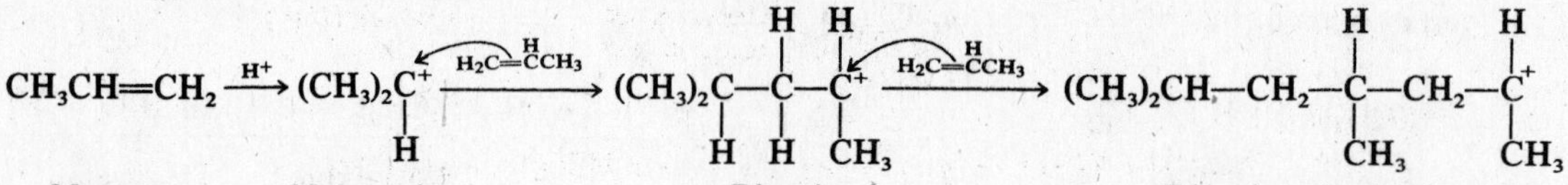

(b)

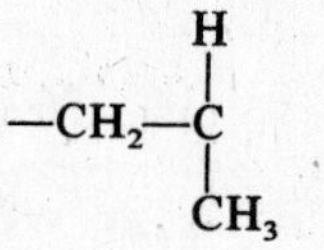

Problem 6.53 List ways in which R^+'s can react. ◄

(a) Combine with a nucleophile.
(b) As a strong acid, give up a vicinal H^+ to form an alkene.
(c) Rearrange by :H or :R shift to a more stable R'^+.
(d) Add to a molecule of alkene to give a larger-molecular-weight R^+ (a type of combination with nucleophile).
(e) Abstract a 3° hydride from an alkane.

Problem 6.54 From propene, prepare (a) CH_3CHDCH_2D, (b) CH_3CHDCH_3, (c) $CH_3CH_2CH_2D$. ◄

(a) Add D_2 in the presence of Pd.

(b) CH_3CH=$CH_2 + HCl \longrightarrow CH_3CHClCH_3 \xrightarrow{Mg} CH_3CHMgClCH_3 \xrightarrow{D_2O} CH_3CHDCH_3$

or propene + $B_2D_6 \longrightarrow (CH_3CHDCH_2)_3B \xrightarrow{CH_3COOH}$ product

(c) CH_3CH=$CH_2 \xrightarrow{B_2H_6} (CH_3CH_2CH_2)_3B \xrightarrow{CH_3COOD} CH_3CH_2CH_2D$

Problem 6.55 Ethylene is alkylated with isobutane in the presence of acid (HF) to give chiefly $(CH_3)_2CHCH(CH_3)_2$, not $(CH_3)_3CCH_2CH_3$. Account for the product. ◄

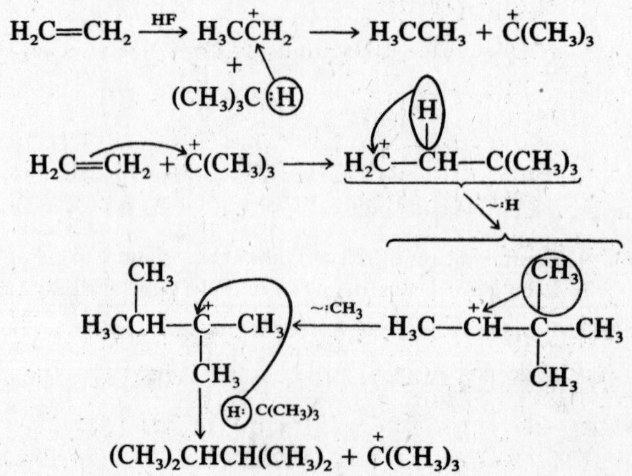

Problem 6.56 Give 4 simple chemical tests to distinguish an alkene from an alkane.

A positive simple chemical test is indicated by one or more detectable events, such as a change in color, formation of a precipitate, evolution of a gas, uptake of a gas, evolution of heat.

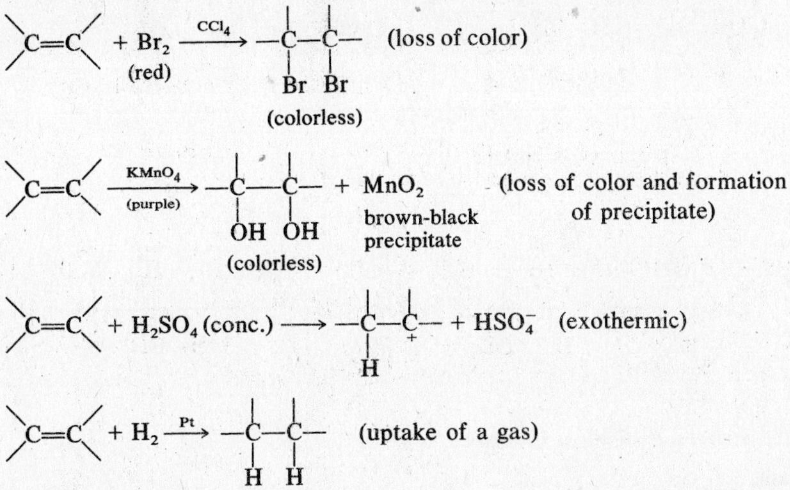

$$\underset{\text{(red)}}{\overset{}{C{=}C}} + Br_2 \xrightarrow{CCl_4} \underset{\underset{Br\ \ Br}{\text{(colorless)}}}{-\overset{|}{C}-\overset{|}{C}-} \quad \text{(loss of color)}$$

$$C{=}C \xrightarrow[\text{(purple)}]{KMnO_4} \underset{\underset{OH\ \ OH}{\text{(colorless)}}}{-\overset{|}{C}-\overset{|}{C}-} + \underset{\substack{\text{brown-black}\\\text{precipitate}}}{MnO_2} \quad \underset{\text{of precipitate)}}{\text{(loss of color and formation}}$$

$$C{=}C + H_2SO_4 \text{(conc.)} \longrightarrow \underset{\overset{|}{H}}{-\overset{|}{C}-\overset{|}{\underset{+}{C}}-} + HSO_4^- \quad \text{(exothermic)}$$

$$C{=}C + H_2 \xrightarrow{Pt} \underset{H\ \ H}{-\overset{|}{\underset{|}{C}}-\overset{|}{\underset{|}{C}}-} \quad \text{(uptake of a gas)}$$

Alkanes give none of these tests.

Problem 6.57 Give the configuration, stereochemical designation and R,S specification for the indicated tetrahydroxy products.

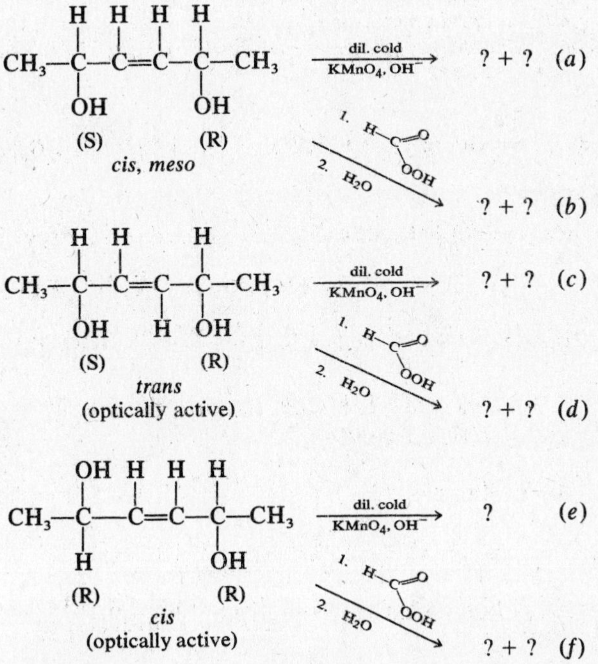

(a) syn Addition of encircled OH's:

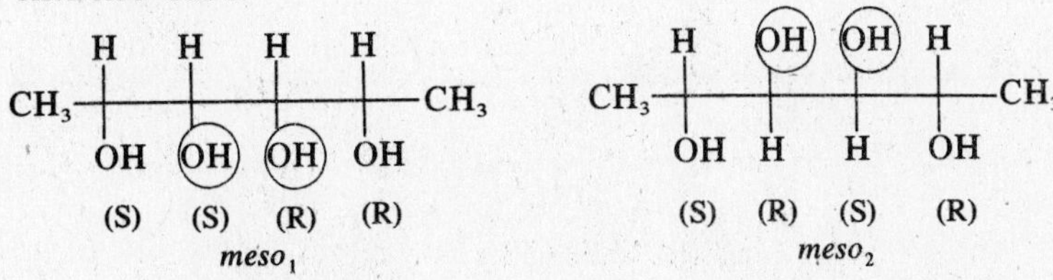

(*b*)   *anti* Addition of encircled OH's:

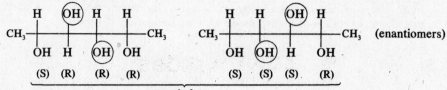

racemic form

(*c*)   *syn* Addition;   same products as part (*b*).
(*d*)   *anti* Addition;   same products as part (*a*).
(*e*)   *syn* Addition:

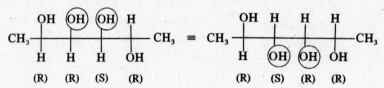

One optically active stereoisomer is formed.

(*f*)   *anti* Addition:

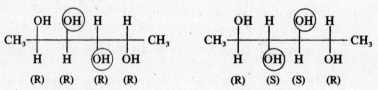

Two optically active stereoisomers.

Chapter 7

Alkyl Halides

7.1 INTRODUCTION

Alkyl halides have the general formula RX, where R is an alkyl or substituted alkyl group and X is any halogen atom.

Problem 7.1 Write structural formulas and IUPAC names for all isomers of: (a) $C_5H_{11}Br$. Classify the isomers as to whether they are tertiary (3°), secondary (2°), or primary (1°). (b) $C_4H_8Cl_2$. Classify the isomers which are *gem*-dichlorides and *vic*-dichlorides. ◄

Take each isomer of the parent hydrocarbon and replace one of each type of equivalent H by X. The correct IUPAC name is written to avoid duplication.

(a) The parent hydrocarbons are the isomeric pentanes. From pentane, $CH_3CH_2CH_2CH_2CH_3$, we get 3 monobromo products, shown with their classification.

$$BrCH_2CH_2CH_2CH_2CH_3 \qquad CH_3CHBrCH_2CH_2CH_3 \qquad CH_3CH_2CHBrCH_2CH_3$$

1-Bromopentane (1°)   2-Bromopentane (2°)   3-Bromopentane (2°)
PRIMARY   SECONDARY   SECONDARY

Classification is based on the structural features: RCH_2Br is 1°, R_2CHBr is 2°, and R_3CBr is 3°.

From isopentane, $(CH_3)_2CHCH_2CH_3$, we get 4 isomers.

$$\underset{\text{1-Bromo-2-methyl-butane (1°)}}{BrCH_2\overset{CH_3}{\underset{|}{C}}HCH_2CH_3} \qquad \underset{\text{2-Bromo-2-methyl-butane (3°)}}{CH_3\overset{CH_3}{\underset{|}{C}}BrCH_2CH_3} \qquad \underset{\text{2-Bromo-3-methyl-butane (2°)}}{CH_3\overset{CH_3}{\underset{|}{C}}HCHBrCH_3} \qquad \underset{\text{1-Bromo-3-methyl-butane (1°)}}{CH_3\overset{CH_3}{\underset{|}{C}}HCH_2CH_2Br}$$

$$\overset{CH_2Br}{\underset{|}{CH_3CHCH_2CH_3}}$$ is also 1-bromo-2-methylbutane; the two CH_3's on C^2 are equivalent.

Neopentane has 12 equivalent H's and has only one monobromo substitution product, $(CH_3)_3CCH_2Br$ (1°), 1-bromo-2,2-dimethylpropane.

(b) For the dichlorobutanes the two Cl's are first placed on each C of the straight chain. These are geminal or *gem*-dichlorides.

$$Cl_2CHCH_2CH_2CH_3 \qquad CH_3CCl_2CH_2CH_3$$

1,1-Dichlorobutane   2,2-Dichlorobutane

Then the Cl's are placed on different C's. The isomers with the Cl's on adjacent C's are vicinal or *vic*-dichlorides.

$$\underset{\substack{\text{1,2-Dichloro-}\\\text{butane (vic)}}}{\overset{Cl}{\underset{|}{ClCH_2CHCH_2CH_3}}} \qquad \underset{\substack{\text{1,3-Dichloro-}\\\text{butane}}}{\overset{Cl}{\underset{|}{ClCH_2CH_2CHCH_3}}} \qquad \underset{\substack{\text{1,4-Dichloro-}\\\text{butane}}}{ClCH_2CH_2CH_2CH_2Cl} \qquad \underset{\substack{\text{2,3-Dichloro-}\\\text{butane (vic)}}}{CH_3CHClCHClCH_3}$$

From isobutane we get

$$\underset{\substack{\text{1,1-Dichloro-2-methyl-}\\\text{propane (gem)}}}{Cl_2CH-\overset{CH_3}{\underset{|}{CH}}-CH_3} \qquad \underset{\substack{\text{1,2-Dichloro-2-methyl-}\\\text{propane (vic)}}}{ClCH_2-\overset{CH_3}{\underset{|}{CCl}}-CH_3} \qquad \underset{\substack{\text{1,3-Dichloro-2-methyl-}\\\text{propane}}}{ClCH_2-\overset{CH_3}{\underset{|}{CH}}-CH_2Cl}$$

Problem 7.2 Give the structural formula and IUPAC name for (*a*) isobutyl chloride, (*b*) *t*-amyl bromide (amyl ≡ pentyl). ◄

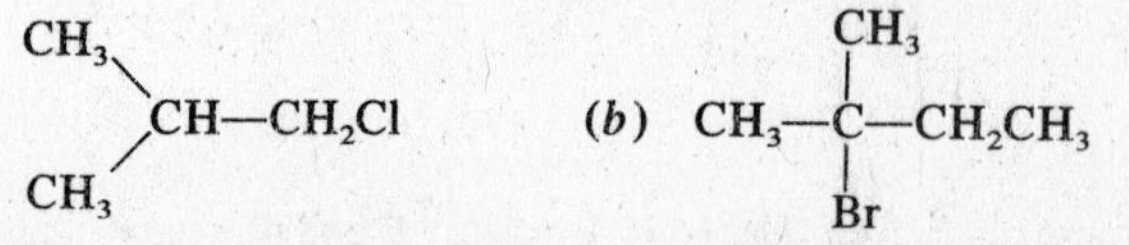

$$(a) \quad \begin{matrix} CH_3 \\ \diagdown \\ \diagup \\ CH_3 \end{matrix} CH-CH_2Cl \qquad (b) \quad CH_3-\underset{\underset{Br}{|}}{\overset{\overset{CH_3}{|}}{C}}-CH_2CH_3$$

1-Chloro-2-methylpropane   2-Bromo-2-methylbutane

Problem 7.3 Compare and account for differences in the (*a*) dipole moment, (*b*) boiling point, (*c*) density and (*d*) solubility in water of an alkyl halide RX and its parent alkane RH. ◄

(*a*) RX has a larger dipole moment because the C—X bond is polar. (*b*) RX has a higher boiling point since it has a larger molecular weight and also is more polar. (*c*) RX is more dense since it has a heavy X atom; the order of decreasing density is RI > RBr > RCl > RF. (*d*) RX, like RH, is insoluble in H_2O, but RX is somewhat more soluble because some H-bonding can occur:

$$\overset{\delta+}{R}-\overset{\delta-}{X}\cdots\overset{\delta+}{H}-\overset{\delta-}{OH}$$

This effect is greatest for RF.

7.2 SYNTHESIS OF RX

1. Halogenation of alkanes with Cl_2 or Br_2 (page 43).
2. From alcohols (ROH) with HX or PX_3 (X = I, Br, Cl); $SOCl_2$.
3. Addition of HX to alkenes (page 80).
4. X_2 (X = Br, Cl) + alkenes give *vic*-dihalides (page 80).
5. $RX + X'^- \longrightarrow RX' + X^-$ (halogen exchange).

Problem 7.4 Give the products of the following reactions:

(*a*) $CH_3CH_2CH_2OH + HI \longrightarrow$ (*b*) $n\text{-}C_4H_9OH + NaBr + H_2SO_4 \longrightarrow$

(*c*) $CH_3CH_2OH + PI_3 (P + I_2) \longrightarrow$ (*d*) $(CH_3)_2CHCH_2OH + SOCl_2 \longrightarrow$

(*e*) $H_2C{=}CH_2 + Br_2 \longrightarrow$ (*f*) $CH_3CH{=}C(CH_3)_2 + HI \longrightarrow$

(*g*) $CH_3CH_2CH_2Br + I^- \longrightarrow$ ◄

(*a*) $CH_3CH_2CH_2I + H_2O$ (*b*) $n\text{-}C_4H_9Br + NaHSO_4 + H_2O$

(*c*) $CH_3CH_2I + H_2PHO_3$ (phosphorous acid)   (*d*) $(CH_3)_2CHCH_2Cl + HCl(g) + SO_2(g)$

(*e*) H_2CBrCH_2Br (*f*) $CH_3CH_2CI(CH_3)_2$ (*g*) $CH_3CH_2CH_2I + Br^-$

Problem 7.5 Which of the following chlorides can be made in good yield by light-catalyzed monochlorination of the corresponding hydrocarbon?

(*a*) CH_3CH_2Cl (*c*) $(CH_3)_3CCH_2Cl$ (*e*)   (*f*) $H_2C{=}CHCH_2Cl$ ◄

(*b*) $CH_3CH_2CH_2CH_2Cl$ (*d*) $(CH_3)_3CCl$

To get good yields, all the reactive H's of the parent hydrocarbon must be equivalent. This is true for

(*a*) CH_3CH_3 (*c*) $(CH_3)_3CCH_3$ (*e*) $H_2C\overset{\overset{\textstyle CH_2}{\diagup\diagdown}}{\underline{\quad\quad}}CH_2$ (*f*) $H_2C{=}CHCH_3$

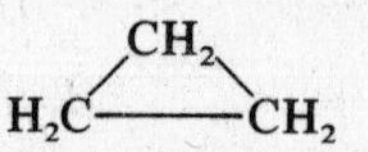

(the allylic H's are much more reactive than the inert vinylic H's). The precursors for (*b*) and (*d*), which are $CH_3CH_2CH_2CH_3$ and $(CH_3)_3CH$, respectively, both have more than one type of equivalent H and would give mixtures.

Problem 7.6 Prepare

$$(a) \ CH_3CHBrCH_3 \qquad (b) \ CH_3CH_2CH_2CH_2I$$

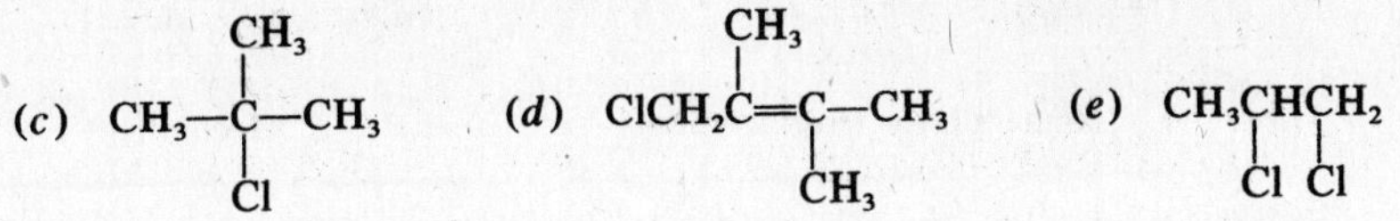

$$(c) \ CH_3-\underset{\underset{Cl}{\overset{\overset{CH_3}{|}}{C}}}{}-CH_3 \qquad (d) \ ClCH_2\underset{\underset{CH_3}{|}}{\overset{\overset{CH_3}{|}}{C}}=C-CH_3 \qquad (e) \ CH_3\underset{\underset{Cl}{|}}{CH}\underset{\underset{Cl}{|}}{CH_2}$$

from a hydrocarbon or an alcohol.

(a)
$$CH_3CH{=}CH_2 + HBr \longrightarrow CH_3CHBrCH_3 \xleftarrow{\ PBr_3\ } CH_3CHOHCH_3$$

(b)
$$CH_3CH_2CH{=}CH_2 + HBr \xrightarrow{\text{peroxide}} CH_3CH_2CH_2CH_2Br \xrightarrow[\text{acetone}]{I^-}$$

$$CH_3CH_2CH_2CH_2I \xleftarrow[(P+I_2)]{PI_3} CH_3CH_2CH_2CH_2OH$$

HI does not undergo an anti-Markovnikov radical addition.

(c)
$$CH_3-\underset{\underset{}{\overset{\overset{CH_3}{|}}{C}}}{}=CH_2 + HCl \longrightarrow CH_3-\underset{\underset{Cl}{|}}{\overset{\overset{CH_3}{|}}{C}}-CH_3 \longleftarrow (CH_3)_3COH + HCl$$

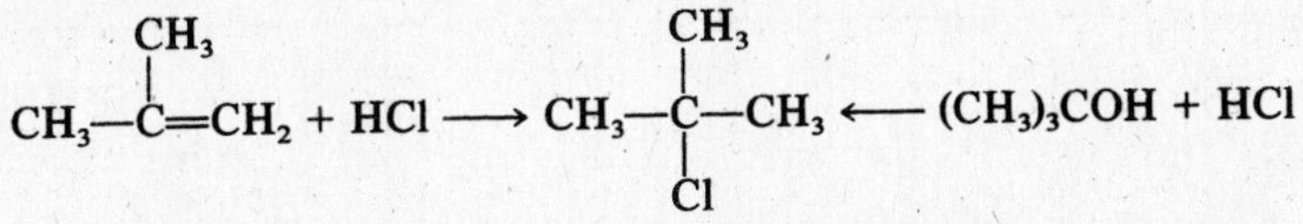

(d)
$$CH_3-\underset{\underset{CH_3}{|}}{\overset{\overset{CH_3}{|}}{C}}=C-CH_3 + Cl_2 \xrightarrow{500\ °C} ClCH_2-\underset{\underset{CH_3}{|}}{\overset{\overset{CH_3}{|}}{C}}=C-CH_3 + HCl$$

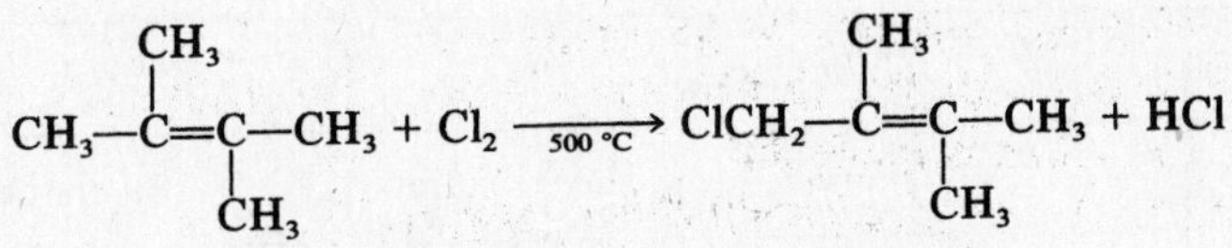

(e)
$$CH_3CH{=}CH_2 + Cl_2 \longrightarrow CH_3\underset{\underset{Cl}{|}}{CH}\underset{\underset{Cl}{|}}{CH_2}$$

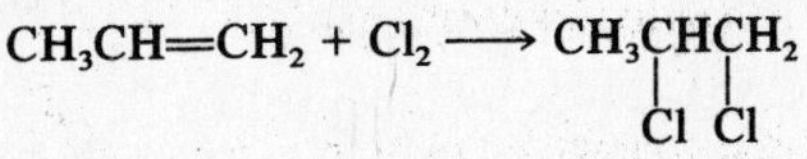

7.3 CHEMICAL PROPERTIES

Alkyl halides react mainly by heterolysis of the polar C—X bond.

NUCLEOPHILIC DISPLACEMENT

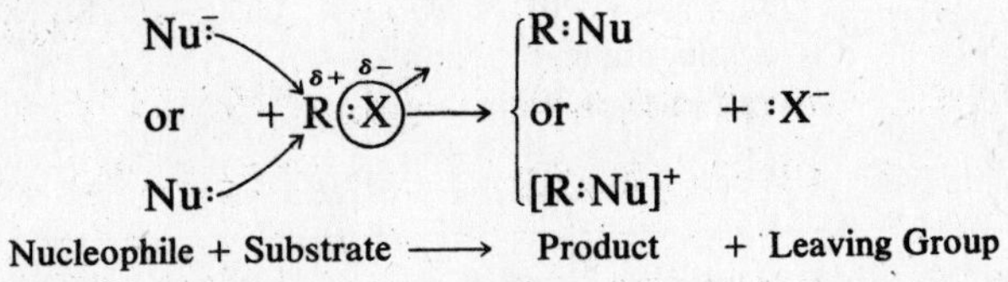

$$\begin{matrix} Nu^{\bar{\cdot}} \\ \text{or} \\ Nu\cdot \end{matrix} \Bigg\} \ + \ R\overset{\delta+}{(\cdot\,}\overset{\delta-}{X)} \longrightarrow \begin{cases} R{:}Nu \\ \text{or} \\ [R{:}Nu]^+ \end{cases} + \ {:}X^-$$

Nucleophile + Substrate $\longrightarrow$ Product + Leaving Group

The order of reactivity is RI > RBr > RCl > RF.

Problem 7.7 Write equations for the reaction of RCH_2X with

$$(a) \ I^- \qquad (b) \ OH^- \qquad (c) \ OR'^- \qquad (d) \ R'^{\bar{\cdot}} \qquad (e) \ RC\underset{O^-}{\overset{\overset{O}{\|}}{\ }} \qquad (f) \ H_3N{\cdot} \qquad (g) \ {:}CN^-$$

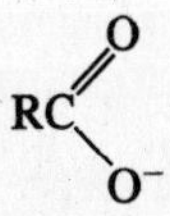

and classify the functional group in each product.

(a) $\qquad \qquad {:}I^- + RCH_2X \longrightarrow RCH_2I + {:}X^- \qquad$ Iodide

(b) $\qquad \qquad {:}OH^- + RCH_2X \longrightarrow RCH_2OH + {:}X^- \qquad$ Alcohol

(c) $\qquad \qquad {:}\bar{O}R' + RCH_2X \longrightarrow RCH_2OR' + {:}X^- \qquad$ Ether

(d) $\qquad \qquad {:}\bar{R}' + RCH_2X \longrightarrow RCH_2R' + {:}X^- \qquad$ Alkane

(e) $:\overline{O}OCR' + RCH_2X \longrightarrow RCH_2OOCR' + :X^-$ Ester

(f) $:NH_3 + RCH_2X \longrightarrow RCH_2NH_3^+ + :X^-$ Ammonium salt

(g) $:CN^- + RCH_2X \longrightarrow RCH_2CN + :X^-$ Nitrile (or Cyanide)

Table 7-1.   Nucleophilic Displacements by S_N1 and S_N2 Mechanisms

	S_N1	S_N2
Steps	Two: (1) $R{:}X \xrightarrow{\text{slow}} R^+ + :X^-$ carbonium ion (2) $R^+ + Nu^- \xrightarrow{\text{fast}} RNu$ or $R^+ + Nu \longrightarrow RNu^+$	One: $R{:}X + :Nu^- \longrightarrow RNu + :X^-$ or $R{:}X + :Nu \longrightarrow RNu^+{:}X^-$
Rate	$= k[RX]$ (1st-order)	$= k[RX][:Nu^-]$ (2nd-order)
TS of slow step	$\overset{\delta+}{\underset{}{C}}{-}{-}{-}{-}{-}\overset{\delta-}{:X}$	$\overset{\delta-}{:Nu}{-}{-}{-}\overset{}{C}{-}{-}{-}\overset{\delta-}{X}$ (with $:Nu^-$)
Molecularity	Unimolecular	Bimolecular
Stereochemistry	Inversion and racemization.	Inversion (backside attack)

Reactivity
   Structure of R
   Determining factor
   Nature of X
   Solvent effect
   on rate
	$3° > 2° > 1° > CH_3$
	 Stability of R^+
	 RI > RBr > RCl > RF
	 Rate increase in
	 polar solvents
		$CH_3 > 1° > 2° > 3°$
		 Steric hindrance in R group
		 RI > RBr > RCl > RF
		 With Nu^- there is a large
		 rate increase in polar
		 aprotic solvents

Effect of nucleophile	R^+ reacts with nucleophilic solvents rather than with $:Nu^-$ (**solvolysis**)	Rate depends on nucleophilicity $I^- > Br^- > Cl^-$; $RS^- > RO^-$ Equilibrium lies towards weaker Brönsted base
Catalysis	Lewis acid, e.g. Ag^+, $AlCl_3$, $ZnCl_2$	None
Competition, reaction	Elimination, rearrangement	Elimination, especially with 3° RX in strong Brönsted base

Problem 7.8 (a) Give an orbital representation for an S_N2 reaction with L-RCHDX and $:Nu^-$, if in the transition state the C on which displacement occurs uses sp^2 hybrid orbitals. (b) How does this representation explain (i) inversion, (ii) the order of reactivity $3° > 2° > 1°$? ◄

(a) See Fig. 7-1.

(b) (i) The reaction is initiated by the nucleophile beginning to overlap with the tail of the sp^3 hybrid orbital holding X. In order for the tail to become the head, the configuration must change; inversion occurs. (ii) As H's on the attacked C are replaced by R's, the TS becomes more crowded and has a higher enthalpy. With a 3° RX, there is a higher $\Delta H^{\ddagger}$ and a lower rate.

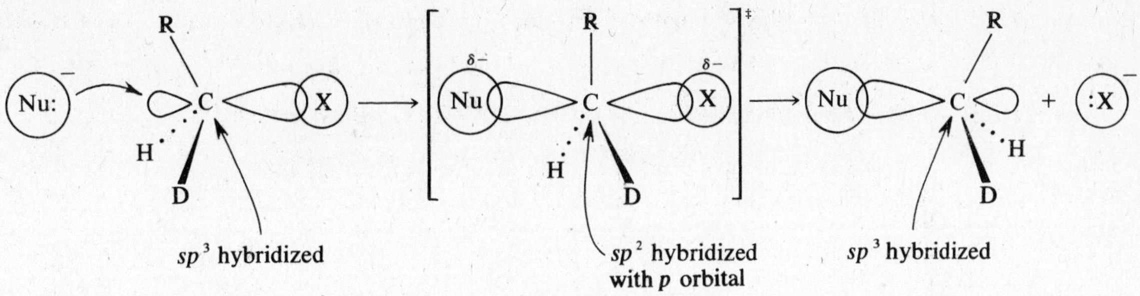

sp^3 hybridized　　　　　　　　　　　sp^2 hybridized　　　　sp^3 hybridized
　　　　　　　　　　　　　　　　　　with p orbital

L Configuration　　　　　　　　　　TS　　　　　　　　　D Configuration

Fig. 7-1

Problem 7.9 (a) Give a representation of an S_N1 TS which assigns a role to the nucleophilic protic solvent molecules (HS⋮) needed to solvate the ion. (b) In view of this representation, explain why (i) the reaction is first-order; (ii) R^+ reacts with solvent rather than with stronger nucleophiles that may be present; (iii) catalysis by Ag^+ takes place; (iv) the more stable the R^+, the less inversion and the more racemization occurs. ◄

(a)

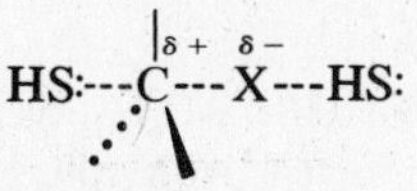

$$HS⋮---\overset{\delta+}{\underset{\Large\diagdown}{C}}---\overset{\delta-}{X}---HS⋮$$

Solvated S_N1 TS

(b) (i) Although solvent HS⋮ appears in the TS, it does not appear in the rate expression. (ii) HS⋮ is already partially bonded with the incipient R^+. (iii) Ag^+ has a stronger affinity for X^- than has a solvent molecule; the dissociation of X^- is accelerated. (iv) The HS⋮ molecule solvating an unstable R^+ is more apt to form a bond, causing inversion. When R^+ is stable, the TS gives an intermediate that reacts with another HS⋮ molecule to give a symmetrically solvated cation,

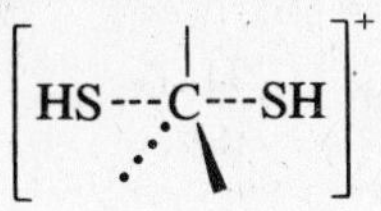

$$\left[HS---\overset{\Large\mid}{\underset{\Large\diagdown}{C}}---SH \right]^+$$

which collapses to a racemic product:

$$H\overset{+}{S}—\overset{\diagup}{\underset{\diagdown}{C}}— \; + \; —\overset{\diagup}{\underset{\diagdown}{C}}—\overset{+}{S}H$$

Problem 7.10 Give differences between S_N1 and S_N2 transition states. ◄

(1) In the S_N1 TS *there is considerable positive charge on* C; there is much weaker bonding between the attacking group and leaving groups with C. There is little or no charge on C in the S_N2 TS.
(2) The S_N1 TS is approached by separation of the leaving group; the S_N2 TS by attack of :Nu$^-$ or :Nu.
(3) The $\Delta H^‡$ of the S_N1 TS (and the rate of the reaction) depends on the stability of the incipient R^+. When R^+ is more stable, $\Delta H^‡$ is lower and the rate is greater. The $\Delta H^‡$ of the S_N2 TS depends on the steric effects. When there are more R's on the attacked C or when the attacking :Nu$^-$ is bulkier, $\Delta H^‡$ is greater and the rate is less.

Problem 7.11 How can the stability of an intermediate R^+ in an S_N1 reaction be assessed from its enthalpy–reaction diagram? ◄

The intermediate R^+ is a trough between two transition-state peaks. More stable R^+'s have a deeper trough and differ less in energy from the reactants and products.

Problem 7.12 Give the solvolysis products for the reaction of $(CH_3)_3CCl$ with   (a) CH_3OH, (b) CH_3COOH.   ◄

   (a) $(CH_3)_3C—OCH_3$ Methyl t-butyl ether.   (b) $(CH_3)_3CO\overset{\displaystyle O}{\overset{\|}{C}}—CH_3$, t-Butyl acetate.

Problem 7.13   (a) Formulate $(CH_3)_3COH + HCl \rightarrow (CH_3)_3CCl + H_2O$ as an S_N1 reaction.   (b) Formulate the reaction $CH_3OH + HI \rightarrow CH_3I + H_2O$ as an S_N2 reaction.   ◄

(a)   **Step 1** $(CH_3)_3COH + HCl \underset{}{\overset{fast}{\rightleftharpoons}} (CH_3)_3C—\overset{\displaystyle H}{\underset{+}{\overset{|}{O}}}H + Cl^-$

 Base$_1$ Acid$_2$ Acid$_1$ Base$_2$

 oxonium ion

Step 2 $(CH_3)_3C\overset{+}{O}H_2 \overset{slow}{\rightleftharpoons} (CH_3)_3C^+ + H_2O$

Step 3 $(CH_3)_3C^+ + Cl^- \xrightarrow{fast} (CH_3)_3CCl$

(b)   **Step 1** $CH_3OH + HI \overset{fast}{\rightleftharpoons} CH_3\overset{+}{O}H_2 + I^-$

Step 2 $:I^- + H_3C\overparen{OH_2^+} \xrightarrow{slow} ICH_3 + H_2O$

Problem 7.14   Optically pure (S)-(+)-CH_3CHBr-n-C_6H_{13} has $[\alpha]_D^{25} = +36.0°$.   A partially racemized sample having a specific rotation of $+30°$ is reacted with dilute NaOH to form (R)-(−)-$CH_3CH(OH)$-n-C_6H_{13} ($[\alpha]_D^{25} = -5.97°$), whose specific rotation is $-10.3°$ when optically pure.   (a) Write an equation for the reaction using projection formulas.   (b) Calculate the percent optical purity of reactant and product.   (c) Calculate percentages of racemization and inversion.   (d) Calculate percentages of frontside and backside attack. (e) Draw a conclusion concerning the reactions of 2° alkyl halides.   (f) What change in conditions would increase inversion?   ◄

(a) $HO:^- + CH_3—\overset{\displaystyle C_6H_{13}}{\underset{\displaystyle H}{\overset{|}{\underset{|}{C}}}}—\ddot{B}r: \longrightarrow HO—\overset{\displaystyle C_6H_{13}}{\underset{\displaystyle H}{\overset{|}{\underset{|}{C}}}}—CH_3 + :\ddot{B}r:^-$

 (S)   (R)

(b)   The percentage of optically active enantiomer (optical purity) is calculated by dividing the observed specific rotation by that of pure enantiomer and multiplying the quotient by 100.   The optical purities are:

$$Bromide = \frac{+30°}{+36°}(100) = 83\% \qquad Alcohol = \frac{-5.97°}{-10.3°}(100) = 58\%$$

(c)   The percentage of inversion is calculated by dividing the percentage of optically active alcohol of opposite configuration by that of reacting bromide.   The percentage of racemization is the difference between this percentage and 100%.

$$Percentage\ inversion = \frac{58\%}{83\%}(100) = 70\%$$

$$Percentage\ racemization = 100\% - 70\% = 30\%$$

(d)   Inversion involves only backside attack, while racemization results from equal backside and frontside attack.   The percentage of backside reaction is the sum of the inversion and one-half of the racemization; the percentage of frontside attack is the remaining half of the percentage of racemization.

$$Percentage\ backside\ reaction = 70\% + \tfrac{1}{2}(30\%) = 85\%$$

$$Percentage\ frontside\ reaction = \tfrac{1}{2}(30\%) = 15\%$$

(e)   The large percentage of inversion indicates chiefly S_N2 reaction, while the smaller percentage of racemization indicates some S_N1 pathway.   This duality of reaction mechanism is typical of 2° alkyl halides.

(f)   The S_N2 rate is increased by raising the concentration of the nucleophile—in this case, OH^-.

Problem 7.15   Account for the following stereochemical results:

$$C_6H_5-\underset{\underset{H}{|}}{\overset{\overset{CH_3}{|}}{C}}-Br$$

$\xrightarrow{H_2O}$ $HO-\underset{\underset{H}{|}}{\overset{\overset{CH_3}{|}}{C}}-C_6H_5 + HBr$ 2% inversion
98% racemization

$\xrightarrow{CH_3OH}$ $CH_3O-\underset{\underset{H}{|}}{\overset{\overset{CH_3}{|}}{C}}-C_6H_5 + HBr$ 27% inversion
73% racemization   ◄

H_2O is more nucleophilic and polar than CH_3OH.   It is better able to react to give $HS\overset{\delta+}{\cdots}R\cdots:SH$ (see Problem 7.9), leading to racemization.

Problem 7.16   ROH does not react with NaBr, but adding H_2SO_4 forms RBr.   Explain.   ◄

Br^-, an extremely weak Brönsted base, cannot displace the strong base OH^-.   In acid, $R\overset{+}{O}H_2$ is first formed.   Now, Br^- displaces H_2O, which is a very weak base and a good leaving group.

Problem 7.17 NH_3 reacts with RCH_2X to form an ammonium salt, $RCH_2NH_3^+X^-$.   Show the transition state, indicating the partial charges.   ◄

$$H_3\overset{\delta+}{N}\cdots\underset{\underset{H}{\overset{\overset{H}{|}}{\cdots}}}{C}\cdots\overset{\delta-}{:X}$$
$$\underset{H\quad R}{}$$

N gains $\delta+$ as it begins to form a bond.

Problem 7.18   Explain why neopentyl chloride, $(CH_3)_3CCH_2Cl$, a 1° RCl, does not participate in typical S_N2 reactions.   ◄

The bulky $(CH_3)_3C$ group sterically hinders backside attack by a nucleophile.

Problem 7.19   In terms of transition-state theory, account for the following solvent effects: (a) The rate of solvolysis of a 3° RX increases as the polarity of the protic nucleophilic solvent (:SH) increases, e.g.

$$H_2O > HCOOH > CH_3OH > CH_3COOH$$

(b) The rate of the S_N2 reaction $:Nu^- + RX \rightarrow RNu + :X^-$ decreases slightly as the polarity of protic solvent increases.   (c) The rate of the S_N2 reaction $:Nu + RX \rightarrow RNu^+ + :X^-$ increases as the polarity of the solvent increases.   (d) The rate of reaction in (b) is greatly increased in a polar aprotic solvent.   ◄

Polar solvents stabilize and lower the enthalpies of charged reactants and charged transition states.   The more diffuse the charge on the species, the less effective the stabilization by the polar solvent.   See Table 7-2.

Table 7-2

	Ground State (GS)	TS	Relative Charge	Effect of Solvent Change	$\Delta H^{\ddagger}$	Rate
(a)	RX + HS:	$HS\overset{\delta+}{\cdots}R\overset{\delta-}{\cdots}X\cdots HS:$	none in GS charge in TS	lower H of TS	decreases	increases
(b)	$RX + Nu^-$	$\overset{\delta-}{Nu}\cdots R\cdots\overset{\delta-}{X}$	full in GS diffuse in TS	lower H of GS $> H$ of TS	increases	decreases
(c)	RX + Nu	$\overset{\delta-}{Nu}\cdots R\cdots\overset{\delta-}{X}$	none in GS charge in TS	lower H of TS	decreases	increases
(d)	same as (b)			raise H of GS* $> H$ of TS	decreases	increases

*Aprotic solvents do not solvate anions.

Problem 7.20 Compare the nucleophilicity (rate of S_N2 reactivity) of

$$(a) \quad H_2O, \ OH^-, \ CH_3O^- \ \text{ and } \ CH_3\overset{O}{\overset{\|}{C}}{-}O^-$$

(b) NH_3 and PH_3 ◄

(a)   When the nucleophilic site is the same atom (here an O), nucleophilicity parallels basicity.   Therefore,
      $CH_3O^- > OH^- > CH_3COO^- > H_2O$.
(b)   When the attacking atoms are different but in the same periodic family, the one with the largest atomic
      weight is the most reactive.   Therefore, $PH_3 > NH_3$.   This order is the reverse of basicity.

Problem 7.21 $H_2C{=}CHCH_2Cl$ is solvolyzed faster than $(CH_3)_3CCl$.   Explain.   ◄

      Solvolyses go by an S_N1 mechanism.   Relative rates of different reactants in S_N1 reactions depend on the
stabilities of intermediate carbonium ions. $H_2C{=}CHCH_2Cl$ is more reactive because

$$[H_2C{\cdots}C{\cdots}CH_2]^+$$

is more stable than $(CH_3)_3C^+$.   (See Problem 6.39 for a corresponding explanation of stability of an allyl radical.)

Problem 7.22   In terms of (a) the inductive effect and (b) steric factors, account for the decreasing stability of
R^+:

$$Me_3\overset{+}{C} > Me_2\overset{+}{C}H > Me\overset{+}{C}H_2 > \overset{+}{C}H_3$$ ◄

(a)   Compared to H, R has an electron-releasing inductive effect.   Replacing H's on the positive C by CH_3's
      disperses the positive charge and thereby stabilizes R^+.
(b)   Steric acceleration also contributes to this order of R^+ stability.   Some steric strain of the three Me's in
      $Me_3C{-}Br$ separated by a 109° angle (sp^3) is relieved upon going to a 120° separation in R^+ with a C using
      sp^2 hybrid orbitals.

Problem 7.23   The following are relative rates observed for the formation of alcohols from the listed alkyl
halides in 80% water and 20% ethanol at 25 °C.   Account for the minimum value for $(CH_3)_2CHBr$.

Compound	CH_3Br	CH_3CH_2Br	$(CH_3)_2CHBr$	$(CH_3)_3CBr$
Relative Rate	2140	171	4.99	1010

◄

      The first three halides react mainly by the S_N2 pathway.   Reactivity decreases as CH_3's replace H's on the
attacked C, because of increased steric hindrance.   A change to the S_N1 mechanism accounts for the sharp rise
in the reactivity of $(CH_3)_3CBr$.

Problem 7.24   Account for the observation that RCl is hydrolyzed to ROH slowly but reaction is rapid if
catalytic amounts of KI are added to the reaction mixture.   ◄

      I^- is a powerful nucleophile which reacts rapidly with RCl to form RI. I^- is also a better leaving group than
Cl^-, and RI is therefore hydrolyzed rapidly to form ROH and regenerate I^-, which recycles in the reaction.

$$RCH_2Cl + H_2O \xrightarrow{\text{slow}} RCH_2OH + Cl^-$$

$$\underset{+I^- \text{ (fast)}}{\searrow} \quad RCH_2I \xrightarrow[\ ]{+H_2O \text{ (fast)}} \Big]^{-I^-}$$

Problem 7.25   Compare the effectiveness of acetate (CH_3COO^-), phenoxide ($C_6H_5O^-$) and benzenesulfonate
($C_6H_5SO_3^-$) anions as leaving groups if the acid strengths of their conjugate acids are given by the pK_a values
4.5, 10.0 and 2.6 respectively.   ◄

      The best leaving group is the weakest base, $C_6H_5SO_3^-$; the poorest is $C_6H_5O^-$, which is the strongest base.

Table 7-3. Elimination Reactions by E1 and E2 Mechanisms

	E1	E2
Steps	Two: 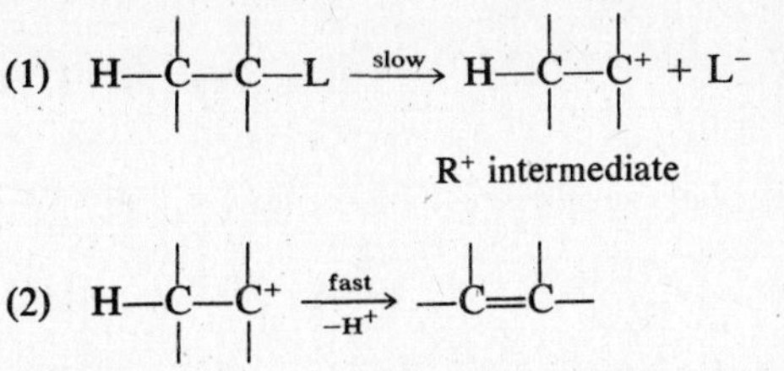 (1) $H-\overset{\mid}{\underset{\mid}{C}}-\overset{\mid}{\underset{\mid}{C}}-L \xrightarrow{\text{slow}} H-\overset{\mid}{\underset{\mid}{C}}-\overset{\mid}{\underset{\mid}{C}}{}^{+} + L^-$ R^+ intermediate (2) $H-\overset{\mid}{\underset{\mid}{C}}-\overset{\mid}{\underset{\mid}{C}}{}^{+} \xrightarrow[-H^+]{\text{fast}} \overset{\mid}{C}{=}\overset{\mid}{C}$	One: $B\!:^- + H-\overset{\mid}{\underset{\mid}{C}}-\overset{\mid}{\underset{\mid}{C}}-L \longrightarrow$ $B\!:\!H + \ \ \diagdown C{=}C\diagup \ \ + :L^-$ 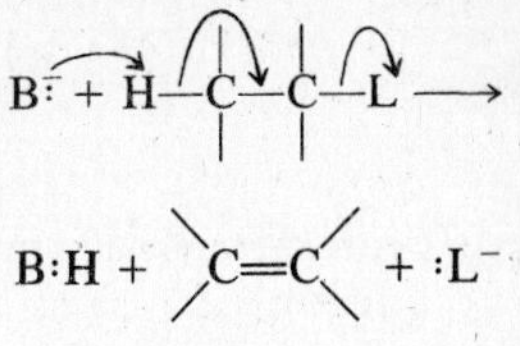
Transition states	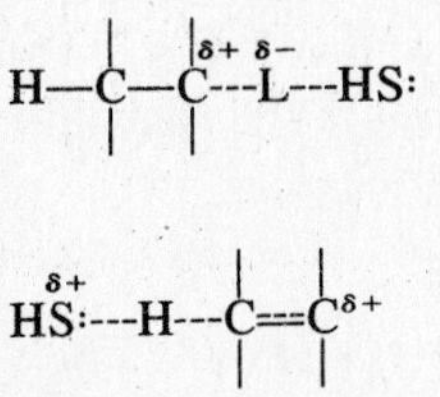 $H-\overset{\mid}{\underset{\mid}{C}}-\overset{\mid}{\underset{\mid}{C}}{}^{\delta+}\text{---}L^{\delta-}\text{---}HS:$ $HS:^{\delta+}\text{---}H\text{---}\overset{\mid}{\underset{\mid}{C}}{=}\overset{\mid}{C}{}^{\delta+}$	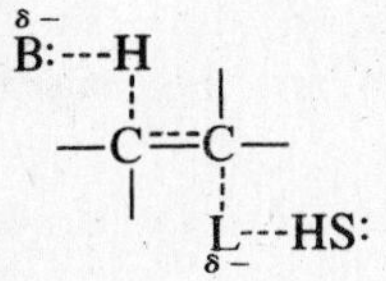 $\overset{\delta-}{B:}\text{---}H$ $-\overset{\mid}{C}{=}\overset{\mid}{C}-$ $\underset{\delta-}{L}\text{---}HS:$
	INDICATES E1	**INDICATES E2**
Kinetics	First-order Rate = $k[RL]$ Ionization determines rate Unimolecular	Second-order Rate = $k[RL][:B^-]$ Bimolecular
Stereochemistry	Nonstereospecific	*anti* Elimination (*syn* when *anti* impossible)
Reactivity order factor	$3° > 2° > 1°$ RX Stability of R^+	$3° > 2° > 1°$ RX Stability of alkenes (Saytzeff rule)
Rearrangements	Common	None
Deuterium isotope effect	None	Observed
Competing reaction	S_N1	S_N2
	FAVORS E1	**FAVORS E2**
Alkyl group	$3° > 2° > 1°$	$3° > 2° > 1°$
Loss of H	No effect	Increased acidity
Base strength concentration	Weak Low	Strong High
Leaving group	Weak base $I^- > Br^- > Cl^- > F^-$	Weak base $I^- > Br^- > Cl^- > F^-$

Problem 7.26   RBr reacts with $AgNO_2$ to give RNO_2 and $RONO$.   Explain.   ◄

The nitrite ion,

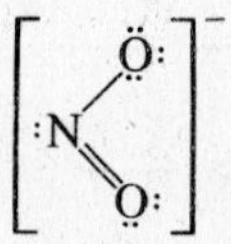

$$\left[\begin{array}{c} \ddot{O}: \\ :N \\ \ddot{O}: \end{array} \right]^{-}$$

has two different nucleophilic sites: the N and either O.   Reaction with the unshared pair on N gives RNO_2, while RONO is formed by reaction at O.   Anions with two nucleophilic sites are called **ambident** anions.

ELIMINATION REACTIONS

In an elimination (dehydrohalogenation) reaction a halogen and a hydrogen atom are removed from adjacent carbon atoms to form a double bond between the 2 C's.   This is sometimes called β-elimination, and the reagent commonly used to remove HX is the strong base KOH in ethanol (cf. Section 6.2).

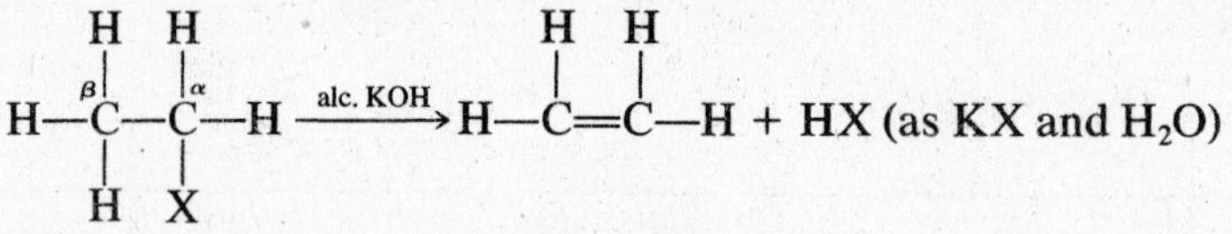

$$H-\underset{\underset{H}{|}}{\overset{\overset{H}{|}}{\underset{\beta}{C}}}-\underset{\underset{X}{|}}{\overset{\overset{H}{|}}{\underset{\alpha}{C}}}-H \xrightarrow{\text{alc. KOH}} H-\overset{\overset{H}{|}}{C}=\overset{\overset{H}{|}}{C}-H + HX \text{ (as KX and } H_2O)$$

Ethyl halide   Ethylene

n-Propyl bromide $CH_3CH_2CH_2Br \xrightarrow{\text{alc. KOH}} CH_3CH{=}CH_2$ Propene

sec-Butyl chloride $CH_3CHClCH_2CH_3 \xrightarrow{\text{alc. KOH}} CH_3CH{=}CHCH_3$ 2-Butene (cis and trans)

Elimination reactions compete with substitution reactions.   (See Tables 7-3, page 105, and 7-4.)

Table 7-4.   E vs S

	FAVORS E2	FAVORS S_N2
Structure of alkyl	3° > 2° > 1°	1° > 2° > 3°
Reagent	Strong bulky Brönsted base	Strong nucleophile
Temperature	High	Low
Low-polarity solvent	Yes	No
	FAVORS E2	FAVORS S_N1
Structure of R	3° > 2° > 1°	3° ≫ 2° > 1°
Base strength concentration	Strong High	Very Weak Low

Problem 7.27 Assuming that *anti* elimination is favored, illustrate the stereospecificity of the E2 dehydrohalogenation by predicting the products formed from (*a*) *meso-* and (*b*) either of the enantiomers of 2,3-dibromobutane.   Use the wedge-sawhorse and Newman projections.   ◄

(*a*)

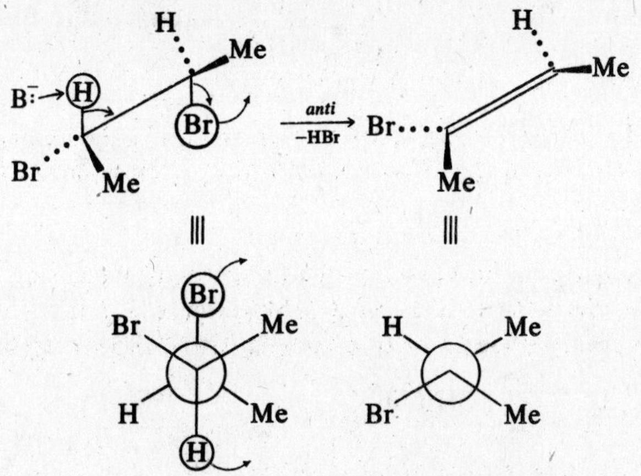

meso   *cis-* or (*E*)-2-Bromo-2-butene

(*b*)

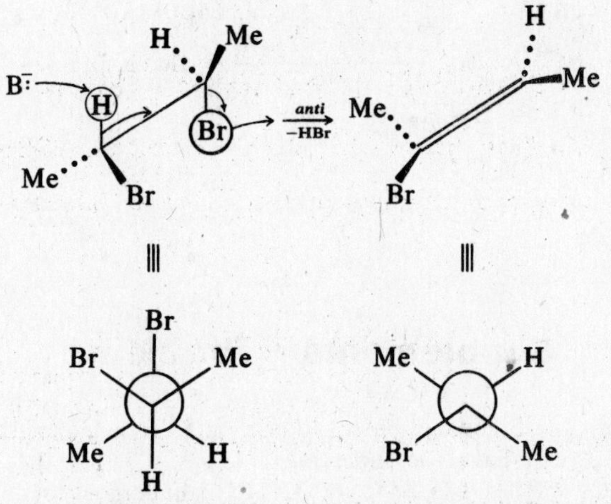

Enantiomer (R,R)   *trans-* or (*Z*)-2-Bromo-2-butene

Problem 7.28 Dehalogenation of *vic*-dihalides with active metals (Mg or Zn) is also an *anti* elimination. Predict the products from   (*a*) *meso-* and   (*b*) either enantiomer of 2,3-bromobutane.   ◄

(*a*)   Mg +

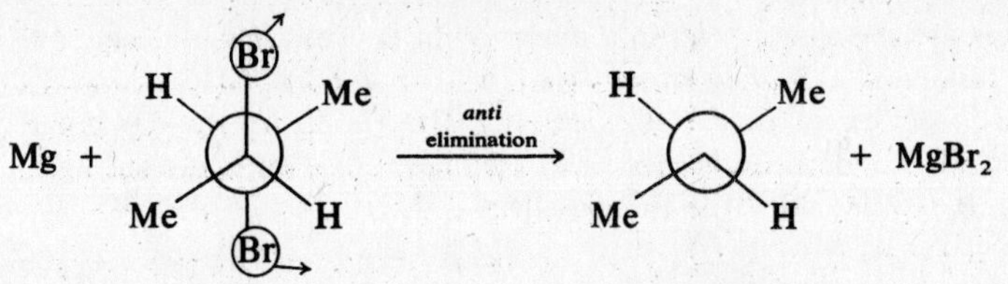

meso   *trans*-2-Butene

(*b*)   Mg +

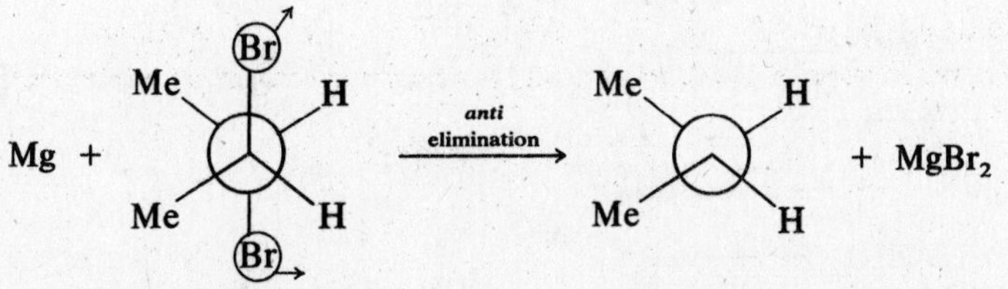

Enantiomer (S,S)   *cis*-2-Butene

7.4 SUMMARY OF ALKYL HALIDE CHEMISTRY

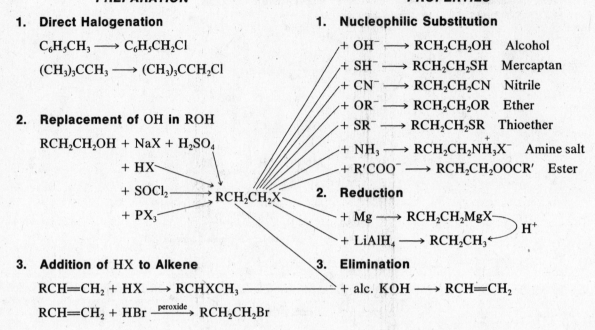

PREPARATION

1. Direct Halogenation

$$C_6H_5CH_3 \longrightarrow C_6H_5CH_2Cl$$

$$(CH_3)_3CCH_3 \longrightarrow (CH_3)_3CCH_2Cl$$

2. Replacement of OH in ROH

$$RCH_2CH_2OH + NaX + H_2SO_4$$

$$+ HX$$

$$+ SOCl_2$$

$$+ PX_3$$

$$\longrightarrow RCH_2CH_2X$$

3. Addition of HX to Alkene

$$RCH{=}CH_2 + HX \longrightarrow RCHXCH_3$$

$$RCH{=}CH_2 + HBr \xrightarrow{peroxide} RCH_2CH_2Br$$

PROPERTIES

1. Nucleophilic Substitution

$+ OH^- \longrightarrow RCH_2CH_2OH$ Alcohol

$+ SH^- \longrightarrow RCH_2CH_2SH$ Mercaptan

$+ CN^- \longrightarrow RCH_2CH_2CN$ Nitrile

$+ OR^- \longrightarrow RCH_2CH_2OR$ Ether

$+ SR^- \longrightarrow RCH_2CH_2SR$ Thioether

$+ NH_3 \longrightarrow RCH_2CH_2\overset{+}{N}H_3X^-$ Amine salt

$+ R'COO^- \longrightarrow RCH_2CH_2OOCR'$ Ester

2. Reduction

$+ Mg \longrightarrow RCH_2CH_2MgX$

$+ LiAlH_4 \longrightarrow RCH_2CH_3$ $\Big\rangle H^+$

3. Elimination

$+$ alc. KOH $\longrightarrow RCH{=}CH_2$

Supplementary Problems

Problem 7.29 Account for the following observations: (*a*) In a polar solvent such as water the S_N1 and E1 reactions of a 3° RX have the same rate. (*b*) $(CH_3)_3CI + H_2O \longrightarrow (CH_3)_3COH + HI$ but $(CH_3)_3CI + OH^- \longrightarrow (CH_3)_2C{=}CH_2 + H_2O + I^-$. (*c*) CH_3CH_2I undergoes loss of HI with strong base faster than CD_3CH_2I loses DI. ◄

(*a*) The rate-controlling step both for E1 and S_N1 reactions is the same:

$$\overset{\delta+}{R}{-}\overset{\delta-}{X} \xrightarrow{slow} R^+ + X^-$$

and therefore the rates are the same. (*b*) In a nucleophilic solvent in the absence of a strong base, a 3° RX undergoes an S_N1 solvolysis. In the presence of a strong base (OH^-) a 3° RX undergoes mainly E2 elimination. (*c*) This reaction is an E2 elimination. The C—H (or C—D) bond is being broken in the rate-controlling step. The C—H bond is broken at a higher rate than is the stronger C—D bond. This ratio of the rate constants, k_H/k_D, is called the **isotope effect**.

Problem 7.30 Give the organic product in the following substitution reactions. The solvent is given above the arrow.

(*a*) $HS^- + CH_3CH_2CHBrCH_3 \xrightarrow{CH_3OH}$

(*b*) $I^- + (CH_3)_3CBr \xrightarrow{HCOOH}$

(*c*) $CH_3CH_2Br + AgCN \longrightarrow$ 2 products

(*d*) $CH_3CHBrCH_3 + CH_3\ddot{N}H_2 \longrightarrow$

(*e*) $CH_3CHBrCH_3 + (CH_3)_2\ddot{S}{:} \longrightarrow$

(f) $CH_3CH_2Br + :P(C_6H_5)_3 \longrightarrow$

(g) $CH_3CH_2Br + \begin{bmatrix} & O \\ & \| \\ S - & S - O \\ & \| \\ & O \end{bmatrix}^{2-}$ (thiosulfate ion) $\longrightarrow$

(a) $CH_3CH_2CHSHCH_3$ (a mercaptan).

(b) $(CH_3)_3CO\overset{\displaystyle O}{\overset{\|}{C}}H$; 3° RX undergoes S_N1 solvolysis.

(c) $CH_3CH_2CN + CH_3CH_2NC$; $:C\equiv N:^-$ is an ambident anion (Problem 7.26).

(d) $\begin{bmatrix} & H \\ & | \\ (CH_3)_2CHNCH_3 \\ & | \\ & H \end{bmatrix}^+ Br^-$; an ammonium salt.

(e) $[(CH_3)_2CHS(CH_3)_2]^+Br^-$; a sulfonium salt.

(f) $[CH_3CH_2P(C_6H_5)_3]^+Br^-$; a phosphonium salt.

(g) $CH_3CH_2-S-\overset{\displaystyle O}{\underset{\displaystyle O}{\overset{\|}{\underset{|}{S}}}}-O^-$; S is a more nucleophilic site than is O.

Problem 7.31 Indicate the products of the following reactions and point out the mechanism as S_N1, S_N2, E1 or E2.

(a)	$CH_3CH_2CH_2Br + LiAlH_4$	(d)	$BrCH_2CH_2Br + Mg$ (ether)
(b)	$(CH_3)_3CBr + C_2H_5OH$, heat at 60 °C	(e)	$BrCH_2CH_2CH_2Br + Mg$ (ether)
(c)	$CH_3CH{=}CHCl + NaNH_2$	(f)	$CH_3CHBrCH_3 + NaOCH_3$ in CH_3OH

(a) $CH_3CH_2CH_3$; an S_N2 reaction, $:H^-$ of AlH_4^- replaces Br^-.

(b)
$$(CH_3)_3CBr \longrightarrow Br^- + (CH_3)_3C^+ \xrightarrow[-\,H^+]{CH_3CH_2OH} (CH_3)_3COCH_2CH_3 + CH_3\overset{\displaystyle CH_3}{\overset{|}{C}}{=}CH_2$$
$\quad\quad$ 3° RX $\qquad\qquad\qquad\qquad\qquad\qquad\qquad\qquad$ Major (S_N1) $\quad$ Very minor; (E1) in absence of base

(c)
$$CH_3CH{=}CHCl + NaNH_2 \longrightarrow CH_3C\equiv CH + NH_3 + NaCl \qquad (E2)$$

Vinyl halides are quite inert toward S_N2 reactions.

(d)
$$BrCH_2CH_2Br + Mg \longrightarrow H_2C{=}CH_2 + MgI_2$$

This is an E2 type of elimination via an alkyl magnesium iodide.

$$Mg + BrCH_2CH_2Br \longrightarrow Br\overset{+}{M}g^-:CH_2{\frown}CH_2{\overset{\frown}{\,}}Br \longrightarrow MgBr_2 + H_2C{=}CH_2$$

(e) This reaction resembles that in (d) and is an internal S_N2 reaction.

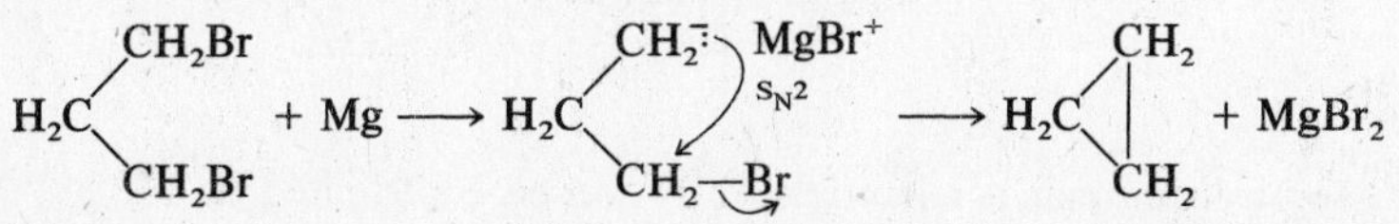

$$H_2C\overset{\displaystyle CH_2Br}{\underset{\displaystyle CH_2Br}{\big<}} + Mg \longrightarrow H_2C\overset{\displaystyle CH_2^-\,MgBr^+}{\underset{\displaystyle CH_2-Br}{\big<}} \overset{S_N2}{\longrightarrow} H_2C\overset{\displaystyle CH_2}{\underset{\displaystyle CH_2}{\big|}} + MgBr_2$$

(f) This 2° RBr undergoes both E2 and S_N2 reactions to form propylene and isopropyl methyl ether.

$$CH_3CHBrCH_3 + Na\overset{+}{O}CH_3\,(CH_3OH) \longrightarrow CH_3CH{=}CH_2 + CH_3\underset{\displaystyle OCH_3}{\overset{|}{C}}HCH_3$$

Problem 7.32 Draw structural formulas for the most important organic products of the following reactions and point out how they are formed.

(a) 2,3-Dimethyl-2-bromobutane + $NaNH_2$ in decane (heat)

(b) $CH_3I + NaNO_2$ (c) $CHCl_3 + (CH_3)_3CO^-K^+ + H_2C{=}CH_2$ ◀

(a)

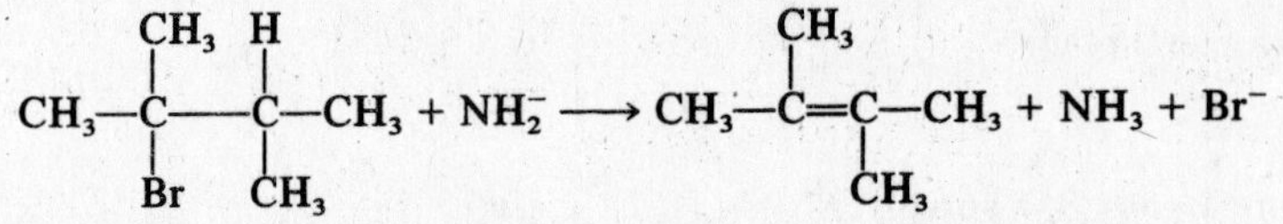

$$CH_3-\underset{\underset{Br}{|}}{\overset{\overset{CH_3}{|}}{C}}-\underset{\underset{CH_3}{|}}{\overset{\overset{H}{|}}{C}}-CH_3 + NH_2^- \longrightarrow CH_3-\underset{}{\overset{\overset{CH_3}{|}}{C}}{=}\underset{\underset{CH_3}{|}}{C}-CH_3 + NH_3 + Br^- \qquad (E2)$$

(b) CH_3NO_2 + some CH_3ONO; S_N2 reaction.   (NO_2^- is an ambident ion.)

(c) Strong base abstracts H^+.   The carbanion $:CCl_3^-$ loses Cl^- to form the reactive dichlorocarbene which adds to $C{=}C$.

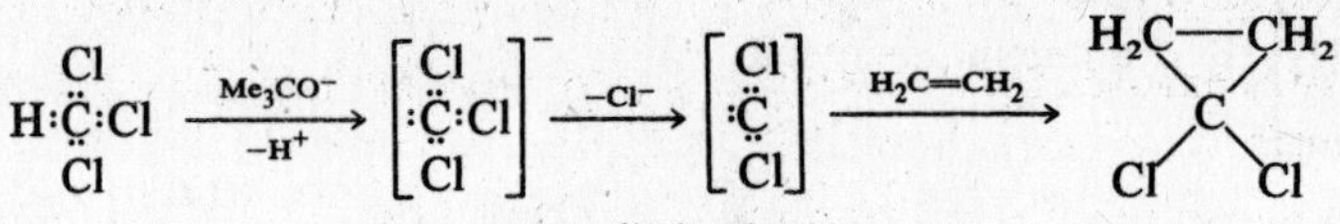

$$H\overset{\overset{Cl}{\cdot\cdot}}{\underset{\underset{Cl}{\cdot\cdot}}{C}}:Cl \xrightarrow[-H^+]{Me_3CO^-} \left[\overset{\overset{Cl}{|}}{\underset{\underset{Cl}{|}}{:C:}}Cl\right]^- \xrightarrow{-Cl^-} \left[\overset{\overset{Cl}{|}}{\underset{\underset{Cl}{|}}{:C}}\right] \xrightarrow{H_2C{=}CH_2} \underset{\underset{Cl \qquad Cl}{}}{\overset{\overset{H_2C{-}CH_2}{}}{C}}$$

$$\textit{an } \alpha\textit{-elimination}$$

Problem 7.33 Account for the inertness of the following rigid bicyclic compound to nucleophilic substitution reactions:

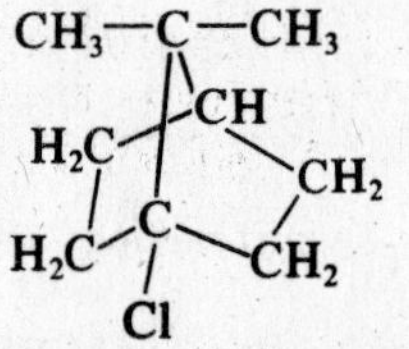

$$\begin{array}{c} CH_3-C-CH_3 \\ |CH \\ H_2C \quad \quad CH_2 \\ H_2C \quad C \quad CH_2 \\ | \\ Cl \end{array}$$ ◀

Although this is a 3° RCl, it does not undergo an S_N1 reaction because the bridgehead C bonded to Cl is part of a rigid structure and therefore cannot form a planar R^+.   The bicyclic ring structure does not permit a backside nucleophilic attack on C, ruling out the S_N2 mechanism.

Problem 7.34 Is each of the following R^+'s stabilized or destabilized by the adjacent atom or group?   Explain.

(a) $F_3C-\overset{+}{\underset{|}{C}}-$ (b) $:\ddot{F}_3C^+$ (c) $H_2\ddot{N}-\overset{+}{\underset{|}{C}}-$ (d) $H_3\overset{+}{N}-\overset{+}{\underset{|}{C}}-$ ◀

(a) Destabilized.   The strongly electron-withdrawing F's place a $\delta+$ on the atom adjacent to C^+.

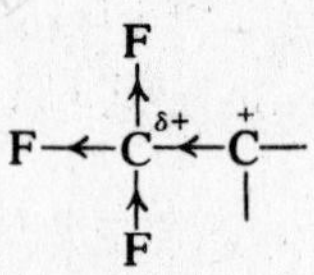

$$F{\leftarrow}\overset{\overset{F}{\uparrow}}{\underset{\underset{F}{\uparrow}}{C}}\overset{\delta+}{\leftarrow}\overset{+}{\underset{|}{C}}-$$

(Arrows indicate withdrawn electron density.)

(b) Stabilized.   Each F has an unshared pair of electrons in a p orbital which can be shifted to $-\overset{+}{\underset{|}{C}}-$ via p-p orbital overlap.

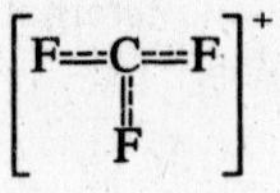

$$\left[\overset{}{\underset{\underset{F}{||}}{F{=}C{=}F}}\right]^+$$

(c) Stabilized.   The unshared pair of electrons on N can be contributed to C^+.

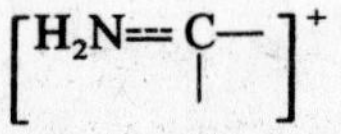

$$\left[H_2N{=}\underset{|}{C}-\right]^+$$

(d) Destabilized.   The adjacent N has a + charge.

Problem 7.35 As a rule a $1°$ RX is the least reactive in S_N1 solvolysis reactions. Account for the high reactivity observed in S_N1 displacements with $ClCH_2OCH_2CH_3$ in ethanol. ◄

The rapid S_N1 reaction is attributed to the stability of a C^+ adjacent to $—\overset{..}{\overset{..}{O}}—$. The empty p orbital on the C^+ can overlap with the p orbital on O, thereby delocalizing the + charge. The positive ion then reacts with the nucleophilic alcohol, giving an ether.

$$CH_3CH_2—O—CH_2Cl \longrightarrow Cl^- + \begin{array}{c} CH_3CH_2—\overset{..}{\underset{..}{O}}\overset{\frown}{—}CH_2^+ \\ \updownarrow \\ CH_3CH_2—\overset{+}{\underset{..}{O}}{\underset{\smile}{=}}CH_2 \end{array} \xrightarrow[-H^+]{+C_2H_5OH} CH_3CH_2—O—CH_2OC_2H_5$$

Problem 7.36 Account for the following observations when (S)-$CH_3CH_2CH_2CHID$ is heated in acetone solution with NaI: (a) The enantiomer is racemized. (b) If radioactive $^*I^-$ is present in excess, the rate of racemization is twice the rate at which the radioactive $^*I^-$ is incorporated into the compound. ◄

(a) Since enantiomers have identical energy, reaction proceeds in both directions until a racemic equilibrium mixture is formed.

$$I^- + CH_3CH_2CH_2—\overset{\displaystyle D}{\underset{\displaystyle H}{\vert\!\!\!\!-\!\!\!\!\vert}}—I \underset{\text{inversion}}{\overset{\text{inversion}}{\rightleftharpoons}} I—\overset{\displaystyle D}{\underset{\displaystyle H}{\vert\!\!\!\!-\!\!\!\!\vert}}—CH_2CH_2CH_3 + I^- \quad (S_N2 \text{ reaction})$$

$$\quad\quad\quad\quad\quad\quad\quad\quad (R) \quad\quad\quad\quad\quad\quad (S)$$

(b) Each radioactive $^*I^-$ incorporated into the compound forms one molecule of enantiomer. Now one unreacted molecule and one molecule of its enantiomer, resulting from reaction with $^*I^-$, form a racemic modification. Since two molecules are racemized when one $^*I^-$ reacts, the rate of racemization will be twice that at which $^*I^-$ reacts.

Problem 7.37 Indicate the effect on the rate of S_N1 and S_N2 reactions of the following: (a) Doubling the concentration of substrate (RL) or Nu^-. (b) Using a mixture of ethanol and H_2O or only acetone as solvent. (c) Increasing the number of R groups on the C bonded to the leaving group, L. (d) Using a strong Nu^-. ◄

(a) Doubling either [RL] or [Nu^-] doubles the rate of the S_N2 reaction. For S_N1 reactions the rate is doubled only by doubling [RL] and is not affected by any change in [Nu^-].
(b) A mixture of ethanol and H_2O has a high dielectric constant and therefore enhances the rate of S_N1 reactions. This usually has little effect on S_N2 reactions. Acetone has a low dielectric constant and is aprotic and favors S_N2 reactions.
(c) Increasing the number of R's on the reaction site enhances S_N1 reactivity through electron release and stabilization of R^+. The effect is opposite in S_N2 reactions because bulky R's sterically hinder formation of, and raise $\Delta H^‡$ for, the transition state.
(d) Strong nucleophiles favor S_N2 reactions and do not affect S_N1 reactions.

Problem 7.38 List the following alkyl bromides in order of decreasing reactivity in the indicated reactions.

$$\begin{array}{ccc} \text{(I)} & \text{(II)} & \text{(III)} \end{array}$$

$$\begin{array}{ccc} \overset{\displaystyle CH_3}{\underset{\displaystyle Br}{CH_3\overset{\vert}{\underset{\vert}{C}}—CH_2CH_3}} & CH_3CH_2CH_2CH_2CH_2Br & CH_3CH_2\underset{\displaystyle Br}{\overset{\vert}{CH}}CH_2CH_3 \end{array}$$

(a) S_N1 reactivity, (b) S_N2 reactivity, (c) reactivity with alcoholic $AgNO_3$. ◄

(a) Reactivity for the S_N1 mechanism is $3°$ (I) > $2°$ (III) > $1°$ (II).
(b) The reverse reactivity order for S_N2 reactions gives $1°$ (II) > $2°$ (III) > $3°$ (I).
(c) Ag^+ catalyzes S_N1 reactions and the reactivities are $3°$ (I) > $2°$ (III) > $1°$ (II).

Problem 7.39 Give structures of all alkenes formed and underline the major product expected from E2 elimination of: (a) 1-chloropentane, (b) 2-chloropentane. ◄

(a)
$$CH_3CH_2CH_2CH_2CH_2Cl \longrightarrow CH_3CH_2CH_2CH{=}CH_2$$

A 1° alkyl halide, therefore one alkene.

(b)
$$CH_3CH_2CH_2\underset{\underset{Cl}{|}}{CH}{-}CH_3 \longrightarrow CH_3CH_2CH_2CH{=}CH_2 + \underline{CH_3CH_2CH{=}CHCH_3}$$

A 2° alkyl halide flanked by 2 R's; therefore 2 alkenes are formed. The more substituted alkene is the major product because of its greater stability.

Problem 7.40 How is conformational analysis used to explain the 6:1 ratio of *trans-* to *cis*-2-butene formed on dehydrochlorination of 2-chlorobutane? ◄

For either enantiomer there are two conformers in which the H and Cl eliminated are *anti* to each other.

(I) ... *trans*-2-Butene

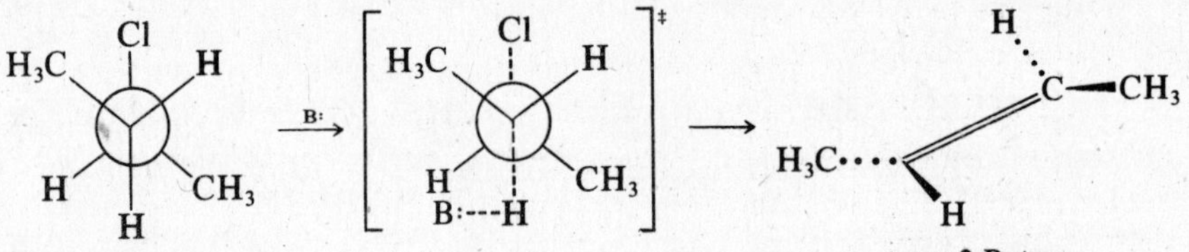

(II) ... *cis*-2-Butene

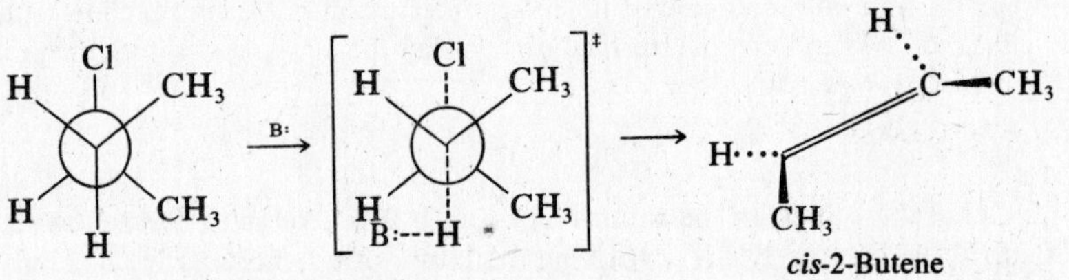

Conformer I has a less crowded, lower-enthalpy transition state than conformer II. Its $\Delta H^{\ddagger}$ is less and reaction rate greater; this accounts for the greater amount of *trans* isomer obtained from conformer I and the smaller amount of *cis* isomer from conformer II.

Problem 7.41 Potassium *tert*-butoxide, $K^+\bar{O}CMe_3$, is used as a base in E2 reactions. (a) How does it compare in effectiveness with ethylamine, $CH_3CH_2NH_2$? (b) Compare its effectiveness in the solvents *tert*-butyl alcohol and dimethylsulfoxide (DMSO). ◄

(a) $K^+\bar{O}CMe_3$ is more effective because it is more basic. Its larger size also precludes S_N2 reactions.
(b) Its reactivity is greater in aprotic DMSO because its basic anion is not solvated. Me_3COH reduces the effectiveness of Me_3CO^- by H-bonding.

Problem 7.42 Indicate and explain why the following may or may not be used as synthetic reactions.

(a)
$$CH_3{-}\underset{\underset{CH_3}{|}}{\overset{\overset{CH_3}{|}}{C}}{-}O^- + CH_3{-}\underset{\underset{CH_3}{|}}{\overset{\overset{CH_3}{|}}{C}}{-}Cl \xrightarrow{S_N2} CH_3{-}\underset{\underset{CH_3}{|}}{\overset{\overset{CH_3}{|}}{C}}{-}O{-}\underset{\underset{CH_3}{|}}{\overset{\overset{CH_3}{|}}{C}}{-}CH_3 + Cl^-$$

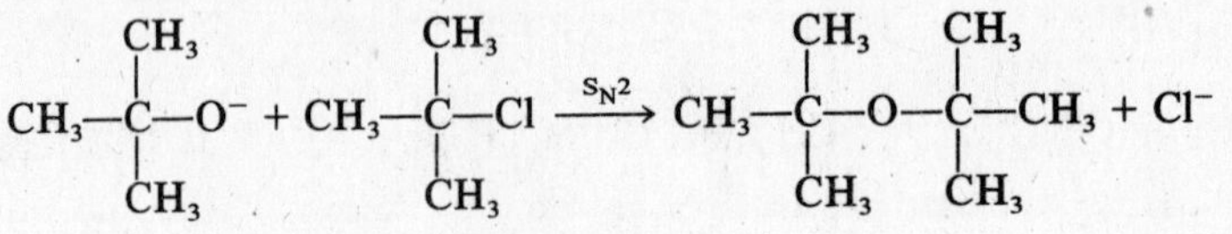

(b)
$$H_2O + CH_3CHClCH_3 \xrightarrow{100\ °C} CH_3CH{=}CH_2$$ ◄

(a) A 3° RX does not undergo S_N2 displacement. Instead, strong bases induce E2 elimination to give $CH_2{=}C(CH_3)_2$.
(b) H_2O is too weakly basic to effect E2 elimination. The E1 reaction does not occur because $\Delta H^{\ddagger}$ for the transition state leading to an incipient R_2CH^+ is very large and the rate is extremely low.

Problem 7.43 Account for the formation of

$$CH_3CH{=}CH{-}CH_2CN \quad \text{and} \quad CH_3{-}\overset{\displaystyle CN}{\underset{\displaystyle |}{CH}}{-}CH{=}CH_2$$

from the reaction with CN^- of 1-chloro-2-butene, $CH_3CH{=}CH{-}CH_2Cl$. ◄

Formation of 1-cyano-2-butene results from S_N2 reaction at the terminal C.

$$CH_3{-}CH{=}CH{-}CH_2{\overset{\frown}{-}}Cl + CN^- \longrightarrow CH_3{-}CH{=}CH{-}CH_2{-}CN + Cl^-$$

Attack by CN^- can also occur at C^3 with the π electrons of the double bond acting as nucleophile to displace Cl^- in an allylic rearrangement:

$$N{\equiv}C{:}\,\overset{\frown}{CH_3}{-}CH{=}CH{-}CH_2{-}\overset{\frown}{Cl} \longrightarrow CH_3{-}\underset{\displaystyle CN}{\underset{\displaystyle |}{CH}}{-}CH{=}CH_2 + Cl^- \quad \text{(an } S_N2' \text{ reaction)}$$

Problem 7.44 Calculate the rate for the S_N2 reaction of 0.1 M C_2H_5I with 0.1 M CN^- if the reaction rate for 0.01 M concentrations is 5.44×10^{-9} moles per liter per second. ◄

The rates are proportional to the products of the concentrations,

$$\frac{\text{Rate}}{5.44 \times 10^{-9}\,\text{mole/l} \cdot \text{sec}} = \frac{[0.1][0.1]}{[0.01][0.01]}$$

$$\text{Rate} = 100 \times 5.44 \times 10^{-9}\,\text{mole/l} \cdot \text{sec} = 5.44 \times 10^{-7}\,\text{mole/l} \cdot \text{sec}$$

Problem 7.45 Give reactions for tests that can be carried out rapidly in a test tube to differentiate the following compounds: hexane, $CH_3CH{=}CHCl$, $H_2C{=}CHCH_2Cl$ and $CH_3CH_2CH_2Cl$. ◄

Hexane is readily distinguished from the other three compounds because there is a negative test for Cl^- after Na fusion and treatment with acidic $AgNO_3$. The remaining three compounds are differentiated by their reactivity with alcoholic $AgNO_3$ solution. $CH_3CH{=}CHCl$ is a vinylic chloride and does not react even on heating. $H_2C{=}CHCH_2Cl$ is most reactive (allylic) and precipitates AgCl in the cold, while $CH_3CH_2CH_2Cl$ gives a precipitate of AgCl on warming with the reagent.

Problem 7.46 Will the following reactions be primarily displacement or elimination?

    (a) $CH_3CH_2CH_2Cl + I^- \longrightarrow$ (b) $(CH_3)_3CBr + CN^-$ (ethanol) $\longrightarrow$

    (c) $CH_3CHBrCH_3 + OH^-$ (H_2O) $\longrightarrow$ (d) $CH_3CHBrCH_3 + OH^-$ (ethanol) $\longrightarrow$

               (e) $(CH_3)_3CBr + H_2O \longrightarrow$ ◄

(a) S_N2 displacement. I^- is a good nucleophile, and a poor base.
(b) E2 elimination. A 3° halide and a fairly strong base.
(c) Mainly S_N2 displacement.
(d) Mainly E2 elimination. A less polar solvent than that in (c) favors E2.
(e) S_N1 displacement. H_2O is not basic enough to remove a proton to give elimination.

Chapter 8

Alkynes and Dienes

8.1 ALKYNES

NOMENCLATURE AND STRUCTURE

Alkynes or acetylenes (C_nH_{2n-2}) have a —C≡C— and are isomeric with alkadienes, which have 2 double bonds. In IUPAC, a —C≡C— is indicated by the suffix -yne.

Acetylene, C_2H_2, is a linear molecule in which each C uses sp hybrid orbitals to form two σ bonds with a 180° angle. The unhybridized p orbitals form two π bonds.

Problem 8.1 Name the structures below by the IUPAC system:

(a) $CH_3C≡CCH_3$

(b) $CH_3C≡CCH_2CH_3$

(c)
$$CH_3-\underset{\underset{H}{|}}{\overset{\overset{CH_3}{|}}{C}}-C≡C-\underset{\underset{CH_3}{|}}{\overset{\overset{CH_3}{|}}{C}}-CH_3$$

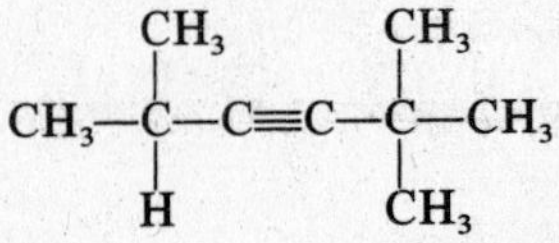

(d) $HC≡C-CH_2CH=CH_2$

(e) $HC≡C-CH_2CH_2Cl$

(f) $CH_3CH=CH-C≡C-C≡CH$ ◄

(a) 2-Butyne

(b) 2-Pentyne

(c) 2,2,5-Trimethyl-3-hexyne

(d) 1-Penten-4-yne
C=C has priority over C≡C and gets the smaller number.

(e) 4-Chloro-1-butyne

(f) 5-Hepten-1,3-diyne

Problem 8.2 Supply structural formulas and IUPAC names for all alkynes with the molecular formula (a) C_5H_8, (b) C_6H_{10}. ◄

(a) Insert a triple bond when possible in n-pentane, isopentane and neopentane. Placing a triple bond in an n-pentane chain gives $H—C≡C—CH_2CH_2CH_3$ (1-pentyne) and $CH_3—C≡C—CH_2CH_3$ (2-pentyne). Isopentane gives one compound,

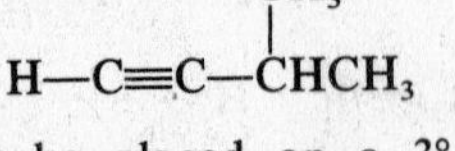

$$H-C≡C-\underset{}{\overset{\overset{CH_3}{|}}{C}}HCH_3 \quad \text{3-Methyl-1-butyne}$$

because a triple bond cannot be placed on a 3° C. No alkyne is obtainable from neopentane, $(CH_3)_2C(CH_3)_2$.

(b) Inserting a triple bond in n-hexane gives:

$$H—C≡C—CH_2CH_2CH_2CH_3 \qquad CH_3—C≡C—CH_2CH_2CH_3 \qquad CH_3CH_2—C≡C—CH_2CH_3$$

1-Hexyne   2-Hexyne   3-Hexyne

Isohexane yields two alkynes, and 3-methylpentane and 2,2-dimethylbutane one alkyne each.

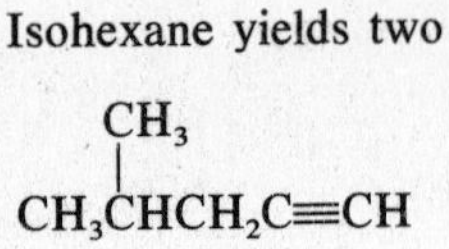

$$\underset{}{\overset{\overset{CH_3}{|}}{CH_3}}CHCH_2C≡CH \qquad \underset{}{\overset{\overset{CH_3}{|}}{CH_3}}CH—C≡C—CH_3 \qquad HC≡C—\underset{}{\overset{\overset{CH_3}{|}}{C}}HCH_2CH_3 \qquad CH_3—\underset{\underset{CH_3}{|}}{\overset{\overset{CH_3}{|}}{C}}—C≡C—H$$

4-Methyl-1-
pentyne

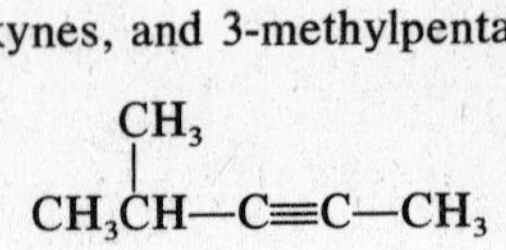

4-Methyl-2-
pentyne

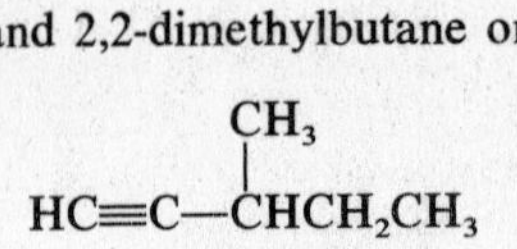

3-Methyl-1-
pentyne

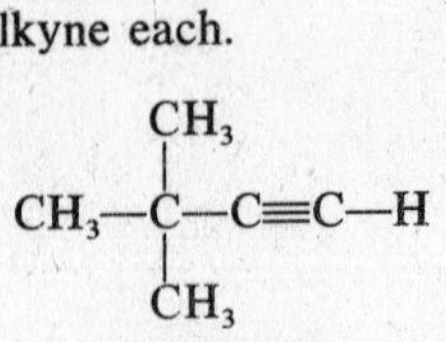

3,3-Dimethyl-1-
butyne

Problem 8.3   Draw models of   (a) sp hybridized C   and (b) C_2H_2 to show bonds formed by orbital overlap. ◄

(a) See Fig. 8-1(a). Only one of three p orbitals of C is hybridized. The two unhybridized p orbitals (p_z and p_y) are at right angles to each other and also to the axis of the sp hybrid orbitals.

(b) See Fig. 8-1(b). Sidewise overlap of the p_y and p_z orbitals on each C forms the π_y and π_z bonds respectively.

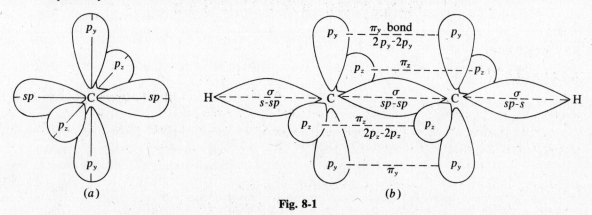

Fig. 8-1

Problem 8.4   Why is the C≡C distance (1.24 Å) shorter than the C=C (1.32 Å) and C—C (1.53 Å)?   ◄

The carbon nuclei in C≡C are shielded by 6 electrons (from 3 bonds) rather than by 4 or 2 electrons as in C=C or C—C, respectively. With more shielding electrons present, the C's of —C≡C— can get closer, thereby affording more orbital overlap and stronger bonds.

Problem 8.5   Explain how the orbital picture of —C≡C— accounts for   (a) the absence of geometric isomers in $CH_3C≡CC_2H_5$;   (b) the acidity of an acetylenic H, e.g.

$$HC≡CH + NH_2^- \longrightarrow HC≡C^- + NH_3 \quad (pK_a = 25)$$ ◄

(a) The sp hybridized bonds are linear, ruling out cis-trans isomers in which substituents must be on different sides of the multiple bond.

(b) We apply the principle: "The more s character in the C—H bond, the more acidic is the H." Therefore the order of acidity of hydrocarbons is

$$≡C—H > =\overset{|}{C}—H > —\overset{|}{\underset{|}{C}}—H$$
$$\quad\; sp \qquad\; sp^2 \qquad\; sp^3$$

LABORATORY METHODS OF PREPARATION

1. Dehydrohalogenation of vic-Dihalides or gem-Dihalides

$$
\begin{array}{c}
\overset{H}{\underset{H}{\overset{|}{\underset{|}{-C}}}}\overset{X}{\underset{X}{\overset{|}{\underset{|}{-C-}}}} \quad \text{or} \quad \overset{H}{\underset{X}{\overset{|}{\underset{|}{-C}}}}\overset{H}{\underset{X}{\overset{|}{\underset{|}{-C-}}}} \xrightarrow{\text{alc. KOH}} KX + H_2O + \overset{H}{\underset{X}{\overset{|}{\underset{|}{-C=C-}}}} \xrightarrow{\text{NaNH}_2} NaX + NH_3 + —C≡C—
\end{array}
$$

A gem-dihalide   A vic-dihalide   A vinyl halide   An acetylene

The vinyl halide requires the stronger base sodamide ($NaNH_2$).

2. Dehalogenation of vic-Tetrahalogen Compounds

$$CH_3—CBr_2—CBr_2—CH_3 + 2Zn \xrightarrow[\text{heat}]{\text{EtOH}} CH_3—C≡C—CH_3 + 2ZnBr_2$$
$$\text{2,2,3,3-Tetrabromobutane} \qquad\qquad \text{2-Butyne}$$

This reaction has the drawback that the halogen compound is itself prepared by halogen addition to alkynes.

3. Alkyl Substitution in Acetylene; Acidity of $\equiv$C—H

$$R\text{—}C\equiv C\text{—}H + \left\{ \begin{array}{c} NaNH_2 \\ or \\ Na \end{array} \right. \longrightarrow R\text{—}C\equiv C^{\text{-}}Na^+ + \left\{ \begin{array}{c} NH_3 \\ or \\ \frac{1}{2}H_2 \end{array} \right.$$

$$R\text{—}C\equiv C^{\text{-}} + 1°\,R'\text{—}CH_2(\text{:}X) \longrightarrow R\text{—}C\equiv C\text{—}CH_2\text{—}R' + X^- \quad (S_N2 \text{ mechanism})$$

Problem 8.6 Why is this last reaction not used with 2° or 3° RX's? ◄

Acetylides are strongly basic (conjugate bases of very weak acids) and form alkenes with 2° and 3° RX's by E2 *elimination*.

$$Na^+Br^- + CH_3C\equiv CH + CH_3CH\text{=}CH_2 \xleftarrow{(CH_3)_2CHBr} CH_3C\equiv C^{\text{-}}Na^+ \xrightarrow{(CH_3)_3CBr}$$

Propylene

$$\underset{\text{Isobutylene}}{CH_3\text{—}\overset{\overset{\displaystyle CH_3}{|}}{C}\text{=}CH_2} + CH_3C\equiv CH + Na^+Br^-$$

Problem 8.7 Outline a synthesis of propyne from isopropyl or propyl bromide. ◄

The needed *vic*-dihalide is formed from propene, which is prepared from the alkyl halides.

$$\left.\begin{array}{l} CH_3CH_2CH_2Br \\ \textit{n}\text{-Propyl bromide} \\ \\ \text{or} \\ \\ CH_3CHBrCH_3 \\ \text{Isopropyl bromide} \end{array}\right\} \xrightarrow[\text{KOH}]{\text{alc.}} \underset{\text{Propene}}{CH_3\text{—}CH\text{=}CH_2} \xrightarrow{Br_2} \underset{\substack{\text{Propylene} \\ \text{bromide}}}{CH_3CHBrCH_2Br} \xrightarrow[\text{KOH}]{\text{alc.}} \underset{\substack{\text{Propenyl} \\ \text{bromide}}}{CH_3CH\text{=}CHBr} \xrightarrow{NaNH_2} \underset{\text{Propyne}}{CH_3C\equiv CH}$$

Problem 8.8 Synthesize the following compounds from HC$\equiv$CH and any other organic and inorganic reagents (do not repeat steps): (*a*) 1-pentyne, (*b*) 2-hexyne. ◄

(*a*) $H\text{—}C\equiv C\text{—}H \xrightarrow{NaNH_2} H\text{—}C\equiv \overset{+}{C}\text{:}Na \xrightarrow{CH_3CH_2CH_2I} H\text{—}C\equiv C\text{—}CH_2CH_2CH_3$

(*b*) $Na\text{:}\overset{+}{C}\equiv C\text{—}H \xrightarrow{CH_3I} CH_3\text{—}C\equiv C\text{—}H \xrightarrow{NaNH_2} CH_3\text{—}C\equiv \overset{+}{C}\text{:}Na \xrightarrow{CH_3CH_2CH_2I} CH_3\text{—}C\equiv C\text{—}CH_2CH_2CH_3$

Problem 8.9 Industrially, acetylene is made from calcium carbide, $CaC_2 + H_2O \longrightarrow HC\equiv CH + Ca(OH)_2$. Formulate the reaction as a Brönsted acid-base reaction. ◄

The carbide anion C_2^{2-} is the base formed when HC$\equiv$CH loses two H$^+$'s.

$$\underset{\text{Base}}{\text{:}C\equiv C\text{:}^{-2}} + \underset{\text{Acid}}{2HOH} \longrightarrow \underset{\text{Acid}}{H\text{—}C\equiv C\text{—}H} + \underset{\text{Base}}{2OH^-} \quad (Ca^{2+} \text{ precipitates as } Ca(OH)_2)$$

8.2 CHEMICAL PROPERTIES OF ACETYLENES

ADDITION REACTIONS AT THE TRIPLE BOND

Nucleophilic π electrons of alkynes add *electrophiles* in reactions similar to additions to alkenes. Alkynes can add two moles of reagent but are less reactive (except to H_2) than alkenes.

1. Hydrogen

(*a*) $CH_3\text{—}C\equiv C\text{—}CH_2CH_3 + 2H_2 \xrightarrow{Pt} CH_3CH_2CH_2CH_2CH_3$

(*b*)

cis-2-Pentene   2-Pentyne   *trans*-2-Pentene

stereospecific reductions

2. HX (HCl, HBr, HI)

$$CH_3-C\equiv C-H \xrightarrow{HBr} CH_3-CBr=CH_2 \xrightarrow{HBr} CH_3-CBr_2-CH_3 \quad \text{(Markovnikov addition)}$$

$$CH_3-C\equiv C-H + HBr \xrightarrow{\text{peroxide}} CH_3-CH=CHBr \quad \text{(anti-Markovnikov)}$$

3. Halogen (Br_2, Cl_2)

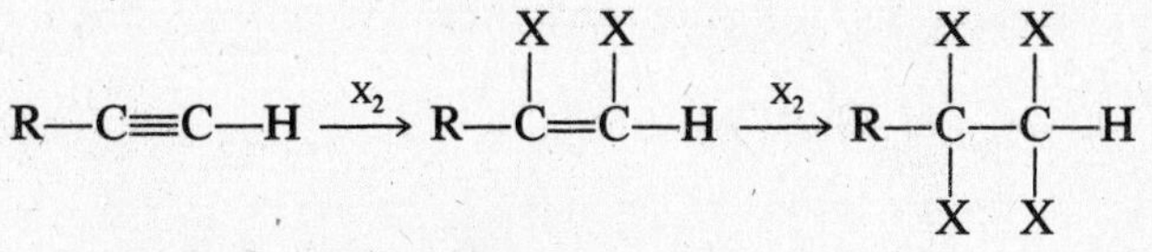

$$R-C\equiv C-H \xrightarrow{X_2} R-\overset{\overset{\displaystyle X}{|}}{C}=\overset{\overset{\displaystyle X}{|}}{C}-H \xrightarrow{X_2} R-\overset{\overset{\displaystyle X}{|}}{\underset{\underset{\displaystyle X}{|}}{C}}-\overset{\overset{\displaystyle X}{|}}{\underset{\underset{\displaystyle X}{|}}{C}}-H$$

4. H_2O (Hydration to Carbonyl Compounds)

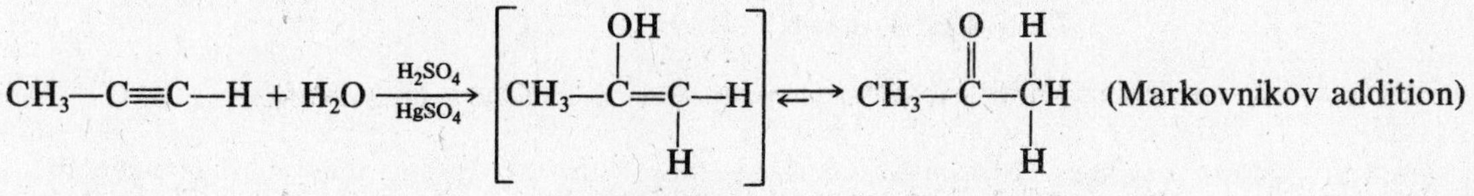

$$CH_3-C\equiv C-H + H_2O \xrightarrow[HgSO_4]{H_2SO_4} \left[CH_3-\overset{\overset{\displaystyle OH}{|}}{C}=\overset{\overset{\displaystyle}{}}{\underset{\underset{\displaystyle H}{|}}{C}}-H \right] \rightleftharpoons CH_3-\overset{\overset{\displaystyle O}{\|}}{C}-\overset{\overset{\displaystyle H}{|}}{\underset{\underset{\displaystyle H}{|}}{C}}H \quad \text{(Markovnikov addition)}$$

Propyne　　　　　　　　　　　　　a vinyl alcohol　　　　　　　Acetone
　　　　　　　　　　　　　　　　　(unstable)

5. Boron Hydride

$$R'-C\equiv C-H + R_2BH \longrightarrow \overset{R'}{\underset{H}{>}}C=C\overset{H}{\underset{BR_2}{<}}$$

$$\xrightarrow[\text{oxidation}]{H_2O_2,\ NaOH} R'CH_2CHO$$

$$\xrightarrow[\text{hydrolysis}]{CH_3COOH} R'CH=CH_2$$

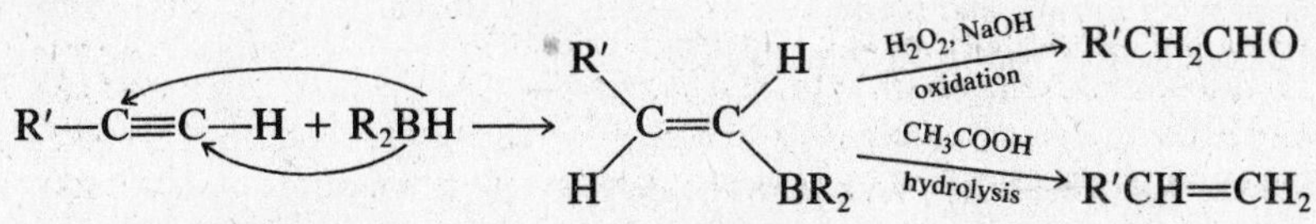

a dialkylborane　　　　　　　　a vinylborane

With dialkylacetylenes, the products of hydrolysis and oxidation are *cis*-alkenes and ketones, respectively.

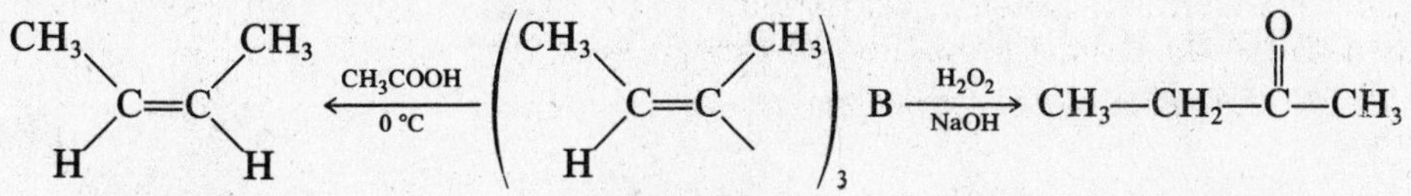

$$\overset{CH_3}{\underset{H}{>}}C=C\overset{CH_3}{\underset{H}{<}} \xleftarrow[0\ ^\circ C]{CH_3COOH} \left(\overset{CH_3}{\underset{H}{>}}C=C\overset{CH_3}{\underset{}{<}} \right)_3 B \xrightarrow[NaOH]{H_2O_2} CH_3-CH_2-\overset{\overset{\displaystyle O}{\|}}{C}-CH_3$$

cis-2-Butene　　　　　　　　a vinylborane　　　　　　　　2-Butanone

6. Dimerization

$$2\ H-C\equiv C-H \xrightarrow[H_2O]{Cu(NH_3)_2^+Cl^-} H_2C=CH-C\equiv C-H$$
　　　　　　　　　　　　　　　　　　　　Vinylacetylene

7. Oxidation to Carboxylic Acids

$$3\ CH_3C\equiv CH + 8KMnO_4 + KOH \longrightarrow 3CH_3COOK + 3K_2CO_3 + 8MnO_2 + 2H_2O$$

$$CH_3CH_2C\equiv C-CH_3 + 2KMnO_4 \longrightarrow CH_3CH_2COOK + CH_3COOK + 2MnO_2 + 2H_2O$$

8. Ozonolysis-Hydrolysis

$$CH_3C\equiv CCH_2CH_3 \xrightarrow[\text{2. hydrolysis}]{\text{1. } O_3} CH_3COOH + HOOCCH_2CH_3$$

2-Pentyne　　　　　　　　　　　Acetic　　　Propionic
　　　　　　　　　　　　　　　　acid　　　　acid

9. Nucleophiles

$$CH_3C\equiv CCH_3 + CN^-, HCN \longrightarrow CH_3CH=C(CN)CH_3$$

ACIDITY AND SALTS OF 1-ALKYNES [see Problem 8.5(b)]

$$CH_3C\equiv CH + Ag^+ \xrightarrow[-H^+]{NH_3} CH_3C\equiv CAg \xrightarrow{HNO_3} CH_3C\equiv CH + Ag^+$$

Problem 8.10 Alkynes differ from alkenes in adding nucleophiles such as CN^-. Explain. ◄

The intermediate carbanion from addition of CN^- to an alkyne has the unshared electron pair on an sp^2 hybridized C. It is more stable and is formed more readily than the sp^3 hybridized carbanion formed from a nucleophile and an alkene.

$$H-C\equiv C-H \xrightarrow{:CN^-} H-\overset{..}{C}=C:CN \xrightarrow{HCN} H_2C=C-CN$$

Acrylonitrile

$$H_2C=CH_2 + :C\equiv N^- \xrightarrow{\times\!\!\!\!\rightarrow} H_2\overset{..}{C}-CH_2C\equiv N$$
$$sp^3$$

Problem 8.11 Outline a synthesis of *trans*-2,3-dibromo-2-hexene from C_2H_2 and point out the conditions for optimum yield of product. ◄

trans-Dibromoalkenes can be prepared by the *anti* addition of 1 mole of Br_2 to an alkyne. The resulting vinyl Br's decrease the nucleophilic reactivity of the double bond. Addition of a second mole of Br_2 is slower. To further prevent tetrabromo formation we slowly add Br_2 to an excess of alkyne. The preparation of 2-hexyne from acetylene is given in Problem 8.8(b).

Problem 8.12 Dehydrohalogenation of 3-bromohexane gives a mixture of *cis*-2-hexene and *trans*-2-hexene. How can this mixture be converted to pure (a) *cis*-2-hexene? (b) *trans*-2-hexene? ◄

Relatively pure alkene geometric isomers are prepared by stereoselective reduction of alkynes.

(a) Hydrogenation of 2-hexyne with Lindlar's catalyst gives 98% *cis*-2-hexene.

$$CH_3CH=CHCH_2CH_2CH_3 \xrightarrow{Br_2} CH_3CH-CHCH_2CH_2CH_3 \xrightarrow{NaNH_2}$$

cis and *trans*-2-Hexene
$$\underset{Br}{}\;\underset{Br}{}$$

$$CH_3C\equiv CCH_2CH_2CH_3 \xrightarrow{H_2/Pt(Pb)}$$

2-Hexyne   *cis*-2-Hexene

(b) Reduction with Na in liquid NH_3 gives the *trans* product.

$$CH_3C\equiv CCH_2CH_2CH_3 \xrightarrow[NH_3]{Na}$$

trans-2-Hexene

Problem 8.13 Will the following compounds react? Give any products and the reason for their formation.

(a) $CH_3-C\equiv C-H + aq. Na^+OH^- \longrightarrow$

(b) $CH_3CH_2C\equiv C-MgI + CH_3OH \longrightarrow$

(c) $CH_3C\equiv C:^- Na^+ + NH_4^+ \longrightarrow$ ◄

(a) No. The products would be the stronger acid H_2O and the stronger base $CH_3C\equiv C:^-$.
(b) Yes. The products are the weaker acid $CH_3CH_2C\equiv C:H$ and the weaker base $MgI(OCH_3)$.
(c) Yes. The products are the weaker acid propyne and the weaker base NH_3.

8.3 ALKADIENES

Problem 8.14 Name by the IUPAC method and classify as *cumulated, conjugated* or *isolated*:

$$(a) \quad H_2C{=}CH{-}CH{=}CHCH_3 \qquad (b) \quad H_2C{=}CHCH_2\overset{\displaystyle CH_2CH_3}{\underset{\displaystyle |}{C}}{=}CHCH_2CH_3$$

$$(c) \quad H_2C{=}C{=}CH_2 \qquad\qquad (d) \quad H_2C{=}CH{-}CH{=}CHCH{=}CH_2 \qquad \blacktriangleleft$$

(a) 1,3-Pentadiene. *Conjugated* diene since it has alternating double and single bonds, i.e., $-C{=}C{-}C{=}C-$. (b) 4-Ethyl-1,4-heptadiene. *Isolated* diene since the double bonds are separated by at least one sp^3 hybridized C, i.e., $-C{=}C{-}(CH_2)_n{-}C{=}C-$. (c) 1,2-Propadiene (allene). *Cumulated* diene since 2 double bonds are on the same C, i.e., $-C{=}C{=}C-$. (d) 1,3,5-Hexatriene. *Conjugated* since it has alternating single and double bonds.

Problem 8.15 Give steps for the conversion $HC{\equiv}CCH_2CH_2CH_3 \longrightarrow H_2C{=}CH{-}CH{=}CHCH_3$. $\blacktriangleleft$

$$HC{\equiv}CCH_2CH_2CH_3 \xrightarrow{\text{H}_2/\text{Pt(Pb)}} H_2C{=}CHCH_2CH_2CH_3 \xrightarrow[\substack{500\ {}^\circ C \\ \text{(allylic substitution)}}]{\text{Cl}_2}$$

$$H_2C{=}CHCHClCH_2CH_3 \xrightarrow{\text{alc. KOH}} H_2C{=}CHCH{=}CHCH_3$$

Problem 8.16 Compare the stabilities of the 3 types of dienes from the following heats of hydrogenation, ΔH_h (in kcal/mole). (For comparison, ΔH_h for 1-pentene is -30.1.)

Conjugated	$H_2C{=}\overset{\displaystyle H}{\underset{\displaystyle |}{C}}{-}CH{=}CH{-}CH_3$ 1,3-Pentadiene	-54.1
Isolated	$H_2C{=}CH{-}CH_2{-}CH{=}CH_2$ 1,4-Pentadiene	-60.2
Cumulated	$H_2C{=}C{=}CH{-}CH_2CH_3$ 1,2-Pentadiene	-71

The calculated ΔH_h, assuming no interaction between the double bonds, is $2(-30.1) = -60.2$. The more negative the observed value of ΔH_h compared to -60.2, the less stable the diene; the less negative the observed value, the more stable the diene. Conjugated dienes are most stable and cumulated dienes are least stable.

Problem 8.17 Predict the product of the reaction

$$\text{alc. KOH} + CH_3{-}\overset{\displaystyle CH_3}{\underset{\displaystyle H}{C}}{-}\overset{\displaystyle H}{\underset{\displaystyle Cl}{C}}{-}CH_2{-}CH{=}CH_2 \longrightarrow \qquad \blacktriangleleft$$

The more stable conjugated diene

$$(CH_3)_2\overset{\displaystyle H}{\underset{}{C}}{-}\overset{\displaystyle H}{\underset{}{C}}{=}\overset{\displaystyle H}{\underset{}{C}}{-}CH{=}CH_2$$

is formed, even though it is less substituted than the isolated diene $(CH_3)_2C{=}CHCH_2CH{=}CH_2$, which *is not formed*.

Problem 8.18 Account for the stability of conjugated dienes by (a) extended π bonding, (b) molecular orbital theory, (c) resonance theory. ◄

(a) The four p orbitals of conjugated dienes are adjacent and parallel (Fig. 8-2) and overlap to form an extended π system involving all four C's. This results in *greater stability* and decreased energy.

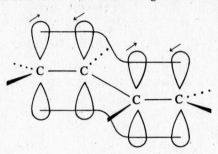

Fig. 8-2

(b) Four atomic p orbitals interact laterally to give four π molecular orbitals, the nature of which can be seen by using $+$ and $-$ signs for the individual p orbitals, as in Fig. 8-3. For simplicity, only the upper lobe is shown. A comparison of relative energies is made with ethylene and the distribution shown as standing waves indicating node points. The energy of $\pi_1 + \pi_2$ for 1,3-butadiene is less than twice π for two ethylenes (or an isolated diene).

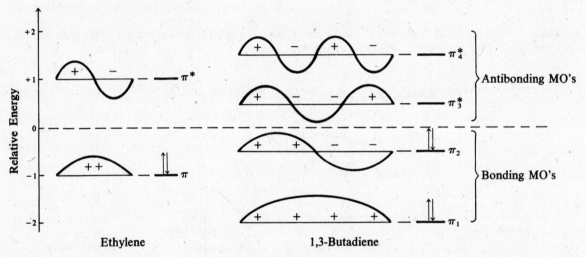

Fig. 8-3

(c) A conjugated diene is a resonance hybrid:

$$-\overset{|}{C}=\overset{|}{C}-\overset{|}{C}=\overset{|}{C}- \longleftrightarrow -\overset{|}{\ddot{C}}-\overset{|}{C}=\overset{|}{C}-\overset{|}{\overset{+}{C}}- \longleftrightarrow -\overset{|}{\overset{+}{C}}-\overset{|}{C}=\overset{|}{C}-\overset{|}{\ddot{C}}- \longleftrightarrow -\overset{|}{C}-\overset{|}{C}=\overset{|}{C}-\overset{|}{C}-$$

(i)

Structure (i) has 11 bonds and makes a more significant contribution than the other 3 structures, which have only 10 bonds. Since the contributing structures are not equivalent, the resonance energy is small.

Problem 8.19 Typical of conjugated dienes, 1,3-butadiene undergoes 1,2- and 1,4-addition, as illustrated with HBr:

$$H_2C-CH=CHCH_2 \xleftarrow[\text{addition}]{1,4-} \boxed{H_2C=CHCH=CH_2 + HBr} \xrightarrow[\text{addition}]{1,2-} H_2C-CH-CH=CH$$
$$\;\;\;|\qquad\qquad\qquad\qquad\qquad\qquad\qquad\qquad\qquad\qquad\qquad\;\;\;|\;\;\;\;|$$
$$\;\;\;H\qquad\qquad Br\qquad\qquad\qquad\qquad\qquad\qquad\qquad\qquad H\;\;Br$$

Explain 1,4-addition in terms of the mechanism of electrophilic addition. ◄

The electrophile (H^+) adds to form an allylic carbonium ion with positive charge delocalized at C^2 and C^4 (resonance forms II and III).   This cation adds the nucleophile at C^2 to form the 1,2-addition product or at C^4 to form the 1,4-addition product.

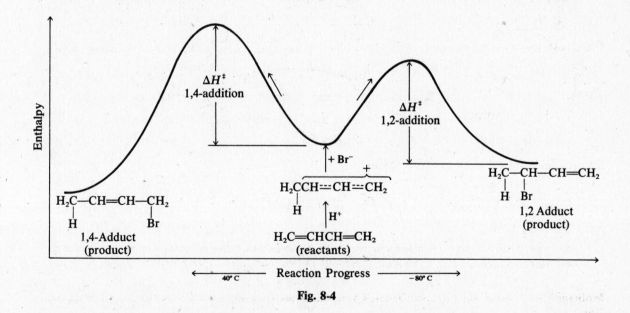

Problem 8.20   Use an enthalpy–reaction diagram to explain the following observations.   Start from the allylic carbonium ion, the common intermediate.

The different products arise from enthalpy variations in the second step, the reaction of Br^- and the allyl R^+.   See Fig. 8-4.

Fig. 8-4

At −80 °C the 1,2-adduct is favored because its formation has the lower $\Delta H^{\ddagger}$.   It is said to be the *kinetic-controlled* or *rate-controlled* product.   1,2-Adduct formation is reversible, and at 40 °C the more stable product, the 1,4-adduct, results.   The 1-4 adduct is called the *thermodynamic-controlled* or *equilibrium-controlled* product and is more stable because it has more R's on the C=C.

Problem 8.21 (*a*) Which additional C's in the following R^+ bear some + charge?

$$-\overset{|}{\underset{7}{C}}=\overset{|}{\underset{6}{C}}-\overset{|}{\underset{5}{C}}=\overset{|}{\underset{4}{C}}-\overset{|}{\underset{3}{C}}=\overset{|}{\underset{2}{C}}-\overset{+}{\underset{1}{\overset{|}{C}}}-$$

(*b*) With which of these C's will an A^- react to give the thermodynamic-controlled product?   ◄

(*a*) C^3, C^5 and C^7.   These are alternating sites.

$$\underset{7}{C}=\underset{6}{C}-\underset{5}{C}=\underset{4}{C}-\overset{+}{\underset{3}{C}}-\underset{2}{C}=\underset{1}{C} \longleftrightarrow \underset{7}{C}=\underset{6}{C}-\overset{+}{\underset{5}{C}}-\underset{4}{C}=\underset{3}{C}-\underset{2}{C}=\underset{1}{C} \longleftrightarrow \overset{+}{\underset{7}{C}}-\underset{6}{C}=\underset{5}{C}-\underset{4}{C}=\underset{3}{C}-\underset{2}{C}=\underset{1}{C}$$

(*b*) A^- adds to equivalent C^1 or C^7 to give the conjugated triene

$$-\overset{|}{C}=\overset{|}{C}-\overset{|}{C}=\overset{|}{C}-\overset{|}{C}=\overset{|}{C}-\overset{|}{\underset{|}{C}}-A \quad \text{(more stable)}$$

Addition at C^3 (or C^5) gives a triene with only 2 conjugated C=C's,

$$-\overset{|}{C}=\overset{|}{C}-\overset{|}{\underset{|}{C}}A-\overset{|}{C}=\overset{|}{C}-\overset{|}{C}=\overset{|}{C}- \quad \text{(less stable)}$$

Problem 8.22 Write structural formulas for major and minor products from acid-catalyzed dehydration of H_2C=CHCH$_2$CH(OH)CH$_3$.   ◄

Dehydration can occur by removal of H from either C^3 or C^5.

$$\underset{\substack{\displaystyle | \;\; | \;\; | \\ \displaystyle H \;\; OH \;\; H}}{H_2C-CH-CH-CH=CH_2} \longrightarrow \underset{\substack{\displaystyle | \\ \displaystyle H}}{H_2C-CH=CH-CH=CH_2} + \underset{\substack{\displaystyle | \\ \displaystyle H}}{H_2C=CH-CH-CH=CH_2}$$

4-Hydroxy-1-pentene   1,3-Pentadiene (major)   1,4-Pentadiene (minor)

conjugated diene   isolated diene

Problem 8.23 Write the structures of the intermediate R^+'s and the two products obtained from the reaction of H_2C=C(CH$_3$)CH=CH$_2$ with   (*a*) HBr,   (*b*) Cl$_2$.   ◄

(*a*) H^+ adds to C^1 to form a more stable allylic 3° R^+, rather than to C^2 or C^3 to form a nonallylic 1° R^+
($\overset{+}{H_2C}-CH(CH_3)CH=CH_2$ and $H_2C=C(CH_3)CH_2-\overset{+}{CH_2}$, respectively), or to C^4 to yield a 2° allylic R^+
($H_2C=C(CH_3)\overset{+}{C}HCH_3$).

3-Methyl-3-bromo-   1-Bromo-3-methyl-

1-butene   2-butene

(b) Cl$^+$ also adds to C^1 to form a hybrid allylic R$^+$.

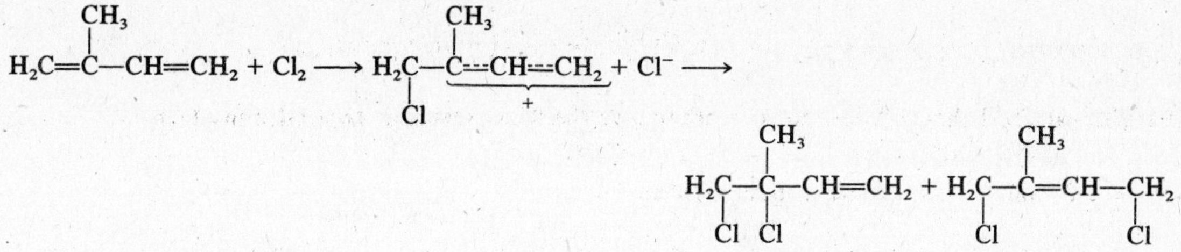

$$H_2C=\overset{\underset{\displaystyle |}{CH_3}}{C}-CH=CH_2 + Cl_2 \longrightarrow H_2C-\underset{\underset{\displaystyle |}{Cl}}{\overset{\underset{\displaystyle |}{CH_3}}{\underbrace{C\text{---}CH\text{==}CH_2}_{+}}} + Cl^- \longrightarrow$$

3-Methyl-3,4-   2-Methyl-1,4-
dichloro-1-butene   dichloro-2-butene

Problem 8.24 Consider addition of one mole of Br$_2$ to one mole of 1,3,5-hexatriene. (a) Deduce structures of possible products. (b) Which products will be obtained under thermodynamic-control conditions? ◄

(a) First, Br$^+$ adds at C^1 to form a 2° allylic-type R$^+$ having + charge on C^2, C^4 and C^6. Then Br$^-$ adds to form three dibromo addition products.

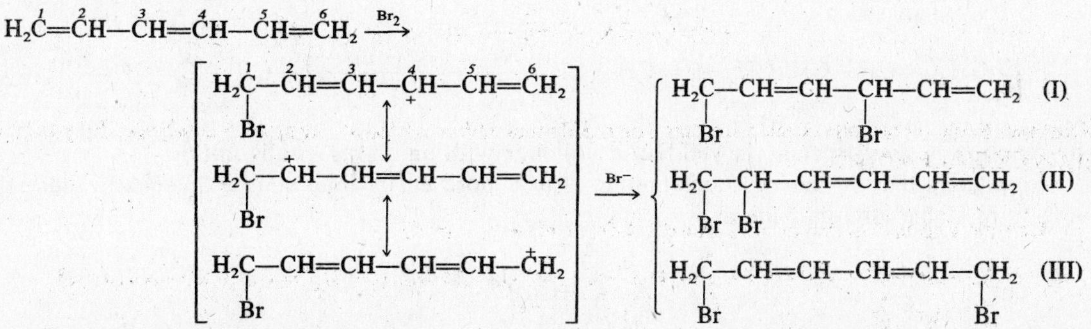

A fourth product (IV) is possible from addition at C^3 and C^4, H$_2$C=CHCHBrCHBrCH=CH$_2$.

(b) Thermodynamic-control conditions favor the most stable product. Compounds II and III are conjugated dienes and hence more stable than the isolated dienes, I and IV. III is more stable than II because its double bonds have more substituents.

Problem 8.25 Write initiation and propagation steps in radical-catalyzed addition of BrCCl$_3$ to 1,3-butadiene and show how the structure of the intermediate accounts for: (a) greater reactivity of conjugated dienes than alkenes, (b) orientation in addition.

Initiation

$$F\cdot\ +\ Br\text{:}CCl_3 \longrightarrow F\text{:}Br + \cdot CCl_3$$

Propagation

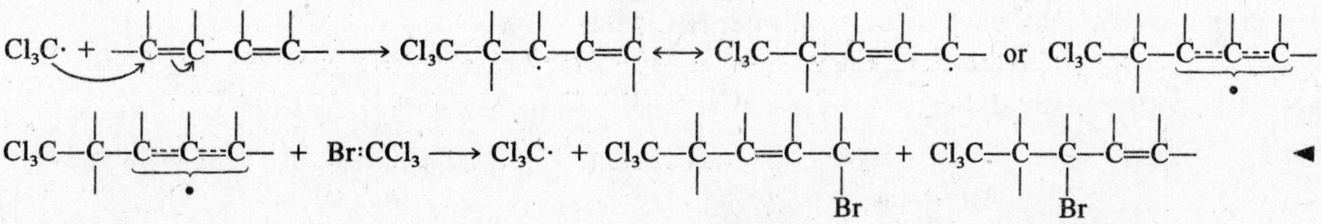

1,4-Adduct   1,2-Adduct
(major)   (minor)

(a) The allyl radical formed in the first propagation step is more stable and requires a lower $\Delta H^{\ddagger}$ than the alkyl free radical from alkenes. The order of free-radical stability is allyl > 3° > 2° > 1°.

(b) The orientation is similar to that in ionic addition because of the relative stabilities of the 2 products.

8.4 POLYMERIZATION OF DIENES

ELECTROPHILIC CATALYSIS

$$E^+ + H_2C\overset{1}{=}\overset{2}{CH}-\overset{3}{CH}\overset{4}{=}CH_2 \longrightarrow E\overset{1}{CH_2}-\overset{2}{CH}\overset{3}{=}CH-\overset{4}{CH_2}\overset{+}{\underset{4}{}} \xrightarrow{\;H_2C\overset{5}{=}\overset{6}{CH}\overset{7}{CH}\overset{8}{=}CH_2\;}$$

$$E\overset{1}{CH_2}-\overset{2}{CH}\overset{3}{=}\overset{4}{CH}\overset{5}{CH_2}\overset{6}{CH_2}\overset{7}{CH}\overset{8}{=}CH\underset{8}{CH_2}^{+} \xrightarrow{n(CH_2=CHCH=CH_2)} [-CH_2CH=CHCH_2-]_{n+2}$$
$$\text{Mer of polymer}$$

NUCLEOPHILIC OR ANIONIC POLYMERIZATION

$$Nu\overset{..}{:} + H_2\overset{CH_3}{\underset{|}{C}}=\overset{|}{C}-CH=CH_2 \longrightarrow Nu:CH_2-\overset{CH_3}{\underset{|}{C}}=CH-\overset{..}{CH_2} \xrightarrow{\;H_2C=\overset{CH_3}{\underset{|}{C}}-CH=CH_2\;}$$

$$Nu:CH_2\overset{CH_3}{\underset{|}{C}}=CH-CH_2:CH_2-\overset{CH_3}{\underset{|}{C}}=CH-\overset{..}{CH_2} \xrightarrow{n(CH_2=\overset{CH_3}{\underset{|}{C}}-CH=CH_2)} \left[-CH_2-\overset{CH_3}{\underset{|}{C}}=CH-CH_2-\right]_{n+2}$$
$$\text{Dimeric anion} \qquad\qquad\qquad\qquad\qquad\qquad\qquad\qquad\qquad \text{Mer of polymer}$$

The reaction is stereospecific in yielding a polymer with an all-*cis* configuration.

Conjugated dienes undergo nucleophilic attack more easily than simple alkenes because they form more stable allyl carbanions,

$$A-\overset{|}{\underset{|}{C}}-\underset{\underline{}}{\overset{|}{C}\text{:::}\overset{|}{C}}-\overset{|}{\underset{|}{C}}-$$

Like the allyl cation, the allylic anion is stabilized by charge delocalization through extended π bonding.

RADICAL POLYMERIZATION

$$Z\cdot + -\overset{|}{\underset{|}{C}}=\overset{|}{\underset{|}{C}}-\overset{|}{\underset{|}{C}}=\overset{|}{\underset{|}{C}}- \longrightarrow Z:\overset{|}{\underset{|}{C}}-\overset{|}{\underset{|}{C}}=\overset{|}{\underset{|}{C}}-\overset{|}{\underset{|}{C}}\cdot \xrightarrow[\text{Monomer}]{-\overset{|}{C}=\overset{|}{C}-\overset{|}{C}=\overset{|}{C}-} Z:\overset{|}{\underset{|}{C}}-\overset{|}{\underset{|}{C}}=\overset{|}{\underset{|}{C}}-\overset{|}{\underset{|}{C}}-\overset{|}{\underset{|}{C}}-\overset{|}{\underset{|}{C}}=\overset{|}{\underset{|}{C}}-\overset{|}{\underset{|}{C}}\cdot$$
$$\text{Monomer} \qquad\qquad \text{Monomeric} \qquad\qquad\qquad\qquad\qquad \text{Dimeric}$$
$$\text{free radical} \qquad\qquad\qquad\qquad\qquad \text{free radical}$$

Conjugated diene polymers are modified and improved by copolymerizing them with other unsaturated compounds, such as acrylonitrile, $H_2C=CH-C\equiv N$.

$$H_2C=CH-CH=CH_2 + H_2\underset{\underset{C\equiv N}{|}}{C}=CH + H_2C=CH-CH=CH_2 + H_2C=CH-CH=CH_2 \longrightarrow$$
$$\text{Acrylonitrile}$$

$$\left[-H_2C-CH=CH-CH_2-CH_2-\underset{\underset{C\equiv N}{|}}{CH}-CH_2-CH=CH-CH_2-CH_2-CH=CH-CH_2-\right]_n$$
$$\text{Buna N rubber unit}$$

8.5 SUMMARY OF ALKYNE CHEMISTRY

PREPARATION

1. Industrial

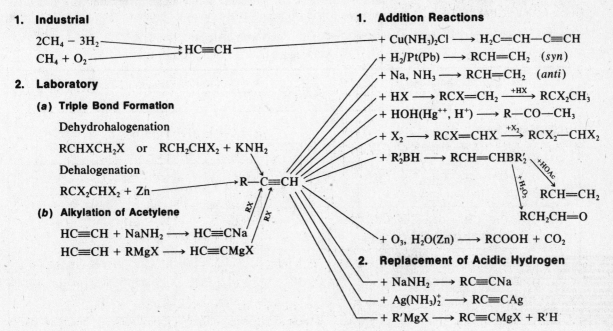

$2CH_4 - 3H_2$
$CH_4 + O_2$ $\longrightarrow$ HC≡CH

2. Laboratory

(a) Triple Bond Formation

Dehydrohalogenation

$RCHXCH_2X$ or $RCH_2CHX_2 + KNH_2$

Dehalogenation

$RCX_2CHX_2 + Zn$ $\longrightarrow$ R—C≡CH

(b) Alkylation of Acetylene

HC≡CH + $NaNH_2$ $\longrightarrow$ HC≡CNa

HC≡CH + RMgX $\longrightarrow$ HC≡CMgX

PROPERTIES

1. Addition Reactions

+ $Cu(NH_3)_2Cl$ $\longrightarrow$ H_2C=CH—C≡CH

+ $H_2/Pt(Pb)$ $\longrightarrow$ RCH=CH_2 *(syn)*

+ Na, NH_3 $\longrightarrow$ RCH=CH_2 *(anti)*

+ HX $\longrightarrow$ RCX=CH_2 $\xrightarrow{+HX}$ RCX_2CH_3

+ $HOH(Hg^{++}, H^+)$ $\longrightarrow$ R—CO—CH_3

+ X_2 $\longrightarrow$ RCX=CHX $\xrightarrow{+X_2}$ RCX_2—CHX_2

+ $R_2'BH$ $\longrightarrow$ RCH=$CHBR_2'$ $\xrightarrow[+H_2O_2]{+HOAc}$ $\begin{array}{c} RCH=CH_2 \\ RCH_2CH=O \end{array}$

+ O_3, $H_2O(Zn)$ $\longrightarrow$ RCOOH + CO_2

2. Replacement of Acidic Hydrogen

+ $NaNH_2$ $\longrightarrow$ RC≡CNa

+ $Ag(NH_3)_2^+$ $\longrightarrow$ RC≡CAg

+ $R'MgX$ $\longrightarrow$ RC≡CMgX + R'H

8.6 SUMMARY OF DIENE CHEMISTRY

PREPARATION

1. Dehydration of Diols

$HOCH_2$—CH_2—CH_2—CH_2—OH

2. Dehydrogenation

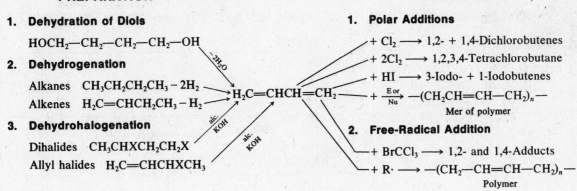

Alkanes $CH_3CH_2CH_2CH_3 - 2H_2$

Alkenes H_2C=$CHCH_2CH_3 - H_2$ $\longrightarrow$ H_2C=CHCH=CH_2

3. Dehydrohalogenation

Dihalides $CH_3CHXCH_2CH_2X$

Allyl halides H_2C=$CHCHXCH_3$

PROPERTIES

1. Polar Additions

+ Cl_2 $\longrightarrow$ 1,2- + 1,4-Dichlorobutenes

+ $2Cl_2$ $\longrightarrow$ 1,2,3,4-Tetrachlorobutane

+ HI $\longrightarrow$ 3-Iodo- + 1-Iodobutenes

+ $\frac{E \text{ or}}{Nu}$ $\longrightarrow$ —$(CH_2CH$=CH—$CH_2)_n$—

Mer of polymer

2. Free-Radical Addition

+ $BrCCl_3$ $\longrightarrow$ 1,2- and 1,4-Adducts

+ R· $\longrightarrow$ —$(CH_2$—CH=CH—$CH_2)_n$—

Polymer

Supplementary Problems

Problem 8.26 For the conjugated and isolated dienes of molecular formula C_6H_{10} tabulate (a) structural formula and IUPAC name, (b) possible geometric isomers, (c) ozonolysis products. ◄

In Table 8-1 a box is placed about the C=C associated with geometric isomers.

Table 8-1

(a) Formula and Name	(b) Geometric Isomers	(c) Ozonolysis Products
(1) H_2C=CH—[CH=CH]—CH_2—CH_3 1,3-Hexadiene	2	H_2C=O, O=CH—CH=O, O=CHCH$_2$CH$_3$
(2) H_2C=CH—CH_2—[CH=CH]—CH_3 1,4-Hexadiene	2	H_2C=O, O=CHCH$_2$CH=O, O=CHCH$_3$
(3) H_2C=CH—CH_2—CH_2—CH=CH_2 1,5-Hexadiene	None	H_2C=O, O=CHCH$_2$CH$_2$CH=O, O=CH$_2$
(4) H_2C=C(CH$_3$)—[CH=CH]—CH_3 2-Methyl-1,3-pentadiene	2	H_2C=O, O=C(CH$_3$)—CH=O, O=CHCH$_3$
(5) H_2C=C(CH$_3$)—CH_2—CH=CH_2 2-Methyl-1,4-pentadiene	None	H_2C=O, O=C(CH$_3$)—CH$_2$CH=O, O=CH$_2$
(6) H_2C=CH—[C(CH$_3$)=CH]—CH_3 3-Methyl-1,3-pentadiene	2	H_2C=O, O=CH—C(CH$_3$)=O, O=CHCH$_3$
(7) H_2C=CH—CH=C(CH$_3$)—CH_3 4-Methyl-1,3-pentadiene	None	H_2C=O, O=CH—CH=O, O=C(CH$_3$)—CH$_3$
(8) CH_3—[CH=CH]—[CH=CH]—CH_3 2,4-Hexadiene	3 *cis, cis*; *cis, trans*; *trans, trans*	CH_3CH=O, O=CH—CH=O, O=CHCH$_3$
(9) H_2C=C(CH$_3$)—C(CH$_3$)=CH_2 2,3-Dimethyl-1,3-butadiene	None	H_2C=O, O=C(CH$_3$)—C(CH$_3$)=O, O=CH$_2$
(10) H_2C=C(CH$_2$CH$_3$)—CH=CH_2 2-Ethyl-1,3-butadiene	None	H_2C=O, O=C(CH$_2$CH$_3$)—CH=O, O=CH$_2$
(11) H_2C=CH—CH(CH$_3$)CH=CH_2 3-Methyl-1,4-pentadiene	None	H_2C=O, O=C—CH(CH$_3$)—CH=O, O=CH$_2$

Problem 8.27 Show reagents and reactions needed to prepare the following compounds from the indicated starting compounds. (a) Acetylene to ethylidene iodide (1,1-diiodoethane). (b) Propyne to isopropyl bromide. (c) 2-Butyne to racemic 2,3-dibromobutane. (d) 2-Bromobutane to trans-2-butene. (e) n-Propyl bromide to 2-hexyne. (f) 1-Pentene to 2-pentyne.   ◄

(a)
$$H-C\equiv C-H \xrightarrow{HI} H_2C=CHI \xrightarrow{HI} CH_3CHI_2$$

(b)
$$CH_3C\equiv C-H \xrightarrow{H_2/Pt} CH_3-CH=CH_2 \xrightarrow{HBr} CH_3CHBrCH_3$$

Add H_2 first; the reaction can be stopped after 1 mole is added.

(c)
$$CH_3C\equiv CCH_3 \xrightarrow{H_2/Pt(Pb)} cis\text{-}CH_3CH=CHCH_3 \xrightarrow{Br_2} (\pm)\text{-}CH_3CHBrCHBrCH_3 \quad (trans \text{ addition})$$

(d)* $CH_3CHBrCH_2CH_3 \xrightarrow{alc.\ KOH} cis+trans\text{-}CH_3CH=CHCH_3 \xrightarrow{Br_2}$

$$CH_3CHBrCHBrCH_3 \xrightarrow{KNH_2} CH_3C\equiv CCH_3 \xrightarrow{Na,\ NH_3} trans\text{-}CH_3CH=CHCH_3$$
$$\text{meso and } rac$$

(e) $CH_3CH_2CH_2Br \xrightarrow{alc.\ KOH} CH_3CH=CH_2 \xrightarrow{Br_2} CH_3CHBrCH_2Br \xrightarrow{KNH_2}$

$$CH_3C\equiv CH \xrightarrow{Na} CH_3C\equiv C^-Na^+ \xrightarrow{n\text{-}C_3H_7Br} CH_3C\equiv C-CH_2CH_2CH_3$$

(f) $H_2C=CHCH_2CH_2CH_3 \xrightarrow{HBr} CH_3CHBrCH_2CH_2CH_3 \xrightarrow{alc.\ KOH}$

$$CH_3CH=CHCH_2CH_3 \xrightarrow{Br_2} CH_3CHBrCHBrCH_2CH_3 \xrightarrow{KNH_2}$$
(Saytzeff product)

$$CH_3C\equiv CCH_2CH_3 \quad (\text{not the less stable allene,}$$
$$CH_3CH=CH=CHCH_3)$$

Problem 8.28 Write a structural formula for organic compounds (A) through (N):

(a)
$$HC\equiv CCH_2CH_2CH_3 \xrightarrow{Ag(NH_3)_2^+} (A) \xrightarrow{HNO_3} (B)$$

(b)
$$CH_3C\equiv CH \xrightarrow{CH_3MgBr} (C,\ a\ gas) + (D) \xrightarrow{CH_3I} (E)$$

(c)
$$CH_3CH_2C\equiv CH + Na^+NH_2^- \longrightarrow (F) \xrightarrow{C_2H_5I} (G) \xrightarrow{H_3O^+,\ Hg^{++}} (H)$$

(d)
$$CH_3C\equiv CH + BH_3 \longrightarrow (I) \xrightarrow{CH_3COOH} (J) \xrightarrow[KMnO_4]{dil.\ aq.} (K)$$

(e)
$$\overset{\displaystyle CH_3}{\underset{}{ClCH_2-\overset{|}{C}HCHClCH_3}} + alc.\ KOH \longrightarrow (L) \xrightarrow[peroxide]{BrCCl_3} (M) + (N)$$
◄

(a) $AgC\equiv CCH_2CH_2CH_3$ (A) $HC\equiv CCH_2CH_2CH_3$ (B)

(b) CH_4 (C) + $CH_3C\equiv CMgBr$ (D) $CH_3C\equiv CCH_3$ (E)

(c) $CH_3CH_2C\equiv C^-Na^+$ (F) $CH_3CH_2C\equiv CCH_2CH_3$ (G) $\longrightarrow$ $CH_3CH_2-\overset{\displaystyle O}{\overset{\|}{C}}-CH_2CH_2CH_3$ (H)

(d) $(CH_3CH=CH)_3B$ (I) $CH_3CH=CH_2$ (J) $CH_3CHOHCH_2OH$ (K)

(e) $H_2C=\overset{\displaystyle CH_3}{\underset{}{\overset{|}{C}}}-CH=CH_2$ (L) $Cl_3CCH_2-\overset{\displaystyle CH_3}{\underset{\displaystyle Br}{\overset{|}{\underset{|}{C}}}}-CH=CH_2$ (M) $Cl_3CCH_2-\overset{\displaystyle CH_3}{\overset{|}{C}}=CH-CH_2-Br$ (N)

1,2-addition product   1,4-addition product

*Trans- and cis-alkenes are made by stereospecific reductions of corresponding alkynes.

Problem 8.29 Give possible products, with IUPAC names, of the following addition reactions: (a) 1,3-butadiene and 1,4-pentadiene, each with 1 and 2 moles of HBr; (b) 2-methyl-1,3-butadiene and 1,3-pentadiene with 1 mole of HI. ◄

(a) $H_2C=CH-CH=CH_2$ + HBr $\longrightarrow$

$\begin{cases} \text{(1,4-Adduct)} \\ CH_3-CH=CH-CH_2Br \xrightarrow{HBr} CH_3CH_2CHBrCH_2Br + CH_3-CHBr-CH_2-CH_2Br \\ \text{1-Bromo-2-butene} \qquad\qquad \text{1,2-Dibromobutane} \qquad \text{1,3-Dibromobutane} \\ + \\ \text{(1,2-Adduct)} \\ CH_3-CHBr-CH=CH_2 \xrightarrow{HBr} CH_3-CHBr-CHBr-CH_3 \\ \text{3-Bromo-1-butene} \qquad\qquad \text{2,3-Dibromobutane} \end{cases}$

$H_2C=CH-CH_2-CH=CH_2$ + HBr $\longrightarrow$ $CH_3-CHBrCH_2-CH=CH_2 \xrightarrow{HBr} CH_3CHBrCH_2CHBrCH_3$
                                           4-Bromo-1-pentene   2,4-Dibromopentane

(b) $H_2C=\overset{\overset{\displaystyle CH_3}{|}}{C}-CH=CH_2$ + HI $\longrightarrow$

$CH_3-\overset{\overset{\displaystyle CH_3}{|}}{C}=CH-CH_2I$ + $CH_3-\overset{\overset{\displaystyle CH_3}{|}}{C}ICH=CH_2$ + $ICH_2-\overset{\overset{\displaystyle CH_3}{|}}{C}=CHCH_3$ + $H_2C=\overset{\overset{\displaystyle CH_3}{|}}{C}-CHICH_3$

  1-Iodo-3-methyl-   3-Iodo-3-methyl-   1-Iodo-2-methyl-   3-Iodo-2-methyl-
    2-butene   1-butene   2-butene   1-butene

$\underbrace{\qquad\qquad\qquad\qquad\qquad\qquad}$ $\underbrace{\qquad\qquad\qquad\qquad\qquad\qquad}$
from more stable $(CH_3)_2\overset{+}{C}CH=CH_2$ (3°)   from less stable $H_2C=C-\overset{+}{C}HCH_3$ (2°)
                                                                              $|$
                                                                            CH_3

$H_2C=CH-CH=CH-CH_3$ + HI $\longrightarrow$

$CH_3CHI-CH=CH-CH_3$ + $H_2C=CHCHICH_2CH_3$ + $H_2CICH=CHCH_2CH_3$
  4-Iodo-2-pentene   3-Iodo-1-pentene   1-Iodo-2-pentene
  (1,2- and 1,4-adduct)   (1,2-adduct)   (1,4-adduct)

Problem 8.30 Assign numbers from 1 for LEAST to 5 for MOST to indicate the relative reactivity on HBr addition to the following compounds:

(a) $H_2C=CH-CH_2CH_3$ (b) $CH_3-CH=CH-CH_3$ (c) $H_2C=CH-CH=CH_2$

(d) $CH_3-CH=CH-CH=CH_2$ (e) $H_2C=\overset{\overset{\displaystyle CH_3}{|}}{C}-\underset{\underset{\displaystyle CH_3}{|}}{C}=CH_2$ ◄

Conjugated dienes form the more stable allyl R^+'s and therefore are more reactive than alkenes. Alkyl groups on the unsaturated C's increase reactivity. Relative reactivities are: (a) 1, (b) 2, (c) 3, (d) 4, (e) 5.

Problem 8.31 For the reaction of propyne with (a) HOBr, (b) Br_2 + NaOH, give the structures of the products and the mechanisms of their formation. ◄

(a)
$$H-O-Br + H^+ \rightleftharpoons H-\overset{\overset{\displaystyle H}{|}}{\underset{+}{O}}-Br$$

$CH_3-C{\equiv}C-H + Br-\overset{\overset{\displaystyle H}{|}}{\underset{\underset{\displaystyle H}{|}}{O}} \longrightarrow CH_3-\overset{+}{C}=CH + OH_2 \xrightarrow{-H^+}$
                                                            $|$
                                                           Br

$\left[\begin{array}{c} CH_3-C=CH \\ | \qquad | \\ O \quad Br \\ | \\ H \end{array}\right] \xrightarrow{\text{tautomerize}} CH_3-\overset{}{\underset{\underset{\displaystyle O}{\|}}{C}}-\overset{\overset{\displaystyle H}{|}}{\underset{\underset{\displaystyle H}{|}}{C}}Br$

                                                            Bromoacetone

(b) Propyne reacts with strong bases to form a nucleophilic carbanion which displaces $:\ddot{B}r:$ from Br_2 by attacking $\overset{\delta+}{Br}$ to form 1-bromopropyne.

$$CH_3-C\equiv C:H + :\ddot{O}:H^- \rightleftharpoons H_2O + CH_3-C\equiv C:^- \xrightarrow{:\ddot{B}r:\ddot{B}r:} CH_3-C\equiv C:\ddot{B}r: + :\ddot{B}r:^-$$

1-Bromopropyne

Problem 8.32 $^{14}CH_3CH=CH_2$ is subjected to allylic free-radical bromination. Will the reaction product be exclusively labeled $H_2C=CH^{14}CH_2Br$? Explain. ◄

No. The product consists of an equal number of $H_2C=CH^{14}CH_2Br$ and $^{14}CH_2=CHCH_2Br$ molecules. H-abstraction produces a resonance hybrid of two contributing structures having both ^{12}C and ^{14}C as equally reactive, free-radical sites that attack Br_2.

$$Br\cdot + {}^{14}CH_3CH=CH_2 \longrightarrow HBr + {}^{14}\dot{C}H_2CH=CH_2$$

$$\updownarrow$$

$${}^{14}CH_2=CH\dot{C}H_2$$

$$\xrightarrow{Br_2} \begin{cases} Br^{14}CH_2CH=CH_2 \\ + \\ {}^{14}CH_2=CHCH_2Br \end{cases} + Br\cdot$$

Problem 8.33 (a) Write a schematic structure for the mer of the polymer from head-to-tail reaction of 2-methyl-1,3-butadiene. (b) Account for this orientation in polymerization. (c) Show how the structure is deduced from the product

$$CH_3-\overset{\overset{\displaystyle O}{\|}}{C}-CH_2-CH_2-CH=O$$

obtained from ozonolysis of the polymer. ◄

(a) 1,4-Addition with regular head-to-tail orientation produces a polymer with the following repeating unit (mer):

$$-CH_2-\overset{\overset{\displaystyle CH_3}{|}}{C}=CH-CH_2-$$

(b) This orientation results from more rapid formation of the more stable intermediate free radical.

$$R\cdot + CH_2=\overset{\overset{\displaystyle CH_3}{|}}{C}-CH=CH_2 \longrightarrow R:CH_2\overset{\overset{\displaystyle CH_3}{|}}{C}-CH-CH_2 \ (3°) \longleftrightarrow RCH_2\overset{\overset{\displaystyle CH_3}{|}}{C}=CH\dot{C}H_2 \ (1°)$$

$$[RCH_2\overset{\overset{\displaystyle CH_3}{|}}{C}=CH-CH_2]\cdot + CH_2=\overset{\overset{\displaystyle CH_3}{|}}{C}-CH=CH_2 \longrightarrow [RCH_2\underset{mer}{\underbrace{\overset{\overset{\displaystyle CH_3}{|}}{C}=CHCH_2}}-CH_2-\overset{\overset{\displaystyle CH_3}{|}}{C}=CH=CH_2]\cdot$$

head   tail   formed bond

The 1° allylic site is more reactive than the 3° allylic site. Attack at the other terminal $=CH_2$ gives the less stable free radical.

$$CH_2=\overset{\overset{\displaystyle CH_3}{|}}{C}-CH=CH_2 + \cdot R \longrightarrow CH_2=\overset{\overset{\displaystyle CH_3}{|}}{C}-\dot{C}HCH_2R \ (2°) \longleftrightarrow \dot{C}H_2-\overset{\overset{\displaystyle CH_3}{|}}{C}=CHCH_2R \ (1°)$$

(c) Write the ozonolysis products with the O's pointing at each other. Now erase the O's and join the C's by a double bond.

$$=\overset{\overset{\displaystyle CH_3}{|}}{C}-\underset{\text{piece of mer}}{\underbrace{CH_2CH_2}}-\underset{\text{mer}}{\underbrace{CH=\overset{\overset{\displaystyle CH_3}{|}}{C}-CH_2}}-\underset{\text{mer}}{\underbrace{CH_2-CH=\overset{\overset{\displaystyle CH_3}{|}}{C}-CH_2}}-\underset{\text{piece of next mer}}{\underbrace{CH_2-CH=}} \xrightarrow[\text{2. Zn, H}_2\text{O}]{\text{1. O}_3}$$

$$O=\overset{\overset{\displaystyle CH_3}{|}}{C}-CH_2CH_2CH=O + O=\overset{\overset{\displaystyle CH_3}{|}}{C}-CH_2CH_2CH=O + O=\overset{\overset{\displaystyle CH_3}{|}}{C}-CH_2CH_2CH=O$$

Problem 8.34 (a) Calculate the heat of hydrogenation, ΔH_h, of acetylene to ethylene if the ΔH_h's to ethane are -32.8 kcal/mole for ethylene and -75.0 kcal/mole for acetylene. (b) Use these data to compare the ease of hydrogenation of acetylene to ethylene with that of ethylene to ethane. ◄

(a) Write the reaction as the algebraic sum of two other reactions whose terms cancel out to give wanted reactants, products and enthalpy. These are the hydrogenation of acetylene to ethane and the dehydrogenation of ethane to ethylene (reverse of hydrogenation of $H_2C\!=\!CH_2$).

$$
\begin{array}{lll}
\text{(Eq. 1)} & H\!-\!C\!\equiv\!C\!-\!H + 2H_2 \longrightarrow CH_3\!-\!CH_3 & -75.0 \text{ kcal/mole} \\
\text{(Eq. 2)} & CH_3\!-\!CH_3 \longrightarrow H_2C\!=\!CH_2 + H_2 & +32.8 \text{ kcal/mole} \\
\hline
& H\!-\!C\!\equiv\!C\!-\!H + H_2 \longrightarrow H_2C\!=\!CH_2 & -42.2 \text{ kcal/mole}
\end{array}
$$

Equation 2 (dehydrogenation) is the reverse of hydrogenation ($\Delta H_h = -32.8$ kcal/mole). Hence the ΔH_h of Eq. 2 has a + value.

(b) Acetylene is less stable thermodynamically relative to ethylene than ethylene is to ethane because ΔH_h for acetylene → ethylene is -42.2 kcal/mole, while for ethylene → ethane it is -32.8 kcal/mole. Therefore acetylene is more easily hydrogenated and the process can be stopped at the ethylene stage. In general, hydrogenation of alkynes can be stopped at the alkene stage.

Problem 8.35 Deduce the structural formula of a compound of molecular formula C_6H_{10} which adds 2 moles of H_2 to form 2-methylpentane, forms a carbonyl compound in aqueous H_2SO_4-$HgSO_4$ solution and does not react with ammoniacal $AgNO_3$ solution, $[Ag(NH_3)_2]^+NO_3^-$. ◄

There are two degrees of unsaturation since the compound C_6H_{10} lacks four H's from being an alkane. The addition of 2 moles of H_2 excludes a cyclic compound. It may be either a diene or an alkyne, and the latter functional group is established by hydration to a carbonyl compound. The skeleton must be

$$
\begin{array}{c}
C \\
| \\
C\!-\!C\!-\!C\!-\!C\!-\!C
\end{array}
$$

as established by the reduction product. The two possible alkynes with this skeleton are

$$(CH_3)_2CHCH_2C\!\equiv\!CH \quad \text{and} \quad (CH_3)_2CH\!-\!C\!\equiv\!C\!-\!CH_3$$

The negative test for a 1-alkyne with Ag^+ establishes the second structure, 4-methyl-2-pentyne.

Problem 8.36 Set up a table to show convenient chemical reactions, and the visible sign of these reactions, which will differentiate n-pentane, 1-pentene and 1-pentyne. ◄

See Table 8-2. Positive and negative tests are indicated by + and − signs.

Table 8-2

Compound \ Test	Decolorization of Br_2 (Red) in CCl_4	White Precipitate with $Ag(NH_3)_2^+$
n-Pentane	−	−
1-Pentene	+	−
1-Pentyne	+	+

Problem 8.37 What is the structure of a hydrocarbon of molecular formula C_8H_{12} if 1 mole adds 2 moles of H_2 and also undergoes reductive ozonolysis to 2 moles of $O\!=\!CHCH_2CH_2CH\!=\!O$? ◄

The compound C_8H_{12} lacks 6 H's from being an alkane, C_8H_{18}. It has 3 degrees of unsaturation; two are in multiple bonds, as shown by the uptake of two moles of H_2, and one is ascribed to a ring. The cyclic compound has either 1 triple or 2 double bonds. Since a four-carbon dicarbonyl compound is formed on ozonolysis, the hydrocarbon (a) cannot be a cycloalkyne because a dicarboxylic acid would be formed, (b) must have both olefin bonds as part of the ring and (c) must be a symmetrical cyclooctadiene with two CH_2 groups separating the $C=C$ groups. The compound is 1,5-cyclooctadiene.

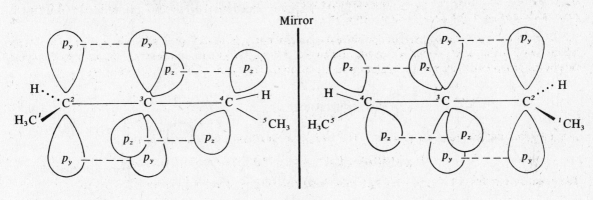

Cyclooctane   1,5-Cyclooctadiene

Problem 8.38 The allene 2,3-pentadiene ($\overset{1}{C}H_3\overset{2}{C}H=\overset{3}{C}=\overset{4}{C}H\overset{5}{C}H_3$) does not have a chiral C but is resolved into enantiomers. Draw an orbital picture that accounts for the chirality [see Problem 5.18(d)]. ◄

C^3 is sp hybridized and forms two σ bonds by sp-sp^2 overlap with the orbitals of C^2 and C^4. The two remaining p orbitals of C^3 form two π bonds, one with C^2 and one with C^4. These π bonds are at right angles to each other. The H and CH_3 on C^2 are in a plane at right angles to the plane of the H and CH_3 on C^4.

Because there is no free rotation about the two π bonds, the two H's and two CH_3's have a fixed spatial relationship. Whenever the two substituents on C^2 are different and the two substituents on C^4 are different, the molecule lacks symmetry and is chiral.

Fig. 8-5

Problem 8.39 Heating C_4H_9Br (A) with alcoholic KOH forms an alkene, C_4H_8 (B), which reacts with bromine to give $C_4H_8Br_2$ (C). (C) is transformed by KNH_2 to a gas, C_4H_6 (D), which forms a precipitate when passed through ammoniacal CuCl. Give the structures of compounds (A) through (D). ◄

The precipitate with ammoniacal CuCl indicates that (D) is a 1-alkyne, which can only be 1-butyne. The reactions and compounds are:

$$H-C{\equiv}CCH_2CH_3 \overset{KNH_2}{\longleftarrow} BrCH_2CHBrCH_2CH_3 \overset{Br_2}{\longleftarrow} H_2C{=}CHCH_2CH_3 \overset{alc.}{\underset{KOH}{\longleftarrow}} BrCH_2CH_2CH_2CH_3$$

(D)   (C)   (B)   (A)

(A) cannot be $CH_3CHBrCH_2CH_3$, which would give mainly $H_3CCH{=}CHCH_3$ and finally $CH_3C{\equiv}CCH_3$.

Problem 8.40 $H_2C{=}CHCH_2C{\equiv}CH$ (A) adds HBr to give $H_3CCHBrCH_2C{\equiv}CH$. $HC{\equiv}CCH{=}CH_2$ (B) adds HBr to give $H_2C{=}CBrCH{=}CH_2$. Explain. ◄

An isolated double bond is more reactive than an isolated triple bond in electrophilic additions, which accounts for the behavior of (A). If HBr had added to the double bond in (B), a butyne would have resulted. By HBr adding as it does, a more stable conjugated diene is formed.

Problem 8.41 In butadiene one can consider 3 individual pairs of p orbital overlaps, C^1-C^2, C^2-C^3 and C^3-C^4. In terms of these pairings, justify the assignment of the relative energies to the four molecular orbitals of 1,3-butadiene. See Problem 8.18(b). ◄

π_1: 3 bonding situations: ++, ++, ++; lowest energy.
π_2: 2 bonding, 1 antibonding: ++, +−, ++; next-lowest energy.
π_3: 1 bonding, 2 antibonding: +−, −−, −+; next-to-highest energy.
π_4: 3 antibonding: +−, −+, +−; highest energy.

Problem 8.42 (a) The allyl system (carbonium ion, free radical, or carbanion) has three overlapping p orbitals and hence has three π molecular orbitals. In terms of the signs of the overlapping p's, indicate the relative energies of the molecular orbitals and state if they are bonding, antibonding or nonbonding. A nonbonding MO is one whose energy is the same as the individual noninteracting atomic p orbitals'. Such an orbital can be recognized if the number of bonding pairs equals the number of antibonding pairs (see Problem 8.41) or if there is no overlap. (b) Insert the electrons for $C_3H_5^+$, $C_3H_5\cdot$, and $C_3H_5^-$. ◄

(a) See Fig. 8-6. A zero means that an atomic orbital is at the node point and therefore does not overlap with a p orbital at either side.

$$\pi_2\uparrow \qquad\qquad \pi_2\,\text{⇅}$$

(b) R^+ $\pi_1\,\text{⇅}$ $R\cdot$ $\pi_1\,\text{⇅}$ $R^{\bar{}}$ $\pi_1\,\text{⇅}$

The electrons in the π_2 orbital do not appreciably affect the stability of the species. Therefore all three species are more stable than the corresponding alkyl systems $C_3H_7^+$, $C_3H_7\cdot$, and $CH_3H_7^{\bar{}}$. The extra electrons do increase the repulsive forces between electrons slightly, so the order of stability is $C_3H_5^+ > C_3H_5\cdot > C_3H_5^-$.

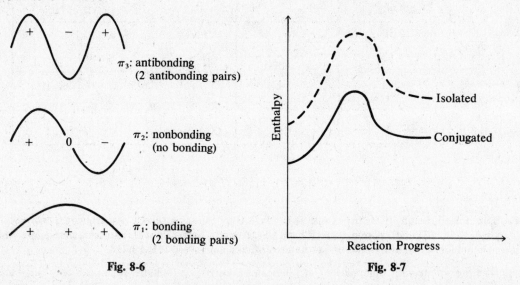

π_3: antibonding (2 antibonding pairs)

π_2: nonbonding (no bonding)

π_1: bonding (2 bonding pairs)

Fig. 8-6   Fig. 8-7

Problem 8.43 Is the fact that conjugated dienes are more stable and more reactive than isolated dienes an incongruity? ◄

No. Reactivity depends on the relative $\Delta H^{\ddagger}$ values. Although the ground-state enthalpy for the conjugated diene is lower than that of the isolated diene, the transition-state enthalpy for the conjugated system is lower by a greater amount (see Fig. 8-7).

$$\Delta H^{\ddagger} \text{ conjugated} < \Delta H^{\ddagger} \text{ isolated} \quad \text{and} \quad \text{rate}_{\text{conjugated}} > \text{rate}_{\text{isolated}}$$

Problem 8.44 Explain why 1,3-butadiene and O_2 do not react unless irradiated by uv light to give the 1,4-adduct ◄

Ordinary ground-state O_2 is a diradical,

$$:\ddot{O}—\ddot{O}:$$
$$\uparrow \quad \uparrow$$

One bond could form, but the intermediate has 2 electrons with the same spin and a second bond cannot form.

$$O—O + H_2C \overset{\text{⇈}}{—} CH—CH \overset{\text{⇅}}{—} CH_2 \longrightarrow O—O—CH_2—CH=CH—CH_2$$
$$\uparrow \quad \uparrow \qquad\qquad \uparrow \qquad\qquad\qquad\qquad\qquad \uparrow \qquad\qquad\qquad\qquad \uparrow$$

When irradiated, O_2 is excited to the singlet spin-paired state

$$:\ddot{O}—\ddot{O}:$$
$$\text{⇅}$$

Singlet O_2 reacts by a concerted mechanism to give the product.

Problem 8.45 (*a*) Relate the observed C—H and C—C bond lengths and bond energies given in Table 8-3 in terms of the hybrid orbitals used by the C's involved. (*b*) Predict the relative C—C bond lengths in CH_3CH_3, $CH_2=CH—CH=CH_2$, and $H—C\equiv C—C\equiv C—H$.

Table 8-3

Compound	Bond	Bond Length, Å	Bond Energy, kcal/mole
(1) $CH_3—CH_3$	—C—H	1.10	98
(2) $CH_2=CH_2$	=C—H	1.08	101
(3) $H—C\equiv C—H$	$\equiv$C—H	1.06	110
(4) $CH_3—CH_3$	C—C—	1.54	85
(5) $CH_3—CH=CH_2$	C—C=	1.51	90
(6) $CH_3—C\equiv C—H$	C—C$\equiv$	1.46	101

Bond energy increases as bond length decreases; the shorter bond length makes for greater orbital overlap and a stronger bond.

(*a*) The hybrid nature of C is: (1) C_{sp^3}-H_s, (2) C_{sp^2}-H_s, (3) C_{sp^3}-H_s, (4) C_{sp^3}-C_{sp^3}, (5) C_{sp^3}-C_{sp^2} and (6) C_{sp^3}-C_{sp}. In going from (1) to (3) the C—H bond length decreases as the *s* character of the hybrid orbital used by C increases. The same situation prevails for the C—C bond in going from (4) to (6). Bonds to C therefore become shorter as the *s* character of the hybridized orbital used by C increases.

(*b*) The hybrid character of the C's in the C—C bond is: for $CH_3—CH_3$, C_{sp^3}-C_{sp^3}; $H_2C=CH—CH=CH_2$, C_{sp^2}-C_{sp^2}; and $H—C\equiv C—C\equiv C—H$, C_{sp}-C_{sp}. Bond length becomes shorter as *s* character increases and hence relative C—C bond lengths should decrease in the order

$$CH_3CH_3 > CH_2=CH—CH=CH_2 > HC\equiv C—C\equiv CH$$

The observed bond lengths are respectively 1.54 Å, 1.49 Å and 1.38 Å.

Problem 8.46 Is there any inconsistency between the facts that the C—H bond in acetylene has the greatest bond energy of all C—H bonds and that it is also the most acidic? ◄

No. Bond energy is a measure of homolytic cleavage, $\equiv C:H \longrightarrow \equiv C\cdot + \cdot H$. Acidity is due to a heterolytic cleavage, $\equiv C:H + Base \longrightarrow \equiv C:^- + H^+(Base)$.

Chapter 9

Alicyclic Compounds

9.1 NOMENCLATURE AND STRUCTURE

Alicyclic (<u>ali</u>phatic <u>cyclic</u>) compounds and their derivatives are named by combining the prefix **cyclo-** with the name of the alkane having the same number of C's as the ring.

Alicyclic compounds are often shown by the topological formulation (Problem 4.27). The cycle can also be designated as a substituent on a chain, as in Problem 9.2(*b*).

Cyclopentene	1,4-Dimethyl-cyclohexane	Ethylcycloheptane	1,3,5,7-Cyclooctatetraene

For bicyclic compounds, the prefix **bicyclo-** is combined with a pair of brackets enclosing numbers separated by periods, which is followed by a name showing the total number of atoms in the rings. The numbers in brackets indicate how many atoms are in each bridge, and are in order of decreasing ring size.

The inability of atoms in rings to rotate completely about their bonds leads to *cis-trans* (geometric) isomerism in alicyclic compounds, as shown in 1,2-dimethylcyclopropanes. The *cis* isomer has both Me's on the same side of the ring; the *trans* isomer has them on opposite sides.

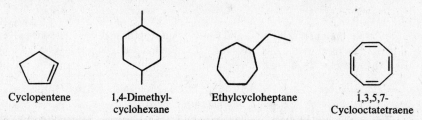

cis-1,2-Dimethylcyclopropane   *trans*-1,2-Dimethylcyclopropane

In topological notation the solid dot denotes an H projecting toward the viewer. The planar representation of polysubstituted alicyclic compounds is the best means for identifying geometric and optical isomers.

Cyclopropane and, to a lesser extent, cyclobutane are strained rings because the normal tetrahedral bond angle of 109.5° is constrained to 60° for the equilateral triangle and to 90° for the square structure (see Problem 9.13).

Problem 9.1 Draw topological structural formulas for (*a*) bromocycloheptane, (*b*) 1-ethylcyclo-pentene, (*c*) *trans*-1-chloro-2-bromocyclobutane, (*d*) bicyclo[3.1.0]hexane. ◄

(*a*)   (*b*)   (*c*)   (*d*)   (must be *cis*)

134

Problem 9.2 Name the compounds shown in Fig. 9-1.

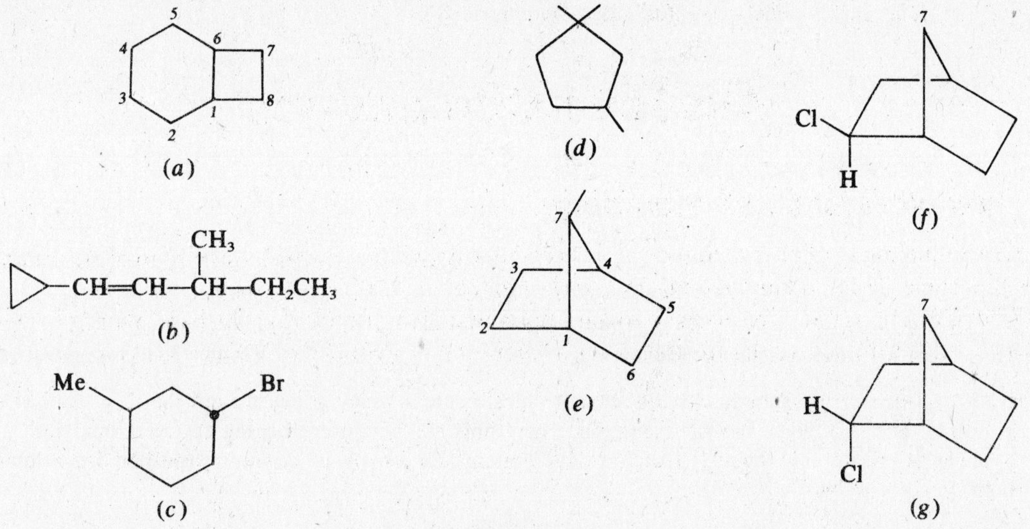

(a)

(b)

(c)

(d)

(e)

(f)

(g)

Fig. 9-1

(a) Bicyclo[4.2.0]octane.   (b) 1-Cyclopropyl-3-methyl-1-pentene.   (c) *trans*-3-Bromo-1-methylcyclo-hexane.   (d) 1,1,3-Trimethylcyclopentane.   (e) 7-Methylbicyclo[2.2.1]heptane   (7-Methylnorbornane). (f) *exo*-2-Chloronorbornane (Cl is *cis* to C^7 bridge, the shortest bridge).   (g) *endo*-2-Chloronorbornane (Cl is *trans* to C^7 bridge, the shortest bridge).

Problem 9.3 Use the solid-dot topological convention to indicate the geometric isomers, if any, for: (a) 1,1,2-trimethylcyclopropane, (b) 1,3-dimethylcyclobutane, (c) 1,2,3-trimethylcyclohexane, (d) 1,2,4-trimethylcyclopentane, (e) 1-methyl-2-propenylcyclopentane.

(a) None:

(b)

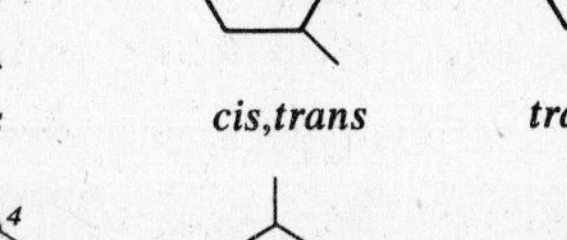

cis   trans

(c) To designate *cis,trans*, start at C^1 and go around the ring increasing the numbers.

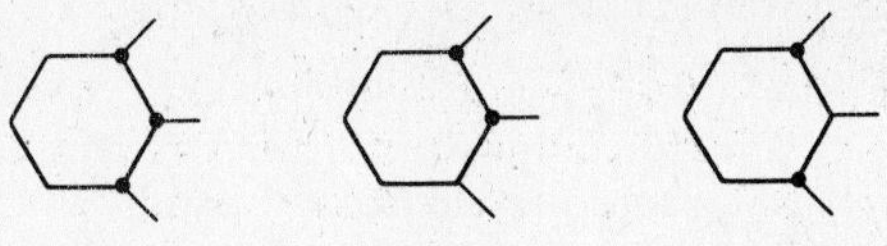

cis,cis   cis,trans   trans,trans

(d)

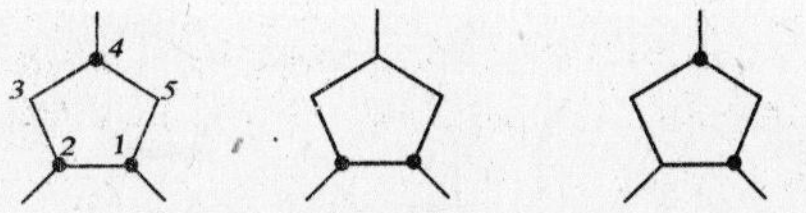

cis,cis   cis,trans   trans,trans

(e)

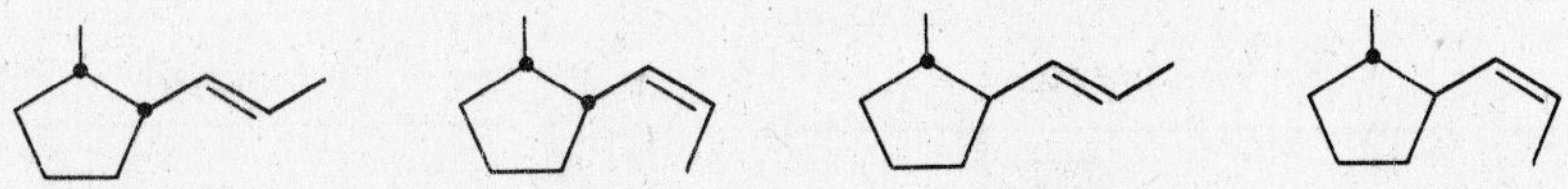

cis,trans   cis,cis   trans,trans   trans,cis

There is geometric isomerism on the ring as well as on the double bond.

Problem 9.4 Give the name, structural formula and stereochemical designation of the isomers of (*a*) bromochlorocyclobutane, (*b*) dichlorocyclobutane, (*c*) bromochlorocyclopentane, (*d*) diiodocyclopentane, (*e*) dimethylcyclohexane. Indicate asymmetric C's. ◀

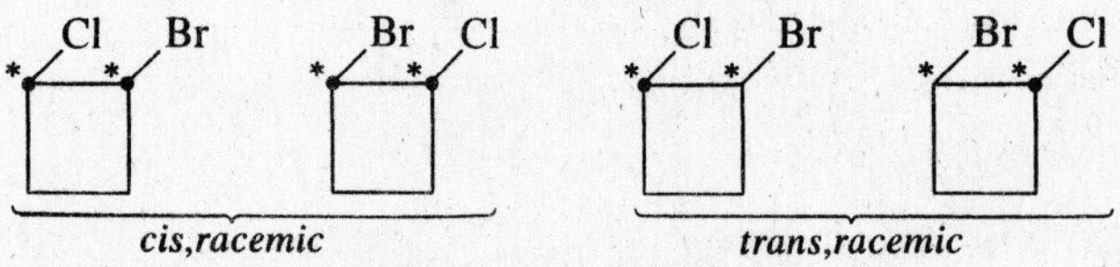

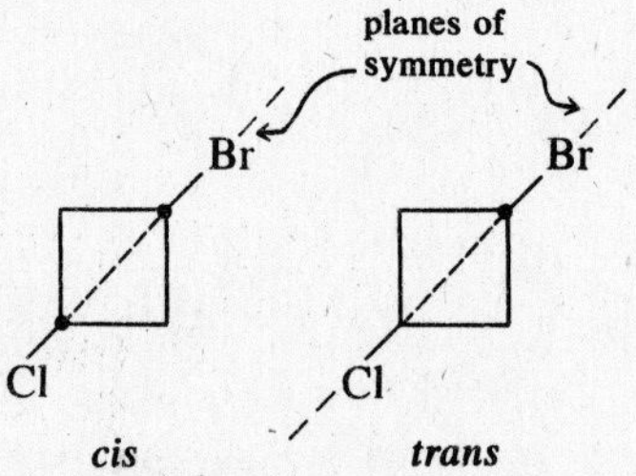

(*a*)

1-Bromo-1-chlorocyclobutane

cis,racemic *trans,racemic*

1-Bromo-2-chlorocyclobutane

cis *trans*

1-Bromo-3-chlorocyclobutane

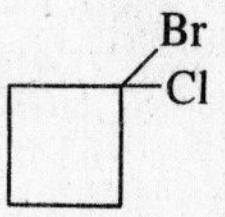

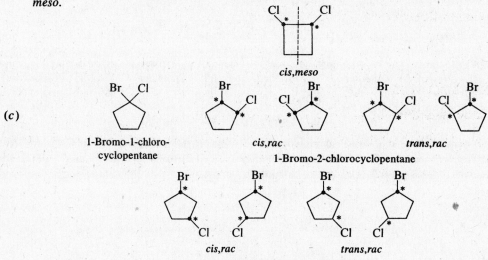

In the 1,3-isomer the substituted C's are not chiral centers, since going around the ring clockwise from either of these C's gives the same sequence of atoms as encountered going counterclockwise.

(*b*) Same as (*a*) except that the *cis*-1,2-dichlorocyclobutane has a plane of symmetry (dashed line below) and is *meso.*

cis,meso

(*c*)

1-Bromo-1-chloro-
cyclopentane

cis,rac *trans,rac*

1-Bromo-2-chlorocyclopentane

cis,rac *trans,rac*

1-Bromo-3-chlorocyclopentane

(*d*) The diiodo compound is similar to the bromochloro derivative except that both the *cis*-1,2- and the *cis*-1,3-diiodo derivatives are *meso.*

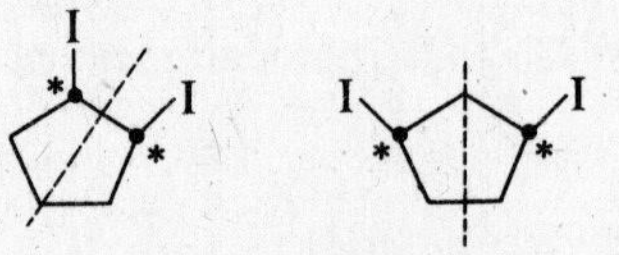

(*e*) Nine isomeric disubstituted cyclohexanes are similarly derived.

1,1-Dimethyl

1,2-Dimethyl

cis,meso *trans,rac*

1,3-Dimethyl

cis,meso *trans,rac*

1,4-Dimethyl

cis *trans*

No optical isomers

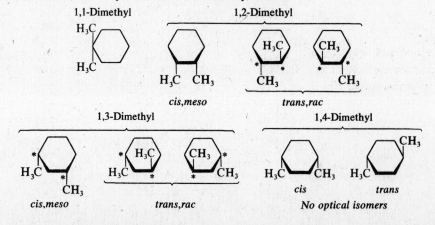

9.2 METHODS OF PREPARATION

FROM BENZENE DERIVATIVES (FOR CYCLOHEXANES)

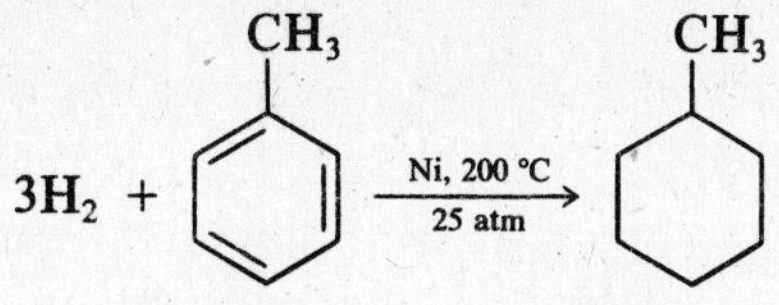

Toluene
from coal tar

Methylcyclohexane

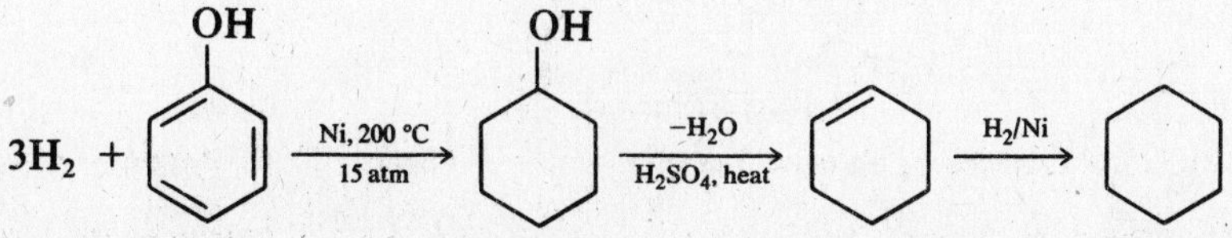

$3H_2 +$ ... $\xrightarrow[\text{25 atm}]{\text{Ni, 200 °C}}$ ...

$3H_2 +$ Phenol $\xrightarrow[\text{15 atm}]{\text{Ni, 200 °C}}$ Cyclohexanol $\xrightarrow[\text{H}_2\text{SO}_4,\ \text{heat}]{-\text{H}_2\text{O}}$ Cyclohexene $\xrightarrow{\text{H}_2/\text{Ni}}$ Cyclohexane

FROM OPEN-CHAIN COMPOUNDS

1. From Dihalogen Compounds (Freund Reaction)

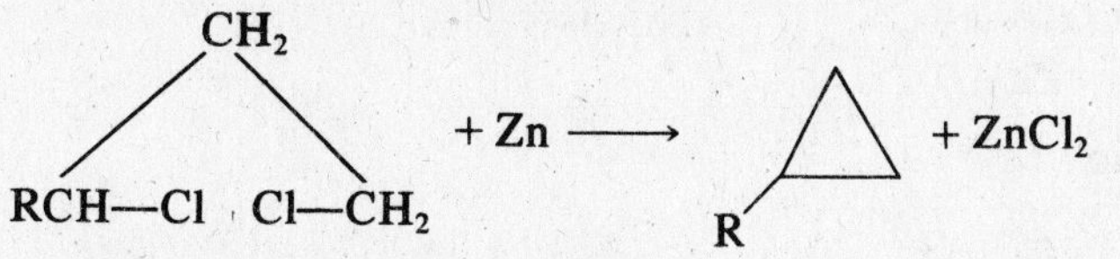

$+ Zn \longrightarrow$ $+ ZnCl_2$ (Best for cyclopropanes)

2. Cycloaddition Reactions of Alkenes

(*a*) [2 + 2].   Ultraviolet-light-catalyzed dimerization of alkenes yields cyclobutanes in one step.

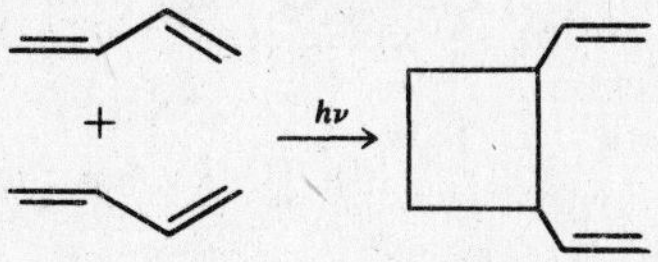

1,3-Butadiene

1,2-Divinylcyclobutane
(one of several products)

Thermal dimerizations occur especially with fluoroalkenes in two steps via an intermediate diradical.

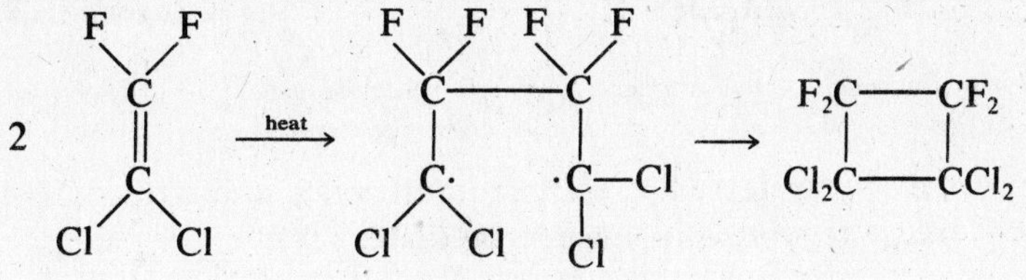

*more stable diradical
intermediate*

(*b*) [2 + 4] (the **Diels-Alder reaction**). A conjugated diene and an alkene form a cyclohexene. Reactive alkenes, called **dienophiles**, have electron-attracting groups on their unsaturated C's.

1,3-Butadiene
+
Ethylene

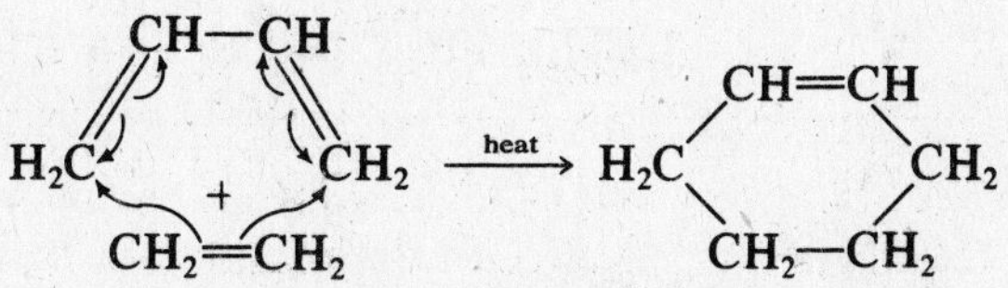

Cyclohexene

1,3-Butadiene
+
Acrylic aldehyde
a dienophile

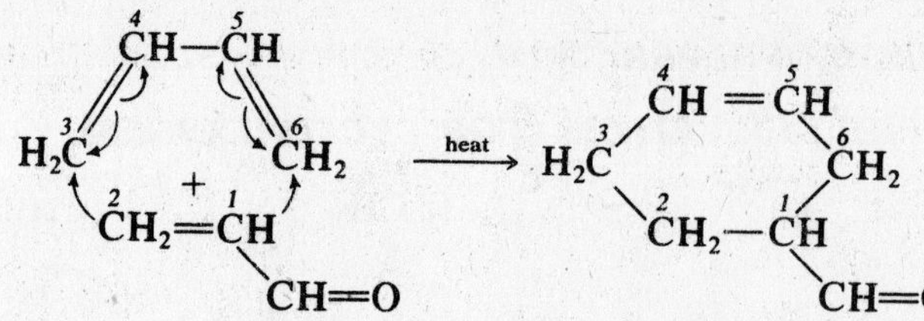
CH—CH
H$_2$C
CH$_2$
+
CH$_2$=CH
CH=O
heat
CH =CH
H$_2$C
CH$_2$
CH$_2$—CH
CH=O

1,2,3,6-Tetrahydrobenzaldehyde

3. Electrocyclic Reactions

These are intramolecular cyclizations of polyenes.

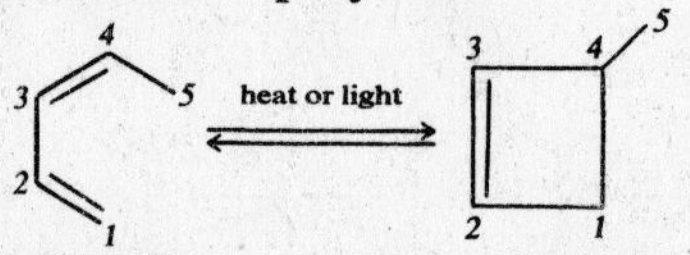
heat or light

1,3-Pentadiene (a diene) 3-Methylcyclobutene

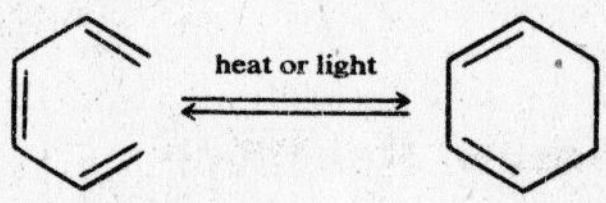
heat or light

1,3,5-Hexatriene (a triene) 1,3-Cyclohexadiene

4. Tetramerization of Acetylene

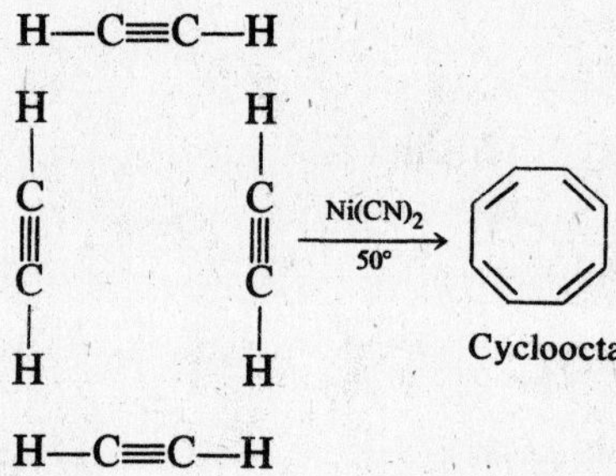
H—C≡C—H
Ni(CN)$_2$
50°
H—C≡C—H

(access to cyclooctane)

Cyclooctatetraene

5. Carbene Addition to Alkenes

$$H{:}\ddot{C}{:}H + H_3CCH{=}CH_2 \longrightarrow$$

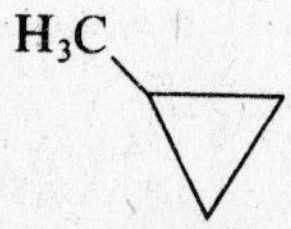
H$_3$C

+ Insertion Products

Carbene (See page 86.) Propylene Methylcyclopropane (See page 46.)

Methylene can be transferred directly from the reagent mixture, CH_2I_2 + Zn-Cu alloy, to the alkene without being generated as an intermediate.

$$CH_2I_2 + CH_3CH{=}CH_2 \xrightarrow[\text{Zn-Cu}]{}$$

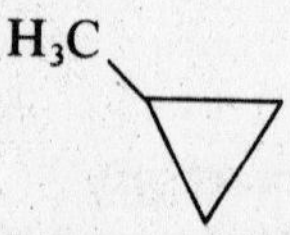
H$_3$C

+ ZnI_2 (no insertion)

Methylene diiodide Methylcyclopropane

$$[{:}CCl_2] + CH_3CH{=}CH_2 \longrightarrow$$

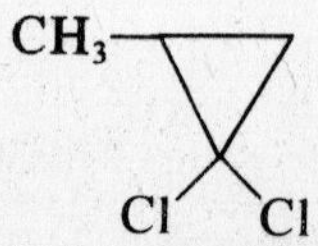
CH$_3$
Cl
Cl

Dichlorocarbene [see Problem 7.32(*c*)]

1,1-Dichloro-2-methylcyclopropane

Problem 9.5 Which dihalogen compounds can be used to prepare 1,1,2-trimethylcyclopropane by the Freund reaction? ◄

Each ring C—C bond in

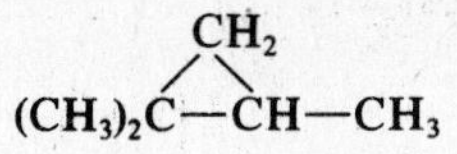

$$\underset{\text{CH}_2}{\overset{\text{CH}_2}{(\text{CH}_3)_2\text{C}-\text{CH}-\text{CH}_3}}$$

is different, and so the ring can be formed using three different starting materials.

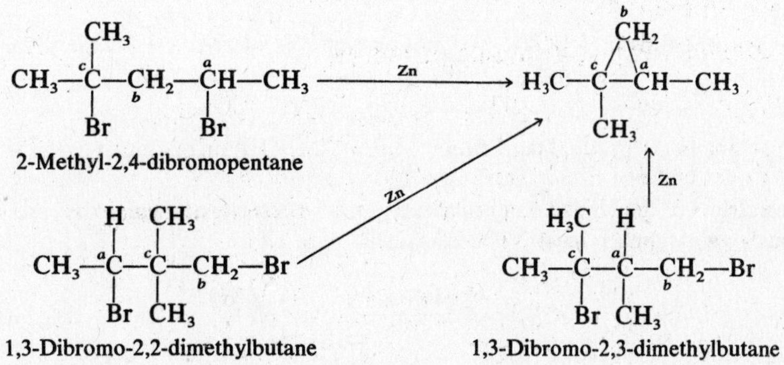

$$\underset{\substack{\text{Br}\quad\text{Br}\\ \text{2-Methyl-2,4-dibromopentane}}}{\text{CH}_3-\overset{c}{\underset{b}{\text{C}}}-\text{CH}_2-\overset{a}{\text{CH}}-\text{CH}_3} \xrightarrow{\text{Zn}} \underset{\substack{\text{CH}_3}}{\text{H}_3\text{C}-\overset{c}{\text{C}}\overset{b}{\underset{a}{\triangle}}\text{CH}-\text{CH}_3}$$

1,3-Dibromo-2,2-dimethylbutane
$$\text{CH}_3-\overset{a}{\text{C}}-\overset{c}{\text{C}}-\text{CH}_2-\text{Br}$$

1,3-Dibromo-2,3-dimethylbutane
$$\text{CH}_3-\overset{c}{\text{C}}-\overset{a}{\text{C}}-\text{CH}_2-\text{Br}$$

Problem 9.6 Outline a synthesis of the following alicyclic compounds from acyclic compounds without using the Freund reaction.

(a) 1,1-Dimethylcyclopropane   (b) ⬡—CH=CH₂   (c) Cyclooctane   ◄

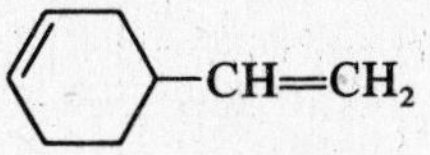

(a)
$$\text{CH}_3-\underset{\text{CH}_3}{\text{C}}=\text{CH}_2 + \text{CH}_2\text{I}_2 \xrightarrow{\text{Zn-Cu}} \triangle$$

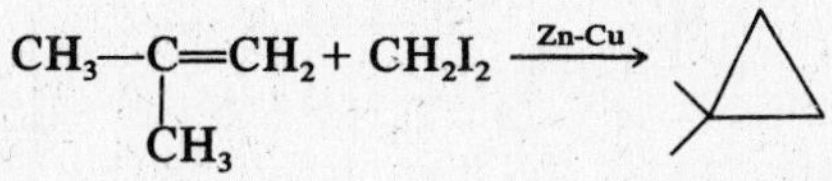

(b)
$$\underset{\text{CH}_2}{\text{HC}} + \underset{\text{CH}_2}{\overset{\text{H}}{\text{C}}}-\text{CH}=\text{CH}_2 \xrightarrow{\text{heat}} \text{⬡}-\text{CH}=\text{CH}_2$$

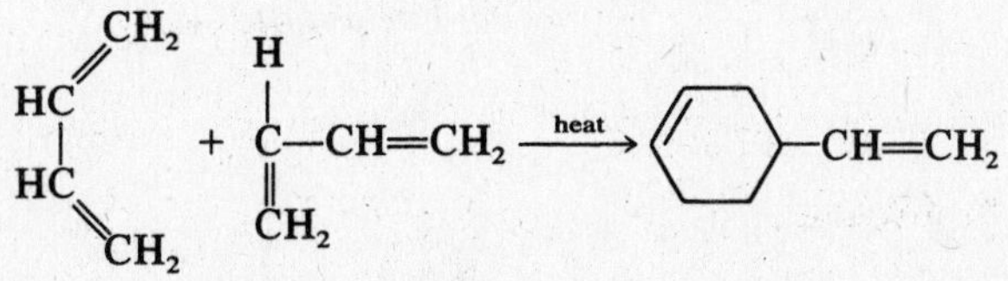

(c)
$$\text{Acetylene} \xrightarrow[\text{50°}]{\text{Ni(CN)}_2} \text{Cyclooctatetraene} \xrightarrow{\text{H}_2/\text{Pd}} \text{Cyclooctane}$$

Problem 9.7 Starting with cyclopentanol, show the reactions and reagents needed to prepare (a) cyclopentene, (b) 3-bromocyclopentane, (c) 1,3-cyclopentadiene, (d) *trans*-1,2-dibromocyclopentene, (e) cyclopentane. ◄

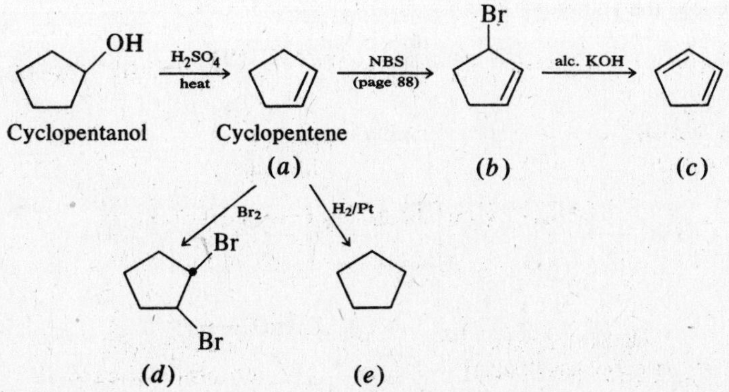

Cyclopentanol $\xrightarrow[\text{heat}]{\text{H}_2\text{SO}_4}$ Cyclopentene (a) $\xrightarrow[\text{(page 88)}]{\text{NBS}}$ (b) $\xrightarrow{\text{alc. KOH}}$ (c)

(d) Br₂   (e) H₂/Pt

Problem 9.8 (*a*) Use the following observations to discuss stereospecificity of carbene addition. (*b*) Suggest mechanisms for addition of (i) singlet and (ii) triplet carbenes to alkenes.

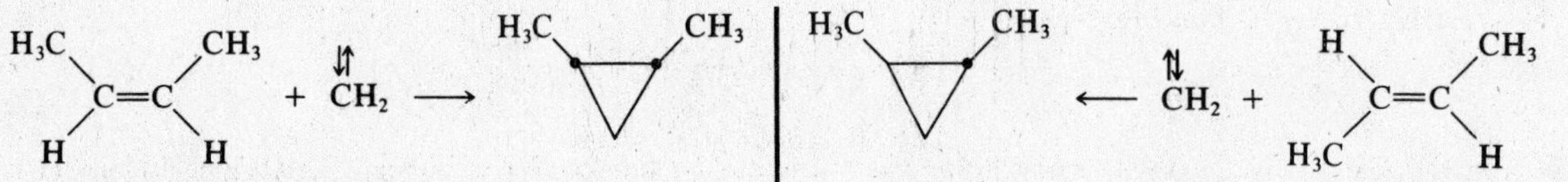

cis-2-Butene (singlet) cis trans (singlet) trans-2-Butene

cis- or trans-2-Butene + ↑ CH₂ ⟶ mixture of cis- and trans-1,2-Dimethylcyclopropane ◀
(triplet)

(*a*) Since *cis* reactant → *cis* product, and *trans* → *trans*, the addition of singlet carbene is stereospecific. The addition of triplet carbene is *not* stereospecific.

(*b*) (i) The π electrons of the alkene and unshared pair of electrons of singlet CH_2 are properly paired to form simultaneously two sigma bonds. The transition state is

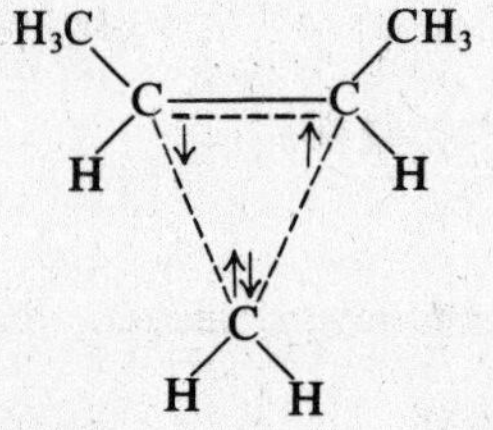

(ii) The electrons are not matched when triplet carbene adds. The sequence of steps is:

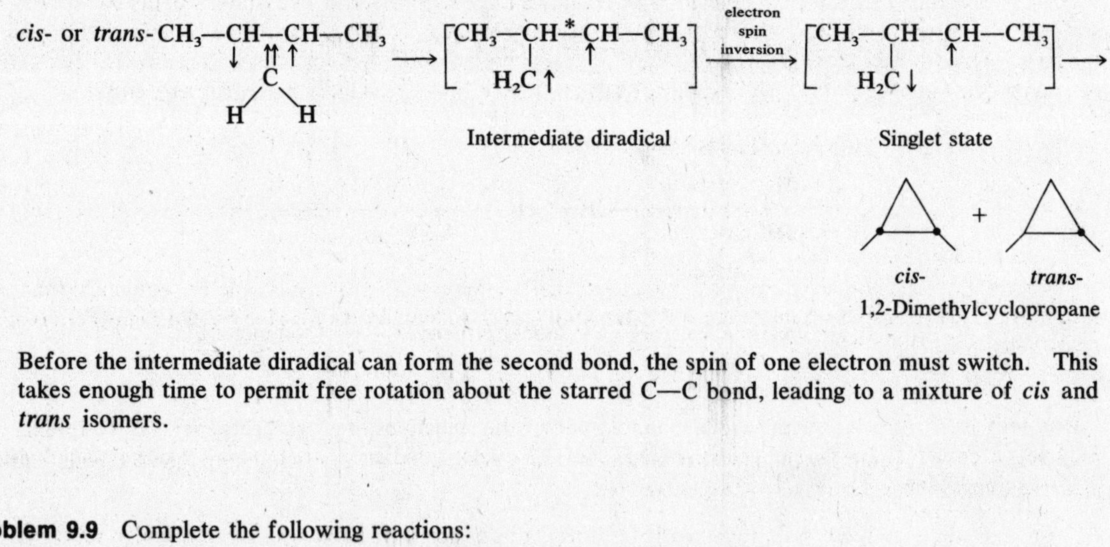

Intermediate diradical   Singlet state

cis-   trans-
1,2-Dimethylcyclopropane

Before the intermediate diradical can form the second bond, the spin of one electron must switch. This takes enough time to permit free rotation about the starred C—C bond, leading to a mixture of *cis* and *trans* isomers.

Problem 9.9 Complete the following reactions:

(*a*) 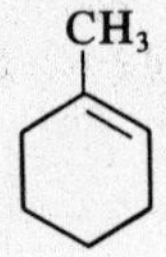   + HBr ⟶   (*b*) ozonolysis of 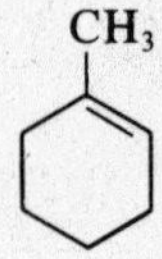   (*c*)   + H₂ $\xrightarrow[\text{Ni}]{120\ °C}$

(*d*)   + Br₂ ⟶   (*e*)   + H₂ $\xrightarrow[200\ °C]{\text{Ni}}$ (*f*)   + H₂ $\xrightarrow[200\ °C]{\text{Ni}}$

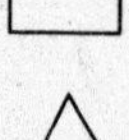

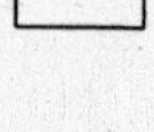

(*g*)   + KMnO₄ ⟶   ◀

Cycloalkenes behave chemically like alkenes.

(a) (a Markovnikov addition)   (b)

(c) $CH_3CH_2CH_3$. Under these conditions the strained 3-membered ring opens. (d) $BrCH_2CH_2CH_2Br$. Again, the 3-membered ring opens. (e) $CH_3CH_2CH_2CH_3$. The strained 4-membered ring opens, but a higher temperature is needed than in part (c). (f) No reaction. The 5-membered ring has no ring strain. (g) No reaction. Even the strained rings are stable toward oxidation.

Problem 9.10 The heat of combustion per —CH_2— unit in alkanes and in cyclopentane and larger rings is approximately 157.5 kcal/mole. For cyclopropane and cyclobutane the values are 166.6 and 164.0 respectively. Explain these data.   ◄

The higher the heat of combustion, the more unstable the substance (Problem 6.10). Cyclopropane has the highest value because it has the greatest ring strain. Cyclobutane has the next-highest value; it is less strained than cyclopropane but more strained than cyclopentane. All other cycloalkanes have approximately the same value as alkanes because they are relatively strain-free.

Problem 9.11 Cycloalkanes with more than 6 C's are difficult to synthesize by intramolecular ring closures, yet they are stable. On the other hand, cyclopropanes are synthesized this way, yet they are the least stable cycloalkanes. Are these facts incompatible? Explain.   ◄

No. The relative ease of synthesis of cycloalkanes by intramolecular cyclizations depends on both ring stability and the probability of bringing the two ends of the chain together to form a C-to-C bond, thereby closing the ring. This probability is greatest for smallest rings and decreases with increasing ring size. The interplay of ring stability and this probability factor are summarized below (numbers represent ring sizes).

Probability of Ring Closure	$3 > 4 > 5 > 6 > 7 > 8 > 9$
Thermal Stability	$6 > 7, 5 > 8, 9 \geqslant 4 > 3$
Ease of Synthesis	$5 > 3, 6 > 4, 7, 8, 9$

The high yield of cyclopropane indicates that a favorable probability factor outweighs the ring instability. For rings with more than 6 C's the ring stability effect is outweighed by the highly unfavorable probability factor.

Problem 9.12 Account for the fact that intramolecular cyclizations of rings with more than 6 C's are effected at extremely low concentrations (**Ziegler method**).   ◄

Chains can also react intermolecularly to form longer chains. Although intramolecular reactions are ordinarily faster than intermolecular reactions, the opposite is true in the reaction of chains leading to rings with more than 6 C's. This side reaction from collisions between different chains is minimized by carrying out the reaction in extremely dilute solutions.

Very Dilute Solution   Concentrated Solution

$(CH_2)_n$ | $\xleftarrow{\text{-AB}}$ $ACH_2(CH_2)_nCH_2B$ $\xrightarrow{\text{-AB}}$ $ACH_2(CH_2)_nCH_2$—$CH_2(CH_2)_nCH_2B$

Ring closure   Chain lengthening

Problem 9.13   Account for the ring strain of cyclopropane in terms of orbital overlap.   ◄

The strongest chemical bonds are formed by the greatest overlap of atomic orbitals.   For sigma bonding, maximum overlap is achieved when the orbitals point directly at each other along the bond axis, as in Fig. 9-2(*a*).   This type of overlap in cyclopropane could not lead to ring closure for sp^3 hybridized C's because it would demand bond angles of 109.5°.   Hence the overlap must be off the bond axis to give a *bent* bond, as shown in Fig. 9-2(*b*).

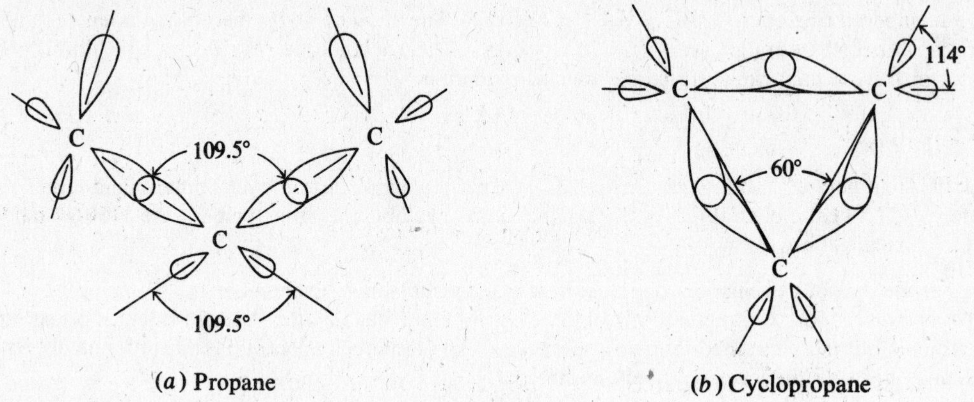

(*a*) Propane   (*b*) Cyclopropane

Fig. 9-2

In order to minimize the angle strain the C's assume more *p* character in the orbitals forming the ring and more *s* character in the external bonds, in this case the C—H bonds.   Additional *p* character narrows the expected angle, while more *s* character expands the angle.   The observed H—C—H bond angle of 114° confirms this suggestion.   Clearly, there are deviations from pure *p*, *sp*, sp^3 and sp^2 hybridizations.   The situation in cyclobutane is similar but not so pronounced.

9.3   CONFORMATIONS OF CYCLOALKANES

Cyclohexane maintains normal sp^3 tetrahedral bond angles of 109° by being puckered rather than flat (a planar structure has internal angles of 120°).   The two extreme conformations are the more stable **chair** and the less stable **boat**.   The **twist-boat** conformer is less stable than the chair by about 5.5 kcal/mole but is more stable than the boat.   It is formed from the boat by moving one "flagpole" to the left and the other to the right.   See Fig. 9-3.

Problem 9.14   In terms of eclipsing interactions explain why (*a*) the chair conformation is more stable than the boat conformation and (*b*) the twist-boat conformation is more stable than the boat conformation.   ◄

(*a*)   The H's on C^2—C^3 and C^5—C^6 ("gunwale" H's) in the boat form are eclipsing.   (This kind of strain is called **torsional**.)   Additional strain arises from the crowding of the "flagpole" H's on C^1 and C^4, as indicated in Fig. 9-3.   (This is an example of **steric** strain.)   All H's in the chair form are staggered.   The C—C bonds in the chair form are *gauche*.   (See C^1—C^6 and C^5—C^4 in the Newman projection structures.)

(*b*)   Twisting of the boat reduces eclipsing of "gunwale" H's and crowding of "flagpole" H's.

Problem 9.15   Account for the fact that (*a*) even though planar cyclopentane has stable internal bond angles of 109°, at any given instant one of the five C's is puckered out of the plane; (*b*) the C's of cyclobutane are not coplanar.   ◄

(*a*)   Planar cyclopentane has 5 pairs of eclipsed H's and has considerable torsional strain.   This strain is reduced at the expense of some increase in angle strain by puckering one CH_2 out of the plane of the ring (Fig. 9-4).   The puckering is not fixed but alternates around the ring.

(*b*)   The torsional strain from eclipsing of 4 pairs of H's causes the ring to fold.   Cyclobutane is an equilibrium mixture of two equivalent, nonplanar conformers (Fig. 9-5).

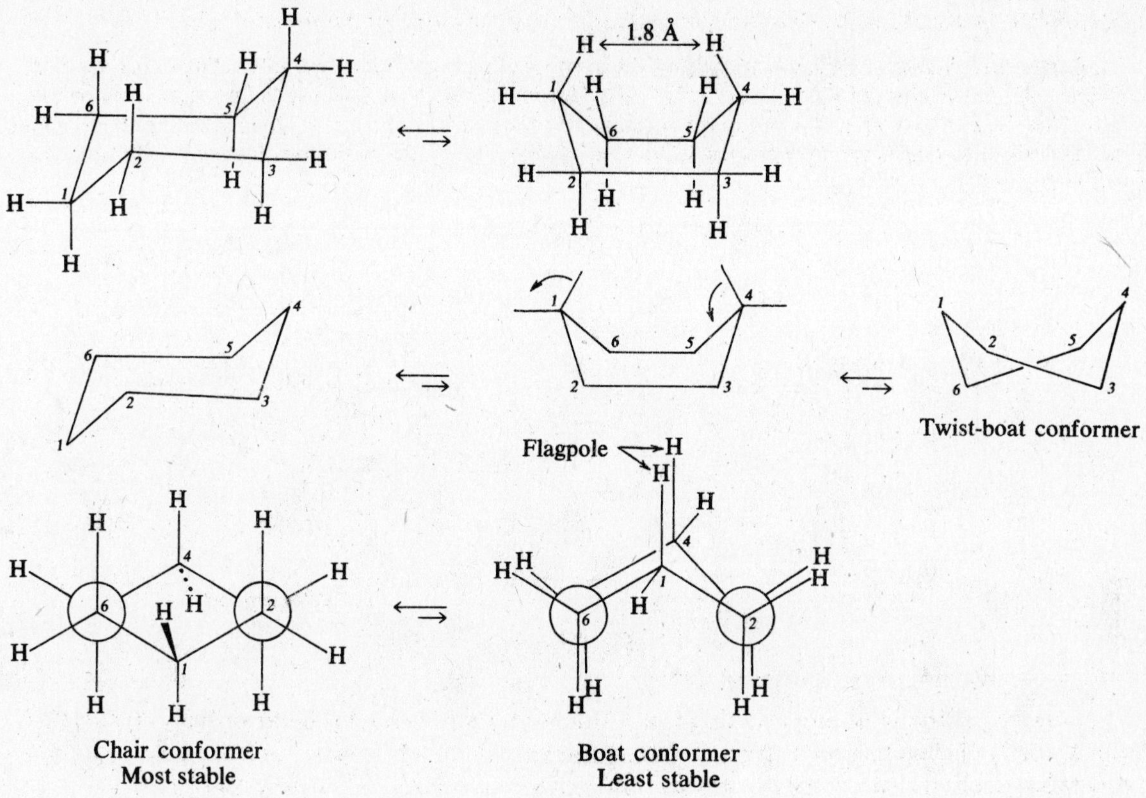

Fig. 9-3

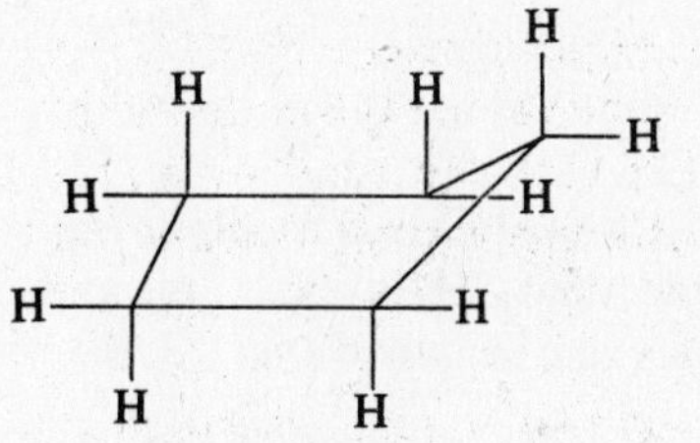

Fig. 9-4

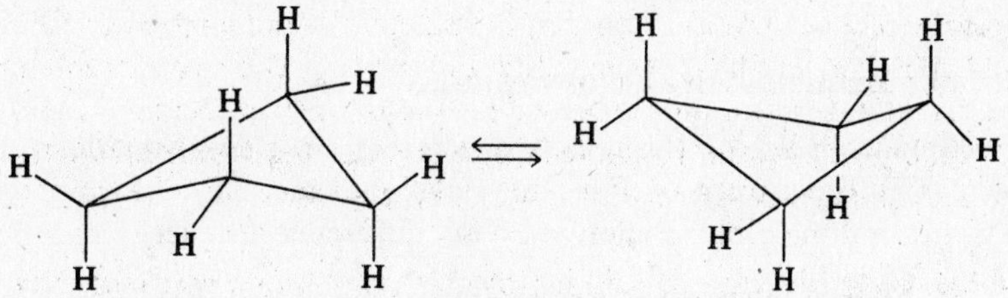

Fig. 9-5

SUBSTITUTED CYCLOHEXANES

1. Axial and Equatorial Bonds

Six of the twelve H's of cyclohexane are **equatorial** (e); they project from the ring, forming a belt around the ring perimeter as in Fig. 9-6(a). The other 6 H's, shown in Fig. 9-6(b), are **axial** (a); they are perpendicular to the plane of the ring and parallel to each other. Three of these axial H's on alternate C's extend up and the other 3 point down.

Converting one chair conformer to the other also changes the axial bonds, shown as heavy lines in Fig. 9-6(c), to equatorial bonds in Fig. 9-6(d). The equatorial bonds of Fig. 9-6(c) similarly become axial bonds in Fig. 9-6(d).

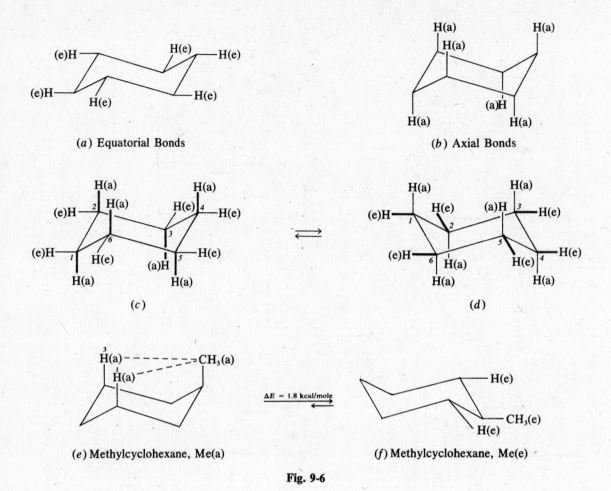

(a) Equatorial Bonds

(b) Axial Bonds

(c)   ⇌   (d)

(e) Methylcyclohexane, Me(a)

ΔE = 1.8 kcal/mole

(f) Methylcyclohexane, Me(e)

Fig. 9-6

2. Monosubstituted Cyclohexanes

Replacing H by CH_3 gives two different chair conformations; in Fig. 9-6(e) CH_3 is axial, in Fig. 9-6(f) CH_3 is equatorial.   For methylcyclohexane, the conformer with the axial CH_3 is less stable and has 1.8 kcal/mole more energy.   This difference in energy can be analyzed in two ways:

1,3-Diaxial interactions (transannular effect).   In Fig. 9-6(e) the axial CH_3 is closer to the two axial H's than is the equatorial CH_3 to the adjacent equatorial H's in Fig. 9-6(f).   The steric strain for each CH_3—H 1,3-diaxial interaction is 0.9 kcal/mole and the total is 1.8 kcal/mole for both.

Gauche interaction (Fig. 9-7).   An axial CH_3 on C^1 has a gauche interaction with the C^2—C^3 bond of the ring.   One gauche interaction is also 0.9 kcal/mole; for the two the difference in energy is 1.8 kcal/mole.   The equatorial CH_3 indicated as (CH_3) is *anti* to the C^2—C^3 ring bond.

In general, a *given substituent prefers the less crowded equatorial position to the more crowded axial position.*

Problem 9.16 (a) Draw the possible chair conformational structures for the following pairs of dimethylcyclohexanes: (i) *cis-* and *trans-*1,2-;   (ii) *cis-* and *trans-*1,3-;   (iii) *cis-* and *trans-*1,4-.   (b) Compare the stabilities of the more stable conformers for each pair of geometric isomers.   (c) Determine which of the isomers of dimethylcyclohexane are chiral.   ◄

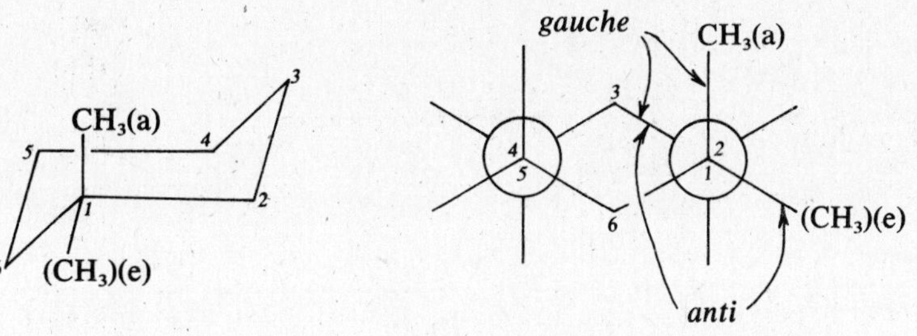

1,1-Dimethylcyclohexane

Fig. 9-7

The best way to determine if groups are *cis* or *trans*, when using chair conformations, is to look at the axial groups.

(*a*)   (i) In the 1,2-isomer, since one axial bond is up and one is down, they are *trans* (Fig. 9-8).   The equatorial bonds are also *trans* although this is not obvious from the structure.   In the *cis*-1,2-isomer an H and CH_3 are *trans* to each other (Fig. 9-9).

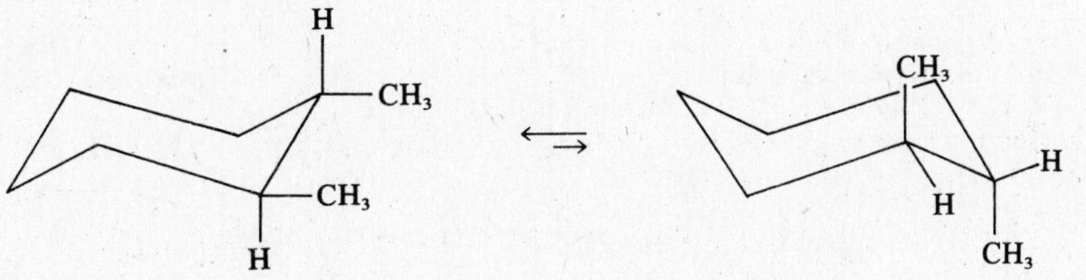

trans-1,2- (CH_3's ee); more stable   *trans*-1,2- (CH_3's aa); less stable

Fig. 9-8

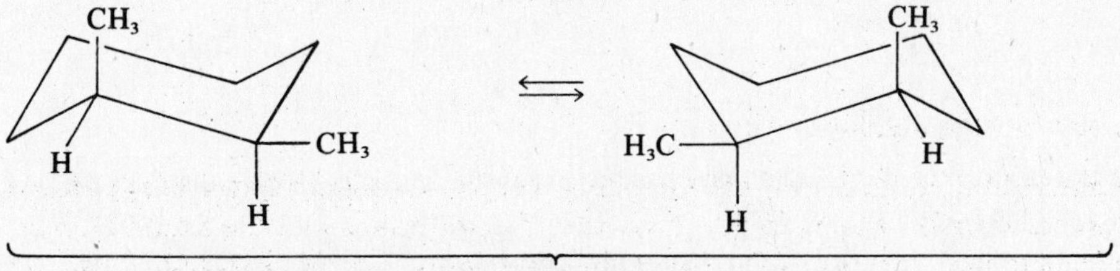

cis-1,2- (CH_3's ea); (conformational enantiomers)

Fig. 9-9

(ii) In the 1,3-isomer both axial bonds are up (or down) and *cis* (Fig. 9-10).   In the more stable conformer [Fig. 9-10(*a*)] both CH_3's are equatorial.   In the *trans* isomer, one CH_3 is axial and one equatorial (Fig. 9-11).

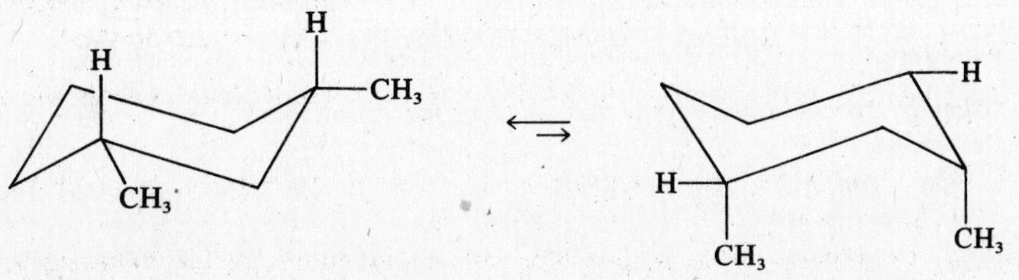

(*a*) *cis*-1,3- (CH_3's ee); more stable   (*b*) *cis*-1,3- (CH_3's aa); less stable

Fig. 9-10

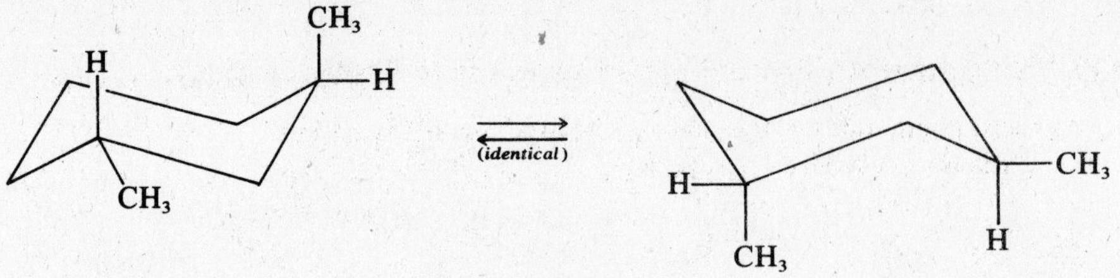

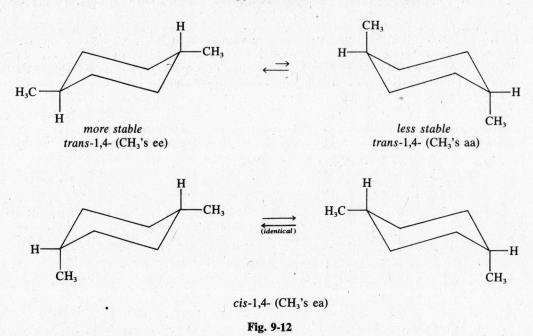

trans-1,3- (CH₃'s ea)

Fig. 9-11

(iii) In the 1,4-isomer the axial bonds are in opposite directions and *trans* (Fig. 9-12).

more stable
trans-1,4- (CH₃'s ee)

less stable
trans-1,4- (CH₃'s aa)

cis-1,4- (CH₃'s ea)

Fig. 9-12

(*b*)   Since an (e) substituent is more stable than an (a) substituent, in each case the (CH₃'s ee) isomer is more stable than the (CH₃'s ea) isomer.

(i) *trans* > *cis*   (ii) *cis* > *trans*   (iii) *trans* > *cis*

(*c*)   The best way to detect chirality in cyclic compounds is to examine the flat structures as in Problem 9.4(*e*); *trans*-1,2- and *trans*-1,3- are the chiral isomers.

Problem 9.17   Give your reasons for selecting the isomers of dimethylcyclohexane shown in Figs. 9-8 to 9-12 that exist as: (*a*) a pair of configurational enantiomers, each of which exists in one conformation;   (*b*) a pair of conformational diastereomers;   (*c*) a pair of configurational enantiomers, each of which exists as a pair of conformational diastereomers;   (*d*) a single conformation;   (*e*) a pair of conformational enantiomers.   ◄

(*a*)   *trans*-1,3-Dimethylcyclohexane is chiral and exists as two enantiomers.   Each enantiomer is (ae) and has only one conformer.

(*b*)   Both *cis*-1,3- and *trans*-1,4-dimethylcyclohexane have conformational diastereomers, the stable (ee) and unstable (aa).   Neither has configurational isomers.

(*c*)   *trans*-1,2-Dimethylcyclohexane is a racemic form of a pair of configurational enantiomers.   Each enantiomer has (ee) and (aa) conformational diastereomers.

(*d*)   *cis*-1,4-Dimethylcyclohexane has no chiral C's and has only a single (ae) conformation.

(*e*)   *cis*-1,2-Dimethylcyclohexane has two (ae) conformers that are nonsuperimposable mirror images.

Problem 9.18 Write planar structures for optical isomers of *cis*-2-chlorocyclohexanol. Indicate chiral C's.◄

This is not a *meso* compound because the two chiral C's are different and there are two enantiomers, as shown on the right.

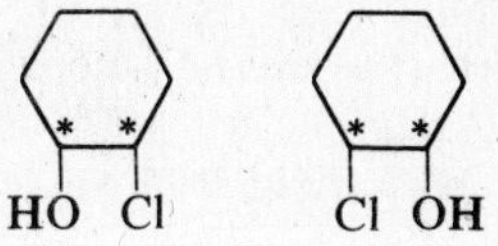

Problem 9.19 Use 1,3- interactions and *gauche* interactions, when needed, to find the difference in energy between (*a*) *cis*- and *trans*-1,3-dimethylcyclohexane; (*b*) (ee) *trans*-1,2- and (aa) *trans*-1,2-dimethylcyclohexane. ◄

Each CH_3/H 1,3-interaction and each CH_3/CH_3 *gauche* interaction imparts 0.9 kcal/mole of instability to the molecule.

(*a*) In the *cis*-1,3-isomer (Fig. 9-10) the more stable conformer has (ee) CH_3's and thus has no 1,3-interactions. The *trans* isomer has (ea) CH_3's. The axial CH_3 has two CH_3/H 1,3-interactions, accounting for 2(0.9) = 1.8 kcal/mole instability. The *cis* isomer is *more* stable than the *trans* isomer by 1.8 kcal/mole.
(*b*) See Fig. 9-13; (ee) is more stable than (aa) by 3.6 − 0.9 = 2.7 kcal/mole.

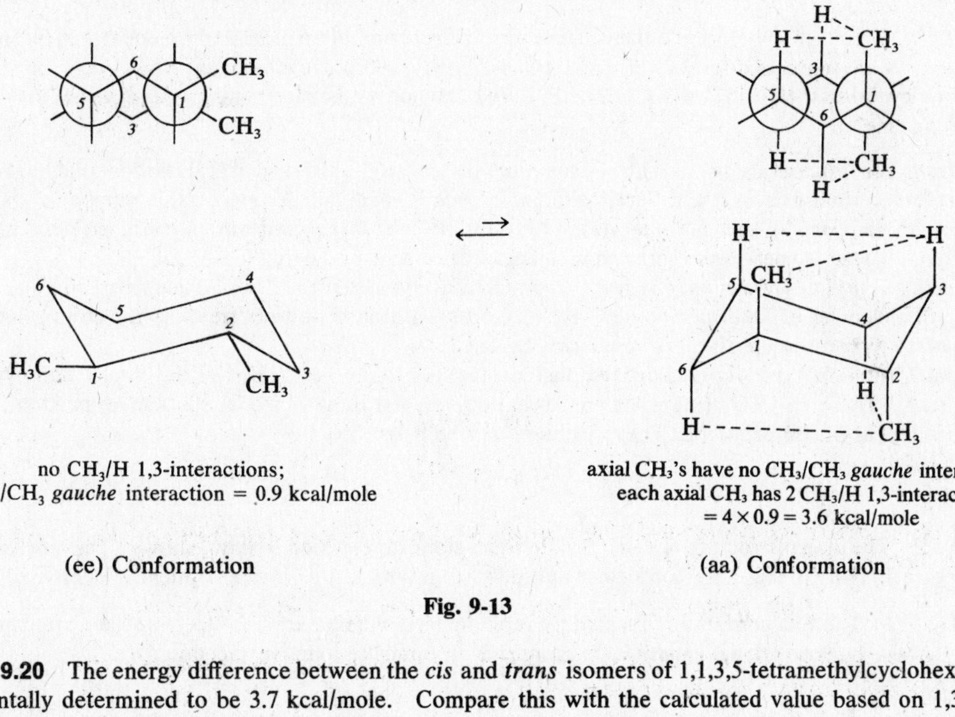

no CH_3/H 1,3-interactions;
1 CH_3/CH_3 *gauche* interaction = 0.9 kcal/mole

axial CH_3's have no CH_3/CH_3 *gauche* interactions;
each axial CH_3 has 2 CH_3/H 1,3-interactions
= 4 × 0.9 = 3.6 kcal/mole

(ee) Conformation　　　　　　　　　　(aa) Conformation

Fig. 9-13

Problem 9.20 The energy difference between the *cis* and *trans* isomers of 1,1,3,5-tetramethylcyclohexane was experimentally determined to be 3.7 kcal/mole. Compare this with the calculated value based on 1,3-diaxial interactions of 0.9 and 3.6 kcal/mole for H/CH_3 and CH_3/CH_3 respectively. Consider only the more stable conformers. ◄

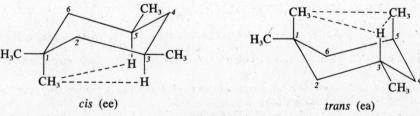

cis (ee)　　　　　　　　　　*trans* (ea)

Fig. 9-14

See Fig. 9-14. *Cis* (ee) has 1.8 kcal/mole of strain energy from two CH_3/H 1,3-interactions. *Trans* (ea) has 5.4 kcal/mole, 3.6 kcal/mole from one CH_3/CH_3 and 1.8 kcal/mole from two CH_3/H. The difference, 5.4 − 1.8 = 3.6 kcal/mole, compares favorably with the experimental value of 3.7 kcal/mole.

Problem 9.21   Write the structure of the preferred conformation of   (*a*) *trans*-1-ethyl-3-isopropylcyclo-hexane,   (*b*) *cis*-2-chloro-*cis*-4-chlorocyclohexyl chloride.   ◄

(*a*)   The *trans*-1,3-isomer is (ea); the bulkier group, in this case *i*-propyl, is equatorial, and the smaller group, in this case ethyl, is axial.   See Fig. 9-15(*a*).

(*b*)   See Fig. 9-15(*b*).

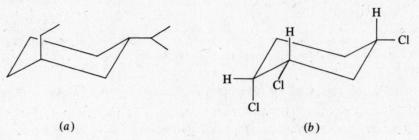

              (*a*)   (*b*)

Fig. 9-15

Problem 9.22   You wish to determine the relative rates of reaction of an axial and an equatorial Br in an S_N2 displacement.   Can you compare   (*a*) *cis*- and *trans*-1-methyl-4-bromocyclohexane?   (*b*) *cis*- and *trans*-1-*t*-butyl-4-bromocyclohexane?   (*c*) *cis*-3,5-dimethyl-*cis*-1-bromocyclohexane   and   *cis*-3,5-dimethyl-*trans*-1-bromocyclohexane?   ◄

(*a*)   The *trans* substituents are (ee).   The *cis* substituents are (ea).   Although CH_3 is bulkier and has a greater (e) preference than has Br, the difference in preference is small and an appreciable number of molecules exist with the Br (e) and the CH_3 (a).   At no time are there conformers with Br only in an (a) position.   These isomers, therefore, cannot be used for this purpose.

(*b*)   The bulky *t*-butyl group can only be (e).   In practically all molecules of the *cis* isomer, Br is forced to be (a).   All molecules of the *trans* isomer have an (e) Br.   Because *t*-butyl "freezes" the conformation and prevents interconversion, these isomers *can* be used.

(*c*)   The *cis*-3,5-dimethyl groups are almost exclusively (ee) to avoid severe CH_3/CH_3 1,3-interactions were they to be (aa).   These *cis*-CH_3's freeze the conformation.   When Br at C^1 is *cis*, it has an (e) position; when it is *trans*, it has an (a) position.   These isomers can be used.

Problem 9.23   The planar structure of *cis*-1,2-dimethylcyclohexane, which is *meso*, shows a plane of symmetry [Problem 9.4(*e*)].   (*a*) Is the chair conformer achiral?   (*b*) Why is this isomer optically inactive?   ◄

(*a*) No.   (*b*) The two conformers formed by rapid interconversion are nonsuperimposable mirror images (Fig. 9-9).   These **conformational enantiomers** comprise an optically inactive racemic form.

Problem 9.24   Are the following compounds stable?

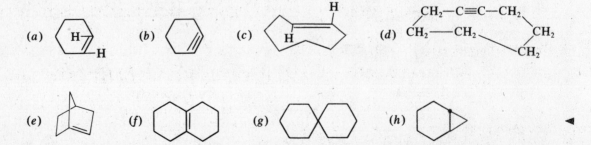

(*a*) No.   A *trans*-cyclohexene is too strained.   The *trans* unit C—C=C—C cannot be bridged by two more C's.   *trans*-Cycloalkenes are stable for eight-membered, (*c*), and larger rings.   (*b*) No.   Cycloalkynes of less than 8 C's are too strained.   The triple bond imposes linearity on 4 C's, C—C≡C—C, which cannot be bridged by 2 more C's but can be bridged by 4 C's, as in (*d*).   (*e*) No.   A molecule cannot have a double bond at a bridgehead C (a C common to 3 or more sides of rings) if there is at least one C in each bridge and the bridges are not too large.   Such a bridgehead C and the 3 atoms bonded to it cannot assume the flat, planar structure required of an sp^2 hybridized C.   This is known as **Bredt's rule**.   (*f*) Yes.   This exists because one of the bridges has no C, and these bridgehead C's can easily use sp^2 hybridized orbitals to form triangular, planar sigma bonds.   (*g*) Yes.   Compounds (**spiranes**) having a single C which is a junction for two separate rings are known for all size rings.   However, the rings must be at right angles.

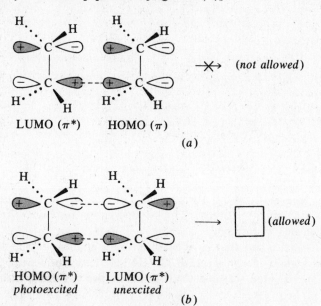

(*h*) No.   Three- and six-membered rings cannot be fused *trans*, since there is too much strain.

9.4 MOLECULAR-ORBITAL INTERPRETATION OF CONCERTED CYCLOADDITION REACTIONS: WOODWARD-HOFFMANN RULES

INTERMOLECULAR REACTIONS

The rule states that reactions can occur only when all overlaps between the *highest occupied π molecular orbital* (HOMO) of one reactant and the *lowest unoccupied π molecular orbital* (LUMO) of the other reactant occur so that (see Figs. 9-16, 9-17, 9-18) a *positive* (shaded) *lobe overlaps only with a positive* (shaded) *lobe, and a negative* (unshaded) *lobe only with another negative* (unshaded) *lobe*.   We consider only the end atomic orbitals comprising the MO because these overlap to form two new σ bonds and a ring compound.

   1.   **Ethylene dimerization (2 + 2) to cyclobutane.**   Without ultraviolet light we have the situation indicated in Fig. 9-16(*a*).   Irradiation with uv causes a $\pi \rightarrow \pi^*$ transition (Fig. 8-3) and now the proper orbital symmetry for overlap prevails [Fig. 9-16(*b*)].

Fig. 9-16

2. **Diels-Alder reaction** (2 + 4). See Fig. 9-17.

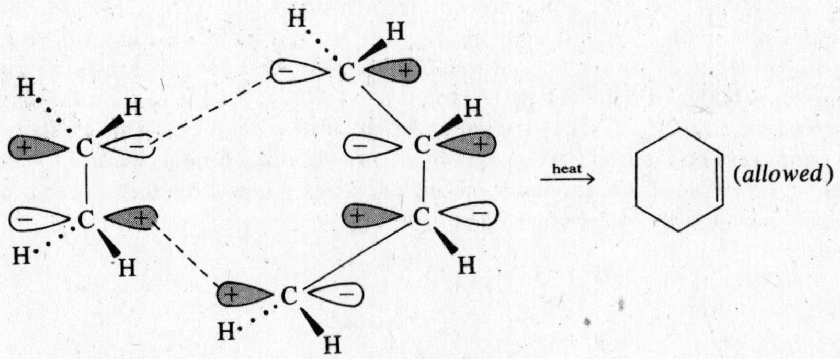

LUMO (π^*)   HOMO (π_2; see Fig. 8-3)
dienophile   *diene*

Fig. 9-17

ELECTROCYCLIC (INTRAMOLECULAR) REACTIONS

In electrocyclic reactions of acyclic conjugated polyenes, one double bond is lost and a single bond is formed between the terminal C's to give a ring.

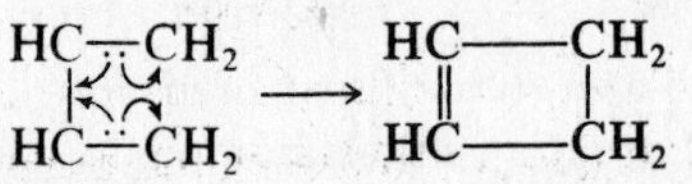

a diene   *a cyclobutene*

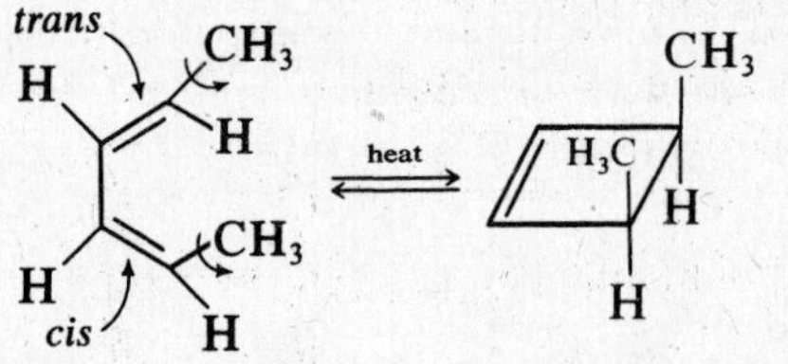

cis,trans-2,4-Hexadiene   *cis*-3,4-Dimethylcyclobutane

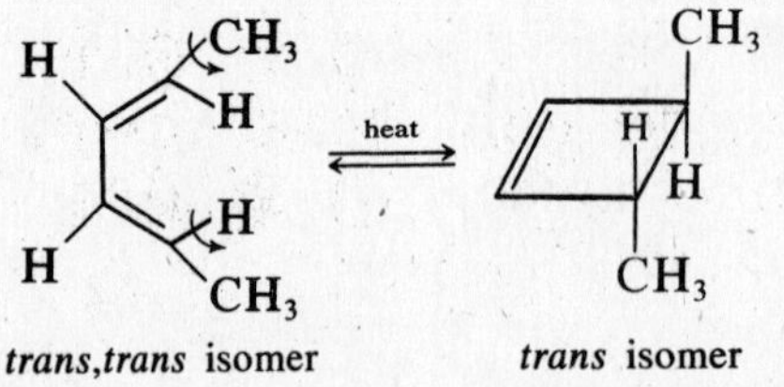

trans,trans isomer   *trans* isomer

To achieve this stereospecificity, both terminal C's rotate 90° in the *same* direction, called a **conrotatory motion**. Movement of these C's in opposite directions (one clockwise and one counterclockwise) is termed **disrotatory**.

The **Woodward-Hoffmann rule** that permits the proper analysis of the stereochemistry is: **The orbital symmetry of the highest-energy occupied MO (HOMO) must be considered**, and rotation occurs to permit overlap of two shaded (or unshaded) lobes of the p orbitals to form the σ bond after rehybridization.

The HOMO for the thermal reaction then requires a conrotatory motion [Fig. 9-18(a)]. Irradiation causes a disrotatory motion by exciting an electron from $\pi_2 \rightarrow \pi_3^*$, which now becomes HOMO [Fig. 9-18(b)].

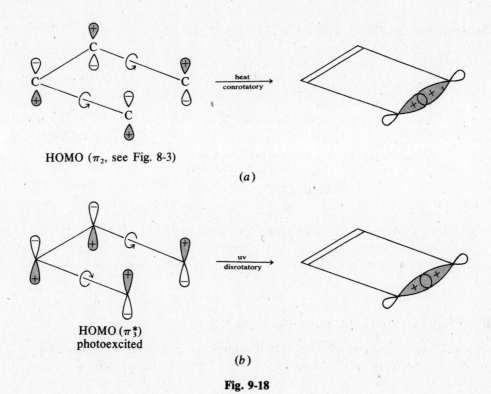

HOMO (π_2, see Fig. 8-3)

(a)

HOMO (π_3^*)
photoexcited

(b)

Fig. 9-18

Problem 9.25 When applying the Woodward-Hoffmann rule to the Diels-Alder reaction, (*a*) would the same conclusion be drawn if we reacted the highest occupied MO of the dienophile with the lowest unoccupied MO of the diene?   (*b*) Would the reaction be light-catalyzed?   ◄

(*a*) Yes; see Fig. 9-19(*a*).   (*b*) No; see Fig. 9-19(*b*).

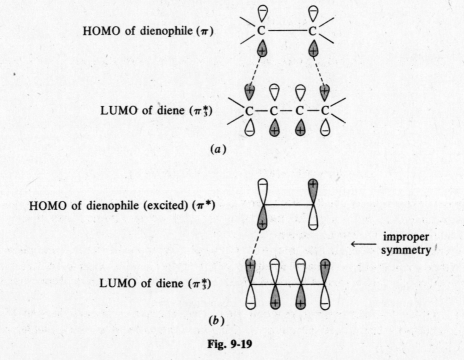

HOMO of dienophile (π)

LUMO of diene (π_3^*)

(a)

HOMO of dienophile (excited) (π^*)

improper symmetry

LUMO of diene (π_3^*)

(b)

Fig. 9-19

Problem 9.26 Use Woodward-Hoffmann rules to predict whether the following reaction would be expected to occur thermally or photochemically.

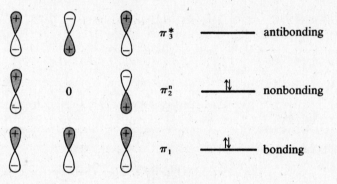

The MO energy levels of the allyl carbanion π system showing the distribution of the four π electrons (2 from π double bond and 2 unshared) are indicated in Fig. 9-20. The 0 is used whenever a node point is at an atom. The allowed reaction occurs thermally as shown in Fig. 9-21(a). The photoreaction is forbidden [Fig. 9-21(b)].

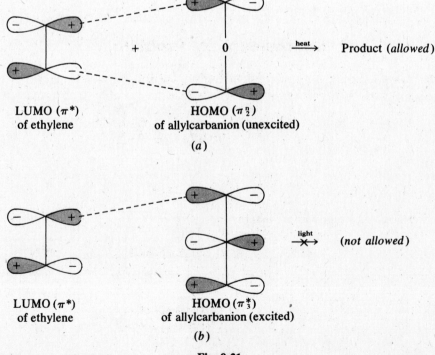

Fig. 9-20

Fig. 9-21

9.5 TERPENES AND THE ISOPRENE RULE

The carbon skeleton

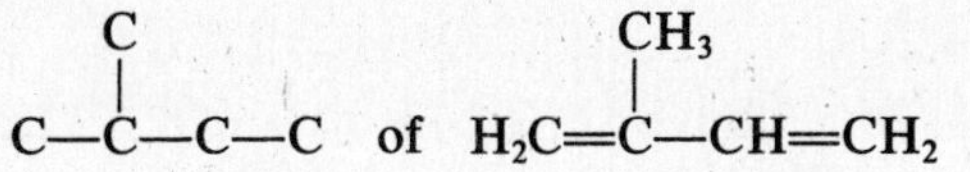

$$C-\overset{\overset{\displaystyle C}{|}}{C}-C-C \quad \text{of} \quad H_2C=\overset{\overset{\displaystyle CH_3}{|}}{C}-CH=CH_2$$

Isoprene

is the structural unit of many naturally occuring compounds, among which are the **terpenes**, whose generic formula is $(C_5H_8)_n$.

Problem 9.27 Pick out the isoprene units in the terpenes limonene, myrcene and α-phellandrene, and in vitamin A, shown below. ◀

In the structures below, dashed lines separate the isoprene units.

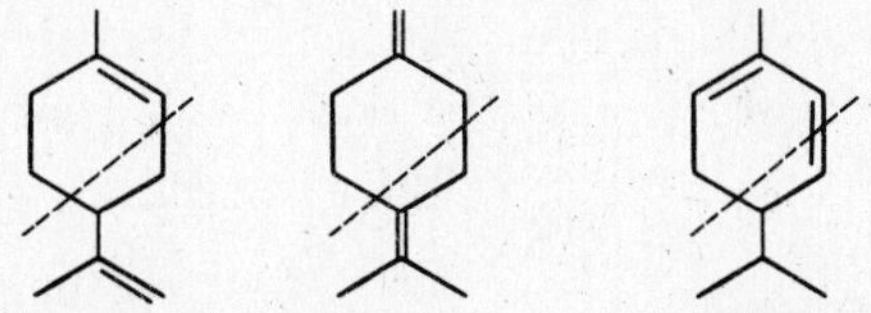

Limonene   Myrcene   α-Phellandrene

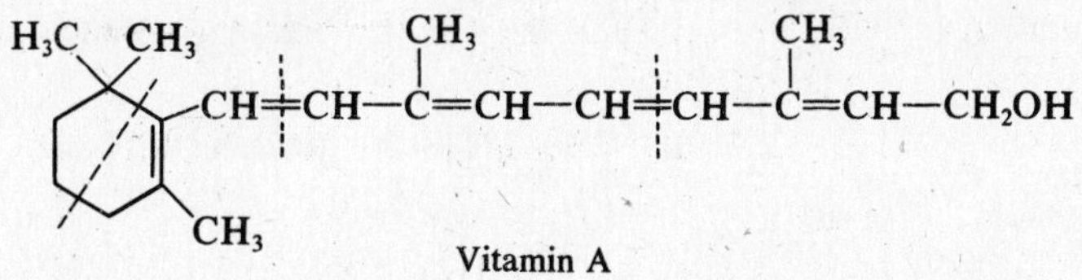

Vitamin A

Supplementary Problems

Problem 9.28 Write structural formulas for the organic compounds designated by a ?. Indicate the stereochemistry where necessary and account for the products.

(a)

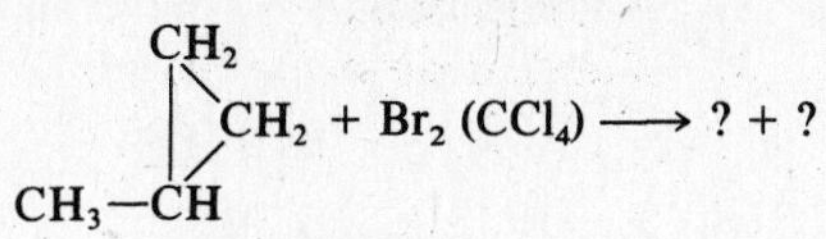

$$\underset{CH_3-CH}{\overset{CH_2}{\diagup}}\!\!\!\diagdown CH_2 + Br_2\ (CCl_4) \longrightarrow ?\ +\ ?$$

(b)

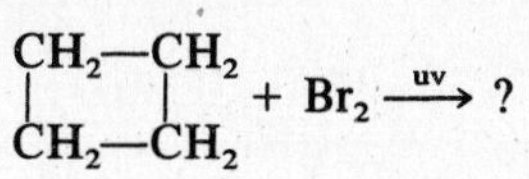

$$\underset{CH_2-CH_2}{\overset{CH_2-CH_2}{|\qquad|}} + Br_2 \xrightarrow{uv} ?$$

(c)

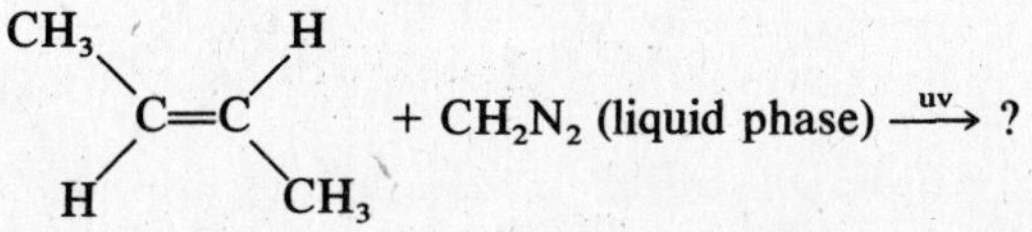

$$\underset{H}{\overset{CH_3}{\diagdown}}C=C\underset{CH_3}{\overset{H}{\diagup}} + CH_2N_2 \text{ (liquid phase)} \xrightarrow{uv} ?$$

(d)

$$\underset{H}{\overset{CH_3}{\diagdown}}C=C\underset{H}{\overset{CH_3}{\diagup}} + CH_2N_2 \ \underset{\text{inert gas, argon}}{\overset{\text{(in presence of an}}{}} \xrightarrow{uv} ?\ +\ ?$$

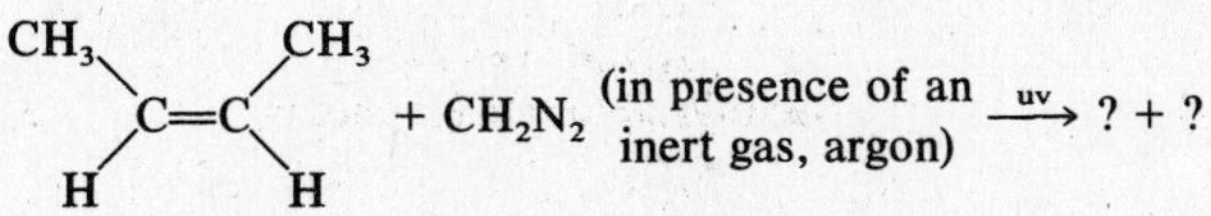

(e)   Same as (d) but with added $O_2 \longrightarrow$?

(f)
$$\begin{array}{c} CH(CH_3)—CH \\ CH_2 \quad\quad | \\ CH_2————CH \end{array} + CHCl_3 \xrightarrow{(CH_3)_3CO^-K^+} ? + ?$$

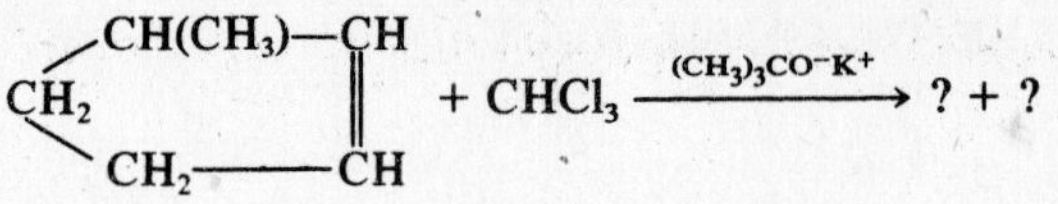

(g)
$$\begin{array}{c} CH_2—C—CH_3 \\ CH_2 \quad\quad || \\ CH_2—CH \end{array} + CHClBr_2 \xrightarrow{(CH_3)_3CO^-K^+} ? + ?$$

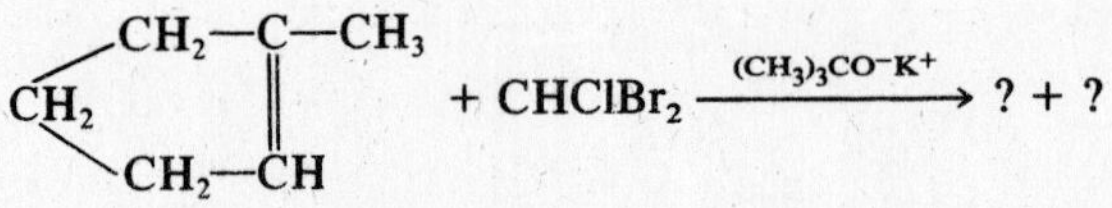

(h)
$$\begin{array}{c} CH_2 \\ CH_2 \quad CH_2 \\ CH_3—C=CH \end{array} + Br_2 \xrightarrow{CCl_4} ?$$

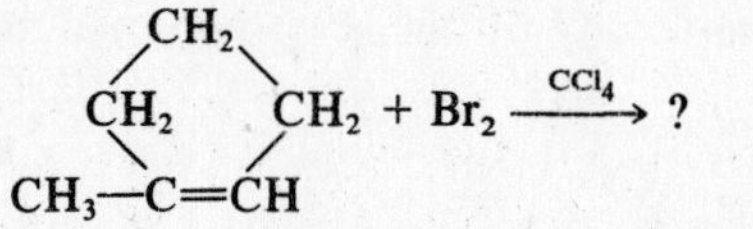

(i)
$$\begin{array}{c} CH_2—CH_2 \\ CH_2 \quad\quad CH_2 \\ CH_2—CHOH \end{array} \xrightarrow[heat]{H_2SO_4} ? \xrightarrow[KMnO_4]{dil.\ aq.} ?$$
◀

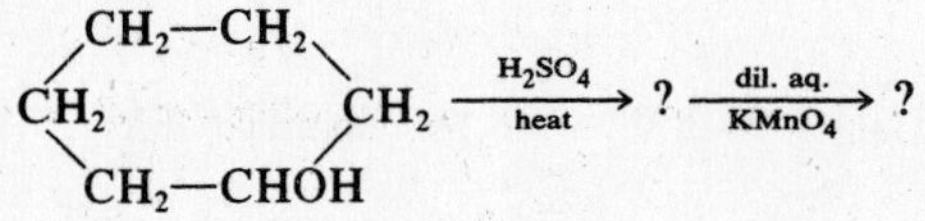

(a) Two Br's add to each of the two nonequivalent single bonds I and II of

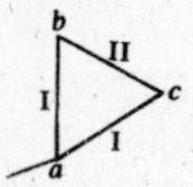

to form two products, (±)-$\overset{c}{C}H_2Br—\overset{b}{C}H_2—\overset{a}{C}HBr—CH_3$ by breaking (I) and

$\overset{b}{C}H_2Br—\overset{a}{C}H—\overset{c}{C}H_2Br$ by breaking (II)
$\quad\quad\quad | $
$\quad\quad\quad CH_3$

(b)
$$\begin{array}{c} CH_2—CHBr \\ | \quad\quad | \\ CH_2—CH_2 \end{array}$$ by radical substitution; unlike the 3-membered ring, the 4-membered ring is stable.

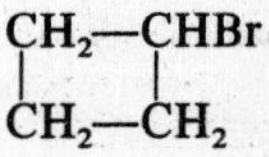

(c) In the liquid phase we get singlet CH_2 which adds cis; trans-2-butene forms trans-1,2-dimethyl-cyclopropane:

$$\begin{array}{c} CH_3 \quad H \\ H \quad CH_3 \end{array}$$

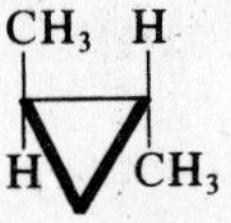

(d) Some initially formed singlet CH_2 collides with inert-gas molecules and changes to triplet

$$\uparrow CH_2 \uparrow$$

which adds nonstereospecifically. cis-2-Butene yields a mixture of cis- and trans-1,2-dimethyl-cyclopropane:

$$\begin{array}{c} CH_3 \quad CH_3 \\ H \quad\quad H \end{array} \quad and \quad \begin{array}{c} CH_3 \quad H \\ H \quad\quad CH_3 \end{array}$$

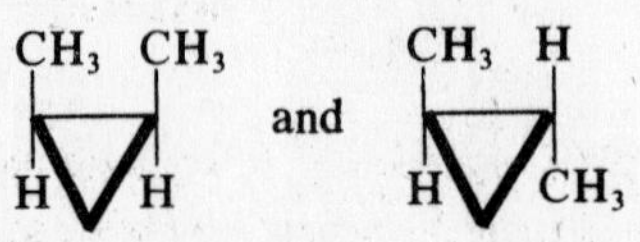

(e) O_2 is a diradical which combines with triplet carbenes, leaving the singlet species to react with cis-2-butene to give cis-1,2-dimethylcyclopropane.

(f)

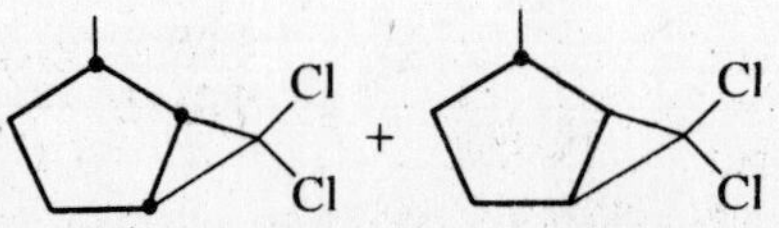

Dichlorocarbene (see page 138) adds cis to C=C but either cis or trans to the Me.

(*g*) $CClBr_2$ loses Br^-, the better leaving group, rather than Cl^- to give ClBrC:, which then adds so that either Cl or Br can be *cis* to CH_3.

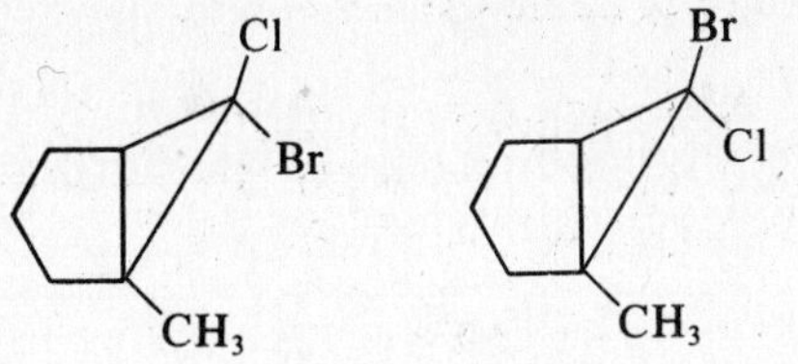

(*h*)

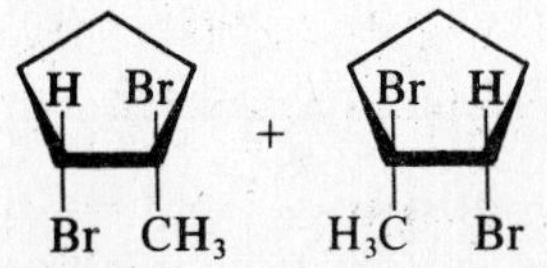

trans addition; 2 enantiomers (*rac*)

(*i*)

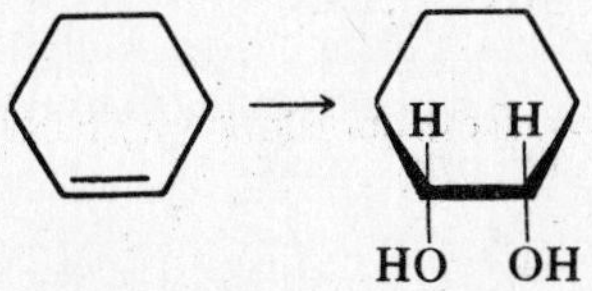

Cyclohexanol is dehydrated to cyclohexene, which forms a *meso* glycol by *cis* addition of two OH's.

Problem 9.29 Explain why (*a*) a carbene is formed by dehydrohalogenation of $CHCl_3$ but not from methyl, ethyl or *n*-propyl chlorides; (*b*) *cis*-1,3- and *trans*-1,4-di-*tert*-butylcyclohexane exist in chair conformations, but their geometric isomers, *trans*-1,3- and *cis*-1,4-, do not. ◄

(*a*) Carbene is formed from $CHCl_3$ because the three strongly electronegative Cl's make this compound sufficiently acidic to have its proton abstracted by a base. CH_3Cl has only one Cl and is considerably less acidic. Carbene formation is an α-elimination of HCl from the same C; it does not occur with ethyl or propyl chlorides because protons are more readily eliminated from the β C's to form alkenes.

(*b*) Both *cis*-1,3 and *trans*-1,4 compounds exist in the chair form because of the stability of their (ee) conformers. *Trans*-1,3- and *cis*-1,4- are (ea). An axial *t*-butyl group is very unstable, so that a twist-boat with a quasi-(ee) conformation (Fig. 9-22) is more stable than the chair.

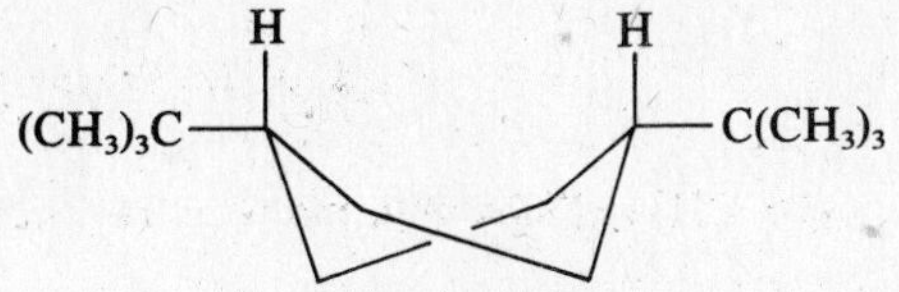

Twist-boat

Twisting reduces eclipsed and "flagpole" interactions

Fig. 9-22

Problem 9.30 Assign structures or configurations for A through D. (*a*) Two isomers, A and B, with formula C_8H_{14} differ in that one adds one mole and the other two moles of H_2. Ozonolysis of A gives only one product, $O{=}CH(CH_2)_6CH{=}O$, while the same reaction with one mole of B produces two moles of $CH_2{=}O$ and one mole of $O{=}CH(CH_2)_6CH{=}O$. (*b*) Two stereoisomers, C and D, of 3,4-dibromocyclopentane-1,1-dicarboxylic acid undergo decarboxylation as shown:

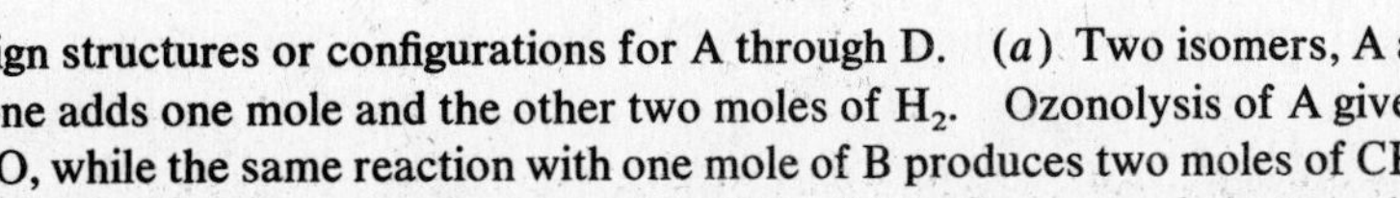

a *gem*-dicarboxylic acid a monocarboxylic acid

C gives one, while D yields two, monocarboxylic acids.

(a) Both compounds have two degrees of unsaturation (see Problem 6.37). B absorbs two moles of H_2 and has 2 multiple bonds. A absorbs one mole of H_2 and has a ring and a double bond; it is a cycloalkene. As a cycloalkene, A can form only a single product, a dicarbonyl compound, on ozonolysis.

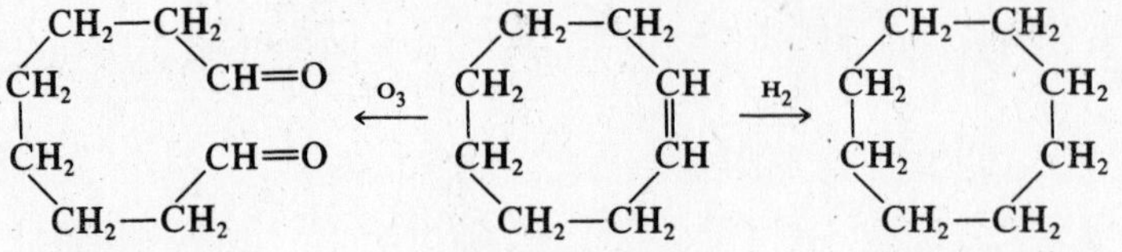

Octane-1,8-dial   Cyclooctene (A)   Cyclooctaene

Since one molecule of B gives 3 carbonyl molecules, it must be a diene and not an alkyne.

$$H_2C{=}O + O{=}HC(CH_2)_4CH{=}O + O{=}CH_2 \xleftarrow{O_3} H_2C{=}CH(CH_2)_4CH{=}CH_2 \xrightarrow{2H_2} CH_3CH_2(CH_2)_4CH_2CH_3$$

1,6-Hexanedial   1,7-Octadiene (B)   n-Octane

(b) The Br's of the dicarboxylic acid may be *cis* or *trans*. Decarboxylation of the *cis* isomer yields two isomeric products in which both Br's are *cis* (E) or *trans* (F) with respect to COOH. The *cis* isomer is D. In the monocarboxylic acid G formed from C (*trans* isomer), one Br is *cis* and the other *trans* with respect to COOH and there is only one isomer.

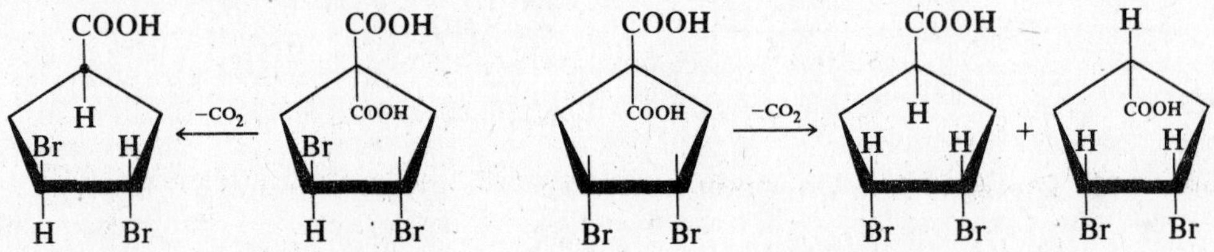

(G)   (C)   (D)   (F)   +   (E)

            trans   *cis*

Problem 9.31 Outline the reactions and reagents needed to synthesize the following from any acyclic compounds having up to 4 C's and any needed inorganic reagents: (a) *cis*-1-methyl-2-ethylcyclopropane; (b) *trans*-1,1-dichloro-2-ethyl-3-*n*-propylcyclopropane; (c) 4-cyanocyclohexene; (d) bromocyclobutane.◄

(a) The *cis*-disubstituted cyclopropane is prepared by stereospecific addition of singlet carbene to *cis*-2-pentene.

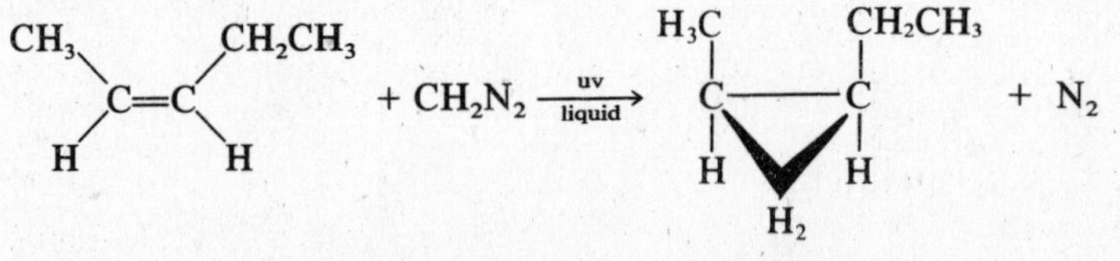

The alkene, having 5 C's, is best formed from 1-butyne, a compound with 4 C's.

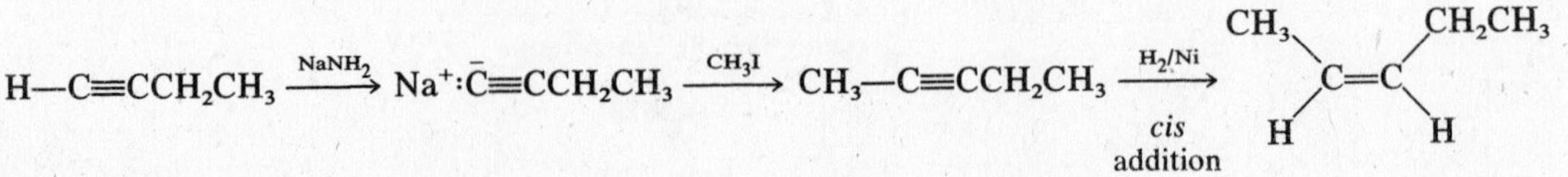

$$H{-}C{\equiv}CCH_2CH_3 \xrightarrow{NaNH_2} Na^+{:}\bar{C}{\equiv}CCH_2CH_3 \xrightarrow{CH_3I} CH_3{-}C{\equiv}CCH_2CH_3 \xrightarrow{H_2/Ni}$$

cis addition

(b) Add dichlorocarbene to *trans*-3-heptene, which is formed from 1-butyne.

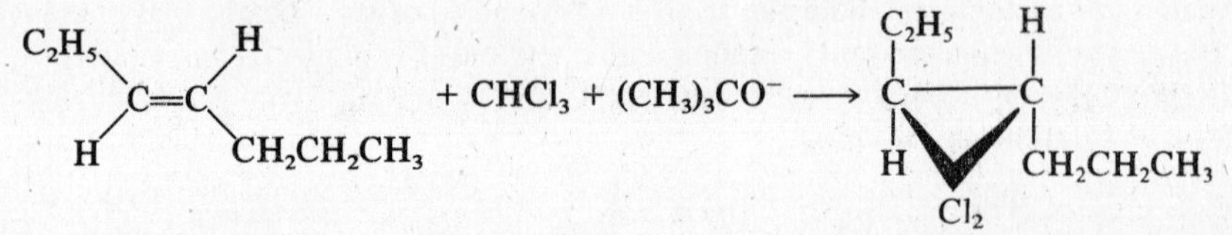

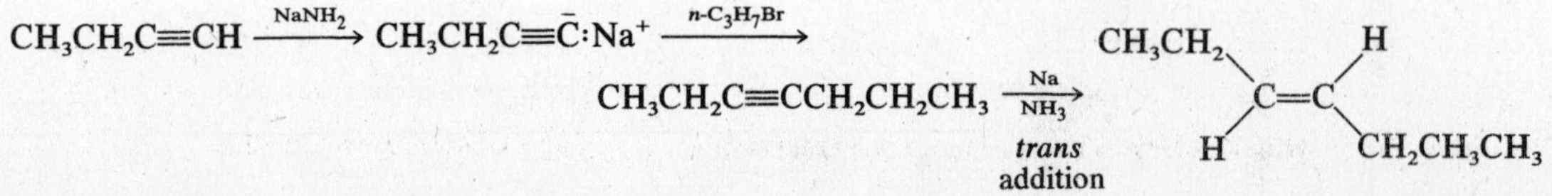

$$CH_3CH_2C{\equiv}CH \xrightarrow{NaNH_2} CH_3CH_2C{\equiv}\bar{C}{:}Na^+ \xrightarrow{n\text{-}C_3H_7Br}$$

$$CH_3CH_2C{\equiv}CCH_2CH_2CH_3 \xrightarrow[NH_3]{Na}$$

trans addition

(*c*) Cyclohexenes are best made by Diels-Alder reactions. The CN group is strongly electron-withdrawing and when attached to the C=C engenders a good dienophile.

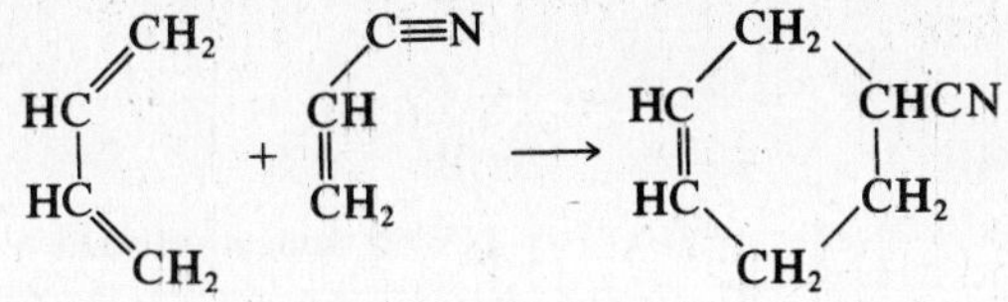

1,3-Butadiene   Acrylonitrile   (a Diels-Alder reaction)

(*d*)

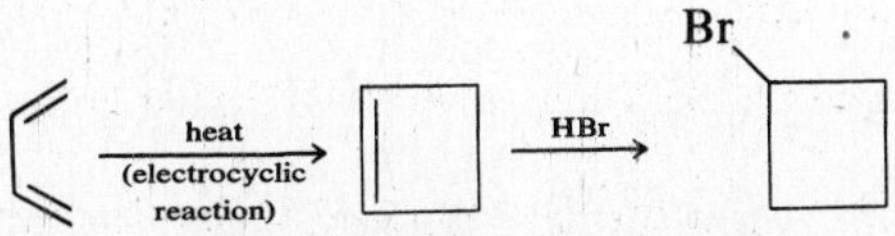

Cyclobutene

Problem 9.32 Write planar structures for the cyclic derivatives formed in the following reactions, and give their stereochemical labels.

(*a*)   3-Cyclohexenol + dil. aq. KMnO$_4$ $\longrightarrow$

(*b*)   3-Cyclohexenol + HCO$_3$H and then H$_2$O $\longrightarrow$

(*c*)   (R,R)-*trans*-1,2-Dibromocyclopropane + Br$_2$ $\xrightarrow{\text{light}}$

(*d*)   *meso-cis*-1,2-Dibromocyclopropane + Br$_2$ $\xrightarrow{\text{light}}$

(*e*)   1-Methylcyclohexene + HBr (peroxide) $\longrightarrow$ (an *anti* addition)   ◄

(*a*)   meso   rac   (*b*)   meso   rac

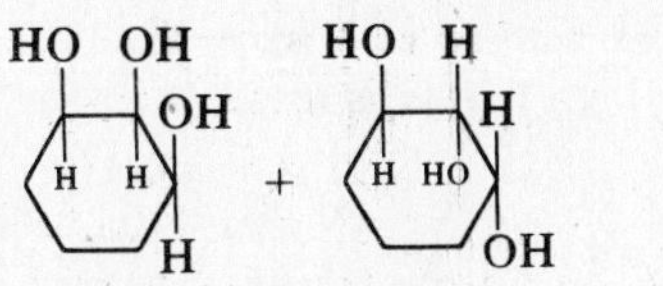

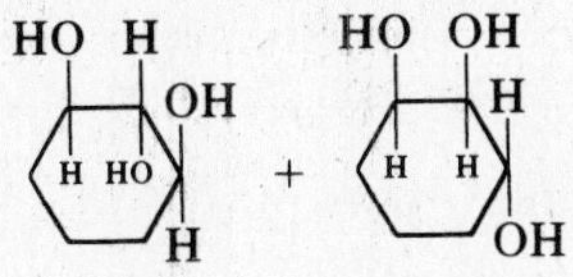

(*c*)   (*d*)

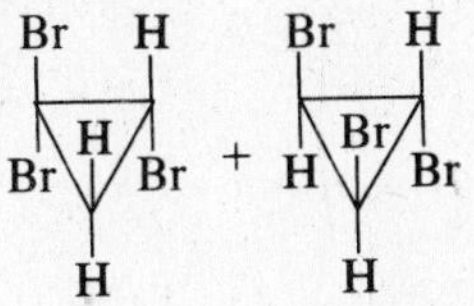

cis,trans

Optically   Optically   racemate   all *cis*   cis,trans
active, R   inactive   Optically inactive

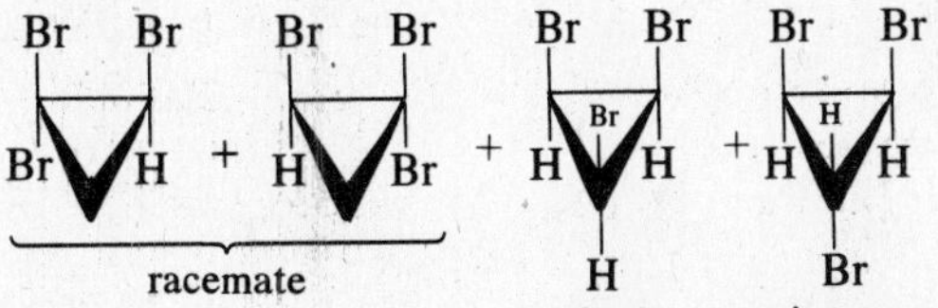

(Note that in absence of chiral catalysts an optically inactive reactant gives only optically inactive products).

(*e*)

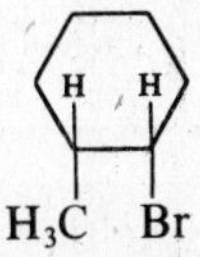

H$_3$C   Br

cis,rac (*erythro*)

Problem 9.33 Use cyclohexanol and any inorganic reagents to synthesize (*a*) *trans*-1,2-dibromo-cyclohexane, (*b*) *cis*-1,2-dibromocyclohexane, (*c*) *trans*-1,2-cyclohexanediol.   ◄

(*a*)

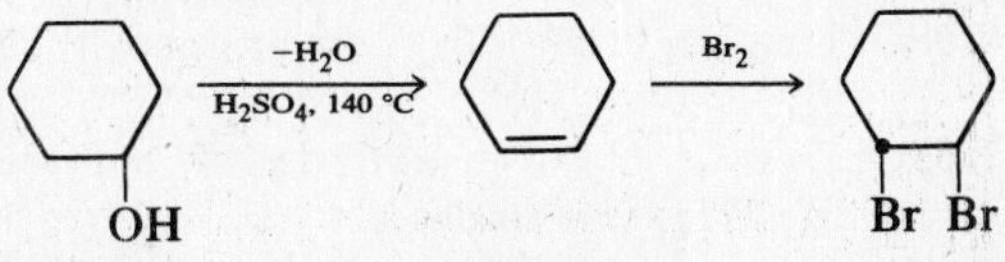

Cyclohexanol   Cyclohexene   *trans*-1,2-Dibromocyclohexane

(b) See Problem 9.32(e).

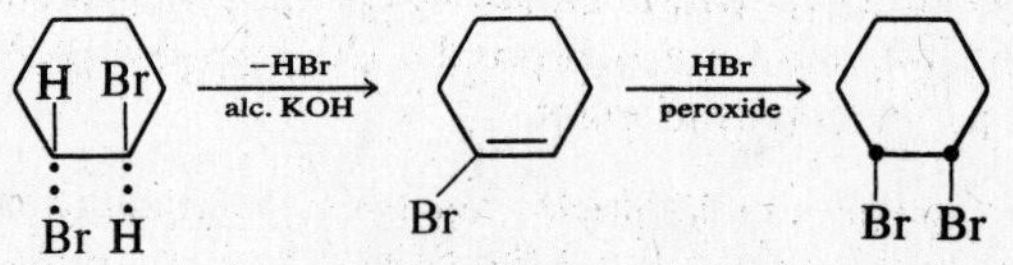

1-Bromocyclohexene   cis-1,2-Dibromocyclohexane

(c)

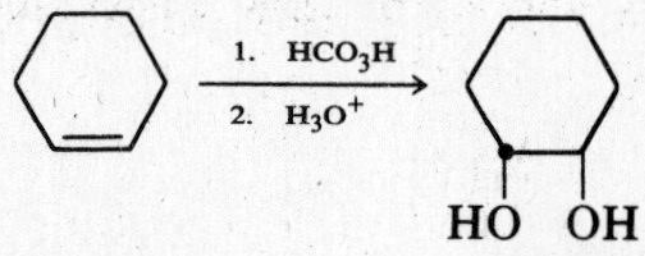

Cyclohexene   trans-1,2-Cyclohexanediol

Problem 9.34 For the two alkylated cyclohexanes

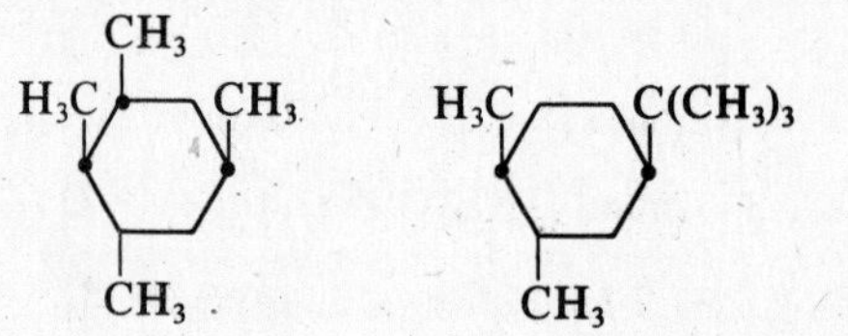

(a) indicate whether the compounds are chiral and draw the possible chair conformations; (b) explain which is the more stable conformer. ◄

(a) Both compounds are chiral. The chiral centers are marked with an asterisk in Fig. 9-23. To properly assign (e) and (a) conformations, consider pairs of groups, e.g., 1,2-cis is (ea), etc. (See Problem 9-16.)

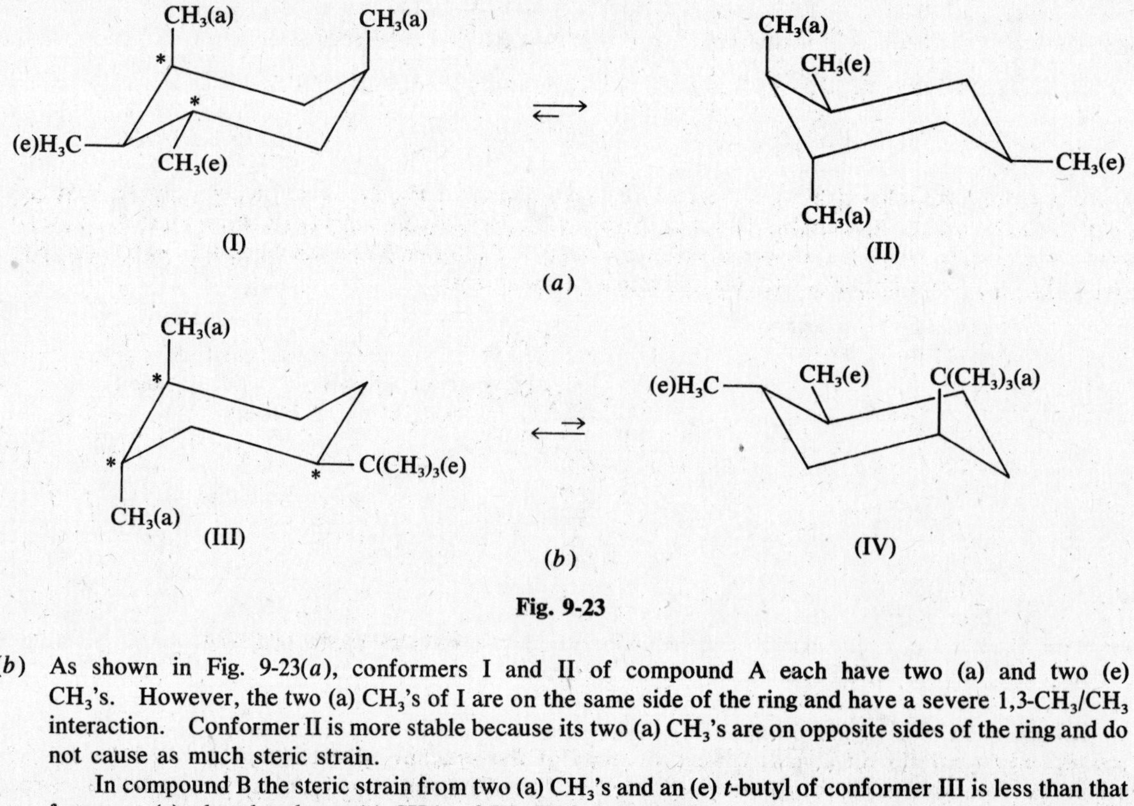

Fig. 9-23

(b) As shown in Fig. 9-23(a), conformers I and II of compound A each have two (a) and two (e) CH$_3$'s. However, the two (a) CH$_3$'s of I are on the same side of the ring and have a severe 1,3-CH$_3$/CH$_3$ interaction. Conformer II is more stable because its two (a) CH$_3$'s are on opposite sides of the ring and do not cause as much steric strain.

In compound B the steric strain from two (a) CH$_3$'s and an (e) t-butyl of conformer III is less than that from one (a) t-butyl and two (e) CH$_3$'s of IV; IV is less stable.

Problem 9.35 Account for the apparent anomalies in the following substances whose diaxial conformations are more stable than the diequatorial: (a) *trans*-1,2-Dibromocyclohexane in nonpolar solvents exists 50% in the (aa) conformation, but in polar solvents the (ee) form is favored. (b) *cis*-1,3-Cyclohexanediol in CCl₄ is shown by infrared measurement to exist as the (aa) conformer. ◄

(a) The two equatorial Br's introduce some dipole-dipole repulsion, tending to destabilize the (ee) conformer. However, polar solvent molecules surround the Br's and minimize the dipole interactions, and the sterically favored (ee) conformer now predominates. See Fig. 9-24(a).

(b) H-bonding stabilizes the two (aa) OH's and more than compensates for the destabilizing 1,3-interactions. See Fig. 9-24(b).

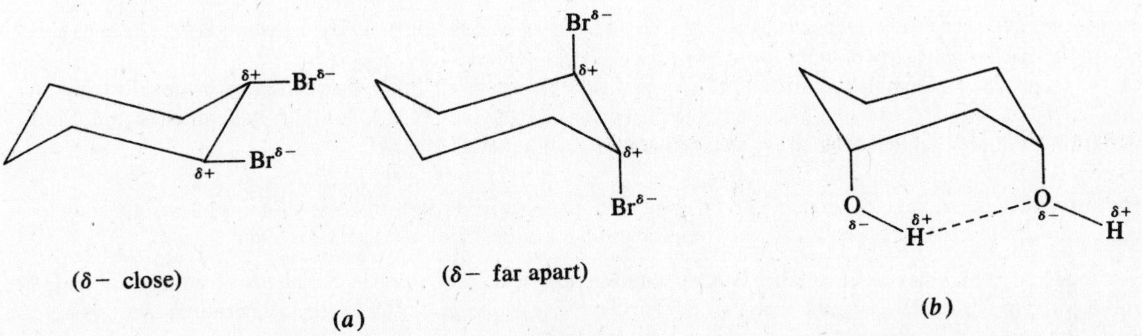

$(\delta - $ close$)$ $(\delta - $ far apart$)$

(a) (b)

Fig. 9-24

Problem 9.36 Decalin, $C_{10}H_{18}$, has two fused cyclohexane rings that share two C's (a common side). There are *cis* and *trans* isomers that differ in the configurations about the two shared C's as shown below. Draw their conformational structural formulas.

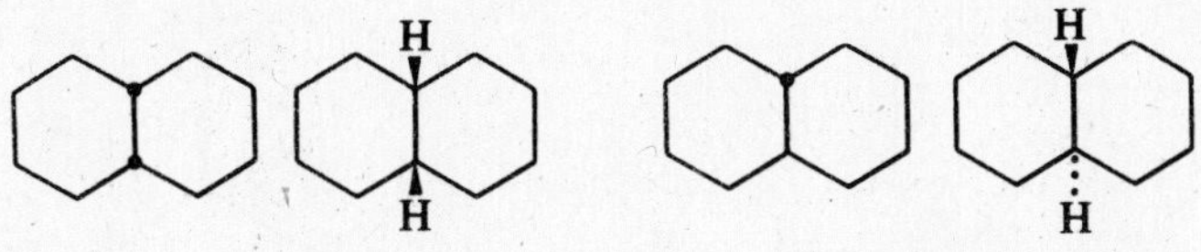

cis   *trans*   ◄

For each ring the other ring can be viewed as 1,2-substituents. For the *trans* isomer, only the rigid (ee) conformation is possible structurally. As shown in Fig. 9-25, diaxial bonds point 180° away from each other and cannot be bridged by only 4 C's to complete the second ring. *Cis* fusion is (ea) and the bonds can be twisted to reverse the (a) and (e) positions, yielding conformation enantiomers.

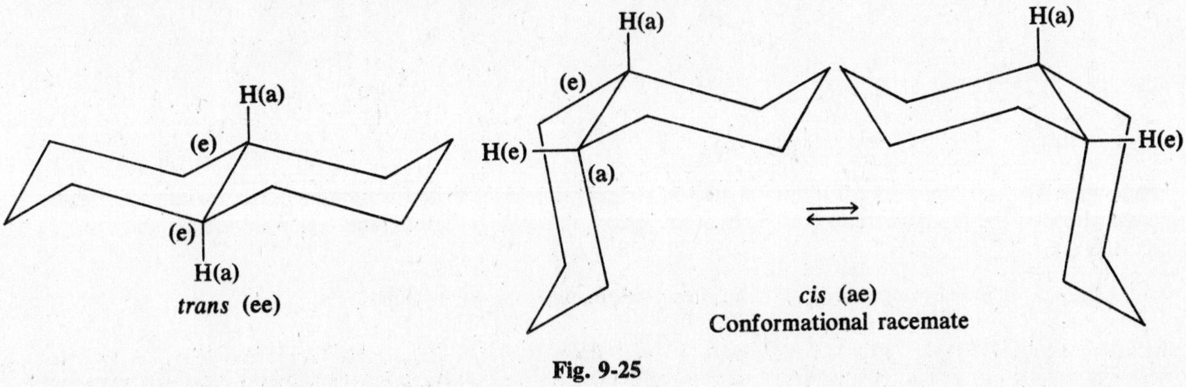

trans (ee)   *cis* (ae)
                               Conformational racemate

Fig. 9-25

Problem 9.37 Explain the following facts in terms of the structure of cyclopropane. (a) The H's of cyclopropane are more acidic than those of propane. (b) The Cl of chlorocyclopropane is less reactive toward S_N2 and S_N1 displacements than the Cl in $CH_3CHClCH_3$. ◄

(a) The external C—H bonds of cyclopropane have more s character than those of an alkane. The more s character in the C—H bond, the more acidic the H.

(b) The C—Cl bond of chlorocyclopropane also has more s character, which diminishes the reactivity of the Cl. Remember that vinyl chlorides are inert in S_N2 and S_N1 reactions. The R^+ formed during the S_N1 reaction would have very high energy, since the C would have to use sp^2 hybrid orbitals needing a bond angle of 120°. The angle strain of the R^+ is much more severe $(120° - 60°)$ than in cyclopropane itself $(109° - 60°)$.

Problem 9.38 Use quantitative and qualitative tests to distinguish between (a) cyclohexane, cyclohexene and 1,3-cyclohexadiene; (b) cyclopropane and propene. ◀

(a) Cyclohexane does not decolorize Br_2 in CCl_4. The uptake of H_2, measured quantitatively, is 2 moles for 1 mole of the diene, but 1 mole for 1 mole of the cycloalkene.

(b) Cyclopropane resembles alkenes and alkynes, and differs from other cycloalkanes in decolorizing Br_2 slowly, adding H_2, and reacting readily with H_2SO_4. However, it is like other cycloalkanes and differs from multiple-bonded compounds in not decolorizing aqueous $KMnO_4$.

Problem 9.39 Contrary to what might be expected from considering proximity of bulky groups, cis-1,3-dimethylcyclobutane has a smaller heat of combustion than the trans isomer. Explain. ◀

Since the cis isomer has a smaller heat of combustion, it is more stable. In terms of conformation [see Problem 9.15(b)], the cis-1,3-isomer has two conformers; one is the more stable (ee), the other is the less stable (aa). The trans-1,3-isomer has one conformer, which is (ea). The cis-(ee) is more stable than the trans-(ea).

Problem 9.40 Draw a conclusion about the stereochemistry of the Diels-Alder reaction from Fig. 9-26. ◀

The stereochemistry of the dienophile is preserved in the product, the substituted cyclohexene. This fact is consistent with a one-step concerted mechanism.

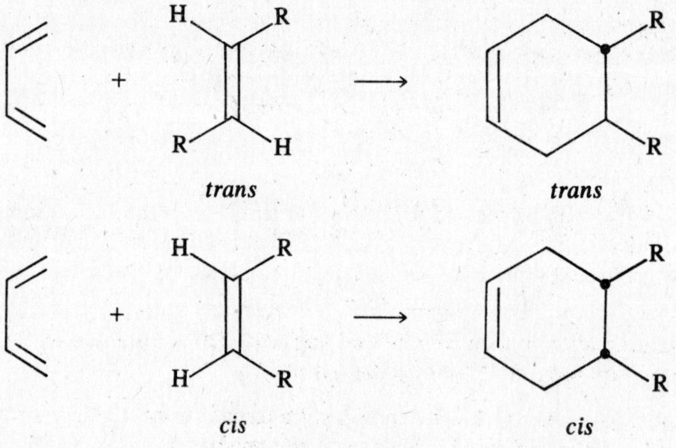

Fig. 9-26

Problem 9.41 (a) Give the structure of the major product, A, whose formula is C_5H_8, resulting from the dehydration of cyclobutylmethanol. On hydrogenation, A yields cyclopentane. (b) Give a mechanism for this reaction. ◀

(a) Compound A is cyclopentene, which gives cyclopentane on hydrogenation.

(b)

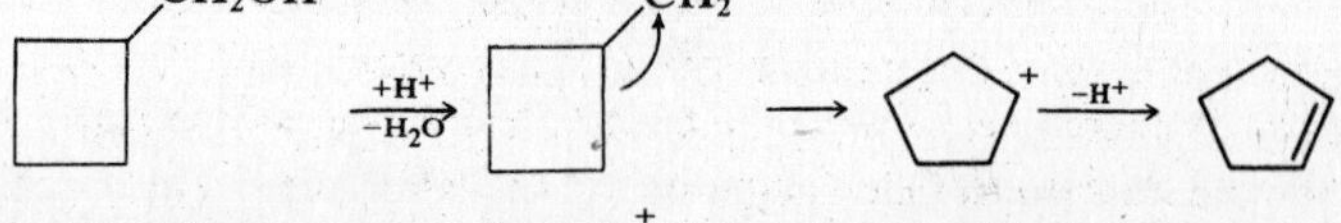

The side of a ring migrates, thereby converting a $R\overset{+}{C}H_2$ having a strained 4-membered ring to a much more stable R_2CH^+ with a strain-free 5-membered ring.

Chapter 10

Benzene and Aromaticity

10.1 INTRODUCTION

Benzene, C_6H_6, is the prototype of **aromatic** compounds, which are unsaturated compounds showing a low degree of reactivity. The so-called **Kekulé structure** (1865) for benzene,

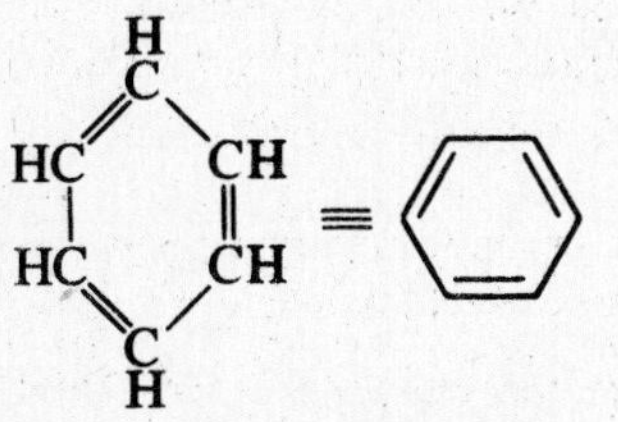

has only one monosubstituted product (C_6H_5Y) since all 6 H's are equivalent. There are 3 disubstituted benzenes—the 1,2-, 1,3-, and 1,4-position isomers—designated as *ortho*, *meta* and *para* respectively.

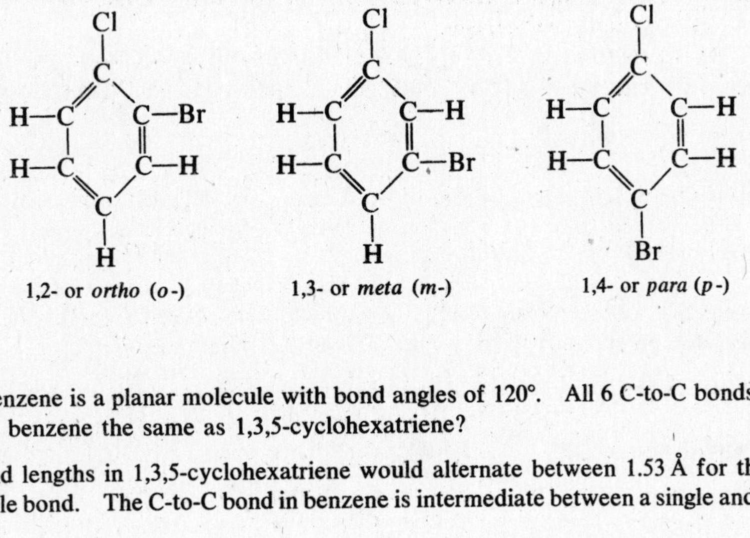

| 1,2- or *ortho* (*o-*) | 1,3- or *meta* (*m-*) | 1,4- or *para* (*p-*) |

Problem 10.1 Benzene is a planar molecule with bond angles of 120°. All 6 C-to-C bonds have the identical length, 1.39 Å. Is benzene the same as 1,3,5-cyclohexatriene? ◄

No. The bond lengths in 1,3,5-cyclohexatriene would alternate between 1.53 Å for the single bond and 1.32 Å for the double bond. The C-to-C bond in benzene is intermediate between a single and double bond.

Problem 10.2 (*a*) How do the following heats of hydrogenation (ΔH_h, kcal/mole) show that benzene is not the ordinary triene 1,3,5-cyclohexatriene? Cyclohexene, -28.6; 1,4-cyclohexadiene, -57.2; 1,3-cyclohexadiene, -55.4; and benzene, -49.8. (*b*) Calculate the resonance (conjugation) energy of benzene. ◄

In computing the first column of Table 10-1, we assume that in the absence of any orbital interactions each double bond should contribute -28.6 kcal/mole to the total ΔH_h of the compound, since this is the ΔH_h of an isolated C=C (in cyclohexene). Any difference between such a calculated ΔH_h value and the observed value is the conjugation (resonance) energy (page 15). Since ΔH_h for 1,3-cyclohexadiene is 1.8 kcal/mole less than that for 1,4-cyclohexadiene, conjugation stabilizes the 1,3-isomer.

161

Table 10-1

	Calculated ΔH_h	Observed ΔH_h	Resonance (Conjugation) Energy
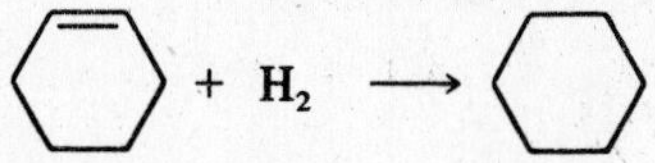 Cyclohexene + H_2 → Cyclohexane		-28.6	
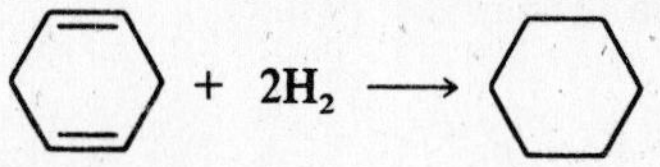 1,4-Cyclohexadiene + $2H_2$ →	$2(-28.6)$ $=-57.2$	-57.2	0
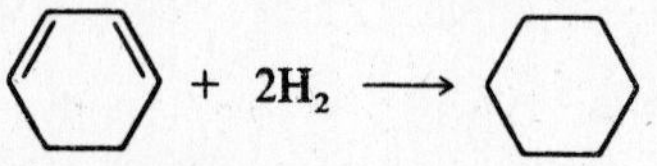 1,3-Cyclohexadiene + $2H_2$ →	$2(-28.6)$ $=-57.2$	-55.4	-1.8
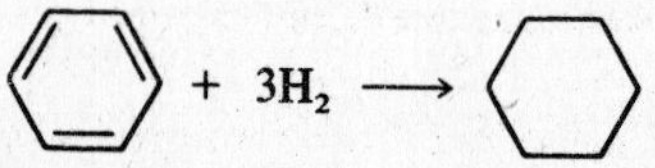 Benzene + $3H_2$ →	$3(-28.6)$ $=-85.8$	-49.8	-36.0

(a)   1,3,5-Cyclohexatriene should behave as a typical triene and have $\Delta H_h = -85.8$. The observed ΔH_h for benzene is -49.8. Benzene is *not* 1,3,5-cyclohexatriene; in fact, the latter does not exist.

(b)   See Table 10-1.

Problem 10.3   (a) Use the ΔH_h's given in Problem 10.2 for complete hydrogenation of cyclohexene, 1,3-cyclohexadiene and benzene to calculate ΔH_h for the addition of 1 mole of H_2 to (i) 1,3-cyclohexadiene, (ii) benzene. (b) What conclusion can you draw from these values about the rate of adding 1 mole of H_2 to these 3 compounds? (The ΔH of a reaction step is not necessarily related to $\Delta H^{\ddagger}$ of the step. However, in the cases being considered in this problem, $\Delta H_{reaction}$ is directly related to $\Delta H^{\ddagger}$.) (c) Can cyclohexadiene and cyclohexene be isolated on controlled hydrogenation of benzene?   ◀

Equations are written for the reactions so that their algebraic sum gives the desired reactant, products and enthalpy.

(a)   (i)   Add reactions (1) and (2):

(1) 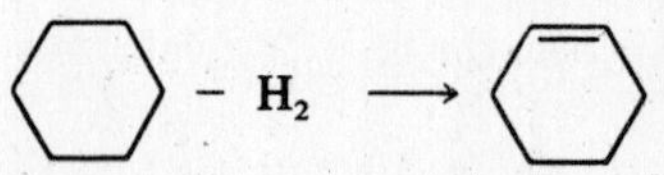   Cyclohexane $- H_2$ → Cyclohexene $\Delta H = +28.6$

(2) 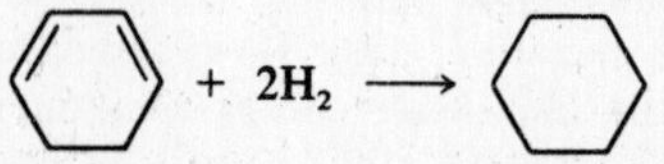   1,3-Cyclohexadiene $+ 2H_2$ → Cyclohexane $\Delta H_h = -55.4$

(3)   1,3-Cyclohexadiene $+ H_2$ → Cyclohexene $\Delta H_h = -26.8$

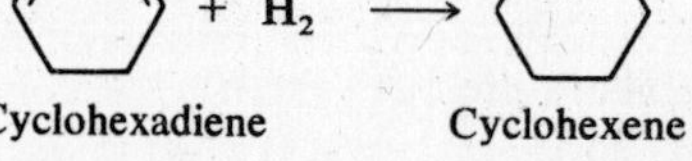

Note that reaction (1) is a dehydrogenation (reverse of hydrogenation) and the sign of its ΔH is positive.

(ii) Add the following 2 reactions:

$$\text{Cyclohexane} - 2H_2 \longrightarrow \text{1,3-Cyclohexadiene} \qquad \Delta H = +55.4$$

$$\text{Benzene} + 3H_2 \longrightarrow \text{Cyclohexane} \qquad \Delta H_h = -49.8$$

$$\text{Benzene} + H_2 \longrightarrow \text{1,3-Cyclohexadiene} \qquad \Delta H_h = +5.6$$

(b) The reaction with the largest negative ΔH_h value is the most exothermic and, in this case, also has the fastest rate. The ease of addition of 1 mole of H_2 is:

$$\text{cyclohexene } (-28.6) > \text{1,3-cyclohexadiene } (-26.8) \gg \text{benzene } (+5.6)$$

(c) No. When one molecule of benzene is converted to the diene, the diene is reduced all the way to cyclohexane by 2 more molecules of H_2 before more molecules of benzene react. If one mole each of benzene and H_2 are reacted, the product is 1/3 mole of cyclohexane and 2/3 mole of unreacted benzene.

Problem 10.4 The observed heat of combustion (ΔH_c) of C_6H_6 is -789.1 kcal/mole.* Theoretical values are calculated for C_6H_6 by adding the contributions from each bond obtained experimentally from other compounds; these are (in kcal/mole) -117.7 for C=C, -49.3 for C—C and -54.0 for C—H. Use these data to calculate the heat of combustion for C_6H_6 and the difference between this and the experimental value. Compare the difference with that from heats of hydrogenation. ◄

The contribution is calculated for each bond and these are totaled for the molecule.

$$\text{Six C—H bonds} \quad = 6(-54.0) \; = -324.0 \text{ kcal/mole}$$
$$\text{Three C—C bonds} = 3(-49.3) \; = -147.9$$
$$\text{Three C=C bonds} = 3(-117.7) = \underline{-353.1}$$
$$\text{TOTAL} = -825.0 \text{ (calculated } \Delta H_c \text{ for } C_6H_6)$$
$$\text{Experimental} = \underline{-789.1}$$
$$\text{Difference} = \;\;-35.9$$

This difference is the resonance energy of C_6H_6. The same value is obtained from ΔH_h (Problem 10.2).

Problem 10.5 How is the structure of benzene explained by (a) resonance, (b) the orbital picture, (c) molecular orbital theory? ◄

(a) Benzene is a hybrid of two equal-energy (Kekulé) structures differing only in the location of the double bonds:

(b) Each C is sp^2 hybridized and is σ bonded to 2 other C's and one H (Fig. 10-1). These σ bonds comprise the skeleton of the molecule. Each C also has 1 electron in a p orbital at right angles to the plane of the ring. These p orbitals overlap *equally* with each of the 2 adjacent p orbitals to form a π system parallel to and above and below the plane of the ring (Fig. 10-2). The 6 p electrons in the π system are associated with all 6 C's. They are therefore more *delocalized* and this accounts for the great stability and large resonance energy of aromatic rings.

---

*Some books define heat of combustion as $-\Delta H_c$ and values are given as positive numbers.

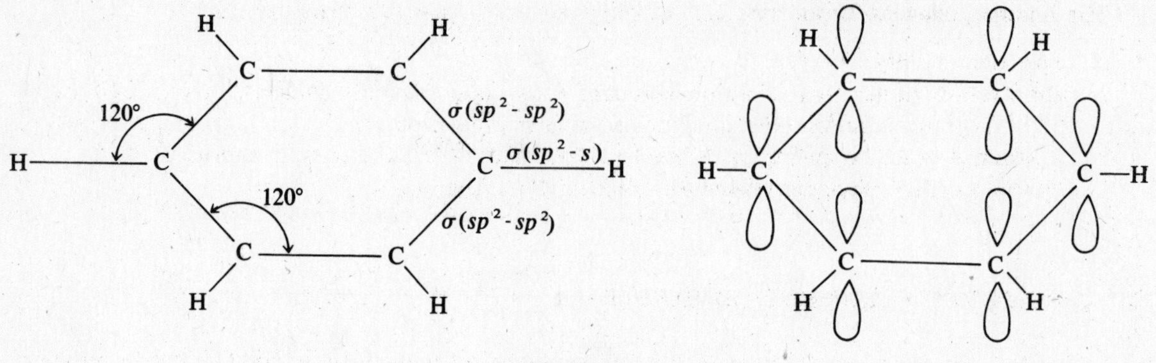

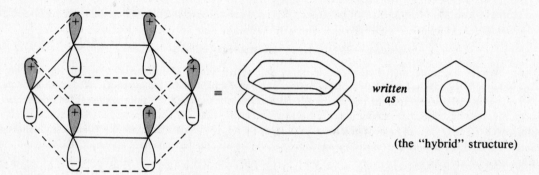

Fig. 10-1

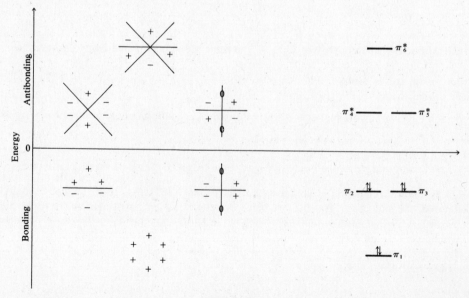

≡   *written as*   (the "hybrid" structure)

Fig. 10-2

(c)   The 6 p AO's discussed in part (b) interact to form 6 π MO's. These are indicated in Fig. 10-3, which gives the signs of the upper lobes (cf. Fig. 8-3 for butadiene). Since benzene is cyclic, the stationary waves representing the electron clouds are cyclic and have nodal planes, shown as lines, instead of nodal points. See Problem 9.26 for the significance of a 0 sign. The 6 p electrons fill the 3 bonding MO's, thereby accounting for the stability of C_6H_6.

Fig. 10-3

The unusual benzene properties collectively known as **aromatic character** are:

1. **Thermal stability.**
2. **Substitution rather than addition reactions** with polar reagents such as HNO_3, H_2SO_4 and Br_2. In these reactions the unsaturated ring is preserved.
3. **Resistance to oxidation** by aq. $KMnO_4$, HNO_3 and all but the most vigorous oxidants.
4. **Unique nuclear magnetic resonance spectra.** (See page 202.)

10.2 AROMATICITY AND HÜCKEL'S RULE

Hückel's rule (1931) for aromaticity states that **if the number of π electrons is equal to $2 + 4n$, where n equals zero or a whole number, the system is aromatic.** The rule applies to carbon-containing monocyclics in which each C is capable of being sp^2 hybridized to provide a p orbital for extended π bonding. It has been extended to unsaturated heterocyclic compounds and fused ring compounds. All atoms participating in π bonding must line in the same plane. The system can be an ion.

The numbers of π electrons that fit Hückel's rule are $2\,(n = 0), 6\,(n = 1), 10\,(n = 2), 14\,(n = 3),$ $18\,(n = 4)$, etc; benzene has 6.

Problem 10.6 Account for aromaticity observed in: (*a*) 1,3-cyclopentadienyl anion but not 1,3-cyclopentadiene; (*b*) 1,3,5-cycloheptatrienyl cation but not 1,3,5-cycloheptatriene; (*c*) cyclopropenyl cation; (*d*) the heterocycles pyrrole, furan and pyridine. ◄

(*a*) 1,3-Cyclopentadiene has an sp^3 hybridized C, making cyclic p orbital overlap impossible. Removal of H^+ from this C leaves a carbanion whose C is now sp^2 hybridized and has a p orbital capable of overlapping to give a cyclic π system. The 4 π electrons from 2 double bonds plus the 2 unshared electrons total 6 π electrons; the anion is aromatic.

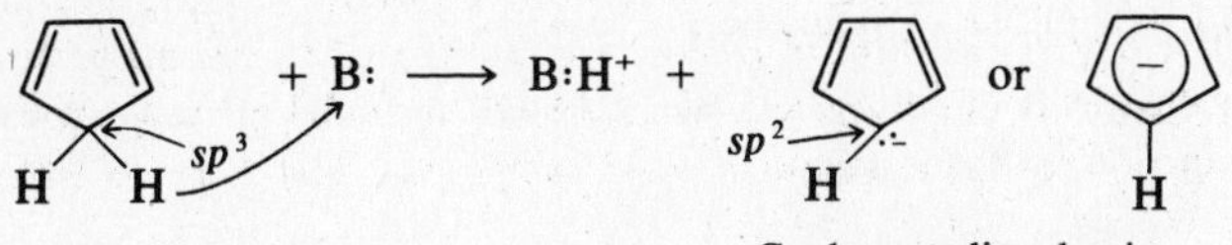

Cyclopentadienyl anion

(*b*) Although the triene has 6 p electrons in 3 C=C bonds, the lone sp^3 hybridized C prevents cyclic overlap of p orbitals.

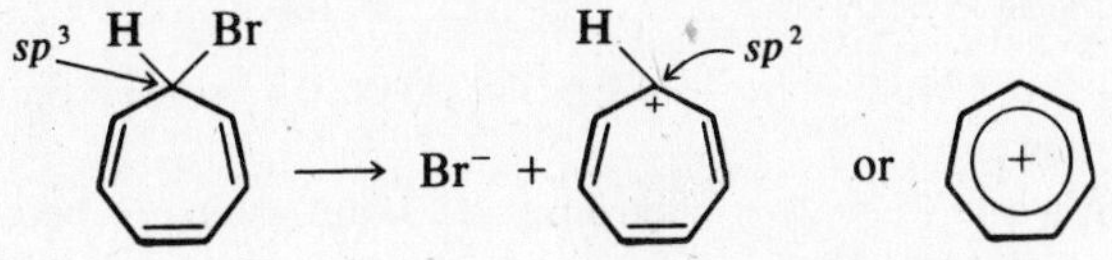

Cycloheptatrienyl cation

Generation of a carbonium ion as shown above permits cyclic overlap of p orbitals on each C, and with 6 π electrons the cation is aromatic.

(*c*) Cyclopropenyl cation has 2 π electrons and $n = 0$.

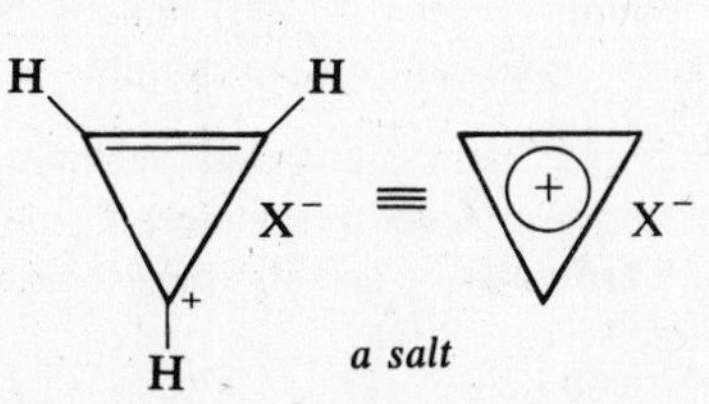

a salt

The ions in parts (*a*), (*b*) and (*c*) are reactive but they are much more stable than the corresponding acyclic ions.

(d) Hückel's rule is extended to heterocyclic compounds as follows:

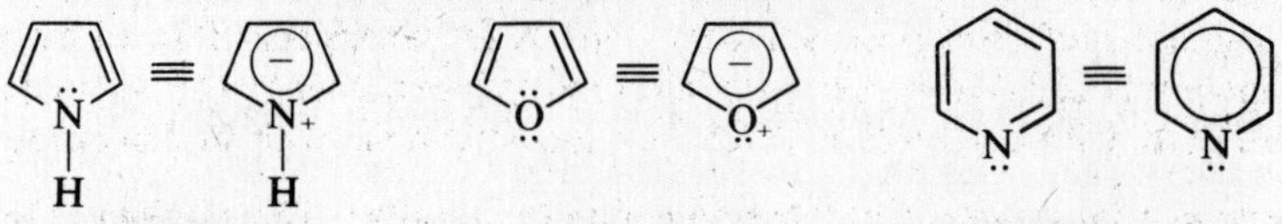

Pyrrole	Furan	Pyridine
(6 π electrons;	(6 π electrons;	(6 π electrons;
2 unshared	only 2 electrons	(the unshared pair
electrons on N	on O participate in	of electrons on N
overlap in the	the π system)	does not participate
π system)		in the π overlap)

Note that dipoles are generated in pyrrole and furan because of delocalization of electrons from the heteroatoms.

10.3 ANTIAROMATICITY

Planar cyclic conjugated species less stable than corresponding acyclic unsaturated species are called **antiaromatic**. They have $4n$ π electrons. 1,3-Cyclobutadiene ($n = 1$), for which one can write two equivalent contributing structures, is an extremely unstable antiaromatic molecule. This shows that the ability to write equivalent contributing structures is not sufficient to predict stability.

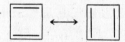

Problem 10.7 Cyclooctatetraene (C_8H_8), unlike benzene, is not aromatic; it decolorizes both dil. aq. $KMnO_4$ and Br_2 in CCl_4. Its experimentally determined heat of combustion is -1095.0 kcal/mole. (a) Use the Hückel rule to account for the differences in chemical properties of C_8H_8 from those of benzene. (b) Use thermochemical data in Problem 10.4 to calculate the resonance energy. (c) Compare the bond lengths and shape of this compound with those of benzene. ◄

(a) C_8H_8 has 8 rather than 6 p electrons. Since it is not aromatic it undergoes addition reactions.
(b) The calculated heat of combustion is:

$$8 \text{ C—H bonds} = 8(-54.0) = -432.0 \text{ kcal/mole}$$
$$4 \text{ C—C bonds} = 4(-49.3) = -197.2$$
$$4 \text{ C}=\text{C bonds} = 4(-117.7) = \underline{-470.8}$$
$$\text{TOTAL: } -1100.0$$

The difference $-1100.0 - (-1095.0) = -5.0$ kcal/mole shows little resonance energy and no aromaticity.

(c) Although the molecule has $4n$ ($n = 2$) π electrons it is *not* antiaromatic, because unlike benzene it is not planar. It exists chiefly in a "tub" conformation (Fig. 10-4). It also differs from benzene in having typical alternate single and double bond lengths.

Fig. 10-4

Problem 10.8 The deep blue compound azulene ($C_{10}H_8$) has a 5- and 7-membered ring fused through two adjacent C's. It is aromatic and has a significant dipole moment of 1.0 D. Explain. ◄

Azulene can be written as fused cyclopentadiene and cycloheptatriene rings, neither of which alone is aromatic. However, some of its resonance structures have a fused cyclopentadienyl anion and cycloheptatrienyl cation, which accounts for its aromaticity and its dipole moment of 1.0 D.

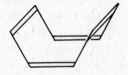

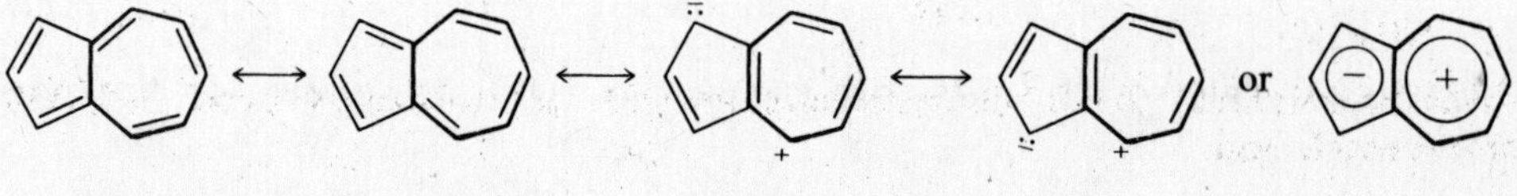

Problem 10.9 Deduce the structure and account for the stability of the following substances which are insoluble in nonpolar but soluble in polar solvents. (*a*) A red compound formed by reaction of 2 moles of AgBF$_4$ with one mole of 1,2,3,4-tetraphenyl-3,4-dibromocyclobut-1-ene. (*b*) A stable compound from the reaction of 2 moles of K with one mole of 1,3,5,7-cyclooctatetraene with no liberation of H$_2$. ◄

The solubility properties suggest that these compounds are salts. The stability of the organic ions formed indicates that they conform to the Hückel rule and are aromatic.

(*a*) Two Br$^-$ are abstracted by two Ag$^+$ to form two AgBr and a tetraphenylcyclobutenyl dication.

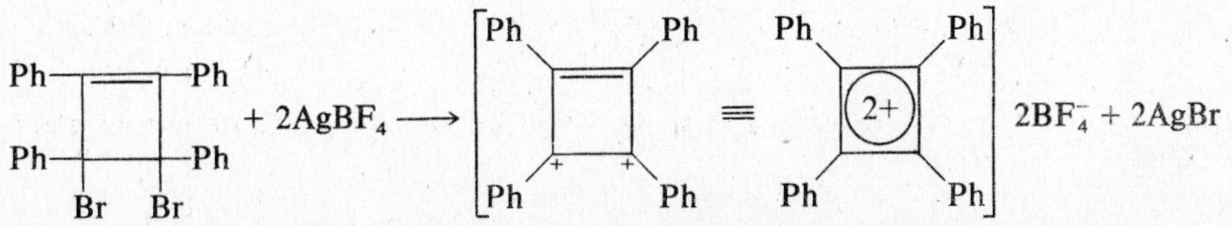

an aromatic cation ($n = 0$)

(*b*) Since K· is a strong reductant and no H$_2$ is evolved, 2 K's supply 2 electrons to form a cyclooctatetraenyl dianion (Fig. 10-5). This planar conjugated unsaturated monocycle has 10 electrons, conforms to the Hückel rule ($n = 2$) and is aromatic.

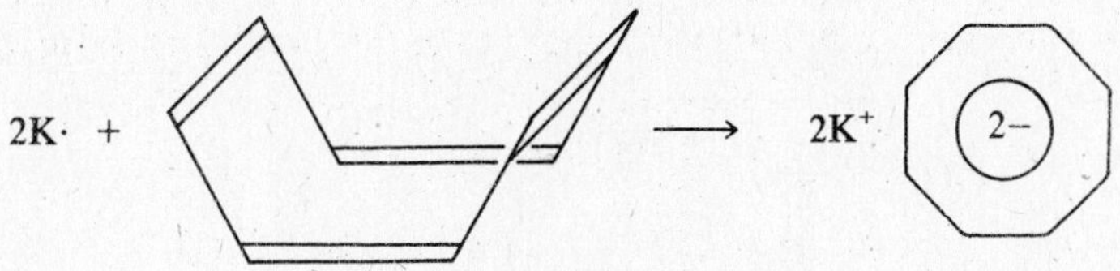

Fig. 10-5

10.4 NOMENCLATURE

Some common names are **toluene** (C$_6$H$_5$CH$_3$), **xylene** (C$_6$H$_4$(CH$_3$)$_2$), **phenol** (C$_6$H$_5$OH), **aniline** (C$_6$H$_5$NH$_2$), **benzaldehyde** (C$_6$H$_5$CHO), **benzoic acid** (C$_6$H$_5$COOH), **benzenesulfonic acid** (C$_6$H$_5$SO$_3$H), **styrene** (C$_6$H$_5$CH=CH$_2$), and **mesitylene** (1,3,5-(CH$_3$)$_3$C$_6$H$_3$).

Derived names combine the name of the substituent as a prefix with the word *benzene*. Examples are **nitrobenzene** (C$_6$H$_5$NO$_2$), **ethylbenzene** (C$_6$H$_5$CH$_2$CH$_3$) and **fluorobenzene** (C$_6$H$_5$F).

Some **aryl** (Ar—) groups are: C$_6$H$_5$— (**phenyl**), *p*-CH$_3$C$_6$H$_4$— (*p*-tolyl) and (CH$_3$)$_2$C$_6$H$_3$— (**xylyl**). Some **arylalkyl** groups are: C$_6$H$_5$CH$_2$— (**benzyl**), C$_6$H$_5$CH— (**benzal**), C$_6$H$_5$C≡ (**benzo**), (C$_6$H$_5$)$_2$CH— (**benzhydryl**), (C$_6$H$_5$)$_3$C— (**trityl**).

The order of decreasing priorities of common substituents is: COOH, SO$_3$H, CHO, CN, C=O, OH, NH$_2$, R, NO$_2$, X. For disubstituted benzenes with a group giving the ring a common name, *o-*, *p-* or *m-* is used to designate the position of the second group. Otherwise positions of groups are designated by the lowest combination of numbers.

Problem 10.10 Name the compounds:

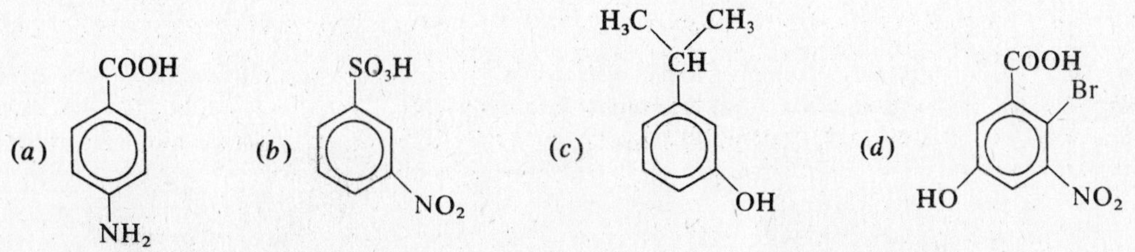

(*a*) *p*-Aminobenzoic acid. (*b*) *m*-Nitrobenzenesulfonic acid. (*c*) *m*-Isopropylphenol. (*d*) 2-Bromo-3-nitro-5-hydroxybenzoic acid.

Problem 10.11 Give the structural formulas for   (*a*) 2,4,6-tribromoaniline,   (*b*) *m*-toluenesulfonic acid, (*c*) *p*-bromobenzalbromide,   (*d*) di-*o*-tolylmethane,   (*e*) trityl chloride. ◀

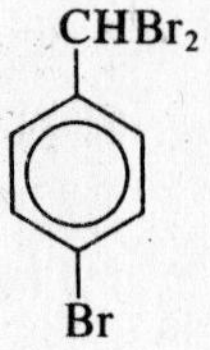

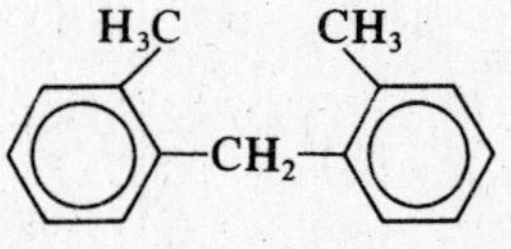

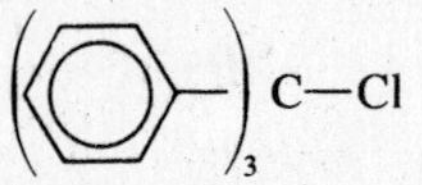

Problem 10.12 Which xylene gives   (*a*) 1,   (*b*) 2,   (*c*) 3 monobromo derivatives?   Name the derivatives. (Numbers are not used with *o*-, *m*- and *p*-designations.) ◀

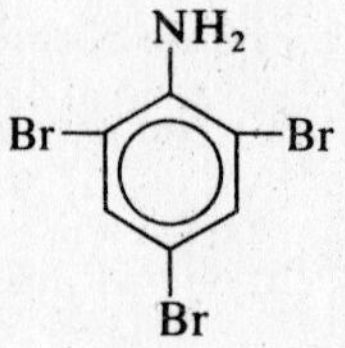

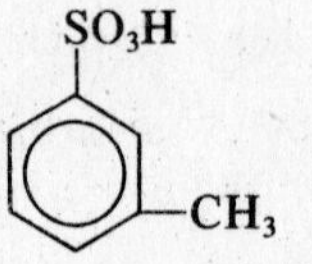

(*a*)   Monobromination of *p*-xylene yields 1 monobromo derivative because the 4 available positions, labeled *a*, are equivalent.

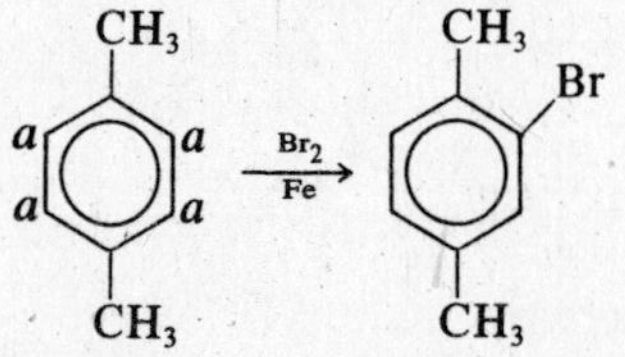

p-Xylene   2-Bromo-1,4-dimethylbenzene

(*b*)   *o*-Xylene has 2 types of positions, labeled *a* and *b*.

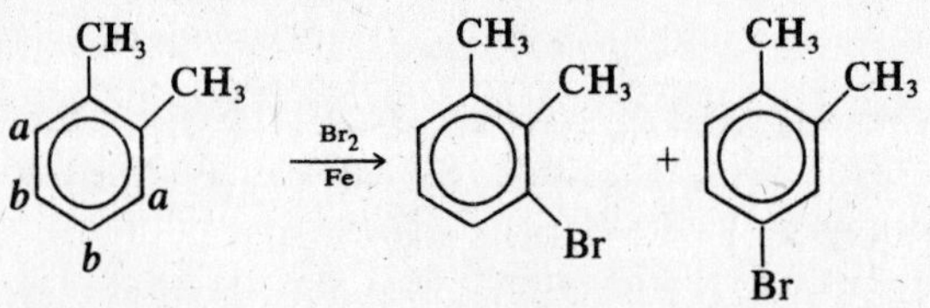

o-Xylene   3-Bromo-1,2-   4-Bromo-1,2-dimethylbenzene
                dimethylbenzene

(*c*)   *m*-Xylene has 3 types of positions, designated *a*, *b*, and *c*.

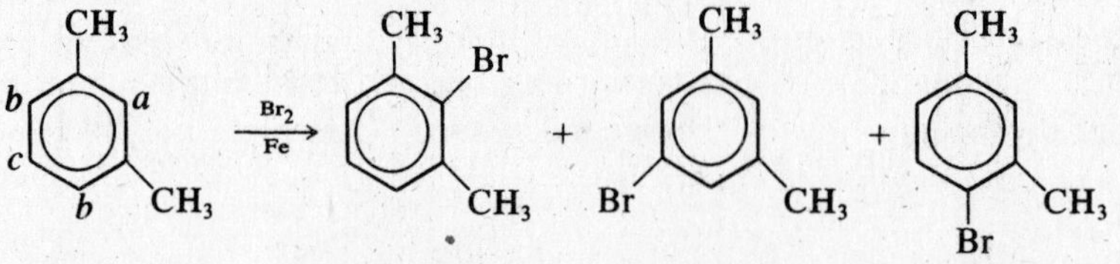

m-Xylene   2-Bromo-1,3-   5-Bromo-1,3-   4-Bromo-1,3-dimethylbenzene
                dimethylbenzene   dimethylbenzene

Problem 10.13 Give structural formulas and names for all isomers of   (*a*) nitroanilines,   (*b*) tribromobenzenes,   (*c*) diaminobenzoic acids,   (*d*) tribromochlorobenzenes. ◀

(*a*)

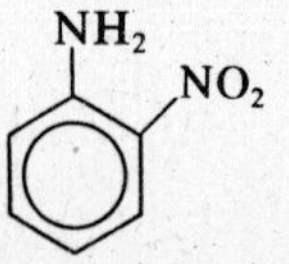

o-Nitroaniline 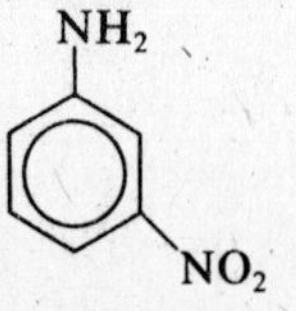   m-Nitroaniline 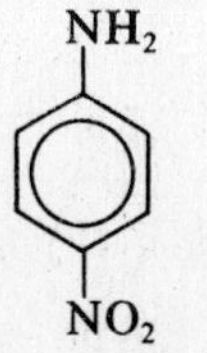   p-Nitroaniline

(b) [Indicated letters are for use in part (d).]

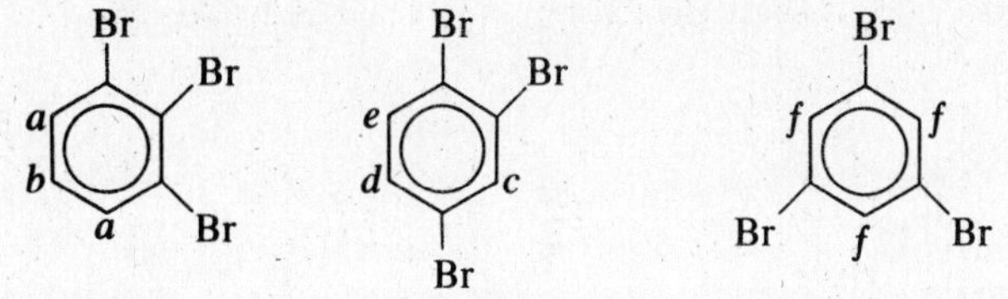

1,2,3-   1,2,4-   1,3,5-Tribromobenzene

(c) Six isomeric diaminobenzoic acids are derived when COOH is introduced into p-, o- and m-diaminobenzenes.

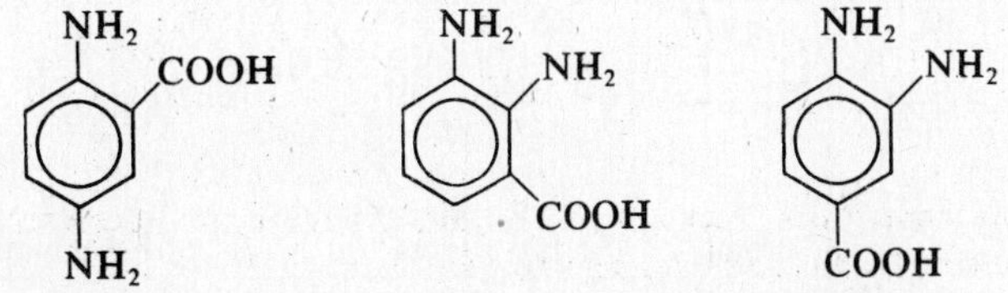

2,5-   2,3-   3,4-

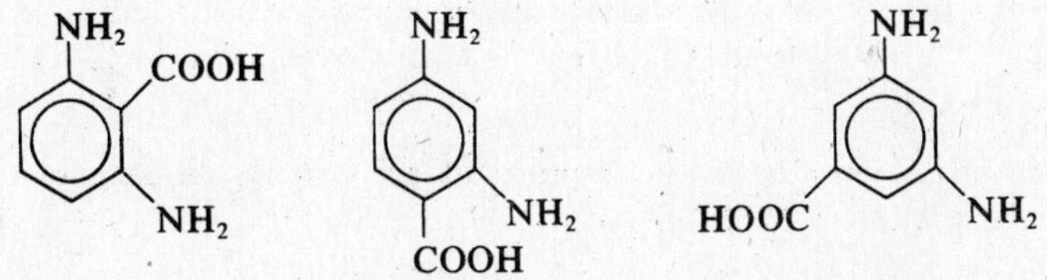

2,6-   2,4-   3,5-Diaminobenzoic acid

(d) See part (b) for the different positions that can be monochlorinated in the three tribromobenzenes.

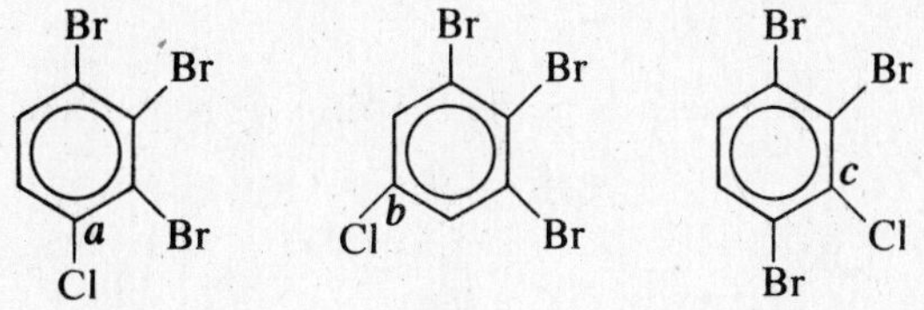

2,3,4-   3,4,5-   2,3,6-

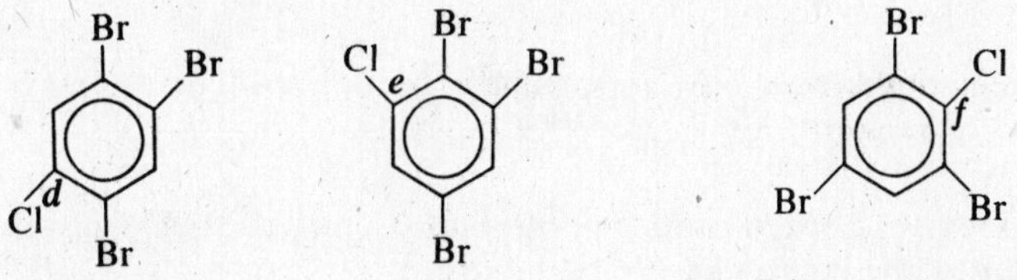

2,4,5-   2,3,5-   2,4,6-Tribromochlorobenzene

Problem 10.14 Give structural formulas and names for all possible isomers formed from (a) monochlorination of p-nitrotoluene, (b) sulfonation of o-chlorobromobenzene (introduce SO₃H group), (c) nitration of m-bromotoluene (introduce NO₂ group). ◀

Letters are used to show different positions in the reactants.

(a)

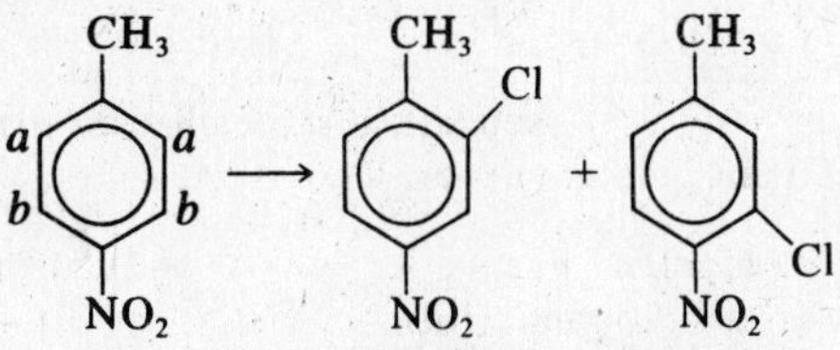

2-Chloro-4-   3-Chloro-4-
nitrotoluene   nitrotoluene

(b)

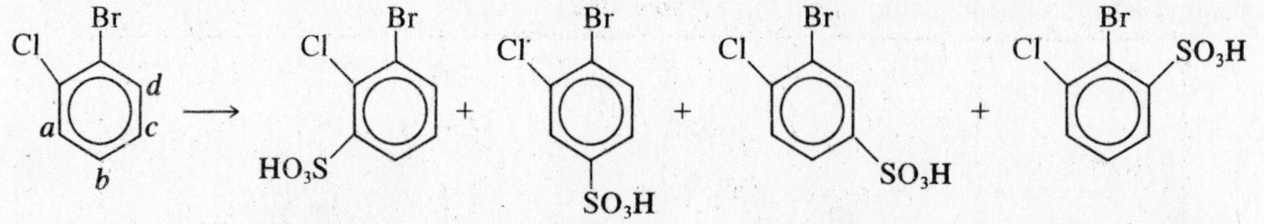

2-Chloro-3-bromo-benzenesulfonic acid + 3-Chloro-4-bromo-benzenesulfonic acid + 3-Bromo-4-chloro-benzenesulfonic acid + 2-Bromo-3-chloro-benzenesulfonic acid

(c)

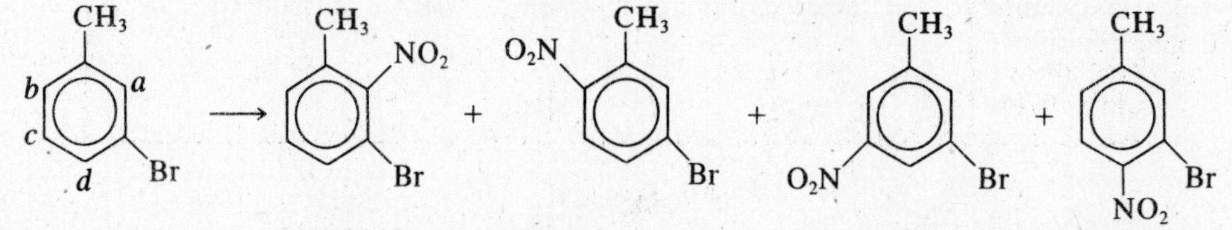

2-Nitro-3-bromo-toluene + 2-Nitro-5-bromo-toluene + 3-Nitro-5-bromo-toluene + 3-Bromo-4-nitro-toluene

Problem 10.15 Give the structures of the **alkylbenzenes** that can have the indicated numbers of monosubstituted derivatives. (a) 1, 2, or 3 isomeric monoiodo derivatives of C_8H_{10}. (b) 1, 2, 3, or 4 isomeric monofluoro derivatives of C_9H_{12}. ◄

Isomeric alkylbenzenes having the molecular formula are written with equivalent positions on the ring designated by the letters a, b, etc.

(a) Six C's are needed for the benzene ring, which leaves two C's attached to the ring either as two CH_3's or as CH_2CH_3. There are four C_8H_{10} alkylbenzenes, as shown in Table 10-2.

(b) Here three C's are attached to the ring either as an n-Pr, an i-Pr, an Et and a Me, or as three Me's. See Table 10-3.

Problem 10.16 Use the Diels-Alder reaction in the synthesis of benzoic acid, C_6H_5COOH. ◄

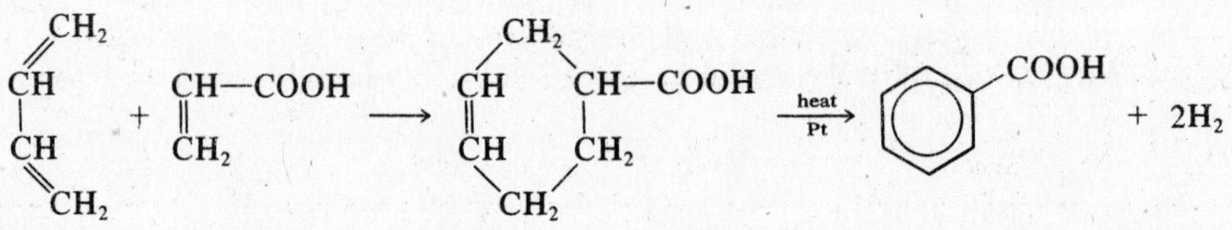

The stability of the benzene ring causes dehydrogenation to be favored (2nd step). S and Se can be used in place of Pt, and H_2S and H_2Se are then the respective products.

Problem 10.17 Sondheimer synthesized a series of interesting conjugated cyclopolyalkenes that he designated the [n]-**annulenes**, where n is the number of C's in the ring.

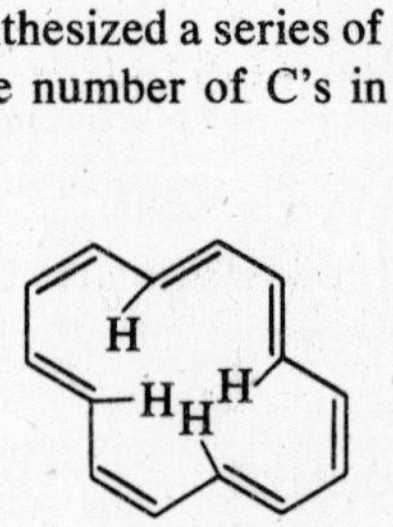

[14]-Annulene 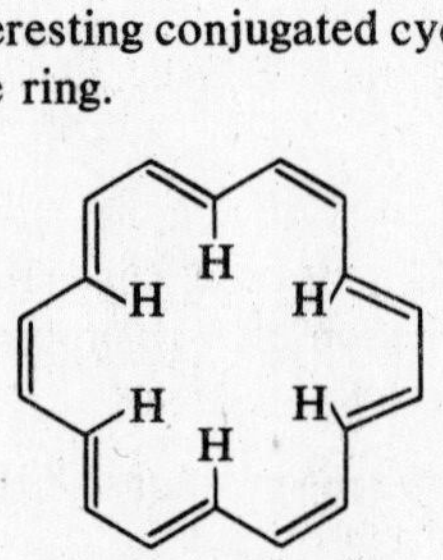   [18]-Annulene

Account for his observations that (a) [18]-annulene is somewhat aromatic, [16]- and [20]-annulene are not; (b) [18]-annulene is more stable than [14]-annulene. ◄

(a) The somewhat aromatic [18]-annulene has $4n + 2$ ($n = 4$) pi electrons; there are $4n$ pi electrons in the nonaromatic, nonplanar [16]- and [20]-annulenes.

(b) [14]-Annulene is somewhat strained because the H's in the center of the ring are crowded. This steric strain prevents a planar conformation, which diminishes aromaticity.

Table 10-2

Isomer	CH₃ ... CH₃ *p*-Xylene	CH₃ CH₃ ... *o*-Xylene	CH₃ ... CH₃ *m*-Xylene	CH₂CH₃ ... Ethylbenzene
Number of Monoiodo Derivatives	1	2	3	3

Table 10-3

Isomer	CH₂CH₂CH₃ *n*-	CH(CH₃)₂ *iso*-	CH₂CH₃ ... CH₃ *para*	CH₂CH₃ CH₃ ... *ortho*
Number of Monofluoro Derivatives	3	3	2	4
Isomer	CH₂CH₃ ... CH₃ *meta*	CH₃ CH₃ ... CH₃ 1,2,3-	CH₃ CH₃ ... CH₃ 1,2,4-	CH₃ ... H₃C ... CH₃ 1,3,5-
Number of Monofluoro Derivatives	4	2	3	1

Problem 10.18 Use the Hückel rule to indicate whether the following planar species are aromatic or antiaromatic:

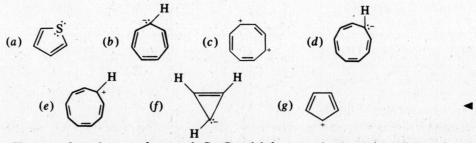

(*a*) Aromatic. There are 2 π electrons from each C=C and 2 from an electron pair on S to make an aromatic sextet. (*b*) Antiaromatic. There are $4n$ ($n = 2$) π electrons. (*c*) Aromatic. There are 6 π electrons. (*d*) Aromatic. There are 10 π electrons and this anion conforms to the $(4n + 2)$ rule ($n = 2$). (*e*) Antiaromatic. The cation has $4n$ ($n = 2$) π electrons. (*f*) and (*g*) Antiaromatic. They have $4n$ ($n = 1$) π electrons.

Supplementary Problems

Problem 10.19 The relative energies of the MO's of conjugated cyclic polyenes can be determined by the following simple method instead of using nodal planes as in Problem 10.5(c). Inscribe a regular polygon in a circle, with one vertex at the bottom of the circle and with the total number of vertices equal to the number of MO's. Then the height of a vertex is proportional to the energy of the associated MO. Vertices below the horizontal diameter are bonding π, those above are antibonding π^*, and those on the diameter are nonbonding π^n. Apply the method to 3-, 4-, 5-, 6-, 7- and 8-carbon systems and indicate the character of the MO's. ◄

     See Fig. 10-6.

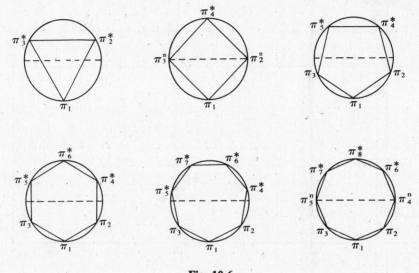

Fig. 10-6

Problem 10.20 Explain aromaticity and antiaromaticity in terms of MO energy levels. ◄

     Aromaticity is observed when all bonding MO's are filled and nonbonding MO's, if present, are empty or completely filled. Hückel's rule arises from this requirement. A species is antiaromatic if it has electrons in antibonding MO's or if it has half-filled bonding or nonbonding MO's, provided it is planar.

Problem 10.21 Design a table showing the structure, number of π electrons, energy levels of π MO's showing electron distribution, and state of aromaticity of:

(a)   cyclopropenyl cation
(b)   cyclopropenyl anion
(c)   cyclobutadiene
(d)   cyclobutadienyl dication
(e)   cyclopentadienyl anion
(f)   cyclopentadienyl cation
(g)   benzene
(h)   cycloheptatrienyl anion
(i)   cyclooctatetraene
(j)   cyclooctatetraenyl dianion. ◄

     See Table 10-4. (H's are understood to be attached to each doubly bonded C.)

Table 10-4

Structure	Number of π Electrons	π MO's	Aromaticity
(a) H	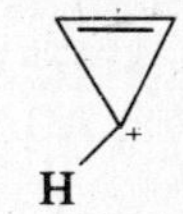 2	π_3^*— —π_2^* $\underset{\uparrow\downarrow}{}\pi_1$	Aromatic
(b) H	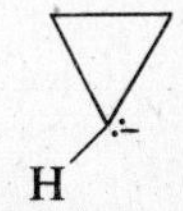 4	$\pi_3^*\uparrow$ $\uparrow\pi_2^*$ $\underset{\uparrow\downarrow}{}\pi_1$	Antiaromatic
(c)	 4	—π_4^* $\pi_3^n\uparrow$ $\uparrow\pi_2^n$ $\underset{\uparrow\downarrow}{}\pi_1$	Antiaromatic
(d) H H	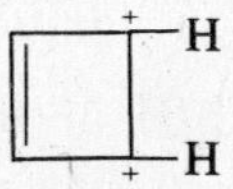 2	—π_4^* π_3^n— —π_2^n $\underset{\uparrow\downarrow}{}\pi_1$	Aromatic
(e) H	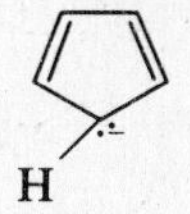 6	π_5^*— —π_4^* $\pi_3\underset{\uparrow\downarrow}{}$ $\underset{\uparrow\downarrow}{}\pi_2$ $\underset{\uparrow\downarrow}{}\pi_1$	Aromatic
(f) H	 4	π_5^*— —π_4^* $\pi_3\uparrow$ $\uparrow\pi_2$ $\underset{\uparrow\downarrow}{}\pi_1$	Antiaromatic
(g)	 6	—π_6^* π_5^*— —π_4^* $\pi_3\underset{\uparrow\downarrow}{}$ $\underset{\uparrow\downarrow}{}\pi_2$ $\underset{\uparrow\downarrow}{}\pi_1$	Aromatic
(h) H	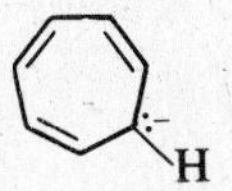 8	π_7^*— —π_6^* $\pi_5^*\uparrow$ $\uparrow\pi_4^*$ $\pi_3\underset{\uparrow\downarrow}{}$ $\underset{\uparrow\downarrow}{}\pi_2$ $\underset{\uparrow\downarrow}{}\pi_1$	Antiaromatic
(i)	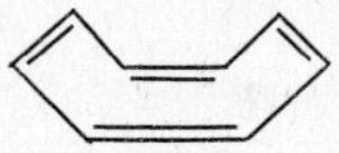 8	—π_8^* π_7^*— —π_6^* $\pi_5^n\uparrow$ $\uparrow\pi_4^n$ $\pi_3\underset{\uparrow\downarrow}{}$ $\underset{\uparrow\downarrow}{}\pi_2$ $\underset{\uparrow\downarrow}{}\pi_1$	Nonaromatic[†] (nonplanar)
(j) $[\quad]^{2-}$	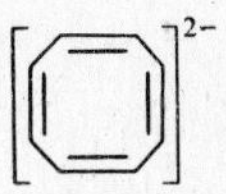 10	—π_8^* π_7^*— —π_6^* $\pi_5^n\underset{\uparrow\downarrow}{}$ $\underset{\uparrow\downarrow}{}\pi_4^n$ $\pi_3\underset{\uparrow\downarrow}{}$ $\underset{\uparrow\downarrow}{}\pi_2$ $\underset{\uparrow\downarrow}{}\pi_1$	Aromatic

†If it were planar it would be antiaromatic; to avoid this, (i) is nonplanar.

Chapter 11

Aromatic Substitution. Arenes

11.1 AROMATIC SUBSTITUTION BY ELECTROPHILES (LEWIS ACIDS, E^+ OR E)

MECHANISM

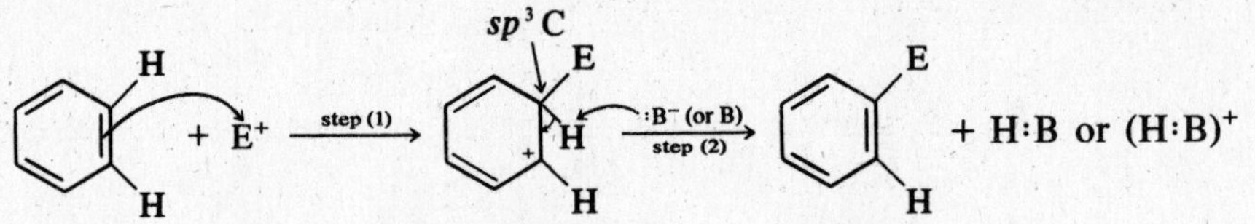

| Benzene | Lewis acid | Benzenonium ion (strong Brönsted acid) | Substituted benzene |

or

$$C_6H_6 + E^+ \longrightarrow \left[C_6H_5 \begin{array}{c} E \\ H \end{array} \right]^+ \xrightarrow[\text{or B}]{:B^-} C_6H_5E + B \cdot H \text{ or } B \cdot H^+$$

Problem 11.1 Account for the relative stability of the benzenonium ion by (a) resonance theory and (b) charge delocalization. ◄

(a)

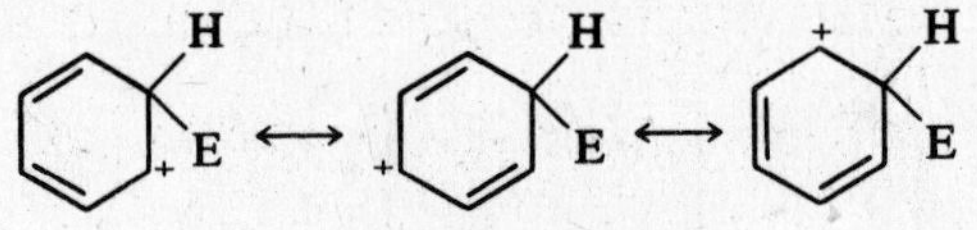

contributing structures

Note that + is at C's *ortho* and *para* to sp^3 hybridized C, which is the one bonded to E^+.

(b) The benzenonium ion is a type of allylic cation (see Problem 8.21). The 5 remaining C's using sp^2 hybridized orbitals each have a p orbital capable of overlapping laterally to give a delocalized π structure as indicated.

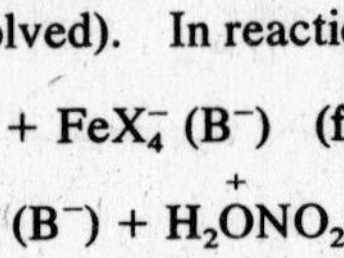

delocalized (hybrid) structure

The $\delta +$ indicates positions where + charge exists in delocalized structure.

Problem 11.2 For each electrophilic aromatic substitution in Table 11-1, give equations for formation of E^+ and indicate what is B^- or B (several bases may be involved). In reaction (c) the electrophile is a molecule, E. ◄

(a)
$$X_2 + FeX_3 \longrightarrow X^+ (E^+) + FeX_4^- (B^-) \quad \text{(forms HX + FeX_3)}$$

(b)
$$H_2SO_4 + HONO_2 \longrightarrow HSO_4^- (B^-) + H_2\overset{+}{O}NO_2 \longrightarrow H_2O + NO_2^+ (E^+)$$

(c)
$$2H_2SO_4 \longrightarrow H_3O^+ + HSO_4^- (B^-) + SO_3 (E)$$

$$C_6H_6 + SO_3 \longrightarrow \overset{+}{C_6H_5} \begin{array}{c} H \\ SO_3^- (B^-) \end{array} \longrightarrow C_6H_5SO_3H$$

174

Table 11-1

Reaction	Reagent	Catalyst	Product
(a) Halogenation	X_2 (X = Cl, Br)	FeX_3 (from Fe + X_2)	ArCl, ArBr
(b) Nitration	HNO_3	H_2SO_4	$ArNO_2$
(c) Sulfonation	H_2SO_4 or $H_2S_2O_7$	none	$ArSO_3H$
(d) Friedel-Craft Alkylation	RX, $ArCH_2X$ ROH $\overset{H\ \ H}{RC=CH}$	$AlCl_3$ HF, H_2SO_4 or BF_3 H_3PO_4 or HF	Ar—R, Ar—CH_2Ar Ar—R Ar—$\underset{R}{CHCH_3}$
(e) Friedel-Craft Acylation	RCOCl	$AlCl_3$	Ar—$\overset{\overset{O}{\|\|}}{C}$—R
(f) Thallation	$Tl(OCOCF_3)_3$	CF_3COOH	$ArTl(OCOCF_3)_2$

(d) $$RX + AlX_3 \longrightarrow R^+ (E^+) + AlX_4^- (B^-) \quad (forms\ HAlX_4)$$

$$ROH + HF \longrightarrow R^+ (E^+) + H_2O (B) + F^- (B^-)$$

$$-\overset{\|}{C}=\overset{\|}{C}- + H_3PO_4 \longrightarrow -\overset{\|}{C}-\overset{\|}{C}H (E^+) + H_2PO_4^- (B^-)$$

(e) $$RCOCl + AlCl_3 \longrightarrow \underbrace{R\overset{+}{C}=\overset{..}{O}: \longleftrightarrow RC\equiv\overset{+}{O}:}_{Oxycarbonium\ ion} (E^+) + AlCl_4^- (B^-)$$

(f) $$Tl(OCOCF_3)_3 \xrightarrow{CF_3COOH} \overset{+}{Tl}(OCOCF_3)_2 (E^+) + \bar{O}COCF_3 (B^-)$$

Problem 11.3   How does the absence of a primary isotope effect prove experimentally that the first step in aromatic electrophilic substitution is rate-determining?   ◄

A C—H bond is broken faster than is a C—D bond.   This rate difference (isotope effect, k_H/k_D) is observed only if the C—H (or C—D) bond is broken in the rate-determining step.   If no difference is observed, as is the case for most aromatic electrophilic substitutions, C—H bond-breaking must occur in a fast step (in this case the second step).   Therefore, the first step, involving no C—H bond-breaking, is rate-determining.   This slow step requires the loss of aromaticity, the fast second step restores the aromaticity.

Problem 11.4   How is E^+ generated, and what is the base, in the following reactions?   (a) Nitration of reactive aromatics with HNO_3 alone.   (b) Chlorination with HOCl using HCl as catalyst.   (c) Nitrosation (introduction of a NO group) of reactive aromatics with HONO in strong acid.   (d) Deuteration with DCl.   ◄

(a) $$HNO_3 + H-O-NO_2 \longrightarrow NO_3^- + \left[H-\overset{H}{\underset{+}{O}}-NO_2 \right] \longrightarrow H_2O\ (Base) + \overset{+}{N}O_2\ (E^+)$$
$$\underset{unstable}{} \qquad \underset{Nitronium\ ion}{}$$

(b) $$H^+ + H-O-Cl \longrightarrow H-\overset{H}{\underset{+}{O}}-Cl \longrightarrow H_2O\ (Base) + Cl^+ (E^+)$$

(c) $H-O-N=O + H^+ \longrightarrow H-\overset{H}{\underset{|}{\overset{+}{O}}}-N=O \longrightarrow H_2O \text{ (Base)} + NO^+ (E^+)$

                                                                    Nitrosonium ion

(d) D^+, transferred by DCl to benzene. Base is Cl^-.

$$C_6H_6 + DCl \longrightarrow \left[C_6H_5 \overset{H}{\underset{D}{\diagdown\!\!\diagup}} \right]^+ + Cl^- \longrightarrow C_6H_5D + HCl$$

        Base$_1$ Acid$_2$ Acid$_1$ Base$_2$

Problem 11.5   Mercuric acetate,

$$Hg(O-\overset{\overset{\displaystyle O}{\|}}{C}-CH_3)_2$$

a covalent compound, reacts with benzene in the presence of H^+ (from $HClO_4$) to give

$$C_6H_5HgO\overset{\overset{\displaystyle O}{\|}}{C}CH_3$$

(a mercuration reaction). (a) Suggest a mechanism for this electrophilic substitution. (b) The reaction has a significant isotope effect of 6.   Explain in terms of the mechanism. ◀

(a) $CH_3\underset{\underset{\displaystyle O}{\|}}{C}OHgO\underset{\underset{\displaystyle O}{\|}}{C}CH_3 + H^+ \longrightarrow CH_3\underset{\underset{\displaystyle O}{\|}}{C}-\overset{+}{\underset{\underset{\displaystyle H}{|}}{O}}HgO\underset{\underset{\displaystyle O}{\|}}{C}CH_3 \longrightarrow CH_3\underset{\underset{\displaystyle O}{\|}}{C}OH + \overset{+}{Hg}O\underset{\underset{\displaystyle O}{\|}}{C}CH_3 \,(E^+)$

$$\overset{+}{Hg}-O-\overset{\overset{\displaystyle O}{\|}}{C}-CH_3 + C_6H_6 \rightleftharpoons \left[C_6H_5 \overset{H}{\underset{HgOCOCH_3}{\diagdown\!\!\diagup}} \right]^+ \xrightarrow[\text{slow}]{-H^+} C_6H_5HgO\overset{\overset{\displaystyle O}{\|}}{C}CH_3$$

      $Hg(OCOCH_3)_2$ is the base in the last step.

(b)   A positive isotope effect means that a C—H bond is broken in the rate-determining step; in this case, the second step.

ORIENTATION AND ACTIVATION OF SUBSTITUENTS

The 5 ring H's of monosubstituted benzenes, C_6H_5G, are not equally reactive.   Introduction of E into C_6H_5G rarely gives the statistical distribution of 40% *ortho*, 40% *meta* and 20% *para* disubstituted benzenes.   The substituent(s) determines (a) the orientation of E (*meta* or a mixture of *ortho* and *para*) and (b) the reactivity of the ring toward substitution.

Problem 11.6   (a) Give the delocalized structure (Problem 11.1) for the 3 benzenonium ions resulting from the common ground state for electrophilic substitution, $C_6H_5G + E^+$. (b) Give resonance structures for the *para*-benzenonium ion when G is (i) OH, (ii) CH_3.   (c) Which ions have G attached to a positively charged C? (d) If the products from this reaction are usually determined by kinetic control (Problem 8.20), how can the **Hammond principle** be used to predict the relative yields of *op* (i.e. the mixture of *ortho* and *para*) as against *m* (*meta*) products?   (e) In terms of electronic effects, what kind of G is a (i) *op*-director, (ii) *m*-director? (f) Classify G in terms of its structure and its electronic effect.   ◀

(a)

        ortho   *para*   *meta*

(b) (i)

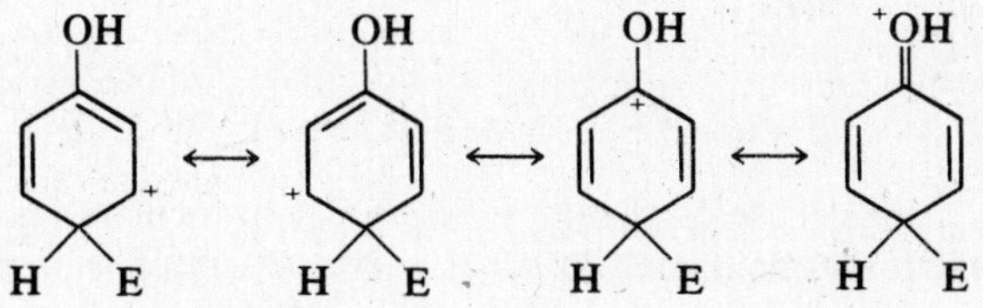

(major contributor
because all atoms
obey octet rule)

(ii)

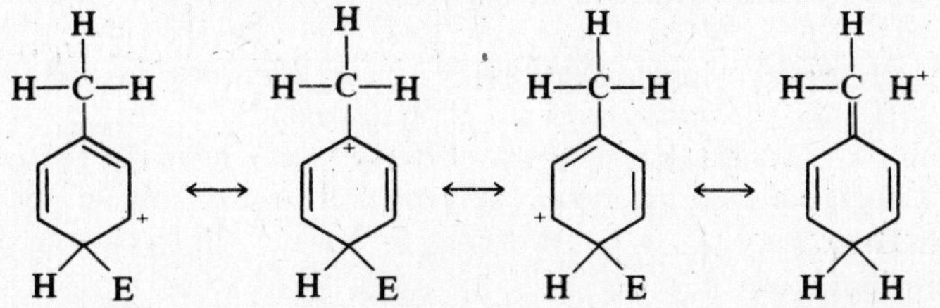

hyperconjugated structure

(c) The *ortho* and *para*.   This is why G is either an *op*- or an *m*-director.

(d) Because of kinetic control, the intermediate with the lowest-enthalpy transition state (TS) is formed in the greatest amount.   Since this step is endothermic, the Hammond principle says that the intermediate resembles the TS.   We then evaluate the relative energies of the intermediates (*op* vs. *m*) and predict that the one with the lowest enthalpy is formed in the greatest yield.

(e) (i) An electron-donating G can better stabilize the intermediate when it is attached directly to positively charged (*op*) C's.   Such G's are *op*-directing.   (ii) An electron-withdrawing G destabilizes the ion to a greater extent when attached directly to positively charged (*op*) C's.   They destabilize less when attached *meta* and are thus *m*-directors.   Such G's have, attached to the benzene ring, an atom with a full + charge, e.g. $-\overset{+}{N}R_3$, or with a partial + charge as the result of unequal electron distribution in a π bond, e.g.

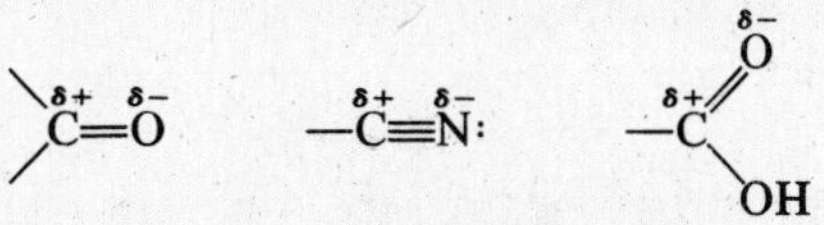

or as the result of being attached to electron-withdrawing groups,

$$\rightarrow \overset{\delta+ \ \delta-}{CCl_3}$$

(f) **Electron-donating (*op*-directors):** (i) Those that have an unshared pair of electrons on the atom bonded to the ring, which can be delocalized to the ring by extended π bonding.

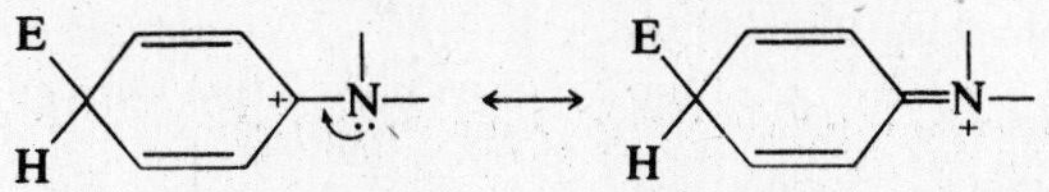

Other examples are $-\overset{..}{\underset{..}{O}}-$, $-\overset{..}{\underset{..}{X}}:$ (halogen) and $-\overset{..}{\underset{..}{S}}-$.   (ii) Those without an unshared pair, which are electron-donating by induction or by **hyperconjugation** (absence of bond resonance), e.g. alkyl groups.   In hyperconjugation the sigma C—H bond on an alpha C is delocalized with the empty p orbital of a C=C or a carbocation.   The resulting H^+ does not change its position.

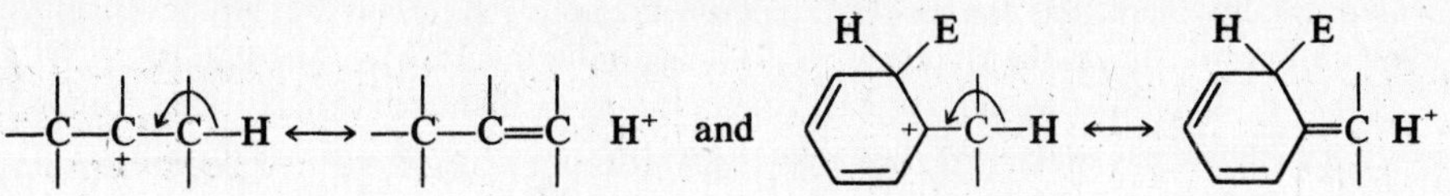

(iii) Those with an attached atom participating in an electron-rich π bond, e.g.

Electron-withdrawing (*m*-directors): The attached atom has no unshared pair of electrons and has some positive charge, e.g.

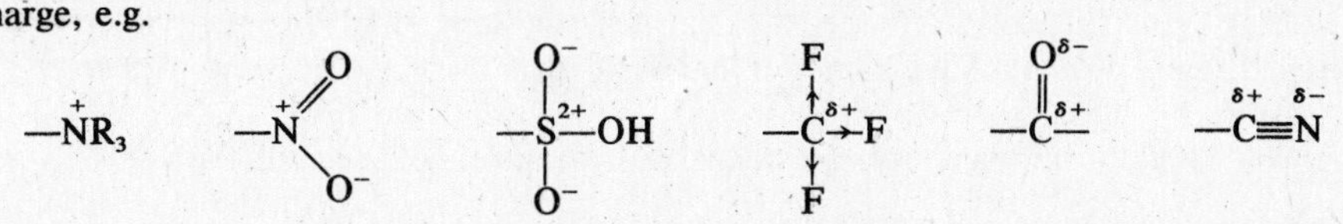

Problem 11.7   Explain: (*a*) All *m*-directors are deactivating.   (*b*) Most *op*-directing substituents make the ring more reactive than benzene itself—they are activating.   (*c*) As exceptions, the halogens are *op*-directors but are deactivating.   ◄

(*a*)   All *m*-directors are electron-attracting and destabilize the incipient benzenonium ion in the TS.   They therefore diminish the rate of reaction as compared to the rate of reaction of benzene.

(*b*)   Most *op*-directors are, on balance, electron-donating.   They stabilize the incipient benzenonium ion in the TS, thereby increasing the rate of reaction as compared to the rate of reaction of benzene.   For example, the ability of the —ÖH group to donate electrons by extended *p* orbital overlap (resonance) far outweighs the ability of the OH group to withdraw electrons by its inductive effect.

(*c*)   In the halogens, unlike the OH group, the electron-withdrawing inductive effect predominates and consequently the halogens are deactivating.   The *o*-, *p*- and *m*-benzenonium ions each have a higher $\Delta H^{\ddagger}$ than does the cation from benzene itself.   However, on demand, the halogens contribute electron density by extended π bonding,

$$\underset{H}{\overset{E}{\diagdown}}\!\!\!\!\bigcirc\!\!\!=\!\ddot{X}\!:\quad \text{(showing delocalization of } + \text{ to X)}$$

and thereby lower the $\Delta H^{\ddagger}$ of the *ortho* and *para* intermediates but not the *meta* cation.   Hence the halogens are *op*-directors, but deactivating.

Problem 11.8   Compare the activating effects of the following *op*-directors:

$$(a) \quad -\ddot{\text{O}}\text{H}, -\ddot{\text{O}}\overset{\cdot\cdot}{:}^{-} \text{ and } -\underset{\underset{\text{O}}{\parallel}}{\ddot{\text{O}}\text{C}}-\text{CH}_3 \qquad (b) \quad -\ddot{\text{N}}\text{H}_2 \text{ and } -\ddot{\text{N}}\text{H}-\underset{\underset{\text{O}}{\parallel}}{\text{C}}-\text{CH}_3$$

Explain your order.   ◄

(*a*)   The order of activation is —O⁻ > —OH > —OCOCH₃.   The —O⁻, with a full negative charge, is best able to donate electrons, thereby giving the very stable uncharged intermediate

$$\underset{H}{\overset{E}{\diagdown}}\!\!\!\!\bigcirc\!\!\!=\!\text{O}$$

In —OCOCH₃ the C of the $\overset{\delta+}{\text{C}}\!\!=\!\!\overset{\delta-}{\text{O}}$ group has + charge and makes demands on the —Ö— for electron density, thereby diminishing the ability of this —Ö— to donate electrons to the benzenonium ion.

(*b*)   The order is —NH₂ > —NHCOCH₃ for the same reason that OH is a better activator than —OCOCH₃.

Problem 11.9   Account for the following.   (*a*) Nitration of PhC(CH₃)₃ gives only 16% *ortho* product, whereas PhCH₃ gives 50%.   (*b*) Of all the aryl halides, PhF gives the least amount of *ortho* product on nitration.◄

(*a*) Steric hindrance, which is more pronounced with the bulky —C(CH₃)₃ group, inhibits formation of the *ortho* isomer, thereby increasing the yield of the *para* isomer.   (*b*) Even though the X can donate electrons to stabilize the *op*-intermediates, the electron-withdrawing inductive effect is substantial.   The inductive effect is strongest at the closer *ortho* position and weakest at the more distant *para* position.   Since F has the greatest inductive effect among halogens, PhF has the smallest percentage of *ortho* substitution.

Problem 11.10   (*a*) Draw enthalpy–reaction diagrams for the first step of electrophilic attack on benzene, toluene (*meta* and *para*) and nitrobenzene (*meta* and *para*).   Assume all ground states have the same energy. (*b*) Where would the *para* and *meta* substitution curves for C₆H₅Cl lie on this diagram?   ◄

(*a*)   Since CH₃ is an activating group, the intermediates and TS's from PhCH₃ have less enthalpy than those from benzene.   The *para* intermediate has less enthalpy than the *meta* intermediate.   The TS and intermediates for PhNO₂ have higher enthalpies than those for C₆H₆, with the *meta* at a lower enthalpy than the *para*.   See Fig. 11-1.

(*b*)   They would both lie between those for benzene and *p*-nitrobenzene, with the *para* lower than the *meta*.

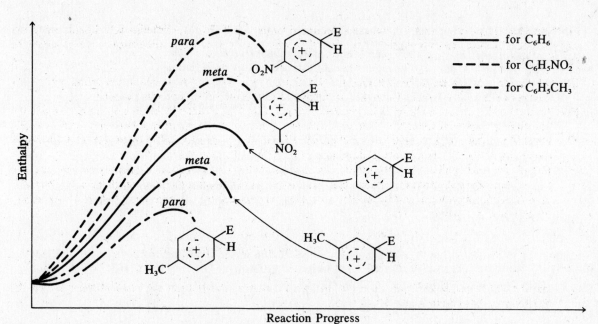

Fig. 11-1

Problem 11.11 (*a*) Explain in terms of the Selectivity-Reactivity principle (page 44) the following yields of *meta* substitution observed with toluene: Br_2 in CH_3COOH, 0.5%; HNO_3 in CH_3COOH, 3.5%; CH_3CH_2Br in $GaBr_3$, 21%. (*b*) In terms of kinetic vs. thermodynamic control, explain the following effect of temperature on isomer distribution in sulfonation of toluene: at 0 °C, 43% *o*- and 53% *p*-; at 100 °C, 13% *o*- and 79% *p*-. ◄

(*a*) The most reactive electrophile is least selective and gives the most *meta* isomer. The order of reactivity is:

$$CH_3CH_2^+ > NO_2^+ > Br_2 \ (Br^+)$$

(*b*) Sulfonation is one of the few reversible electrophilic substitutions and therefore kinetic and thermodynamic products can result. At 100 °C the thermodynamic product predominates; this is the *para* isomer. The *ortho* isomer is somewhat more favored by kinetic control at 0 °C.

Problem 11.12 Draw contributing resonance structures for the benzenonium ion formed on nitration in (*a*) the *para* position of anisole, $C_6H_5OCH_3$, (*b*) the *ortho* position of toluene, $C_6H_5CH_3$. ◄

Problem 11.13 $PhNO_2$, but not C_6H_6, is used as a solvent for the Friedel-Craft alkylation of PhBr. Explain. ◀

C_6H_6 is more reactive than PhBr and would preferentially undergo alkylation. $—NO_2$ is so strongly deactivating that $PhNO_2$ does not undergo Friedel-Craft alkylations or acylations.

Problem 11.14 $—OCH_3$ strongly activates the *op*-positions but weakly deactivates the *m*-position. $—CH_3$ activates all positions, but mainly the *op*. Explain. ◀

$—OCH_3$ is electron-donating and activating by extended π bonding [Problems 11.6(*b*) and 11.12(*a*)] only when it is attached to the positively charged (*op*) positions of the benzenonium ion. It is also electron-withdrawing by induction and this factor prevails in the *m*-position, which is deactivated.

$—CH_3$ is electron-donating by induction and hyperconjugation [Problems 11.6(*b*)(ii) and 11.12(*b*)] and activates all positions. Hyperconjugation is effective only in the *op*-positions, which therefore are more activated than the *m*-position.

Problem 11.15 Account for the percentages of *m*-orientation in the following compounds: (*a*) $C_6H_5CH_3$ (4.4%), $C_6H_5CH_2Cl$ (15.5%), $C_6H_5CHCl_2$ (33.8%), $C_6H_5CCl_3$ (64.6%); (*b*) $C_6H_5N^+(CH_3)_3$ (100%), $C_6H_5CH_2N^+(CH_3)_3$ (88%), $C_6H_5(CH_2)_2N^+(CH_3)_3$ (19%). ◀

(*a*) Substitution of the CH_3 H's by Cl's causes a change from electron-release ($\leftarrow CH_3$) to electron-attraction

$$\rightarrow \overset{\delta+}{C} \xrightarrow{} \overset{\overset{\delta-}{Cl}}{\underset{\underset{\delta-}{Cl}}{\overset{\delta-}{Cl}}}$$

and *m*-orientation increases.

(*b*) $^+NMe_3$ has a strong electron-attracting inductive effect and is *m*-orienting. When CH_2 groups are placed between this N^+ and the ring, this inductive effect falls off rapidly, as does the *m*-orientation. When two CH_2's intercede, the electron-releasing effect of the CH_2 bonded directly to the ring prevails, and chiefly *op*-orientation is observed.

Problem 11.16 Predict and explain the reaction, if any, of (*a*) phenol (PhOH), (*b*) PhH and (*c*) benzenesulfonic acid with D_2SO_4 in D_2O. ◀

(*a*) D_2SO_4 transfers D^+, an electrophile, to form 2,4,6-trideuterophenol. Reaction is rapid because of the activating $—OH$ group. The *meta* positions are deactivated. (*b*) PhH reacts slowly to give hexadeuterobenzene. (*c*) The sulfonic acid does not react, because $—SO_3H$ is too deactivating.

Problem 11.17 Write structural formulas for the principal monosubstitution products of the indicated reactions from the following monosubstituted benzenes. For each write an S or F to show whether reaction is SLOWER or FASTER than with benzene. (*a*) Monobromination, $C_6H_5CF_3$. (*b*) Mononitration, $C_6H_5COOCH_3$. (*c*) Monochlorination, $C_6H_5OCH_3$. (*d*) Monosulfonation, C_6H_5I. (*e*) Mononitration, $C_6H_5C_6H_5$. (*f*) Monochlorination, C_6H_5CN. (*g*) Mononitration, $C_6H_5NHCOCH_3$. (*h*) Monosulfonation, $C_6H_5CH(CH_3)CH_2CH_3$. ◀

(*a*) F_3C—⟨benzene ring with Br⟩   S   (*b*) $CH_3O\overset{O}{\overset{\|}{C}}$—⟨benzene ring with NO_2⟩   S   (*c*) CH_3O—⟨benzene ring⟩—Cl   F

(*d*) I—⟨benzene ring⟩—SO_3H S   (*e*) ⟨benzene ring⟩—⟨benzene ring⟩—NO_2 F   (*f*) $\overset{\delta-}{N}\equiv\overset{\delta+}{C}$—⟨benzene ring with Cl⟩   S

(*g*) CH_3CONH—⟨benzene ring⟩—NO_2 F   (*h*) $CH_3CH_2\underset{\underset{CH_3}{|}}{CH}$—⟨benzene ring⟩—$SO_3H$ F

Problem 11.18 Explain why *p*-nitrotoluene has a larger dipole moment (4.40 D) than does *p*-chloronitrobenzene (2.40 D).◄

See Fig. 11-2.

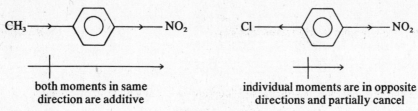

both moments in same   individual moments are in opposite
direction are additive   directions and partially cancel

Fig. 11-2

Problem 11.19 From C_6H_6 (PhH) or $PhCH_3$ synthesize: (*a*) *p*-$ClC_6H_4NO_2$, (*b*) *m*-$ClC_6H_4NO_2$, (*c*) *p*-$O_2NC_6H_4COOH$ and (*d*) *m*-$O_2NC_6H_4COOH$.◄

In the synthesis of disubstituted benzenes, the first substituent present determines the position of the incoming second. Therefore the order of introducing substituents must be carefully planned to yield the desired isomer.

(*a*) Since the two substituents are *para*, it is necessary to introduce the *op*-directing Cl first.

$$PhH \xrightarrow[Cl_2]{Fe} PhCl \xrightarrow[H_2SO_4]{HNO_3} p\text{-}ClC_6H_4NO_2$$

(*b*) Since the substituents are *meta*, the *m*-directing NO_2 is introduced first.

$$PhH \xrightarrow[H_2SO_4]{HNO_3} PhNO_2 \xrightarrow[Fe]{Cl_2} m\text{-}ClC_6H_4NO_2$$

(*c*) The COOH group is formed by oxidation of CH_3. Since *p*-$O_2NC_6H_4COOH$ has two *m*-directing groups, the NO_2 must be added while the *op*-directing CH_3 is still present.

$$PhCH_3 \xrightarrow[H_2SO_4]{HNO_3} p\text{-}CH_3C_6H_4NO_2 + o\text{-}CH_3C_6H_4NO_2$$

The *para* isomer is usually easily separated from the *op* mixture.

$$p\text{-}O_2NC_6H_4CH_3 \xrightarrow[H^+]{KMnO_4} p\text{-}O_2NC_6H_4COOH$$

(*d*) Now the substituents are *meta*, and NO_2 is introduced when the *m*-directing COOH is present.

$$PhCH_3 \xrightarrow[H^+]{KMnO_4} PhCOOH \xrightarrow[H_2SO_4]{HNO_3} m\text{-}O_2NC_6H_4COOH$$

RULES FOR PREDICTING ORIENTATION IN DISUBSTITUTED BENZENES

1. If the groups *reinforce* each other, there is no problem.
2. If an *op*-director and an *m*-director are *not reinforcing*, the *op*-director controls the orientation. (The incoming group goes mainly *ortho* to the *m*-director.)
3. A *strongly activating* group, competing with a *weakly activating* group, controls the orientation.
4. When *two weakly activating or deactivating* groups or *two strongly activating or deactivating* groups compete, substantial amounts of both isomers are obtained; there is little preference.
5. Very little substitution occurs in the *sterically hindered* position between *meta* substituents.

Problem 11.20 Indicate by an arrow the position(s) most likely to undergo electrophilic substitution in each of the following compounds. List the number of the above rule(s) used in making your prediction. (*a*) *m*-xylene, (*b*) *p*-nitrotoluene, (*c*) *m*-chloronitrobenzene, (*d*) *p*-methoxytoluene, (*e*) *p*-chlorotoluene, (*f*) *m*-nitrotoluene, (*g*) *o*-methylphenol (*o*-cresol).◄

(a) very little (b) (c)

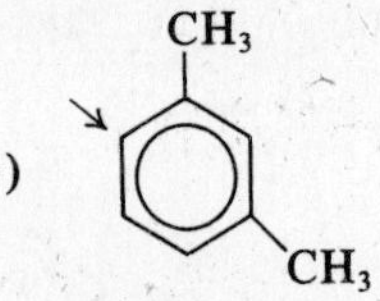

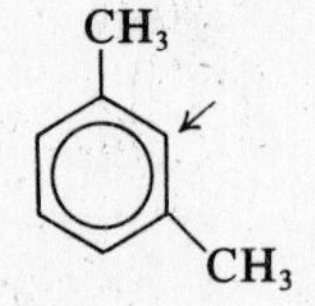

(Rule 1)   (Rule 5)   (Rule 1)   (Rule 2)

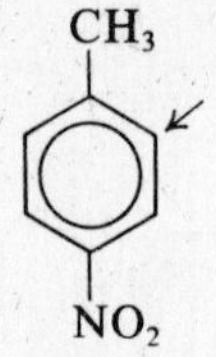

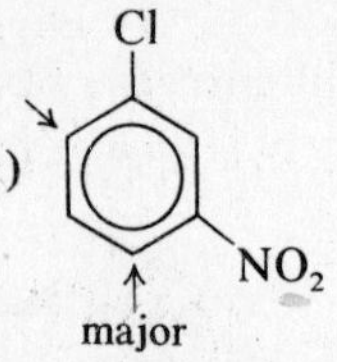

(d)   (e) or   (f)   (g)

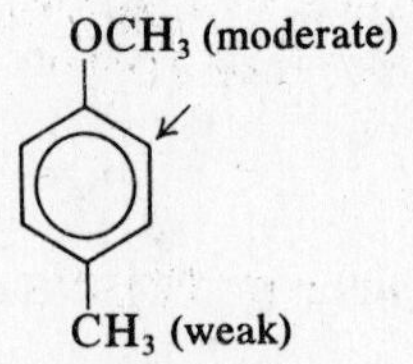

(Rule 3)   (Rule 4)   (Rule 2)   (Rule 3)

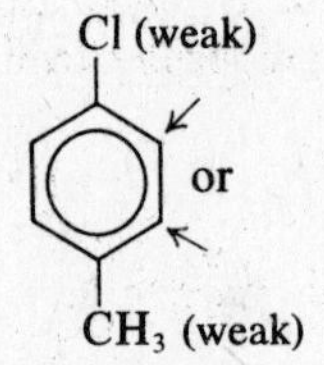

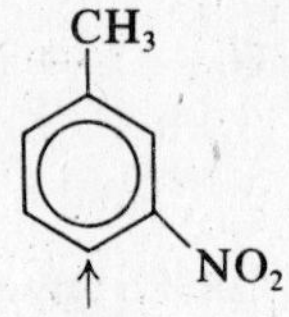

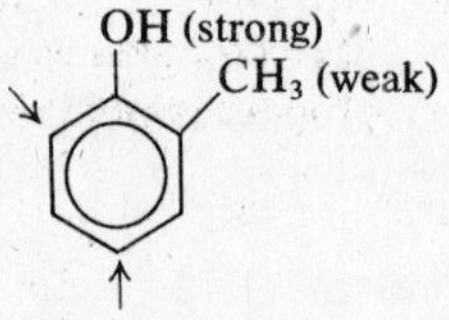

Problem 11.21 Which xylene is most easily sulfonated?   ◄

m-Xylene is most reactive and sulfonates at C^4 because its CH_3's reinforce each other [Rule 1; Problem 11.20(a)].

Problem 11.22 Write structures for the principal mononitration products of   (a) o-cresol (o-methylphenol), (b) p-$CH_3CONHC_6H_4SO_3H$,   (c) m-cyanotoluene (m-toluonitrile).   ◄

(a)   (b) 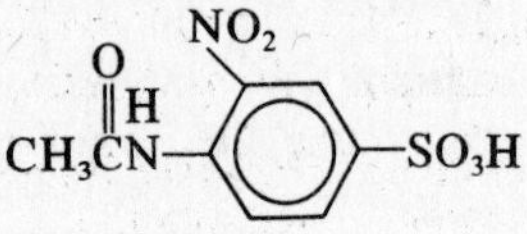   (c)

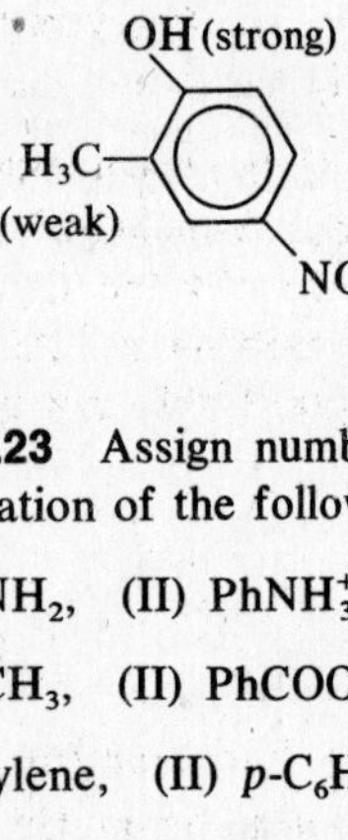

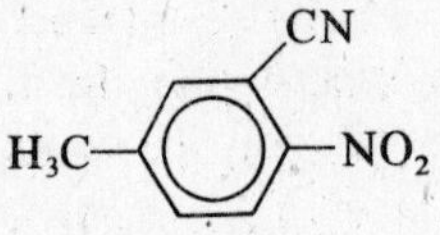

Problem 11.23 Assign numbers from 1 for LEAST to 5 for MOST to designate relative reactivity to ring monobromination of the following groups.

(a) (I) $PhNH_2$, (II) $PhNH_3^+Cl^-$, (III) $PhNHCOCH_3$, (IV) $PhCl$, (V) $PhCOCH_3$.

(b) (I) $PhCH_3$, (II) $PhCOOH$, (III) PhH, (IV) $PhBr$, (V) $PhNO_2$.

(c) (I) p-xylene, (II) p-$C_6H_4(COOH)_2$, (III) $PhMe$, (IV) p-$CH_3C_6H_4COOH$, (V) m-xylene.   ◄

See Table 11-2.

Table 11-2

	(I)	(II)	(III)	(IV)	(V)
(a)	5	1	4	3	2
(b)	5	2	4	3	1
(c)	4	1	3	2	5

Problem 11.24 Use PhH, PhMe and any aliphatic or inorganic reagents to prepare the following compounds in reasonable yields: (a) m-bromobenzenesulfonic acid, (b) 3-nitro-4-bromobenzoic acid, (c) 3,4-dibromonitrobenzene, (d) 2,6-dibromo-4-nitrotoluene.   ◄

(a) (*m*-director is added first)

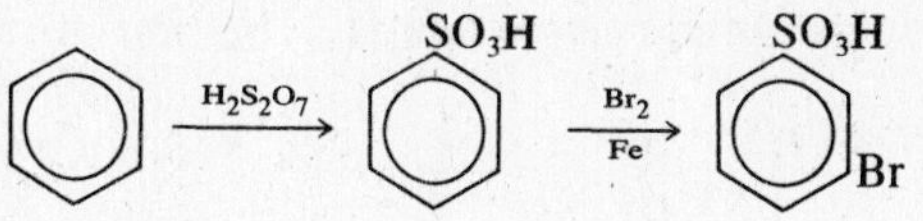

(b)

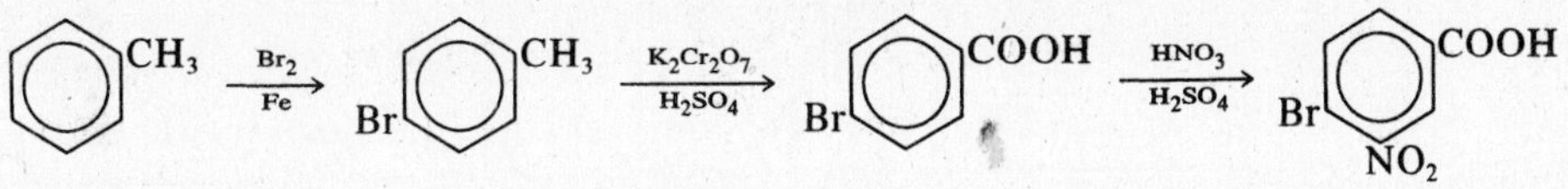

Nitration of *p*-BrC$_6$H$_4$CH$_3$ would have given about a 50–50 mixture of two products; 2-nitro-4-bromotoluene would be unwanted. When oxidation precedes nitration, an excellent yield of the desired product is obtained.

(c) (only product)

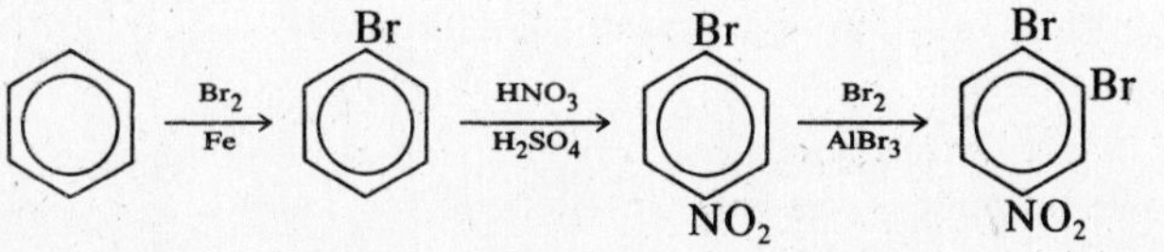

Nitration followed by dibromination would give as the major product 2,5-dibromonitrobenzene (see Rule 2, page 181).

(d)

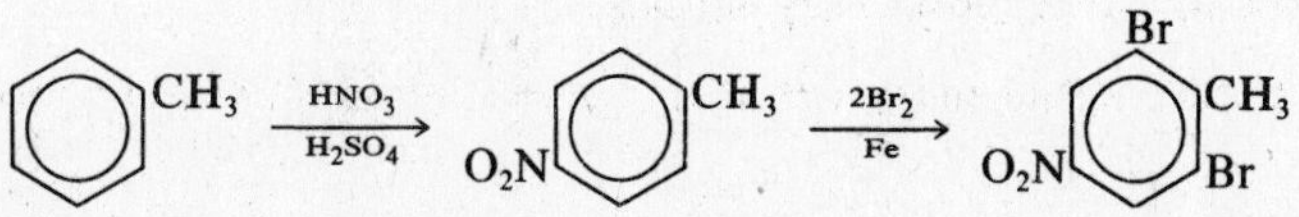

Problem 11.25 Available organic starting materials for synthesizing the pairs of compounds below are PhH, PhMe and PhNO$_2$. Select the compound in each pair whose synthesis is more expensive and explain your choice.   (a) *p*-C$_6$H$_4$(NO$_2$)$_2$ and *m*-C$_6$H$_4$(NO$_2$)$_2$,   (b) *m*-nitrotoluene and *p*-nitrotoluene,   (c) 2,4,6-trinitrotoluene (TNT) and 1,3,5-trinitrobenzene,   (d) *m*-C$_6$H$_4$Cl$_2$ and *p*-C$_6$H$_4$Cl$_2$.   ◄

(a) *p*-C$_6$H$_4$(NO$_2$)$_2$. NO$_2$ is *meta*-directing. The *meta* isomer is easily made directly by nitrating PhNO$_2$. The *para* isomer cannot be prepared directly and requires a more circuitous, expensive route. (b) *m*-Nitrotoluene. Nitration of PhMe gives the *para* isomer. PhNO$_2$ is too deactivated to undergo *meta* alkylation. (c) 1,3,5-Trinitrobenzene. Two NO$_2$'s deactivate the ring so that insertion of a third NO$_2$ is very difficult. This extreme deactivation is countered somewhat by Me. Therefore, PhMe can be more easily trinitrated than PhH. (d) *m*-C$_6$H$_4$Cl$_2$. Cl is *op*-directing.

11.2 NUCLEOPHILIC AND FREE-RADICAL SUBSTITUTIONS

Nucleophilic aromatic substitutions of H are rare. The intermediate benzenanion in aromatic nucleophilic substitution is analogous to the intermediate benzenonium ion in aromatic electrophilic substitution; negative charge is dispersed to the *op*-positions.

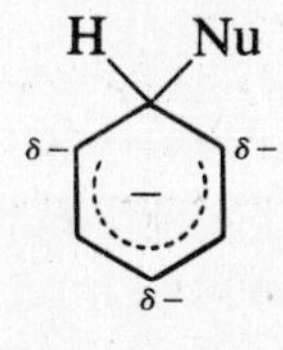

Benzenanion

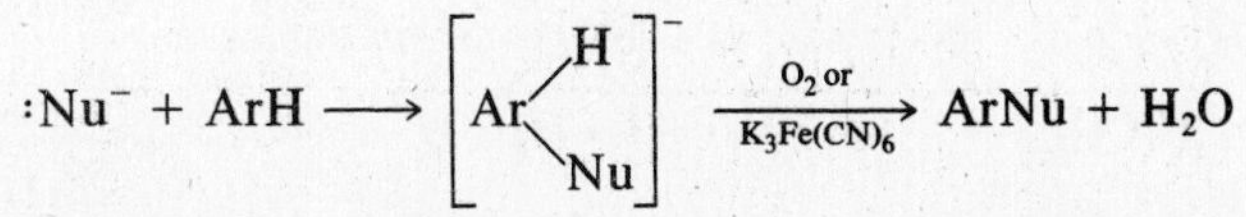

$$:Nu^- + ArH \longrightarrow \left[Ar \underset{Nu}{\overset{H}{\diagdown}} \right]^- \xrightarrow[K_3Fe(CN)_6]{O_2 \text{ or}} ArNu + H_2O$$

Oxidants such as O$_2$ and K$_3$Fe(CN)$_6$ facilitate the second step, which may be rate-controlling, by oxidizing the ejected :H$^-$, a powerful base and a very poor leaving group, to H$_2$O.

Free-radical substitutions proceed by a similar intermediate, having free-radical character distributed to the *op*-positions.

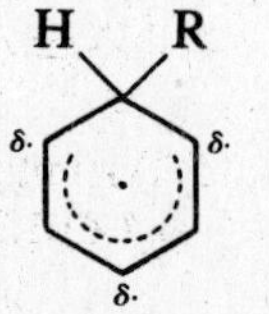

The effects of substituents are summarized for **aromatic radical substitutions**:

1. Substituents have **much less effect** than in electrophilic or nucleophilic substitutions.
2. Both electron-withdrawing and electron-donating groups **increase** reactivity at *ortho* and *para* positions.   Except when large groups cause steric hindrance, the *ortho* is somewhat more reactive than the *para* position.

Problem 11.26 Account for the product in the following reactions:

(a)

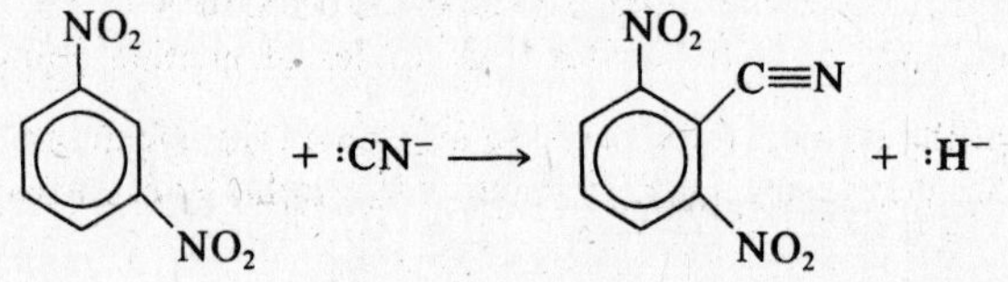

(b) $PhN{=}N{-}OCOCH_3 + PhH \xrightarrow{\Delta} Ph{-}Ph + HOOCCH_3 + N_2$ ◀

(a) CN^- is a nucleophile. NO_2's activate the ring toward nucleophilic substitution at *op*-positions by withdrawing the electron density and placing charge on the O's of NO_2:

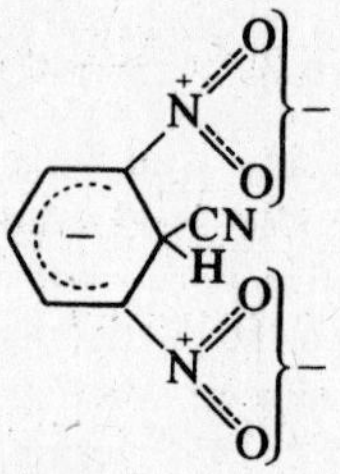

When not sterically hindered, the *ortho* position may be more reactive. CN^- is a "thin" nucleophile and its insertion *ortho* to each NO_2 is not hindered.

(b) $PhN{=}NOCOCH_3$ undergoes homolysis to give $N_2 + Ph\cdot + \cdot OCOCH_3$, then

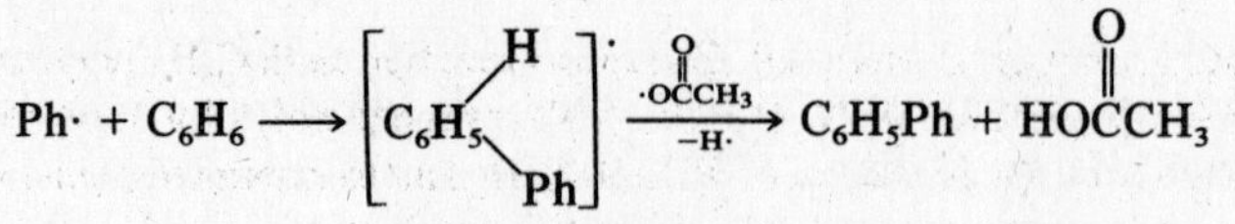

11.3 ARENES

Benzene derivatives with saturated or unsaturated C-containing side chains are **arenes**. Examples are cumene or isopropylbenzene, $C_6H_5CH(CH_3)_2$, and styrene or phenylethene, $C_6H_5CH{=}CH_2$.

Problem 11.27 Supply systematic and, where possible, common names for:

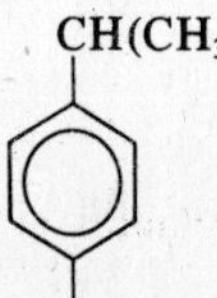

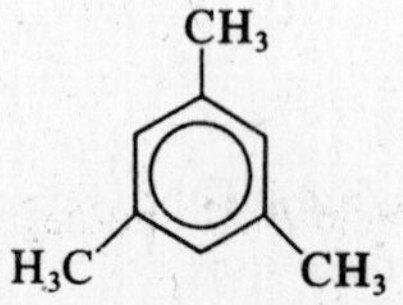

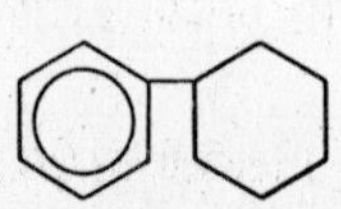

(d) $\langle\!\!\bigcirc\!\!\rangle$—$CH_2$—$C\equiv C$—$CH_2$—$\langle\!\!\bigcirc\!\!\rangle$　　　(e) $\begin{array}{c}Ph\\ \diagdown\\ \end{array}C=C\begin{array}{c}Ph\\ \diagup\\ \end{array}$
　　　　　　　　　　　　　　　　　　　　　　　　$\begin{array}{c}\diagup\\ H\end{array}\qquad\begin{array}{c}\diagdown\\ H\end{array}$　　◀

(a) *p*-Isopropyltoluene (*p*-cymene).　(b) 1,3,5-Trimethylbenzene (mesitylene).　(c) Cyclohexylbenzene.
(d) 1,4-Diphenyl-2-butyne (dibenzylacetylene).　(e) (Z)-1,2-Diphenylethene (*cis*-stilbene).

Problem 11.28 Write structural formulas for　(a) *p*-methylstyrene,　(b) *m*-chlorophenylacetylene,
(c) 1,3-diphenyl-1,4-pentadiene.　　　　　　　　　　　　　　　　　　　　　　　　◀

(a) CH_3—$\langle\!\!\bigcirc\!\!\rangle$—$CH=CH_2$　　(b) $\langle\!\!\bigcirc\!\!\rangle\begin{array}{c}C\equiv C-H\\ \\ Cl\end{array}$　　(c) C_6H_5—$CH=CH$—$\underset{\underset{C_6H_5}{\vert}}{CH}$—$CH=CH_2$

Problem 11.29　Arrange the isomeric tetramethylbenzenes, prehnitene (1,2,3,4-) and durene (1,2,4,5-), in order
of decreasing melting point and verify this order from tables of melting points.　　　◀

The more symmetrical the isomer, the closer the molecules are packed in the crystal and the higher is the
melting point.　The order of decreasing symmetry, durene > prehnitene, corresponds to that of their respective
melting points, +80 °C > −6.5 °C.

Problem 11.30　Explain the following observations about the Friedel-Craft alkylation reaction.　(a) In
monoalkylating C_6H_6 with RX in AlX_3 an excess of C_6H_6 is used.　(b) The alkylation of PhOH and $PhNH_2$ gives
poor yields.　(c) Ph—Ph *cannot* be prepared by the reaction

$$PhH + PhCl \xrightarrow{\text{AlCl}_3} Ph\text{—}Ph + HCl$$

(d)　At 0 °C

$$PhH + 3CH_3Cl \xrightarrow{\text{AlCl}_3} 1,2,4\text{-trimethylbenzene}$$

but at 100° one gets 1,3,5-trimethylbenzene (mesitylene).　(e) The reaction

$$PhH + CH_3CH_2CH_2Cl \xrightarrow{\text{AlCl}_3} PhCH_2CH_2CH_3 + HCl$$

gives poor yield, whereas

$$PhH + CH_3CHClCH_3 \xrightarrow{\text{AlCl}_3} PhCH(CH_3)_2 + HCl$$

gives very good yield.　　　　　　　　　　　　　　　　　　　　　　　　　　　◀

(a)　The monoalkylated product, C_6H_5R, which is more reactive than C_6H_6 itself since R is an activating group,
　　will react to give $C_6H_4R_2$ and some $C_6H_3R_3$.　To prevent polyalkylation an excess of C_6H_6 is used to
　　increase the chance for collision between R^+ and C_6H_6 and to minimize collision between R^+ and C_6H_5R.
(b)　OH and NH_2 groups react with and inactivate the catalyst.
(c)　　　　　　　　　　　　$PhCl + AlCl_3 \xrightarrow{\hspace{0.3cm}\not\rightarrow\hspace{0.3cm}} Ph^+ + AlCl_4^-$

　　Ph^+ has a very high enthalpy and doesn't form.
(d)　The alkylation reaction is *reversible* and therefore gives the kinetic-controlled product at 0° and the
　　thermodynamic-controlled product at 100°.
(e)　The R^+ intermediates, especially the 1° RCH_2^+ can undergo rearrangements.　With $CH_3CH_2CH_2Cl$ we get

$$CH_3CH_2CH_2^+ \xrightarrow{\sim H\cdot} CH_3\overset{+}{C}HCH_3$$

and the major product is $PhCH(CH_3)_2$.

Problem 11.31　Prepare $PhCH_2CH_2CH_3$ from PhH and any acyclic compound.　　　◀

$$PhH + ClCH_2CH=CH_2 \xrightarrow{\text{AlCl}_3} PhCH_2CH=CH_2 \xrightarrow{\text{H}_2/\text{Pt}} PhCH_2CH_2CH_3$$

Problem 11.32 Give the structural formula and the name for the major alkylation product:

(a) $C_6H_6 + (CH_3)_2CHCH_2Cl \xrightarrow{AlCl_3}$

(b) $C_6H_5CH_3 + (CH_3)_3CCH_2OH \xrightarrow{BF_3}$

(c) $C_6H_6 + CH_3CH_2CH_2CH_2Cl \xrightarrow[100\ °C]{AlCl_3}$

(d) m-xylene + $(CH_3)_3CCl \xrightarrow[100\ °C]{AlCl_3}$ ◀

(a)

$$CH_3\overset{\overset{CH_3}{|}}{C}HCH_2Cl \xrightarrow{AlCl_3} CH_3\overset{\overset{CH_3}{|}}{C}HCH_2^+ \xrightarrow{\sim H} \underset{+}{CH_3\overset{\overset{CH_3}{|}}{C}CH_3} \xrightarrow{C_6H_6} CH_3\overset{\overset{CH_3}{|}}{\underset{\underset{C_6H_5}{|}}{C}}CH_3$$

| Isobutyl | Isobutyl | *tert*-Butyl | *tert*-Butyl- |
| alcohol | cation (1°) | cation (3°) | benzene |

(b)

$$CH_3\overset{\overset{CH_3}{|}}{\underset{\underset{CH_3}{|}}{C}}CH_2OH \xrightarrow{BF_3} CH_3\overset{\overset{CH_3}{|}}{\underset{\underset{CH_3}{|}}{C}}CH_2^+ \xrightarrow{\sim :CH_3} \underset{+}{CH_3\overset{\overset{CH_3}{|}}{C}CH_2CH_3} \xrightarrow{C_6H_5CH_3} p\text{-}CH_3C_6H_4\overset{\overset{CH_3}{|}}{\underset{\underset{CH_3}{|}}{C}}CH_2CH_3$$

| Neopentyl | Neopentyl | *tert*-Pentyl | *p*-*tert*-Pentyl- |
| alcohol | cation (1°) | cation (3°) | toluene |

(c) $CH_3CH_2CH_2CH_2Cl \xrightarrow{AlCl_3} \begin{bmatrix} CH_3CH_2CH_2CH_2^+ \\ \Big\downarrow {\sim H} \\ CH_3CH_2\underset{+}{C}HCH_3 \end{bmatrix} \xrightarrow{C_6H_6}$ 34% $PhCH_2CH_2CH_2CH_3$ + 66% $C_6H_5\overset{}{\underset{\underset{CH_3}{|}}{C}}HCH_2CH_3$

n-Butyl chloride   *n*-Butylbenzene   *s*-Butylbenzene

(d)

Thermodynamic product; it has less steric strain and is more stable than the kinetic-controlled isomer having a bulky *t*-butyl group *ortho* to a CH_3.

Problem 11.33 Take advantage of the reversibility of the Friedel-Crafts alkylation reaction to prepare 1,2,3-trimethylbenzene from toluene. ◀

Methylation of $PhCH_3$ gives mainly *p*-xylene. Therefore it is necessary to block the *para* position with a group that later can readily be removed. This group is $-C(CH_3)_3$.

In the reaction with HF, the electrophile H^+ replaces $C(CH_3)_3^+$, which forms $(CH_3)_2C=CH_2$.

Problem 11.34 $PhCH_3$ reacts with Br_2 and Fe to give a mixture of 3 monobromo products. With Br_2 in light only 1 compound, a 4th monobromo isomer, is isolated. What are the 4 products? Explain the formation of the light-catalyzed product. ◀

With Fe, the products are *o*-, *p*- and some *m*-$BrC_6H_4CH_3$. In light the product is benzyl bromide, $PhCH_2Br$. Like allylic halogenation (page 88), the latter reaction is a free-radical substitution:

(1) $Br_2 \xrightarrow{uv} 2Br\cdot$ (2) $Br\cdot + PhCH_3 \longrightarrow Ph\dot{C}H_2 + HBr$

(3) $Ph\dot{C}H_2 + Br_2 \longrightarrow PhCH_2Br + Br\cdot$

Steps (2) and (3) are the propagating steps.

Problem 11.35 Account for the following:

$$PhCH_2CH(CH_3)_2 + Br_2 \xrightarrow{uv} PhCHBrCH(CH_3)_2$$

with little or no $PhCH_2CBr(CH_3)_2$ formed.   ◄

The H's on the C attached to the ring (the benzylic H's), although they are in this case 2°, are nevertheless more reactive toward Br· than are ordinary 3° H's.   Like a C=C group in the allylic system, the Ph group can stabilize the free radical by electron donation through extended p orbital overlap.

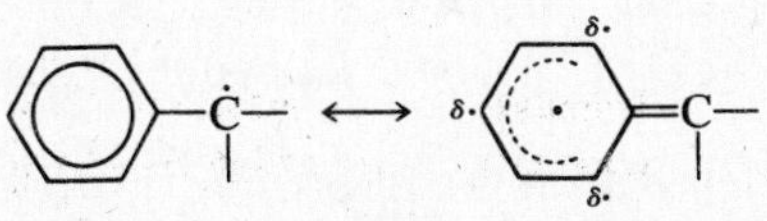

Problem 11.36 Irradiation of an equimolar mixture of cyclohexane, toluene and Br_2 in CCl_4 gives almost exclusively benzyl bromide.   A similar reaction with Cl_2 gives mainly cyclohexyl chloride.   Explain.   ◄

This is a *competitive* reaction that compares the reactivities of toluene and cyclohexane.   Br· is less reactive and more selective than Cl·, and reactivity differences of the respective H's determine the product (benzyl > 2°).   With the more reactive and less selective Cl·, the statistical advantage of cyclohexane (12 H's) over toluene (3 H's) controls the product formation.

Problem 11.37 Which is more reactive to radical halogenation, $PhCH_3$ or p-xylene?   Explain.   ◄

p-Xylene reactivity depends on the rate of formation of the benzyl-type radical.   Electron-releasing groups such as CH_3 stabilize the transition state, producing the benzyl radical on the other CH_3 and thereby lowering the $\Delta H^{\ddagger}$ and increasing the reaction rate.

Problem 11.38 Outline a synthesis of 2,3-dimethyl-2,3-diphenylbutane from benzene, propylene and any inorganic reagents.   ◄

The symmetry of this hydrocarbon makes possible a self-coupling reaction with 2-bromo-2-phenylpropane.

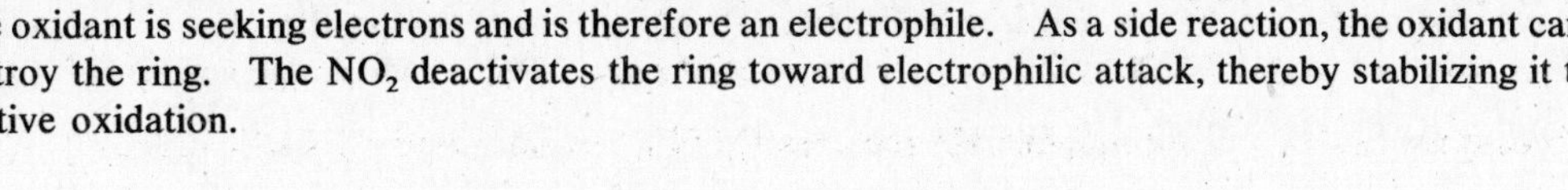

Problem 11.39 Explain why the oxidation of $PhCH_3$ to $PhCOOH$ by $KMnO_4$ or $K_2Cr_2O_7$ and H^+ goes in poor yield, whereas the same oxidation of p-$O_2NC_6H_4CH_3$ to p-$O_2NC_6H_4COOH$ goes in good yield.   ◄

The oxidant is seeking electrons and is therefore an electrophile.   As a side reaction, the oxidant can attack and destroy the ring.   The NO_2 deactivates the ring toward electrophilic attack, thereby stabilizing it towards degradative oxidation.

Problem 11.40 Give all possible products of the following reactions and underline the major product.

(a) $PhCH_2CHOHCH(CH_3)_2 \xrightarrow{H_2SO_4}$ (b) $PhCH_2CHBrCH(CH_3)_2 \xrightarrow{\text{alc.}{KOH}}$

(c) $PhCH=CHCH_3 + HBr \longrightarrow$ (d) $PhCH=CHCH_3 + HBr \xrightarrow{peroxide}$

(e) $PhCH=CHCH=CH_2 + Br_2$ (equimolar amounts) $\longrightarrow$ ◄

(a) $PhCH_2CH=C(CH_3)_2 + \underline{PhCH=CHCH(CH_3)_2}$. The major product has the C=C conjugated with the benzene ring and therefore, even though it is a disubstituted alkene, it is more stable than the minor product, which is a trisubstituted nonconjugated alkene.

(b) Same as part (a) and for the same reason.

(c) $PhCH_2CHBrCH_3$ + __PhCHBrCH_2CH_3__. H^+ adds to C=C to give the more stable benzyl-type $Ph\overset{+}{C}HCH_2CH_3$. Reaction with Br^- gives the major product. The benzyl-type cation $Ph\overset{+}{C}HR$ (like $CH_2=CH\overset{+}{C}H_2$) can be stabilized by delocalizing the + to the op-positions of the ring:

(d) __PhCH_2CHBrCH_3__ + $PhCHBrCH_2CH_3$. Br· adds to give the more stable benzyl-type $Ph\overset{.}{C}HCHBrCH_3$ rather than $PhCHBr\overset{.}{C}HCH_3$. We have already discussed the stability of benzylic free radicals (Problem 11.35).

(e) $PhCHBrCHBrCH=CH_2$ + __PhCH=CHCHBrCHBr__ + $PhCHBrCH=CHCH_2Br$. The major product is the conjugated alkene, which is more stable than the other two products [see part (a)].

Problem 11.41 Explain the following observations. (a) A yellow color is obtained when Ph_3COH (trityl alcohol) is reacted with concentrated H_2SO_4, or when Ph_3CCl is treated with $AlCl_3$. On adding H_2O, the color disappears and a white solid is formed. (b) Ph_3CCl is prepared by the Friedel-Crafts reaction of benzene and CCl_4. It does not react with more benzene to form Ph_4C. (c) A deep-red solution appears when Ph_3CH is added to a solution of $NaNH_2$ in liquid NH_3. The color disappears on adding water. (d) A red color appears when Ph_3CCl reacts with Zn in C_6H_6. O_2 decolorizes the solution. ◄

(a) The yellow color is attributed to the stable Ph_3C^+, whose + is delocalized to the op-positions of the 3 rings.

$$Ph_3COH + H_2SO_4 \longrightarrow Ph_3C^+ + H_3O^+ + HSO_4^-$$

$$Ph_3CCl + AlCl_3 \longrightarrow Ph_3C^+ + AlCl_4^-$$

$$Ph_3C^+ + 2H_2O \longrightarrow Ph_3COH + H_3O^+$$

Lewis	Lewis	white
acid	base	solid

(b) With $AlCl_3$, Ph_3CCl forms a salt, $Ph_3C^+AlCl_4^-$, whose carbonium ion is too stable to react with benzene. Ph_3C^+ may also be too sterically hindered to react further.

(c) The strong base $:NH_2^-$ removes H^+ from Ph_3CH to form the stable, deep red-purple carbanion $Ph_3C^{\bar{:}}$, which is then decolorized on accepting H^+ from the feeble acid H_2O.

$$Ph_3CH + :NH_2^- \longrightarrow H:NH_2 + Ph_3C^{\bar{:}}$$

$$\text{Acid}_1 \quad\quad \text{Base}_2 \quad\quad \text{Acid}_2 \quad \text{Base}_1 \text{ (deep red)}$$

$$Ph_3C^{\bar{:}} + H_2O \longrightarrow Ph_3CH + OH^-$$

$$\text{Base}_1 \quad \text{Acid}_2 \quad\quad \text{Acid}_1 \quad \text{Base}_2$$

The $Ph_3C^{\bar{:}}$ is stabilized because the − can be delocalized to the op-positions of the 3 rings (as in the corresponding carbonium ion and free radicals).

(d) Cl· is removed from Ph_3CCl by Zn to give the colored radical Ph_3C·, which decolorizes as it forms the peroxide in the presence of O_2.

$$2Ph_3CCl + Zn \longrightarrow 2Ph_3C\cdot + ZnCl_2$$

$$2Ph_3C\cdot + \cdot\ddot{O}-\ddot{O}\cdot \longrightarrow Ph_3C:\ddot{O}:\ddot{O}:CPh_3$$

Problem 11.42 In the following hydrocarbons the alkyl H's are designated by the Greek letters α, β, γ, etc. Assign each letter an Arabic number, beginning with 1 for LEAST, in order of *increasing* ease of abstraction by a Br·.

(b) $CH_3\underset{\alpha}{-}CH_2\underset{\beta}{-}CH_2\underset{\gamma}{-}CH\underset{\delta}{=}CH\underset{\delta}{-}CH_2\underset{\epsilon}{-}$⬡   (c) ⬡ with $\underset{\alpha}{CH_3}$, $\underset{\beta}{CH_3}$, $\underset{\gamma}{CH_3}$ ◄

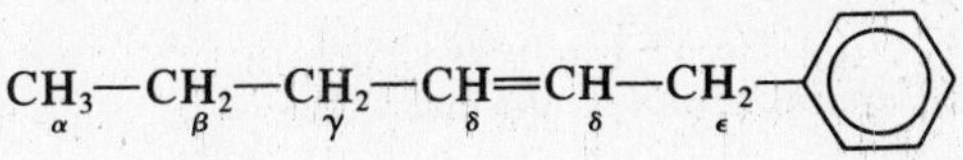

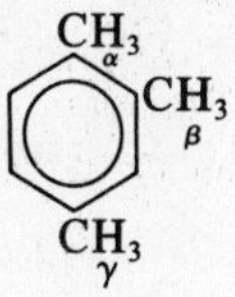

See Table 11-3.

Table 11-3

	α	β	γ	δ	ε
(a)	1 (1°)	2 (2°)	4 (2°, benzylic)	5 (2°, dibenzylic)	3 (1°, benzylic)
(b)	2 (1°)	3 (2°)	4 (allylic)	1 (vinylic)	5 (allylic, benzylic)
(c)	3 (op to other CH₃'s)	1 (mp to 2 other CH₃'s; more hindered)	2 (less sterically hindered than β; also mp to 2 other CH₃'s)		

Problem 11.43 Use + and − signs for positive and negative tests in tabulating rapid chemical reactions that can be used to distinguish among the following compounds: (a) chlorobenzene, benzyl chloride and cyclohexyl chloride;   (b) ethylbenzene, styrene, and phenylacetylene.   ◄

See Tables 11-4(a) and 11-4(b).

Table 11-4(a)

Reaction	Chloro-benzene	Benzyl chloride	Cyclohexyl chloride
Ring sulfonation is exothermic	+	+	−
Alc. AgNO₃ (forms AgCl, a white precipitate)	−	+ (very fast)	+ (much slower)

(Ag⁺ induces an S_N1 reaction; $PhCH_2^+ > C_6H_{11}^+$.)

Table 11-4(b)

Reactions	Ethyl-benzene	Styrene	Phenyl-acetylene
Br₂ in CCl₄ (is decolorized)	−	+	+
Ag(NH₃)₂⁺ (forms a precipitate)	−	−	+

Supplementary Problems

Problem 11.44   Supply structures for organic compounds (A) through (O).

(a) $PhCH_2CH_2CH_3 + Br_2 \xrightarrow{\text{uv}} (A) \xrightarrow[\text{KOH}]{\text{alc.}} (B) \xrightarrow[\text{KMnO}_4]{\text{cold dil.}} (C) \xrightarrow[\text{KMnO}_4]{\text{hot}} (D)$

(b) $PhBr + Mg \xrightarrow{\text{Et}_2\text{O}} (E) \xrightarrow{\text{H}_2\text{C=CHCH}_2\text{Br}} (F) \xrightarrow[\text{heat}]{\text{KOH}} (G) \xrightarrow[\text{300 °C}]{\text{Br}_2} (H)$

(c) $Ph—C≡CH + CH_3MgX \longrightarrow (I) \xrightarrow{\text{ArCH}_2\text{Cl}} (J) \xrightarrow{\text{Li, NH}_3} (K)$

(d) $p\text{-}CH_3C_6H_4C≡CPh + H_2/Pt \longrightarrow (L) \xrightarrow{\text{HBr}} (M)$

(e)

$$\begin{array}{c} Ph \qquad\qquad H \\ \diagdown\qquad\diagup \\ C=C \\ \diagup\qquad\diagdown \\ H \qquad\qquad CH_3 \end{array} + Br_2 \text{ (Fe)} \longrightarrow (N) \xrightarrow[\text{peroxide}]{\text{HBr}} (O)$$ ◄

(a)   (A) $PhCHBrCH_2CH_3$ (B) $PhCH=CHCH_3$ (C) $PhCHOHCHOHCH_3$

(D) $PhCOOH + CH_3COOH$

(b)   (E) $PhMgBr$ (F) $PhCH_2CH=CH_2$ (G) $PhCH=CHCH_3$ (H) $PhCH=CHCH_2Br$
                                                              (conjugated alkene
                                                              is more stable)

(c)   (I) $PhC≡CMgX (+CH_4)$ (J) $PhC≡CCH_2Ar$ (K)

$$\begin{array}{c} Ph \qquad\qquad H \\ \diagdown\qquad\diagup \\ C=C \\ \diagup\qquad\diagdown \\ H \qquad\qquad CH_2Ar \end{array} \quad (trans)$$

(d)   (L)

$$\begin{array}{c} p\text{-}CH_3C_6H_4 \qquad\qquad Ph \\ \diagdown\qquad\qquad\diagup \\ C=C \\ \diagup\qquad\qquad\diagdown \\ H \qquad\qquad\quad H \end{array} \quad (cis)$$ (M) $p\text{-}CH_3C_6H_4CH—CH_2Ph$
                                                                                                                                $\overset{\displaystyle |}{Br}$

In (M), H^+ adds to give more stable R^+, which is $p\text{-}CH_3C_6H_4\overset{+}{C}HCH_2Ph$ rather than $p\text{-}CH_3C_6H_4CH_2\overset{+}{C}HPh$, because of electron release by $p\text{-}CH_3$.

(e)   (N)

$$\begin{array}{c} p\text{-}BrC_6H_4 \qquad\qquad H \\ \diagdown\qquad\qquad\diagup \\ C=C \\ \diagup\qquad\qquad\diagdown \\ H \qquad\qquad\quad CH_3 \end{array}$$ (O) $p\text{-}BrC_6H_4—CH_2—\underset{\underset{\displaystyle Br}{|}}{C}HCH_3$

In (O), Br· adds to give more stable R·, $p\text{-}BrC_6H_4\overset{\cdot}{C}HCHBrCH_3$, which is benzylic.

Problem 11.45   Assign numbers from 1 for LEAST to 3 for MOST to the Roman numerals for the indicated compounds to show their relative reactivities in the designated reactions.

(a)   HBr addition to   (I) $PhCH=CH_2$,   (II) $p\text{-}CH_3—C_6H_4CH=CH_2$,   (III) $p\text{-}O_2NC_6H_4CH=CH_2$.

(b)   Dehydration of   (I) $p\text{-}O_2NC_6H_4CHOHCH_3$,   (II) $p\text{-}H_2NC_6H_4CHOHCH_3$,   (III) $C_6H_5CHOHCH_3$.

(c)   Dehydration of   (I) $Ph—\underset{\underset{\displaystyle OH}{|}}{\overset{\overset{\displaystyle CH_3}{|}}{C}}—CH_2CH_3$,   (II) $Ph—\overset{\overset{\displaystyle CH_3}{|}}{C}H—CH_2OH$,   (III) $PhCHOHCH_2CH_2CH_3$.

(d)   Solvolysis of   (I) $C_6H_5CH_2Cl$,   (II) $p\text{-}O_2NC_6H_4CH_2Cl$,   (III) $p\text{-}CH_3OC_6H_4CH_2Cl$.   ◄

See Table 11-5.

Table 11-5

	I	II	III
(a)	2	3 (p-Me stabilizes the benzylic R^+)	1 (p-NO_2 destabilizes the benzylic R^+)
(b)	1 (p-NO_2 destabilizes benzylic R^+)	3 (p-NH_2 stabilizes the benzylic R^+)	2
(c)	3 (get 3° benzylic R^+)	1 (get 1°, non-benzylic R^+)	2 (get 2° benzylic R^+)
(d)	2	1 (p-NO_2 destabilizes benzylic R^+)	3 (p-CH_3O stabilizes benzylic R^+)

Problem 11.46 Show the syntheses of the following compounds from benzene, toluene and any inorganic reagents or aliphatic compounds having up to three C's:

(a) p-BrC$_6$H$_4$CH$_2$Cl   (b) p-ClC$_6$H$_4$CH=CH$_2$ (c) Ph—$\overset{\overset{\displaystyle CH_3}{|}}{\underset{\underset{\displaystyle OH}{|}}{C}}$—CH$_3$

(d) p-O$_2$NC$_6$H$_4$—CH$_2$Ph   (e) Ph$_2$CHCH$_3$　　　　◀

(a)　　　　C$_6$H$_5$CH$_3$ $\xrightarrow{\text{Br}_2, \text{Fe}}$ p-BrC$_6$H$_4$CH$_3$ $\xrightarrow{\text{Cl}_2, \text{uv}}$ p-BrC$_6$H$_4$CH$_2$Cl

(b)　　　　C$_6$H$_6$ $\xrightarrow{\underset{\text{AlCl}_3}{\text{C}_2\text{H}_5\text{Cl}}}$ C$_6$H$_5$CH$_2$CH$_3$ $\xrightarrow{\underset{\text{Fe}}{\text{Br}_2}}$ p-BrC$_6$H$_4$CH$_2$CH$_3$ $\xrightarrow{\underset{\text{uv}}{\text{Cl}_2}}$

p-BrC$_6$H$_4$CHClCH$_3$ $\xrightarrow{\underset{\text{KOH}}{\text{alc.}}}$ p-BrC$_6$H$_4$CH=CH$_2$

The C=C must be introduced by a base-induced reaction. If acid is used, such as in the dehydration of an alcohol, the product would undergo polymerization.

(c)　　　PhH + (CH$_3$)$_2$CHCl $\xrightarrow{\text{AlCl}_3}$ Ph—CH(CH$_3$)$_2$ $\xrightarrow{\underset{\text{uv}}{\text{Br}_2}}$ Ph—$\underset{\underset{\displaystyle Br}{|}}{C}$(CH$_3$)$_2$ $\xrightarrow{\underset{(\text{S}_\text{N}\text{I})}{\text{H}_2\text{O}}}$ Ph—$\underset{\underset{\displaystyle OH}{|}}{C}$(CH$_3$)$_2$

(d)　　　C$_6$H$_5$CH$_3$ $\xrightarrow{\underset{\text{H}_2\text{SO}_4}{\text{HNO}_3}}$ p-O$_2$NC$_6$H$_4$CH$_3$ $\xrightarrow{\underset{\text{uv}}{\text{Cl}_2}}$ p-O$_2$NC$_6$H$_4$CH$_2$Cl $\xrightarrow{\underset{\text{AlCl}_3}{\text{C}_6\text{H}_6}}$ p-O$_2$NC$_6$H$_4$CH$_2$C$_6$H$_5$

The deactivated C$_6$H$_5$NO$_2$ cannot be alkylated with PhCH$_2$Cl.

(e)　　　PhH $\xrightarrow{\underset{\text{AlCl}_3}{\text{C}_2\text{H}_5\text{Cl}}}$ PhCH$_2$CH$_3$ $\xrightarrow{\underset{\text{light}}{\text{Br}_2}}$ PhCHBrCH$_3$ $\xrightarrow{\underset{\text{KOH}}{\text{alc.}}}$ PhCH=CH$_2$ $\xrightarrow{\underset{\text{HF}}{\text{PhH}}}$ Ph$_2$CHCH$_3$

Problem 11.47 Deduce the structural formulas of the following arenes. (a) (i) Compound A (C$_{16}$H$_{16}$) decolorizes both Br$_2$ in CCl$_4$ and cold aqueous KMnO$_4$. It adds an equimolar amount of H$_2$. Oxidation with hot KMnO$_4$ gives a dicarboxylic acid, C$_6$H$_4$(COOH)$_2$, having only one monobromo substitution product. (ii) What structural feature is uncertain? (b) Arene B (C$_{10}$H$_{14}$) has 5 possible monobromo derivatives (C$_{10}$H$_{13}$Br). Vigorous oxidation of B yields an acidic compound, C$_8$H$_6$O$_4$, having only one mononitro substitution product, C$_8$H$_5$O$_4$NO$_2$.　　　　◀

(a) (i) Compound A has one C=C since it adds one H$_2$. The other 8 degrees of unsaturation mean the presence of 2 benzene rings. Since oxidative cleavage gives a dicarboxylic acid, C$_6$H$_4$(COOH)$_2$, each benzene ring must be disubstituted. Since C$_6$H$_4$(COOH)$_2$ has only 1 monobromo derivative, the COOH's must be para to each other.

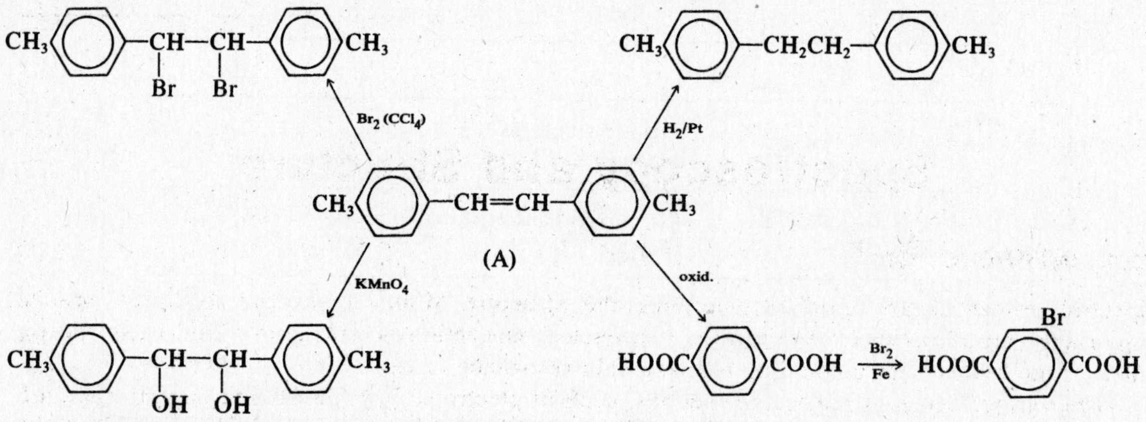

(A)

(ii) It may be *cis* or *trans*.

(b) $C_8H_6O_4$ must be a dicarboxylic acid, $C_6H_4(COOH)_2$, and, as in part (a), has *para* COOH's.   Compound B must therefore be a *p*-dialkylbenzene.

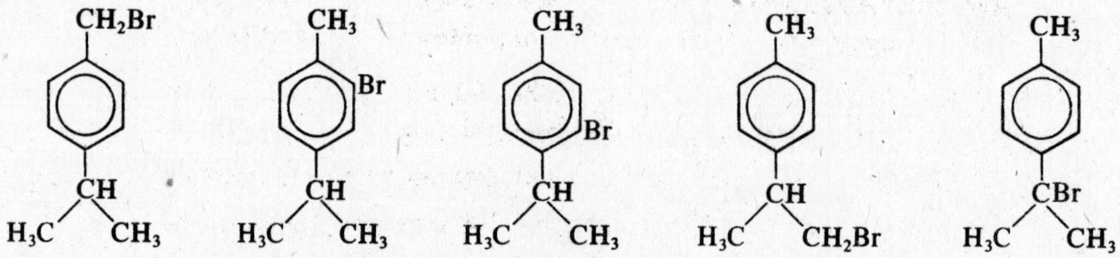

Monobromo substitution products

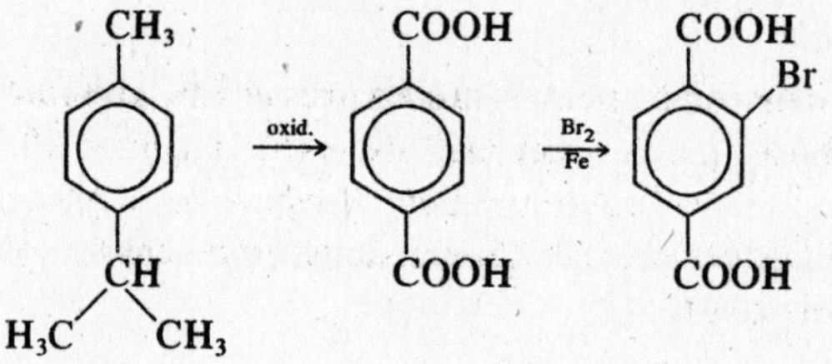

(B)

Problem 11.48   Is the benzenonium ion

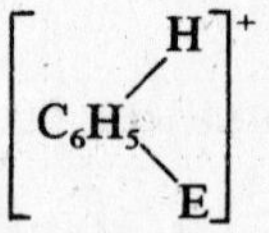

$$\left[C_6H_5 \overset{H}{\underset{E}{\diagdown}} \right]^+$$

flat?   ◄

No; the C bonded to E and H is tetrahedral.

Problem 11.49   Which isomer of the arene $C_{10}H_{14}$ resists vigorous oxidation to an aryl carboxylic acid?   ◄

For a side chain R to undergo oxidation to COOH, it must have at least 1 benzylic H.   To resist oxidation, the benzylic C must be 3°.   The arene is $C_6H_5C(CH_3)_3$.

Chapter 12

Spectroscopy and Structure

12.1 INTRODUCTION

Spectral properties are used to determine the structure of molecules and ions. Of special importance are ultraviolet (uv), infrared (ir), nuclear magnetic resonance (nmr) and mass spectra (ms). Free radicals are studied by electron spin resonance (esr).

The various types of molecular energies, such as electronic, vibrational and nuclear spin, are quantized. That is, only certain energy states are permitted. The molecule can be raised from its lowest energy state (**ground state**) to a higher energy state (**excited state**) by a photon (quantum of energy) of electromagnetic radiation of the correct wavelength.

Region of electromagnetic spectrum	Wavelength of photon	Type of excitation
far ultraviolet (uv)	100–200 nm	electronic
near ultraviolet (uv)	200–350 nm	electronic
visible	350–800 nm	electronic
infrared (ir)	1–300 μm	vibrational
radio	1 m	electron spin and nuclear spin

(left margin, vertical): Increasing Energy

Wavelengths (λ) for ultraviolet spectra are expressed in **nanometers** (1 nm $= 10^{-9}$ m); for the infrared, **micrometers** (formerly called **microns**) are used (1 μm $= 10^{-6}$ m). Frequencies (ν) in the infrared are often specified by the **wave number** $\bar{\nu}$, where $\bar{\nu} = 1/\lambda$. A common unit for $\bar{\nu}$ is the **reciprocal centimeter** (1 cm$^{-1} = 100$ m^{-1}). The basic SI units for frequency and energy are the **hertz** (Hz) and the **joule** (J), respectively.

Problem 12.1 (*a*) Calculate the frequencies of violet and red light if their wavelengths are 400 nm and 750 nm respectively. (*b*) Calculate and compare the energies of their photons. ◄

(*a*) The wavelengths are substituted into the equation $\nu = c/\lambda$, where $c =$ speed of light $= 3.0 \times 10^8$ m s^{-1}. Thus

$$\text{Violet:} \quad \nu = \frac{3.0 \times 10^8 \text{ m s}^{-1}}{400 \times 10^{-9} \text{ m}} = 7.5 \times 10^{14} \text{ s}^{-1} = 750 \text{ THz}$$

$$\text{Red:} \quad \nu = \frac{3.0 \times 10^8 \text{ m s}^{-1}}{750 \times 10^{-9} \text{ m}} = 4.0 \times 10^{14} \text{ s}^{-1} = 400 \text{ THz}$$

where 1 THz $= 10^{12}$ Hz $= 10^{12}$ s^{-1}. Violet light has the shorter wavelength and higher frequency.

(*b*) The frequencies from part (*a*) are substituted into the equation $E = h\nu$, where $h = 6.624 \times 10^{-34}$ J s (**Planck's constant**). Thus

$$\text{Violet:} \quad E = (6.624 \times 10^{-34} \text{ J s})(7.5 \times 10^{14} \text{ s}^{-1}) = 5.0 \times 10^{-19} \text{ J}$$

$$\text{Red:} \quad E = (6.624 \times 10^{-34} \text{ J s})(4.0 \times 10^{14} \text{ s}^{-1}) = 2.7 \times 10^{-19} \text{ J}$$

Photons of violet light have more energy than those of red light.

Problem 12.2 Express 10 microns (a) in centimeters, (b) in angstroms ($1 \text{ Å} = 10^{-10}$ m), (c) in nanometers, (d) as a wave number. ◀

(a)
$$10 \ \mu m = (10 \times 10^{-6} \text{ m})\left(\frac{100 \text{ cm}}{1 \text{ m}}\right) = 10^{-3} \text{ cm}$$

(b)
$$10 \ \mu m = (10 \times 10^{-6} \text{ m})\left(\frac{1 \text{ Å}}{10^{-10} \text{ m}}\right) = 10^5 \text{ Å}$$

(c)
$$10 \ \mu m = (10 \times 10^{-6} \text{ m})\left(\frac{10^9 \text{ nm}}{1 \text{ m}}\right) = 10^4 \text{ nm}$$

(d)
$$\bar{\nu} = \frac{1}{10 \times 10^{-6} \text{ m}} = 10^5 \text{ m}^{-1} = (10^5 \text{ m}^{-1})\left(\frac{1 \text{ cm}^{-1}}{100 \text{ m}^{-1}}\right) = 10^3 \text{ cm}^{-1}$$

In a typical spectrophotometer, a compound is exposed to electromagnetic radiation with a continuous spread in wavelength. The radiation passing through or absorbed is recorded on a chart against the wavelength or wave number. Absorption peaks are plotted as *minima* in *infrared*, and usually as *maxima* in *ultraviolet*, spectroscopy.

At a given wavelength, the percentage of radiant energy absorbed (the **absorbance**) depends on:

1. Nature of compound absorbing.
2. Concentration (C) of molecules (moles/liter of solution).
3. Length, ℓ (in cm), of path of solution through which light passes.

The expression for the absorbance is

$$A = \epsilon C \ell$$

where ϵ, the **molar extinction coefficient**, is an inherent property of the compound. The wavelengths of maximum absorption, λ_{max}, and the corresponding ϵ_{max} are specific physical properties used to identify compounds.

12.2   ULTRAVIOLET AND VISIBLE SPECTROSCOPY

Ultraviolet (uv) and visible light cause an electron to be excited from a lower-energy, occupied MO to a higher-energy, unoccupied MO*.

There are three kinds of electrons: those in σ bonds, those in π bonds, and unshared electrons, which are designated by the letter n for nonbonding. These are illustrated in formaldehyde:

$$\begin{array}{c} H \\ \diagdown \sigma \\ \quad\quad C \overset{\pi}{\underset{\sigma}{=\!=\!=}} \overset{n}{O} \colon n \\ \diagup \sigma \\ H \end{array}$$

On absorbing energy, any of these electrons can enter excited states, which are either antibonding σ^* or π^*. All molecules have σ and σ^* orbitals, but only those with π orbitals have π^* orbitals.

Only the $n \to \pi^*$, $\pi \to \pi^*$, and more rarely the $n \to \sigma^*$ excitations occur in the near ultraviolet and visible regions, which are the available regions for ordinary spectrophotometers. Species which absorb in the visible region are colored, and black is observed when all visible light is absorbed.

Problem 12.3 The relative energy for various electronic states (MO's) is:

List the 3 electronic transitions detectable by uv spectrophotometers in order of increasing ΔE. ◀

$$n \to \pi^* < \pi \to \pi^* < n \to \sigma^*$$

Problem 12.4 List all the electronic transitions possible for   (a) CH_4,   (b) CH_3Cl,   (c) $H_2C=O$.   ◄

(a) $\sigma \to \sigma^*$.   (b) $\sigma \to \sigma^*$ and $n \to \sigma^*$ (there are no π or π^* MO's).   (c) $\sigma \to \sigma^*$, $\sigma \to \pi^*$, $\pi \to \sigma^*$, $n \to \sigma^*$, $\pi \to \pi^*$ and $n \to \pi^*$.

Problem 12.5 The uv spectrum of acetone shows two peaks of $\lambda_{max} = 280$ nm, $\epsilon_{max} = 15$ and $\lambda_{max} = 190$, $\epsilon_{max} = 100$. (a) Identify the electronic transition for each. (b) Which is more intense?   ◄

(a) The longer wavelength (280 nm) is associated with the smaller-energy ($n \to \pi^*$) transition. $\pi \to \pi^*$ occurs at 190 nm.

(b) $\pi \to \pi^*$ has the larger ϵ_{max} and is the more intense peak.

Problem 12.6 Draw conclusions about the relationship of λ_{max} to the structure of the absorbing molecule from the following λ_{max} values (in nm): ethylene (170), 1,3-butadiene (217), 2,3-dimethyl-1,3-butadiene (226), 1,3-cyclohexadiene (256), and 1,3,5-hexatriene (274).   ◄

1. Conjugation of π bonds causes molecules to absorb at longer wavelengths. 2. As the number of conjugated π bonds increases, λ_{max} increases. 3. Cyclic polyenes absorb at higher wavelengths than do acyclic polyenes. 4. Substitution of alkyl groups on C=C causes a shift to longer wavelength (**red shift**).

Problem 12.7 Account for the following variations in λ_{max} (nm) of CH_3X: X = Cl (173), Br (204), and I (258). ◄

The transition must be $n \to \sigma^*$ [Problem 12.4(b)]. On going from Cl to Br to I the n electrons (a) are found in higher principal energy levels (the principal quantum numbers are 3, 4, 5 respectively), (b) are further away from the attractive force of the nucleus, and (c) become more easily excited. Hence absorption occurs at progressively higher λ_{max}.

Problem 12.8 Use MO theory (see Fig. 8.3) to explain why λ_{max} for 1,3-butadiene is higher than for ethylene. (See Problem 12.6 for data.)   ◄

For 1,3-butadiene, the excitation is from π_2, the highest occupied MO (HOMO), to π_3^*, the lowest unoccupied MO (LUMO). The ΔE for this transition is less than the ΔE for $\pi \to \pi^*$ for ethylene. Therefore λ_{max} for 1,3-butadiene is greater than λ_{max} for ethylene.

Problem 12.9 Identify the 2 geometric isomers of stilbene, $C_6H_5CH=CHC_6H_5$, from their λ_{max} values, 294 nm and 278 nm.   ◄

The higher-energy *cis* isomer has the shorter wavelength. Steric strain prevents full coplanarity of the *cis* phenyl groups, and the conjugative effect is attenuated.

Problem 12.10 Each of three C_6H_{10} isomers absorbs 2 moles of H_2 to form *n*-hexane. CO_2 and RCOOH are not formed on ozonolysis from any of the compounds. Deduce possible structures for the 3 compounds if their uv absorption maxima are 176, 211 and 216 nm.   ◄

The isomers are hexadienes. An allene or alkyne is not possible since CO_2 and RCOOH are not ozonolysis products. Absorption at 175 nm is attributed to the isolated 1,4-hexadiene. The conjugated *cis*- and *trans*-1,3-hexadienes absorb at 211 and 216 nm, respectively.

Problem 12.11 The complementary color pairs are: violet–yellow, blue–orange, and green–red. Given a red, an orange and a yellow polyene, which is most and which is least conjugated?   ◄

The orange polyene absorbs *blue*, the red absorbs *green* and the yellow absorbs *purple*. The most conjugated polyene, in this case the red one, absorbs the color of longest wavelength, in this case green. Purple has the shortest wavelength and therefore the yellow polyene is the least conjugated.

12.3 INFRARED SPECTROSCOPY

Atoms in a diatomic molecule, e.g. H—H and H—Cl, vibrate in only one way; they move, as though attached to a coiled spring, toward and away from each other. This mode is called **bond stretching**. Triatomic molecules, such as CO_2 (O=C=O), possess two different stretching modes. In the **symmetrical stretch**, each O moves away from the C at the same time. In the **antisymmetrical stretch**, one O moves toward the C while the other O moves away.

Molecules with more than 2 atoms have continuously changing bond angles. These **bending modes** are indicated in Fig. 12-1.

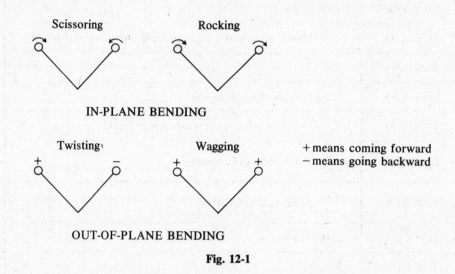

Scissoring   Rocking

IN-PLANE BENDING

Twisting   Wagging   + means coming forward
                                    − means going backward

OUT-OF-PLANE BENDING

Fig. 12-1

In a molecule, each bond such as O—H and each group of three or more atoms such as NH_2 and CH_3 absorbs ir radiation at certain wave numbers to give quantized excited stretching and bending vibrational states. See Tables 12-1 and 12-2. Only vibrations that cause a change in dipole moment give rise to an absorption band. The absorptions *depend only slightly on the molecular environment* of the bond or group. An observed absorption band at a specific wavelength proves the identity of a particular bond or group of bonds in a molecule. Conversely, the absence of a certain band in the spectrum usually rules out the presence of the bond that would produce it.

Between 1400 and 800 cm^{-1} there are many peaks which are difficult to interpret. However, this range, called the **fingerprint region**, is useful for determining whether compounds are identical.

Problem 12.12 Use Tables 12-1 and 12-2 to answer the following questions. (*a*) Which types of bonds appear to be the strongest? (*b*) How do the stretching frequencies of corresponding single, double, and triple bonds compare? (*c*) How does the hybridized orbital state of C influence the stretching frequency of the C—H bond? (*d*) What is the effect of H-bonding on the stretching frequency of O—H?   ◄

(*a*) The stretching bands for the strongest bonds are found at the highest frequencies or shortest wavelengths, where most energy is provided. The strongest bonds are those between an H and another element, such as C, N or O. (*b*) The order of decreasing frequency and bond strength are triple (C≡C, 2100 cm^{-1}), double (C=C, 1620–1680 cm^{-1}) and single (C—C, 700–1200 cm^{-1}; very weak and of little use in identification). (*c*) The strongest bond, H—C_{sp}, absorbs at 3300 cm^{-1}; the weakest, H—C_{sp^3}, absorbs at about 2900 cm^{-1}; the H—C_{sp^2} stretch is at 3100 cm^{-1}. (*d*) H-bonding causes a shift of the O—H stretch to lower frequencies (3600 cm^{-1} → 3300 cm^{-1}). The band also becomes broader and less intense.

Table 12-1. Infrared Absorption Peaks (mostly stretching)

$\bar{\nu}$, cm^{-1}	Structure
1050–1400	C—O (in ethers, alcohols and esters)
1150–1360	SO$_2$ (in sulfonic acid derivatives)
1315–1475	C—H (in alkanes)
1340, 1500	NO$_2$
1450–1600	C=C bond in aromatic ring (usually shows several peaks)
1620–1680	C=C
1630–1690	C=O (in amides O=C—N)
1690–1750	C=O (in carbonyl compounds and esters)
1700–1725	$\begin{matrix}\text{OH}\\\text{C=O}\end{matrix}$ (in carboxylic acids)
1770–1820	$\begin{matrix}\text{Cl}\\\text{C=O}\end{matrix}$ (in acid chlorides)
2100–2200	C≡C
2210–2260	C≡N
2500	S—H
2700–2800	C—H (of aldehyde group)
2500–3000	O—H in COOH
3000–3100	C—H (C is part of aromatic ring)
3300	C—H (C is acetylenic)
3020–3080	C—H (C is ethylenic)
2800–3000	C—H (in alkanes)
3300–3500	N—H (in amines and amides)
3200–3600	O—H (in H-bonded ROH and ArOH)
3600–3650	O—H
2100	O—D

Table 12-2. Bending Frequencies (cm^{-1}) of Hydrocarbons

Alkanes	CH$_3$ 1420–1470 1375	=CH$_2$ 1430–1470	CH(CH$_3$)$_2$ Doublet of equal intensities at 1370, 1385. Also 1170	C(CH$_3$)$_3$ Doublet at 1370 (strong) 1395 (moderate)
Alkenes Out-of-Plane	RCH=CH$_2$ 910–920 990–1000	R$_2$C=CH$_2$ 880–900	RCH=CHR cis 675–730 (variable) trans 965–975	
Aromatic C—H Out-of-Plane	Monosubstituted 690–710 730–770		Disubstituted ortho meta para 735–770 690–710 810–840 750–810	

Problem 12.13   Identify the peaks marked by Roman numerals in Fig. 12-2, the ir spectrum of ethyl acetate, CH$_3$COCH$_2$CH$_3$. ◄

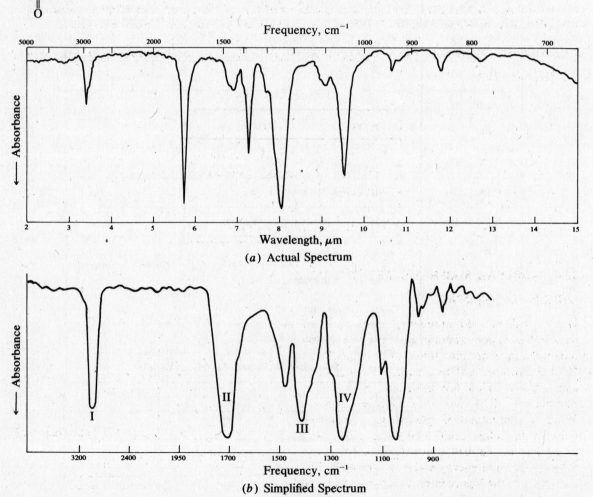

(a) Actual Spectrum

(b) Simplified Spectrum

Fig. 12-2

The deep valleys represent wave numbers of low transmission and are therefore absorption peaks or bands. At about 2800 cm^{-1}, peak I is due to H—C$_{sp^3}$ stretching. The peak at 1700 cm^{-1}, II, is due to stretching of the

$$\text{\Large $>$}\!\!C\!\!=\!\!O$$

group. The two peaks at 1400–1500 cm^{-1}, III, are again due to C—H bonds. The one at 1250 cm^{-1}, IV, is due to the C—O stretch. It is extremely difficult and impractical to attempt an interpretation of each band in the spectrum.

Problem 12.14 Which of the following vibrational modes show no ir absorption bands? (a) symmetrical CO$_2$ stretch, (b) antisymmetrical CO$_2$ stretch, (c) symmetrical O=C=S stretch, (d) C=C stretch in o-xylene, (e) C=C stretch in p-xylene and (f) C=C stretch in p-bromotoluene. ◄

Those vibrations which do not result in a change in dipole moment show no band. These are (a) and (e), which are symmetrical about the axis of the stretched bonds.

Problem 12.15 Explain the following observation. A concentrated solution of C$_2$H$_5$OH in CCl$_4$, as well as one of CH$_2$OHCH$_2$OH, has a broad O—H stretch near 3350 cm^{-1}. On dilution with CCl$_4$, the spectrum of CH$_2$OHCH$_2$OH does not change, but that of C$_2$H$_5$OH shows a sharp O—H stretch at 3600 cm^{-1} instead of the broad band at 3350 cm^{-1}. ◄

H-bonding in HOCH$_2$CH$_2$OH is intramolecular (Section 2.6) and is not disturbed by dilution. H-bonding in C$_2$H$_5$OH is intermolecular. On dilution the molecules move too far apart for H-bonding to occur.

Problem 12.16 Where in the ir spectrum of p-tolunitrile, p-CH$_3$C$_6$H$_4$CN, would absorption peaks be expected? ◄

See Table 12-3.

Table 12-3

$\bar{\nu}$, cm^{-1}	Structure
2800–2900	alkane C—H bonds
3000–3100	aromatic C—H bonds
1450–1600	aromatic ring
2210–2260	C≡N
810–840	p-substituted

12.4 NUCLEAR MAGNETIC RESONANCE

ORIGIN OF SPECTRA

Nuclei with an odd number of protons or of neutrons have permanent magnetic moments and quantized nuclear spin states. For example, an H in a molecule has two equal-energy nuclear spin states, which are assigned the quantum numbers $+\frac{1}{2}$ (↑) and $-\frac{1}{2}$ (↓) [Fig. 12-3(a)]. When a compound is placed in a magnetic field its H's align their own fields either *with* or *against* the applied magnetic field, giving rise to two separated energy states, as shown in Fig. 12-3(b). In the higher energy state the fields are aligned *against* each other; in the lower energy state they are aligned *with* each other. The

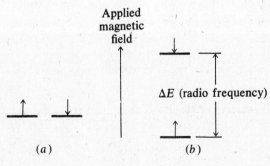

Fig. 12-3

difference in energy between the two states corresponds to the frequency of radiowaves. For this reason, radio-frequency photons "flip" H nuclei from lower to higher energy states. The required frequency is directly proportional to the magnetic field.

In practice it is easier to fix the frequency and vary the applied magnetic field. The plot for the nmr spectrum is transmittance vs. magnetic field strength. When the radiowaves are removed the excited nuclei quickly return to the lower-energy spin state. The same sample can then be used for obtaining repeated spectra.

Nmr spectroscopy is useful because not all H's change spin at the same applied magnetic field, for the energy absorbed depends on the bonding environment of the H. The magnetic field "sensed" by an H is not necessarily that which is applied by the magnet, because the electrons in the bond to the H and the electrons in nearby π bonds *induce* their own magnetic fields. This induced field may oppose or reinforce the applied field. The field felt by the H is then the resultant of the applied and induced fields, as shown in Fig. 12-4.

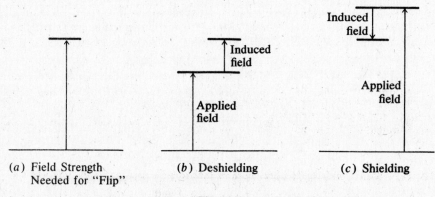

(a) Field Strength   (b) Deshielding   (c) Shielding
Needed for "Flip"

Fig. 12-4

When the two fields reinforce, as in Fig. 12-4(b), a smaller field must be applied to flip the proton. Such a proton is **deshielded** and absorbs more **downfield** (i.e. at lower magnetic field strength). When the fields oppose, as in Fig. 12-4(c), a stronger field must be applied. The proton is now **shielded** and absorbs more **upfield**. Such shifts in position of nmr absorption caused by the specific environment of H's are called **chemical shifts**.

Each nonequivalent H has a unique and characteristic chemical shift resulting in an individual peak or grouping of peaks. To decide if any two H's are equivalent, replace each alternately by a group X. If the X derivatives are the same, the H's are equivalent. If replacing each H in a CH_2 gives 2 diastereomers, the H's are *not* equivalent.

Problem 12.17 Which of the following atoms do not exhibit nuclear magnetic resonance? ^{12}C, ^{16}O, ^{14}N, ^{15}N, ^{2}H, ^{19}F, ^{31}P, ^{13}C and ^{32}S. ◀

Atoms with odd numbers of protons and/or neutrons are nmr active. The inactive atoms are: ^{12}C ($6p, 6n$), ^{16}O ($8p, 8n$) and ^{32}S ($16p, 16n$). To detect the nmr activity of atoms other than ^{1}H requires alteration of the nmr spectrometer. The ordinary spectrometer selects the range of radiowave frequency that excites only ^{1}H.

Problem 12.18 How many kinds of H's are there in (a) CH_3CH_3, (b) $CH_3CH_2CH_3$, (c) $(CH_3)_2$-$CHCH_2CH_3$, (d) $H_2C{=}CH_2$, (e) $CH_3CH{=}CH_2$, (f) $C_6H_5NO_2$, (g) $C_6H_5CH_3$? ◀

(a) One (all equivalent).
(b) 2: $CH_3^a CH_2^b CH_3^a$.
(c) 4: $(CH_3^a)_2 CH^b CH_2^c CH_3^d$.

(d) One (all equivalent).
(e) 4:

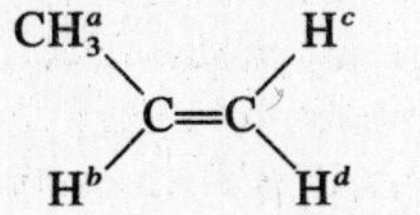

$$CH_3^a \quad\quad H^c$$
$$\diagdown C = C \diagup$$
$$H^b \diagup \quad\quad \diagdown H^d$$

The $=CH_2$ H's are not equivalent since one is *cis* to the CH_3 and the other is *trans*. Replacement of H^c by X gives the *cis*-diastereomer. Replacement of H^d gives the *trans*-diastereomer.
(f) 3: 2 *ortho*, 2 *meta* and one *para*.
(g) Theoretically there are 3 kinds of aromatic H's, as in (f). Actually the ring H's are little affected by alkyl groups and are equivalent. There are 2 kinds: $C_6H_5^a CH_3^b$.

Problem 12.19 How many kinds of equivalent H's are there in the following?

(a) $CH_3CHClCH_2CH_3$　　　　(b) $p\text{-}CH_3CH_2\text{—}C_6H_4\text{—}CH_2CH_3$　　◀

(a) *Five* as shown:

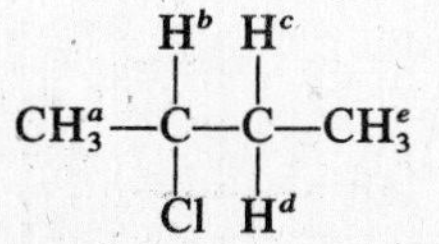

$$\begin{array}{ccc} & H^b \; H^c & \\ CH_3^a\text{—} & \!\!\!\underset{\underset{Cl}{|}}{C}\text{—}\underset{\underset{H^d}{|}}{C} & \!\!\!\text{—}CH_3^e \end{array}$$

The 2 H's of CH_2 are *not* equivalent, because of the presence of a chiral C. Replacing H^c and H^d separately by X gives 2 diastereomers.
(b) *Three*. All 4 aromatic H's are equivalent, as are the 6 in each CH_3 and the 4 in each CH_2.

Problem 12.20 How many kinds of H's are there in the isomers of dimethylcyclopropane?　　◀

Dimethylcyclopropane has 3 isomers, shown with labeled H's to indicate differences and equivalencies.

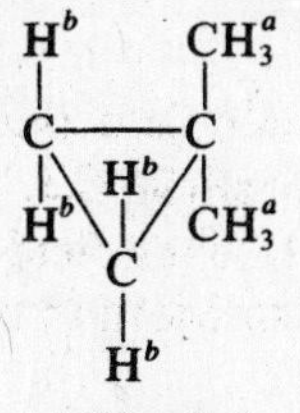

1,1-Dimethyl-
cyclopropane
(I)

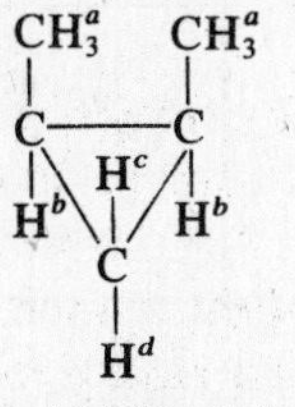

cis-1,2-Dimethyl-
cyclopropane
(II)

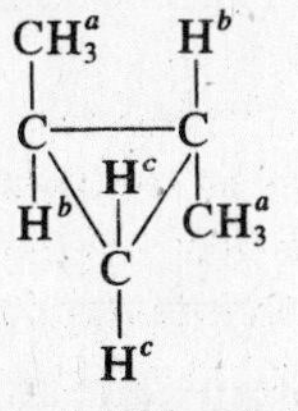

trans-1,2-Dimethyl-
cyclopropane
(III)

In II, H^c and H^d are different since H^c is *cis* to the CH_3's and H^d is *trans*. In III the CH_2 H's are equivalent; they are each *cis* to a CH_3 and *trans* to a CH_3.

CHEMICAL SHIFT

The positions (chemical shifts) of peaks are measured from a reference point produced by the H's in tetramethylsilane, $(CH_3)_4Si$. This reference point is given a value of $\delta = 0$ ppm (parts per million) or $\tau = 10$ ppm. Both scales are used; $\delta + \tau = 10$. The larger the δ value or the smaller the τ value, the more downfield is the peak.

A few important generalizations can be made about molecular structure and chemical shift (see Table 12-4):

1. Electronegative atoms such as N, O and X deshield H's. The extent of deshielding is directly proportional to the electronegativity of the N, O or X atom and the proximity of the atom to the H.

Table 12-4. Proton Chemical Shifts

δ, ppm	τ, ppm	Character of Underlined Proton	δ, ppm	τ, ppm	Character of Underlined Proton
0.2	9.8	Cyclopropane: ▷—H	4–2.5	6–7.5	Bromide: Br—C—H
0.9	9.1	Primary: R—CH₃	4–3	6–7	Chloride: Cl—C—H
1.3	8.7	Secondary: R₂CH₂			
1.5	8.5	Tertiary: R₃—CH	4–3.4	6–6.6	Alcohol: HO—C—H
1.7	8.3	Allylic: —C=C—CH₃	4–4.5	6–5.5	Fluoride: F—C—H
2.0–4.0	8.0–6.0	Iodide: α H I—C—H	4.1–3.7	5.9–6.3	Ester (I): α H to alkyl O R—C=O O—C—H
2.2–2.0	7.8–8.0	Ester (II): α H to C=O H—C—C=O OR	5.0–1.0	5.0–9.0	Amine: R—NH₂
2.6–2.0	7.4–8.0	Carboxylic acid: α H H—C—C=O OH	5.5–1.0	4.5–9.0	Hydroxyl: RO—H
2.7–2.0	7.3–8.0	Carbonyl: α H —C=O —C—H	5.9–4.6	4.1–5.4	Olefinic: —C=C—H
3–2	7–8	Acetylenic: —C≡C—H	8.5–6.0	1.5–4.0	Aromatic: ⬡—H (Ar—H)
3–2.2	7–7.8	Benzylic: ⬡—C—H	10.0–9.0	0–1.0	Aldehyde: —C=O H
3.3–4.0	6.7–6.0	Ether: α H R—O—C—H	12.0–10.5	(−2)–(−0.5)	Carboxyl: R—C=O O—H
			12.0–4.0	(−2)–6	Phenolic: ⬡—O—H
			15.0–17.0	(−5)–(−7)	Enolic: —C=C—O—H

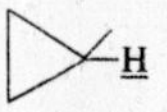

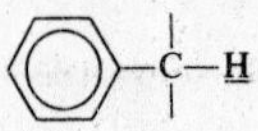

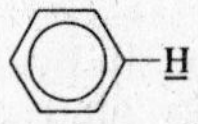

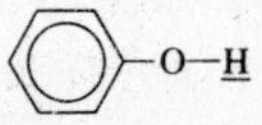

2. Electrons of an aromatic ring, —C≡C—,

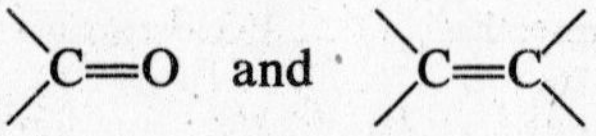

$$\diagup_{\diagdown}C=O \quad \text{and} \quad \diagup_{\diagdown}C=C\diagdown^{\diagup}$$

deshield an attached H. The order of deshielding is:

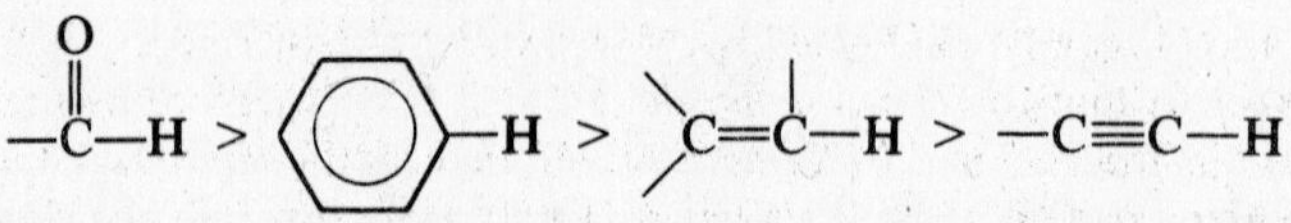

$$-\overset{O}{\overset{\|}{C}}-H \; > \; \text{⬡}-H \; > \; \diagup_{\diagdown}C=\overset{|}{C}-H \; > \; -C\equiv C-H$$

3. The C of a

$$\diagup_{\diagdown}C=O$$

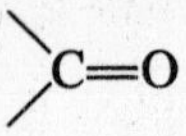

group is electron-withdrawing and deshields an H on an adjacent C, as in

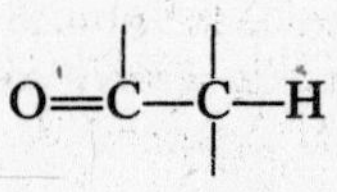

$$O=\overset{|}{C}-\overset{|}{C}-H$$

4. 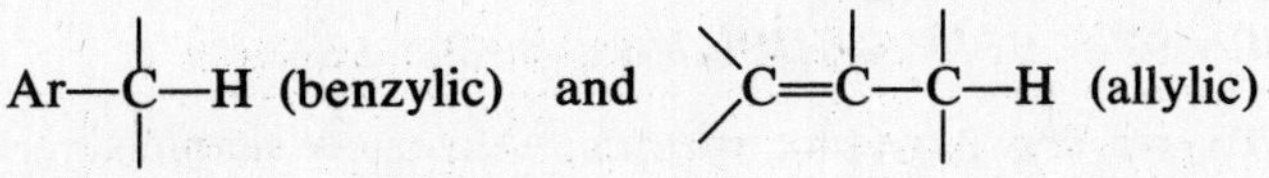 $$Ar-\overset{|}{\underset{|}{C}}-H \text{ (benzylic)} \quad \text{and} \quad \overset{|}{\underset{}{>}}C=\overset{|}{C}-\overset{|}{\underset{|}{C}}-H \text{ (allylic)}$$

H's are deshielded because Ar— and

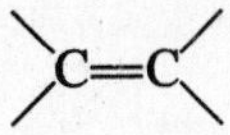

$$\overset{}{\underset{}{>}}C=C\overset{}{\underset{}{<}}$$

are electron-withdrawing by induction.

5.   For alkyl groups, the order of being downfield is:

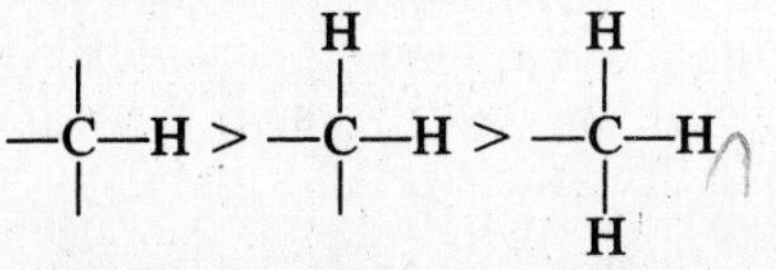

$$-\overset{|}{\underset{|}{C}}-H > -\overset{H}{\underset{|}{C}}-H > -\overset{H}{\underset{H}{C}}-H$$

Methine   Methylene   Methyl

6.   Electropositive atoms, such as Si, shield H's. The H's in $(CH_3)_4Si$ are so shielded and upfield that they are usually fully isolated from all other kinds of H's. For this reason $(CH_3)_4Si$ is used as an internal reference.

7.   H's attached to a cyclopropane ring and those situated *in* the π cloud of an aromatic system are strongly shielded. Some may have negative δ values.

RELATIVE PEAK AREAS; COUNTING H's

The area under a signal graph is directly proportional to the number of equivalent H's giving the signal. For example, the compound $C_6H_5^aCH_2^bC(CH_3^c)_3$ has 5 aromatic protons (a), 2 benzylic protons (b), and 9 equivalent CH_3 protons (c). Its nmr spectrum shows 3 peaks for the 3 different kinds of H, which appear at: (a) $\delta = 7.1$ (aromatic H), (b) $\delta = 2.2$ (benzylic H), (c) $\delta = 0.9 (1° H)$. The relative areas under the peaks are $a:b:c = 5:2:9$. The nmr instrument integrates the areas as follows: When no signal is present, it draws a horizontal line. When the signal is reached the line ascends and levels off when the signal ends. The relative distance from plateau to plateau gives the relative area. See Problem 12.62 for a typical nmr spectrum showing integration.

Problem 12.21   Suggest a structure for a compound (C_9H_{12}) showing nmr peaks at δ values of 7.1, 2.2, 1.5 and 0.9 ppm.   ◀

The value 7.1 indicates H's on a benzene ring. The formula shows 3 more C's, which might be attached to the ring as shown below (assuming that, since this is an alkylbenzene, all aromatic H's are equivalent):

(1)   3 CH_3's in trimethylbenzene, $(CH_3^a)_3C_6H_3^b$ (2)   a CH_3 and a CH_2CH_3 in $CH_3^aC_6H_4^bCH_2^cCH_3^d$

(3)   a $CH_2CH_2CH_3$ in $C_6H_5^aCH_2^bCH_2^cCH_3^d$ (4)   a $CH(CH_3)_2$ in $C_6H_5^aCH^b(CH_3^c)_2$

Compounds (1) and (4) can be eliminated because they would give 2 and 3 signals respectively rather than the 4 observed signals. Although (2) has 4 signals, H^a and H^c are different benzylic H's and the compound should have 2 signals in the region 3.0–2.2 ppm rather than the single observed signal. Hence (2) can be eliminated. Only (3) can give the 4 observed signals with the proper chemical shifts.

Problem 12.22   What compound (C_7H_8O) has nmr signals at $\delta = 7.3, 4.4$ and 3.7 ppm, with relative intensities of $7:2.9:1.4$, respectively?   ◀

The relative intensities of the 3 different kinds of H become $5:2:1$ on dividing by 1.4. That is, 5 H's contribute to the $\delta = 7.3$, 2 H's to $\delta = 4.4$ and 1 H to $\delta = 3.7$, for a total of 8 H's, which is consistent with the formula. The 5 H's at $\delta = 7.2$ are aromatic, indicating a C_6H_5 compound. The remaining portion of the formula comprises the CH_2OH group. The H at $\delta = 3.7$ is part of OH. The 2 H's at $\delta = 4.4$ are benzylic and alpha to OH. The compound is $C_6H_5CH_2OH$, benzyl alcohol.

PEAK SPLITTING; SPIN-SPIN COUPLING

Because of **spin-spin coupling**, most nmr spectra do not show simple single peaks but rather *groups of peaks* that tend to cluster about certain δ values. To see how this coupling arises we examine the molecular fragment

$$-\overset{|}{C}H^a-\overset{|}{C}H_2^b$$

present in a very large number of like molecules. The signal for H^b is shifted slightly upfield or downfield depending on whether the spin of H^a is aligned against or with the applied field. Since in about half of the molecules the H^a are spinning ↑ and in half ↓, H^b appears as a doublet instead of a singlet. The effect is reciprocal: the two H^b split the signal of H^a. There are 4 spin states of approximately equal probability for the two H^b:

$$↑↑; \quad \underbrace{↑↓, \ ↓↑}; \quad ↓↓$$

Because the middle 2 spin states have the same effect, the signal of H^a is split into a triplet with relative intensities 1:2:1. In the molecular fragment

$$-\overset{|}{C}H^a-CH_3^b$$

the three H^b appear as a *doublet* due to the effect of the single H^a. H^a, however, appears as a *quartet* due to the effects of the three H^b, which may be spinning as follows:

$$↑↑↑; \quad \underbrace{↑↑↓, \ ↑↓↑, \ ↓↑↑}; \quad \underbrace{↑↓↓, \ ↓↓↑, \ ↓↑↓}; \quad ↓↓↓$$

Intensities   1   3   3   1

The entire quartet integrates for one H.

Spin-spin coupling usually, but not always, occurs between nonequivalent H's on adjacent atoms. In general, if n equivalent H's are affecting the peak of H's on an adjacent C, the peak is split into $n + 1$ peaks. A symmetrical multiplet is an ideal condition and not always observed in practice.

Problem 12.23 In which of the following molecules does spin-spin coupling occur? If splitting is observed give the multiplicity of each kind of H.

(a) $ClCH_2CH_2Cl$ (b) $ClCH_2CH_2I$ (c) $CH_3-\overset{\overset{\displaystyle CH_3}{|}}{\underset{\underset{\displaystyle CH_3}{|}}{C}}-CH_2Br$

(d) $\begin{array}{c} H \\ Br \end{array}\!\!>\!C=C\!<\!\!\begin{array}{c} H \\ Br \end{array}$ (e) $\begin{array}{c} H \\ Br \end{array}\!\!>\!C=C\!<\!\!\begin{array}{c} Cl \\ H \end{array}$ (f) $\begin{array}{c} I \\ Cl \end{array}\!\!>\!C=C\!<\!\!\begin{array}{c} H \\ H \end{array}$ ◄

Splitting is not observed in (a) or (d), which each have only equivalent H's, or in (c), which has no nonequivalent H's on *adjacent* C's. The H's of CH_2 in (b) are nonequivalent and each is split into a triplet ($n = 2$; $2 + 1 = 3$). In (e) the 2 H's are not equivalent and each is split into a doublet. The vinyl H's in (f) are nonequivalent since one is *cis* to Cl and the other is *cis* to I. Each is split into a doublet. In this case the interacting H's are on the same C.

Problem 12.24 Give schematic coupling patterns showing relative chemical shifts observed for the following common alkyl groups: (a) ethyl, $-CH_2CH_3$; (b) isopropyl $-CH(CH_3)_2$; (c) t-butyl, $-C(CH_3)_3$. ◄

See Fig. 12-5.

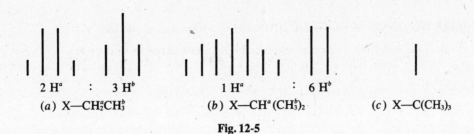

$2\,H^a$: $3\,H^b$ $1\,H^a$: $6\,H^b$

(a) X—CH$_2^a$CH$_3^b$ (b) X—CH^a(CH$_3^b$)$_2$ (c) X—C(CH$_3$)$_3$

Fig. 12-5

Problem 12.25 Sketch the expected nmr spectrum (as in Problem 12.24), showing the integration, for (a) 1,1-dichloroethane, (b) 1,1,2-trichloroethane, (c) 1,1,2,2-tetrachloroethane and (d) 1-bromo-2-chloroethane. ◄

See Fig. 12-6. In (d), H^a is more downfield than H^b because Cl is more electron-withdrawing than Br.

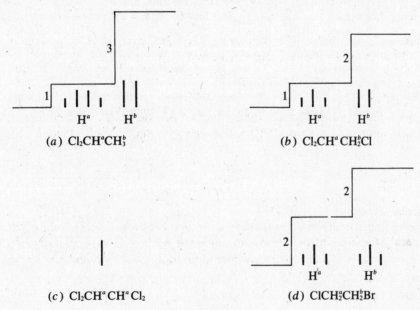

(a) Cl$_2$CH^{a}CH$_3^b$ (b) Cl$_2$CH^{a}CH$_2^b$Cl

(c) Cl$_2$CH^{a}CH^{a}Cl$_2$ (d) ClCH$_2^a$CH$_2^b$Br

Fig. 12-6

Problem 12.26 Why is splitting observed in 2-methylpropene but not in 1-chloro-2,2-dimethylpropane? ◄

In (CH$_3^a$)$_3$C—CH$_2^b$Cl, H^a and H^b do not couple because they are not on adjacent C's; they are too far away from one another. In

$$\underset{CH_3^a}{\overset{CH_3^a}{\diagup}}C=C\underset{H^b}{\overset{H^b}{\diagdown}}$$

although H^a and H^b are not on adjacent C's, they are close enough to couple because of the shorter C=C bond.

Problem 12.27 F's couple H's in the same way as do other H's. Predict the splitting in the nmr spectrum of 2,2-difluoropropane. ◄

In CH$_3^a$CF$_2$CH$_3^a$ the two F's split the H^a into a 1:2:1 triplet. The F signal, when detected by a special probe, would be a septet.

Problem 12.28 Deuterium does not give a signal in the proton nmr spectrum nor does it split nearby protons. Thus D's might just as well not be there. What is the difference between nmr spectra of CH$_3$CH$_2$Cl and CH$_3$CHDCl? ◄

See Problem 12.24(a) for the spectrum of CH$_3$CH$_2$Cl. CH$_3^a$CH^{b}DCl has a *doublet* for H^a and, more downfield, a *quartet* for H^b.

Problem 12.29 The stable *anti* conformer of CH_3CH_2Cl shows a nonequivalency of the CH_3 H's:

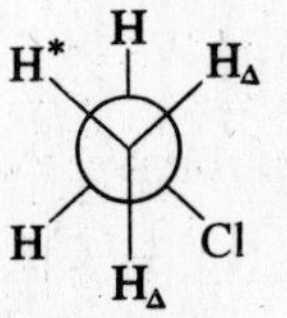

H* is *anti* to the Cl while the H_Δ's are *gauche*. Why does H* *not* give a signal different from the H_Δ's? (Instead, the 3 H's produce an equivalent triplet.) ◄

Rotation around the C—C bond is rapid. Detection by the nmr spectrometer is slower. The spectrometer therefore detects the average condition, which is the same for each H: 1/3 *anti* and 2/3 *gauche*.

Problem 12.30 What information can you deduce from the fact that one signal in the nmr spectrum of 2,2,6,6-tetradeuterobromocyclohexane changes to two smaller signals when the spectrum is taken at low temperatures? ◄

As the ring changes its conformation from one chair form to another, the Br—C—H proton changes its position from axial to equatorial (Fig. 12-7). An axial H and an equatorial H have different chemical shifts. But at room temperature the ring "flips" too fast for the instrument to detect the difference; it senses the average condition. At low temperatures this process becomes slow enough so that the instrument can pick up the two different H_{ax} and H_{eq} signals. D's are used to ensure that the H under study is a singlet.

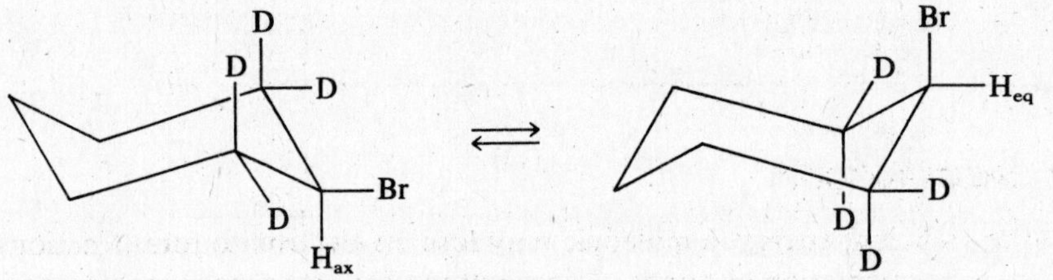

Fig. 12-7

COUPLING CONSTANTS

Figure 12-8 summarizes the nmr spectrum of CH_3CH_2Cl by using a vertical axis line for each peak.

The spacing between lines within a multiplet is typically constant; furthermore, the *spacing in each coupled multiplet* is constant. This constant distance is called the **coupling constant**, *J*, and is expressed in hertz, Hz. The value of the coupling constant depends on the structural relationship of the coupled H's, and becomes a valuable tool for structure proof. Some typical values are given in Table 12-5.

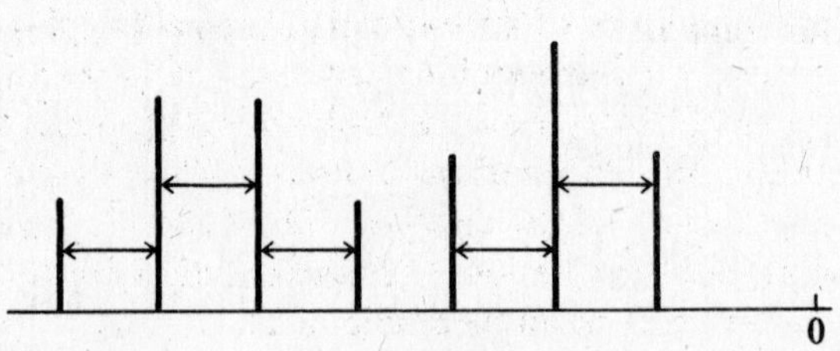

Fig. 12-8

Problem 12.31 A compound, C_2H_2BrCl, has two doublets, $J = 16\,Hz$. Suggest a structure. ◄

The three possibilities showing *J* values for two doublets are:

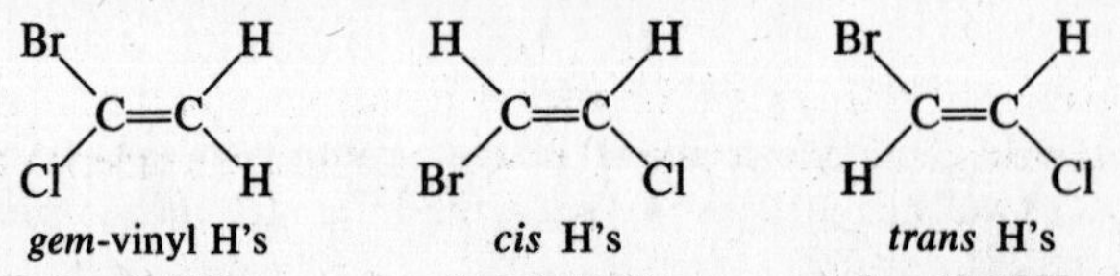

gem-vinyl H's	*cis* H's	*trans* H's
$J = 0-3\,Hz$	$J = 7-12\,Hz$	$J = 13-18\,Hz$

The *trans* isomer fits the data.

Table 12-5

Type of H's	J, Hz
H—C—C—H (free rotation of C—C)	~7
$\begin{array}{c}\text{H}\\ \diagdown\\ C{=}C\\ \diagup \diagdown\\ \text{H}\end{array}$ *trans*	13–18
$\begin{array}{c}\text{H} \text{H}\\ \diagdown \diagup\\ C{=}C\\ \diagup \diagdown\end{array}$ *cis*	7–12
$\begin{array}{c}\text{H}\\ \diagup\\ ={=}C\\ \diagdown\\ \text{H}\end{array}$	0–3
Phenyl H's *ortho* *meta* *para*	 6–9 1–3 0–1

12.5 MASS SPECTROSCOPY

When exposed to sufficient energy, a molecule may lose an electron to form a cation-radical, which then may undergo fragmentation of bonds.   These processes make mass spectroscopy a useful tool for structure proof.   The parent molecules (RS) in the vapor state are ionized by a beam of energetic electrons (e^-),

$$\underset{\substack{\text{Parent}\\\text{molecule}}}{R{:}S} + \overset{\cdot}{e}{}^- \longrightarrow 2e^- + \underset{\substack{\text{Parent}\\\text{cation-radical}}}{R{\cdot}S^+} \longrightarrow \underset{\substack{\text{Fragment}\\\text{ions}}}{R{\cdot} + S^+ \text{ or } R^+ + S{\cdot}}$$

and a number of parent cation-radicals ($R{\cdot}S^+$) may then fragment to give other cations and neutral species.   Fragment ions can undergo further bond-cleavage to give even smaller cations and neutral species.   In a mass spectrogram, sharp peaks appear at the values of m/e (the mass divided by the charge) for the various cations.   The relative heights (intensities) of the peaks represent relative abundances of the cations.   Most cations have a charge of $+1$ and therefore most peaks record the masses of the cations.   Fragmentation tends to give the more stable cations.   The most abundant cations are the most stable ones.

Unless all the parent ions fragment, which rarely happens, the largest observed m/e value is the molecular weight of the parent (RS) molecule.   This generalization overlooks the presence of naturally occurring isotopes in the parent.   Thus, the chances of finding a ^{13}C atom in an organic molecule are 1.11%; the chances of finding two are negligible.   Therefore, the instrument detects a small peak at $m_{RS} + 1$, owing to the ^{13}C-containing parent.   The chances of finding ^2H in a molecule are negligible.

The masses and possible structures of fragment cations, especially the more stable ones, are clues to the structure of the original molecule.   However, rearrangement of cations complicates the interpretation.

Mass spectra, like uv and ir spectra, are unique properties used to identify known and unknown compounds.

Problem 12.32   (a) What structural formulas containing only C and H can be assigned to a cation with m/e equal to   (i) 43,   (ii) 65,   (iii) 91?   (Assume that $e = +1$.)   (b) What combination of C, H and N can account for an m/e of   (i) 43,   (ii) 57?   (Assume that $e = +1$.)   ◀

(a)   Divide by 12 to get the number of C's; the remainder of the weight is due to H's.   (i) $C_3H_7^+$,   (ii) $C_5H_5^+$, (iii) $C_7H_7^+$.

(b)   (i) If one N is present, subtracting 14 leaves a mass of 29, which means 2 C's (mass of 24) are present.   Therefore the formula is $C_2H_5N^+$.   If 2 N's are present, it is $CH_3N_2^+$.   (ii) $CH_3N_3^+$, $C_2H_5N_2^+$ or $C_3H_7N^+$.

Problem 12.33   (a) Do parent (molecular) ions, RS^+, of hydrocarbons ever have odd m/e values?   (b) If an RS^+ contains only C, H and O, may its m/e value be either odd or even?   (c) If an RS^+ contains only C, H and N, may its m/e value be either odd or even?   (d) Why can an ion, $m/e = 31$, not be $C_2H_7^+$?   What might it be?◀

(a)   No.   Hydrocarbons, and their parent ions, must have an even number of H's: $C_nH_{2n+2}, C_nH_{2n}, C_nH_{2n-2}$, C_nH_{2n-6}, etc.   Since the atomic weight of C is even (12), the m/e values must be even.

(b)   The presence of O in a formula does not change the ratio of C to H.   Since the mass of O is even (16), the mass of RS^+ with C, H and O *must be even*.

(c)   The presence of each N ($m = 14$) requires an additional H ($C_nH_{2n+3}N$, $C_nH_{2n+1}N$, $C_nH_{2n-1}N$).   Therefore, if the number of N's is odd, an odd number of H's and an odd m/e value result.   An even number of N's requires an even number of H's and an even m/e value.   These statements apply *only* to parent ions, *not* to fragment ions.

(d)   The largest number of H's for two C's is six (C_2H_6).   Some possibilities are CH_3O^+ and CH_5N^+.

Problem 12.34   Which electron is most likely to be lost in the ionization of the following compounds?   Write an electronic structure for RS^+.   (a) CH_4,   (b) CH_3CH_3,   (c) $H_2C{=}CH_2$,   (d) CH_3Cl,   (e) $H_2C{=}O$.   ◀

The electron in the highest-energy MO is most likely to be lost.   The relative energies for electrons are: $n > \pi > \sigma$ (Problem 12.3).   In a compound with no n or π electrons, the electron most likely comes from the highest-energy σ bond.

(a)   H:Ċ₊:H   (+ is the charge due to the missing electron)

(b)   H:Ċ₊:Ċ:H   (the C—C bond is weaker than the C—H bond)

(c) $H_2C{\overset{+}{\underset{\cdot}{=}}}CH_2$ (electron comes from the π bond)

(d) $H_3C{:}\ddot{Cl}^{+\cdot}$ (an n electron is lost)

(e) $H_2C{=}\ddot{O}^{+\cdot}$ (an n electron is lost)

Problem 12.35   Write equations involving the electron-dot formulas for each fragmentation used to explain the following.   (a) Isobutane, a typical branched-chain alkane, has a lower-intensity RS^+ peak than does n-butane, a typical unbranched alkane.   (b) All 1° alcohols, RCH_2CH_2OH, have a prominent fragment cation at $m/e = 31$. (c) All $C_6H_5CH_2R$-type hydrocarbons have a prominent fragment cation at $m/e = 91$.   (d) Alkenes of the type $H_2C{=}CHCH_2R$ have a prominent fragment cation at $m/e = 41$.   (e) Aldehydes,

$$\overset{\displaystyle H}{\underset{\displaystyle |}{R{-}C{=}O}}$$

show intense peaks at $(m/e)_{RS} - 1$ and $m/e = 29$.   ◀

(a)   Cleavage of C—C is more likely than cleavage of C—H.   Fragmentation of RS^+ for isobutane,

$$(H_3C)_2\overset{H}{\underset{|}{C^+}}CH_3 \longrightarrow (H_3C)_2\overset{H}{\underset{|}{C^+}} + \cdot CH_3$$

gives a 2° R$^+$, which is more stable than the 1° R$^+$ from *n*-butane,

$$H_3CC^+CCH_3 \longrightarrow H_3CC^+ + \cdot CH_2CH_3$$

Hence RS$^+$ of isobutane undergoes fragmentation more readily than does RS$^+$ of *n*-butane, and fewer RS$^+$ fragments of isobutane survive. Consequently, isobutane, typical of branched-chain alkanes, has a lower-intensity RS$^+$ peak than has *n*-butane.

(b)

$$R-\overset{\beta}{C}\overset{\alpha}{CH_2\ddot{O}H} \longrightarrow R-\overset{\beta}{C}\cdot + \overset{+}{\underset{\alpha}{CH_2\ddot{O}H}} \longleftrightarrow H_2C=\overset{+}{\ddot{O}H}$$

$$m/e = 31$$

A —C$\overset{+}{-}$ next to an O is stabilized by extended π bonding (resonance). The RS$^+$ species of alcohols generally undergoes cleavage of the bond

$$\overset{\beta}{C}\!\!\mid\!\!\overset{\alpha}{C}-OH$$

(c)

$$C_6H_5:\overset{+}{C}R \longrightarrow R\cdot + C_6H_5:CH_2^+ \longrightarrow \quad \text{a more stable aromatic cycloheptatrienyl cation, } m/e = 91$$

stable benzyl R$^+$

(d)

$$H_2C\overset{+}{-}CH-\overset{}{C}R \longrightarrow H_2C=CH-\overset{+}{CH_2} + \cdot R$$

stable allylic
cation, $m/e = 41$

(e)

$$R:\overset{+}{C}=\ddot{O}: \longrightarrow \quad R-\overset{+}{C}=\ddot{O}: \longleftrightarrow R-\overset{+}{C}\equiv O \quad (m/e = (m/e)_{RS} - 1)$$

$$\overset{+}{C}=\ddot{O}: \longleftrightarrow H-\overset{+}{C}\equiv O \quad \text{stable acylonium ions} \quad (m/e = 29)$$

Problem 12.36 Why is less than 1 mg of parent compound needed for mass spectral analysis? ◀

A relatively small number of molecules are taken to prevent collision and reaction between fragments. Combination of fragments might lead to ions with larger masses than RS$^+$, making it impossible to determine the molecular weight. The fragmentation pattern would also become confusing.

Problem 12.37 Suggest structures for three of the peaks (86, 43 and 42) in the mass spectrum of *n*-hexane. ◀

The RS of *n*-hexane, $CH_3CH_2CH_2CH_2CH_2CH_3$, is of mass 86. A split in half would give a fragment $CH_3CH_2\overset{+}{C}H_2$ of mass 43. The 42 fragment may be made up of three CH$_2$'s coming from the middle of the chain: $\cdot CH_2CH_2CH_2^+$.

Problem 12.38 Give the structure of a compound, $C_{10}H_{12}O$, whose mass spectrum shows m/e values of 15, 43, 57, 91, 105, and 148. ◀

The 15 value hints at a CH$_3$. Because $43 - 15 = 28$ is the mass of a C=O group, the value of 43 could mean an acetyl, CH$_3$CO, group in the compound. The highest value, 148, gives the molecular weight. Cleaving an acetyl group ($m/e = 43$) from 148 gives 105, which is an observed peak. Next below 105 is 91, a difference of 14; this suggests a CH$_2$ attached to CH$_3$CO. So far we have CH$_3$COCH$_2$ adding up to 57, leaving $148 - 57 = 91$ to be accounted for. This peak is likely to be $C_7H_7^+$, whose precursor is the stable benzyl cation, $C_6H_5CH_2$. The structure is $CH_3-\overset{O}{\underset{\|}{C}}-CH_2-CH_2-C_6H_5$.

Problem 12.39   How could mass spectroscopy distinguish among the three deuterated forms of ethyl methyl ketone?

$$(1) \quad DCH_2CH_2COCH_3 \qquad (2) \quad CH_3CH_2COCH_2D \qquad (3) \quad CH_3CHDCOCH_3 \qquad \blacktriangleleft$$

The expected peaks for each compound are shown in Table 12-6; each has a different combination of peaks.

Table 12-6

m/e	$DCH_2CH_2COCH_3$	$CH_3CH_2COCH_2D$	$CH_3CHDCOCH_3$
15	CH_3^+	CH_3^+	CH_3^+
16	DCH_2^+	DCH_2^+	—
29	—	$CH_3CH_2^+$	—
30	$DCH_2CH_2^+$	—	CH_3CHD^+
43	CH_3CO^+	—	CH_3CO^+
44	—	DCH_2CO^+	—

Problem 12.40   A prominent peak in the mass spectrum of 2-methyl-2-phenylbutane is at $m/e = 119$.   To what fragment cation is this peak due?   Show how cleavage occurs to form this ion. $\blacktriangleleft$

The RS is

$$C_6H_5 \overset{\overset{\displaystyle CH_3}{|}}{\underset{\underset{\displaystyle CH_3}{|}}{\overset{\alpha}{C}}} \overset{}{\underset{\beta}{}} CH_2CH_3$$

Cleavage between C^α and C^β gives $C_6H_5\overset{+}{C}(CH_3)_2^+$, $m/e = 119$.

Supplementary Problems

Problem 12.41   Match the type of spectrometer with the kind of information which it can provide the chemist.

1.	Mass	A.	functional groups
2.	Infrared	B.	molecular weights
3.	Ultraviolet	C.	proton environment
4.	Nuclear magnetic resonance	D.	conjugation

1 B,   2 A,   3 D,   4 C

Problem 12.42   Which peaks in the ir spectra distinguish cyclohexane from cyclohexene? $\blacktriangleleft$

One of the C—H stretches in cyclohexene is above 3000 cm^{-1} (C_{sp^2}—H); in cyclohexane the C—H stretches are below 3000 cm^{-1} (C_{sp^3}—H).   Cyclohexene has a C=C stretch at about 1650 cm^{-1}.

Problem 12.43   The mass spectrum of a compound containing C, H, O and N gives a maximum m/e of 121.   Its ir spectrum shows peaks at 700, 750, 1520, 1685 and 3100 cm^{-1}, and a twin peak at 3440 cm^{-1}.   What is a reasonable structure for the compound? $\blacktriangleleft$

The molecular weight is 121.   Since the mass is odd, there must be an odd number of N's [Problem 12.33(c)].   The ir data indicate the following groups to be present:

1520 cm^{-1}:   aromatic ring (1450–1600 cm^{-1} range)

1685 cm^{-1}:   C=O stretch of amide structure —CO—N= (1630–1690 cm^{-1} range)

3100 cm^{-1}:   aromatic C—H bond (3000–3100 cm^{-1} range)

3440 cm^{-1}:   —N—H in amine or amide (3300–3500 cm^{-1} range)

700, 750 cm^{-1}:   monosubstituted phenyl

A twin peak due to symmetric and antisymmetric N—H stretches means an NH$_2$ group.   By putting the pieces together we find that the compound is benzamide,

$$C_6H_5-\overset{\overset{\textstyle O}{\|}}{C}-NH_2$$

Problem 12.44   The ir spectrum of methyl salicylate, o-HOC$_6$H$_4$COOCH$_3$, has peaks at 3300, 1700, 3050, 1540, 1590, and 2990 cm^{-1}.   Attribute these peaks to the following structures: (a) CH$_2$,   (b) C=O,   (c) OH group on ring,   and (d) aromatic ring.   ◄

   (a) 2990 cm^{-1};   (b) 1700 cm^{-1};   (c) 3300 cm^{-1};   (d) 3050, 1540, 1590 cm^{-1}.

Problem 12.45   Calculate ϵ_{max} for a compound whose maximum absorbance (A) is 1.2.   The cell length (ℓ) is 1.0 cm and the concentration is 1.9 mg per 25.0 ml of solution.   The mass spectrum of the compound has the largest m/e value at 100.   ◄

   The molecular weight is 100 g/mole.   The concentration, C, in mole/liter is:

$$\left(\frac{1.9 \text{ mg}}{25.0 \text{ ml}}\right)\left(\frac{1000 \text{ ml}}{1 \text{ liter}}\right)\left(\frac{1 \text{ g}}{1000 \text{ mg}}\right)\left(\frac{1 \text{ mole}}{100 \text{ g}}\right) = 7.6 \times 10^{-4} \frac{\text{mole}}{\text{liter}}$$

Hence
$$\epsilon_{max} = \frac{A}{C\ell} = \frac{1.2}{(7.6 \times 10^{-4})(1.0)} = 1.6 \times 10^3$$

Problem 12.46   Methanol is a good solvent for uv but not for ir determinations.   Why?   ◄

   Methanol absorbs in the uv at 183 nm, which is below 190 nm, the cutoff for most spectrophotometers, and therefore it doesn't interfere.   Its ir spectrum has bands in most regions and therefore it cannot be used.   Solvents such as CCl$_4$ and CS$_2$ have few interfering bands and are preferred for ir determinations.

Problem 12.47   Draw structural formulas for compounds with the following molecular formulas, that have only one nmr signal: (a) C$_5$H$_{12}$,   (b) C$_3$H$_6$,   (c) C$_2$H$_6$O,   (d) C$_3$H$_4$,   (e) C$_2$H$_4$Br$_2$,   (f) C$_4$H$_6$,   (g) C$_8$H$_{18}$.   ◄

   All H's are equivalent since there is only one signal.   (a) The 12 H's must be in 4 equivalent CH$_3$'s; (CH$_3$)$_4$C. (b) There must be 3 equivalent CH$_2$'s as found only in cyclopropane;

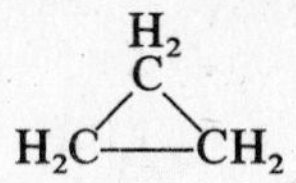

$$\overset{\textstyle \overset{H_2}{C}}{\underset{H_2C \!-\!\!-\!\!-\! CH_2}{\diagup \diagdown}}$$

(c) 2 equivalent CH$_3$'s; CH$_3$OCH$_3$.   (d) 2 equivalent CH$_2$'s and a C with no H's; H$_2$C=C=CH$_2$.   (e) 2 equivalent CH$_2$'s; BrCH$_2$CH$_2$Br.   (f) 2 equivalent CH$_3$'s and 2 C's with no H's; CH$_3$C≡CCH$_3$.   There is no way to have 3 equivalent CH$_2$'s with a fourth C.   (g) 6 equivalent CH$_3$'s and 2 other C's; (CH$_3$)$_3$C—C(CH$_3$)$_3$.

Problem 12.48   Write structural formulas for compounds with the following molecular formulas, having 2 singlets: (a) C$_3$H$_5$Cl$_3$,   (b) C$_2$H$_5$OCl,   (c) C$_3$H$_8$O$_2$,   (d) C$_3$H$_6$O$_2$,   (e) C$_5$H$_{10}$Cl$_2$.   ◄

   Since there is no coupling, dissimilar H's cannot be on adjacent C's.

   (a) CH$_3^a$CCl$_2$CH$_2^b$Cl   (b) ClCH$_2^a$OCH$_3^b$ (c) CH$_3^a$OCH$_2^b$OCH$_3^a$

   (d) CH$_3^a$C—OCH$_3^b$ (e) ClCH$_2^a$C(CH$_3^b$)$_2$CH$_2^a$Cl
        $\overset{\|}{\underset{O}{}}$

Problem 12.49 How can nmr distinguish between p-xylene and ethylbenzene?   ◄

p-$CH_3^a C_6H_4^b CH_3^a$ has 2 kinds of H's, and its nmr spectrum has 2 singlets. $C_6H_5^a CH_2^b CH_3^c$ has 3 kinds of H's, and its nmr spectrum has a singlet for H^a, a quartet for H^b and a triplet for H^c.

Problem 12.50 A compound, C_3H_6O, contains a

$$\diagdown C=O$$

group. How could nmr establish whether this compound is an aldehyde or a ketone?   ◄

If an aldehyde, the compound is CH_3CH_2CHO, with 3 multiplet peaks and a downfield signal for

$$\overset{H}{\underset{|}{-C}}=O$$

$(\delta = 9\text{--}10)$. If a ketone, it is $(CH_3)_2C=O$, with one singlet.

Problem 12.51 Compare the nmr spectra of p-xylene and mesitylene (1,3,5-trimethylbenzene).   ◄

Both show 2 singlets; one for ring H's and one for CH_3 H's. The relative intensities of the peaks are different. In p-xylene, $C_6H_4(CH_3)_2$, the ratio of CH_3 to ring H is $6:4$ or $3:2$, while in mesitylene, $C_6H_3(CH_3)_3$, it is $9:3$ or $3:1$.

Problem 12.52 Describe the expected appearance of the nmr spectrum of n-propylbenzene,

$$\text{〈◯〉}-CH_2^a-CH_2^b-CH_3^c$$ ◄

(All chemical shifts are given as δ in ppm.) At about 7.2 there is a signal due to the aromatic ring H's. At 2.3–3 there is a triplet for the benzylic H^a. If the coupling effects of H^a and H^c with H^b were similar, the multiplet for H^b would be a sextet $(5 + 1)$. If the effects were different, H^b would be split into a triplet by an H^a, and each peak of the triplet would be further split into a quartet. Some of these peaks might coincide. At best, H^b is a complex multiplet at about 1.3. The H^c signal is a triplet at about 0.9. The relative peak intensities will be: ring $H:H^a:H^b:H^c = 5:2:2:3$.

Problem 12.53 A compound, $C_{10}H_{14}$, shows 2 nmr singlets: A at $\delta = 8.0$ ppm and B at $\delta = 1.0$ ppm. The ratio of their intensities is $A:B = 5:9$. What structure is consistent with these data?   ◄

A is caused by aromatic ring H's and B by alkyl H's. The 5 ring H's suggests a monosubstituted benzene. The singlet accounting for 9 H's must arise from 3 equivalent CH_3's bonded to the 10th C. The structure is $C_6H_5C(CH_3)_3$.

Problem 12.54 The nmr spectrum of a compound, $C_3H_5ClF_2$, has 2 noncoupled triplets. Triplet A is 1.5 times the intensity of the more downfield triplet B. Suggest a structure.   ◄

Since A is 1.5 times as intense as B and the molecular formula has 5 H's, A must come from CH_3 and B from CH_2. A and B are not coupling each other; if they did, we would find a triplet and a quartet. A and B must each be coupled to 2 F's. B is more downfield than A, indicating a $-CH_2Cl$ group. The compound is

$$Cl\overset{B}{C}H_2CF_2\overset{A}{C}H_3$$

See Problem 12.27 for H/F coupling.

Problem 12.55 A compound, C_3H_7Cl, gives an nmr triplet A at about $\delta = 0.9$ and another triplet B, with 2/3 the intensity and more downfield. A complex multiplet C appears between the two triplets. Is this compound *normal*- (I) or *isopropyl* chloride (II)?   ◄

The possible structures are:

$$\text{(I)} \quad CH_3^a—CH_2^c—CH_2^b Cl \qquad \text{(II)} \quad CH_3^d—CH^e—CH_3^d$$
$$|$$
$$Cl$$

For I, H^a and H^b are split into individual triplets by the 2 H^c's. The intensity ratio of $B:A$ is 2:3 and H^b is more downfield. H^c's give a complex multiplet because they are split by H^a and H^b. For II, H^d has an upfield doublet and H^e has a more downfield septet $(6 + 1)$. The compound is $CH_3CH_2CH_2Cl$.

Problem 12.56 The nmr spectrum of a dichloropropane shows a quintuplet and, downfield, a triplet of about twice the intensity. Is the isomer 1,1-, 1,2-, 1,3-, or 2,2-dichloropropane? ◄

We would expect the following signals:

1,1-Dichloropropane, $Cl_2CH^a CH_2^b CH_3^c$: a triplet (H^c), a complex multiplet more downfield (H^b), and a triplet still more downfield (H^a).

1,2-Dichloropropane, $ClCH_2^a CH^b CH_3^c$: a doublet (H^c), another doublet more downfield (H^a), a complex
$$|$$
$$Cl$$

multiplet most downfield (H^b).

1,3-Dichloropropane, $ClCH_2^a CH_2^b CH_2^a Cl$: a quintuplet (H^b), and, downfield, a triplet (H^a).

$$Cl$$
$$|$$
2,2-Dichloropropane, $CH_3^a—C—CH_3^a$: a singlet (H^a).
$$|$$
$$Cl$$

The compound is 1,3-dichloropropane.

Problem 12.57 Assign a structure for a compound, $C_3H_5O_2Cl$, showing a signal at δ between 10.5 and 12, a doublet around $\delta = 1.5$ and a quartet at about $\delta = 4.2$. ◄

The 10.5–12 signal is from a COO<u>H</u>. Subtracting COOH from $C_3H_5O_2Cl$ gives C_2H_4Cl, which is arranged as $CH_3^a CH^b Cl$. H^a gives the upfield doublet and H^b the downfield quartet. The structure is $CH_3—CHCl—COOH$.

Problem 12.58 Give the signals, their multiplicities and relative intensities for the nmr spectrum of $CH_3COCH_2C\equiv CCH_3$. ◄

$$CH_3^a—C—CH_2^b—C\equiv C—CH_3^c$$
$$\|$$
$$O$$

H^a, H^b and H^c each give rise to a singlet. The relative downfield chemical shifts are $H^b > H^a > H^c$, with relative intensities of 2:3:3 respectively.

Problem 12.59 How is nmr used to determine on which C monochlorination of methyl ethyl ether occurs? ◄

The 3 possible products with their spectra described are:

$CH_3^t CH_2^q—O—CH_2^s Cl$ (a triplet, a more downfield quartet, and a most downfield singlet)

$CH_3^d CH^a Cl—O—CH_3^s$ (a doublet, a quartet downfield, and a singlet between them)

$CH_2^{t_2}Cl—CH_2^{t_1}—O—CH_3^s$ (a singlet and two triplets; the downfield order is $t_1 < s < t_2$)

Problem 12.60 In CCl_4 as solvent, the nmr spectrum of CH_3OH shows two singlets. In $(CH_3)_2SO$, there is a doublet and a quartet. Explain in terms of the "slowness" of nmr detection. ◄

In CCl_4, CH_3OH H-bonds intermolecularly, leading to a rapid interchange of the H of O—H. The instrument senses an average situation and therefore there is no coupling between CH_3 and OH protons. In $(CH_3)_2SO$, H-bonding is with solvent, and the H stays on the O of OH. Now coupling occurs. This technique can be used to distinguish among RCH_2OH, R_2CHOH and R_3COH, whose signals for H of OH are a triplet, a doublet and a singlet, respectively.

Problem 12.61 Indicate whether the following statements are true or false and give a reason in each case. (a) The ir spectra are identical for

$$Br\!-\!\!\overset{\displaystyle C_2H_5}{\underset{\displaystyle H}{|}}\!\!-\!CH_3 \quad \text{and} \quad CH_3\!-\!\!\overset{\displaystyle C_2H_5}{\underset{\displaystyle H}{|}}\!\!-\!Br$$

(b) The nmr spectra of the compounds in (a) are also identical. (c) The ir spectrum of 1-hexene has more peaks than the uv spectrum. (d) Compared to CH_3CH_2CHO, the $n \to \pi^*$ for $H_2C{=}CHCHO$ has shifted to a shorter wavelength (**blue shift**). ◄

(a) and (b) True. The compounds are enantiomers which have identical vibrational modes and proton resonances.

(c) True. The ir spectrum has peaks for stretching and bending of all bonds, while the uv spectrum has only 1 peak for excitation of a pi electron.

(d) False. The shift is to longer wavelength (red shift), since

$$H_2C{=}CHC\!\!\overset{\displaystyle }{\underset{\displaystyle H}{|}}\!\!{=}O$$

is a conjugated system.

Problem 12.62 Assign the nmr spectra shown in Fig. 12-9 to the appropriate monochlorination products of 2,4-dimethylpentane $(C_7H_{15}Cl)$ and justify your assignment. Note the integration assignments drawn in the spectra. ◄

The three possible structures are:

$$\overset{1}{Cl}CH_2-\overset{2}{\underset{\displaystyle }{\overset{\displaystyle CH_3}{|}}}CH-\overset{3}{CH_2}-\overset{4}{\underset{\displaystyle }{\overset{\displaystyle CH_3}{|}}}CH-\overset{5}{CH_3}$$

1-Chloro-2,4-
dimethylpentane (I)

$$\overset{1}{CH_3}-\overset{2}{\underset{\displaystyle Cl}{\overset{\displaystyle CH_3}{|}}}C-\overset{3}{CH_2}-\overset{4}{\underset{\displaystyle }{\overset{\displaystyle CH_3}{|}}}CH-\overset{5}{CH_3}$$

2-Chloro-2,4-
dimethylpentane (II)

$$\overset{1}{CH_3}-\overset{2}{\underset{\displaystyle }{\overset{\displaystyle CH_3}{|}}}CH-\overset{3}{\underset{\displaystyle Cl}{CH}}-\overset{4}{\underset{\displaystyle }{\overset{\displaystyle CH_3}{|}}}CH-\overset{5}{CH_3}$$

3-Chloro-2,4-
dimethylpentane (III)

The best clue is the most downfield signal arising from the H's closest to Cl. In spectrum (a) the signal with the highest δ value is a *doublet*, integrating for 2 H's, that corresponds only to structure I ($ClCH_2$—). This is confirmed by the 9 H's of the 3 CH_3's that are most upfield and the four 2° and 3° H's with signals between these.

In spectrum (b) the most downfield signal is a triplet, for one H, which arises from the

$$-\overset{\displaystyle H}{\underset{\displaystyle }{|}}C-\overset{\displaystyle Cl}{\underset{\displaystyle H}{|}}C-\overset{\displaystyle H}{\underset{\displaystyle }{|}}C-$$

grouping in III. In addition, the most upfield signal is a doublet, integrating for 12 H's, which is produced by the H's of the 4 CH_3's split by the 3° H.

This leaves II for spectrum (c). The most downfield group of irregular signals, integrating for 3 H's, comes from the two 2° and one 3° H on C^3 and C^4, respectively. The most upfield doublet, integrating for 6 H's, arises from the 2 equivalent CH_3's on C^4 split by the C^4 3° H. The 2 CH_3's on C^2 give rise (6 H's) to the singlet of median δ value.

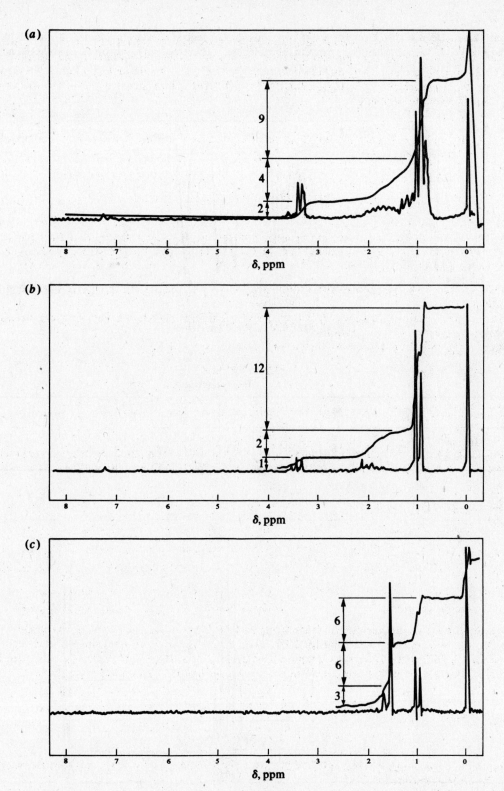

Fig. 12-9

Problem 12.63 Deduce structures for the compound whose spectral data are presented in Fig. 12-10, Table 12-7, and Fig. 12-11. Assume an O is present in the molecule. There was no uv absorption above 180 nm. ◀

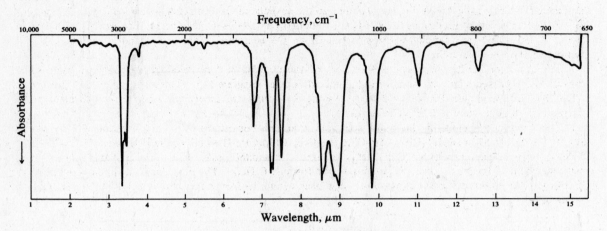

Fig. 12-10. Infrared Spectrum

Table 12-7. Mass Spectrum

m/e	26	27	29	31	39	41	42	43	44	45	59	87	102
Relative Intensity, % of base peak	3	18	6	4	11	17	6	61	4	100	11	21	0.63

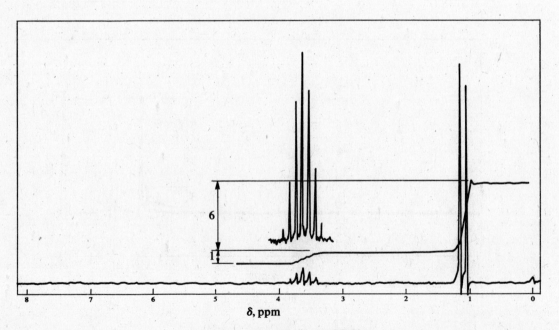

Fig. 12-11. nmr Spectrum (CDCl₃)

From the mass spectrum, we know that RS^+ has molecular weight 102. The C, H portion of the compound has a mass of $102 - 16$ (for O) = 86. Dividing by 12 gives 6 C's, leaving a remainder of 14 for 14 H's. The molecular formula is $C_6H_{14}O$. (Had we chosen 7 C's we would have had C_7H_2O, which is an impossible formula.) The compound has no degrees of unsaturation. This fact is consistent with (but not proven by) the absence of uv absorption above 180 nm. Note also the absence of C—H stretch above 3000 cm^{-1} (H—C$_{sp^2}$ or H—C$_{sp}$); the O must be present as C—O—H (an alcohol) or as C—O—C (an ether). The absence of a peak in the ir at 3300–3600 cm^{-1} precludes the presence of an O—H group. The strong band at about 1110 cm^{-1} represents the C—O stretch.

The structure of the alkyl groups R—O—R of the ether is best revealed by the nmr spectrum. The downfield septet (see blown-up signal) and the upfield doublet integrate 1:6. This arrangement is typical for a CH(CH$_3$)$_2$ grouping [see Problem 12.24(b)]. Both R groups are isopropyl, since no other signals are present. The compound is (CH$_3$)$_2$CHOCH(CH$_3$)$_2$.

The significant peaks in the mass spectrum that are consistent with our assignment are: $m/e = 102 - 15$ (CH$_3$) = 87, which is (CH$_3$)$_2$CHOCHCH$_3$; 43 is (CH$_3$)$_2$CH; 41 would be the allyl cation H$_2$C=CHCH$_2^+$ formed from fragmentation of (CH$_3$)$_2$CH. The most abundant peak, $m/e = 45$, probably comes from rearrangement of fragment ions. It could have the formula C$_2$H$_5$O. The presence of this peak convinces us that O was indeed present in the compound. The peak cannot be from a fragment with only C and H; C$_3$H$_9^+$ is impossible.

Chapter 13

Alcohols

13.1 GENERAL

ROH is an **alcohol**, ArOH is a **phenol** (Chapter 21). Alcohols are named by: (1) **common names** such as ethyl alcohol (C_2H_5OH); (2) **carbinol method**, in which alcohols are derivatives of methyl alcohol (carbinol) and wherein one or more of the CH_3 H's are replaced by other groups; and (3) **IUPAC**, in which the suffix **-ol** replaces the **-e** of the alkane to indicate the OH.

Problem 13.1 Give a common name for each of the following alcohols and classify them as 1°, 2° or 3°:

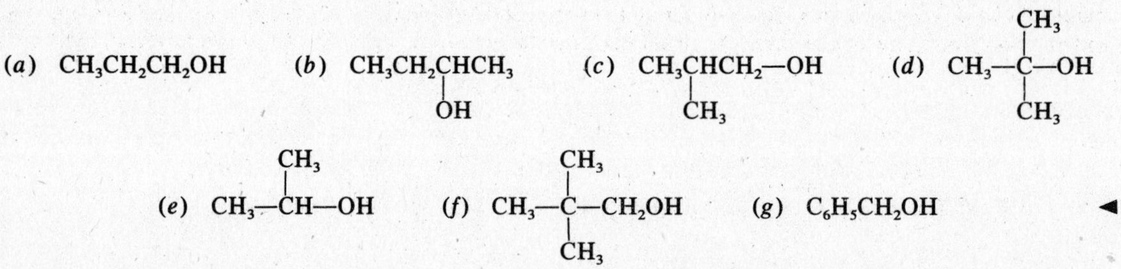

(a) $CH_3CH_2CH_2OH$

(b) $CH_3CH_2CHCH_3$
        |
        OH

(c) CH_3CHCH_2—OH
        |
        CH_3

(d) CH_3—C—OH, with CH_3 above and CH_3 below

(e) CH_3—CH—OH, with CH_3 above

(f) CH_3—C—CH_2OH, with CH_3 above and CH_3 below

(g) $C_6H_5CH_2OH$ ◄

(a) *n*-propyl alcohol, 1°; (b) *sec*-butyl alcohol, 2°; (c) isobutyl alcohol, 1°; (d) *t*-butyl alcohol, 3°; (e) isopropyl alcohol, 2°; (f) neopentyl alcohol, 1°; (g) benzyl alcohol, 1°.

Problem 13.2 Name the following alcohols by carbinol and by IUPAC methods:

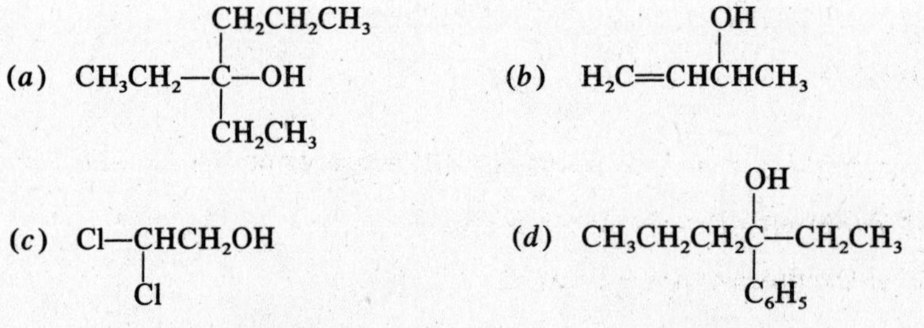

(a) CH_3CH_2—C—OH, with $CH_2CH_2CH_3$ above and CH_2CH_3 below

(b) H_2C=CHCHCH$_3$, with OH above

(c) Cl—CHCH$_2$OH, with Cl below

(d) $CH_3CH_2CH_2$C—CH_2CH_3, with OH above and C_6H_5 below ◄

	Carbinol	IUPAC
(a)	Diethyl-*n*-propylcarbinol	3-Ethyl-3-hexanol
(b)	Methylvinylcarbinol	3-Buten-2-ol
(c)	Dichloromethylcarbinol	2,2-Dichloroethanol
(d)	Ethyl-*n*-propylphenylcarbinol	3-Phenyl-3-hexanol

Note that in IUPAC the OH is given a lower number than C=C or Cl.

Problem 13.3 Explain why (a) propanol boils at a higher temperature than the corresponding hydrocarbon; (b) propanol, unlike propane or butane, is soluble in H_2O; (c) *n*-hexanol is not soluble in H_2O; (d) dimethyl ether (CH_3OCH_3) and ethyl alcohol (CH_3CH_2OH) have the same molecular weight, yet dimethyl ether has a lower boiling point (−24 °C) than ethyl alcohol (78 °C). ◄

(a) Propanol can H-bond intermolecularly: C_3H_7—O---H—O—C_3H_7.
$$\overset{\displaystyle |}{H}$$

(b) Propanol can H-bond with H_2O: C_3H_7—O---H—O—H.
$$\overset{\displaystyle |}{H}$$

(c) As the R group becomes larger, ROH resembles the hydrocarbon more closely. There is little H-bonding between H_2O and *n*-hexanol. When the ratio of C to OH is more than 4, alcohols have little solubility in water.

(d) The ether CH_3OCH_3 has no H on O and cannot H-bond.

Problem 13.4 The ir spectra of *trans*- and *cis*-1,2-cyclopentanediol show a broad band in the region 3450–3570 cm^{-1}. On dilution with CCl_4, this band of the *cis* isomer remains unchanged, but the band of the *trans* isomer shifts to a higher frequency and becomes sharper. Account for this difference in behavior. ◄

The OH's of the *cis* isomer participate in *intramolecular* H-bonding, Fig. 13-1(*a*), which is not affected by dilution. In the *trans* isomer, the H-bonding is *intermolecular*, Fig. 13-1(*b*), and dilution breaks these bonds, causing disappearance of the broad band and its replacement by a sharp band at higher frequency.

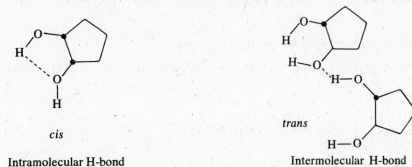

cis

Intramolecular H-bond

trans

Intermolecular H-bond

Fig. 13-1

13.2 PREPARATION

1. $RX + OH^- \xrightarrow{H_2O} ROH + X^-$ **(S_N2 or S_N1 Displacement**, Table 7-1)

2. Hydration of Alkenes [see Problem 6.22(*d*)]

3. Hydroboration-Oxidation [see Problem 6.22(*f*)]

Treatment of the alkylboranes with H_2O_2 in OH^- replaces —B— with OH.
$$\overset{\displaystyle |}{}$$

$$RCH{=}CH_2 + (BH_3)_2 \longrightarrow (R\underset{\textcircled{H}}{CH}CH_2)_3{-}B \xrightarrow[OH^-]{H_2O_2} R\underset{\textcircled{H}\ \textcircled{OH}}{CH}CH_2$$

trialkylborane

The net addition of H—OH to alkenes is *cis*, anti-Markovnikov and free from rearrangement.

4. Oxymercuration-demercuration

$$RCH{=}CH_2 \xrightarrow[H_2O,\,25\,°C]{Hg(OCOCH_3)_2} RCH{-}CH_2 \xrightarrow{NaBH_4} R\underset{\textcircled{HO}\ \textcircled{H}}{CH}CH_2$$
$$\phantom{RCH{=}CH_2 \xrightarrow{}} \underset{OH\ HgOCOCH_3}{}$$

not isolated

The net addition of H—OH is Markovnikov and is free from rearrangement.

Problem 13.5 Give structures and IUPAC names of the alcohols formed from $(CH_3)_2CHCH{=}CH_2$ by reaction with (*a*) dilute H_2SO_4; (*b*) B_2H_6, then H_2O_2, OH^-; (*c*) $Hg(OCOCH_3)_2$, H_2O, then $NaBH_4$. ◄

(*a*) The expected product is 3-methyl-2-butanol, $(CH_3)_2CHCHOHCH_3$, from a Markovnikov addition of H_2O. However, the major product is likely to be 2-methyl-2-butanol, $(CH_3)_2COHCH_2CH_3$, formed by rearrangement of the intermediate R^+.

$$(CH_3)_2CHCH{=}CH_2 \xrightarrow[(H_2O)]{+H^+} \underset{(2^\circ)}{(CH_3)_2CH\overset{+}{C}HCH_3} \xrightarrow{\sim H:} \underset{(3^\circ)}{(CH_3)_2\overset{+}{C}CH_2CH_3} \xrightarrow[-H^+]{+H_2O} (CH_3)_2COHCH_2CH_3$$

(*b*) Anti-Markovnikov HOH addition forms $(CH_3)_2CH{-}CH_2CH_2OH$, 3-methyl-1-butanol.
(*c*) Markovnikov HOH addition with no rearrangement gives $(CH_3)_2CHCHOHCH_3$, 3-methyl-2-butanol.

Problem 13.6 Give the structure and IUPAC name of the product formed on hydroboration-oxidation of 1-methylcyclohexene. ◄

H and OH add *cis* and therefore CH_3 and OH are *trans*.

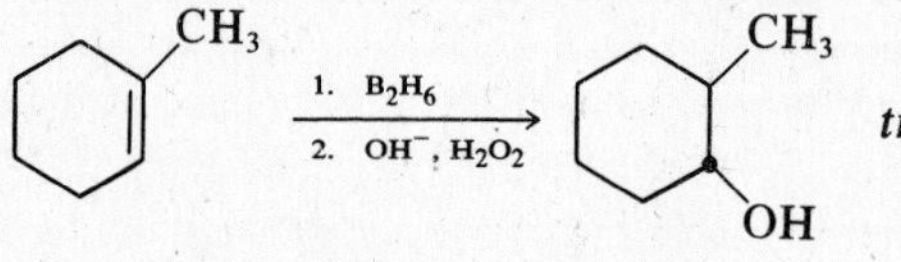

trans-2-Methylcyclohexanol

5. Carbonyl Compounds and Grignard Reagents

Grignard reagents, RMgX, are reacted with aldehydes or ketones and the intermediate products hydrolyzed to alcohols.

$$\boxed{R}MgX + \underset{\text{Formaldehyde}}{H{-}\overset{H}{\overset{|}{C}}{=}O} \longrightarrow \boxed{R}{-}CH_2{-}\bar{O}(MgX)^+ \xrightarrow{H_2O} \boxed{R}{-}CH_2{-}OH \quad (1^\circ\ \text{alcohol})$$

$$\boxed{R}MgX + \underset{\text{aldehyde}}{R'{-}\overset{H}{\overset{|}{C}}{=}O} \longrightarrow R'{-}\overset{H}{\underset{\boxed{R}}{C}}{-}\bar{O}(MgX)^+ \xrightarrow{H_2O} R'{-}\overset{H}{\underset{\boxed{R}}{C}}{-}OH \quad (2^\circ\ \text{alcohol})$$

$$\boxed{R}MgX + \underset{\text{ketone}}{R'{-}\overset{R''}{\overset{|}{C}}{=}O} \longrightarrow \underset{\text{alkoxide salts}}{R'{-}\overset{R''}{\underset{\boxed{R}}{C}}{-}\bar{O}(MgX)^+} \xrightarrow{H_2O} R'{-}\overset{R''}{\underset{\boxed{R}}{C}}{-}OH \quad (3^\circ\ \text{alcohol})$$

The boxed group in the alcohol comes from the Grignard; the remainder comes from the carbonyl compound.

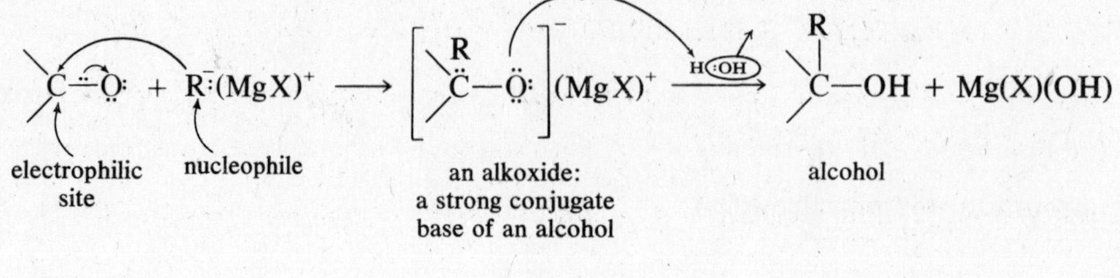

Problem 13.7 Give 3 combinations of RMgX and a carbonyl compound that could be used to prepare

$$C_6H_5CH_2\overset{CH_3}{\underset{OH}{C}}{-}CH_2CH_3$$ ◄

This 3° alcohol is made from RMgX and a ketone, R'COR''. The possibilities are:

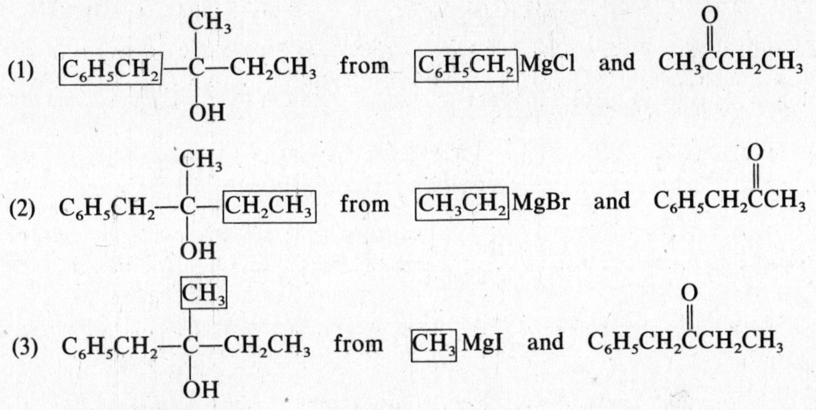

(1) $\boxed{C_6H_5CH_2}$—$\underset{\underset{OH}{|}}{\overset{\overset{CH_3}{|}}{C}}$—$CH_2CH_3$ from $\boxed{C_6H_5CH_2}MgCl$ and $CH_3\overset{O}{\overset{||}{C}}CH_2CH_3$

(2) $C_6H_5CH_2$—$\underset{\underset{OH}{|}}{\overset{\overset{CH_3}{|}}{C}}$—$\boxed{CH_2CH_3}$ from $\boxed{CH_3CH_2}MgBr$ and $C_6H_5CH_2\overset{O}{\overset{||}{C}}CH_3$

(3) $C_6H_5CH_2$—$\underset{\underset{OH}{|}}{\overset{\overset{\boxed{CH_3}}{|}}{C}}$—$CH_2CH_3$ from $\boxed{CH_3}MgI$ and $C_6H_5CH_2\overset{O}{\overset{||}{C}}CH_2CH_3$

The best combination is usually the one in which the two reactants share the C content as equally as possible. In this case, (1) is best.

Problem 13.8 Prepare ethyl p-chlorophenylcarbinol by a Grignard reaction. ◀

Prepare this 2° alcohol, $p\text{-}ClC_6H_4CHOHCH_2CH_3$, from RCHO and R'MgX. Since the groups on the carbinol C are different, there are two combinations possible:

(1) $p\text{-}ClC_6H_4\overset{\overset{H}{|}}{C}=O$ $\xrightarrow[\text{2. }H_3O^+]{\text{1. }CH_3CH_2MgCl}$ $p\text{-}ClC_6H_4CHOHCH_2CH_3$

(ring Cl has not interfered)

(2) $p\text{-}ClC_6H_4MgBr$ $\xrightarrow[\text{2. }H_3O^+]{\text{1. }CH_3CH_2CHO}$ $p\text{-}ClC_6H_4CHOHCH_2CH_3$

Br is more reactive than Cl when making a Grignard of $p\text{-}ClC_6H_4Br$.

Problem 13.9 Prepare 1-butanol from   (a) an alkene,   (b) 1-chlorobutane,   (c) 1-chloropropane and (d) ethyl bromide. ◀

(a) $CH_3CH_2CH=CH_2$ $\xrightarrow[\text{2. }H_2O_2,\,OH^-]{\text{1. }(BH_3)_2}$ $CH_3CH_2CH_2CH_2OH$

(b) $HO^- + CH_3CH_2CH_2CH_2Cl \xrightarrow{H_2O} CH_3CH_2CH_2CH_2OH + Cl^-$ (S_N2)

(c) 1-Chloropropane has one less C than the needed 1° alcohol. The Grignard reaction is used to lengthen the chain by adding $H_2C=O$.

$CH_3CH_2CH_2Cl \xrightarrow[\text{ether}]{Mg} CH_3CH_2CH_2MgCl \xrightarrow[\text{2. }H_3O^+]{\text{1. }HCH=O} CH_3CH_2CH_2CH_2OH$

(d) 1-Butanol is a 1° alcohol with two C's more than CH_3CH_2Br. Reaction of CH_3CH_2MgBr with ethylene oxide and hydrolysis gives 1-butanol.

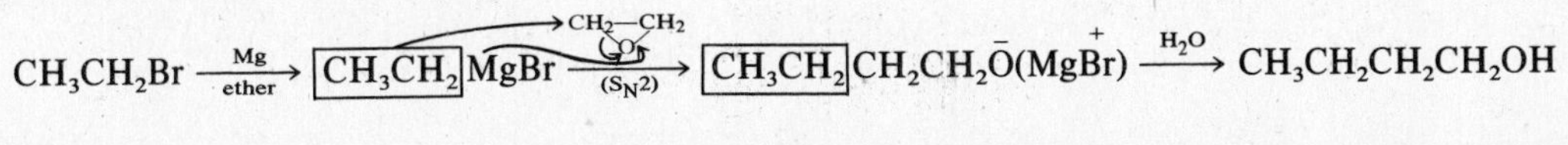

$CH_3CH_2Br \xrightarrow[\text{ether}]{Mg} \boxed{CH_3CH_2}MgBr \xrightarrow[(S_N2)]{\overset{CH_2-CH_2}{\underset{O}{\diagdown\diagup}}} \boxed{CH_3CH_2}CH_2CH_2\bar{O}(MgBr)^+ \xrightarrow{H_2O} CH_3CH_2CH_2CH_2OH$

Problem 13.10 For the following pairs of halides and carbonyl compounds, give the structure of each alcohol formed by the Grignard reaction. (a) Bromobenzene and acetone. (b) p-Chlorophenol and formaldehyde. (c) Isopropyl chloride and benzaldehyde. (d) Chlorocyclohexane and methyl phenyl ketone. ◀

(a)
$$C_6H_5Br \xrightarrow[\text{ether}]{Mg} \boxed{C_6H_5}MgBr \xrightarrow[\text{2. } H_2O]{\text{1. } CH_3-\overset{O}{\underset{\|}{C}}-CH_3} \boxed{C_6H_5}\overset{CH_3}{\underset{CH_3}{\overset{|}{\underset{|}{C}}}}-OH$$

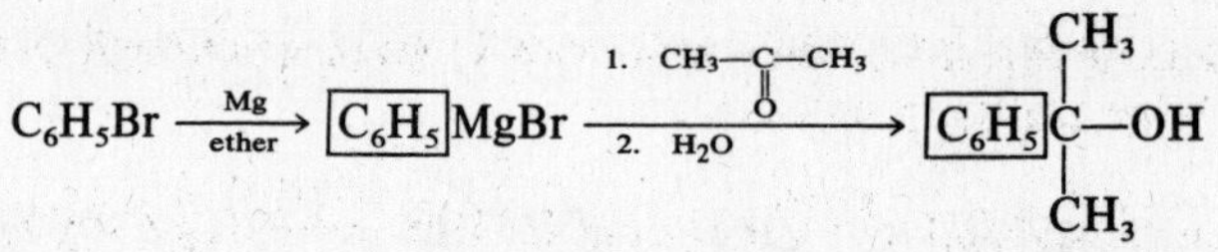

(b) The weakly acidic OH in p-chlorophenol prevents formation of the Grignard reagent.

(c)
$$C_6H_5\overset{H}{\underset{}{\overset{|}{C}}}=O \xrightarrow[\text{2. } H_2O]{\text{1. } \boxed{(CH_3)_2CH}MgCl} C_6H_5\overset{H}{\underset{OH}{\overset{|}{\underset{|}{C}}}}-\boxed{CH(CH_3)_2}$$

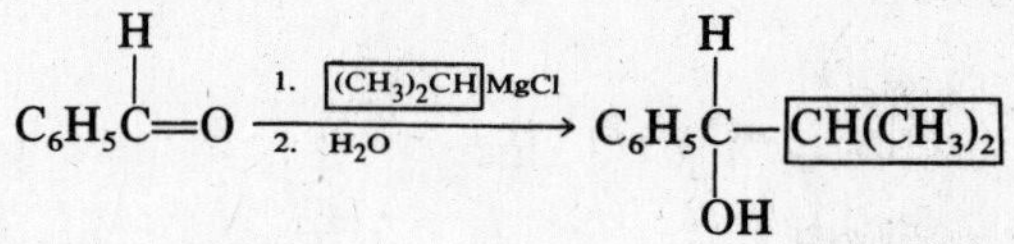

(d)

$$\text{(cyclohexyl-Cl)} \xrightarrow[\text{ether}]{Mg} \text{(cyclohexyl-MgCl)} \xrightarrow[\text{2. } H_2O]{\text{1. } C_6H_5\overset{O}{\underset{\|}{C}}CH_3} C_6H_5\overset{CH_3}{\underset{OH}{\overset{|}{\underset{|}{C}}}}\text{(cyclohexyl)} + Mg(OH)Cl$$

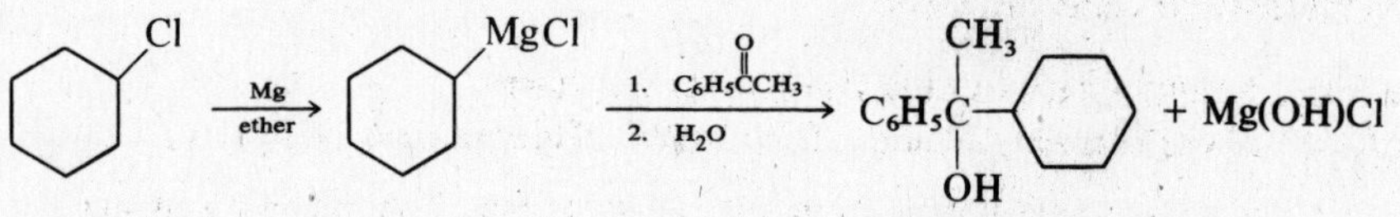

Methylcyclohexyl-
phenylcarbinol

6. Addition of :H⁻ to —C̶=O

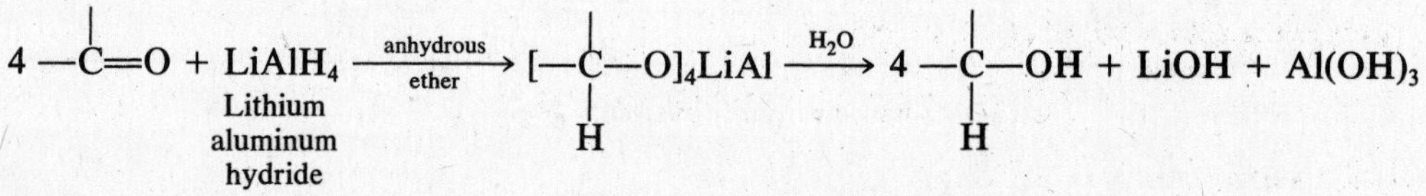

$$4 -\overset{|}{C}=O + LiAlH_4 \xrightarrow[\text{ether}]{\text{anhydrous}} [-\overset{|}{\underset{H}{\overset{|}{C}}}-O]_4LiAl \xrightarrow{H_2O} 4 -\overset{|}{\underset{H}{\overset{|}{C}}}-OH + LiOH + Al(OH)_3$$

Lithium
aluminum
hydride

Sodium borohydride, $NaBH_4$, can be used in protic solvents such as ROH and H_2O.

7. Hydrogenation of —C̶=O

Problem 13.11 How do the alcohols from $LiAlH_4$ or catalytic reduction of ketones differ from those derived from aldehydes? ◄

Ketones yield 2° alcohols while aldehydes give 1° alcohols.

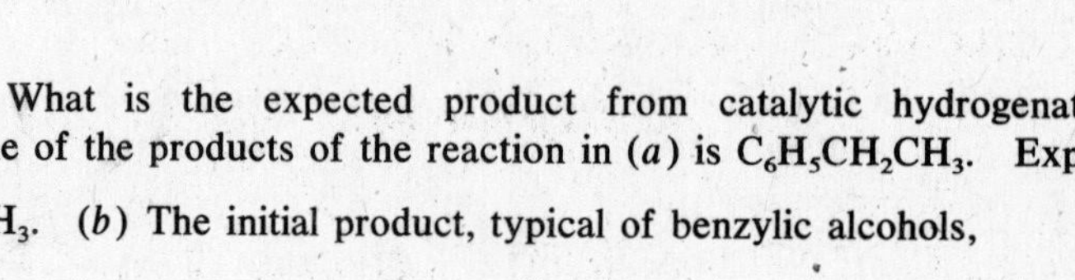

Ketone:
$$R-\overset{O}{\underset{\|}{C}}-R' \xrightarrow[\text{2. } H_2O]{\text{1. } LiAlH_4} R-\overset{H}{\underset{OH}{\overset{|}{\underset{|}{C}}}}-R' \xleftarrow{H_2/Pt} RCR'$$

Aldehyde:
$$R-\overset{O}{\underset{\|}{C}}-H \xrightarrow[\text{2. } H_2O]{\text{1. } LiAlH_4} R-\overset{H}{\underset{OH}{\overset{|}{\underset{|}{C}}}}-H \xleftarrow{H_2/Pt} RCH$$

Problem 13.12 (a) What is the expected product from catalytic hydrogenation of acetophenone, $C_6H_5COCH_3$? (b) One of the products of the reaction in (a) is $C_6H_5CH_2CH_3$. Explain its formation. ◄

(a) $C_6H_5CHOHCH_3$. (b) The initial product, typical of benzylic alcohols,

$$C_6H_5\overset{|}{\underset{|}{C}}-OH$$

can be further reduced with H_2. This reaction is a **hydrogenolysis** (bond-breaking by H_2).

$$C_6H_5CHOHCH_3 + H_2 \xrightarrow{Pd} C_6H_5CH_2CH_3 + H_2O$$

Problem 13.13 Reduction of $H_2C{=}CHCHO$ with $NaBH_4$ gives a different product than does catalytic hydrogenation (H_2/Ni). What are the products? ◄

$$H_2C{=}CHCH_2OH \xleftarrow{\ NaBH_4\ } H_2C{=}CHCHO \xrightarrow{\ H_2/Ni\ } H_3CCH_2CH_2OH$$

selective   Acrolein   *nonselective*
reduction of   *reduction*

$\rangle C{=}O$

13.3 REACTIONS OF ALCOHOLS

1. **The electron pairs on O make alcohols Lewis bases.**
2. **H of OH is very weakly acidic.** The order of decreasing acidity is

$$H_2O > ROH\,(1°) > ROH\,(2°) > ROH\,(3°) > RC{\equiv}CH \gg RCH_3$$

3. **$1°$ and $2°$ alcohols have at least one H on the carbinol C and are oxidized to carbonyl compounds.** They also lose H_2 in the presence of Cu (300 °C) to give carbonyl compounds.
4. **Formation of alkyl halides** (Problem 7.4).
5. **Intramolecular dehydration to alkenes** (Problems 6.13 through 6.16).
6. **Intermolecular dehydration to ethers.**

$$2ROH \xrightarrow[130°]{H_2SO_4} ROR + H_2O$$

7. **Ester formation.**

$$ROH + R'\underset{\|\atop O}{C}OH \xrightarrow{H_2SO_4} R'\underset{\|\atop O}{C}OR + H_2O$$

ester

With cold conc. H_2SO_4, sulfate esters are formed.

$$ROH + H_2SO_4 \longrightarrow H_2O + ROSO_3H \xrightarrow{ROH} ROSO_2OR$$

alkyl acid   *dialkyl*
sulfate   *sulfate*

Problem 13.14 Give (a) S_N2, (b) S_N1 mechanisms for formation of $C_2H_5OC_2H_5$ from C_2H_5OH in conc. H_2SO_4. ◄

(a)   (1) $C_2H_5OH + H_2SO_4 \longrightarrow \left[C_2H_5\overset{H}{\underset{+}{O}}{-}H \right] + HSO_4^-$

           $Base_1$ $Acid_2$ $Acid_1$ $Base_2$

     (2) $C_2H_5\ddot{O}H + CH_3CH_2\overset{+}{O}H_2 \longrightarrow \left[CH_3CH_2\overset{H}{\underset{+}{O}}C_2H_5 \right] + H_2O$

conjugate acid of ether

     (3) $\left[CH_3CH_2\overset{H}{\underset{+}{O}}C_2H_5 \right] + HSO_4^- \text{ (or } C_2H_5OH) \longrightarrow CH_3CH_2OC_2H_5 + H_2SO_4 \text{ (or } C_2H_5\overset{+}{O}H_2)$

(b) (1) Same as (a)   (3) $C_2H_5\ddot{O}H + H_2\overset{+}{C}CH_3 \longrightarrow \left[C_2H_5\overset{H}{\underset{+}{O}}CH_2CH_3 \right]$

    (2) $CH_3CH_2\overset{+}{O}H_2 \longrightarrow CH_3CH_2^+ + H_2O$ (4) Same as (3) in part (a)

Problem 13.15   Give the products for the reaction of isopropanol with conc. H_2SO_4 at   (a) 0 °C,   (b) room temperature,   (c) 130 °C,   (d) 180 °C.   ◄

(a) $(CH_3)_2CH\overset{+}{O}H_2 + HSO_4^-$
     oxonium ion

(b) $(CH_3)_2CHOSO_3H$ Isopropyl hydrogen sulfate
(c) $(CH_3)_2CHOCH(CH_3)_2$ Diisopropyl ether
(d) $CH_3CH{=}CH_2$ (high temperatures favor intramolecular dehydration)

Problem 13.16   Write equations to show why alcohols cannot be used as solvents with Grignard reagents or with $LiAlH_4$.   ◄

Strongly basic R^- and H^- react with weakly acidic alcohols.

$$CH_3O\overset{\frown}{H} + CH_3\overset{-}{C}H_2\overset{+}{M}gCl \longrightarrow CH_3CH_3 + (CH_3O)^-(MgCl)^+$$

$$4CH_3OH + LiAlH_4 \longrightarrow 4H_2 + LiAl(OCH_3)_4$$

Problem 13.17   Why may Na metal be used to remove the last traces of H_2O from benzene but not from ethanol?   ◄

Ethanol is sufficiently acidic to react with Na, although not as vigorously as does H_2O.

$$2C_2H_5OH + 2Na \longrightarrow 2C_2H_5O^-Na^+ + H_2$$

Problem 13.18   Give the main products of reaction of 1-propanol with   (a) alkaline aq. $KMnO_4$ solution during distillation;   (b) hot Cu shavings;   (c) CH_3COOH, H^+.   ◄

(a) CH_3CH_2CHO. Since aldehydes are oxidized further under these conditions, CH_3CH_2COOH is also obtained.
(b) CH_3CH_2CHO. The aldehyde can't be oxidized further.
(c) $CH_3\overset{\|}{\underset{O}{C}}-OCH_2CH_2CH_3$, an ester.

Problem 13.19   Explain the relative acidity of 1°, 2° and 3° alcohols.   ◄

The order of decreasing acidity of alcohols, $CH_3OH > 1° > 2° > 3°$, is attributed to electron-releasing R's. These intensify the charge on the conjugate base, RO^-, and destabilize this ion, making the acid weaker.

Problem 13.20   Write steps in the acidic degradative oxidation of $(CH_3)_3COH$ to CH_3COOH and CO_2.   ◄

$$CH_3{-}\underset{\underset{CH_3}{|}}{\overset{\overset{CH_3}{|}}{C}}{-}OH \xrightarrow{H^+} H_2O + CH_3{-}\underset{\underset{CH_2}{\|}}{\overset{\overset{CH_3}{|}}{C}} \xrightarrow{\text{oxid.}} CO_2 + CH_3{-}\underset{\underset{O}{\|}}{\overset{\overset{CH_3}{|}}{C}} \xrightarrow{\text{oxid.}} CH_3{-}\underset{\underset{O}{\|}}{C}{-}OH + CO_2$$

Problem 13.21   Give simple chemical tests to distinguish   (a) 1-pentanol and n-hexane;   (b) n-butanol and t-butanol;   (c) 1-butanol and 2-buten-1-ol;   (d) 1-hexanol and 1-bromohexane.   ◄

(a) Alcohols such as 1-pentanol dissolve in cold H_2SO_4 [see Problem 13.15(a)]. Alkanes such as n-hexane are insoluble.   (b) Unlike t-butanol (a 3° alcohol), n-butanol (a 1° alcohol) can be oxidized under mild conditions. The analytical reagent is chromic anhydride in H_2SO_4. A positive test is signaled when this orange-red solution turns a deep green because of the presence of Cr^{3+}.   (c) 2-Buten-1-ol decolorizes Br_2 in CCl_4 solution; 1-butanol does not.   (d) 1-Hexanol reduces orange-red CrO_3 to green Cr^{3+}; alkyl halides such as 1-bromohexane do not. The halide on warming with $AgNO_3$ (EtOH) gives AgBr.

Problem 13.22 How can the difference in reactivity of 1°, 2° and 3° alcohols with HCl be used to distinguish among these kinds of alcohols, provided the alcohols have 6 or less C's? ◄

The **Lucas test** uses conc. HCl and $ZnCl_2$ (to increase the acidity of the acid).

$$3° \; ROH + HCl \longrightarrow RCl + H_2O$$

soluble if	insoluble
it has 6 or	liquid
less C's	

The above reaction is immediate; a 2° ROH reacts within 5 minutes; a 1° ROH does not react at all at room temperature.

Problem 13.23 H of OH has a chemical shift that varies with the extent of H-bonding. Why can we detect an nmr signal from H of ROH by shaking the sample with D_2O and then retaking the spectrum? ◄

Through H-bonding the exchange reaction $ROH + D_2O \rightarrow ROD + DOH$ occurs. D signals are not detected by a proton nmr spectrometer. If the signal in question disappears, it came from H of ROH. A new signal appears at about $\delta = 5$ due to DOH.

Problem 13.24 Methyl ketones give the **haloform test**:

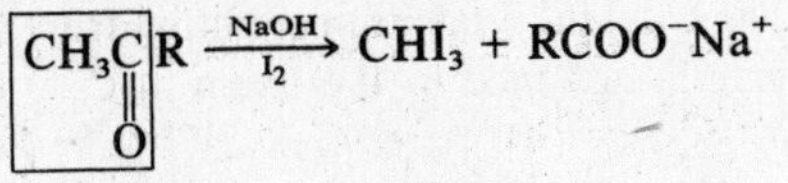

$$\boxed{CH_3CR \atop \quad \| \atop \quad O} \xrightarrow[I_2]{NaOH} CHI_3 + RCOO^-Na^+$$

a methyl	yellow
ketone	precipitate

I_2 can also oxidize 1° and 2° alcohols to carbonyl compounds. Which butyl alcohols give a positive haloform test? ◄

Alcohols with the

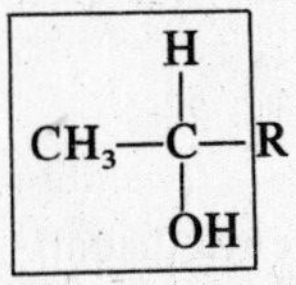

$$\boxed{\begin{array}{c} H \\ | \\ CH_3-C-R \\ | \\ OH \end{array}}$$

group are oxidized to

$$CH_3C-R \atop \quad \| \atop \quad O$$

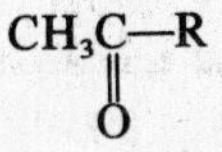

and give a positive test. The only butyl alcohol giving a positive test is

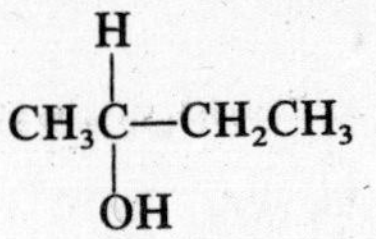

$$\begin{array}{c} H \\ | \\ CH_3C-CH_2CH_3 \\ | \\ OH \end{array}$$

Supplementary Problems

Problem 13.25 Give the IUPAC name for each of the following alcohols. Which are 1°, 2° and 3°?

(a) $CH_3—CH_2—\underset{\underset{CH_3}{|}}{CH}—CH_2—OH$ (b) $HO—CH_2—\underset{\underset{CH_3}{|}}{CH}—CH_2C_6H_5$ (c)

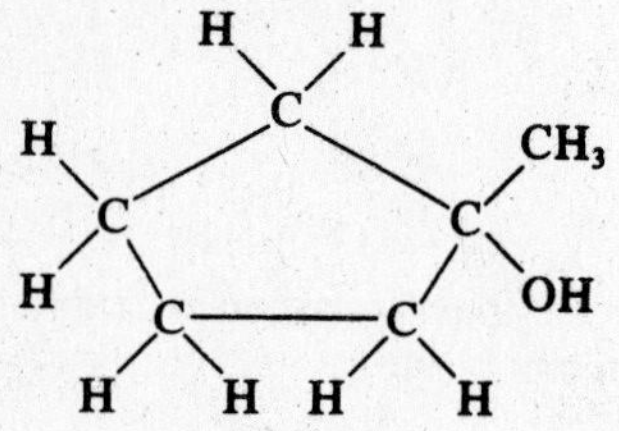

(d) $CH_3\underset{\underset{OH}{|}}{\overset{\overset{CH_2CH_3}{|}}{C}}—CH_2CH_3$ (e) $Cl—CHCH_2CH{=}CHCH_2OH$
$\qquad\qquad\qquad\qquad\qquad\quad CH_3—\overset{|}{\underset{|}{C}}—CH_3$
$\qquad\qquad\qquad\qquad\qquad\quad C_6H_4\text{-}m\text{-}Cl$

   (a) 2-Methyl-1-butanol, 1°.   (b) 2-Methyl-3-phenyl-1-propanol, 1°.   (c) 1-Methyl-1-cyclopentanol, 3°.
(d) 3-Methyl-3-pentanol, 3°.   (e) 5-Chloro-6-methyl-6-(3-chlorophenyl)-2-hepten-1-ol. (The longest chain
with OH has 7 C's and the prefix is **hept-**. Numbering begins at the end of the chain with OH; therefore,
-1-ol. The aromatic ring substituent has Cl at the 3 position counting from the point of attachment, and is put in
parentheses to show that the entire ring is attached to the chain at C^6. Cl on the chain is at C^5.) 1°.

Problem 13.26 Write condensed structural formulas and give IUPAC names for   (a) vinylcarbinol,
(b) diphenylcarbinol,   (c) dimethylethylcarbinol,   (d) benzylcarbinol.

(a) $H_2C{=}CHCH_2OH$ 2-Propen-1-ol (Allyl alcohol)

(b) $(C_6H_5)_2CH—OH$ Diphenylmethanol (Benzhydrol)

(c) $(CH_3)_2\underset{\underset{CH_2CH_3}{|}}{C}—OH$ 2-Methyl-2-butanol

(d) $C_6H_5—\overset{\beta}{C}H_2\overset{\alpha}{C}H_2OH$ 2-Phenylethanol (β-Phenylethanol)

Problem 13.27 Dihydroxy compounds are called **glycols**. Give the structural formulas for   (a) ethylene
glycol,   (b) propylene glycol   and (c) trimethylene glycol.

   (a) $HOCH_2CH_2OH$,   (b) $CH_3CHOHCH_2OH$,   (c) $HOCH_2CH_2CH_2OH$.

Problem 13.28 Why does the reaction of $(CH_3)_3CCl$ with OH^- give no $(CH_3)_3COH$?

   3° Halides react with bases to eliminate HCl by an E2 reaction, forming alkenes. They do not undergo
an S_N2 displacement to give alcohols. Under S_N1 conditions, such as heating with H_2O in dioxane, $(CH_3)_3CCl$
forms the alcohol in good yield.

Problem 13.29 Give the hydroboration-oxidation product from   (a) cyclohexene,   (b) *cis*-2-phenyl-2-
butene,   (c) *trans*-2-phenyl-2-butene.

   Addition of H_2O is *cis* anti-Markovnikov. See Fig. 13-2. In Fig. 13-2(c) the second pair of conformations
show the eclipsing of the H's with each other and of the Me's with each other. The more stable staggered
conformations are not shown.

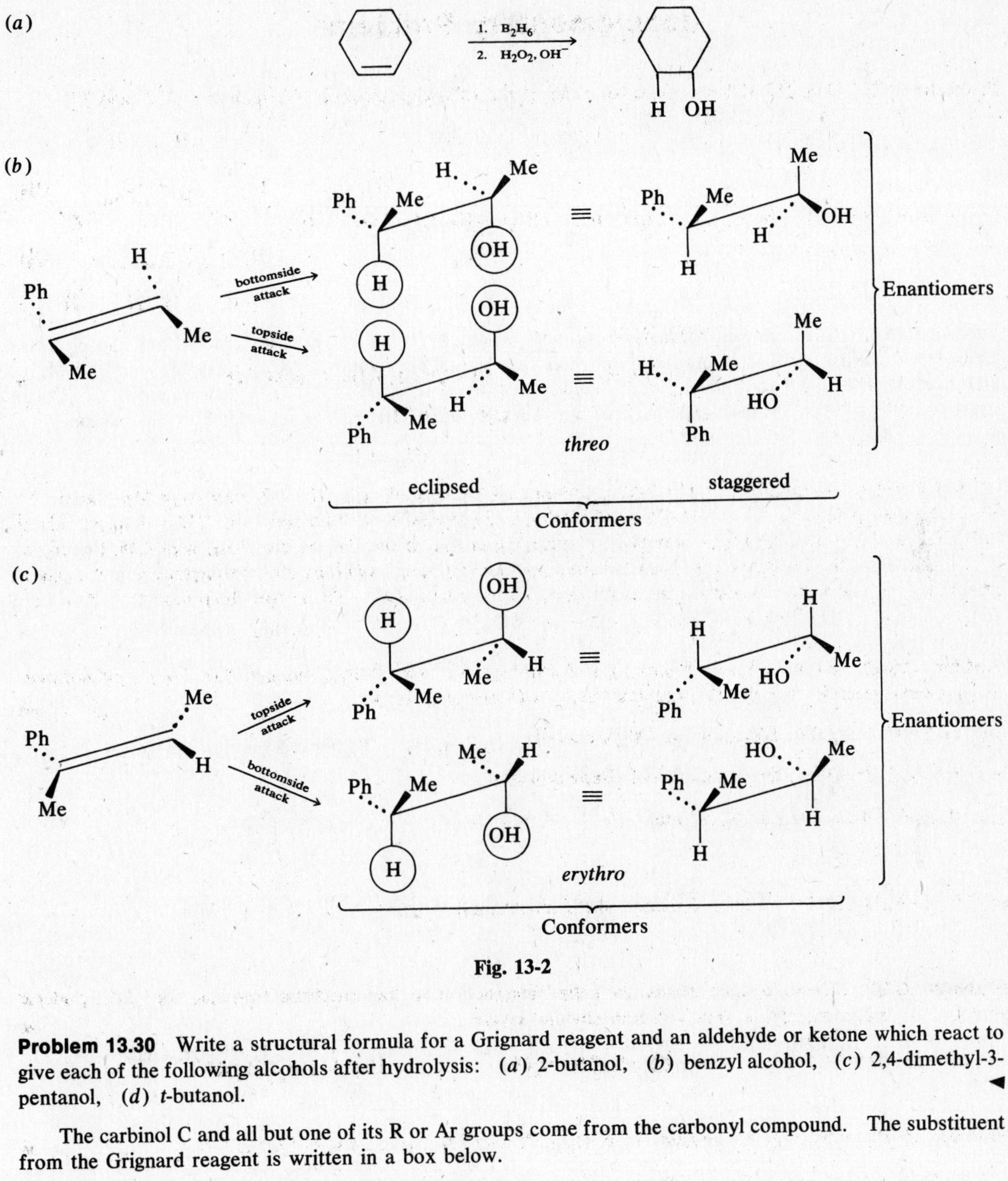

(a)

(b)

(c)

Fig. 13-2

Problem 13.30 Write a structural formula for a Grignard reagent and an aldehyde or ketone which react to give each of the following alcohols after hydrolysis: (a) 2-butanol, (b) benzyl alcohol, (c) 2,4-dimethyl-3-pentanol, (d) t-butanol.

The carbinol C and all but one of its R or Ar groups come from the carbonyl compound. The substituent from the Grignard reagent is written in a box below.

(a) Two possible combinations of reactants are:

(b)
$$H_2C=O \xrightarrow[\text{2. } H_3O^+]{\text{1. } \boxed{C_6H_5}MgBr} \boxed{C_6H_5}CH_2OH$$

(c)
$$(CH_3)_2CH-\overset{H}{\underset{}{C}}=O \xrightarrow[\text{2. } H_3O^+]{\text{1. } \boxed{(CH_3)_2CH}MgBr} \boxed{(CH_3)_2CH}-\overset{H}{\underset{OH}{C}}-CH(CH_3)_2$$

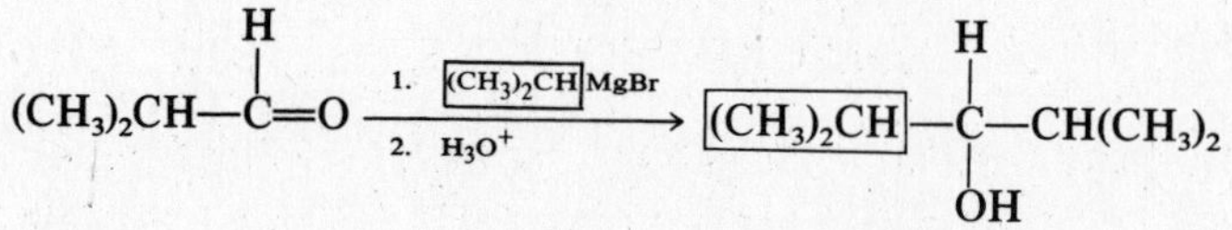

(d)
$$CH_3-\overset{}{\underset{O}{C}}-CH_3 \xrightarrow[\text{2. } H_3O^+]{\text{1. } \boxed{CH_3}MgBr} CH_3-\overset{\boxed{CH_3}}{\underset{OH}{C}}-CH_3$$

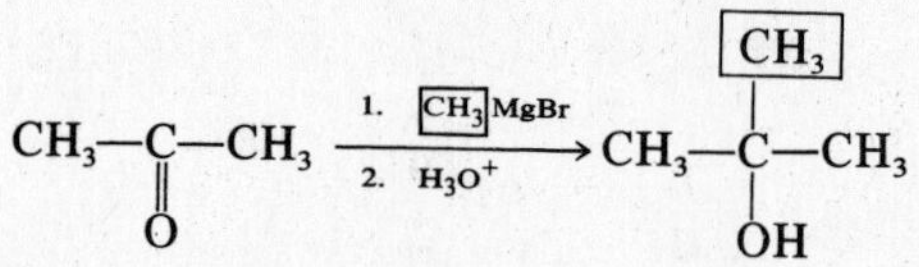

Problem 13.31 The following Grignard reagents and aldehydes or ketones are reacted and the products hydrolyzed. What alcohol is produced in each case? (a) Benzaldehyde ($C_6H_5CH=O$) and C_2H_5MgBr. (b) Acetaldehyde and benzyl magnesium bromide. (c) Acetone and benzyl magnesium bromide. (d) Formaldehyde and cyclohexyl magnesium bromide. (e) Acetophenone ($C_6H_5\overset{}{\underset{O}{C}}CH_3$) and ethyl magnesium bromide. ◄

(a) 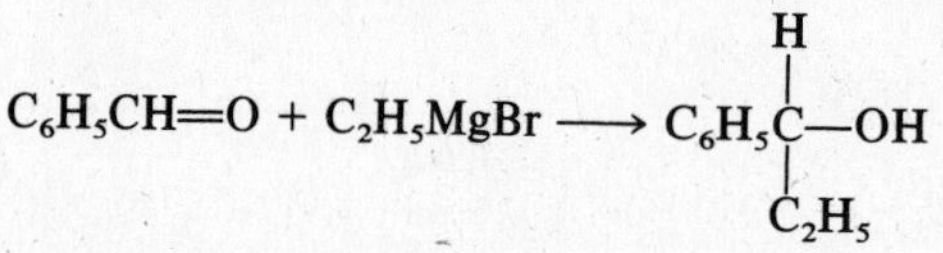
$$C_6H_5CH=O + C_2H_5MgBr \longrightarrow C_6H_5\overset{H}{\underset{C_2H_5}{C}}-OH \qquad \text{1-Phenyl-1-propanol}$$

(b) 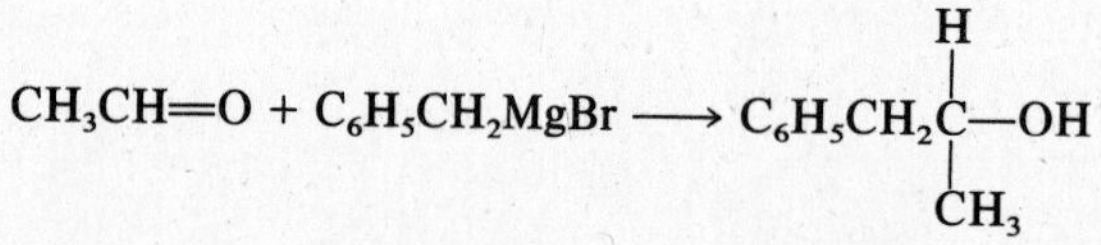
$$CH_3CH=O + C_6H_5CH_2MgBr \longrightarrow C_6H_5CH_2\overset{H}{\underset{CH_3}{C}}-OH \qquad \text{1-Phenyl-2-propanol}$$

(c) 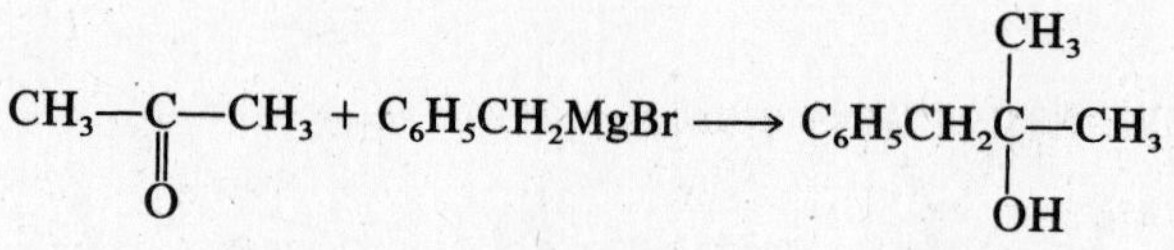
$$CH_3-\overset{}{\underset{O}{C}}-CH_3 + C_6H_5CH_2MgBr \longrightarrow C_6H_5CH_2\overset{CH_3}{\underset{OH}{C}}-CH_3 \qquad \text{1-Phenyl-2-methyl-2-propanol}$$

(d)
$$H_2C=O + C_6H_{11}MgBr \longrightarrow C_6H_{11}CH_2OH \qquad \text{Cyclohexylcarbinol}$$

(e) 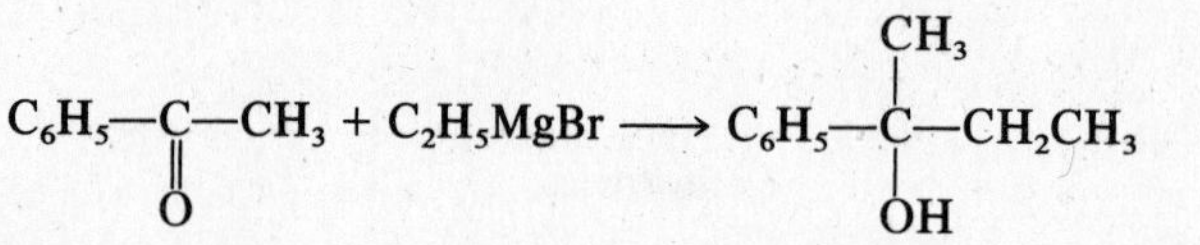
$$C_6H_5-\overset{}{\underset{O}{C}}-CH_3 + C_2H_5MgBr \longrightarrow C_6H_5-\overset{CH_3}{\underset{OH}{C}}-CH_2CH_3 \qquad \text{2-Phenyl-2-butanol}$$

Problem 13.32 Use C_2H_5MgBr in a single Grignard reaction to prepare n-butanol. ◄

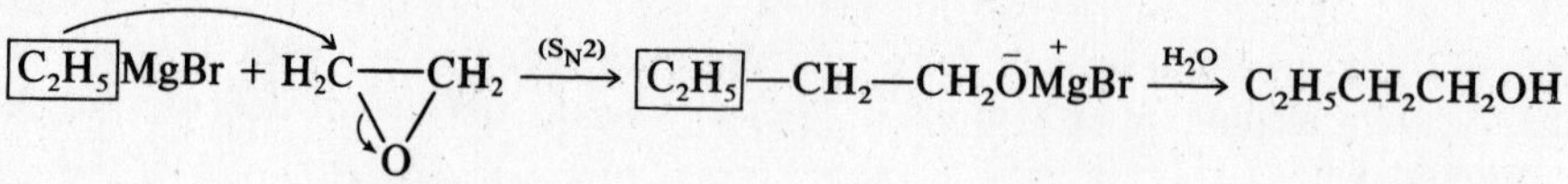

$$\boxed{C_2H_5}MgBr + H_2C-CH_2 \xrightarrow{\text{(}S_N2\text{)}} \boxed{C_2H_5}-CH_2-CH_2\overset{-}{O}\overset{+}{M}gBr \xrightarrow{H_2O} C_2H_5CH_2CH_2OH$$

Ethylene
oxide

Problem 13.33 Give the mechanism in each case:

(a) $CH_3CH_2CH_2CH_2CH_2OH \xrightarrow{HCl} CH_3CH_2CH_2CH_2CH_2Cl$

(b) $CH_3\overset{CH_3}{\underset{OH}{CH}}-CHCH_3 \xrightarrow{HCl} CH_3\overset{CH_3}{\underset{Cl}{C}}CH_2CH_3$

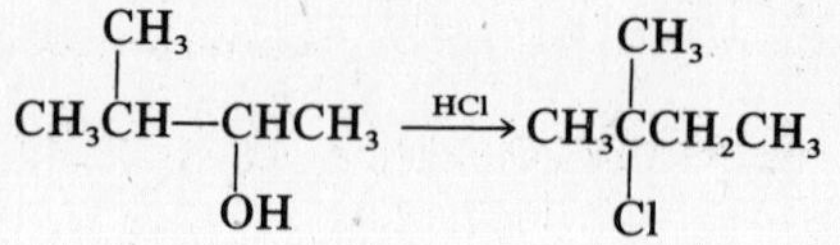

(c) $CH_3-\overset{CH_3}{\underset{OH}{C}}-CH_2CH_3 \xrightarrow{HCl} CH_3-\overset{CH_3}{\underset{Cl}{C}}-CH_2CH_3$

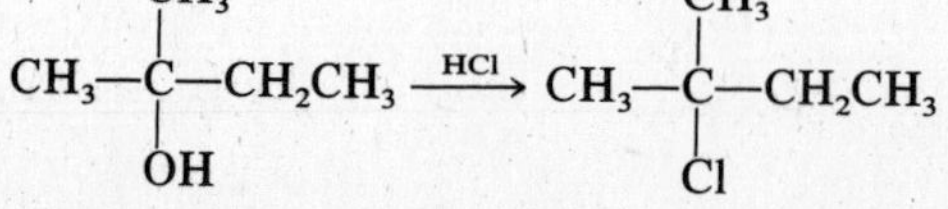

Why did rearrangement occur only in (b)? ◄

(a) The mechanism is S_N2 since we are substituting Cl for H_2O from 1° ROH_2^+.
(b) The mechanism is S_N1.

$$CH_3-\overset{\overset{\displaystyle CH_3}{|}}{C}H-\overset{\overset{}{\underset{\underset{\displaystyle OH}{|}}{}}}{C}H-CH_3 \xrightarrow[-H_2O]{H^+} CH_3-\overset{\overset{\displaystyle CH_3}{|}}{\underset{\underset{\displaystyle H}{|}}{C}}-\overset{+}{C}H-CH_3 \xrightarrow{\sim H:} CH_3-\overset{\overset{\displaystyle CH_3}{|}}{\overset{+}{C}}-CH_2-CH_3 \xrightarrow{Cl^-} CH_3-\overset{\overset{\displaystyle CH_3}{|}}{\underset{\underset{\displaystyle Cl}{|}}{C}}-CH_2-CH_3$$

2° R$^+$ (less stable)   3° R$^+$ (more stable)

(c) S_N1 mechanism.   The stable 3° $(CH_3)_2\overset{+}{C}CH_2CH_3$ reacts with Cl$^-$ with no rearrangement.

Problem 13.34 Why does dehydration of 1-phenyl-2-propanol in acid form 1-phenyl-1-propene rather than 1-phenyl-2-propene? ◄

1-Phenyl-1-propene, $PhCH{=}CHCH_3$, is a more substituted alkene and therefore more stable than 1-phenyl-2-propene, $PhCH_2CH{=}CH_2$. Even more important, it is more stable because the double bond is conjugated with the ring.

Problem 13.35 Give the formula for an alcohol used in, and supply conditions suitable for, the preparation of (a) $C_6H_5COCH_3$, (b) CH_3CH_2CHO, (c) $CH_3CH_2CH(CH_3)COOH$. ◄

(a) Ketones are oxidation products of 2° alcohols.

$$C_6H_5-\overset{\overset{}{\underset{\underset{\displaystyle OH}{|}}{}}}{C}H-CH_3 \xrightarrow[KMnO_4 \text{ or } K_2Cr_2O_7]{H^+} C_6H_5-\overset{\overset{}{\underset{\underset{\displaystyle O}{\|}}{}}}{C}-CH_3$$

1-Phenylethanol   Acetophenone

(b) If the aldehyde has a boiling point less than 100 °C, regular oxidants can be used.   Since the aldehyde has a lower boiling point than the alcohol, it is distilled off as it forms, so that further oxidation to a carboxylic acid is minimized.

$$CH_3CH_2CH_2OH \xrightarrow{H^+, K_2Cr_2O_7} CH_3CH_2CH{=}O$$

1-Propanol   Propionaldehyde
(b.p. 98 °C)   (b.p. 50 °C)

(c) A 1° alcohol RCH_2OH gives RCOOH on extended oxidation.

$$CH_3CH_2\overset{\overset{\displaystyle CH_3}{|}}{\underset{\underset{\displaystyle H}{|}}{C}}-CH_2OH \xrightarrow[Cr_2O_7^{2-} \ H^+]{MnO_4^- \text{ or}} CH_3CH_2\overset{\overset{\displaystyle CH_3}{|}}{\underset{\underset{\displaystyle H}{|}}{C}}-COOH$$

2-Methyl-1-butanol

Problem 13.36 Write the structural formula for the alcohol formed by oxymercuration-demercuration from (a) 1-heptene, (b) 1-methylcyclohexene, (c) 3,3-dimethyl-1-butene. ◄

The net addition of H_2O is Markovnikov.

(a) $CH_3(CH_2)_4CHOHCH_3$ 2-Heptanol

(b)   1-Methylcyclohexanol

(c) $(CH_3)_3CCHOHCH_3$ 3,3-Dimethyl-2-butanol (no rearrangement occurs)

Problem 13.37 Starting with isopropyl alcohol as the only available organic compound, prepare 2,3-dimethyl-2-butanol.

This 3° alcohol, $(CH_3)_2C(OH)CH(CH_3)_2$, is prepared from a Grignard reagent and a ketone.

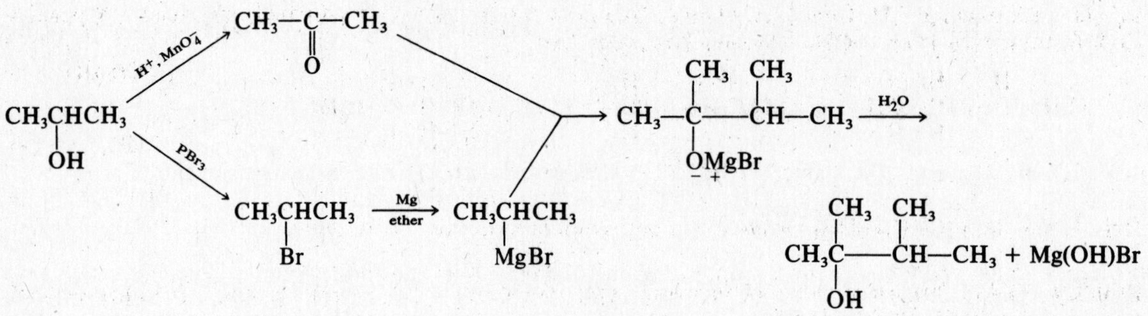

Problem 13.38 Use formaldehyde and 2-butanol to prepare 1,2-dibromo-2-methylbutane.

The product, $BrCH_2CBr(CH_3)CH_2CH_3$, is a *vic*-dibromide formed from the alkene 2-methyl-1-butene,

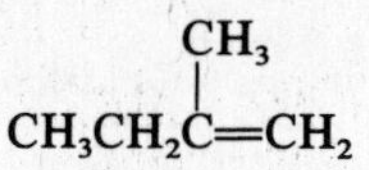

$$\begin{array}{c} CH_3 \\ | \\ CH_3CH_2C{=}CH_2 \end{array}$$

which must be made from

$$\begin{array}{c} CH_3 \\ | \\ CH_3CH_2CHCH_2OH \end{array}$$

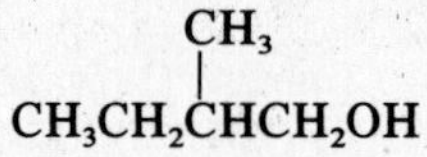

which in turn is prepared as follows:

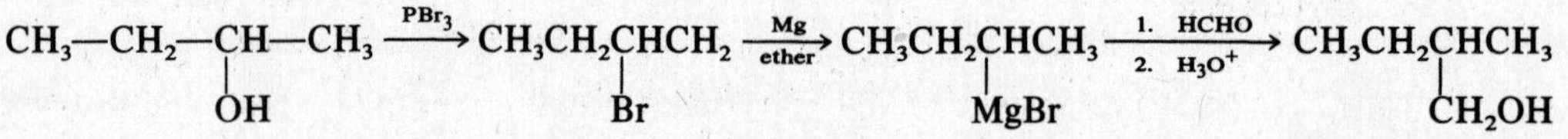

However, dehydration of 2-methyl-1-butanol with H_2SO_4 forms 2-methyl-2-butene by carbonium ion rearrangement.

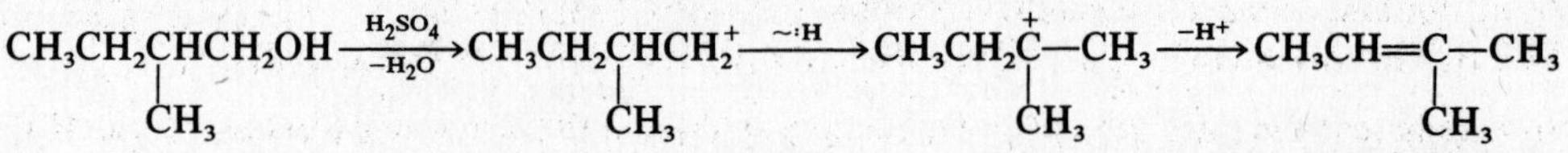

The needed alkene is prepared by dehydrohalogenation.

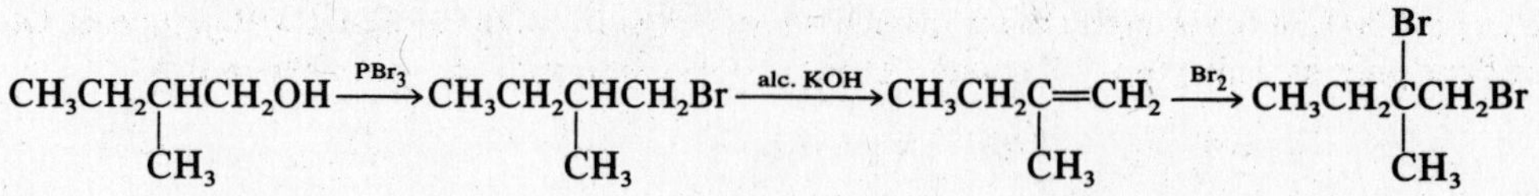

Problem 13.39 Prepare

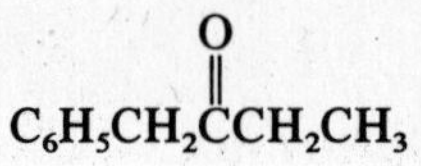

$$\begin{array}{c} O \\ \| \\ C_6H_5CH_2CCH_2CH_3 \end{array}$$

from benzyl bromide and propionaldehyde.

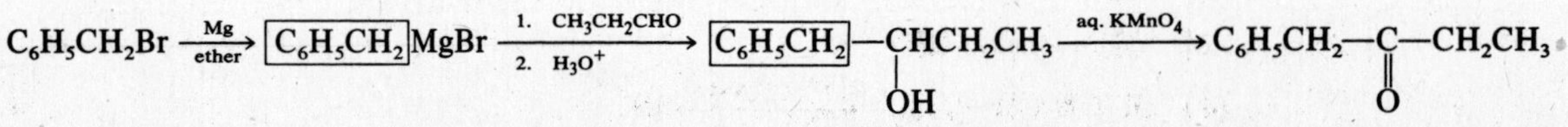

Problem 13.40 A compound, $C_9H_{12}O$, is oxidized under vigorous conditions to benzoic acid. It reacts with CrO_3 and gives a positive iodoform test (Problem 13.24). Is this compound chiral? ◄

Since benzoic acid is the product of oxidation, the compound is a monosubstituted benzene, C_6H_5G. Subtracting C_6H_5 from $C_9H_{12}O$ gives C_3H_7O as the formula for a saturated side chain. A positive CrO_3 test means a 1° or 2° OH. Possible structures are:

Only II has the —CH(OH)CH$_3$ needed for a positive iodoform test. II is chiral.

Problem 13.41 Suggest a possible industrial preparation for (*a*) *t*-butyl alcohol, (*b*) allyl alcohol, (*c*) glycerol (HOCH$_2$CHOHCH$_2$OH). ◄

(*a*)

$$CH_3-\underset{\underset{\displaystyle CH_3}{|}}{CH}-CH_3 \xrightarrow[\text{heat}]{\text{catalyst}} H_2 + CH_3-\underset{\underset{\displaystyle CH_3}{\|}}{C}=CH_2 \xrightarrow{H_3O^+} CH_3-\underset{\underset{\displaystyle OH}{|}}{\overset{\overset{\displaystyle CH_3}{|}}{C}}-CH_3$$

Isobutane   Isobutylene

(*b*)

$$CH_3-CH=CH_2 \xrightarrow[\text{heat}]{Cl_2} ClCH_2-CH=CH_2 \xrightarrow{HOH} HOCH_2CH=CH_2$$

Propene   Allyl chloride   Allyl alcohol

(*c*)

Problem 13.42 Assign numbers, from 1 for the lowest to 5 for the highest, to indicate *relative reactivity* with HBr in forming benzyl bromides from the following benzyl alcohols: (*a*) *p*-Cl—C$_6$H$_4$—CH$_2$OH, (*b*) (C$_6$H$_5$)$_2$CHOH, (*c*) *p*-O$_2$N—C$_6$H$_4$—CH$_2$OH, (*d*) (C$_6$H$_5$)$_3$COH, (*e*) C$_6$H$_5$CH$_2$OH. ◄

Differences in reaction rates depend on the relative abilities of the protonated alcohols to lose H$_2$O to form R$^+$. The stability of R$^+$ affects the $\Delta H^\ddagger$ for forming the incipient R$^+$ in the transition state and determines the overall rate.

Electron-attracting groups such as NO$_2$ and Cl in the *para* position destabilize R$^+$ by intensifying the positive charge. NO$_2$ of (*c*) is more effective since it destabilizes by both resonance and induction, while Cl of (*a*) destabilizes only by induction. The more C$_6$H$_5$'s on the benzyl C, the more stable is R$^+$.

(*a*) 2   (*b*) 4   (*c*) 1   (*d*) 5   (*e*) 3

Problem 13.43 Supply structural formulas and stereochemical designations for the organic compounds (A) through (H).

(*a*)   (A) + conc. H$_2$SO$_4$ $\xrightarrow{\text{cold}}$ CH$_3$—$\underset{\underset{\displaystyle OSO_3H}{|}}{CH}CH_2CH_3$ $\xrightarrow[\text{H}^+]{\text{H}_2\text{O}}$ (B)

(*b*)   (R)-CH$_3$CHOHCH$_2$CH=CH$_2$ $\xrightarrow[\text{H}_2\text{SO}_4]{\text{H}_2\text{O}}$ (C) + (D)

(*c*)   (R)-CH$_3$CH$_2$—$\underset{\underset{\displaystyle OH}{|}}{CH}$—CH=CH$_2$ + HBr $\longrightarrow$ (E) + (F) + (G)

(*d*)   (R)-CH$_3$CH—CH$_2$ $\xrightarrow[\text{2. H}_2\text{O}]{\text{1. CH}_3\text{CH}_2\text{MgBr}}$ (H)
        _O_/
◄

(a) (A) is cis- or trans-$CH_3CH=CHCH_3$ or $H_2C=CHCH_2CH_3$; (B) is rac-$CH_3CHOHCH_2CH_3$. The intermediate $CH_3C^+HCH_2CH_3$ can be attacked from either side by HSO_4^- to give an optically inactive racemic hydrogen sulfate ester which is hydrolyzed to rac-2-butanol.

(b) (C) is (R,R)-CH$_3$—CH$_2$—CH$_3$;   (D) is (R,S)-CH$_3$—CH$_2$—C—CH$_3$

optically active   meso

Hydration follows Markovnikov's rule, and a new similar chiral C is formed. The original chiral C remains R, but the new chiral C may be R or S. The two diastereomers are not formed in equal amounts [see Problem 5.17(b)].

(c) (E) is rac-$CH_3CH_2CHBrCH=CH_2$; (F) is trans- and (G) cis-$CH_3CH_2CH=CHCH_2Br$. The intermediate R^+ in this S_N1 reaction is a resonance-stabilized (charge-delocalized) allylic cation

$$CH_3CH_2\overset{+}{C}HCH=CH_2 \longleftrightarrow CH_3CH_2CH=CH\overset{+}{C}H_2 \text{ or } CH_3CH_2\overset{\delta+}{CH}=\!\!=\!\!CH\overset{\delta+}{=\!\!=}CH_2$$

In the 2nd step, Br^- attacks either of the positively charged C's to give one of the 3 products. Since R^+ is flat the chiral C in (E) can be R or S; (E) is racemic. (F) is the major product because it is most stable (trans and disubstituted).

(d) (H) is (R)-CH$_3$—CH$_2$ CH$_2$CH$_3$ (S_N2 attack at less substituted 1° C)

The configuration of the chiral C does not change.

Problem 13.44 How does the Lewis theory of acids and bases explain the functions of   (a) $ZnCl_2$ in the Lucas reagent?   (b) ether as a solvent in the Grignard reagent?   ◄

(a) $ZnCl_2 + 2HCl \longrightarrow H_2ZnCl_4 \xrightarrow{ROH} ROH_2^+ \xrightarrow{Cl^-} RCl + H_2O$
                       Lewis   (a stronger
                       acid   acid than HCl)

(b) R'MgX acts as a Lewis acid because Mg can coordinate with one unshared electron pair of each O of two ether molecules to form an addition compound,

$$\begin{matrix} & R' & \\ R & | & R \\ \ddot{O}:Mg:\ddot{O} & & \\ R & | & R \\ & X & \end{matrix}$$

that is soluble in ether.

Problem 13.45 Draw Newman projections of the conformers of the following substituted ethanols and predict their relative populations: (a) FCH_2CH_2OH,   (b) $H_2NCH_2CH_2OH$,   (c) $BrCH_2CH_2OH$.   ◄

If the substituents F, H_2N and Br are designated by Z, the conformers may be generalized as anti or gauche.

anti   gauche

For (a) and (b) the gauche is the more stable conformer and has a greater population because of H-bonding with F and N. The anti conformer is more stable in (c) because there is no H-bonding and dipole-dipole repulsion causes OH and Br to lie as far from each other as possible.

Problem 13.46 Deduce the structure of a compound, $C_4H_{10}O$, which gives the following nmr data: $\delta = 0.8$ (doublet, 6 H's), $\delta = 1.7$ (complex multiplet, one H), $\delta = 3.2$ (doublet, 2 H's) and $\delta = 4.2$ (singlet, one H; disappears after shaking sample with D_2O).   ◄

The singlet at $\delta = 4.2$ which disappears after shaking with D_2O is for O$\underline{H}$ (Problem 13.23). The compound must be one of the 4 butyl alcohols. Only isobutyl alcohol, $(CH_3)_2CHCH_2OH$, has 6 equivalent H's (2 CH_3's), accounting for the six-H doublet at $\delta = 0.8$, the one-H multiplet at $\delta = 1.7$, and the two-H doublet which is further downfield at $\delta = 3.2$ because of the electron-attracting O.

Problem 13.47 Write balanced ionic equations for the following redox reactions:

(a) $CH_3CHOHCH_3 + K_2Cr_2O_7 + H_2SO_4 \xrightarrow{\text{heat}} CH_3-CO-CH_3 + Cr_2(SO_4)_3 + H_2O + K_2SO_4$

(b)

Write partial equations for the oxidation and the reduction. Then: (1) **Balance charges** by adding H$^+$ in acid solutions or OH$^-$ in basic solutions. (2) **Balance the number of** O's by adding H_2O's to one side. (3) **Balance the number of** H's by adding H's to one side. The number added is the **number of equivalents** of oxidant or reductant.

(a)   **OXIDATION**   **REDUCTION**

$(CH_3)_2CHOH \longrightarrow (CH_3)_2C{=}O$ $Cr_2O_7^{2-} \longrightarrow 2Cr^{3+}$

(1) In acid balance charges with H$^+$:

(no change) $Cr_2O_7^{2-} + 8H^+ \longrightarrow 2Cr^{3+}$

(2) Balance O with H_2O:

(no change) $Cr_2O_7^{2-} + 8H^+ \longrightarrow 2Cr^{3+} + 7H_2O$

(3) Balance H:

$(CH_3)_2CHOH \longrightarrow (CH_3)_2C{=}O + 2H$ $Cr_2O_7^{2-} + 8H^+ + 6H \longrightarrow 2Cr^{3+} + 7H_2O$

(4) Balance equivalents:

$$3[(CH_3)_2CHOH \longrightarrow (CH_3)_2C{=}O + 2H]$$
$$Cr_2O_7^{2-} + 8H^+ + 6H \longrightarrow 2Cr^{3+} + 7H_2O$$

(5) Add: $3(CH_3)_2CHOH + Cr_2O_7^{2-} + 8H^+ \longrightarrow 3(CH_3)_2C{=}O + 2Cr^{3+} + 7H_2O$

(b)   **OXIDATION**   **REDUCTION**

$MnO_4^- \longrightarrow MnO_2$

(1) In base balance charges with OH$^-$:

$MnO_4^- \longrightarrow MnO_2 + OH^-$

(2) Balance O with H_2O:

$2OH^- + C_6H_{10} + 2H_2O \longrightarrow C_6H_8O_4^{2-}$ $MnO_4^- \longrightarrow MnO_2 + OH^- + H_2O$

(3) Balance H:

$2OH^- + C_6H_{10} + 2H_2O \longrightarrow C_6H_8O_4^{2-} + 8H$ $3H + MnO_4^- \longrightarrow MnO_2 + OH^- + H_2O$

(4) Balance equivalents:

$$3[2OH^- + C_6H_{10} + 2H_2O \longrightarrow C_6H_8O_4^{2-} + 8H]$$
$$8[3H + MnO_4^- \longrightarrow MnO_2 + OH^- + H_2O]$$

(5) Add: $3C_6H_{10} + 8MnO_4^- \longrightarrow 3C_6H_8O_4^{2-} + 8MnO_2 + 2OH^- + 2H_2O$

Problem 13.48 The attempt to remove water from ethanol by fractional distillation gives 95% ethanol, an **azeotrope** that boils at a constant temperature of 78.15 °C. It has a lower boiling point than either water (100 °C) or ethanol (78.3 °C). A liquid mixture is an azeotrope if it gives a vapor of the same composition. How does boiling 95% ethanol with Mg remove the remaining H_2O? ◄

$$Mg + H_2O \longrightarrow H_2 + Mg(OH)_2$$
$$\text{insoluble}$$

The dry ethanol, called **absolute**, is now distilled from the insoluble $Mg(OH)_2$.

Problem 13.49 Explain why the most prominent (base) peak of 1-propanol is at $m/e = 31$, while that of allyl alcohol is at $m/e = 57$. ◄

$CH_3CH_2-CH_2OH^+$ cleaves mainly into

$$CH_3CH_2\cdot + \overset{+}{C}H_2\overset{..}{O}H \longleftrightarrow H_2C=\overset{+}{\overset{..}{O}}H$$

($m/e = 31$) rather than $CH_3CH_2\overset{+}{C}HOH + H\cdot$ because C—C is weaker than C—H. In allyl alcohol,

$$CH_2=CH\overset{\overset{\displaystyle H}{|}}{\underset{\underset{\displaystyle H}{|}}{C}}-OH$$

the C—H bond cleaves to give

$$CH_2=CH\overset{+}{\underset{\underset{\displaystyle H}{|}}{C}}-OH$$

($m/e = 57$). This cation is stabilized by both the $CH_2=CH$ and the O.

Problem 13.50 A compound, $C_5H_{12}O$, in dimethyl sulfoxide shows a singlet peak for the OH proton. What is the compound? ◄

If the signal for the H of the OH group is a singlet, the alcohol must be 3° (Problem 12.60). The compound is $CH_3CH_2C(OH)(CH_3)_2$.

Chapter 14

Ethers, Epoxides
and Glycols

14.1 INTRODUCTION AND NOMENCLATURE

Simple (symmetrical) ethers have the general formula R—O—R or Ar—O—Ar; **mixed (unsymmetrical) ethers** are R—O—R' or Ar—O—Ar' or Ar—O—R. The derived system names R and Ar as separate words and adds the word "ether." In the IUPAC system, ethers (ROR) are named as alkoxy-(RO) substituted alkanes.

Problem 14.1 Give a derived and IUPAC name for the following ethers: (a) $CH_3OCH_2CH_2CH_2CH_3$, (b) $(CH_3)_2CHOCH(CH_3)CH_2CH_3$, (c) $C_6H_5OCH_2CH_3$, (d) $p\text{-}NO_2C_6H_4OCH_3$. ◄

(a) Methyl n-butyl ether, 1-methoxybutane. (b) sec-Butyl isopropyl ether, 2-isopropoxybutane. (Select the longest chain of C's as the alkane root.) (c) Ethyl phenyl ether, ethoxybenzene (commonly called **phenetole**). (d) Methyl p-nitrophenyl ether, p-nitromethoxybenzene (commonly called p-**nitroanisole**).

Problem 14.2 Account for the following. (a) Ethers have significant dipole moments (≈ 1.18 D). (b) Ethers have lower boiling points than isomeric alcohols. (c) The water solubilities of isomeric ethers and alcohols are comparable. ◄

(a) The C—O—C bond angle is about 110° and the dipole moments of the two C—O bonds do not cancel. (b) The absence of OH in ethers precludes H-bonding and therefore there is no strong intermolecular force of attraction between ether molecules as there is between alcohol molecules. The weak polarity of ethers has no appreciable effect. (c) The O of ethers is able to undergo H-bonding with H of H_2O.

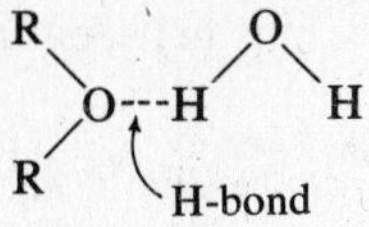

$$\begin{matrix} R & & & O \\ & \diagdown & & \diagup \quad \diagdown \\ & & O\text{-}\text{-}\text{-}H & \quad H \\ & \diagup & & \\ R & & \diagdown_{\text{H-bond}} & \end{matrix}$$

14.2 PREPARATION

SIMPLE ETHERS

1. **Intermolecular Dehydration of Alcohols** (see Section 13.3, page 223)

2. **2° Alkyl Halides with Silver Oxide**

$$2(CH_3)_2CHCl + Ag_2O \longrightarrow (CH_3)_2CHOCH(CH_3)_2 + 2AgCl$$

MIXED ETHERS

1. **Williamson Synthesis**

$$Na^+R':\!\ddot{O}:^- + R:\!X \xrightarrow{(S_N2)} R'OR + Na^+ + X^- \qquad X = Cl, Br, I, OSO_2R, OSO_2Ar$$

$\quad$ alkoxide $\qquad$ 1° or 2°

2. Use of Diazomethane to Form Methyl Ethers

$$n\text{-}C_7H_{13}OH + CH_2N_2 \xrightarrow[\text{(Lewis acid)}]{\text{BF}_3} n\text{-}C_7H_{13}OCH_3 + N_2$$
$$\text{Methyl } n\text{-heptyl ether}$$

$$C_6H_5OH + CH_2N_2 \xrightarrow{\text{BF}_3} C_6H_5OCH_3$$
$$\text{Anisole}$$

3. Alkoxymercuration-demercuration

To form ethers the mercuration of alkenes is performed in an alcohol.

$$RCH{=}CH_2 + R'OH + Hg(OC{-}CF_3)_2 \longrightarrow \underset{\underset{R'O\;\;HgOCOCF_3}{|\quad\quad|}}{RCHCH_2} \xrightarrow{\text{NaBH}_4} \underset{\underset{OR'}{|}}{RCHCH_3}$$

Problem 14.3 Compare the mechanisms for the formation of an ether by intermolecular dehydration of (a) 1°, (b) 3° and (c) 2° alcohols. ◄

(a) For 1° alcohols, such as *n*-butanol, the mechanism is S_N2 with alcohol as the attacking nucleophile and water as the leaving group. There would be no rearrangements.

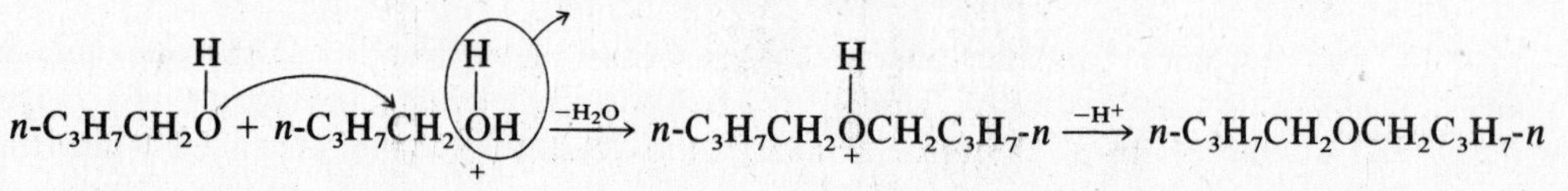

$$n\text{-}C_3H_7CH_2\overset{+}{O} + n\text{-}C_3H_7CH_2\overset{\frown}{OH} \xrightarrow{-H_2O} n\text{-}C_3H_7CH_2\overset{+}{O}CH_2C_3H_7\text{-}n \xrightarrow{-H^+} n\text{-}C_3H_7CH_2OCH_2C_3H_7\text{-}n$$

(b) The mechanism for 3° alcohols is S_N1.

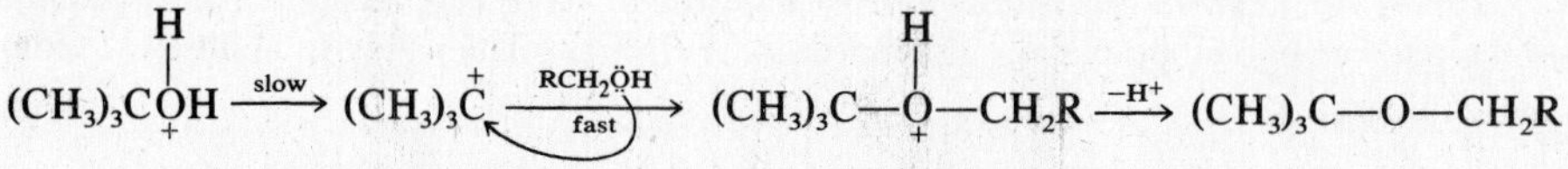

$$(CH_3)_3C\overset{H}{\overset{|}{O}}H \xrightarrow{\text{slow}} (CH_3)_3\overset{+}{C} \xrightarrow[\text{fast}]{RCH_2\overset{\cdot\cdot}{O}H} (CH_3)_3C{-}\overset{H}{\underset{+}{O}}{-}CH_2R \xrightarrow{-H^+} (CH_3)_3C{-}O{-}CH_2R$$

$(CH_3)_3C^+$ cannot react with $(CH_3)_3COH$ or any other 3° ROH because of severe steric hindrance, but it can react with a 1° RCH_2OH as shown above.

(c) 2° alcohols can react either way. Rearrangements may occur when they react by the S_N1 mechanism.

Problem 14.4 List the ethers formed in the reaction between concentrated H_2SO_4, and equimolar quantities of ethanol and (a) methanol, (b) *tert*-butanol. ◄

(a) These 1° alcohols react by S_N2 mechanisms to give a mixture of 3 ethers: $C_2H_5OC_2H_5$ from $2C_2H_5OH$, CH_3OCH_3 from $2CH_3OH$, and $C_2H_5OCH_3$ from C_2H_5OH and CH_3OH.

(b) This is an S_N1 reaction.

$$(CH_3)_3COH \xrightarrow{H^+} (CH_3)_3\overset{+}{C}OH_2 \xrightarrow{-H_2O} (CH_3)_3\overset{+}{C} \xrightarrow[-H^+]{CH_3CH_2OH} (CH_3)_3C{-}O{-}CH_2CH_3$$
$$\text{Ethyl } tert\text{-butyl ether}$$

Reaction between $(CH_3)_3C^+$ and $(CH_3)_3COH$ is sterically hindered and occurs much less readily.

Problem 14.5 Explain why a good yield of a mixed ether can be obtained from a mixture of $H_2C{=}CHCH_2OH$ and *i*-C_3H_7OH. ◄

The ether is $H_2C{=}CHCH_2OCH(CH_3)_2$. $H_2C{=}CHCH_2OH$ forms the stable $H_2C{=}CHCH_2^+$, which then reacts with *i*-C_3H_7OH.

Problem 14.6 Use any needed starting material to synthesize the following ethers, selecting from among intermolecular dehydration, Williamson synthesis or alkoxymercuration-demercuration. Justify your choice of method.

(a) $CH_3(CH_2)_3OCH_2CH_3$ (b) $CH_3CH_2\underset{\underset{CH_3}{|}}{CH}OCH_2CH_2CH_3$ (c) cyclohexyl ether ◄

(a) Use Williamson synthesis; $CH_3(CH_2)_3Cl + C_2H_5O^-Na^+$. Since alkoxymercuration is a Markovnikov addition, it cannot be used to prepare an ether in which both R's are 1°. Unless one of the R's can form a stable R^+, intermolecular dehydration cannot be used to synthesize a mixed ether.

(b)
$$\overset{4}{C}H_3\overset{3}{C}H_2\overset{2}{C}H=\overset{1}{C}H_2 + Hg(OCOCF_3)_2 + n\text{-}C_3H_7OH \longrightarrow \overset{4}{C}H_3\overset{3}{C}H_2\overset{2}{C}H-OC_3H_7\text{-}n \xrightarrow{NaBH_4}$$
$$\overset{1}{C}H_2HgOCOCF_3$$
$$\overset{4}{C}H_3\overset{3}{C}H_2\overset{2}{C}H(CH_3)\overset{1}{O}C_3H_7\text{-}n$$

This is better than Williamson synthesis because there is no competing elimination reaction.

(c) Dehydration; Cyclohexyl—OH $\xrightarrow{H_2SO_4}$ (Cyclohexyl)$_2$O. This is a simple ether.

Problem 14.7 Show steps for the following syntheses: (a) $(CH_3)_2CH-O-CH_2CH(OH)CH_2OH$ from propylene; (b) $CH_3CH_2-O-CH_2C_6H_4NO_2\text{-}p$ from toluene and aliphatic alcohols. ◄

(a) The glycol is prepared by mild oxidation of the unsaturated ether in Problem 14.5.

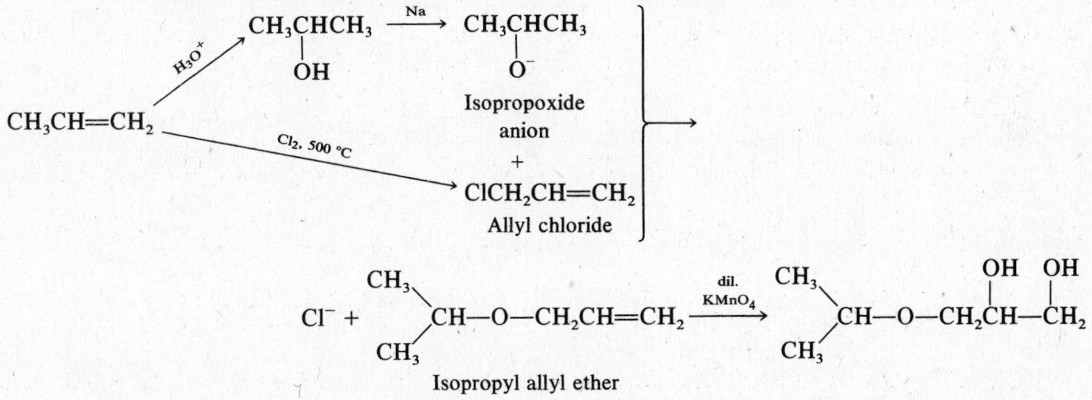

(b)
$$C_6H_5CH_3 \xrightarrow[H_2SO_4]{HNO_3} p\text{-}NO_2C_6H_4CH_3 \xrightarrow[uv]{Cl_2} p\text{-}NO_2C_6H_4CH_2Cl \xrightarrow{C_2H_5O^-Na^+} p\text{-}NO_2C_6H_4CH_2OC_2H_5$$

This method requires fewer steps than does $p\text{-}NO_2C_6H_4CH_2O^- + C_2H_5Br$.

Problem 14.8 (R)-2-octanol and its ethyl ether are levorotatory. Predict the configuration and sign of rotation of the ethyl ether prepared from this alcohol by: (a) reacting with Na and then C_2H_5Br; (b) reacting in a solvent of low dielectric constant with concentrated HBr and then with $C_2H_5O^-Na^+$. ◄

(a) No bond to the chiral C of the alcohol is broken in this reaction; hence the R configuration is unchanged and the ether is levorotatory. (b) These conditions for the reaction of the alcohol with HBr favor an S_N2 mechanism and the chiral C is inverted. Attack by RO^- is also S_N2 and the net result of two inversions is retention of configuration.

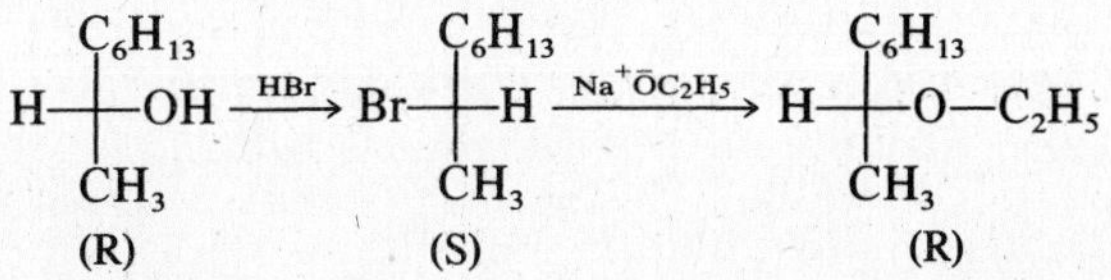

14.3 CHEMICAL PROPERTIES

Ethers are basic because of the unshared electron pairs on O. See Problem 13.14(a).

Ethers are cleaved by concentrated HI (ROR + HI $\longrightarrow$ ROH + RI) and with excess HI (ROH $\xrightarrow{HI}$ RI). They also undergo free-radical substitution at the α carbon.

Problem 14.9 (*a*) Show how the cleavage of ethers with HI can proceed by an S_N2 or an S_N1 mechanism. (*b*) Why is HI a better reagent than HBr for this type of reaction?   ◄

(*a*)

        Step 1 $R\!-\!O\!-\!R' + HI \longrightarrow R\!-\!\overset{+}{\underset{H}{O}}\!-\!R' + I^-$

               Base$_1$ Acid$_2$ Acid$_1$ Base$_2$

        Step 2 for S_N2 $I^- + R\!-\!\overset{\overset{H}{|}}{\underset{+}{O}}\!-\!R' \xrightarrow{\text{slow}} RI + HOR'$ (R is 1°)

        Step 2 for S_N1 $R\overset{\overset{H}{|}}{\underset{+}{O}}R' \xrightarrow{\text{slow}} R^+ + R'OH$ (R is 3°)

        Step 3 for S_N1 $R^+ + I^- \longrightarrow RI$

(*b*) HI is a stronger acid than HBr and gives a greater concentration of the oxonium ion

$$R\!-\!\overset{+}{\underset{H}{O}}\!-\!R$$

I^- is also a better nucleophile in the S_N2 reaction than is Br^-.

Problem 14.10 Account for the following observations:

$$(CH_3)_3COCH_3 \begin{cases} \xrightarrow[\text{ether}]{\text{anhyd. HI,}} CH_3I + (CH_3)_3COH \\ \xrightarrow[\text{(2)}]{\text{aq. HI}} CH_3OH + (CH_3)_3CI \end{cases}$$

    ◄

The high polarity of the solvent (H_2O) in reaction (2) favors an S_N1 mechanism giving the 3° R^+.

$$CH_3\overset{\overset{H}{|}}{\underset{+}{O}}C(CH_3)_3 \longrightarrow CH_3OH + (CH_3)_3C^+ \xrightarrow{I^-} (CH_3)_3CI$$

The low polarity of solvent (ether) in reaction (1) favors the S_N2 mechanism and the nucleophile, I^-, attacks the 1° C of CH_3.

$$I^- + CH_3\!\!\overset{\overset{H}{|}}{\underset{+}{O}}\!C(CH_3)_3 \longrightarrow CH_3I + HOC(CH_3)_3$$

Problem 14.11 Why does ArOR cleave to give RI and ArOH rather than ArI and ROH?   ◄

S_N2 attack by I^- does not occur readily on a C of a benzene ring nor does $C_6H_5^+$ form by an S_N1 reaction. Therefore, ArI cannot be a product.

Problem 14.12 Why do free-radical substitution reactions of ethers occur preferentially on the α carbon? ◄

The intermediate α alkyl radical $R\dot{C}HOR$ is stabilized by delocalization of electron density by the adjacent O through extended π bonding. In resonance terms, we write

$$R\overset{\cdot}{\underset{\underset{H}{|}}{C}}\!-\!\ddot{O}\!-\!R \longleftrightarrow R\underset{\underset{H}{|}}{C}\!=\!\overset{\cdot}{\underset{}{O}}\!-\!R$$

Problem 14.13 (a) Give a mechanism for the formation of the explosive solid hydroperoxides, e.g.

$$\overset{\text{OOH}}{\underset{|}{\text{RCHOCH}_2\text{R}}}$$

from ethers and O_2. (b) Why should ethers be purified before distillation? ◄

(a)   **Initiation Step** $RCH_2OCH_2R + \cdot\ddot{O}{-}\ddot{O}\cdot \longrightarrow R\dot{C}HOCH_2R + H{-}\ddot{O}{-}\ddot{O}\cdot$

Propagation Step 1 $R\dot{C}HOCH_2R + \cdot\ddot{O}{-}\ddot{O}\cdot \longrightarrow \underset{\underset{\text{O}{-}\text{O}\cdot}{|}}{RCHOCH_2R}$

Propagation Step 2 $\underset{\underset{\text{O}{-}\text{O}\cdot}{|}}{RCHOCH_2R} + RCH_2OCH_2R \longrightarrow \underset{\underset{\text{O}{-}\text{OH}}{|}}{RCHOCH_2R} + R\dot{C}HOCH_2R$

(b) An ether may contain hydroperoxides which concentrate as the ether is distilled and which may then explode. Ethers are often purified by mixing with $FeSO_4$ solution, which reduces the hydroperoxides to the nonexplosive alcohols (ROOH → ROH).

Problem 14.14 Does peroxide formation occur more rapidly with $(RCH_2)_2O$ or $(R_2CH)_2O$? ◄

With $(R_2CH)_2O$ because the 2° radical is more stable and forms faster.

Problem 14.15 How can $C_2H_5OC_2H_5$ and $n\text{-}C_4H_9OH$ be distinguished by (a) chemical reactions, (b) spectral methods? ◄

(a) $n\text{-}C_4H_9OH$ gives a positive test with CrO_3 in acid and evolves H_2 when Na is added. Dry ethyl ether is negative to both tests. (b) The ir spectrum of $n\text{-}C_4H_9OH$ shows an O—H stretching bond at about 3500 cm^{-1}.

Problem 14.16 Give a chemical test to distinguish C_5H_{12} from $(C_2H_5)_2O$. ◄

Unlike C_5H_{12}, $(C_2H_5)_2O$ is basic and dissolves in concentrated H_2SO_4.

$$(C_2H_5)_2O + H_2SO_4 \longrightarrow (C_2H_5)_2OH^+ + HSO_4^-$$

14.4 GLYCOLS

PREPARATION OF 1,2-GLYCOLS

1. Oxidation of Alkenes (see Table 6-1 and Problems 6.30 and 6.31)

2. Hydrolysis of Alkylene Dihalides and Halohydrins

$$\underset{\underset{\text{Cl \quad Cl}}{| \quad |}}{R{-}CH{-}CH_2} \xrightarrow{H_2O,\, OH^-} \underset{\underset{\text{OH \quad OH}}{| \quad |}}{R{-}CH{-}CH_2} \xleftarrow{H_2O,\, OH^-} \underset{\underset{\text{Cl \quad OH}}{| \quad |}}{R{-}CH{-}CH_2}$$

$$\textit{vic}\text{-dihalide} \qquad\qquad\qquad\qquad\qquad\qquad \text{1,2-haloalcohol (halohydrin)}$$

3. Hydrolysis of Olefin Oxides

$$\underset{\text{O}}{RCH{-\!\!\!-\!\!\!-}CH_2} \xrightarrow{H_3O^+} \underset{\underset{\text{OH \quad OH}}{| \quad |}}{RCH{-}CH_2}$$

4. Reduction of Carbonyl Compounds

Symmetrical 1,2-glycols, known as **pinacols**, are prepared by bimolecular reduction of aldehydes or ketones.

$$\underset{\underset{\text{O}}{\|}}{R{-}C{-}R'} + \underset{\underset{\text{O}}{\|}}{R{-}C{-}R'} \xrightarrow{\text{Mg in ether}} \underset{\underset{\text{O}^- \text{ Mg}^{2+} \text{ O}^-}{| \qquad\quad |}}{\overset{\overset{\text{R}' \qquad\quad \text{R}'}{| \qquad\quad |}}{R{-}C{-\!\!\!-\!\!\!-\!\!\!-}C{-}R}} \xrightarrow{H_2O} \underset{\underset{\text{OH \quad OH}}{| \qquad |}}{\overset{\overset{\text{R}' \quad\; \text{R}'}{| \qquad |}}{R{-}C{-\!\!\!-\!\!\!-}C{-}R}} + Mg(OH)_2$$

$$\textit{a pinacol}$$

Problem 14.17   What compound would you use to prepare 2,3-diphenyl-2,3-butanediol,

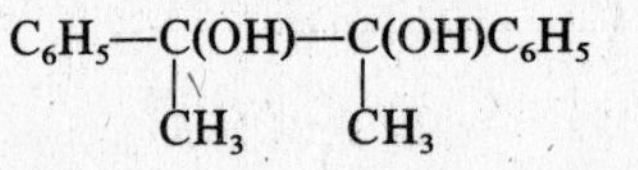

$$C_6H_5-C(OH)-C(OH)C_6H_5$$
$$\quad\quad\quad | \quad\quad\quad | $$
$$\quad\quad\quad CH_3 \quad\quad CH_3$$

by   (*a*) halide hydrolysis?   (*b*) bimolecular reduction of a carbonyl compound?   ◄

  (*a*) $C_6H_5C(CH_3)ClC(CH_3)ClC_6H_5$ or $C_6H_5C(CH_3)ClC(CH_3)OHC_6H_5$,   (*b*) $C_6H_5COCH_3$.

UNIQUE REACTIONS OF GLYCOLS

1.   Periodic Acid (HIO_4) Oxidative Cleavage

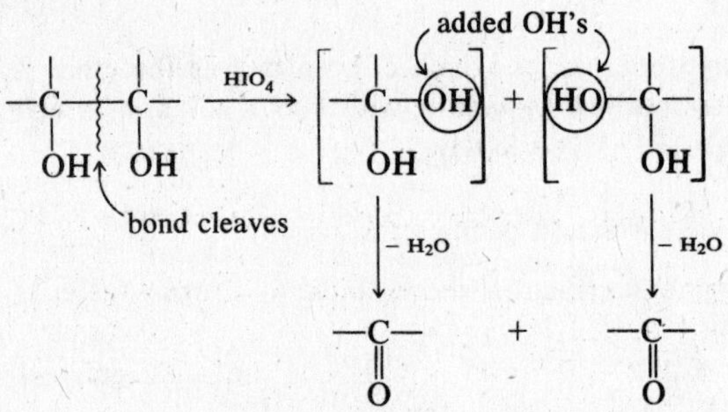

A 1° OH yields $H_2C{=}O$; a 2° OH an aldehyde, $RCHO$; a 3° OH a ketone, $R_2C{=}O$.   In polyols, if 2 vicinal OH's are termed an "adjacency," the number of moles of HIO_4 consumed is the number of such adjacencies.

Problem 14.18   Give the products resulting from periodate cleavage of the following glycols:
  (*a*) H_2COHCH_2OH,   (*b*) $CH_3CHOHCH_2OH$,   (*c*) $(CH_3)_2COHCHOHCH_3$,   (*d*) $HOCH_2CH_2CH_2OH$.   ◄

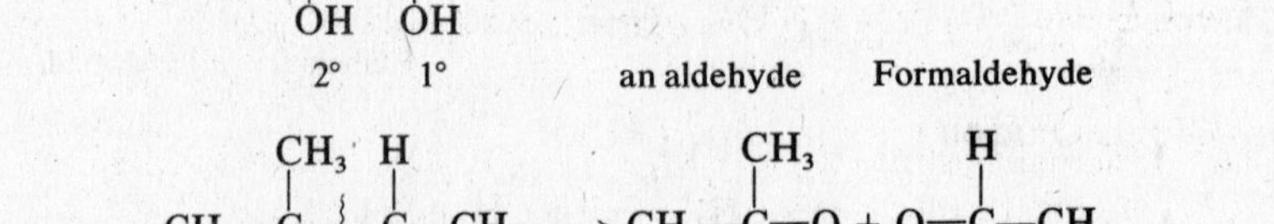

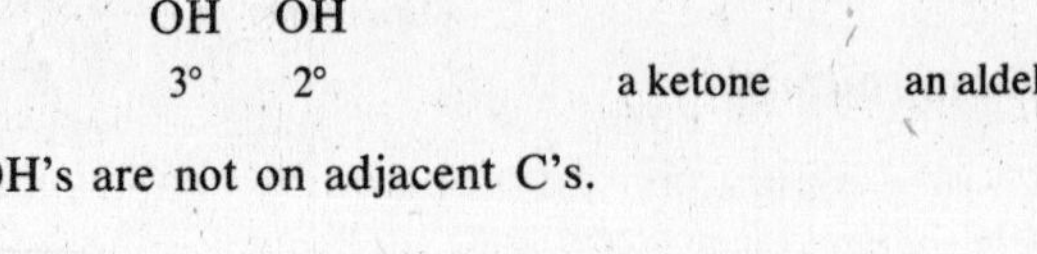

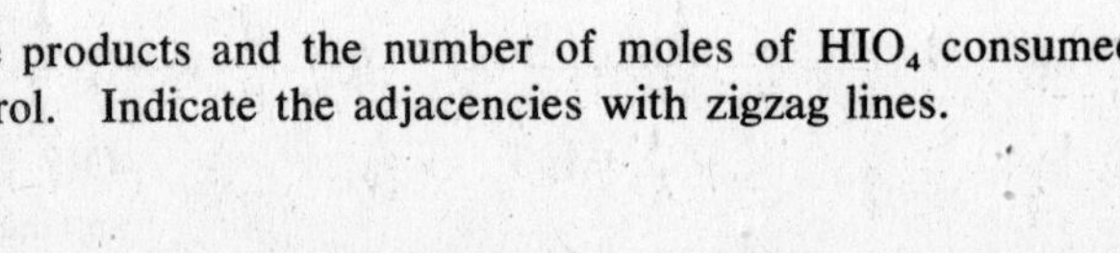

(*d*)   No reaction.   The OH's are not on adjacent C's.

Problem 14.19   Give the products and the number of moles of HIO_4 consumed in the reaction with 2,4-dimethyl-2,3,4,5-hexanetetrol.   Indicate the adjacencies with zigzag lines.   ◄

$$
\begin{array}{c}
\text{H} \quad\ \text{CH}_3 \quad \text{H} \quad\ \text{CH}_3 \\
| \qquad | \qquad | \qquad | \\
\text{CH}_3\!-\!\overset{}{\text{C}}\!\!\Big|\!\!-\!\overset{}{\text{C}}\!\!\Big|\!\!-\!\overset{}{\text{C}}\!\!\Big|\!\!-\!\overset{}{\text{C}}\!-\!\text{CH}_3, \quad \text{3 moles of HIO}_4 \\
| \qquad | \qquad | \qquad | \\
\text{OH} \quad \text{OH} \quad \text{OH} \quad \text{OH}
\end{array}
$$

$$
\text{CH}_3\overset{}{\text{C}} + \left[\ \text{HO}\!-\!\overset{\text{CH}_3}{\underset{\text{OH}}{\text{C}}}\!-\!\text{OH}\ \right] + \left[\ \text{HO}\!-\!\overset{\text{H}}{\underset{\text{OH}}{\text{C}}}\!-\!\text{OH}\ \right] + \overset{\text{CH}_3}{\underset{\text{O}}{\text{C}}}\!-\!\text{CH}_3
$$

| Acetaldehyde | $-\,\text{H}_2\text{O}$ | $-\,\text{H}_2\text{O}$ | Acetone |

$$
\overset{\text{CH}_3}{\underset{\text{OH}}{\overset{|}{\text{C}}=\text{O}}} \qquad\qquad \overset{\text{H}}{\underset{\text{OH}}{\overset{|}{\text{C}}=\text{O}}}
$$

Acetic acid   Formic acid

Note that the middle —C—OH's are oxidized to a —COOH because C—C bonds on both sides are cleaved.

Problem 14.20 Periodate cleavage also occurs with α-hydroxy ketones and α-diketones. Give the products formed from (a) $(\text{CH}_3)_2\text{COHCC}_2\text{H}_5$, (b) $\text{C}_6\text{H}_5\text{C}\!-\!\text{C}\!-\!\text{CH}_3$.
  (with C=O groups below)

(a)
$$
\text{CH}_3\!-\!\overset{\text{CH}_3}{\underset{\text{OH}}{\text{C}}}\!\!\Big|\!\!-\!\overset{}{\underset{\text{O}}{\text{C}}}\!-\!\text{C}_2\text{H}_5 \xrightarrow{\text{HIO}_4} \text{CH}_3\!-\!\overset{\text{CH}_3}{\text{C}}\!=\!\text{O} + \text{HO}\overset{}{\underset{\text{O}}{\text{C}}}\!-\!\text{C}_2\text{H}_5
$$

a ketone   a carboxylic acid

(b)
$$
\text{C}_6\text{H}_5\!-\!\overset{}{\underset{\text{O}}{\text{C}}}\!\!\Big|\!\!-\!\overset{}{\underset{\text{O}}{\text{C}}}\!-\!\text{CH}_3 \xrightarrow{\text{HIO}_4} \text{C}_6\text{H}_5\!-\!\overset{}{\underset{\text{O}}{\text{C}}}\!-\!\text{OH} + \text{H}\!-\!\text{O}\!-\!\overset{}{\underset{\text{O}}{\text{C}}}\!-\!\text{CH}_3
$$

Problem 14.21 Which glycol is oxidized by HIO_4 to (a) $\text{H}_2\text{C}\!=\!\text{O}$ and $\text{CH}_3\text{CH}(\text{CH}_3)\text{CH}_2\text{CHO}$? (b) $(\text{CH}_3)_2\text{C}\!=\!\text{O}$ and $\text{CH}_3\text{CH}_2\text{CHO}$? ◀

Write structures for the products with C=O groups aligned vertically.

$$
\text{CH}_3\!-\!\text{CH}\!-\!\text{CH}_2\!-\!\overset{\text{H}}{\underset{\text{O}}{\text{C}}} \quad \text{and} \quad \overset{\text{H}}{\underset{\text{O}}{\text{C}}\!-\!\text{H}}
$$
  (with CH$_3$ below first CH)

Join the structures through the C=O C's and change the =O to OH to give:

(a) $\text{CH}_3\!-\!\text{CH}\!-\!\text{CH}_2\!-\!\text{CH}\!\!\Big|\!\!\text{CH}_2$
  (with CH$_3$, OH, OH below)

(b) $\text{CH}_3\!-\!\overset{\text{CH}_3}{\underset{\text{OH}}{\text{C}}}\!\!\Big|\!\!-\!\underset{\text{OH}}{\text{CHCH}_2\text{CH}_3}$

4-Methyl-1,2-pentanediol   2-Methyl-2,3-pentanediol

Problem 14.22 Which compound reacts with 2 moles of HIO_4 to give an equimolar mixture of $\text{C}_6\text{H}_5\text{CHO}$, HCOOH, and CH_3CHO? ◀

Oxidation by 2 moles of HIO_4 to produce 3 moles of product indicates a **triol** with two adjacencies. HCOOH comes from oxidation of the middle C, and the aldehydes from those on either side of this C.

$$
\text{C}_6\text{H}_5\!-\!\text{CH}\!\!\Big|\!\!\text{CH}\!\!\Big|\!\!\text{CH}\!-\!\text{CH}_3 \xrightarrow{2\text{HIO}_4} \text{C}_6\text{H}_5\!-\!\overset{}{\underset{\text{O}}{\text{CH}}} + \text{HC}\!-\!\text{OH} + \text{H}\!-\!\overset{}{\underset{\text{O}}{\text{C}}}\!-\!\text{CH}_3
$$
  (with OH OH OH below)

2. Pinacol Rearrangement

Acidification of glycols produces an aldehyde or a ketone by rearrangement.   There are 4 steps:
(1) protonation of an OH;   (2) loss of H_2O to form an R^+;   (3) 1,2-shift of :H, :R or :Ar to form
a more stable cation;   (4) loss of H^+ to give product.

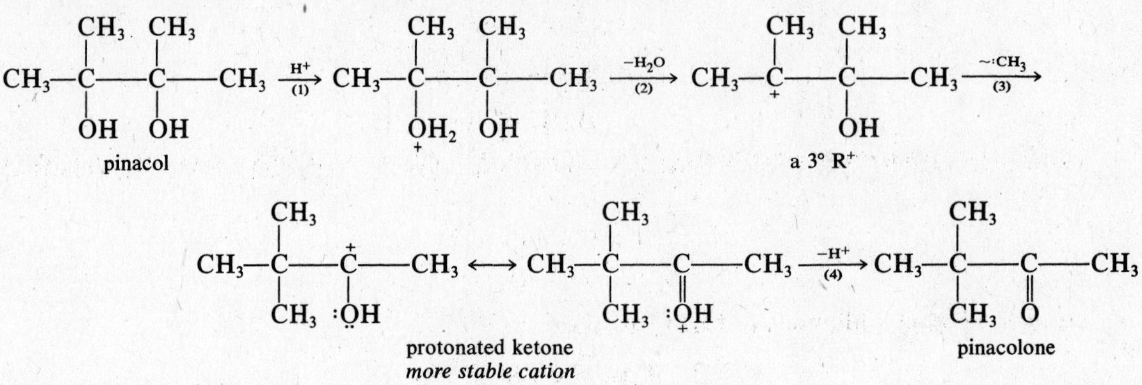

$$
\begin{array}{c}
\underset{\substack{|\\OH}}{\overset{\substack{CH_3\\|}}{CH_3-C}}\underset{\substack{|\\OH}}{\overset{\substack{CH_3\\|}}{-C}}-CH_3 \xrightarrow[(1)]{H^+}
\underset{\substack{|\\\overset{OH_2}{+}}}{\overset{\substack{CH_3\\|}}{CH_3-C}}\underset{\substack{|\\OH}}{\overset{\substack{CH_3\\|}}{-C}}-CH_3 \xrightarrow[(2)]{-H_2O}
\underset{\substack{|\\+}}{\overset{\substack{CH_3\\|}}{CH_3-C}}\underset{\substack{|\\OH}}{\overset{\substack{CH_3\\|}}{-C}}-CH_3 \xrightarrow[(3)]{\sim:CH_3}
\end{array}
$$

pinacol   a 3° R^+

$$
\underset{\substack{|\\CH_3}}{\overset{\substack{CH_3\\|}}{CH_3-C}}\underset{\substack{|\\:OH}}{\overset{\substack{+\\|}}{-C}}-CH_3 \longleftrightarrow
\underset{\substack{|\\CH_3}}{\overset{\substack{CH_3\\|}}{CH_3-C}}\underset{\substack{\|\\\overset{:OH}{+}}}{\overset{\substack{\\}}{-C}}-CH_3 \xrightarrow[(4)]{-H^+}
\underset{\substack{|\\CH_3}}{\overset{\substack{CH_3\\|}}{CH_3-C}}\underset{\substack{\|\\O}}{\overset{\substack{\\}}{-C}}-CH_3
$$

protonated ketone   pinacolone
more stable cation

With unsymmetrical glycols, the product obtained is determined mainly by which OH is lost as H_2O
to give the more stable R^+, and thereafter by which group migrates.   The order of **migratory aptitude**
is Ar > H or R.

Problem 14.23 Give structural formulas for the major products from the pinacol rearrangement of:
(a) 2-methyl-2,3-butanediol, (b) 1,1-diphenyl-2-methyl-1,2-propanediol, (c) 1,1,2-triphenyl-1,2-propanediol.
Indicate the protonated OH and the migrating groups. ◀

(a)

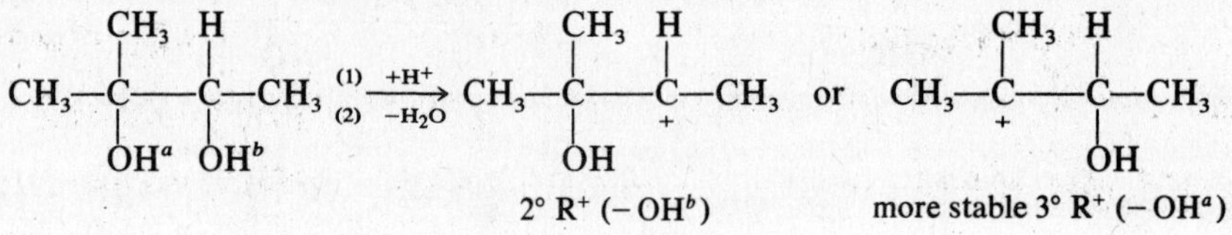

$$
\underset{\substack{|\\OH^a}}{\overset{\substack{CH_3\\|}}{CH_3-C}}\underset{\substack{|\\OH^b}}{\overset{\substack{H\\|}}{-C}}-CH_3 \xrightarrow[(2)\ -H_2O]{(1)\ +H^+}
\underset{\substack{|\\OH}}{\overset{\substack{CH_3\\|}}{CH_3-C}}\underset{\substack{|\\+}}{\overset{\substack{H\\|}}{-C}}-CH_3 \quad or \quad
\underset{\substack{|\\+}}{\overset{\substack{CH_3\\|}}{CH_3-C}}\underset{\substack{|\\OH}}{\overset{\substack{H\\|}}{-C}}-CH_3
$$

2° R^+ $(-OH^b)$ more stable 3° R^+ $(-OH^a)$

The preferred 3° R^+ undergoes rearrangement of :CH_3 to give

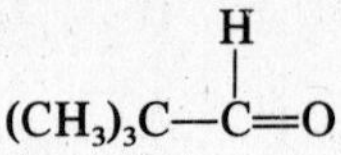

$$(CH_3)_3C-\overset{\overset{\textstyle H}{|}}{C}=O \quad (I)$$

or rearrangement of :H to give

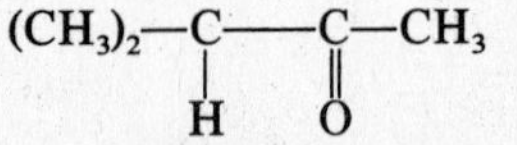

$$(CH_3)_2-\underset{\substack{|\\H}}{\overset{\substack{\\}}{C}}-\underset{\substack{\|\\O}}{\overset{\substack{\\}}{C}}-CH_3 \quad (II)$$

Since there is little difference in migratory aptitudes of H and R, both (I) and (II) are major products.
(b) Protonation of OH^a and H_2O loss gives (I), as shown below. The analogous reaction with OH^b would
give $(C_6H_5)_2C(OH)C^+(CH_3)_2$ (II). Since (I) is stabilized by two C_6H_5's, while (II) is stabilized by two
CH_3's, (I) is more stable than (II). Therefore (I) forms preferentially and yields (III).

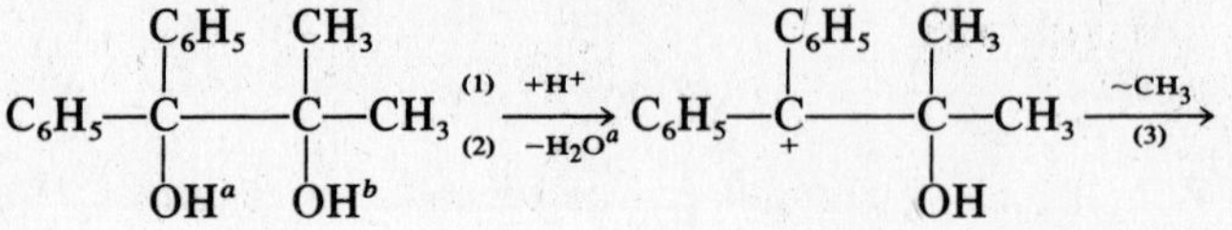

$$
\underset{\substack{|\\OH^a}}{\overset{\substack{C_6H_5\\|}}{C_6H_5-C}}\underset{\substack{|\\OH^b}}{\overset{\substack{CH_3\\|}}{-C}}-CH_3 \xrightarrow[(2)\ -H_2O^a]{(1)\ +H^+}
\underset{\substack{|\\+}}{\overset{\substack{C_6H_5\\|}}{C_6H_5-C}}\underset{\substack{|\\OH}}{\overset{\substack{CH_3\\|}}{-C}}-CH_3 \xrightarrow{\sim CH_3}
$$

(I)

$$
\underset{\substack{|\\CH_3}}{\overset{\substack{C_6H_5\\|}}{C_6H_5-C}}\underset{\substack{|\\OH}}{\overset{\substack{+\\|}}{-C}}-CH_3 \xrightarrow[(4)]{-H^+}
\underset{\substack{|\\CH_3}}{\overset{\substack{C_6H_5\\|}}{C_6H_5-C}}\underset{\substack{\|\\O}}{\overset{\substack{\\}}{-C}}-CH_3
$$

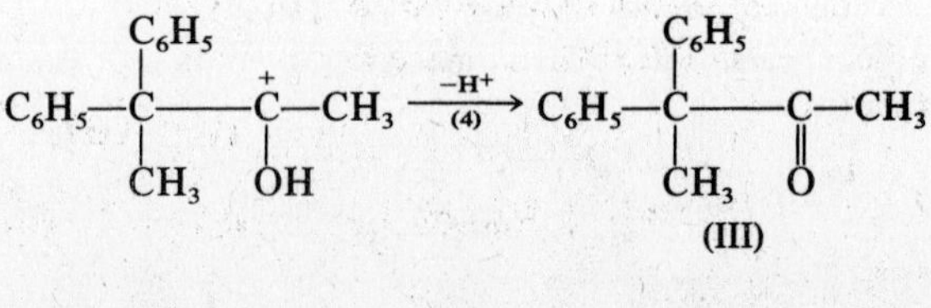

(III)

Loss of OH^b would yield

$$\underset{\underset{O}{\|}}{C_6H_5-C}-C(CH_3)_2C_6H_5 \quad (IV)$$

(c)

$$C_6H_5-\underset{\underset{C_6H_5}{|}}{\overset{\overset{OH^a}{|}}{C}}-\underset{\underset{C_6H_5}{|}}{\overset{\overset{OH^b}{|}}{C}}-CH_3$$

Loss of OH^a yields the more stable $(C_6H_5)_2C^+-C(OH)(C_6H_5)(CH_3)$. C_6H_5 rather than CH_3 migrates to form

$$(C_6H_5)_3C-\underset{\underset{O}{\|}}{C}-CH_3$$

the major product. Migration of CH_3 would give

$$(C_6H_5)_2C(CH_3)\underset{\underset{O}{\|}}{C}-C_6H_5$$

which also arises from the loss of OH^b.

14.5   SUMMARY OF GLYCOL CHEMISTRY

PREPARATION

1. Glycols

(a) Alkene oxidation

$$RCH{=}CHR \xrightarrow[\text{2. H}_2\text{O}]{\text{1. RCO}_3\text{H}} \underset{\underset{OH}{|}}{RCH}-\overset{\overset{OH}{|}}{CHR}$$

$$RCH{=}CHR \xrightarrow{\text{KMnO}_4} \underset{\underset{OH}{|}}{RCH}-\underset{\underset{OH}{|}}{CHR}$$

(b) Hydrolysis

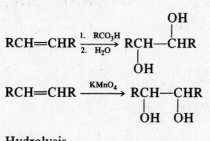

Dihalogen: $\underset{\underset{X}{|}}{RCH}-\underset{\underset{X}{|}}{CHR}$

Halohydrin: $\underset{\underset{OH}{|}}{RCH}-\underset{\underset{X}{|}}{CHR}$ $\xrightarrow{OH^-}$ $\underset{\underset{OH}{|}}{RCH}-\underset{\underset{OH}{|}}{CHR}$

Epoxide: $\underset{\underset{O}{\diagdown/}}{RCH}-CHR$

(c) Bimolecular carbonyl
     reduction

$$R_2C{=}O + Mg\,(HCl) \longrightarrow \underset{\underset{}{}}{R_2}\overset{\overset{HO}{|}}{C}-\overset{\overset{OH}{|}}{C}-R_2$$

PROPERTIES

1. Ester Formation

$$+ R'COX \longrightarrow \underset{\underset{R'COO}{|}}{RCH}-\underset{\underset{OOCR'}{|}}{CHR}$$

2. Oxidation

$$+ KMnO_4 \longrightarrow 2RCOOH$$

$$+ HIO_4 \longrightarrow 2\,RCH{=}O$$

3. Pinacol Rearrangement

$$+ H_2SO_4 \longrightarrow \underset{\underset{R}{|}}{R}\overset{\overset{R}{|}}{C}-\overset{\overset{O}{\|}}{C}-R$$

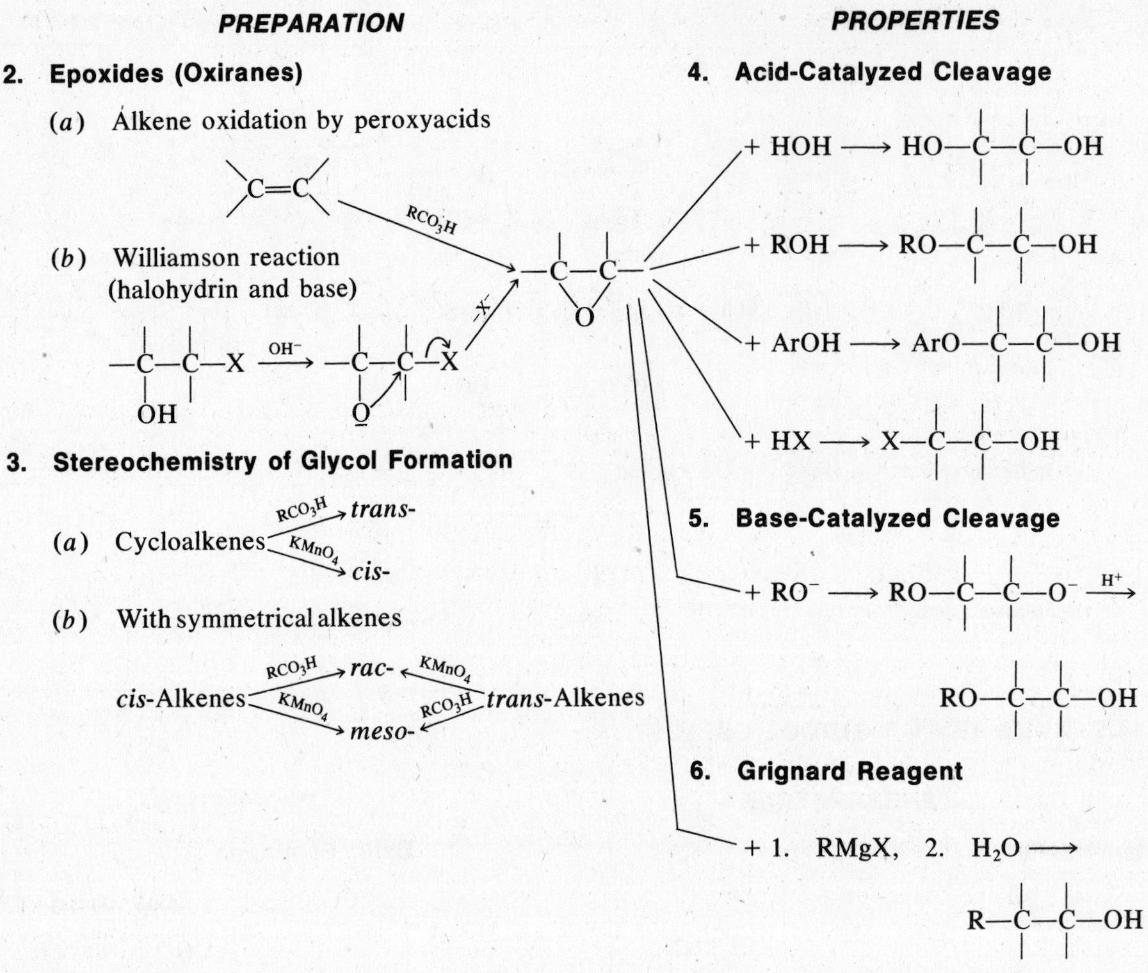

PREPARATION

2. Epoxides (Oxiranes)

(a) Alkene oxidation by peroxyacids

$$\text{C=C} \xrightarrow{RCO_3H}$$

(b) Williamson reaction (halohydrin and base)

$$-\overset{|}{\underset{OH}{C}}-\overset{|}{C}-X \xrightarrow{OH^-} -\overset{|}{C}-\overset{|}{\underset{O}{C}}-X$$

3. Stereochemistry of Glycol Formation

(a) Cycloalkenes $\overset{RCO_3H}{\underset{KMnO_4}{\longrightarrow}} \begin{array}{l} trans- \\ cis- \end{array}$

(b) With symmetrical alkenes

cis-Alkenes $\overset{RCO_3H}{\underset{KMnO_4}{\longrightarrow}} \begin{array}{l} rac- \\ meso- \end{array} \overset{KMnO_4}{\underset{RCO_3H}{\longleftarrow}}$ trans-Alkenes

PROPERTIES

4. Acid-Catalyzed Cleavage

$$-\overset{|}{C}-\overset{|}{C}- \quad \overset{|}{\underset{O}{}}$$

+ HOH $\longrightarrow$ HO—$\overset{|}{C}$—$\overset{|}{C}$—OH

+ ROH $\longrightarrow$ RO—$\overset{|}{C}$—$\overset{|}{C}$—OH

+ ArOH $\longrightarrow$ ArO—$\overset{|}{C}$—$\overset{|}{C}$—OH

+ HX $\longrightarrow$ X—$\overset{|}{C}$—$\overset{|}{C}$—OH

5. Base-Catalyzed Cleavage

+ RO$^-$ $\longrightarrow$ RO—$\overset{|}{C}$—$\overset{|}{C}$—O$^-$ $\xrightarrow{H^+}$

RO—$\overset{|}{C}$—$\overset{|}{C}$—OH

6. Grignard Reagent

+ 1. RMgX, 2. H$_2$O $\longrightarrow$

R—$\overset{|}{C}$—$\overset{|}{C}$—OH

Supplementary Problems

Problem 14.24 Give the structural formula and IUPAC name for (a) *n*-propyl propenyl ether, (b) isobutyl *tert*-butyl ether. ◄

(a) $CH_3CH_2CH_2$—O—CH=CHCH$_3$ 1-(*n*-Propoxy)-1-propene

(b) $CH_3\overset{\overset{\displaystyle CH_3}{|}}{C}HCH_2$—O—$\overset{\overset{\displaystyle CH_3}{|}}{\underset{\underset{\displaystyle CH_3}{|}}{C}}$—CH$_3$ 2-Isobutoxy-2-methylpropane

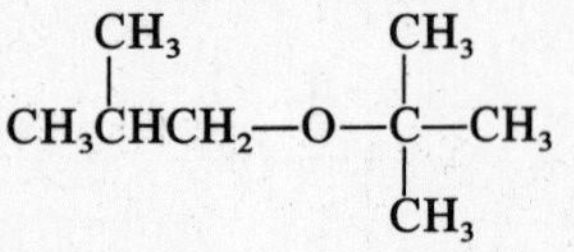

Problem 14.25 Account for the fact that [(CH$_3$)$_3$C]$_2$O cannot be prepared either by a Williamson reaction or by dehydration of (CH$_3$)$_3$COH. ◄

Reaction between (CH$_3$)$_3$CO$^-$K$^+$ and (CH$_3$)$_3$CCl results in an E2 elimination to give isobutylene. This alkene is also obtained from (CH$_3$)$_3$COH and H$_2$SO$_4$. The *tert*-butyl R$^+$ eliminates an H$^+$. Attack by (CH$_3$)$_3$C$^+$ on (CH$_3$)$_3$COH to give the ether is sterically hindered. The instability of *t*-butyl ether in sulfuric acid may also be due to steric crowding of the CH$_3$'s.

Problem 14.26 Outline the mechanism for acid- and base-catalyzed additions to ethylene oxide and give the structural formulas of the products of addition of the following: (a) H_2O, (b) CH_3OH, (c) CH_3NH_2, (d) CH_3CH_2SH. ◄

In acid, O is first protonated.

$$H:A + \underset{O}{-C-C-} \longrightarrow :A^- + \underset{\overset{+}{O}H}{-C-C-} \xrightarrow[or\ S_N1]{S_N2} \underset{A\ \ OH}{-C-C-}$$

The protonated epoxide can also react with nucleophilic solvents such as CH_3OH.

$$\underset{H}{CH_3\ddot{O}:} + \underset{\overset{+}{O}H}{-C-C-} \longrightarrow \underset{\overset{+}{CH_3OH}\ OH}{-C-C-} \xrightarrow{-H^+} \underset{CH_3O\ \ OH}{-C-C-}$$

In base, the ring is cleaved by attack of the nucleophile on the less substituted C to form an alkoxide anion, which is then protonated. Reactivity is attributed to the highly strained three-membered ring, which is readily cleaved.

$$B:^- + \underset{O}{-C-C-} \xrightarrow{S_N2} \underset{B\ \ O^-}{-C-C-} \xrightarrow{H^+} \underset{B\ \ OH}{-C-C-}$$

(a) $HOCH_2CH_2OH$, (b) $CH_3OCH_2CH_2OH$, (c) $CH_3NHCH_2CH_2OH$, (d) $CH_3CH_2SCH_2CH_2OH$.

Problem 14.27 Supply structures for compounds (A) through (F).

$$H_2C=CH_2 + (A) \longrightarrow ClCH_2CH_2OH \xrightarrow[heat]{H_2SO_4} (B) \xrightarrow{alc.\ KOH} (C)$$

$$(CH_3)_3CBr + alc.\ KOH \longrightarrow (D) \xrightarrow{HOCl} (E) \xrightarrow{NaOH} (F)$$ ◄

(A) HOCl   (B) $ClCH_2CH_2-O-CH_2CH_2Cl$ (C) $H_2C=CH-O-CH=CH_2$

(D) $(CH_3)_2C=CH_2$ (E) $\underset{OH}{(CH_3)_2C-CH_2Cl}$ (F) $\underset{O}{(CH_3)_2C-CH_2}$

Formation of (F), isobutylene oxide, is an internal S_N2 reaction.

$$\underset{OH}{(CH_3)_2C-CH_2Cl} \xrightarrow[NaOH]{conc.} \left[\underset{O^-}{(CH_3)_2C-CH_2-Cl}\right] \longrightarrow \underset{O}{(CH_3)_2C-CH_2} + Cl^-$$

Problem 14.28 Are the m/e peaks 102, 87, and 59 (base peak) consistent for n-butyl ethyl ether (A) or methyl n-pentyl ether (B)? Give the structure of the fragments which justify your answer. ◄

$$\overset{\beta'\ \ \alpha'\ \ \ \ \ \alpha\ \ \ \beta}{CH_3CH_2OCH_2CH_2CH_2CH_3},\ (A) \qquad \overset{\alpha'\ \ \ \ \alpha\ \ \ \beta}{CH_3OCH_2CH_2CH_2CH_2CH_3},\ (B)$$

The parent P^+ is $m/e = 102$, the molecular weight of the ether. The other peaks arise as follows:

$$102 - 15\ (CH_3) = 87 \qquad 102 - 43\ (C_3H_7) = 59$$

Fragmentations of P^+ ions of ethers occur mainly at the $C^\alpha-C^\beta$ bonds. (A) fits these data for $H_2\overset{+}{C}-OCH_2CH_2CH_3$ ($m/e = 87$; $C^{\alpha'}-C^{\beta'}$ cleavage) and $CH_3CH_2O-\overset{+}{C}H_2$ ($m/e = 59$; $C^\alpha-C^\beta$ cleavage). Cleavage of the $C^\beta-C^\alpha$ bonds in (B) would give a cation, $CH_3O=\overset{+}{C}H_2$ ($m/e = 45$), but this peak *was not observed*.

Problem 14.29 Prepare the following ethers starting with benzene, toluene, phenol (C_6H_5OH), cyclohexanol, any aliphatic compound of 3 C's or less and any solvent or inorganic reagent: (a) dibenzyl ether, (b) di-n-butyl ether, (c) ethyl isopropyl ether, (d) cyclohexyl methyl ether, (e) p-nitrophenyl ethyl ether, (f) divinyl ether, (g) diphenyl ether. ◄

(a) $$C_6H_5CH_3 \xrightarrow[light]{Cl_2} C_6H_5CH_2Cl \xrightarrow[H_2O]{OH^-} C_6H_5CH_2OH \xrightarrow[-H_2O]{H_2SO_4} (C_6H_5CH_2)_2O$$

(b) $$\boxed{CH_3CH_2}-MgBr + \overset{O}{\overset{\triangle}{CH_2CH_2}} \xrightarrow[2.\ H_2O]{1.\ reaction} \boxed{CH_3CH_2}-CH_2CH_2OH \xrightarrow[-H_2O]{H_2SO_4} (CH_3CH_2CH_2CH_2)_2O$$

(c) $$\underset{(1°)}{CH_3CH_2Br} + Na^+\bar{O}CH(CH_3)_2 \longrightarrow CH_3CH_2OCH(CH_3)_2 + Na^+Br^-$$

Use the 1° RX to minimize the competing E2 elimination reaction or use

$$CH_3CH{=}CH_2 + CH_3CH_2OH \xrightarrow{Hg(OCOCF_3)} CH_3CH(OCH_2CH_3)CH_2HgOCOCF_3 \xrightarrow{NaBH_4} product$$

(d) $$C_6H_{11}OH + CH_2N_2 \xrightarrow{H^+} C_6H_{11}OCH_3 + N_2 \quad or \quad C_6H_{11}OH \xrightarrow{Na} C_6H_{11}O^-Na^+ \xrightarrow{CH_3I} C_6H_{11}OCH_3$$

(e) $$C_6H_5OH \xrightarrow{NaOH} C_6H_5O^-Na^+ \xrightarrow{C_2H_5Br} C_6H_5OC_2H_5 \xrightarrow[H_2SO_4]{HNO_3} p\text{-}NO_2C_6H_4OC_2H_5$$

Williamson synthesis of an aryl alkyl ether requires the Ar to be part of the nucleophile ArO^- and *not the halide*, since ArX does not readily undergo S_N2 displacements. Note that since ArOH is much more acidic than ROH, it is converted to ArO^- by OH^- instead of by Na as required for ROH.

(f) See Problem 14.27, compounds (A), (B) and (C). Vinyl alcohol, $H_2C{=}CHOH$, cannot be used as a starting material because it is not stable and rearranges to CH_3CHO. The double bond must be introduced after the ether bond is formed.

(g) $$C_6H_6 \xrightarrow{Br_2}{Fe} C_6H_5Br \xrightarrow[\substack{Cu\ (> 200\ °C),\\ no\ solvent}]{C_6H_5O^-Na^+} (C_6H_5)_2O$$

Phenols do not undergo intermolecular dehydration. Although aryl halides cannot be used as substrates in typical Williamson syntheses, they do undergo a modified Williamson-type synthesis at higher temperature in the presence of Cu.

Problem 14.30. Prepare ethylene glycol from the following compounds: (a) ethylene, (b) ethylene oxide, (c) 1,2-dichloroethane. ◄

(a) Oxidation: $$H_2C{=}CH_2 \xrightarrow{dil.\ aq.\ KMnO_4} HOCH_2-CH_2OH$$

(b) Acid hydrolysis: $$\overset{CH_2\text{---}CH_2}{\underset{O}{\diagdown\diagup}} \xrightarrow{H_2O,\ H^+} HOCH_2-CH_2OH$$

(c) Alkaline hydrolysis: $ClCH_2-CH_2Cl \xrightarrow{H_2O,\ OH^-} HOCH_2-CH_2OH$

Problem 14.31 What structural unit in a molecule is needed to obtain CO_2 in periodate cleavage? ◄

CO_2 is the acid anhydride of carbonic acid,

$$\underset{\overset{\|}{O}}{HO-C-OH}$$

The OH's signal cleavage of 2 adjacent C—C bonds. The C=O must be present in the original substrate. A possible unit is

$$-\underset{OH}{C}\text{---}\underset{O}{C}\text{---}\underset{OH}{C}-$$

The middle C ends up in CO_2.

Problem 14.32 One mole of compound $C_6H_{14}O_5$ reacts with 4 moles of HIO_4. The moles of products are: 1 H_2CO, 1 $CH_3CH=O$, and 3 HCOOH. Supply a possible structure for $C_6H_{14}O_5$. ◄

There are 4 adjacencies. The aldehydes, $H_2C=O$ and CH_3CHO, are formed by oxidation of the terminal C—OH groups. The C—OH groups in the middle of the polyol sequence have 2 adjacencies and cleave to HCOOH. The sequence

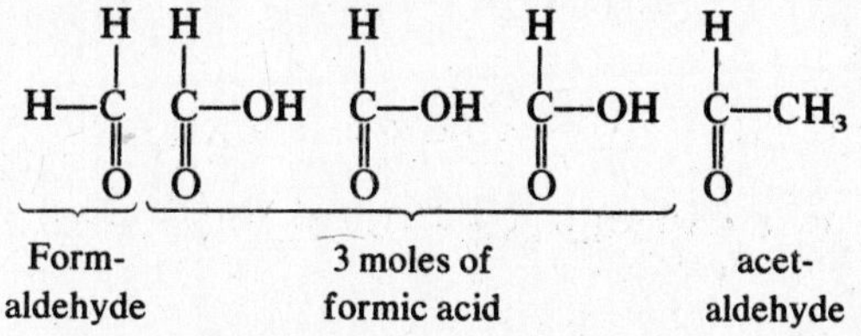

| Form-aldehyde | 3 moles of formic acid | acet-aldehyde |

gives the structure

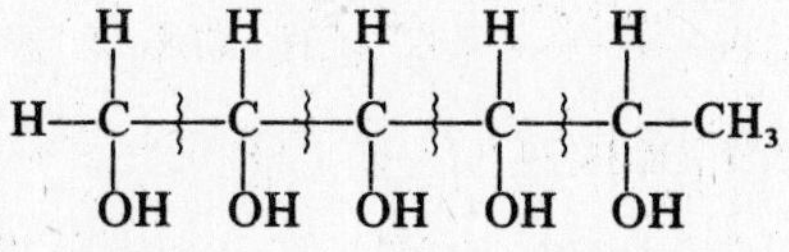

Problem 14.33 Outline the steps and give the product of pinacol rearrangement of: (a) 3-phenyl-1,2-propanediol, (b) 2,3-diphenyl-2,3-butanediol. ◄

(a)

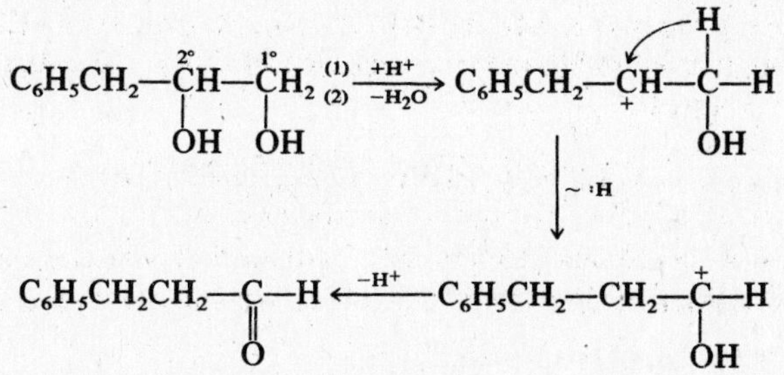

The 2° OH is protonated and lost as H_2O in preference to the 1° OH.

(b)

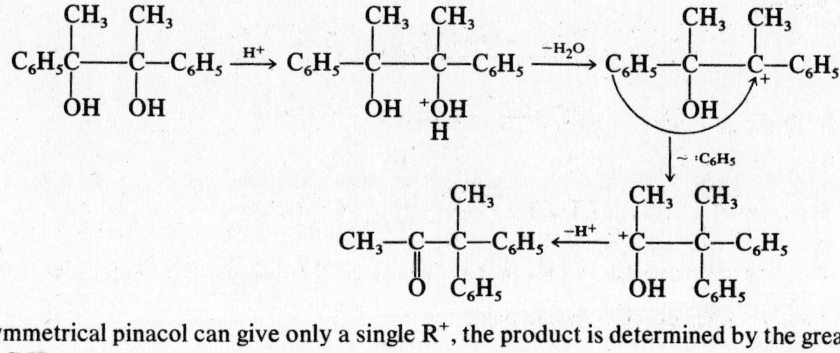

Since the symmetrical pinacol can give only a single R^+, the product is determined by the greater migratory aptitude of C_6H_5.

Problem 14.34 Show how ethylene oxide is used to manufacture the following water soluble organic solvents:

(a) Carbitol ($C_2H_5OCH_2CH_2OCH_2CH_2OH$)   (b) Diethylene glycol ($HOCH_2CH_2OCH_2CH_2OH$)

(c) Diethanolamine ($HOCH_2CH_2NCH_2CH_2OH$ with H on N)   (d) 1,4-Dioxane 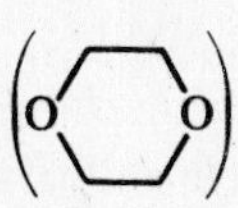 ◄

(a) 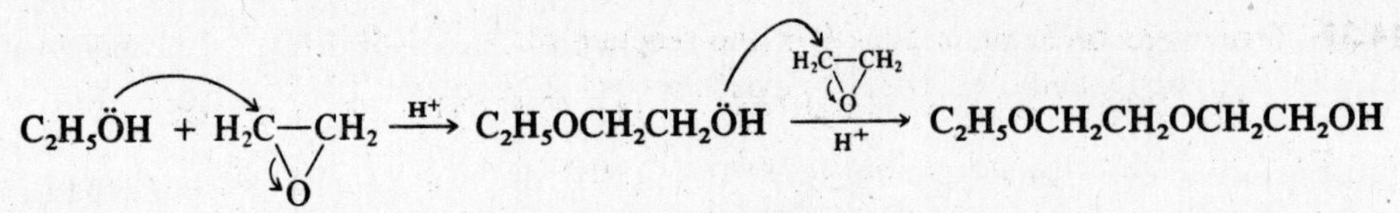　$C_2H_5\ddot{O}H + H_2\overset{\frown}{C}-CH_2 \xrightarrow{H^+} C_2H_5OCH_2CH_2\ddot{O}H \xrightarrow[H^+]{} C_2H_5OCH_2CH_2OCH_2CH_2OH$

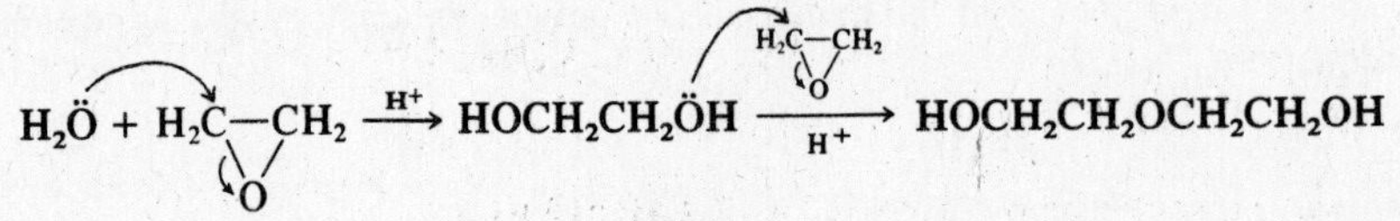

(b)　$H_2\ddot{O} + H_2\overset{\frown}{C}-CH_2 \xrightarrow{H^+} HOCH_2CH_2\ddot{O}H \xrightarrow[H^+]{} HOCH_2CH_2OCH_2CH_2OH$

(c)　$\ddot{N}H_3 + H_2\overset{\frown}{C}-CH_2 \longrightarrow HOCH_2CH_2\ddot{N}H_2 \longrightarrow HOCH_2CH_2-NH-CH_2CH_2OH$

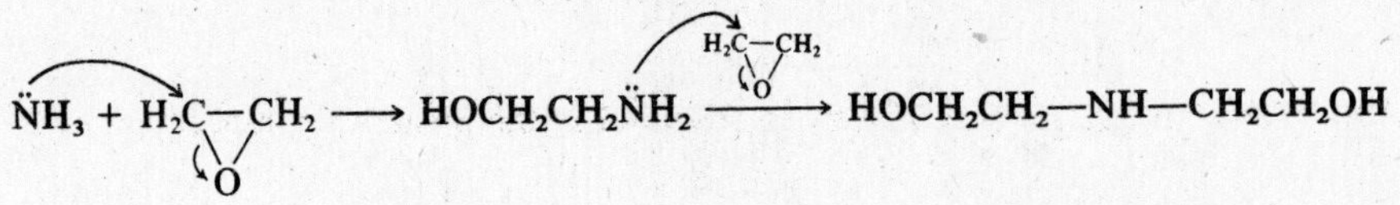

(d) 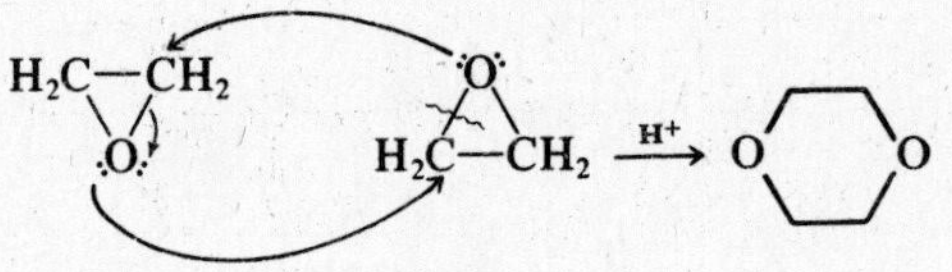　$\xrightarrow{H^+}$ (cyclic dioxane structure)

Problem 14.35 Ethers, especially those with more than one ether linkage, are also named by the **oxa method**. The ether O's are counted as C's in determining the longest hydrocarbon chain. The O is designated by the prefix **oxa-**, and a number indicates its position. Use this method to name:

(a) $(CH_3)_3COCH_2CH(CH_3)_2$　　(b) $C_2H_5OCH_2CH_2OCH_2CH_2OH$　　(c) [THF structure] 　(d) [Dioxane structure] ◄

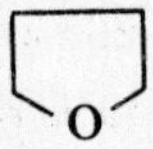

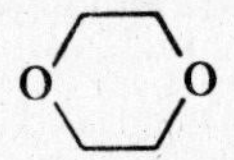

　　　　　　　　　　　　　　　　　　　　　　　　　　　Tetrahydrofuran　　　　Dioxane

(a)　$\overset{1}{CH_3}-\overset{2}{\underset{CH_3}{\overset{CH_3}{C}}}-\overset{3}{O}-\overset{4}{CH_2}-\overset{5}{\underset{CH_3}{\overset{H}{C}}}-\overset{6}{CH_3}$　　2,2,5-Trimethyl-3-oxahexane

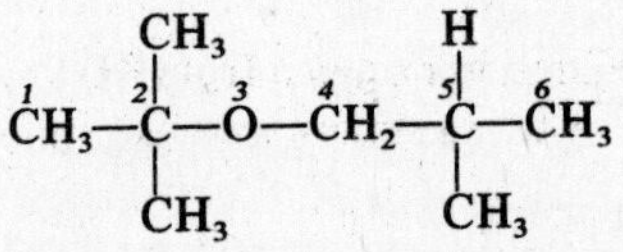

(b)　$CH_3CH_2OCH_2CH_2OCH_2CH_2OH$　　3,6-Dioxa-1-octanol.

(c)　Oxacyclopentane.　　　　(d)　1,4-Dioxacyclohexane.

Problem 14.36 Outline mechanisms to account for the different isomers formed from reaction of

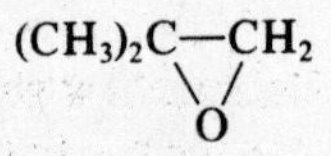

$$(CH_3)_2C-CH_2$$
$$\diagdown O \diagup$$

with CH_3OH in acidic (H^+) and in basic (CH_3O^-) media. ◄

CH_3O^- reacts by an S_N2 mechanism attacking the less substituted C.

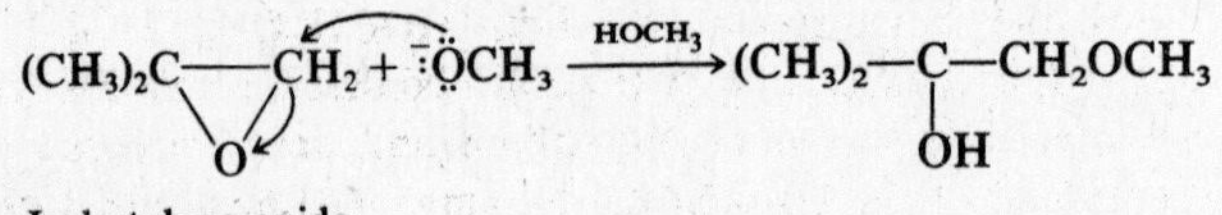

$$(CH_3)_2C-CH_2 + {:}\ddot{O}CH_3 \xrightarrow{HOCH_3} (CH_3)_2-C-CH_2OCH_3$$
$$\diagdown O \diagup \qquad\qquad\qquad\qquad \underset{OH}{|}$$

Isobutylene oxide

In acid, the S_N1 mechanism produces the more stable 3° R^+.

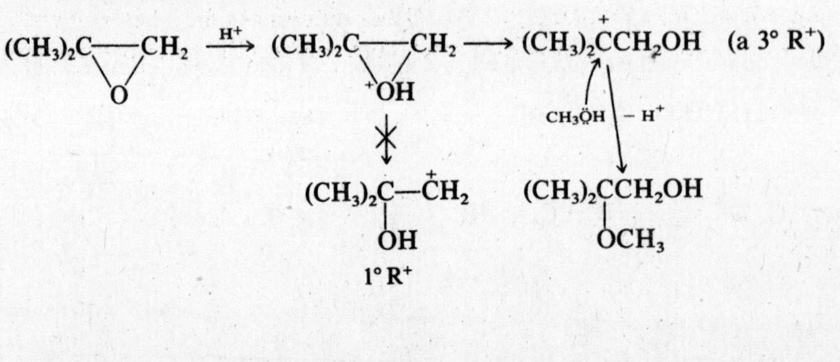

$$(CH_3)_2C-CH_2 \xrightarrow{H^+} (CH_3)_2C-CH_2 \longrightarrow (CH_3)_2\overset{+}{C}CH_2OH \quad \text{(a 3° } R^+)$$
$$\diagdown O \diagup \qquad\qquad \diagdown \overset{+}{OH} \diagup$$

$$\xcancel{\qquad}$$

$$(CH_3)_2C-\overset{+}{C}H_2 \qquad (CH_3)_2CCH_2OH$$
$$\underset{OH}{|} \qquad\qquad \underset{OCH_3}{|}$$

$$1° R^+$$

($CH_3\ddot{O}H$, $-H^+$)

Problem 14.37 Write structures for products of the reaction of:

$$(a)\ \ H_3C—CH—CH_2 + CH_3NH_2 \qquad (b)\ \ C_6H_5CH—CH_2 + dry\ HCl$$
$$\diagdown O \diagup \qquad\qquad\qquad\qquad \diagdown O \diagup$$

$$(a)\ \ CH_3—CH—CH_2\ (by\ S_N2) \qquad (b)\ \ C_6H_5—CH—CH_2OH\ (by\ S_N1)$$
$$\underset{OH\ \ NHCH_3}{} \qquad\qquad\qquad \underset{Cl}{}$$

Reaction (b) goes via the stable benzyl-type R^+, $C_6H_5\overset{+}{C}HCH_2OH$.

Problem 14.38 The following stereochemistry is observed in the pinacol rearrangement:

(R), (S) designate configuration and are used to determine whether there is retention or inversion. The C^* bonded to C^b is chiral. Describe the stereochemistry of the migrating group,

and of C^a, the migration terminus. Suggest a transition state for the migration step.

The migrating group is (S) in both reactant and product, and therefore migrates with retention of configuration. The migrating group begins to bond to the frontside of the terminus C^a before it completely leaves C^b, the migration origin.

The terminus carbon, C^a, is inverted, since its configuration changed from (R) to (S). The migrating group begins to bond to C^a to the rear of the leaving H_2O. The transition state is

Problem 14.39 A compound, $C_3H_8O_2$, gives a negative test with HIO_4. List all possible structures and show how ir and nmr spectroscopy can distinguish among them. (Note that *gem*-diols can be disregarded since they are usually not stable.)

There are no degrees of unsaturation and hence no rings or multiple bonds. The O's must be present as C—O—H and/or C—O—C. The compound can be a diol, a hydroxyether or a diether. A negative test with HIO_4 rules out a *vic*-diol. Possible structures are: a diol, $^1HOCH_2^2CH_2^2CH_2^2OH^1$ (A); two hydroxyethers, $^1HOCH_2^2CH_2^3OCH_3^4$ (B) and $^1HOCH_2^2OCH_2^3CH_3^4$ (C); and a diether (an acetal), $CH_3OCH_2OCH_3$ (D). (D) is pinpointed by ir; it has no OH, there is no O—H stretch and peaks are not observed at greater than 2950 cm^{-1}. (A) can be differentiated from (B) and (C) by nmr. (A) has only 3 kinds of equivalent H's, as labeled, while (B) and (C) each have 4. In dimethyl sulfoxide, the nmr spectrum of (C) shows all H peaks to be split: H^3, a quartet, couples H^4, a triplet; H^2, a doublet, couples H^1, a triplet. The nmr spectrum of (B) in DMSO shows a sharp singlet for H^4 integrating for 3 H's. Other differences may be observed but those described above are sufficient for identification. The DMSO used is deuterated, $(CD_3)_2SO$, to prevent interference with the spectrum.

Chapter 15

The Carbonyl Compounds

15.1 NOMENCLATURE

Carbonyl compounds have H, R, or Ar groups attached to the **carbonyl group**,

$$\text{\Large$>$C=O}$$

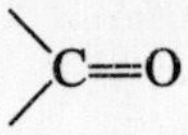

Aldehydes have at least one H bonded to the carbonyl group; ketones have only R's or Ar's.

ALDEHYDES

Common names replace the suffix **-ic** and the word **acid** of the corresponding carboxylic acids by **-aldehyde**. Locations of substituent groups are designated by Greek letters, e.g.

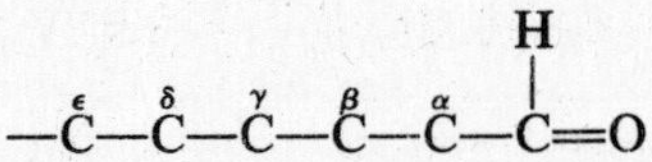

$$-\overset{\epsilon}{C}-\overset{\delta}{C}-\overset{\gamma}{C}-\overset{\beta}{C}-\overset{\alpha}{C}-\overset{H}{\underset{|}{C}}=O$$

IUPAC names use the longest chain with —CH=O and replace **e** of **alkane** by the suffix **-al**. The C of CHO is number 1. —CHO is also called the **formyl** group.

KETONES

Common names use the names of R or Ar as separate words, along with the word **ketone**. The **IUPAC system** replaces the **e** of the name of the longest chain by the suffix **-one**.

In molecules with functional groups, such as —COOH, that have a higher naming priority, the carbonyl group is indicated by the prefix **keto-**. Thus, $CH_3—CO—CH_2—CH_2—COOH$ is 4-ketopentanoic acid.

Problem 15.1 Give the common and IUPAC names for (a) CH_3CHO, (b) $(CH_3)_2CHCH_2CHO$, (c) $CH_3CH_2CH_2CHClCHO$, (d) $(CH_3)_2CHCOCH_3$, (e) $CH_3CH_2COC_6H_5$, (f) $H_2C=CHCOCH_3$. ◄

(a) Acetaldehyde (from acetic acid), ethanal; (b)

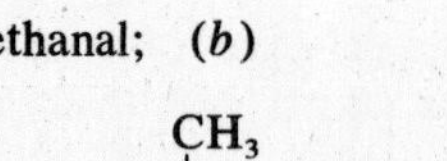

$$\overset{4}{CH_3}—\overset{\gamma}{\underset{3}{CH}}\overset{CH_3}{\overset{|}{\underset{\beta}{}}}\overset{\alpha}{\underset{2}{CH_2}}\overset{}{\underset{1}{CHO}}$$

β-methylbutyraldehyde, 3-methylbutanal; (c) α-chlorovaleraldehyde, 2-chloropentanal; (d) methyl isopropyl ketone, 3-methyl-2-butanone; (e) ethyl phenyl ketone, 1-phenyl-1-propanone (propiophenone); (f) methyl vinyl ketone, 3-buten-2-one. The C=O group has numbering priority over the C=C group.

Problem 15.2 Give structural formulas for (a) methyl isobutyl ketone, (b) phenylacetaldehyde, (c) 2-methyl-3-pentanone, (d) 3-hexenal, (e) β-chloropropionaldehyde. ◄

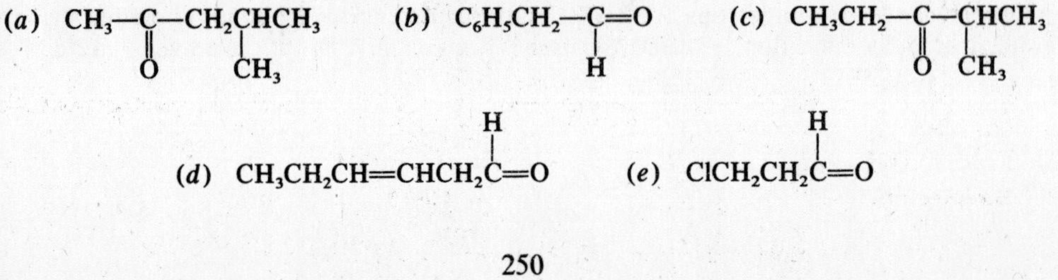

(a) $CH_3—\underset{O}{\overset{\|}{C}}—CH_2\underset{CH_3}{\overset{|}{CH}}CH_3$ (b) $C_6H_5CH_2—\underset{H}{\overset{\|}{C}}=O$ (c) $CH_3CH_2—\underset{O}{\overset{\|}{C}}—\underset{CH_3}{\overset{|}{CH}}CH_3$

(d) $CH_3CH_2CH=CHCH_2\underset{H}{\overset{\|}{C}}=O$ (e) $ClCH_2CH_2\underset{H}{\overset{\|}{C}}=O$

Problem 15.3 Write IUPAC and common names for (a) $(CH_3)_3CCHO$, (b) $C_6H_5CH=CH-CHO$, (c) $HOCH_2CH_2CH=O$, (d) $(CH_3)_2C=CHCOCH_3$, (e) $CH_3CH_2CH(CH_3)COCH_2CH_3$. ◄

(a) 2,2-dimethylpropanal, α,α-dimethylpropionaldehyde or trimethylacetaldehyde; (b) 3-phenyl-propenal, cinnamaldehyde; (c) 3-hydroxypropanal, β-hydroxypropionaldehyde; (d) 4-methyl-3-penten-2-one, mesityl oxide; (e) 4-methyl-3-hexanone, ethyl sec-butyl ketone.

Problem 15.4 n-Butyl alcohol boils at 118 °C and n-butyraldehyde boils at 76 °C, yet their molecular weights are 74 and 72, respectively. Account for the higher boiling point of the alcohol. ◄

H-bonding between alcohol molecules is responsible for the higher boiling point.

15.2 METHODS OF PREPARATION

ALDEHYDES

1. Oxidation

(a) See Problem 13.18(a) and (b).

(b) 1° Alkyl halides with dimethyl sulfoxide in base.

$$CH_3(CH_2)_6I + CH_3\overset{\displaystyle O}{\overset{\|}{S}}CH_3 \xrightarrow{HCO_3^-} CH_3(CH_2)_5CHO + CH_3SCH_3$$

(c) $ArCH_3$ to $ArCHO$.

$$ArCH_3 \xrightarrow[uv]{2Cl_2} ArCHCl_2 \xrightarrow{H_2O} ArCHO + 2HCl$$

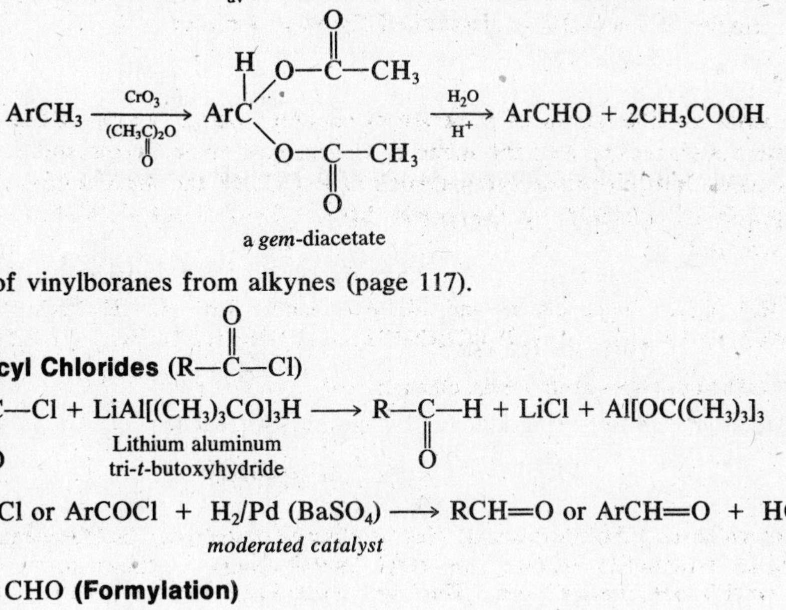

$$ArCH_3 \xrightarrow[(CH_3C)_2O]{CrO_3} ArC\begin{matrix} H \\ \\ \end{matrix}\begin{matrix} O-\overset{O}{\overset{\|}{C}}-CH_3 \\ \\ O-\underset{O}{\underset{\|}{C}}-CH_3 \end{matrix} \xrightarrow[H^+]{H_2O} ArCHO + 2CH_3COOH$$

a gem-diacetate

(d) Oxidation of vinylboranes from alkynes (page 117).

2. Reduction of Acyl Chlorides $(R-\overset{O}{\overset{\|}{C}}-Cl)$

(a) $R-\underset{O}{\underset{\|}{C}}-Cl + LiAl[(CH_3)_3CO]_3H \longrightarrow R-\underset{O}{\underset{\|}{C}}-H + LiCl + Al[OC(CH_3)_3]_3$
Lithium aluminum
tri-t-butoxyhydride

(b) $RCOCl$ or $ArCOCl + H_2/Pd (BaSO_4) \longrightarrow RCH=O$ or $ArCH=O + HCl$
moderated catalyst

3. Introduction of CHO **(Formylation)**

(a) $C_6H_5CH(CH_3)_2 + CO \xrightarrow[CuCl]{HCl, AlCl_3} p\text{-}(CH_3)_2CHC_6H_4CHO$ **(Gatterman-Koch Reaction)**
Isopropylbenzene   p-Isopropylbenzaldehyde

(b) [benzene ring with OH, OH, activated ring] $+ HCN + HCl \xrightarrow[2.\ H_2O]{1.\ ZnCl_2,\ ether}$ [benzene ring with OH, OH, CHO] $+ NH_4Cl$ **(Gatterman Reaction)**
activated ring

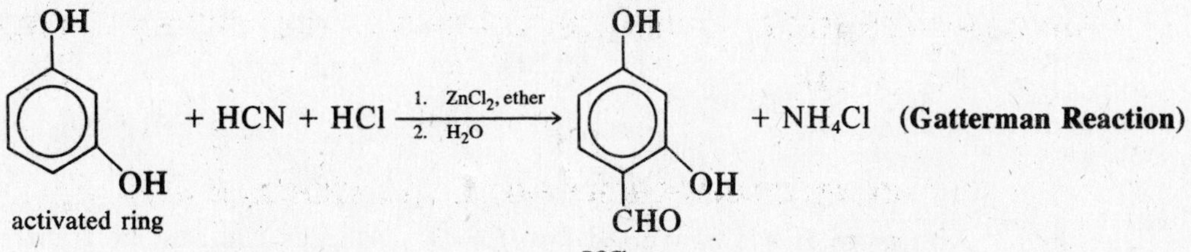

(c) $ArH + H-\underset{O}{\underset{\|}{C}}-N(CH_3)_2 \xrightarrow{POCl_3} ArCHO + (CH_3)_2NH$
activated
ring

KETONES

1. Oxidation of 2° ROH [Problem 13.35(a)]

2. Acylation of Aromatic Rings

$$\text{ArH} + \text{RCOCl} \xrightarrow{\text{AlCl}_3} \text{ArCOR} + \text{HCl}$$

3. Acylation of Alkenes

$$\text{RC}-\text{Cl} + \text{H}_2\text{C}=\text{CHR}' \xrightarrow{\text{BF}_3} \left[\; \overset{O}{\underset{\;}{R-C}}-\text{CH}_2-\overset{R'}{\underset{\;}{\text{CHCl}}} \right] \xrightarrow{-\text{HCl}} \overset{O}{\underset{\;}{R-C}}-\text{CH}=\text{CHR}'$$

This is a Markovnikov addition initiated by $\text{RC}^+{=}\overset{\cdot\cdot}{\text{O}}{:}$, an acylonium cation.

4. With Organometallics

(a) $2\,\underset{O}{R-C}-\text{Cl} + R_2'\text{Cd} \longrightarrow 2\,\underset{O}{R-C}-R' + \text{CdCl}_2$ (R' = Ar or 1° alkyl)

$R_2'\text{Cd}$ is prepared from $R'\text{MgX}$: $2R'\text{MgX} + \text{CdCl}_2 \longrightarrow R_2'\text{Cd} + 2\text{MgXCl}$.

(b) $\boxed{\text{C}_6\text{H}_5}-\text{MgBr} +$ ⬡$-\text{C}{\equiv}\text{N} \longrightarrow$ ⬡$-\underset{^-\text{N(MgBr)}^+}{\overset{\;}{\text{C}}}{-}\boxed{\text{C}_6\text{H}_5} \xrightarrow{\text{H}_3\text{O}^+}$ ⬡$-\underset{O}{\overset{\;}{\text{C}}}{-}\boxed{\text{C}_6\text{H}_5}$

       a nitrile   *an imine salt*   Cyclohexyl phenyl
                                                               ketone

5. Ring Ketones from Dicarboxylic Acids and Their Derivatives

(a) $(\text{CH}_2)_n \overset{\diagup \text{COOH}}{\underset{\diagdown \text{COOH}}{\;}} \xrightarrow[\text{heat}]{\text{BaO}} (\text{CH}_2)_n\;\text{C}{=}\text{O} + \text{BaCO}_3 + \text{H}_2\text{O}$

       dicarboxylic acids
       (n = 4 and 5)

(b) Medium- and large-ring ketones by the **acyloin reaction**.

$(\text{CH}_2)_7 \overset{\diagup \text{COOC}_2\text{H}_5}{\underset{\diagdown \text{COOC}_2\text{H}_5}{\;}} \xrightarrow{\text{Na}} (\text{CH}_2)_7 \overset{\diagup \text{CHOH}}{\underset{\diagdown \text{C}=\text{O}}{\;}} \xrightarrow[\text{HOAc}]{\text{Zn-Hg}} (\text{CH}_2)_8\;\text{C}{=}\text{O}$

   a dicarboxylic ester   *an acyloin*
                              (an α-hydroxyketone)

6. Oxidation of Alkylboranes (see Problem 6.22(f) for hydroboration)

2 (methylcyclohexene) $\xrightarrow{\text{B}_2\text{H}_6}$ $\left[\text{(methylcyclohexyl)} \text{BH}\right]_2 \xrightarrow[\text{H}_3\text{O}^+]{\text{CrO}_3} 2$ (2-methylcyclohexanone)   (78% yield)

The vinyl C with more H's is converted into C=O. Alkenes can also be transformed into dialkyl carbonyls by a carbonylation-oxidation procedure.

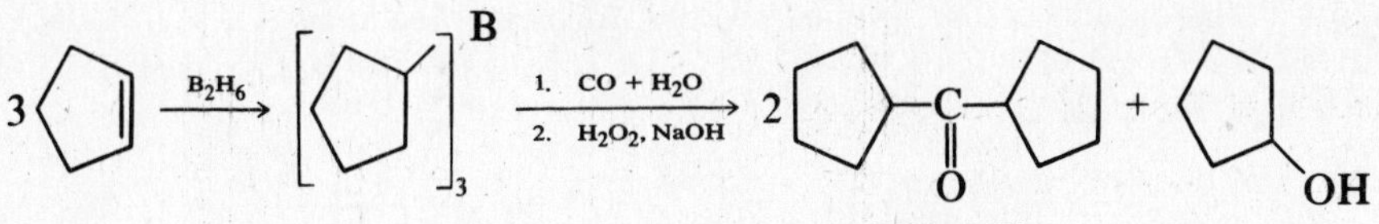

$$3 \; \bigcirc \xrightarrow{B_2H_6} \left[\bigcirc \right]_3^{B} \xrightarrow[\text{2. } H_2O_2, \text{ NaOH}]{\text{1. } CO + H_2O} 2 \; \underset{O}{\bigcirc - \overset{\shortmid}{C} - \bigcirc} + \underset{OH}{\bigcirc}$$

(90% yield)

7. Alkyne Hydration (page 117)

Problem 15.5 What conditions are needed to oxidize a 1° RCH_2OH to an aldehyde (RCHO) by acid dichromate?   ◄

RCHO must be removed from the oxidation mixture as soon as it is formed to prevent its further oxidation to RCOOH, but the alcohol must be kept.   Aldehydes with boiling points below 100 °C, the b.p. of the solvent water, can be removed by distillation.   The corresponding alcohol is always higher-boiling than the aldehyde.

Alcohols can also be oxidized by heating their vapors with hot copper (250–300 °C).

$$H - \overset{\shortmid}{\underset{\shortmid}{C}} - OH \xrightarrow[250\,°C]{Cu} \; \underset{}{}C = O + H_2$$

This method cannot be used for alcohols that are dehydrated by heat.

Problem 15.6 Why can't ketones be prepared from RCOCl and a Grignard, R'MgCl, although they can be prepared from RCOCl and R'_2Cd?   ◄

The Grignard reagent reacts faster with any ketone formed than it does with RCOCl.   The product is a 3° alcohol.

$$RCOCl + \boxed{R'}MgCl \longrightarrow [RCOR'] \xrightarrow[\text{2. } H_3O^+Cl^-]{\text{1. } R'MgCl} R - \overset{\boxed{R'}}{\underset{\boxed{R'}}{C}} - OH + MgCl_2$$

The $\overset{\delta-}{R'} - \overset{\delta+}{Cd}$ bond is more covalent than the $\overset{\delta-}{R} - \overset{\delta+}{Mg}$ bond since Cd is less electropositive (ability to lose electrons) than Mg (Problem 15.50).   R' of R'_2Cd is not nucleophilic enough to add to

$$\underset{}{}C = O$$

The Cd of R'_2Cd is sufficiently electrophilic to initiate the reaction:

$$RCOCl + R'_2Cd \longrightarrow [R'_2CdCl^- + R\overset{+}{C} = \overset{..}{O}] \longrightarrow RCOR' + R'CdCl$$

Problem 15.7 Suggest a mechanism for acylation of ArH with RCOCl in $AlCl_3$.   ◄

The mechanism is similar to that of alkylation:

$$(1) \quad RCOCl + AlCl_3 \longrightarrow R\overset{+}{C} = \overset{..}{O}: + AlCl_4^-$$

Acylonium ion

$$(2) \quad R\overset{+}{C} = \overset{..}{O}: + ArH \longrightarrow \left[\underset{\underset{O}{\overset{|}{C} - R}}{\overset{\nearrow H}{Ar}} \right]^+ \xrightarrow{-H^+} \underset{O}{Ar\overset{\shortmid}{C} - R}$$

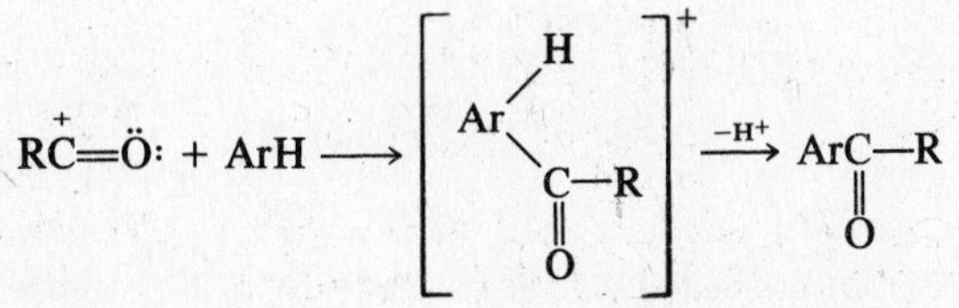

Problem 15.8 What products are formed in the following reactions? (a) CH_3CH_2OH, $Cr_2O_7^{2-}$, H^+; (b) $CH_3CHOHCH_3$, $Cr_2O_7^{2-}$, H^+ (60 °C); (c) CH_3COCl, $LiAl(O—t-C_4H_9)_3H$; (d) CH_3COCl, C_6H_6, $AlCl_3$; (e) CH_3COCl, $C_6H_5NO_2$, $AlCl_3$;

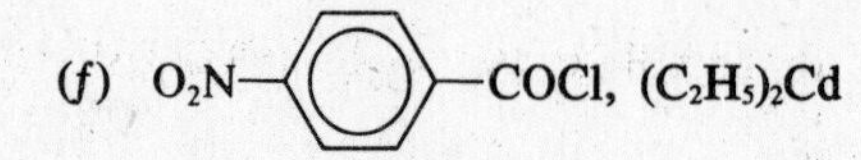

(f) O_2N—⟨○⟩—$COCl$, $(C_2H_5)_2Cd$ ◄

(a) CH_3CHO (some oxidation to CH_3COOH occurs).

$$\text{(b)} \quad CH_3-\overset{\overset{\displaystyle O}{\|}}{C}-CH_3 \qquad \text{(c)} \quad CH_3-\overset{\overset{\displaystyle H}{|}}{C}=O$$

(d) $C_6H_5COCH_3$. (e) No reaction; acylation like alkylation does not occur because NO_2 deactivates the ring.

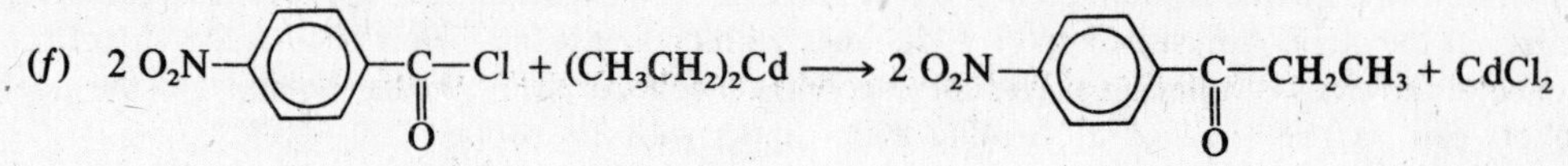

(f) $2\ O_2N$—⟨○⟩—$\overset{\|}{\underset{O}{C}}$—$Cl$ + $(CH_3CH_2)_2Cd$ ⟶ $2\ O_2N$—⟨○⟩—$\overset{\|}{\underset{O}{C}}$—$CH_2CH_3$ + $CdCl_2$

Problem 15.9 Show the substances needed to prepare the following compounds by the indicated reactions:

(a) $CH_3CH_2\overset{\overset{\displaystyle}{\|}}{\underset{O}{C}}CH_2CH_2C_6H_5$ (Grignard)   (b) $C_6H_5CH_2CH=CH—\overset{\overset{\displaystyle}{\|}}{\underset{O}{C}}—CH_2C_6H_5$ (Acylation of an alkene)

(c) $2,4-Cl_2C_6H_3COC_6H_5$ (Friedel-Crafts acylation) ◄

(a) $R'C≡N + RMgX$. The carbonyl C in $RCOR'$ and one alkyl group (R') come from $R'—C≡N$; the other R from $RMgX$. The 2 possible combinations are:

$$CH_3CH_2C≡N + ClMgCH_2CH_2C_6H_5 \quad \text{or} \quad CH_3CH_2MgBr + N≡CCH_2CH_2C_6H_5$$

(b) The R attached to $C=C$ is part of the alkene. $O=\overset{|}{C}R'$ comes from $R'COCl$.

$$C_6H_5CH_2CH=CH_2 + ClCOCH_2C_6H_5 \xrightarrow{BF_3} \text{product}$$

(c) $2,4-Cl_2C_6H_3COCl + C_6H_6 \xrightarrow{AlCl_3} \text{product}$

C_6H_5COCl and $1,3-C_6H_4Cl_2$ cannot react with $AlCl_3$ because the 2 aryl Cl's deactivate the ring.

Problem 15.10 Use hydroboration to prepare (a) cyclohexanone, (b) dicyclohexyl ketone, (c) pentanal, (d) n-butyl cyclopentyl ketone. ◄

(a) $2\ \text{Cyclohexene} + B_2H_6 \longrightarrow (\text{Cyclohexyl})_2BH \xrightarrow[H^+]{CrO_3} \text{Cyclohexanone}$

(b) $(\text{Cyclohexyl})_2BH \xrightarrow[2.\ H_2O_2,\ OH^-]{1.\ CO,\ H_2O} \text{Dicyclohexyl ketone}$

(c) $1\text{-Pentyne} + B_2H_6 \longrightarrow (CH_3CH_2CH_2CH=CH)_3B \xrightarrow[NaOH]{H_2O_2} \text{Pentanal}$

(d) $2\ \langle\text{pentene}\rangle + B_2H_6 \longrightarrow \left[\langle\text{cyclopentyl}\rangle\right]_2 BH \xrightarrow{CH_3CH_2CH=CH_2}$

$\left[\langle\text{cyclopentyl}\rangle\right]_2 B—CH_2CH_2CH_2CH_3 \xrightarrow[2.\ H_2O_2,\ OH^-]{1.\ CO,\ H_2O} CH_3CH_2CH_2CH_2—\overset{\overset{\displaystyle O}{\|}}{C}—\langle\text{cyclopentyl}\rangle$

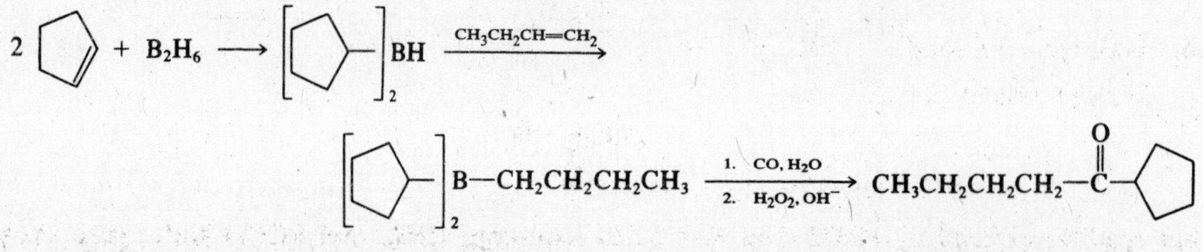

With mixed alkylboranes, the major ketone formed has the 1° R. Likewise, 2° R's predominate over 3°.

Problem 15.11 Prepare the following compounds from benzene, toluene, and alcohols of 4 or fewer C's: (a) 2-methylpropanal (isobutyraldehyde),   (b) p-chlorobenzaldehyde,   (c) p-nitrobenzophenone (p-NO$_2$C$_6$H$_4$COC$_6$H$_5$),   (d) benzyl methyl ketone,   (e) p-methylbenzaldehyde.   ◄

(a)
$$(CH_3)_2CHCH_2OH \xrightarrow[250\ °C]{Cu} (CH_3)_2CHCHO$$

(RCHO is not oxidized further.)

(b)
$$C_6H_5CH_3 \xrightarrow{Cl_2}{Fe} p\text{-}ClC_6H_4CH_3 \xrightarrow[2.\ H_3O^+]{1.\ CrO_3,\ acetic\ anhydride} p\text{-}ClC_6H_4CHO$$

(c) $C_6H_5CH_3 \xrightarrow[H_2SO_4]{HNO_3} p\text{-}O_2NC_6H_4CH_3 \xrightarrow[H^+]{KMnO_4}$

$$p\text{-}O_2NC_6H_4COOH \xrightarrow{SOCl_2} p\text{-}O_2NC_6H_4COCl \xrightarrow[AlCl_3]{C_6H_6} p\text{-}O_2NC_6H_4COC_6H_5$$

We cannot acylate C$_6$H$_5$NO$_2$ with C$_6$H$_5$COCl because NO$_2$ deactivates the ring.

(d) $C_6H_6 \xrightarrow{Br_2}{Fe} C_6H_5Br \xrightarrow{Mg} C_6H_5MgBr \xrightarrow{H_2C-CH_2 \diagdown O \diagup} C_6H_5CH_2CH_2OH \xrightarrow{KMnO_4}$

$$C_6H_5CH_2COOH \xrightarrow{SOCl_2} C_6H_5CH_2COCl \xrightarrow{(CH_3)_2Cd} C_6H_5CH_2COCH_3$$

$$CH_3CH_2OH \xrightarrow{H_2SO_4} H_2C{=}CH_2 \xrightarrow[Ag]{O_2} H_2C\overset{O}{-}CH_2$$

$$CH_3OH \xrightarrow{HBr} CH_3Br \xrightarrow{Mg} CH_3MgBr \xrightarrow{CdCl_2} (CH_3)_2Cd$$

(e)
$$C_6H_5CH_3 + CO,\ HCl \xrightarrow[CuCl]{AlCl_3} p\text{-}CH_3C_6H_4CHO$$

15.3   REACTIONS OF ALDEHYDES AND KETONES

OXIDATION

1.
$$R{-}CH{=}O \xrightarrow{KMnO_4\ or\ K_2Cr_2O_7,\ H^+} R{-}COOH$$

2.   Tollens' Reagent

A specific oxidant for RCHO is Ag(NH$_3$)$_2^+$.

$$R\overset{\displaystyle}{\underset{O}{-}}\overset{}{C}{-}H + 2Ag(NH_3)_2^+ + 3OH^- \longrightarrow R{-}COO^- + 2H_2O + 4NH_3 + 2Ag(c)$$
$$(mirror)$$

3.   Strong Oxidants

Ketones resist mild oxidation, but with strong oxidants at high temperatures they undergo cleavage of C—C bonds on either side of the carbonyl group.

$$RCH_2 \overset{(a)}{-} \underset{O}{C} \overset{(b)}{-} CH_2R' \xrightarrow{oxid.} \underline{RCOOH + R'CH_2COOH} + \underline{RCH_2COOH + R'COOH}$$

from cleavage of bond (a)   from cleavage of bond (b)

4.   Haloform Reaction

Methyl ketones,

$$CH_3\underset{O}{\overset{\|}{C}}{-}R$$

are readily oxidized by NaOI (NaOH + I$_2$) to iodoform, CHI$_3$, and RCOO$^-$Na$^+$. (See Problem 13.24.)

REDUCTION

1. To Alcohols by Metal Hydrides or H_2 (Catalyst). (See Problem 13.11.)

2.

$$\rangle C{=}O \longrightarrow \rangle CH_2 \quad \text{(also see Problem 15.29)}$$

$$R{-}\overset{\overset{\displaystyle O}{\|}}{C}{-}R' \xrightarrow[\text{or } H_2NNH_2 + KOH \text{ (Wolff-Kishner)}]{\text{Zn-Hg + HCl (Clemmensen)}} RCH_2R'$$

3. Disproportionation. Cannizzaro Reaction

Aldehydes with no H on the α C undergo self-redox (disproportionation) in hot concentrated alkali.

$$2HCHO \xrightarrow[\text{heat}]{50\%\ NaOH} CH_3OH + HCOO^-Na^+$$

$$2C_6H_5CHO \xrightarrow[\text{heat}]{50\%\ NaOH} C_6H_5CH_2OH + C_6H_5COO^-Na^+$$

$$\underset{\textit{always used}}{C_6H_5CHO + HCHO} \xrightarrow[\text{heat}]{50\%\ NaOH} C_6H_5CH_2OH + \underset{\textit{always formed}}{HCOO^-Na^+} \quad \textbf{(Crossed-Cannizzaro)}$$

Problem 15.12 Devise a mechanism for the Cannizzaro reaction from the reactions

$$2ArCDO \xrightarrow[H_2O]{OH^-} ArCOO^- + ArCD_2OH \qquad 2ArCHO \xrightarrow[D_2O]{OD^-} ArCOO^- + ArCH_2OH$$ ◄

The D's from OD^- and D_2O (solvent) are not found in the products. The molecule of ArCDO that is oxidized must transfer its D to the molecule that is reduced. A role must also be assigned to OD^-.

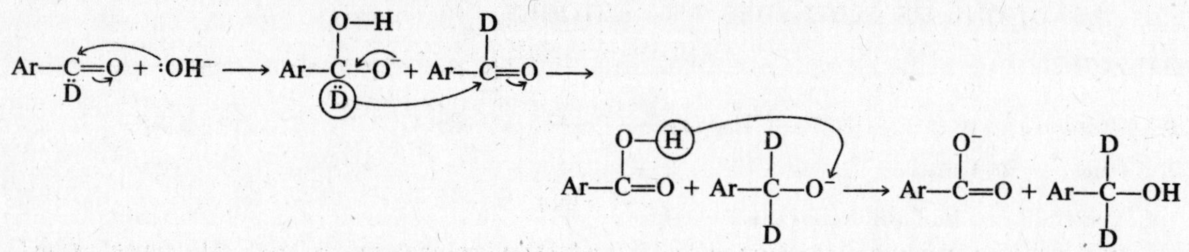

Problem 15.13 Give the products of reaction for (*a*) benzaldehyde + Tollens' reagent; (*b*) cyclohexanone + HNO_3, heat; (*c*) acetaldehyde + dilute $KMnO_4$; (*d*) phenylacetaldehyde + $LiAlH_4$; (*e*) methyl vinyl ketone + H_2/Ni; (*f*) methyl vinyl ketone + $NaBH_4$; (*g*) cyclohexanone + C_6H_5MgBr and then H_3O^+; (*h*) methyl ethyl ketone + strong oxidant; (*i*) methyl ethyl ketone + $Ag(NH_3)_2^+$. ◄

(*a*) $C_6H_5COO^-NH_4^+$, Ag° (*b*) $HOOC(CH_2)_4COOH$, (*c*) CH_3COOH, (*d*) $C_6H_5CH_2CH_2OH$, (*e*) $CH_3{-}CH(OH)CH_2CH_3$ (C=O and C=C are reduced), (*f*) $CH_3{-}CH(OH)CH{=}CH_2$ (only C=O is reduced),

(*g*)

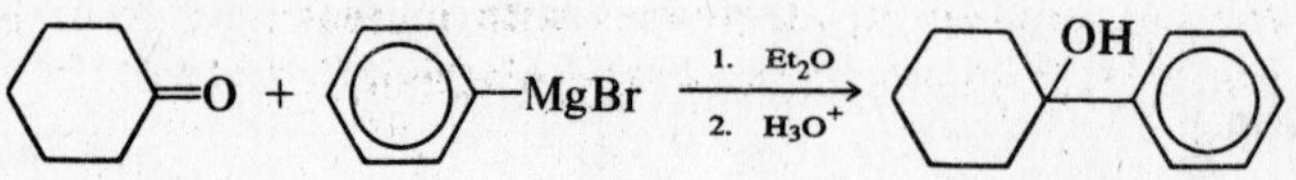

(*h*)

$$CH_3 \overset{(a)}{+} \underset{\underset{\displaystyle O}{\|}}{C} \overset{(b)}{+} CH_2CH_3 \xrightarrow{\text{hot acid } KMnO_4} \underbrace{CO_2 + CH_3CH_2COOH}_{\text{from cleavage at }(a)} + \underbrace{CH_3COOH}_{\text{from cleavage at }(b)}$$

(*i*) no reaction.

Problem 15.14 Convert cinnamaldehyde, C_6H_5—CH=CH—CH=O, to 1-phenyl-1,2-dibromo-3-chloro-propane, $C_6H_5CHBrCHBrCH_2Cl$. ◄

We must add Br_2 to C=C and convert —CHO to —CH_2Cl. Since Br_2 oxidizes —CHO to —COOH, —CHO must be converted to CH_2Cl before adding Br_2.

C_6H_5—CH=CH—CH=O $\xrightarrow[\text{2. } H_2O]{\text{1. } NaBH_4}$ C_6H_5—CH=CH—CH_2OH $\xrightarrow{PCl_3}$

C_6H_5—CH=CH—CH_2Cl $\xrightarrow{Br_2}$ C_6H_5—CH—CH—CH_2Cl
                                                          | |
                                                          Br   Br

ADDITION REACTIONS OF NUCLEOPHILES TO $\diagdown$C=O

The C of the carbonyl group is electrophilic

$$\diagdown C=\ddot{O}: \longleftrightarrow \diagdown \overset{+}{C}-\ddot{O}:^-$$

[Problem 2.23(b)] and initially forms a bond with the nucleophile.

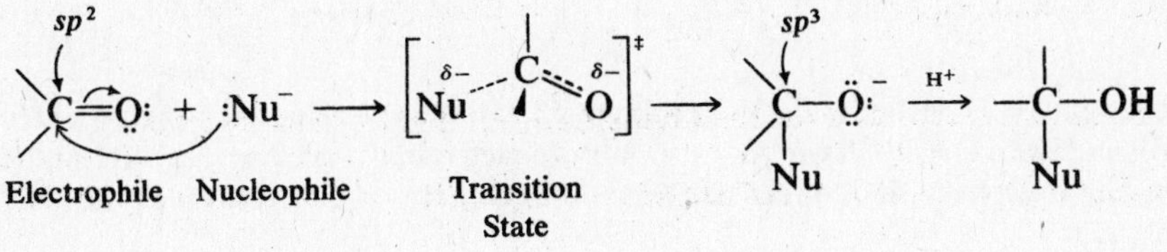

For example, :Nu$^-$ can be :R'$^-$ of R'MgX or :H$^-$ of $LiAlH_4$. With :NuH_2, the addition product

$$\left[HO-\overset{|}{\underset{|}{C}}-NuH \right]$$

dehydrates to

$$-\overset{|}{C}=Nu$$

Acid increases the rate of nucleophilic addition by first protonating the O of C=O, thus making C more electrophilic.

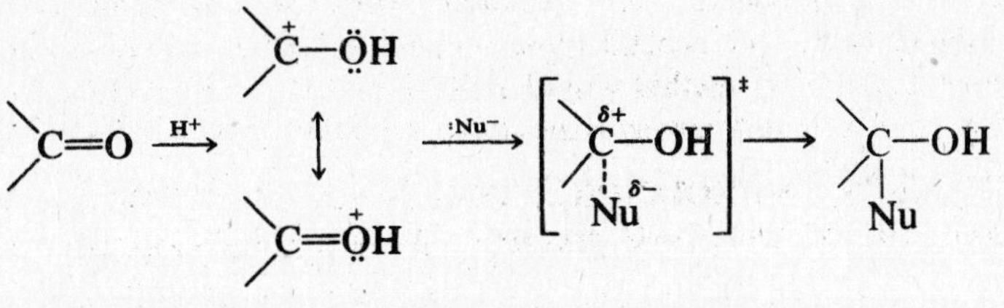

The reactivity of the carbonyl group decreases with increasing size of R's and with electron donation by R. Electron-attracting R's increase the reactivity of C=O.

Problem 15.15 The order of reactivity in nucleophilic addition is

$$CH_2=O > RCH=O > R_2C=O > R\overset{}{\underset{\overset{\|}{O}}{C}}-\ddot{Y}$$

Account for this order in terms of steric and electronic factors. ◄

A change from a trigonal sp^2 to a tetrahedral sp^3 C in the transition state is accompanied by crowding of the 4 groups on C. Crowding and destabilization of the transition state increase in the order

$$CH_2=O > RCH=O > R_2C=O$$

Also, the electron-releasing R's intensify the $-$ charge developing on O, which destabilizes the transition state and decreases reactivity.

In $RC\ddot{Y}$, extended π bonding between $-\ddot{Y}$ and C=O,
$$\underset{O}{\underset{\|}{}}$$

$$R-\underset{\underset{\ddot{O}:}{\|}}{C}-\ddot{Y} \longleftrightarrow R-\underset{\underset{:\ddot{O}\bar{}}{\|}}{C}=\overset{+}{Y} \quad \left(R-\underset{\underset{O^{\delta-}}{\|}}{C}\overset{\delta+}{=}\ddot{Y}\right)$$

lowers the enthalpy of the ground state, raises $\Delta H^{\ddagger}$ and decreases the reactivity of C=O towards nucleophilic attack. Hence, acid derivatives RCOY, in which

$$Y = -\ddot{\underset{..}{X}}:, \ -\ddot{N}H_2, \ -\ddot{O}R, \ -\ddot{O}-\underset{\underset{O}{\|}}{C}-R$$

are less reactive than RCHO or R_2CO.

Problem 15.16 Explain the order of reactivity $ArCH_2COR > R_2C=O > ArCOR > Ar_2CO$ in nucleophilic addition. ◄

When attached to C=O, Ar's, like $-Y:$ (Problem 15.15), are electron-releasing by extended π bonding (resonance) and deactivate C=O. Two Ar's are more deactivating than one Ar. In $ArCH_2COR$ only the electron-withdrawing inductive effect of Ar prevails; consequently, $ArCH_2$ increases the reactivity of C=O.

Problem 15.17 The rate of addition of HCN to $R_2C=O$ to form a cyanohydrin, $R_2C(OH)CN$, is increased by adding a trace of NaCN. Explain. ◄

The rate-controlling step is the addition of CN^-.

$$R_2C\overset{\frown}{=}\ddot{O}\bar{} + CN^- \xrightarrow{\text{slow}} R_2\underset{\underset{CN}{|}}{C}-\ddot{O}\bar{} \xrightarrow{\text{HCN}} R_2\underset{\underset{CN}{|}}{C}-OH + CN^- \quad \text{(regenerated catalyst)}$$

Problem 15.18 NaHSO$_3$ reacts with RCHO in EtOH to give a solid adduct. (a) Write an equation for the reaction. (b) Explain why only RCHO, methyl ketones (RCOCH$_3$) and cyclic ketones react. (c) If the carbonyl compound can be regenerated on treating the adduct with acid or base, explain how this reaction with NaHSO$_3$ can be used to separate RCHO from noncarbonyl compounds such as RCH$_2$OH. ◄

(a) HSO_3^- can protonate RCHO.

$$RC\underset{\underset{H}{|}}{=}O + Na^+HSO_3^- \longrightarrow R-\overset{+}{\underset{\underset{H}{|}}{C}}-OH + Na^+ + :SO_3^{2-} \longrightarrow R-\underset{\underset{H}{|}}{\overset{\overset{SO_3^-Na^+}{|}}{C}}-OH$$

Sodium bisulfite adduct
(*solid*)

A C—S bond is formed because S is a more nucleophilic site than O.

(b) SO_3^{2-} is a large ion and reacts only if C=O is not sterically hindered, as is the case for RCHO, RCOCH$_3$ and cyclic ketones.

(c) The solid adduct is filtered from the ethanolic solution of unreacted RCH$_2$OH and then is decomposed by acid or base:

$$RC\underset{\underset{OH}{|}}{\overset{\overset{H}{|}}{-}}SO_3^-Na^+ \xrightarrow[\text{or}]{\overset{H^+}{\longrightarrow} SO_2} \left.\begin{array}{c} \\ \\ \end{array}\right\} + \underset{\underset{O}{\|}}{RCH} \quad (\textit{extracted with ether})$$

Problem 15.19 Write the formula for the solid derivative formed when an aldehyde or ketone reacts with each of the following ammonia derivatives:

$$(a) \quad \underset{\text{Hydroxylamine}}{H-\overset{\displaystyle H}{\underset{\displaystyle |}{N}}-OH} \qquad (b) \quad \underset{\text{Phenylhydrazine}}{H-\overset{\displaystyle H}{\underset{\displaystyle |}{N}}-NHC_6H_5} \qquad (c) \quad \underset{\text{Semicarbazide}}{H-\overset{\displaystyle H}{\underset{\displaystyle |}{N}}-NHCONH_2} \qquad \blacktriangleleft$$

Since these nucleophiles are of the $:NuH_2$ type, addition is followed by dehydration.

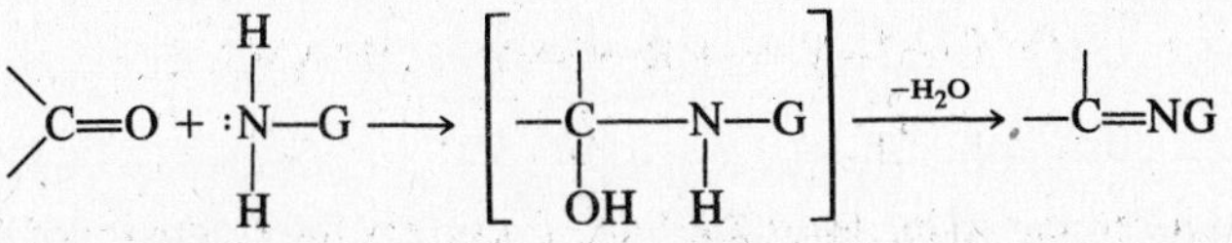

$$\overset{\diagup}{\underset{\diagdown}{}}C=O + \overset{\displaystyle H}{\underset{\displaystyle H}{\overset{\displaystyle |}{\underset{\displaystyle |}{:N}}}}-G \longrightarrow \left[-\overset{\displaystyle |}{\underset{\displaystyle OH}{C}}-\overset{\displaystyle |}{\underset{\displaystyle H}{N}}-G \right] \xrightarrow{-H_2O} , \ -\overset{\displaystyle |}{C}=NG$$

(a) $G = -OH$; $\diagdown C=N-OH$ (Oxime).

(b) $G = -NHC_6H_5$; $\diagdown C=NNHC_6H_5$ (Phenylhydrazone).

(c) $G = -NHCONH_2$; $\diagdown C=NNHCONH_2$ (Semicarbazone).

The melting points of these solid derivatives are used to identify carbonyl compounds.

Problem 15.20 Explain why formation of oximes and other ammonia derivatives requires slightly acidic media (pH ≈ 3.5) for maximum rate, while basic or more highly acid conditions lower the rate. $\blacktriangleleft$

The carbonyl group becomes more electrophilic and reactive when converted by acid to its conjugate acid,

$$\diagdown C\overset{+}{-}OH$$

In more strongly acid solutions (pH < 3.5) the unshared pair of electrons (the nucleophilic site) of N is protonated to give electrophilic H_3NG^+, a species which cannot react. In basic media there is no protonation of $C=O$.

Problem 15.21 Reaction of 1 mole of semicarbazide with a mixture of 1 mole each of cyclohexanone and benzaldehyde precipitates cyclohexanone semicarbazone, but after a few hours the precipitate is benzaldehyde semicarbazone. Explain. $\blacktriangleleft$

The $C=O$ of cyclohexanone is not deactivated by the electron-releasing C_6H_5 and does not suffer from steric hinderance. The semicarbazone of cyclohexanone is the kinetically-controlled product. Conjugation makes $PhCH=NNHCONH_2$ more stable, and its formation is thermodynamically controlled. In such reversible reactions the equilibrium shifts to the more stable product (Fig. 15-1).

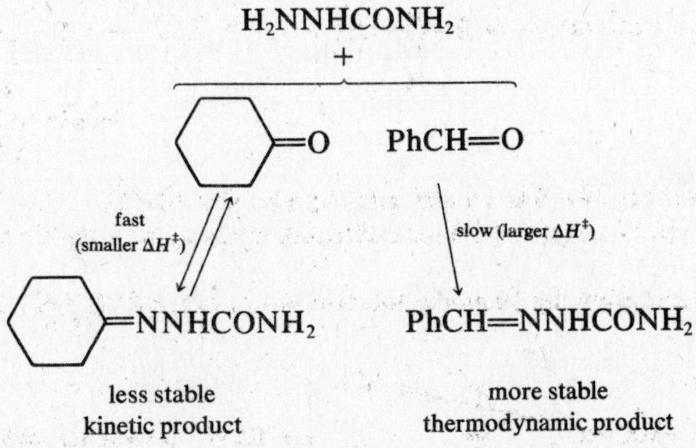

$$H_2NNHCONH_2$$
$$+$$

Fig. 15-1

Problem 15.22 Give structures for the product(s) of the following reactions:

$$(a) \quad \text{1-Butanal} + C_6H_5NHNH_2 \qquad (b) \quad \text{2-Butanone} + \left[H_3\overset{+}{N}OH\right]Cl^-, \ NaC_2H_3O_2 \qquad \blacktriangleleft$$

(a) $CH_3(CH_2)_2CH{=}NNHC_6H_5$. (b) $CH_3CH_2\overset{\displaystyle \overset{NOH}{\|}}{C}{-}CH_3$: $C_2H_3O_2^-$ (acetate) frees H_2NOH from its salt.

Problem 15.23 Symmetrical ketones, $R_2C{=}O$, form a single oxime, but aldehydes and unsymmetrical ketones may form two isomeric oximes. Explain. ◄

The π bond in

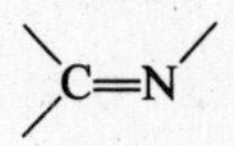

$$\ce{>C=N<}$$

prevents free rotation, and therefore geometric isomerism occurs if the groups on the carbonyl C are dissimilar. The old terms *syn* and *anti* are also used in place of *cis* and *trans*, respectively.

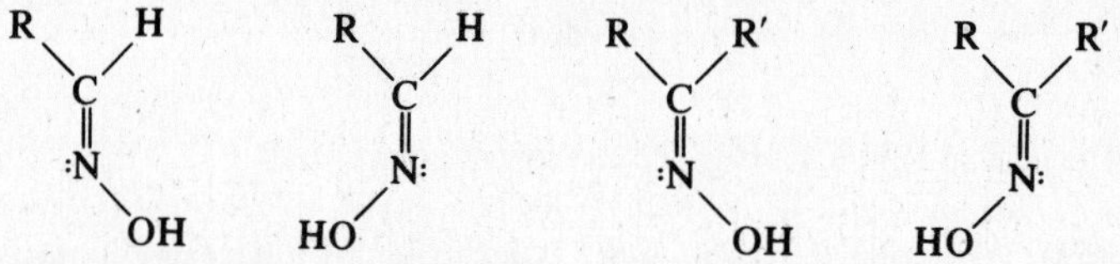

trans (anti)   cis (syn)

Problem 15.24 Why are oximes more *acidic* than hydroxylamine? ◄

Loss of H^+ from H_2NOH gives the conjugate base, H_2NO^-, with the charge localized on O. Delocalization of charge by extended π bonding can occur in the conjugate base of the oxime,

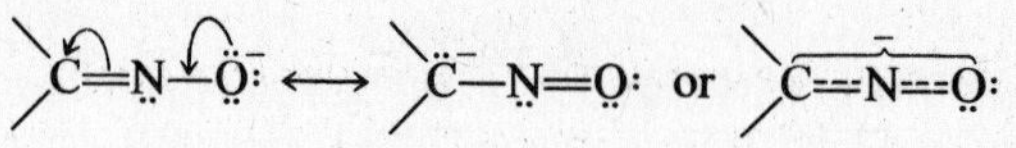

Thus

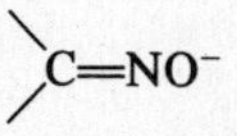

$$\ce{>C=NO^-}$$

is a weaker base, and its conjugate acid, the oxime, is much more acidic.

ACETAL FORMATION

$$\underset{\text{R'}-\overset{\displaystyle \overset{H}{|}}{C}{=}O}{} + 2ROH \ \underset{H_3O^+}{\overset{\text{dry HCl}}{\rightleftharpoons}} \ \underset{\text{R'}-\overset{\displaystyle \overset{H}{|}}{\underset{\displaystyle \underset{OR}{|}}{C}}{-}OR}{} + H_2O$$

an acetal (*gem*-diether)

In H_3O^+, $R'CHO$ is regenerated because acetals undergo acid-catalyzed cleavage much more easily than do ethers. Since acetals are stable in neutral or basic media, they are used to protect the $-CH{=}O$ group. Unhindered ketones form ketals, $R_2C(OR')_2$.

CONVERSION TO DIHALIDES

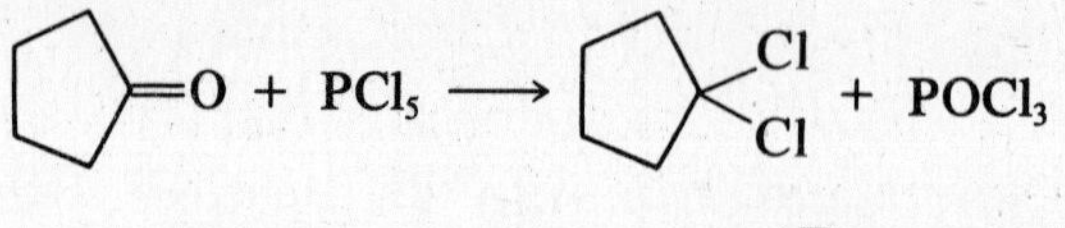

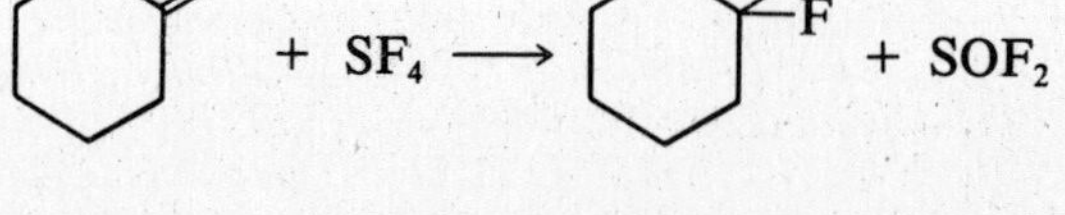

Problem 15.25 Give mechanisms for:

(a)

$$R-\overset{\overset{\displaystyle H}{|}}{C}=O + R'OH \xrightarrow{\text{dry HCl}} \left[R-\overset{\overset{\displaystyle H}{|}}{\underset{\overset{\displaystyle |}{O-R'}}{C}}-O-H \right] \xrightarrow{R'OH} R-\overset{\overset{\displaystyle H}{|}}{\underset{\overset{\displaystyle |}{O-R'}}{C}}-O-R' + H_2O$$

hemiacetal   acetal

(b) hemiacetal formation initiated by basic OR⁻ in ROH.   ◀

(a)

$$R-\overset{\overset{\displaystyle H}{|}}{C}=O \underset{-H^+}{\overset{+H^+}{\rightleftharpoons}} R-\overset{\overset{\displaystyle H}{|}}{\underset{+}{C}}-OH \overset{R'\ddot{O}H}{\rightleftharpoons} R-\overset{\overset{\displaystyle H}{|}}{\underset{\underset{+}{HOR'}}{C}}-OH \underset{+H^+}{\overset{-H^+}{\rightleftharpoons}} \left[R-\overset{\overset{\displaystyle H}{|}}{\underset{\overset{\displaystyle |}{OR'}}{C}}-OH \right] \underset{-H^+}{\overset{+H^+}{\rightleftharpoons}} R-\overset{\overset{\displaystyle H}{|}}{\underset{\overset{\displaystyle |}{OR'}}{C}}-\overset{\overset{\displaystyle H}{|}}{\underset{+}{O}}H$$

hemiacetal

From the protonated hemiacetal the mechanism is similar to that for the formation of ethers from alcohols [Problem 14.3(b)].

$$R-\overset{\overset{\displaystyle H}{|}}{\underset{\overset{\displaystyle |}{OR'}}{C}}-\overset{\overset{\displaystyle H}{|}}{\underset{+}{O}}H \underset{+H_2O}{\overset{-H_2O}{\rightleftharpoons}} R-\overset{\overset{\displaystyle H}{|}}{\underset{\overset{\displaystyle |}{:OR'}}{C^+}} \longleftrightarrow R\overset{\overset{\displaystyle H}{|}}{\underset{\overset{\displaystyle |}{^+OR'}}{C}} \overset{R'OH}{\rightleftharpoons} R-\overset{\overset{\displaystyle H}{|}}{\underset{\overset{\displaystyle |}{OR'}}{C}}-\overset{+}{O}R' \underset{+H^+}{\overset{-H^+}{\rightleftharpoons}} R-\overset{\overset{\displaystyle H}{|}}{\underset{\overset{\displaystyle |}{OR'}}{C}}-OR'$$

protonated   protonated   acetal
hemiacetal   acetal

(b)

$$\overset{\diagdown}{\diagup}C\!\!=\!\!\overset{..}{O}: + {}^-OR' \longrightarrow \overset{\diagdown}{\underset{\overset{\displaystyle |}{OR'}}{\diagup}}C-\overset{..}{O}:^- \xrightarrow{H:OR'} \overset{\diagdown}{\underset{\overset{\displaystyle |}{OR'}}{\diagup}}C-OH + :\overset{..}{O}R'$$

Problem 15.26 In acid, most aldehydes form nonisolable hydrates (*gem*-diols).   Two exceptions are the stable chloral hydrate, $Cl_3CCH(OH)_2$, and ninhydrin,

(a) Given the bond energies 179, 111 and 86 kcal/mole for C=O, O—H and C—O, respectively, show why the equilibrium typically lies toward the carbonyl compound. (b) Account for the exceptions.   ◀

(a) Calculating ΔH for

$$\overset{\diagdown}{\diagup}C\!\!=\!\!O + H\!-\!O\!-\!H \rightleftharpoons \overset{\diagdown}{\diagup}\overset{\overset{\displaystyle OH}{|}}{\underset{\overset{\displaystyle |}{OH}}{C}}$$

we obtain

$$\underset{\substack{\text{(C=O) (O—H)} \\ \text{cleavages} \\ \textit{endothermic}}}{[179 + 2(111)]} + \underset{\substack{\text{(C—O)   (O—H)} \\ \text{formations} \\ \textit{exothermic}}}{[2(-86) + 2(-111)]} = \Delta H$$

or $\Delta H = +7$ kcal/mole.   Hydrate formation is endothermic and not favored.   The carbonyl side is also favored by entropy because 2 molecules,

$$\overset{\diagdown}{\diagup}C\!\!=\!\!O \quad \text{and} \quad H_2O$$

are more random than 1 *gem*-diol molecule [see Problem 3.37(a)].

(b) Strong electron-withdrawing groups on an α C destabilize an adjacent carbonyl group because of repulsion of adjacent + charges.   Hydrate formation overcomes the forces of repulsion.

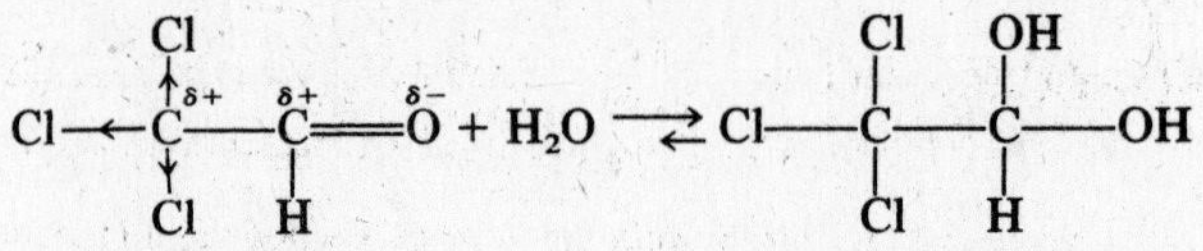

$$Cl\leftarrow\overset{\overset{Cl}{\uparrow}}{\underset{\underset{Cl}{\downarrow}}{C}}\overset{\delta+}{\underset{H}{\_\_}}C\overset{\delta+}{=}\overset{\delta-}{O} + H_2O \rightleftharpoons Cl\_\_\overset{Cl}{\underset{Cl}{C}}\_\_\overset{OH}{\underset{H}{C}}\_\_OH$$

repulsion from
adjacent + charges

Chloral hydrate
less repulsion

Hydration of the middle carbonyl group of ninhydrin removes both pairs of repulsions.

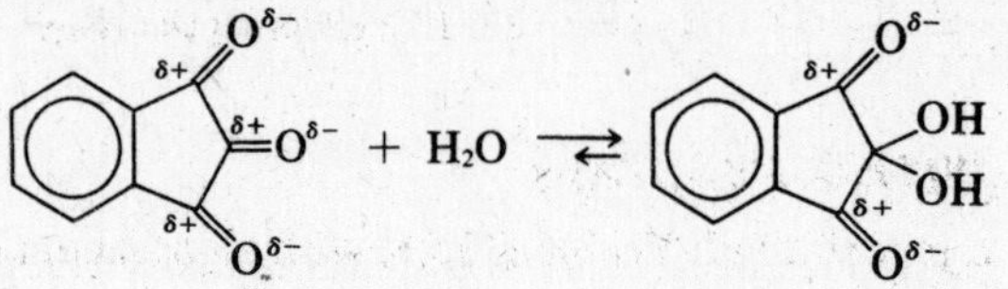

Problem 15.27   Prepare the compound 2,3-dihydroxypropanal ($HOCH_2CHOHCH=O$) from 2-propenal ($H_2C=CH\_CH=O$). ◄

KMnO₄, which converts $C=C$ to $\_C(OH)\_C(OH)\_$, also oxidizes $\_CH=O$ to $\_COOH$. The $\_CH=O$ is protected as an acetal and then regenerated.

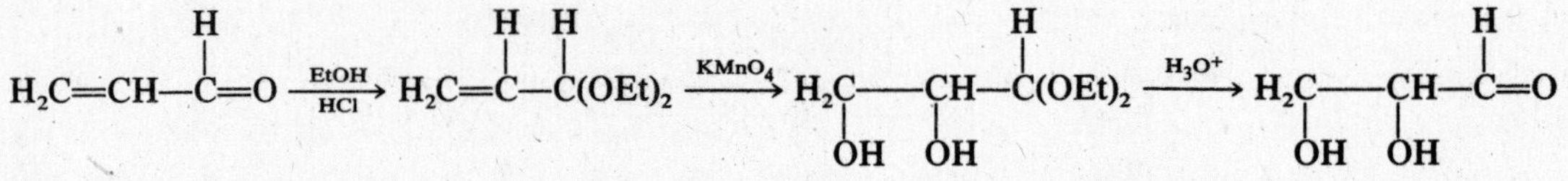

Problem 15.28   Write structures for the cyclic ketals or acetals prepared from (a) butanal + 1,3-propanediol, (b) cyclohexanone + ethylene glycol. ◄

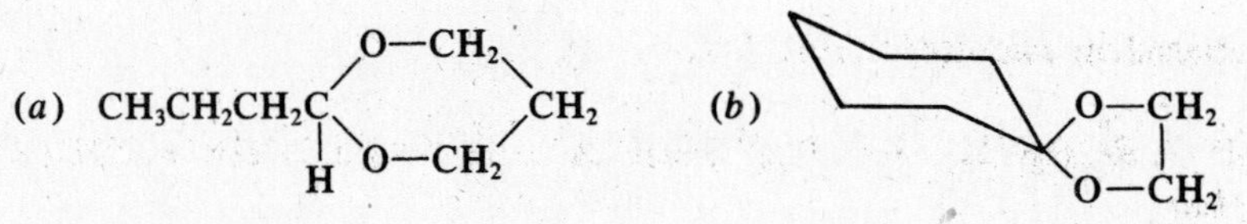

Problem 15.29   The C—S—C bonds of cyclic thioketals (or thioacetals) prepared from $HSCH_2CH_2SH$ are reduced (hydrogenolysis) with Ni to

$$\_\overset{|}{\underset{|}{C}}H_2 + H_2S$$

Use these reactions to convert $\diagdown C=O$ to $\diagdown CH_2$. ◄

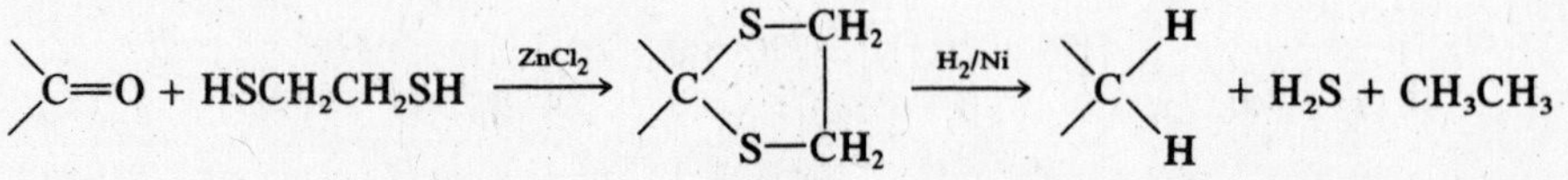

a thioketal

Problem 15.30   Show steps in the synthesis of cyclooctyne from

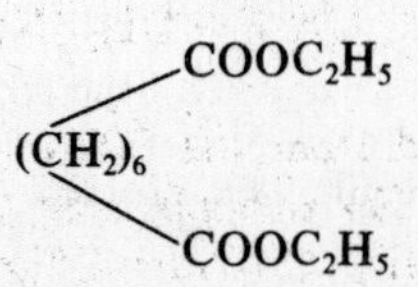

$$(CH_2)_6\diagup\overset{COOC_2H_5}{\diagdown COOC_2H_5}$$

◄

The 1,8-diester is converted to an 8-membered ring acyloin, which is then changed to the alkyne.

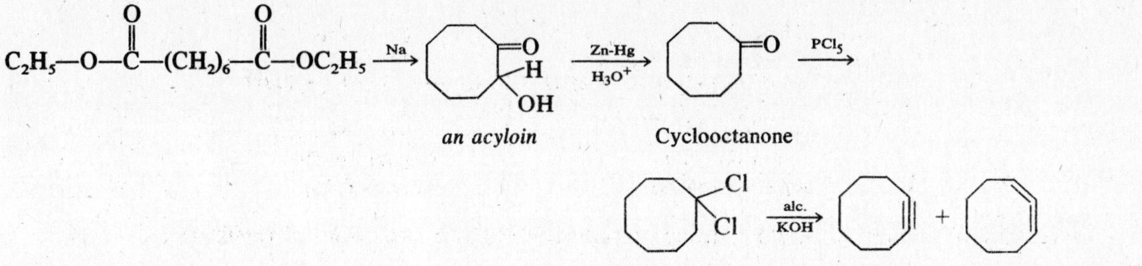

$$C_2H_5-O-\overset{O}{\overset{\|}{C}}-(CH_2)_6-\overset{O}{\overset{\|}{C}}-OC_2H_5 \xrightarrow{Na}$$

an acyloin　　　　Cyclooctanone

1,1-Dichlorocyclooctane　　Cyclooctyne　　1,2-Cyclooctadiene
　　　　　　　　　　　　　　(major)　　　(minor), *an allene*

ATTACK BY YLIDES; WITTIG REACTION

A carbanion C can form a p-d π bond (Problem 3.26) with an adjacent P or S. The resulting charge delocalization is especially effective if P or S, furnishing the empty d orbital, also has a + charge. Carbanions with these characteristics are called **ylides**, e.g.

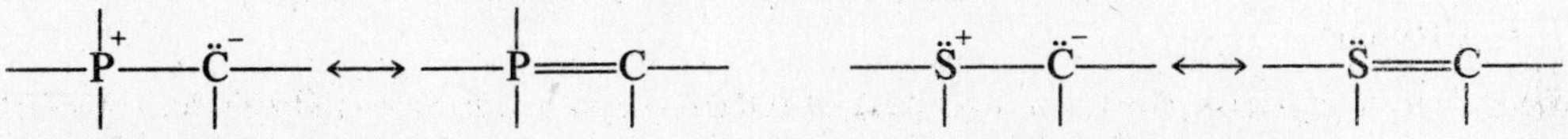

The **Wittig reaction** uses P ylides to change O of the carbonyl group to

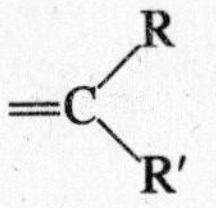

$$=C\overset{R}{\underset{R'}{}}$$

The carbanion portion of the ylide replaces the O.

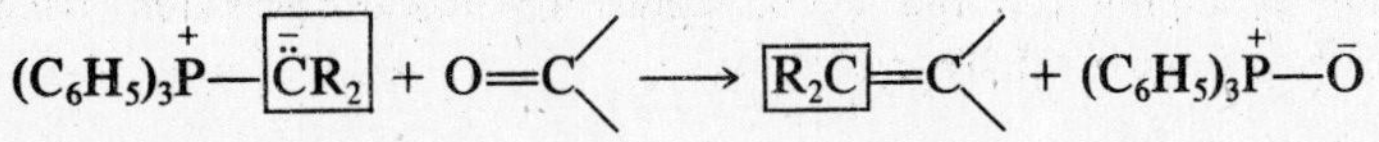

$$(C_6H_5)_3\overset{+}{P}-\boxed{\overset{..}{C}R_2} + O=C\langle \longrightarrow \boxed{R_2C}=C\langle + (C_6H_5)_3\overset{+}{P}-\overset{-}{O}$$

The ylide is prepared in two steps from RX.

$$Ph_3\overset{..}{P}: + RCH_2\overset{\frown}{-X} \xrightarrow{S_N2} \left[Ph_3\overset{+}{P}CH_2R\right]X^- \xrightarrow{C_4H_9Li^+} Ph_3\overset{+}{P}\overset{-}{C}HR + C_4H_{10} + Li\overset{+}{X}$$

a phosphine

Sulfur ylides react with aldehydes and ketones to form epoxides (oxiranes):

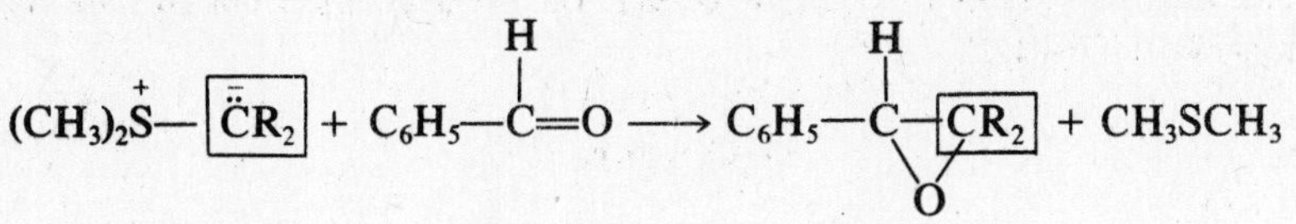

$$(CH_3)_2\overset{+}{S}-\boxed{\overset{..}{C}R_2} + C_6H_5-\overset{H}{\underset{|}{C}}=O \longrightarrow C_6H_5-\overset{H}{\underset{O}{\overset{|}{C}}}\boxed{CR_2} + CH_3SCH_3$$

The sulfur ylide is formed from the sulfonium salt,

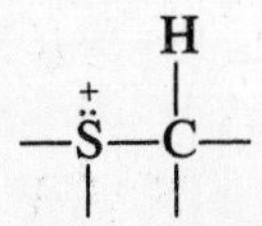

$$-\overset{+}{\overset{..}{S}}-\overset{H}{\underset{|}{C}}-$$

with a strong base, such as sodium dimethyloxosulfonium methylide

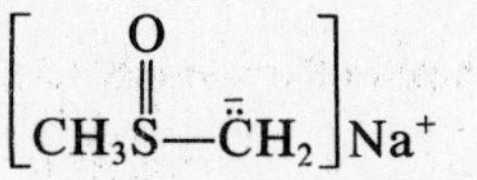

$$\left[\overset{O}{\overset{\|}{CH_3S}}-\overset{-}{C}H_2\right]Na^+$$

Problem 15.31 Which alkenes are formed from the following ylide–carbonyl compound pairs? (*a*) 2-butanone and $CH_3CH_2CH_2CH=P(C_6H_5)_3$, (*b*) acetophenone and $(C_6H_5)_3P=CH_2$, (*c*) benzaldehyde and $C_6H_5-CH=P(C_6H_5)_3$, (*d*) cyclohexanone and $(C_6H_5)_3P=C(CH_3)_2$. (Disregard stereochemistry.) ◄

The boxed portions below come from the ylide.

$$(a) \quad CH_3CH_2\overset{\overset{\displaystyle CH_3}{|}}{C}=\boxed{CHCH_2CH_2CH_3} \qquad (b) \quad C_6H_5-\overset{\overset{\displaystyle CH_3}{|}}{C}=\boxed{CH_2}$$

$$(c) \quad C_6H_5-CH=\boxed{CH-C_6H_5} \qquad (d) \quad \bigcirc=\boxed{C(CH_3)_2}$$

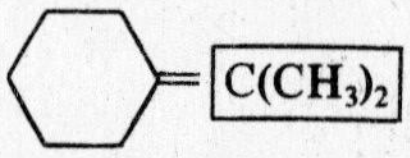

Problem 15.32 Give structures of the ylide and carbonyl compound needed to prepare:

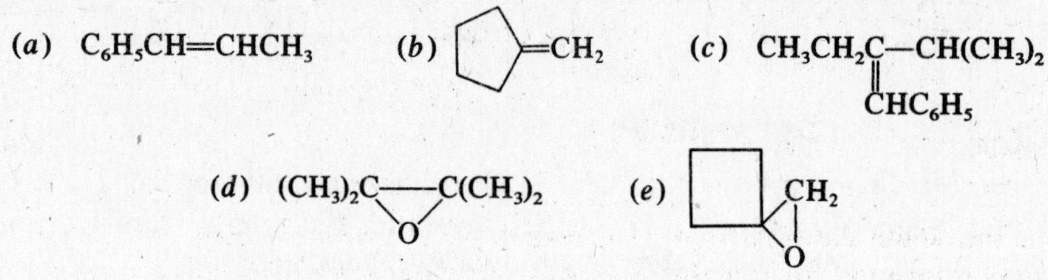

$$(a) \quad C_6H_5CH=CHCH_3 \qquad (b) \quad \bigcirc=CH_2 \qquad (c) \quad CH_3CH_2\overset{\displaystyle C}{\underset{\overset{|}{CHC_6H_5}}{}}-CH(CH_3)_2$$

$$(d) \quad (CH_3)_2C\overset{}{\underset{O}{\diagup\diagdown}}C(CH_3)_2 \qquad (e) \quad \boxed{}\overset{}{\underset{O}{\diagup\diagdown}}CH_2$$

$$(a) \qquad Ph_3\overset{+}{P}\overset{..}{\overset{-}{C}}HCH_3 + C_6H_5\overset{\overset{\displaystyle H}{|}}{C}=O \quad \text{or} \quad Ph_3\overset{+}{P}\overset{..}{\overset{-}{C}}HC_6H_5 + CH_3\overset{\overset{\displaystyle H}{|}}{C}=O$$

The *cis-trans* geometry of the alkene is influenced by the nature of the substituents, solvent and dissolved salts.   Polar protic or aprotic solvents favor the *cis* isomer.

$$(b) \qquad \bigcirc=O + Ph_3\overset{+}{P}\overset{..}{\overset{-}{C}}H_2 \quad \text{or} \quad \bigcirc:\overset{+}{}-PPh_3 + O=CH_2$$

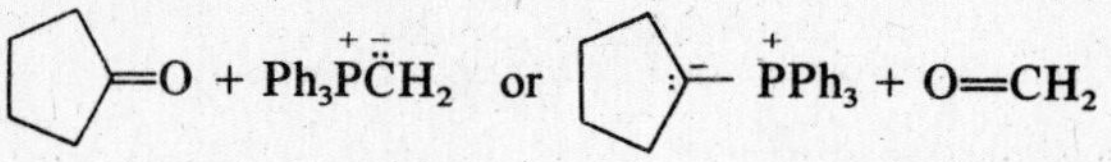

$$(c) \qquad CH_3CH_2\overset{\displaystyle C}{\underset{\displaystyle O}{\|}}CH(CH_3)_2 + Ph_3\overset{+}{P}\overset{..}{\overset{-}{C}}HC_6H_5 \quad \text{or} \quad CH_3CH_2\overset{..}{\overset{-}{C}}CH(CH_3)_2 + C_6H_5CHO$$
$$\qquad\qquad\qquad\qquad\qquad\qquad\qquad\qquad\qquad\qquad\qquad \overset{+}{P}Ph_3$$

$$(d) \qquad (CH_3)_2C=O + Ph_2\overset{+}{S}-\overset{..}{\overset{-}{C}}(CH_3)_2$$

$$(e) \qquad \boxed{}\overset{}{\underset{O}{}} + (CH_3)_2\overset{+}{S}-\overset{..}{\overset{-}{C}}H_2$$

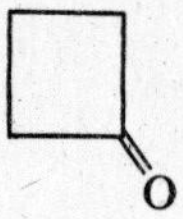

ACIDITY OF α H's; *TAUTOMERISM*

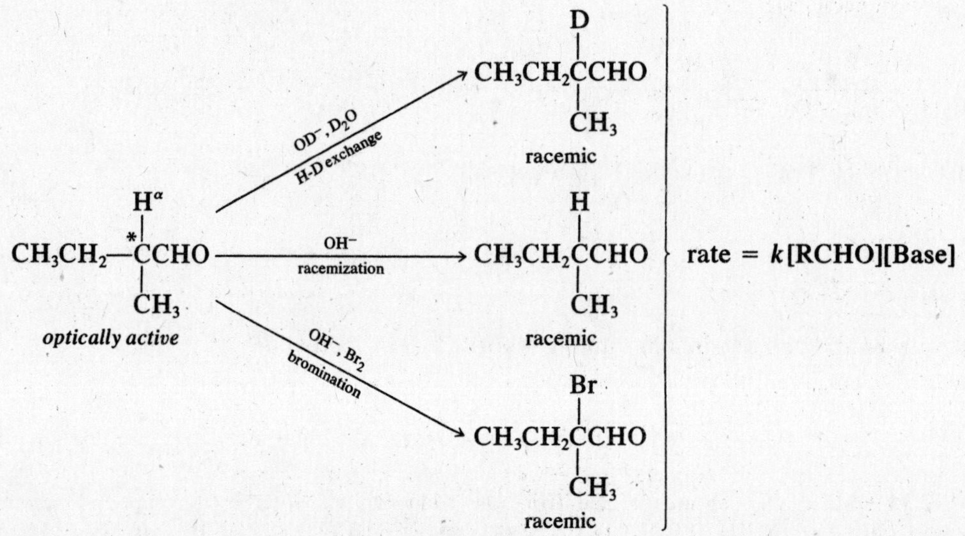

Fig. 15-2

Some base-catalyzed reactions of α H's are shown in Fig. 15-2. Since the 3 reactions have the same rate expression, they have the same rate-determining step: the removal of an α H to form a stabilized carbanion.

$$CH_3CH_2\overset{\underset{\displaystyle H}{|}}{\underset{\underset{\displaystyle H}{\uparrow}}{C}}\overset{CH_3}{\underset{}{|}}C=\ddot{O}: + :B^- \xrightarrow{\text{slow}} \left\{ \begin{array}{c} CH_3CH_2\overset{CH_3}{\underset{\underset{\displaystyle H}{|}}{C}}=C-\ddot{O}:^- \\ \updownarrow \\ CH_3CH_2\overset{CH_3}{\underset{\underset{\displaystyle H}{|}}{C}}-C=\ddot{O}: \end{array} \right. \quad \text{or} \quad CH_3CH_2-\overset{CH_3}{\underset{\underset{\displaystyle H}{|}}{C}}=\overset{-}{\underset{\underset{\displaystyle H}{|}}{C}}=O \left. \right\} + H:B$$

Acid₁   Base₂   Base₁   Acid₂

stable carbanion-enolate ion

Stabilization of the anion causes the α H of carbonyl compounds to be more acidic than H's of alkanes. H^+ may return to C to give the more stable carbonyl compound (the keto structure) or to O to give the less stable enol.

$$\overset{\displaystyle H}{\underset{|}{|}}\overset{}{\underset{|}{-C}}-\overset{}{\underset{|}{C}}=\ddot{O}: \underset{-H^+}{\overset{+H^+}{\rightleftharpoons}} \left\{ \begin{array}{c} -\overset{\cdot\cdot}{\underset{|}{C}}-\overset{}{\underset{|}{C}}=\ddot{O}: \\ \updownarrow \\ -\overset{}{\underset{|}{C}}=\overset{}{\underset{|}{C}}-\ddot{O}:^- \end{array} \right. \underset{+H^+}{\overset{-H^+}{\rightleftharpoons}} -\overset{}{\underset{|}{C}}=\overset{}{\underset{|}{C}}-\ddot{O}H$$

keto (weaker acid)   carbanion-enolate   enol (stronger acid)

With few exceptions (Problem 15.34), the keto structure rather than the enol is isolated from reactions (see hydration of alkynes, page 117).

Structural isomers existing in rapid equilibrium are **tautomers** and the equilibrium reaction is **tautomerism**. The above is a keto-enol tautomerism.

Problem 15.33 (*a*) Give equations for the tautomerism in which each of the following compounds is the more stable tautomer. (1) CH_3CHO, (2) $C_6H_5COCH_3$, (3) CH_3NO_2, (4) $Me_2C=NOH$, and (5) $CH_3CH=NCH_3$. (*b*) Which two enols are in equilibrium with methyl ethyl ketone? Predict which is more stable. ◄

(*a*)   The grouping needed for tautomerism, X=Y—Z—H (a triple bond could also exist between X and Y), is encircled in each case.

(1) (H—C—C=O) ⇌ (C=C—OH)

keto   enol

(2) C_6H_5—(C—C)—H ⇌ C_6H_5—(C=C)

keto   enol

(3) (H—C—N=O) ⇌ (C=N—O—H)

nitro form   *aci* form

(4) Me_2(C=N—O—H) ⇌ Me_2(C—N=O)

oxime   nitroso

(5) (H—C—C=N) ⇌ (C=C—N—H)

imine   enamine

(b)
$$H_2C=\underset{\underset{OH}{|}}{C}-CH_2CH_3 \quad \text{and} \quad CH_3-\underset{\underset{OH}{|}}{C}=CHCH_3$$

The latter is more stable because it has a more substituted double bond.

Problem 15.34 C_6H_5OH, an enol, is much more stable than its keto isomer, cyclohexa-2,4-diene-1-one.

Phenol (enol)   Cyclohexa-2,4-diene-1-one (keto)

Explain this exception.   ◄

Phenol has a stable aromatic ring.

Problem 15.35 Reaction of 1 mole each of Br_2 and $PhCOCH_2CH_3$ in basic solution yields 0.5 mole of $PhCOCBr_2CH_3$ and 0.5 mole of unreacted $PhCOCH_2CH_3$. Explain.   ◄

Substitution by one Br gives $PhCOCHBrCH_3$. The electron-withdrawing Br increases the acidity of the remaining α H, which reacts more rapidly than, and is substituted before, the H's on the unbrominated ketone.

less acidic   more acidic

$$PhCOCH_2CH_3 \xrightarrow[OH^-]{0.5 \text{ mole } Br_2} [PhCOCHBrCH_3] \xrightarrow[OH^-]{0.5 \text{ mole } Br_2} PhCOCBr_2CH_3$$

(0.5 mole)   (0.5 mole)   (0.5 mole)
                              not isolated

Problem 15.36 Describe the formation of CHI_3 from reaction of $C_6H_5COCH_3$ with NaOH and I_2 (NaOI).

First, in this haloform reaction,   ◄

$$C_6H_5\overset{\overset{\displaystyle O}{\|}}{C}-CH_3 \xrightarrow{I_2, NaOH} C_6H_5\overset{\overset{\displaystyle O}{\|}}{C}-CI_3$$

Then, OH^- adds to the carbonyl group, and $I_3C:^-$ is eliminated because this *anion* is stabilized by electron-withdrawal by the 3 I's. Finally, H^+-exchange occurs.

$$R-\overset{\overset{\displaystyle O}{\|}}{C}-CI_3 + :\ddot{O}H^- \longrightarrow R-\underset{\underset{OH}{|}}{\overset{\overset{\displaystyle :\ddot{O}:}{|}}{C}}-CI_3 \longrightarrow \left[R-\underset{\underset{OH}{|}}{\overset{\overset{\displaystyle :O:}{|}}{C}} + :CI_3^- \right] \xrightarrow{H_2O} R-\overset{\overset{\displaystyle O}{\|}}{C} + H:CI_3$$

Problem 15.37 Why does the C—C cleavage in Problem 15.36 not occur with simple aldehydes or ketones? ◄

The basic anions that must be displaced are $:R^-$, $:Ar^-$, or $:H^-$, all of which are more basic than $:OH^-$. Hence the weaker base, OH^-, is ejected and there is no net reaction.

$$HO:^- + (Ar)R-\overset{\overset{\displaystyle O}{\|}}{C}-H \rightleftharpoons (Ar)R-\underset{\underset{OH}{|}}{\overset{\overset{\displaystyle O^-}{|}}{C}}-H$$

$$R-\overset{\overset{\displaystyle O}{\|}}{C}\diagdown_{OH} + :H^-$$

$$H-\overset{\overset{\displaystyle O}{\|}}{C}\diagdown_{OH} + :R^-(:Ar^-)$$

stronger bases than $:OH^-$

reversal is favored

Nucleophilic displacement from the carbonyl group is observed in reactions of carboxylic acid derivatives, RCOY (Chapter 17).

Problem 15.38   (a) Why is (+)-PhCH(CH₃)CHO racemized by base?   (b) Why is

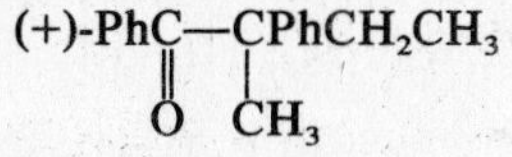

$$(+)\text{-PhC}\!-\!\text{CPhCH}_2\text{CH}_3$$
$$\underset{\text{O}}{\overset{}{}}\ \underset{\text{CH}_3}{\overset{}{}}$$

not racemized by base?   ◄

(a) Racemization occurs when base removes the α H to form an anion.   The α C of the anion is no longer chiral.   Return of an H⁺ gives a racemic keto form.   (b) This ketone has no α H and cannot form the anion.

Problem 15.39   Racemization, D-exchange, and bromination of carbonyl compounds are also acid-catalyzed. (a) Suggest reasonable mechanisms in which enol is an intermediate.   (b) In terms of your mechanisms, are the rate expressions of these reactions the same?   ◄

First we have:

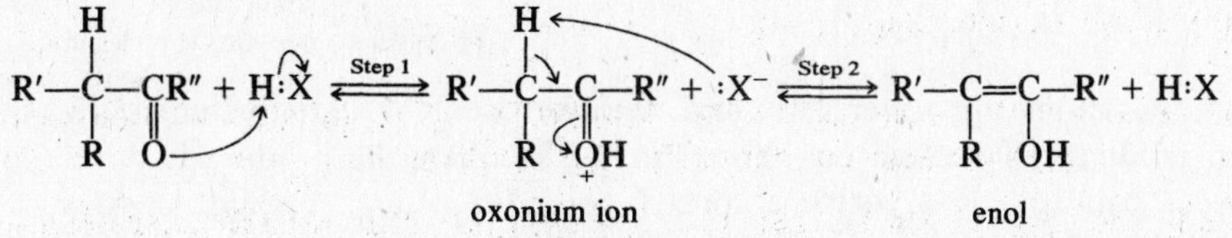

oxonium ion   enol

(a)   **Racemization**

enol   from a   from b

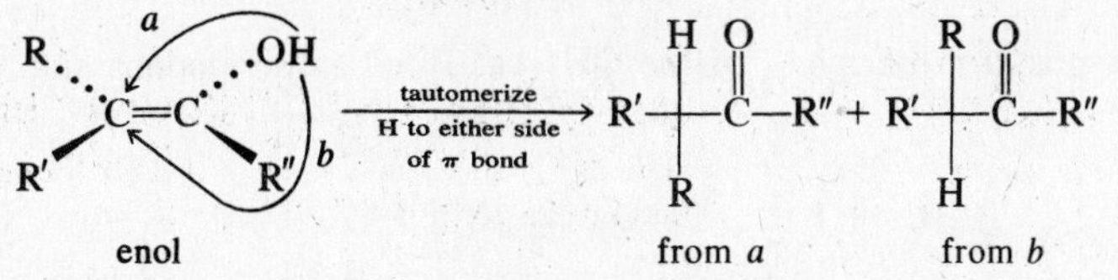

Deuterium Exchange

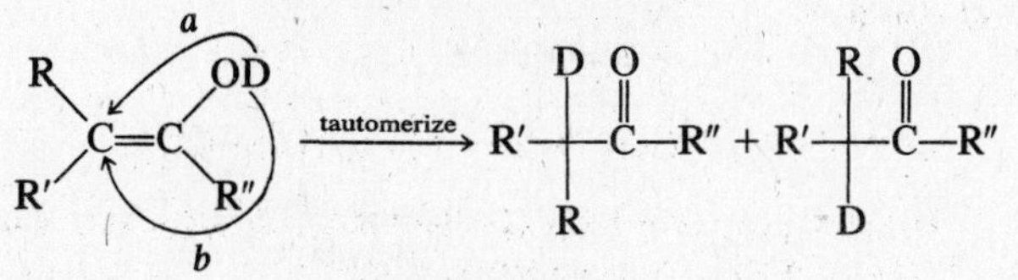

Bromination

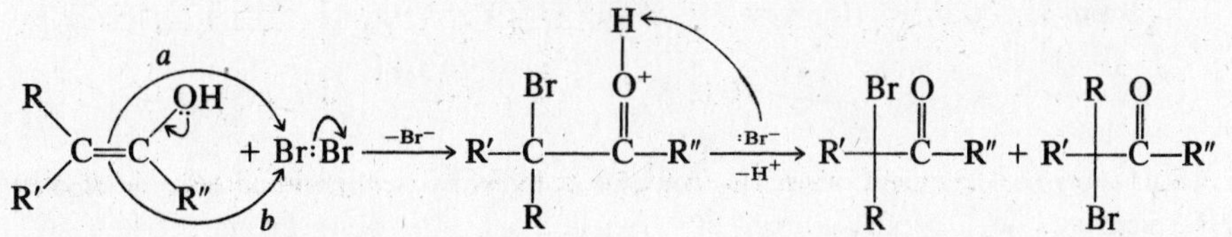

(b)   Since enol formation is rate-determining, these 3 reactions have the same rate expression.

ALDOL CONDENSATION

An α carbanion

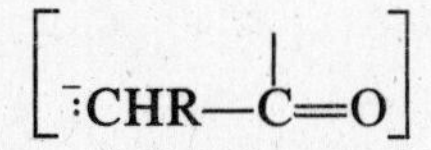

$$\left[\text{:CHR}\!-\!\overset{|}{\text{C}}\!=\!\text{O}\right]$$

is a nucleophile that can add to the carbonyl group of its parent compound.   This is a key step in **aldol condensations** leading to β-hydroxycarbonyl compounds.

Net Reactions

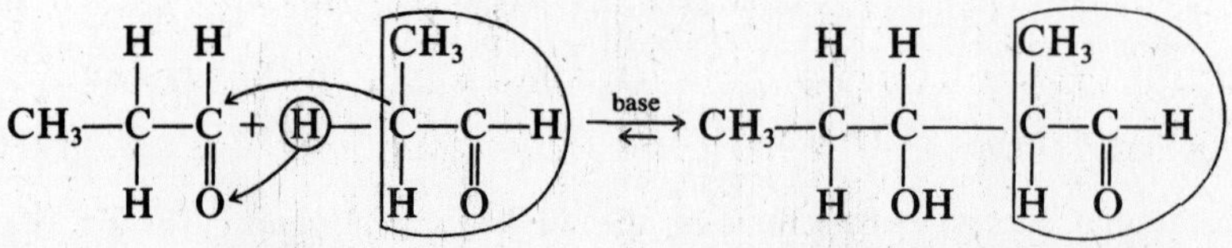

Propanal → 2-Methyl-3-hydroxypentanal

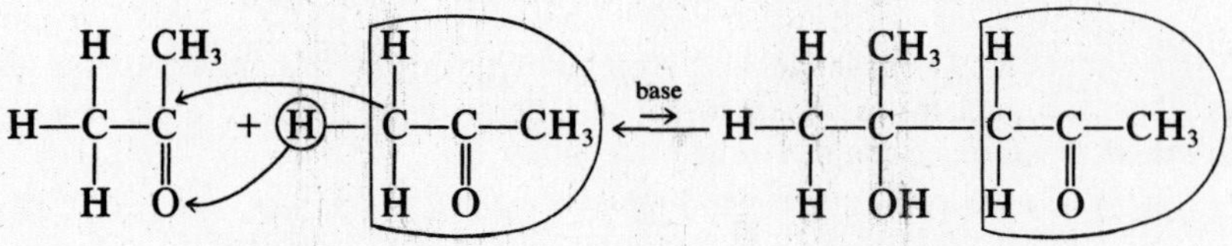

Acetone → Diacetone alcohol
(4-Hydroxy-4-methyl-2-pentanone)

Aldol condensations are reversible, and with ketones the equilibrium is unfavorable for the condensation product. To effect condensations of ketones, the product is continuously removed from the basic catalyst. β-Hydroxycarbonyl compounds are readily dehydrated to give α,β-unsaturated carbonyl compounds. With Ar on the β carbon, only the dehydrated product is isolated.

Problem 15.40 Suggest a mechanism for the OH^--catalyzed aldol condensation of acetaldehyde. ◄

Step 1

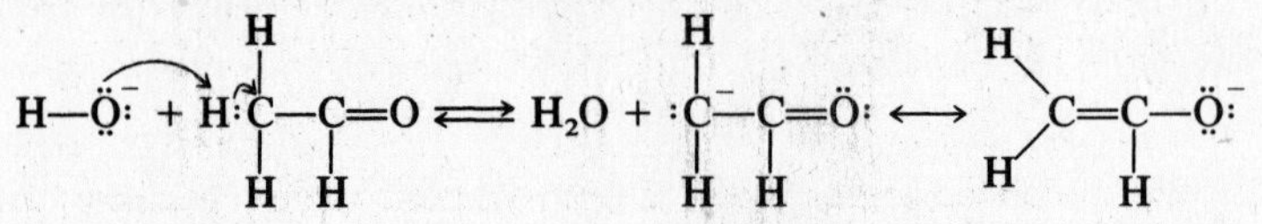

Step 2

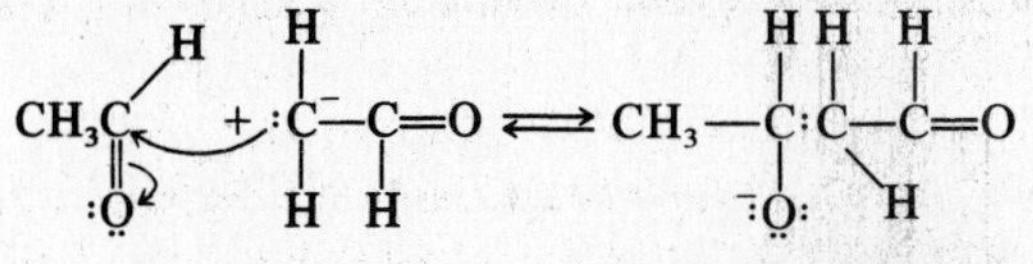

alkoxide ion of the
β-hydroxyaldehyde

Step 3

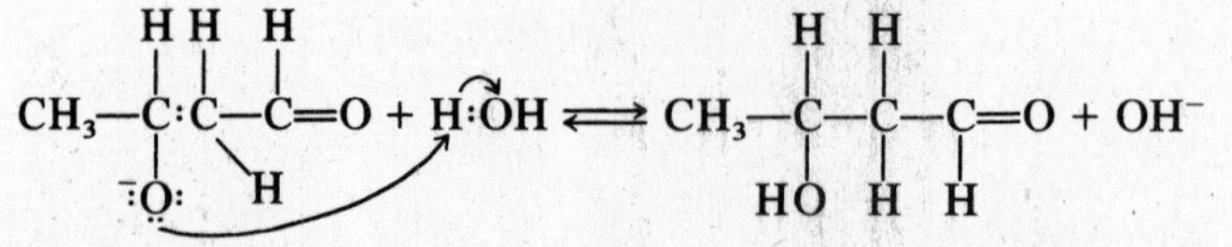

Problem 15.41 Write structural formulas for the β-hydroxycarbonyl compounds and their dehydration products formed by aldol condensations of: (a) butanal, (b) phenylacetaldehyde, (c) diethyl ketone, (d) cyclohexanone, (e) benzaldehyde. ◄

Acceptor   Carbanion Source

(a)

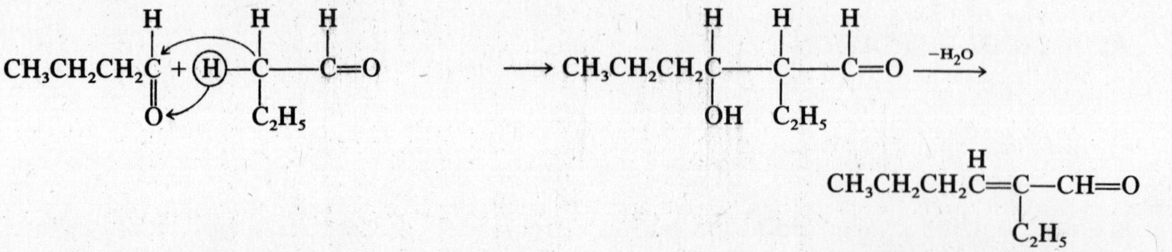

mixture of geometric isomers

Acceptor   Carbanion Source

(b)

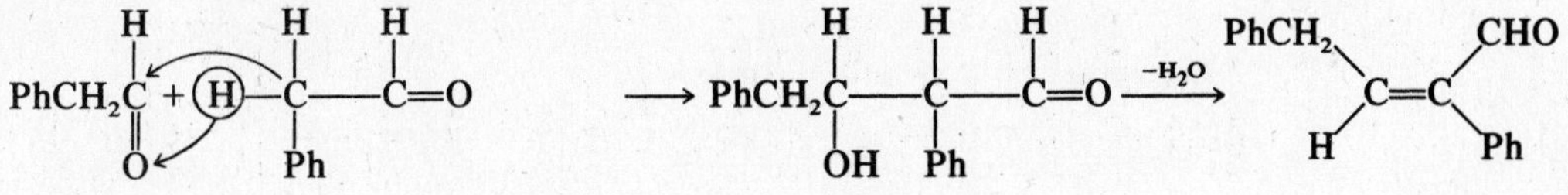

(c)

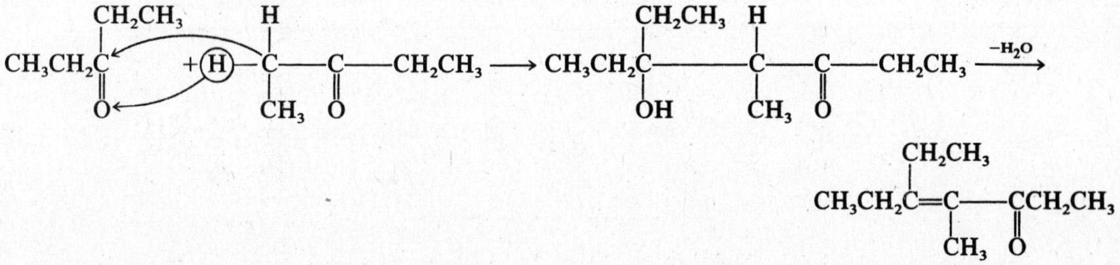

(d)

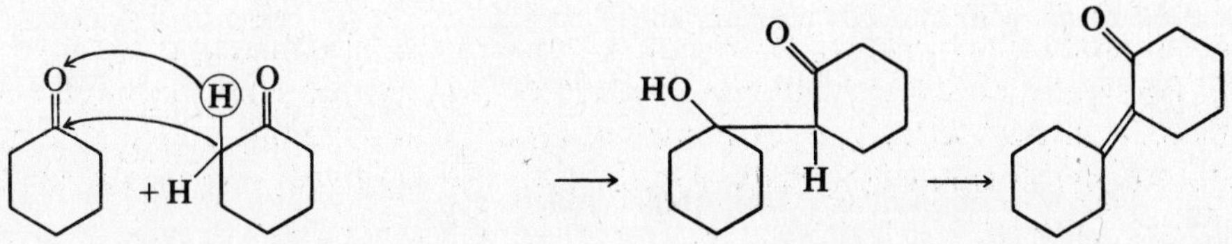

(e) No aldol condensation, because C_6H_5CHO has no α H. Aldol condensations require dilute NaOH at room temperatures. With concentrated NaOH at higher temperatures, C_6H_5CHO undergoes the Cannizzaro reaction (Problem 15.12).

Problem 15.42 Some condensations have little synthetic value because they give mixtures of β-hydroxycarbonyl compounds. Illustrate with (a) a mixture of 2 aldehydes, each with an α H; (b) an unsymmetrical ketone having an H on each α C.   ◄

(a) There are 4 possible products. Each aldehyde reacts with itself to give 2 products. If one aldehyde reacts as carbanion and the other as acceptor, and vice versa, there are 2 more products.

Acceptor   Carbanion Source

Self-aldol

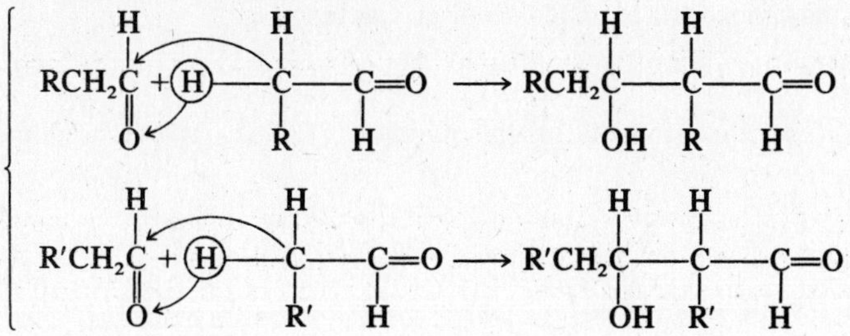

Mixed aldol

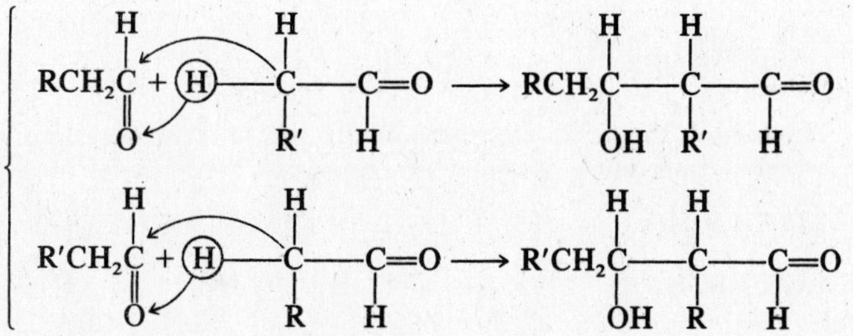

(b) Two products are possible. There is only one acceptor, but 2 different carbanions can be formed—one from each α C.

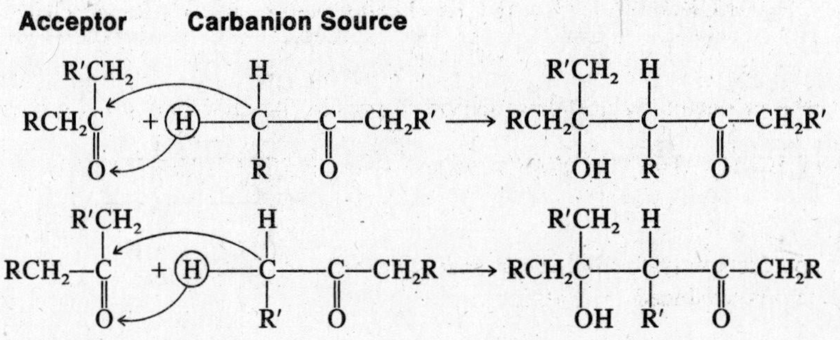

Acceptor Carbanion Source

Problem 15.43 Mixed aldol condensations are useful if (a) one of the 2 aldehydes has no α H, (b) a symmetrical ketone reacts with RCHO. Explain and illustrate. ◀

(a) The aldehyde with no α H, e.g. H_2CO and C_6H_5CHO, is only a carbanion acceptor, so that only 2 products are possible.

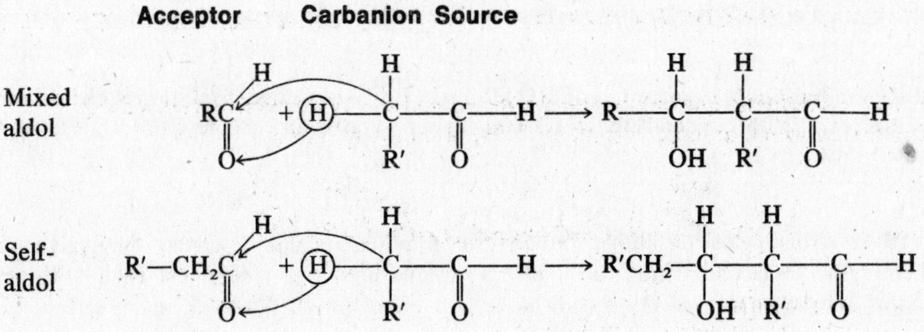

Acceptor Carbanion Source

For reasonably good yields of mixed aldol product, the aldehyde with the α H should be added slowly to a large amount of the one with no α H.

(b) Ketones are poor carbanion acceptors but are carbanion sources. With symmetrical ketones and an RCHO having an α H, 2 products can be formed: (1) the self-aldol of RCHO and (2) the mixed aldol. If RCHO has no α H, only the mixed aldol results. As in part (a), the correct sequence of addition can give a good yield of the mixed aldol products.

Problem 15.44 Give the product for the reaction of $C_6H_5CH{=}O$ in dilute base with (a) CH_3CHO, (b) CH_3COCH_3. ◀

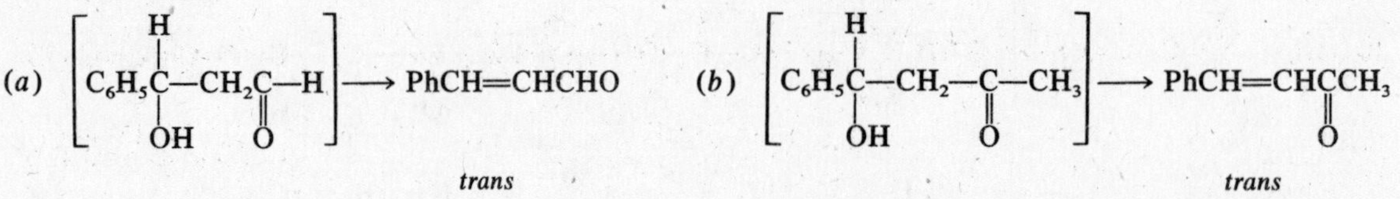

Problem 15.45 Show how the following compounds are made from CH_3CH_2CHO. Do not repeat the synthesis of any compound needed in ensuing syntheses.

(a) $CH_3CH_2CH{=}C(CH_3)CHO$ (b) $CH_3CH_2CH_2CH(CH_3)CHO$ (c) $CH_3CH_2CH{=}C(CH_3)CH_2OH$

(d) $CH_3CH_2CH_2CH(CH_3)CH_2OH$ (e) $CH_3CH_2CH_2CH(CH_3)_2$ (f) $CH_3CH_2CHCH(CH_3)COOH$
                                                                                                   $\overset{|}{OH}$ ◀

Each product has 6 C's, which is twice the number of C's in CH_3CH_2CHO. This suggests an aldol condensation as the first step.

(a)
$$CH_3CH_2CHO \xrightarrow{OH^-} CH_3CH_2\underset{\underset{HO}{|}}{C}H\underset{\underset{CH_3}{|}}{C}HCHO \xrightarrow[-H_2O]{\Delta, H^+} CH_3CH_2CH{=}C(CH_3)CHO \quad (A)$$

(b) —CHO can be protected by acetal formation to prevent its reduction when reducing C=C of (A).

$$(A) \xrightarrow[HCl]{CH_3OH} CH_3CH_2CH{=}C(CH_3)CH(OCH_3)_2 \xrightarrow{H_2/Pt} CH_3CH_2CH_2CH(CH_3)CH(OCH_3)_2 \xrightarrow[H_2O]{HCl}$$
$$CH_3CH_2CH_2CH(CH_3)CHO$$

(There are specific catalysts that permit reduction only of C=C.)

(c) CHO is selectively reduced by $NaBH_4$:

$$(A) \xrightarrow{NaBH_4} CH_3CH_2CH{=}C(CH_3)CH_2OH$$

(d)
$$(A) \xrightarrow{H_2/Pt} CH_3CH_2CH_2CH(CH_3)CH_2OH$$

(e)
$$(A) \xrightarrow[\Delta]{H_2NNH_2, OH^-} CH_3CH_2CH{=}C(CH_3)_2 \xrightarrow{H_2/Pt} CH_3CH_2CH_2CH(CH_3)_2$$

(f) Tollens' reagent, $Ag(NH_3)_2^+$, is a specific oxidant for CHO $\longrightarrow$ COOH.

$$CH_3CH_2CH(OH)CH(CH_3)CHO \xrightarrow[2. \ H^+]{1. \ Ag(NH_3)_2^+} CH_3CH_2CH(OH)CH(CH_3)COOH$$

Problem 15.46 Aldehydes and ketones also undergo acid-catalyzed aldol condensations. Devise a mechanism for this reaction in which an enol (Problem 15.39) is an intermediate. ◄

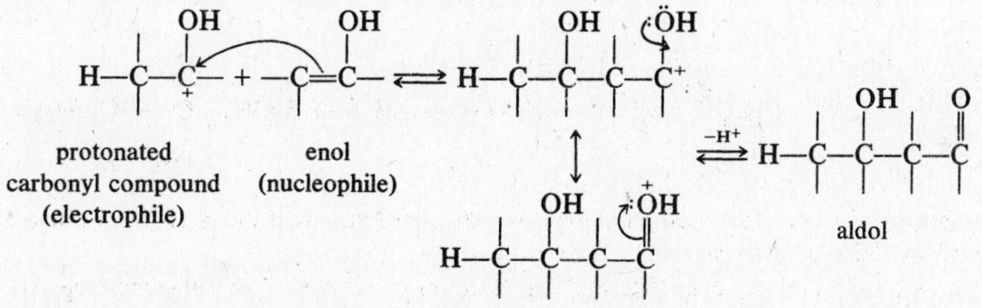

protonated carbonyl compound (electrophile)   enol (nucleophile)   aldol

Problem 15.47 Crotonaldehyde ($\overset{\gamma}{C}H_3\overset{}{C}H{=}\overset{\beta}{C}H\overset{\alpha}{C}H{=}O$) undergoes an aldol condensation with acetaldehyde to form sorbic aldehyde ($CH_3CH{=}CH{-}CH{=}CH{-}CH{=}O$). Explain the reactivity and acidity of the γ H. ◄

Crotonaldehyde has C=C conjugated with C=O. On removal of the γ H by base, the − charge on C is delocalized to O.

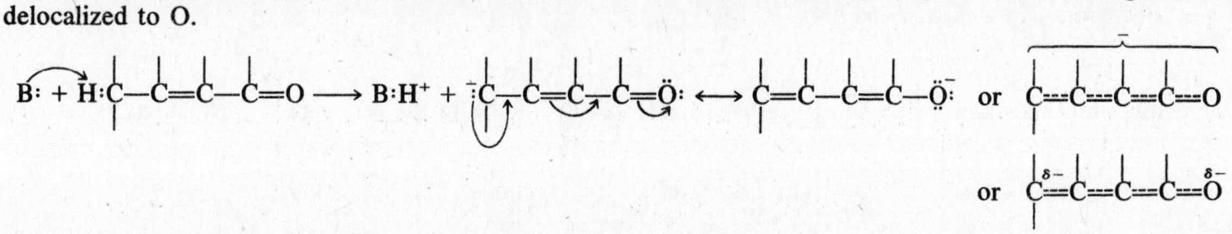

The nucleophilic carbanion adds to the carbonyl group of acetaldehyde.

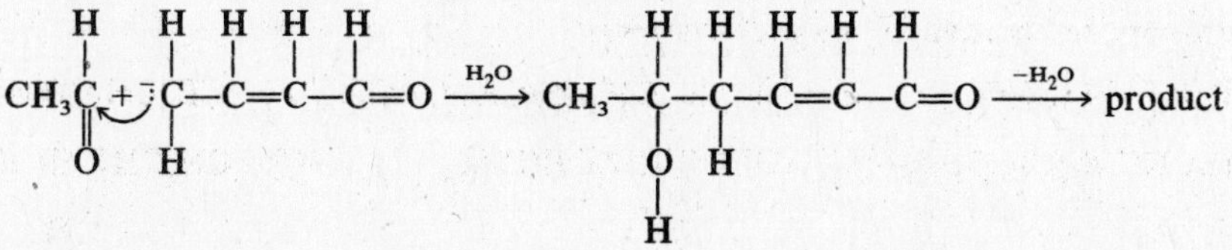

Problem 15.48 Use aldol condensations to synthesize the following useful compounds from cheap and readily available compounds: (a) the food preservative sorbic acid, $CH_3CH=CH—CH=CH—COOH$; (b) 2-ethyl-1-hexanol; (c) 2-ethyl-1,3-hexanediol, an insect repellant; (d) the humectant pentaerythritol, $C(CH_2OH)_4$. ◄

(a) $2\,CH_3CH=O \xrightarrow[-H_2O]{OH^-} CH_3CH=CHCH=O \xrightarrow[OH^-]{CH_3CHO}$

$$CH_3CH=CHCH=CHCH=O \xrightarrow[oxid.]{mild} CH_3(CH=CH)_2COOH$$

(b) $CH_3CH_2CH_2CH_2OH \xrightarrow[heat]{Cu} CH_3CH_2CH_2CH=O \xrightarrow{OH^-} CH_3CH_2CH_2\underset{\underset{OH}{|}}{CH}-\overset{\overset{C_2H_5}{|}}{CH}CH=O \xrightarrow{-H_2O}$

$$CH_3CH_2CH_2CH=\overset{\overset{C_2H_5}{|}}{C}-CH=O \xrightarrow{H_2/Pt} CH_3CH_2CH_2CH_2-\overset{\overset{C_2H_5}{|}}{CH}-CH_2OH$$

(c)   As in (b) to

$$CH_3CH_2CH_2\underset{\underset{OH}{|}}{CH}-\overset{\overset{C_2H_5}{|}}{CH}-CH=O \xrightarrow{H_2/Pt} CH_3CH_2CH_2\underset{\underset{OH}{|}}{CH}-\overset{\overset{C_2H_5}{|}}{CH}-CH_2OH$$

(d)   One mole of CH_3CHO undergoes aldol condensation with 3 moles of H_2CO.   A fourth mole of H_2CO then reacts with the product by a crossed-Cannizzaro reaction.

$$O=CH_2 + H-\overset{\overset{H_2C=O}{\overset{+}{\overset{|}{H}}}}{\underset{\underset{\underset{H_2C=O}{+}}{\underset{|}{H}}}{C}}-CH=O \xrightarrow{Ca(OH)_2} HOCH_2-\overset{\overset{CH_2OH}{|}}{\underset{\underset{CH_2OH}{|}}{C}}-CH=O \xrightarrow[OH^-]{CH_2=O} HOCH_2-\overset{\overset{CH_2OH}{|}}{\underset{\underset{CH_2OH}{|}}{C}}-CH_2OH + H\underset{\underset{O}{||}}{C}O^-$$

Problem 15.49 Which of the following alkanes can be synthesized from a self-aldol condensation product of an aldehyde [see Problem 15.45(a)] or a symmetrical ketone?   (a) $CH_3CH_2CH_2CH_2CH(CH_3)CH_2CH_2CH_3$, (b) $CH_3CH_2CH_2CH_2CH(CH_3)CH_2CH_3$,   (c) $(CH_3)_2CHCH_2CH_2CH_3$,   (d) $(CH_3)_2CHCH_2C(CH_3)_3$, (e) $(CH_3)_2CHCH_2CH_2CH_2CH_2CH_3$,   (f) $(CH_3CH_2)_2CHCH(CH_3)CH_2CH_2CH_3$. ◄

The general formula for the aldol product from $RR'CHCHO$ is

$$R-\boxed{\underset{\underset{R'}{|}}{CH}-\underset{\underset{OH}{|}}{\overset{\overset{H}{|}}{C}}\xleftarrow{} \underset{\underset{R}{|}}{\overset{\overset{R'}{|}}{C}}-CHO}$$

The arrow points to the formed bond, and the α and C=O C's are in the rectangle.   The alkane is

$$R-\boxed{\underset{\underset{R'}{|}}{CHCH_2}\xleftarrow{} \underset{\underset{R'}{|}}{\overset{\overset{R}{|}}{C}}-CH_3}$$

There is always a terminal CH_3 in this four-carbon sequence.   From $RR'CHCOCHRR'$ the products are

$$R\underset{\underset{R'}{|}}{CH}\underset{\underset{RCH}{\underset{R'}{|}}}{\overset{\overset{OH}{|}}{C}}\xleftarrow{}\underset{\underset{RO}{||}}{\overset{\overset{R'}{|}}{C}}\overset{\overset{R'}{|}}{C}CHR \longrightarrow RCH-\underset{\underset{RCH}{\underset{R'}{|}}}{\overset{\overset{R'}{|}}{CH}}\xleftarrow{}\underset{\underset{R}{|}}{\overset{\overset{R'}{|}}{C}}CH_2CHR$$

Each half must have the same skeleton of C's.   Note that R and/or R' can also be Ar or H.   The alkane must always have an even number of C's (twice the number of C's of the carbonyl compound).

(a) No.   There is an odd number of C's in the alkane.

(b)

$$CH_3CH_2\boxed{CH_2CH_2\text{—}\downarrow\text{—}CHCH_3}$$
$$CH_2CH_3$$

Yes.   The four-carbon sequence has a terminal CH_3, and each half has the same sequence of C's.   Use RR'CHCHO, where R = H and R' = CH_2CH_3.

$$CH_3CH_2CH_2CHO \xrightarrow{OH^-} CH_3CH_2CH_2\underset{\underset{CH_2CH_3}{|}}{\overset{\overset{}{|}}{C}}H\underset{\underset{}{|}}{\overset{\overset{HO}{|}}{C}}HCHO \longrightarrow alkane$$

(c)

$$\overset{\overset{CH_3}{|}}{CH_3\text{—}CH}\text{—}\downarrow\text{—}CH_2CH_2CH_3$$

Yes.   Each half has the same skeleton of C's.   A ketone is needed; R = R' = H.

$$(CH_3)_2C{=}O \xrightarrow{OH^-} (CH_3)_2\underset{\underset{OH}{|}}{C}\text{—}CH_2\overset{\overset{O}{\parallel}}{C}\text{—}CH_3 \longrightarrow alkane$$

(d)

$$CH_3\text{—}\boxed{\underset{\underset{H}{|}}{\overset{\overset{CH_3}{|}}{C}}\text{—}CH_2\text{—}\underset{\underset{CH_3}{|}}{\overset{\overset{CH_3}{|}}{C}}\text{—}CH_3}$$

Yes.   Use RR'CHCHO, R = R' = CH_3.

$$(CH_3)_2CHCHO \xrightarrow{OH^-} CH_3\text{—}\underset{\underset{H}{|}}{\overset{\overset{CH_3}{|}}{C}}\text{—}\underset{\underset{HO}{|}}{\overset{\overset{H}{|}}{C}}\text{—}\underset{\underset{CH_3}{|}}{\overset{\overset{CH_3}{|}}{C}}\text{—}CHO \longrightarrow alkane$$

(e)

$$\overset{\overset{CH_3}{|}}{CH_3\text{—}CH}\text{—}CH_2\text{—}\downarrow\text{—}CH_2CH_2CH_2CH_3$$

No.   The formed bond is not part of a four-carbon sequence with a terminal CH_3, and the two halves do not have the same skeleton of C's; one half is branched, the other half is not.

(f)

$$CH_3CH_2\underset{\underset{CH_2CH_3}{|}}{\overset{\overset{H}{|}}{C}}\text{————}\underset{\underset{CH_3}{|}}{\overset{\overset{H}{|}}{C}}\text{—}CH_2CH_2CH_3$$

Yes.   Each half has the same skeleton of C's (5 C's in a row).   Therefore, use a symmetrical ketone with 5 C's in a row (R = H, R' = CH_3).

$$CH_3CH_2\overset{\overset{O}{\parallel}}{C}CH_2CH_3 \xrightarrow{OH^-} CH_3CH_2\underset{\underset{CH_2CH_3}{|}}{\overset{\overset{OH}{|}}{C}}\text{————}\underset{\underset{CH_3}{|}}{\overset{\overset{H}{|}}{C}}\text{—}\overset{\overset{O}{\parallel}}{C}\text{—}CH_2CH_3 \longrightarrow alkane$$

REFORMATSKY REACTION

$$\underset{R'}{\overset{R}{>}}C{=}O + Zn + Br\underset{|}{\overset{|}{C}}\text{—}COOEt \xrightarrow[2.\ H_2O]{1.\ \ dry\ ether} \underset{R'}{\overset{R}{>}}\underset{\underset{OH}{|}}{C}\text{—}\underset{|}{\overset{|}{C}}\text{—}COOEt \quad via \quad \left[BrZn\overset{\delta+}{\text{—}}\underset{|}{\overset{|}{C}}{}^{\delta-}\text{—}COOEt\right]$$

β-hydroxyesters

R' and R may also be H or Ar.

Problem 15.50 Why is Mg or Cd not used in place of Zn in the Reformatsky reaction? ◄

The order of electropositivity is Mg > Zn > Cd. The order of ionic character is

$$Mg-C > Zn-C > Cd-C$$

in $\overset{\delta+}{M}-\overset{\delta-}{R}$. The order of nucleophilicity of R is MgR > ZnR > CdR. Therefore $BrMgCH_2COOC_2H_5$ reacts with the C=O of the ester $BrCH_2COOC_2H_5$. $BrZnCH_2COOC_2H_5$ can only react with RR'C=O, since ketones and aldehydes are more reactive than esters towards nucleophilic addition. CdR_2 does not react with ketones and aldehydes (Problem 15.6).

Problem 15.51 Use the Reformatsky reaction to prepare

   (a) $(CH_3)_2C(OH)CH_2COOC_2H_5$ (b) $PhC(OH)CHCOOC_2H_5$ (c) $PhC{=}CCOOH$ ◄
                                     CH_3 CH_3 CH_3 CH_3

The formed bond is

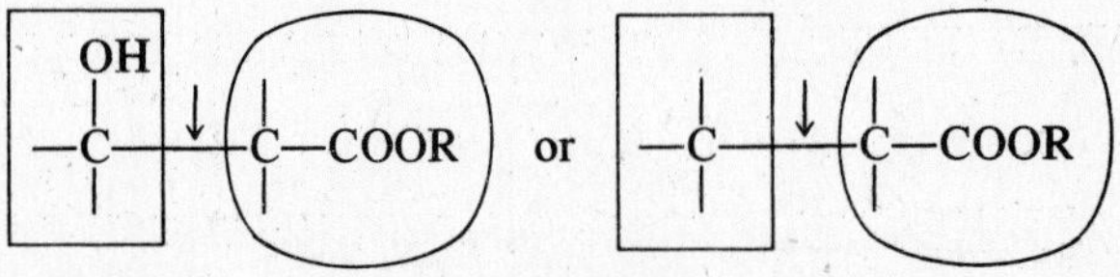

The structure in the box comes from the carbonyl compound (acceptor); the structure in the oval comes from the α-bromoester (carbanion source).

(a)

(b)

(c)   Product from (b) $\xrightarrow[-H_2O]{H^+}$ $PhC{=}C-COOH$
                                                      CH_3 CH_3

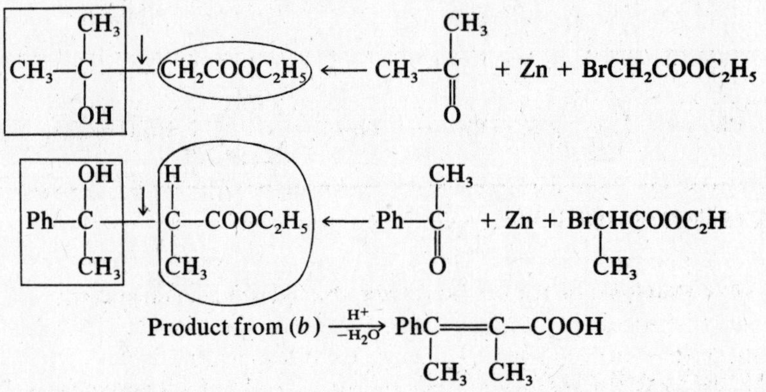

Problem 15.52 For the aldol-type condensations indicated in Table 15-1 give the structure of the stable carbanion, give the product and explain the stability of the carbanion. ◄

Table 15-1

	(a)	(b)	(c)	(d)	(e)	(f)
Acceptor	PhCHO	PhCHO	Me_2CO	Me_2CO	Me_2CO	Ph_2CO
Base	OH^-	OH^-	OH^-	NH_2^-	OH^-	NH_2^-
Carbanion Source		$CH_3C{\equiv}N$	$CHCl_3$	$CH_3C{\equiv}CH$		Ph_2CH_2

See Table 15-2.

Table 15-2

Stable Carbanion	Product	Reason for Stability
(a) $:\overset{-}{C}H_2-\overset{+}{N}\overset{\overset{..}{O}:^-}{\underset{\underset{..}{O}:^-}{}}$　$\longleftrightarrow$　$CH_2=\overset{+}{N}\overset{\overset{..}{O}:^-}{\underset{\underset{..}{O}:^-}{}}$	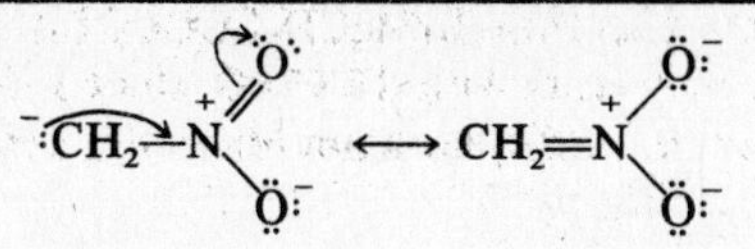 *PhCH=CHNO$_2$	p-p π bond
(b) $:\overset{-}{C}H_2-C\equiv N:$　$\longleftrightarrow$　$CH_2=C=\overset{..}{N}:^-$	*PhCH=CHCN	p-p π bond
(c) $\overset{Cl}{\underset{Cl}{:\overset{\|}{C}-Cl}}$　or　$\left[\overset{Cl}{\underset{Cl}{\overset{\|}{C}=Cl}}\right]^-$ 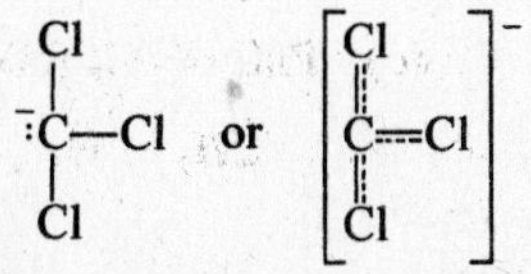	$\underset{OH}{Me_2C-CCl_3}$	p-d π bond
(d) $CH_3C\equiv C:^-$	$\underset{OH}{Me_2C-C\equiv CCH_3}$	sp hybrid
(e) (cyclopentadienyl anion structures)	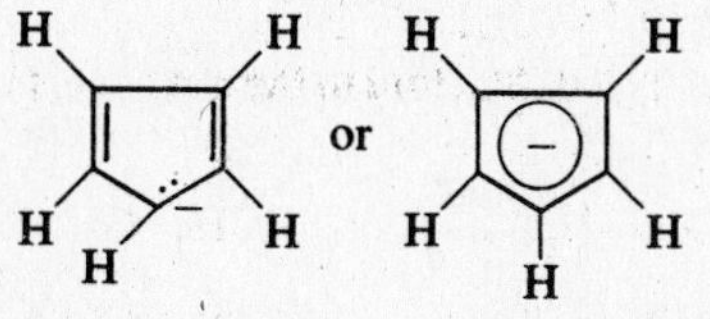 (dimethylfulvene structure) CMe$_2$	 aromaticity
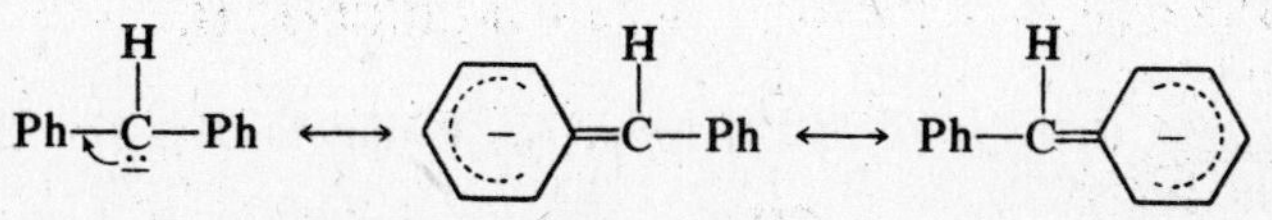 (f) $\overset{H}{Ph-\overset{\|}{\underset{..}{C}}-Ph}$　$\longleftrightarrow$ ... $\longleftrightarrow$ ...	Ph$_2$C=CPh$_2$	p-p π bond

*The more stable *trans* product.

Problem 15.53　Give structures of the products from the following condensations:

(a)　p-CH$_3$C$_6$H$_4$CHO + (CH$_3$CH$_2\overset{\overset{O}{\|}}{C}$)$_2$O $\xrightarrow{CH_3CH_2COO^-Na^+}$　　　(b)　Cyclohexanone + CH$_3$CH$_2$NO$_2$ $\xrightarrow{OH^-}$

(c)　C$_6$H$_5$CHO + C$_6$H$_5$CH$_2$C≡N $\xrightarrow{OH^-}$　　　　　　(d)　Benzophenone + Cyclopentadiene $\xrightarrow{OH^-}$

(e)　CH$_3$COCH$_3$ + 2C$_6$H$_5$CHO $\xrightarrow{OH^-}$　◀

(a)　This is a **Perkin condensation**.

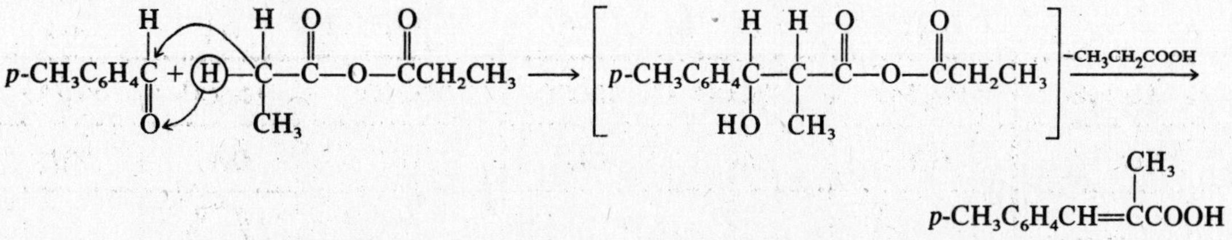

p-CH$_3$C$_6$H$_4\overset{\overset{H}{|}}{\underset{\underset{O}{\|}}{C}}$ + $\overset{H}{(\text{H})}\overset{\overset{H}{|}}{\underset{\underset{CH_3}{|}}{C}}\overset{\overset{O}{\|}}{C}-O-\overset{\overset{O}{\|}}{C}CH_2CH_3$ $\longrightarrow$ $\left[p\text{-CH}_3\text{C}_6\text{H}_4\overset{\overset{H}{|}}{\underset{\underset{HO}{|}}{C}}\overset{\overset{H}{|}}{\underset{\underset{CH_3}{|}}{C}}\overset{\overset{O}{\|}}{C}-O-\overset{\overset{O}{\|}}{C}CH_2CH_3\right]$ $\xrightarrow{-CH_3CH_2COOH}$

p-CH$_3$C$_6$H$_4$CH=$\overset{\overset{CH_3}{|}}{C}$COOH

(b)　(cyclohexanone) =O + $(\text{H})\overset{\overset{H}{|}}{\underset{\underset{CH_3}{|}}{C}}-NO_2$ $\longrightarrow$ $\left[\text{(cyclohexane ring)}\overset{OH}{\underset{\underset{\underset{CH_3}{|}}{CHNO_2}}{|}}\right]$ $\xrightarrow{-H_2O}$ (cyclohexylidene)=$\overset{\overset{CNO_2}{}}{\underset{CH_3}{}}$

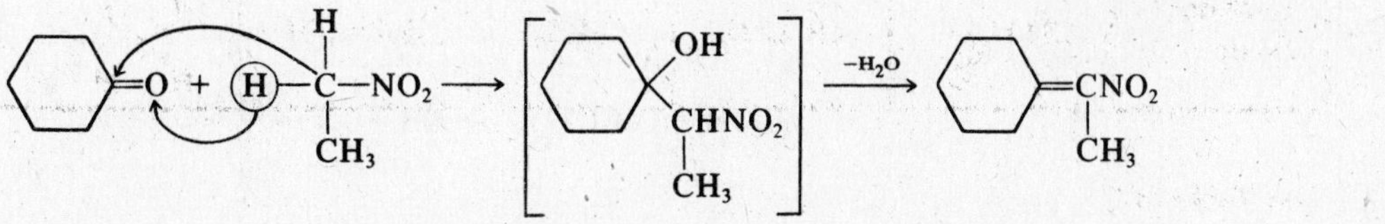

(c)

$$C_6H_5\overset{H}{\underset{O}{C}} + \overset{H}{\underset{C_6H_5}{(H)C}}-CN \longrightarrow \left[C_6H_5\overset{H}{\underset{OH}{C}}-\overset{H}{\underset{C_6H_5}{C}}-CN \right] \xrightarrow{-H_2O} \overset{C_6H_5}{\underset{H}{}}C=C\overset{CN}{\underset{C_6H_5}{}}$$

bulky C_6H_5's are *trans*

(d)

$$Ph_2C=O + \overset{H}{\underset{(H)}{}} \longrightarrow \left[\overset{H}{\underset{Ph_2C}{}} \right] \xrightarrow{-H_2O} Ph_2C$$

Diphenylfulvene

(e) Each CH_3 of $(CH_3)_2CO$ reacts with one PhCHO.

$$\left[\overset{H}{\underset{OH}{PhC}}CH_2\overset{}{\underset{O}{C}}CH_2\overset{H}{\underset{OH}{C}}-Ph \right] \xrightarrow{-2H_2O} PhCH=CH\overset{}{\underset{O}{C}}CH=CHPh$$

Problem 15.54 In the **Knoevenagel reaction**, aldehydes or ketones condense with compounds having a reactive CH_2 between two C=O groups. The cocatalysts are *both* a weak base ($RCOO^-$) *and* a weak acid ($R_2NH_2^+$). Outline the reaction between $C_6H_5CH=O$ and $H_2C(COOEt)_2$. ◀

$$C_6H_5\overset{H}{\underset{O}{C}} + H-\overset{COOC_2H_5}{\underset{COOC_2H_5}{C}}-H \xrightarrow[CH_3COO^-]{Me_2NH_2^+} C_6H_5-\overset{H}{\underset{OH}{C}}-\overset{COOC_2H_5}{\underset{COOC_2H_5}{C}}-H \xrightarrow{-H_2O} C_6H_5-CH=C\overset{COOC_2H_5}{\underset{COOC_2H_5}{}}$$

Ethyl malonate

Problem 15.55 Prepare *trans*-cinnamic acid, $C_6H_5CH=CHCOOH$, by (a) Perkin condensation [Problem 15.53(a)], (b) Reformatsky reaction (Problem 15.51). ◀

(a)

$$\underset{\substack{\text{Benzalde-}\\\text{hyde}}}{PhCHO} + \underset{\substack{\text{Acetic}\\\text{anhydride}}}{(CH_3CO)_2O} \xrightarrow[\text{heat}]{NaOAc} PhCH=CHCOOH$$

(b)

$$PhCHO + BrCH_2COOC_2H_5 \xrightarrow[\substack{2.\ H^+}]{1.\ Zn} PhCH=CHCOOC_2H_5 \xrightarrow[H_2O]{H^+} PhCH=CHCOOH$$

Dehydration of β-hydroxyester occurs on workup because the resulting C=C is conjugated with Ph.

Problem 15.56 The C of the —C≡N group is an electrophilic site capable of being attacked by a carbanion. Show how nitriles like $CH_3CH_2C≡N$ undergo an aldol-type condensation (**Thorpe reaction**) with hindered bases. ◀

$$CH_3-\overset{H}{\underset{H}{C}}-C≡N: + \xrightarrow[-H^+]{R_2N^-Li^+} \left[CH_3-\overset{H}{\underset{}{C}}-C≡N: \longleftrightarrow CH_3\overset{H}{\underset{}{C}}=C=\ddot{N}:^- \right] Li^+ + R_2N\colon H$$

$$CH_3CH_2C≡N: + \left[:\overset{CN}{\underset{}{C}}HCH_3 \right] Li^+ \longrightarrow \left[CH_3CH_2\overset{CN}{\underset{:N^-}{C}}-CHCH_3 \right] Li^+ \xrightarrow{H:OH}$$

$$LiOH + CH_3CH_2\overset{}{\underset{NH}{C}}-\overset{CH_3}{\underset{}{CH}}-CN \xrightarrow{H_3O^+} CH_3CH_2\overset{}{\underset{O}{C}}-\overset{CH_3}{\underset{}{CH}}CN + NH_3$$

an iminonitrile

ALKYLATION OF ENOLATE ANIONS

Enolate anions may act as nucleophiles in S_N2-type reactions with RX.   Since the enolate anion

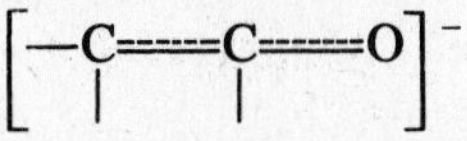

$$\left[-\overset{|}{C}\!=\!=\!=\!\overset{|}{C}\!=\!=\!=\!O \right]^{-}$$

is an ambident ion (Problem 7.26), it can be alkylated at C or at O.

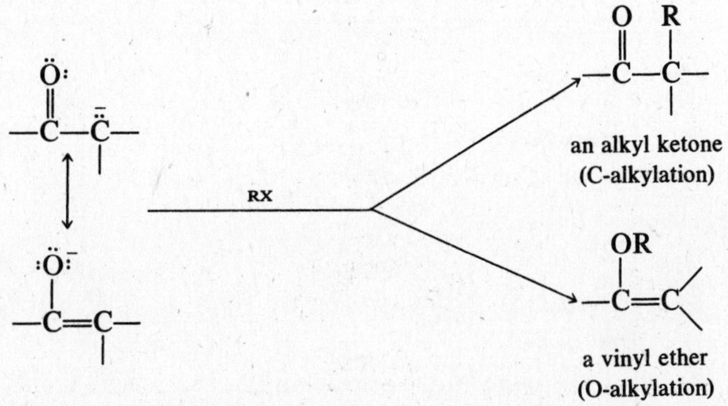

an alkyl ketone
(C-alkylation)

a vinyl ether
(O-alkylation)

If more than one α H is present, further C-alkylation produces mixtures.
   By using their enamines,

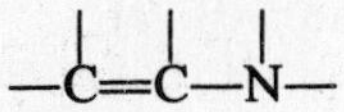

$$-\overset{|}{C}\!=\!\overset{|}{C}\!-\!\overset{|}{N}\!-$$

[Problem 15.33(a)(5)] ketones can be monoalkylated by reactive benzyl and allyl halides in good yield
at the α C.   These N analogs of enol ethers,

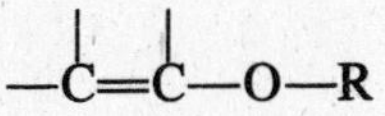

$$-\overset{|}{C}\!=\!\overset{|}{C}\!-\!O\!-\!R$$

are made from the ketone and preferably a 2° amine, R_2NH.   (See page 323.)

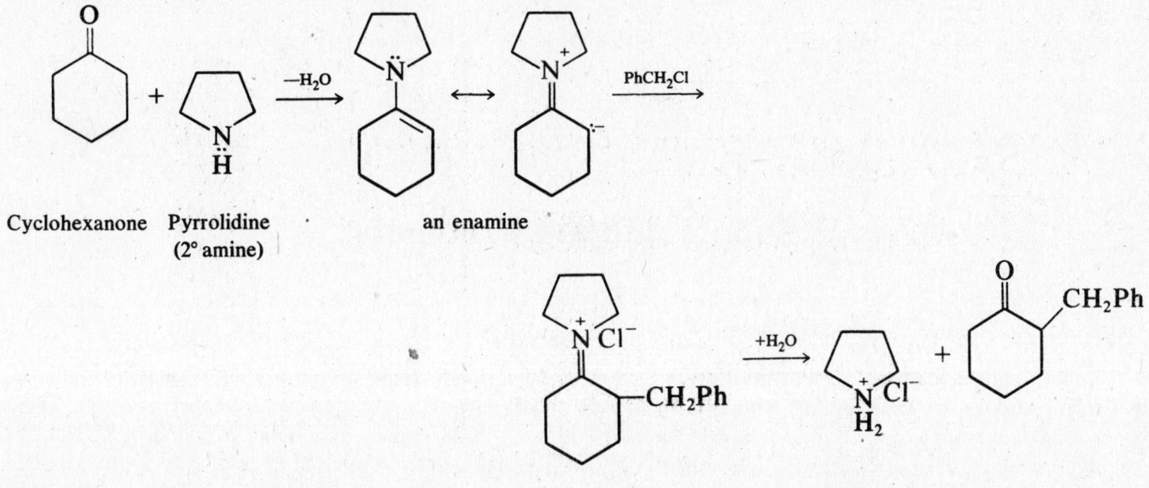

Cyclohexanone   Pyrrolidine   an enamine
                  (2° amine)

NUCLEOPHILIC ADDITION TO CONJUGATED CARBONYL COMPOUNDS; MICHAEL 3,4-ADDITION

α,β-Unsaturated carbonyl compounds add nucleophiles at the β C, leaving a $-$ charge at the α
C.   This intermediate is a stable enolate anion.   These 3,4-Michael additions compete with addition
to the carbonyl group (1,2-addition).

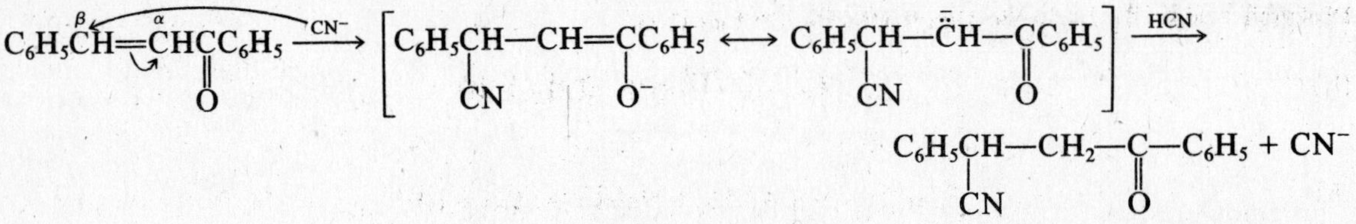

$$C_6H_5\overset{\beta}{C}H \overset{\alpha}{=\!=}CHCC_6H_5 \xrightarrow{CN^-} \left[C_6H_5CH\!-\!CH\!=\!CC_6H_5 \longleftrightarrow C_6H_5CH\!-\!\overset{..}{C}H\!-\!CC_6H_5 \right] \xrightarrow{HCN}$$

$$C_6H_5CH\!-\!CH_2\!-\!\overset{}{C}\!-\!C_6H_5 + CN^-$$

Michael addition product

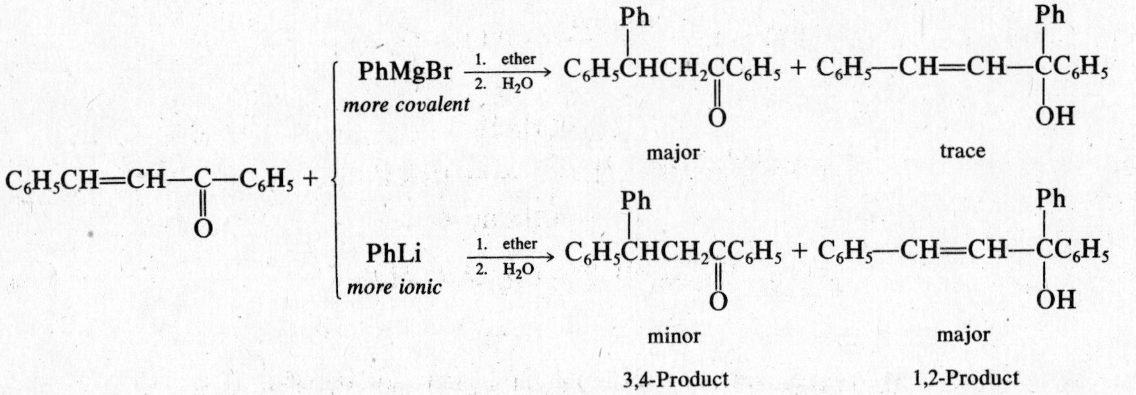

$$C_6H_5CH\!=\!CH\!-\!\overset{}{C}\!-\!C_6H_5 +$$

PhMgBr $\xrightarrow[\text{2. } H_2O]{\text{1. ether}}$ $C_6H_5\overset{Ph}{C}HCH_2CC_6H_5 + C_6H_5\!-\!CH\!=\!CH\!-\!\overset{Ph}{C}C_6H_5$

more covalent

major   trace

PhLi $\xrightarrow[\text{2. } H_2O]{\text{1. ether}}$ $C_6H_5\overset{Ph}{C}HCH_2CC_6H_5 + C_6H_5\!-\!CH\!=\!CH\!-\!\overset{Ph}{C}C_6H_5$

more ionic

minor   major

3,4-Product   1,2-Product

Problem 15.57 Cyanoethylation is the replacement of an acidic α H of a carbonyl compound by a —CH_2CH_2CN group, using acrylonitrile (CH_2=CHCN) and base. Illustrate with cyclohexanone. ◄

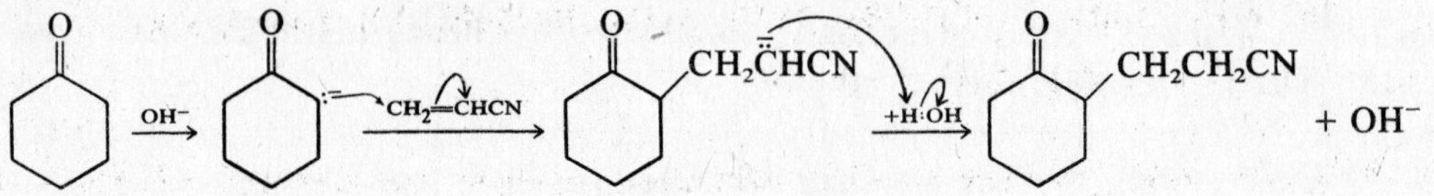

Supplementary Problems

Problem 15.58 (a) What properties identify a carbonyl group of aldehydes and ketones? (b) How can aldehydes and ketones be distinguished? ◄

(a) A carbonyl group (1) forms derivatives with substituted ammonia compounds such as H_2NOH, (2) forms sodium bisulfite adduct with $NaHSO_3$, (3) shows strong ir absorption at 1690–1760 cm^{-1} (C=O stretching frequency), (4) shows weak n-π^* absorption in uv at 289 nm. (b) The H—C bond in RCHO has a unique ir absorption at 2720 cm^{-1}. In nmr the H of CHO has a very downfield peak at δ = 9–10. RCHO gives a positive Tollens' test.

Problem 15.59 What are the similarities and differences between C=O and C=C bonds? ◄

Both undergo addition reactions. They differ in that the C of C=O is more electrophilic than a C of C=C, because O is more electronegative than C. Consequently, the C of C=O reacts with nucleophiles. The C=C is nucleophilic and adds mainly electrophiles.

Problem 15.60 Identify the substances (I) through (VI).

(a)
$$(I) + H_2 \xrightarrow{Pd\,(BaSO_4)} (CH_3)_2CH-CHO$$

(b)
$$CH_3-\underset{\underset{CH_3}{|}}{\overset{\overset{CH_3}{|}}{C}}-\underset{\underset{O}{\|}}{C}-CH_3 + NaOI \longrightarrow (II) + (III)$$

(c)
$$(IV) + H_2O \xrightarrow{HgSO_4,\,H_2SO_4} CH_3CH_2-\underset{\underset{O}{\|}}{C}-CH_3$$

(d)
$$(V) \xrightarrow[-H_2O]{H_2SO_4} CH_3-CH_2-\underset{\underset{CH_2CH_3}{|}}{\overset{\overset{CH_2CH_3}{|}}{C}}-\underset{\underset{O}{\|}}{C}-CH_2CH_3$$

(e)
$$(VI) + CH_3-CHO \xrightarrow[heat]{OH^-} CH_2=CH-CHO$$ ◄

(a) $(CH_3)_2CHC=O$ (I)   (b) $CH_3-\underset{\underset{CH_3}{|}}{\overset{\overset{CH_3}{|}}{C}}-COO^-Na^+$ (II), CHI_3 (III)
$\quad\quad\quad\underset{Cl}{|}$

(c) $H-C\equiv C-CH_2CH_3$ or $CH_3-C\equiv C-CH_3$ (IV)

(d) $(CH_3CH_2)_2\underset{\underset{OH}{|}}{C}-\underset{\underset{OH}{|}}{C}(CH_2CH_3)_2$ (V)   (e) $H-CHO$ (VI) (mixed aldol)

Problem 15.61 By rapid test tube reactions distinguish between (a) pentanal and diethyl ketone, (b) diethyl ketone and methyl n-propyl ketone, (c) pentanal and 2,2-dimethylpropanal, (d) 2-pentanol and 2-pentanone.   ◄

(a) Pentanal, an aldehyde, gives a positive Tollens' test (Ag mirror). (b) Only the methyl ketone gives CHI_3 (yellow precipitate) on treatment with NaOI (iodoform test). (c) Unlike pentanal, 2,2-dimethylpropanal has no α H and so does not undergo an aldol condensation. Pentanal in base gives a colored solution. (d) Only the ketone 2-pentanone gives a solid oxime with H_2NOH. Additionally, 2-pentanol is oxidized by CrO_3 (color change is from orange-red to green). Both give a positive iodoform test.

Problem 15.62 Show steps in the following syntheses: (a) acetaldehyde to n-butyl alcohol, (b) acetyl chloride to acetal $(CH_3CH(OC_2H_5)_2)$, (c) 2-propanol to 2-methyl-2,4-pentanediol, (d) ethyne to 1-butyn-3-ol, (e) allyl chloride to acrolein (propenal), (f) ethanol to 2-butene.   ◄

(a) $2\,CH_3-CHO \xrightarrow{H_3O^+} \left[CH_3-\underset{\underset{OH}{|}}{CH}-CH_2-CHO\right] \xrightarrow{-H_2O} CH_3CH=CH-CHO \xrightarrow{H_2/Pt} CH_3CH_2CH_2CH_2OH$

aldol   Crotonaldehyde

(b)
$$CH_3-\underset{\underset{Cl}{|}}{C}=O \underset{\underset{LiAlH_4}{\searrow}}{\overset{\overset{H_2/Pd}{\nearrow}}{}} \left.\begin{array}{l} CH_3CH=O \\ + \\ CH_3CH_2OH \end{array}\right\} \xrightarrow[dry]{HCl} CH_3-\underset{\underset{OC_2H_5}{|}}{\overset{\overset{OC_2H_5}{|}}{C}}-H$$

(c) $CH_3-\underset{\underset{OH}{|}}{\overset{\overset{CH_3}{|}}{C}}-H \xrightarrow[H^+]{Cr_2O_7^{2-}} CH_3-\underset{\underset{O}{\|}}{\overset{\overset{CH_3}{|}}{C}} \xrightarrow{OH^-} CH_3-\underset{\underset{OH}{|}}{\overset{\overset{CH_3}{|}}{C}}-CH_2-\underset{\underset{O}{\|}}{C}-CH_3 \xrightarrow{H_2/Pt} CH_3-\underset{\underset{OH}{|}}{\overset{\overset{CH_3}{|}}{C}}-CH_2-\underset{\underset{OH}{|}}{\overset{\overset{H}{|}}{C}}-CH_3$

(d)
$$H-C\equiv C-H + H_2O \xrightarrow[H_2SO_4]{HgSO_4} CH_3\overset{\displaystyle}{C}=O \xrightarrow[2.\ H_3O^+]{1.\ H-C\equiv C^-Na^+} CH_3-\overset{\overset{\displaystyle H}{|}}{\underset{\underset{\displaystyle OH}{|}}{C}}-C\equiv C-H$$

(e)
$$H_2C=CHCH_2Cl \xrightarrow[HCO_3^-]{CH_3SOCH_3} H_2C=CHCHO$$

(f)
$$CH_3CH_2OH \xrightarrow{HBr} CH_3CH_2Br \xrightarrow[2.\ OEt^-]{1.\ Ph_3P} CH_3CH=PPh_3 \xrightarrow{CH_3CHO} product$$
$$\overset{|}{\underset{\Delta}{\xrightarrow{Cu}}} CH_3CHO \underline{\hspace{8cm}}$$

Problem 15.63 Use benzene and any aliphatic and inorganic compounds to prepare   (a) 1,1-diphenylethanol, (b) 4,4-diphenyl-3-hexanone. ◄

(a)   The desired 3° alcohol is made by reaction of a Grignard with a ketone by two possible combinations:

$$(C_6H_5)_2CO + CH_3MgBr \quad\quad or \quad\quad C_6H_5COCH_3 + C_6H_5MgBr$$

Since it is easier to make $C_6H_5COCH_3$ than $(C_6H_5)_2CO$ from C_6H_6, the latter pair is used.   Benzene is used to prepare both intermediate products.

$$\left. \begin{array}{l} C_6H_6 + CH_3COCl \xrightarrow{AlCl_3} C_6H_5COCH_3 \\ \\ + \\ \\ C_6H_6 \xrightarrow{Br_2 \atop Fe} C_6H_5Br \xrightarrow{Mg \atop Et_2O} C_6H_5MgBr \end{array} \right\} \xrightarrow[2.\ H_2O]{1.\ reaction} (C_6H_5)_2C(OH)CH_3$$

(b)   The 4° C of $CH_3CH_2\overset{4°}{C}OCPh_2CH_2CH_3$ is adjacent to C=O, and this suggests a pinacol rearrangement of

$$CH_3CH_2\underset{\underset{\displaystyle OH}{|}}{C}Ph-\underset{\underset{\displaystyle OH}{|}}{C}PhCH_2CH_3$$

which is made from CH_3CH_2COPh as follows:

$$C_6H_6 + CH_3CH_2COCl \xrightarrow{AlCl_3} PhCOCH_2CH_3 \xrightarrow{Mg \atop Et_2O} CH_3CH_2-\underset{\underset{\displaystyle OH}{|}}{\overset{\overset{\displaystyle Ph}{|}}{C}}-\underset{\underset{\displaystyle OH}{|}}{\overset{\overset{\displaystyle Ph}{|}}{C}}-CH_2CH_3 \xrightarrow{H^+} product$$

Problem 15.64 Use butyl alcohols and any inorganic materials to prepare 2-methyl-4-heptanone. ◄

The indicated bond

$$CH_3CH_2CH_2\overset{\downarrow}{\underset{\underset{\displaystyle O}{\|}}{C}}CH_2CH(CH_3)_2$$

is formed from 2 four-carbon moieties by a Grignard reaction.

$$CH_3CH_2CH_2CH_2OH \quad\quad CH_3-\overset{\overset{\displaystyle CH_3}{|}}{C}HCH_2Br \xleftarrow{PBr_3} CH_3-\overset{\overset{\displaystyle CH_3}{|}}{C}H-CH_2OH$$

n-Butyl alcohol   Isobutyl alcohol

$$\downarrow Cu,\ heat \quad\quad\quad\quad \downarrow Mg$$

$$CH_2CH_2CH_2CH=O + CH_3-\overset{\overset{\displaystyle CH_3}{|}}{C}HCH_2MgBr \xrightarrow[2.\ H_2O,\ H^+]{1.\ ether} CH_3CH_2CH_2\overset{\downarrow}{\underset{\underset{\displaystyle OH}{|}}{C}}H-CH_2-\overset{\overset{\displaystyle CH_3}{|}}{C}H-CH_3 \xrightarrow{KMnO_4} product$$

2-Methyl-4-heptanol

Problem 15.65 Compound (A), $C_4H_8Cl_2$, is hydrolyzed to compound (B), C_4H_8O, which gives an oxime and a negative Tollens' test. What is the structure of (A)? ◄

(B) is a carbonyl compound because it forms an oxime, and therefore (A) is a *gem*-dihalide,

$$-\overset{|}{\underset{}{C}}Cl_2$$

Since (B) does not reduce Tollens' reagent, (B) is not an aldehyde, but must be a ketone. The only ketone with 4 C's is $CH_3COCH_2CH_3$, and (A) is $CH_3CCl_2CH_2CH_3$.

Problem 15.66 Compound (A), $C_5H_{10}O$, forms a phenylhydrazone, gives negative Tollens' and iodoform tests and is reduced to pentane. What is the compound? ◄

Phenylhydrazone formation indicates a carbonyl compound. Since the negative Tollens' test rules out an aldehyde, (A) must be a ketone. A negative iodoform test rules out the $CH_3C=O$ group, and the reduction product, pentane, establishes the C's to be in a continuous chain. The compound is $CH_3CH_2COCH_2CH_3$.

Problem 15.67 A compound ($C_5H_8O_2$) is reduced to pentane. With H_2NOH it forms a dioxime and also gives positive iodoform and Tollens' tests. Deduce its structure. ◄

Reduction to pentane indicates 5 C's in a continuous chain. The dioxime shows 2 carbonyl groups. The positive CHI_3 test points to

$$CH_3-\overset{\overset{\displaystyle O}{\|}}{C}-$$

while the positive Tollens' test establishes a $-CH=O$. The compound is

$$CH_3-\underset{\underset{\displaystyle O}{\|}}{C}-CH_2CH_2-CHO$$

Problem 15.68 The Grignard reagent of RBr (I) with CH_3CH_2CHO gives a 2° alcohol (II), which is converted to R'Br (III), whose Grignard reagent is hydrolyzed to an alkane (IV). (IV) is also produced by coupling (I). What are the compounds (I), (II), (III), and (IV)? ◄

Since CH_3CH_2CHO reacts with the Grignard of (I) to give (II) after hydrolysis, (II) must be an alkyl ethyl carbinol.

$$RMgBr \xrightarrow{CH_3CH_2CHO} CH_3CH_2\overset{\overset{\displaystyle H}{|}}{\underset{\underset{\displaystyle OMgBr}{|}}{C}}-R \xrightarrow[H^+]{HOH} CH_3CH_2\overset{\overset{\displaystyle H}{|}}{\underset{\underset{\displaystyle OH}{|}}{C}}-R$$

$$\text{Grignard of (I)} \qquad\qquad\qquad\qquad\qquad (II)$$

The conversion of (II) to (IV) is

$$CH_3CH_2\overset{\overset{\displaystyle H}{|}}{\underset{\underset{\displaystyle OH}{|}}{C}}-R \xrightarrow{HBr} CH_3CH_2\overset{\overset{\displaystyle H}{|}}{\underset{\underset{\displaystyle Br}{|}}{C}}R \xrightarrow[\text{ether}]{Mg} CH_3CH_2\overset{\overset{\displaystyle H}{|}}{\underset{\underset{\displaystyle MgBr}{|}}{C}}-R \xrightarrow{HOH} CH_3CH_2CH_2-R$$

$$(II) \qquad\qquad\qquad (III) \qquad\qquad\qquad\qquad\qquad (IV)$$

(IV) must be symmetrical, since it is formed by coupling (I). R is therefore $-CH_2CH_2CH_3$. (I) is $CH_3CH_2CH_2Br$. (IV) is *n*-hexane. (II) is $CH_3CH_2CH(OH)CH_2CH_2CH_3$. (III) is $CH_3CH_2CHBrCH_2CH_2CH_3$.

Problem 15.69 Carry out the following synthesis:

$$C_6H_5-CH=\overset{\overset{\displaystyle H_3C}{|}}{C}-\overset{\overset{\displaystyle O}{\|}}{C}-CH_3 \longrightarrow C_6H_5-CH=\overset{\overset{\displaystyle H_3C}{|}}{C}-\overset{\overset{\displaystyle O}{\|}}{C}-OH \qquad ◄$$

Common oxidants cannot be used, because they oxidize the C=C bond.   Since the starting compound is a methyl ketone, its reaction with NaOI removes CH_3 and converts

$$
\begin{array}{c}
O \\
\parallel \\
-C-CH_3
\end{array}
$$

to $-COO^-$, which is then acidified.

Problem 15.70 Translate the following description into a chemical equation: Friedel-Crafts acylation of resorcinol (1,3-dihydroxybenzene) with $CH_3(CH_2)_4COCl$ produces a compound which on Clemmensen reduction yields the important antiseptic, hexylresorcinol.   ◄

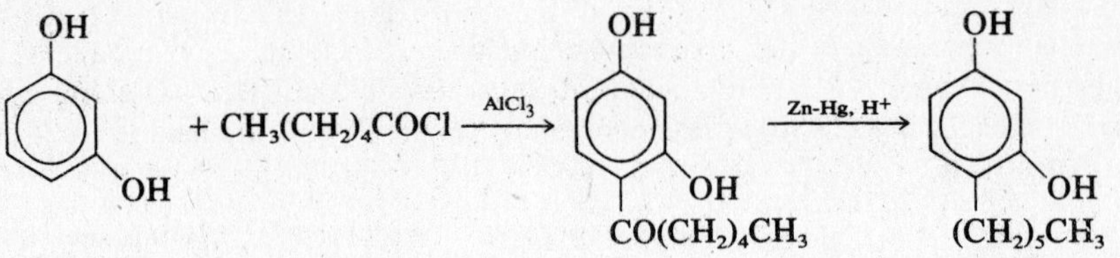

Problem 15.71 Treatment of benzaldehyde with HCN produces a mixture of two isomers that cannot be separated by very careful fractional distillation.   Explain.   ◄

Formation of benzaldehyde cyanohydrin creates a chiral C and produces a racemic mixture, which cannot be separated by fractional distillation.

$$
\begin{array}{c}
H \\
| \\
C_6H_5-C=O + HCN \longrightarrow C_6H_5-\overset{*}{C}-OH \\
| \\
CN
\end{array}
$$

Problem 15.72 Explain the following reaction: ◄

$$
\begin{array}{c}
C_6H_5C-C=O + NaOH \longrightarrow C_6H_5CHC-O^-Na^+ \\
\parallel \ \ | \qquad\qquad\qquad | \ \ \parallel \\
O \ \ H \qquad\qquad\qquad HO \ O
\end{array}
$$

Phenylglyoxal   sodium salt of mandelic acid

This is an internal crossed-Cannizzaro reaction: the keto group is reduced and the $-CHO$ is oxidized.

Problem 15.73 Prepare 1-phenyl-1-(p-bromophenyl)-1-propanol from benzoic acid, bromobenzene and ethanol.   ◄

The compound is a 3° alcohol, conveniently made from a ketone and a Grignard reagent as shown.

$$
C_6H_5COOH \xrightarrow{PCl_5} C_6H_5-\underset{\underset{O}{\parallel}}{C}-Cl \xrightarrow[AlCl_3]{C_6H_5Br} p\text{-}BrC_6H_4-\underset{\underset{O}{\parallel}}{C}-C_6H_5
$$

+

$$
C_2H_5OH \xrightarrow{HBr} C_2H_5Br \xrightarrow[ether]{Mg} C_2H_5MgBr
$$

$$
\Big\downarrow \begin{array}{l} 1. \ \ ether \\ 2. \ \ NH_4^+ \ (mild \ acid, \ prevents \ H_2O \ loss) \end{array}
$$

$$
p\text{-}BrC_6H_4-\underset{\underset{OH}{|}}{\overset{\overset{C_2H_5}{|}}{C}}-C_6H_5
$$

Problem 15.74 Prepare 4-methyl-1-hepten-5-one from $(CH_3CH_2)_2C=O$ and any other needed compounds. ◀

$$CH_3CH_2\overset{\overset{O}{\|}}{C}-\underset{\underset{CH_3}{|}}{CH}(CH_2CH=CH_2)$$

has an allyl group substituted on the α C of diethyl ketone. This substitution is best achieved through the enamine reaction.

$$CH_3CH_2-\underset{\underset{C_2H_5}{|}}{C}=O + HN\Big\rangle \longrightarrow CH_3\underset{\underset{H}{|}}{\overset{..}{C}}-\underset{\underset{C_2H_5}{|}}{C}=N^+\Big\rangle \xrightarrow{ClCH_2CH=CH_2}$$

an enamine

$$\underset{\underset{H}{|}}{\overset{\overset{CH_2CH=CH_2}{|}}{CH_3\overset{|}{C}}}-\underset{\underset{C_2H_5}{|}}{C}\overset{}{=}\overset{+}{N}\Big\rangle Cl^- \xrightarrow[HCl]{H_2O} CH_3-\underset{\underset{H}{|}}{\overset{\overset{CH-CH=CH_2}{|}}{C}}-\underset{\underset{C_2H_5}{|}}{C}=O + \overset{+}{N}\Big\rangle Cl^-$$

Problem 15.75 Compounds "labeled" at various positions by isotopes such as ^{14}C (radioactive), D (deuterium) and ^{18}O are used in studying reaction mechanisms. Suggest a possible synthesis of each of the labeled compounds below, using $^{14}CH_3OH$ as the source of ^{14}C, D_2O as the source of D, and $H_2^{18}O$ as the source of ^{18}O. Once a ^{14}C-labeled compound is made, it can be used in ensuing syntheses. Use any other unlabeled compounds.　　(a) $CH_3^{14}CH_2OH$,　　(b) $^{14}CH_3CH_2OH$,　　(c) $^{14}CH_3CH_2CHO$,　　(d) $C_6H_5^{14}CHO$, (e) $^{14}CD_3CH_2OH$,　(f) $^{14}CH_3CHDOH$,　(g) $CH_3CH^{18}O$. ◀

(a) The 1° alcohol with a labeled carbinol C suggests a Grignard reaction with $H_2^{14}C=O$.

$$^{14}CH_3OH \xrightarrow[300\,°C]{Cu} H_2^{14}C=O \xrightarrow[2.\ H_3O^+]{1.\ CH_3MgBr} CH_3^{14}CH_2OH$$

(b) Now the Grignard reagent is labeled instead of H_2CO.

$$^{14}CH_3OH \xrightarrow{HBr} {}^{14}CH_3Br \xrightarrow[Et_2O]{Mg} {}^{14}CH_3MgBr \xrightarrow[2.\ H_3O^+]{1.\ CH_2O} {}^{14}CH_3CH_2OH$$

(c)　　$^{14}CH_3MgBr$ [see (b)] $+ H_2\overset{\diagdown}{C}\overset{}{-}\overset{\diagup}{CH_2} \xrightarrow[2.\ H_3O^+]{1.\ ether} {}^{14}CH_3CH_2CH_2OH \xrightarrow[heat]{Cu} {}^{14}CH_3CH_2CHO$
　　　　　　　　　　　　　　　　　　$\overset{\diagdown}{O}\overset{\diagup}{}$

(d)　　　　$^{14}CH_3Br + C_6H_6 \xrightarrow{AlCl_3} C_6H_5^{14}CH_3 \xrightarrow[\substack{acetic \\ anhydride \\ 2.\ H_3O^+}]{1.\ CrO_3} C_6H_5^{14}CHO$

(e)　　　　$^{14}CH_3CH_2OH$ [from (b)] $\xrightarrow{Cu} {}^{14}CH_3CHO \xrightarrow[OD^-\ (trace)]{D_2O} {}^{14}CD_3CHO$

(f)　D on carbinol C is best introduced by reduction of a —CHO group with a D-labeled reductant.

$$^{14}CH_3CHO \text{ from } (e) + D_2/Pt \text{ or } LiAlD_4 \longrightarrow {}^{14}CH_3CHDOD \xrightarrow{H_2O} {}^{14}CH_3CHDOH$$

　　D of OD is easily exchanged with excess H_2O.

(g)　Add CH_3CHO to $H_2^{18}O$ with a trace of HCl.

$$H_2^{18}O + CH_3CHO \underset{H^+}{\overset{H^+}{\rightleftarrows}} \left[\underset{\underset{OH}{|}}{CH_3CH^{18}OH}\right] \underset{H^+}{\rightleftarrows} CH_3CH^{18}O + H_2O$$

hydrate

The unstable half-labeled hydrate can lose H_2O to give $CH_3CH^{18}O$.

Problem 15.76 Isopropyl chloride is treated with triphenylphosphine (Ph$_3$P) and then with NaOEt. CH$_3$CHO is added to the reaction product to give a compound, C$_5$H$_{10}$. When C$_5$H$_{10}$ is treated with diborane and then CrO$_3$, a ketone is obtained. Give the structural formula for C$_5$H$_{10}$ and the name of the ketone. ◄

The series of reactions is:

Formation of Ylide

$$(CH_3)_2CHCl + Ph_3P \longrightarrow [(CH_3)_2CH\overset{+}{-}PPh_3]Cl^- \xrightarrow{\text{NaOEt}} (CH_3)_2C=PPh_3 + EtOH + Na^+Cl^-$$

Wittig Reaction

$$(CH_3)_2C=PPh_3 + CH_3-\overset{\overset{\displaystyle H}{|}}{C}=O \longrightarrow (CH_3)_2C=\overset{\overset{\displaystyle H}{|}}{C}-CH_3$$

Anti-Markovnikov Hydroboration-Oxidation

$$(CH_3)_2C=\underset{\underset{\displaystyle H}{|}}{C}-CH_3 \xrightarrow[\text{2. CrO}_3]{\text{1. BH}_3} (CH_3)_2C-\underset{\underset{\displaystyle O}{\|}}{\overset{\overset{\displaystyle H}{|}}{C}}-CH_3 \quad \text{3-Methyl-2-butanone}$$

Problem 15.77 The **Robinson "annelation" reaction** for synthesizing fused rings uses Michael addition followed by intramolecular aldol condensation. Illustrate with cyclohexanone and methyl vinyl ketone, CH$_2$=CHCOCH$_3$. ◄

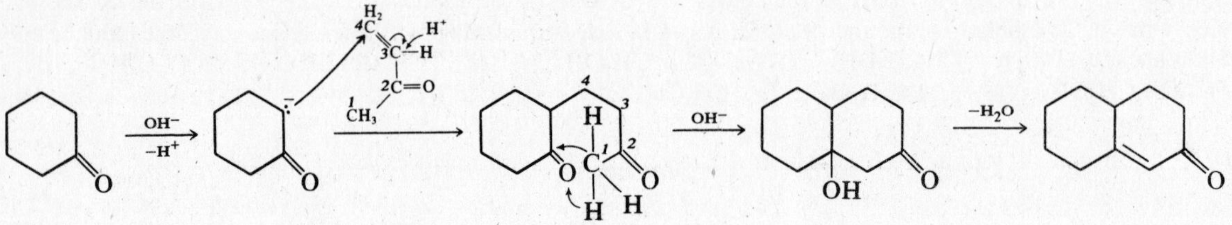

Problem 15.78 Prepare

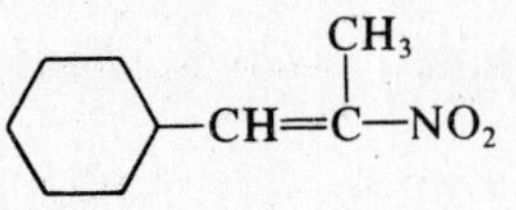

from simple acyclic compounds. ◄

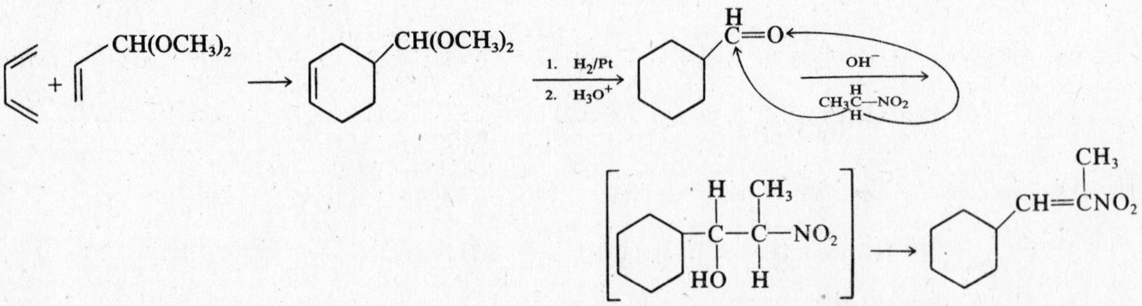

The $\diagup\!\!\!\diagdown$C(OCH$_3$)$_2$ is an electron-withdrawing group which activates the dienophile.

Problem 15.79 Trace the oxidation of 2-propanol to acetone by infrared spectroscopy. ◄

Observe the disappearance of the O—H stretching band at 3600 cm^{-1} and the appearance of the C=O stretching band at about 1720 cm^{-1}.

Problem 15.80 Deduce the structure of a compound, C_4H_6O, with the following spectral data: (a) Electronic absorption at $\lambda_{max} = 213$ nm, $\epsilon_{max} = 7100$ and $\lambda_{max} = 320$ nm, $\epsilon_{max} = 27$. (b) Infrared bands, among others, at 3000, 2900, 1675 (most intense) and 1602 cm^{-1}. (c) Nmr singlet at $\delta = 2.1$ (3 H's), three multiplets each integrating for 1 H at $\delta = 5.0$–6.0.   ◄

The formula C_4H_6O indicates two degrees of unsaturation and may represent an alkyne or some combination of two rings, C=C and C=O groups.

(a) λ_{max} at 213 nm comes from the $\pi \to \pi^*$ transition. It is more intense than the λ_{max} at 320 nm from the $n \to \pi^*$ transition. Both peaks are shifted to higher wavelengths than normal (190 and 280 nm, respectively), thus indicating an α,β-unsaturated carbonyl compound. The two degrees of unsaturation are a C=C and a C=O.

(b) The given peaks and their bonds are 3000 cm^{-1}, sp^2 C—H; 2900 cm^{-1}, sp^3 C—H; 1675 cm^{-1}, C=O (probably conjugated to C=C); 1602 cm^{-1}, C=C. All are stretching vibrations. Absence of a band at 2720 cm^{-1} means no aldehyde H. The compound is probably a ketone.

(c) The singlet at $\delta = 2.1$ is from a

$$\begin{array}{c} O \\ \parallel \\ -C-CH_3 \end{array}$$

There are also three nonequivalent vinylic H's ($\delta = 5.0$–6.0) which intercouple. The compound is

$$\begin{array}{ccc} H^c & & H^b \\ & C{=}C & \\ H^d & & C-CH_3^a \\ & & \parallel \\ & & O \end{array}$$

shown with nonequivalent H's.

Problem 15.81 A compound, $C_5H_{10}O$, has a strong ir band at about 1700 cm^{-1}. The nmr shows no peak at $\delta = 9$–10. The mass spectrum shows the base peak (most intense) at $m/e = 57$ and nothing at $m/e = 43$ or $m/e = 71$. What is the compound?   ◄

The strong ir band at 1700 cm^{-1} indicates a C=O, accounting for the one degree of unsaturation. The absence of a signal at $\delta = 9$–10 means no

$$\begin{array}{c} O \\ \parallel \\ -C-H \end{array}$$

proton. The compound is a ketone, not an aldehyde. Nmr is the best way to differentiate between a ketone and an aldehyde.

Carbonyl compounds undergo fragmentation to give stable acylium ions [see Problem 12.34(e)].

$$\begin{array}{c} R-C{=}\overset{+}{\overset{..}{O}} \\ | \\ R'\ (P^+) \end{array} \longrightarrow R-C{\equiv}\overset{+}{O}{:} + \cdot R' \quad (\text{or } R'-C{\equiv}\overset{+}{O}{:} + \cdot R)$$

an acylium
ion

The possible ketones are

$$\underset{\underset{O}{\parallel}}{CH_3CH_2CCH_2CH_3} \qquad \underset{\underset{O}{\parallel}}{CH_3CCH_2CH_2CH_3} \qquad \underset{\underset{O}{\parallel}}{CH_3CCH(CH_3)_2}$$

$$\qquad\quad A \qquad\qquad\qquad\qquad\quad B \qquad\qquad\qquad\qquad C$$

Compounds B and C would both give some $CH_3C{\equiv}\overset{+}{O}{:}$ ($m/e = 43$) and $C_3H_7C{\equiv}\overset{+}{O}{:}$ ($m/e = 71$). These peaks were absent; therefore A, which fragments to $CH_3CH_2C{\equiv}\overset{+}{O}{:}$ ($m/e = 57$), is the compound.

Problem 15.82 Propanal reacts with 1-butene in the presence of uv to give $CH_3CH_2COCH_2CH_2CH_2CH_3$. Give steps for a likely mechanism. ◄

The net reaction is an addition of CH_3CH_2CHO to $H_2C=CHCH_2CH_3$.

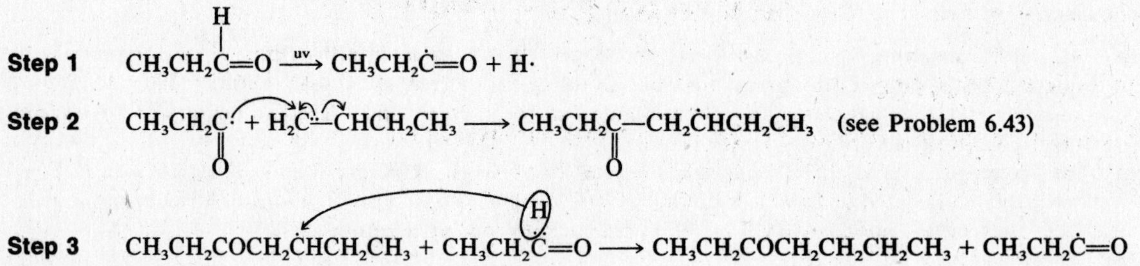

Step 1 $CH_3CH_2\overset{\overset{\displaystyle H}{|}}{C}=O \xrightarrow{uv} CH_3CH_2\dot{C}=O + H\cdot$

Step 2 $CH_3CH_2\overset{|}{\underset{O}{C}}\cdot + H_2C=CHCH_2CH_3 \longrightarrow CH_3CH_2\overset{|}{\underset{O}{C}}-CH_2\dot{C}HCH_2CH_3$ (see Problem 6.43)

Step 3 $CH_3CH_2COCH_2\dot{C}HCH_2CH_3 + CH_3CH_2\dot{C}=O \longrightarrow CH_3CH_2COCH_2CH_2CH_2CH_3 + CH_3CH_2\dot{C}=O$

Step 1 is the initiation step. Steps 2 and 3 propagate the chain.

Problem 15.83 In the biochemical conversion of the sugar glucose (Problem 25.3) to ethanol (**alcoholic fermentation**) a key step is

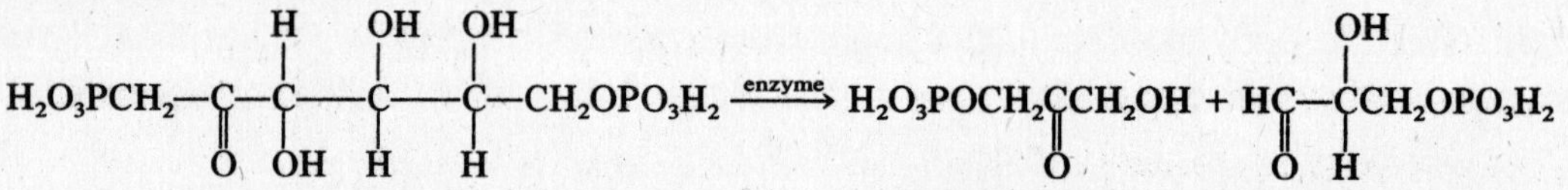

$$H_2O_3PCH_2-\overset{\overset{\displaystyle H}{|}}{\underset{\underset{\displaystyle O}{|}}{C}}-\overset{\overset{\displaystyle OH}{|}}{\underset{\underset{\displaystyle OH}{|}}{C}}-\overset{\overset{\displaystyle OH}{|}}{\underset{\underset{\displaystyle H}{|}}{C}}-\overset{\overset{\displaystyle}{|}}{\underset{\underset{\displaystyle H}{|}}{C}}-CH_2OPO_3H_2 \xrightarrow{enzyme} H_2O_3POCH_2\overset{\overset{\displaystyle}{|}}{\underset{\underset{\displaystyle O}{|}}{C}}CH_2OH + H\overset{\overset{\displaystyle OH}{|}}{\underset{\underset{\displaystyle O}{||} \; H}{C}}-\overset{\overset{\displaystyle}{|}}{\underset{\underset{\displaystyle H}{|}}{C}}CH_2OPO_3H_2$$

Fructose 1,6-diphosphate   Dihydroxyacetone   Glyceraldehyde
(I)   phosphate (II)   3-phosphate (III)

Formulate this reaction as a reversal of an aldol condensation (**retroaldol condensation**). ◄

(I) is a β-hydroxyketone. Loss of a proton from the $C^{\underline{\beta}}$—OH affords an alkoxide (IV) that undergoes a retroaldol condensation by cleavage of the C^{α}—C^{β} bond.

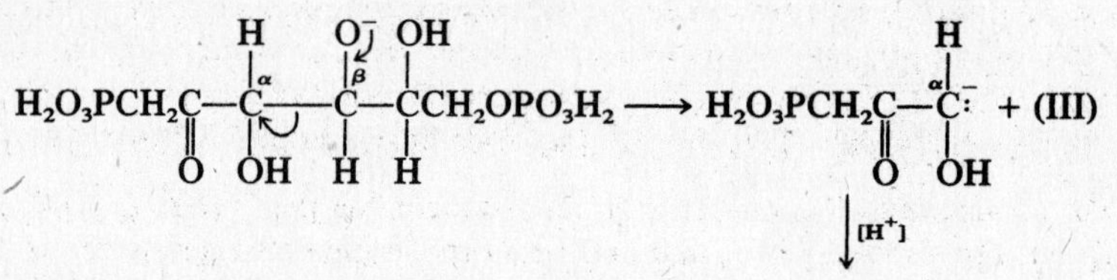

$$H_2O_3PCH_2-\overset{\overset{\displaystyle H}{|}}{\underset{\underset{\displaystyle O}{|}}{\underset{\alpha}{C}}}-\overset{\overset{\displaystyle O^-}{|}}{\underset{\underset{\displaystyle OH}{|}}{\underset{\beta}{C}}}-\overset{\overset{\displaystyle OH}{|}}{\underset{\underset{\displaystyle H}{|}}{C}}CH_2OPO_3H_2 \longrightarrow H_2O_3PCH_2-\overset{\overset{\displaystyle H}{|}}{\underset{\underset{\displaystyle OH}{|}}{\underset{\alpha}{C}}}{:}^- + (III)$$

(IV) $\downarrow$ [H$^+$]
                                    (II)

Chapter 16

Carboxylic Acids

16.1 INTRODUCTION

Carboxylic acids (RCOOH or ArCOOH) have the **carboxyl** group,

$$\overset{\displaystyle O}{\underset{\displaystyle }{\overset{\displaystyle \|}{-C}}}-OH$$

Common names, such as **formic** (ant) and **butyric** (butter) acids, are based on the natural source of the acid. The positions of substituent groups are shown by Greek letters α, β, γ, δ, etc. Some have names derived from acetic acid, e.g. $(CH_3)_3CCOOH$ and $C_6H_5CH_2COOH$ are trimethylacetic acid and phenylacetic acid, respectively. Occasionally they are named as carboxylic acids, e.g.

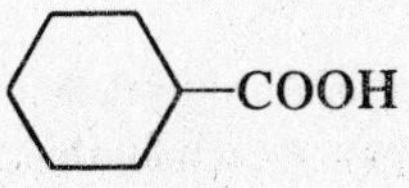

—COOH

is cyclohexanecarboxylic acid.

In the IUPAC system, carboxylic acids are designated by the suffix **-oic** and the word **acid** added to the root name of the alkane; CH_3CH_2COOH is propanoic acid. The C's are numbered; the C of COOH is numbered 1. C_6H_5COOH is benzoic acid. Dicarboxylic acids contain two COOH groups and are named by adding the suffix **-dioic** and the word **acid** to the longest chain with the two COOH's.

Problem 16.1 Give a derived and IUPAC name for the following carboxylic acids. Note the common names. (*a*) $CH_3(CH_2)_4COOH$ (caproic acid); (*b*) $(CH_3)_3CCOOH$ (pivalic acid); (*c*) $(CH_3)_2CHCH_2CH_2COOH$ (γ-methylvaleric acid); (*d*) $C_6H_5CH_2CH_2COOH$ (β-phenylpropionic acid); (*e*) $(CH_3)_2C(OH)COOH$ (α-hydroxyisobutyric acid); (*f*) $HOOC(CH_2)_2COOH$ (succinic acid) (no derived name). ◄

To get the IUPAC name, find the longest chain of C's including the C from COOH, as shown below by a horizontal line. To get the derived name, find and name the groups attached to the α C:

$$R\boxed{\overset{\displaystyle R'}{\underset{\displaystyle R''}{C^\alpha}}-COOH}$$

(*a*) ~~CH₃CH₂CH₂CH₂~~ $\boxed{\text{CH}_2\text{COOH}}$ *n*-Butylacetic acid, Hexanoic acid

   (6 C's in longest chain)

(*b*) $\overset{\displaystyle CH_3}{\underset{\displaystyle CH_3}{\text{~~CH}_3\text{~~}\boxed{\text{CCOOH}}}}$ Trimethylacetic acid, 2,2-Dimethylpropanoic acid

   (3 C's in longest chain)

(*c*) $\underset{\displaystyle CH_3}{\text{~~CH}_3\text{—CHCH}_2\text{~~}\boxed{\text{CH}_2\text{COOH}}}$ Isobutylacetic acid, 4-Methylpentanoic acid

(d) $C_6H_5CH_2CH_2COOH$ Benzylacetic acid, 3-Phenylpropanoic acid

(e)
$$\begin{array}{c} OH \\ | \\ CH_3CCOOH \\ | \\ CH_3 \end{array}$$
Dimethylhydroxyacetic acid, 2-Hydroxy-2-methylpropanoic acid

(f) $HOOCCH_2CH_2COOH$, butanedioic acid.

Problem 16.2   Name the following aromatic carboxylic acids.

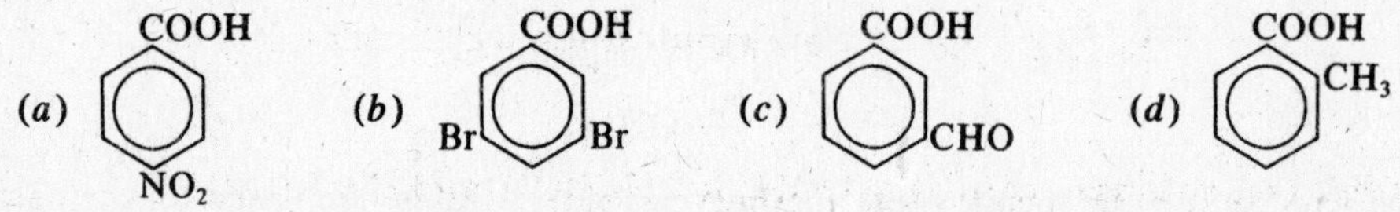

(a) [benzene ring with COOH at top, NO₂ at bottom]   (b) [benzene ring with COOH at top, Br and Br at bottom positions]   (c) [benzene ring with COOH at top, CHO at bottom right]   (d) [benzene ring with COOH at top, CH₃ at top right]   ◄

    (a) p-nitrobenzoic acid;   (b) 3,5-dibromobenzoic acid;   (c) m-formylbenzoic acid (COOH takes priority over CHO, wherefore this compound is named as an acid, not as an aldehyde);   (d) o-methylbenzoic acid, but more commonly called o-toluic acid (from toluene).

Problem 16.3   Account for the following physical properties of carboxylic acids.   (a) Only carboxylic acids with 5 or fewer C's are soluble in H_2O.   (b) Acetic acid in the vapor state shows a molecular weight of 120.   ◄

    (a) RCOOH dissolves because the H of COOH can H-bond with H_2O.   The R portion is nonpolar and lyophobic; this effect predominates as R gets large (over 5 C's).   (b) CH_3COOH typically undergoes dimeric intermolecular H-bonding.

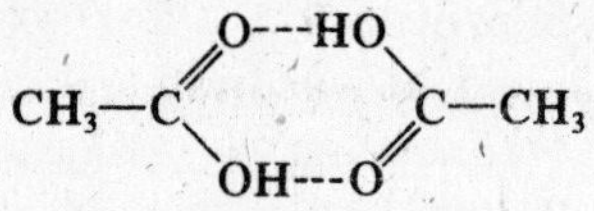

$$CH_3-C \begin{array}{c} O---HO \\ \\ OH---O \end{array} C-CH_3$$

16.2   PREPARATION OF CARBOXYLIC ACIDS

1.   Oxidation of 1° Alcohols, Aldehydes and Arenes

$$R-CH_2OH \xrightarrow{KMnO_4} RCHO \xrightarrow{KMnO_4} RCOOH$$

$$C_6H_5CHR_2 \xrightarrow[H^+]{KMnO_4} C_6H_5COOH$$

2.   Grignard Reagent and CO_2

$$\ddot{R}-MgX + O=C=O \longrightarrow \begin{array}{c} R-C=O \\ | \\ {}^-O(MgX)^+ \end{array} \xrightarrow[H_2O]{HX} \begin{array}{c} R-C=O \\ | \\ OH \end{array} + Mg^{2+} + 2X^-$$

a carboxylate
salt

3.   Hydrolysis of Nitriles

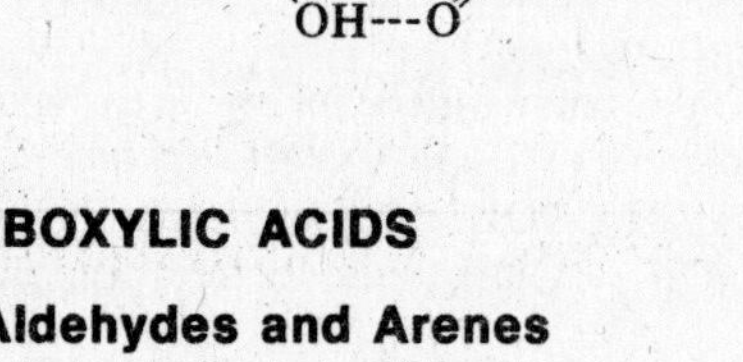

$$RC\equiv N \begin{array}{c} \xrightarrow{H_3O^+} RCOOH + NH_4^+ \\ \\ \xrightarrow[OH^-]{H_2O} RCOO^- + NH_3 \end{array}$$
$$\Big\uparrow H^+$$

Problem 16.4 Prepare the following acids from alkyl halides or dihalides of fewer C's. (a) $C_6H_5CH_2COOH$, (b) $(CH_3)_3CCOOH$, (c) $HOCH_2CH_2CH_2COOH$, (d) $HOOCCH_2CH_2COOH$ (succinic acid). ◄

Replace COOH by X to find the needed alkyl halide. The two methods for $RX \rightarrow RCOOH$ are:

$$RX\ (1°, 2°, \text{or } 3°) \xrightarrow{Mg} RMgX \xrightarrow[\text{2. } H_3O^+]{\text{1. } CO_2} RCOOH \xleftarrow{H_3O^+} RCN \xleftarrow{CN^-} RX\ (1°)$$

(a) Either method can be used starting with $C_6H_5CH_2Br$ (a 1° RX).
(b) With the 3° $(CH_3)_3C$—Br, CN^- *cannot* be used because elimination rather than substitution would occur.
(c) $HOCH_2CH_2CH_2Br$ has an acidic H (O—H); hence the Grignard reaction can't be used.
(d) $BrCH_2CH_2Br$ undergoes dehalogenation with Mg to form an alkene.

$$BrCH_2CH_2Br \xrightarrow{CN^-} N\equiv CCH_2CH_2C\equiv N \xrightarrow{H^+} \text{product}$$

4. Malonic Ester Synthesis of RCH_2COOH, $R_2CHCOOH$ and $RR'CHCOOH$

Step 1   A carbanion is formed.

$$\underset{\substack{\text{Malonic ester}\\ (\textit{Diethyl malonate})}}{EtOOC—CH_2—COOEt} + Na\overset{+}{O}Et \longrightarrow [EtOOC—\overset{..}{C}H—COOEt]Na^+ + EtOH$$

or

stabilized carbanion

Step 2   The carbanion is alkylated by S_N2 reaction.

$$[EtOOC—\overset{..}{C}H—COOEt]Na^+ + R\overset{\frown}{:}X \longrightarrow EtOOC—\overset{\overset{\displaystyle R}{|}}{C}H—COOEt + Na^+X^-$$

For dialkylacetic acids, the second H of the α C is similarly replaced with another R or a different R' group.

$$EtOOC—\underset{\underset{\displaystyle R}{|}}{C}H—COOEt \xrightarrow{Na\overset{+}{O}Et} [EtOOC—\overset{..}{C}R—COOEt]Na^+ \xrightarrow{R'X} EtOOC—\underset{\underset{\displaystyle R}{|}}{\overset{\overset{\displaystyle R'}{|}}{C}}—COOEt + :X^-$$

Step 3   Hydrolysis of the substituted malonic ester gives the malonic acid, which undergoes decarboxylation (loss of CO_2) to form a substituted acetic acid.

$$EtOOC—\underset{\underset{\displaystyle R}{|}}{C}H—COOEt \xrightarrow[\substack{\text{1. } OH^-\\ \text{2. } H_3O^+}]{\text{hydrolysis}} HOOC\underset{\underset{\displaystyle R}{|}}{C}HCOOH \xrightarrow[\text{heat}]{-CO_2} RCH_2COOH$$

$$EtOOC—\underset{\underset{\displaystyle R}{|}}{\overset{\overset{\displaystyle R'}{|}}{C}}—COOEt \xrightarrow{\text{hydrolysis}} HOOC—\underset{\underset{\displaystyle R}{|}}{\overset{\overset{\displaystyle R'}{|}}{C}}—COOH \xrightarrow[\text{heat}]{-CO_2} R'—\underset{\underset{\displaystyle R}{|}}{C}H—COOH$$

Problem 16.5 Use malonic ester to prepare (a) 2-ethylbutanoic acid, (b) 3-methylbutanoic acid, (c) 2-methylbutanoic acid, (d) trimethylacetic acid. ◄

The alkyl groups attached to the α C are introduced by the alkyl halide.

(a) In 2-ethylbutanoic acid

$$\boxed{CH_3CH_2}—CHCOOH$$
$$\boxed{CH_2CH_3}$$

the circled R's are both CH_2CH_3. Therefore each α H is replaced sequentially by an ethyl group ($R = R' = Et$), using CH_3CH_2Br.

$$CH_2(COOC_2H_5)_2 + Na^+OC_2H_5^- \longrightarrow C_2H_5OH + [\bar{C}H(COOC_2H_5)_2] \xrightarrow[-Br^-]{+C_2H_5Br}$$

$$\boxed{C_2H_5}CH(COOC_2H_5)_2 \xrightarrow[\text{alcohol}]{NaOEt} \boxed{C_2H_5}\bar{C}(COOC_2H_5)_2Na^+ \xrightarrow{EtBr} \boxed{(C_2H_5)_2}C(COOC_2H_5)_2 \xrightarrow[\text{heat}]{aq.\ OH^-}$$

$$\boxed{(C_2H_5)_2}C(COO^-)_2 \xrightarrow{H_3O^+} \boxed{(C_2H_5)_2}C(COOH)_2 \xrightarrow{heat} \boxed{(C_2H_5)_2}CHCOOH + CO_2$$

(b) Only $(CH_3)_2CH—Br$ is needed for a single alkylation (see product below).

$$CH_2(COOC_2H_5)_2 \xrightarrow[2.\ i\text{-PrBr}]{1.\ NaOEt} \boxed{\begin{array}{c} CH_3 \\ | \\ CH_3—CH \end{array}}—CH(COOC_2H_5)_2 \xrightarrow[\text{heat}]{aq.\ OH^-}$$

$$\boxed{\begin{array}{c} CH_3 \\ | \\ CH_3—CH \end{array}}—CH(COO^-)_2 \xrightarrow{H_3O^+} \boxed{\begin{array}{c} CH_3 \\ | \\ CH_3—CH \end{array}}—CH(COOH)_2 \xrightarrow[-CO_2]{heat} \boxed{\begin{array}{c} CH_3 \\ | \\ CH_3—CH \end{array}}—CH_2COOH$$

(c) To obtain

$$\boxed{CH_3CH_2}CHCOOH$$
$$\boxed{CH_3}$$

dialkylate stepwise with CH_3CH_2Br and CH_3I. The larger R is introduced first to minimize steric hindrance in the second alkylation step.

$$CH_2(COOEt)_2 \xrightarrow[2.\ EtBr]{1.\ OEt^-} \boxed{CH_3CH_2}CH(COOEt)_2 \xrightarrow[2.\ CH_3I]{1.\ OEt^-}$$

$$\boxed{CH_3CH_2}\underset{\boxed{CH_3}}{C}—(COOEt)_2 \xrightarrow[2.\ H_3O^+,\ heat]{1.\ OH^-} \boxed{CH_3CH_2}\underset{\boxed{CH_3}}{CHCOOH}$$

(d) Trialkylacetic acids cannot be prepared from malonic ester. The product prepared from malonic ester must have at least one α H, which replaces the lost COOH.

Problem 16.6 Prepare an alkyl succinic acid: $\boxed{R}—CH—COOH$ ◀
$$\boxed{CH_2—COOH}$$

Alkylsuccinic acids are disubstituted acetic acids as shown above. Alkylate with RX and then with $BrCH_2COOC_2H_5$ to give

$$R—C(COOC_2H_5)_2$$
$$|$$
$$CH_2COOC_2H_5$$

Hydrolysis and acidification of this ester yields a tricarboxylic acid which is heated and loses CO_2 from one of the *gem*-COOH's to form an alkylsuccinic acid.

$$R—\underset{CH_2COOC_2H_5}{\overset{|}{C}}(COOC_2H_5)_2 \xrightarrow[2.\ H_3O^+]{1.\ NaOH} R—\underset{CH_2COOH}{\overset{|}{C}}(COOH)_2 \xrightarrow[-CO_2]{heat} R—\underset{CH_2—COOH}{\overset{|}{C}H}—COOH$$

16.3 REACTIONS OF CARBOXYLIC ACIDS

H OF COOH IS ACIDIC

$$RCOOH + H_2O \rightleftharpoons RCOO^- + H_3O^+ \qquad pK_a \approx 5 \text{ (page 33)}$$

$$\text{Acid}_1 \qquad \text{Base}_2 \qquad \text{Base}_1 \qquad \text{Acid}_2$$

RCOOH forms carboxylate salts with bases. When R is large these salts are called **soaps**.

$$RCOOH + KOH \longrightarrow RCOO^-K^+ + H_2O$$

$$2RCOOH + Na_2CO_3 \longrightarrow 2RCOO^-Na^+ + H_2O + CO_2$$

Problem 16.7 Use the concept of charge delocalization by extended π bonding (resonance; Problems 3.22 and 3.25) to explain why RCOOH ($pK_a \approx 5$) is more acidic than ROH ($pK_a \approx 15$). ◄

It is usually best to account for relative strengths of acids in terms of relative stabilities of their conjugate bases. The weaker (more stable) base has the stronger acid. Since the electron density in RCOO⁻ is dispersed to both O's,

RCOO⁻ is more stable and a weaker base than RO⁻, whose charge is localized on only one O.

Problem 16.8 Use the inductive effect (page 32) to account for the following differences in acidity.
(a) $ClCH_2COOH > CH_3COOH$, (b) $FCH_2COOH > ClCH_2COOH$, (c) $ClCH_2COOH > ClCH_2CH_2COOH$,
(d) $Me_3CCH_2COOH > Me_3SiCH_2COOH$, (e) $Cl_2CHCOOH > ClCH_2COOH$. ◄

(a) Like all halogens, Cl is electronegative and therefore exerts an electron-withdrawing inductive effect to disperse the electron density from the O's of —COO⁻. The anion

   is therefore a weaker base than $CH_3CO_2^-$, and $ClCH_2COOH$ is the stronger acid.
(b) F is more electronegative than Cl and more effectively withdraws electron density from the O's of —CO_2^-. $FCH_2CO_2^-$ is a weaker base than $ClCH_2CO_2^-$.
(c) The inductive effect diminishes as the number of C's between Cl and the O's increases. $ClCH_2CO_2^-$ is a weaker base than $ClCH_2CH_2CO_2^-$.
(d) Si is electropositive compared to C and therefore has an electron-donating inductive effect,

   which increases the electron density on the O's. Hence, $Me_3SiCH_2CO_2^-$ is a stronger base than $Me_3CCH_2CO_2^-$; Me_3CCH_2COOH is more acidic.
(e) Two Cl's are more electron-withdrawing than one Cl. Cl_2CHCOO^- is the weaker base and $Cl_2CHCOOH$ is the stronger acid.

Problem 16.9 The ionization constants of benzoic acid, p-nitrobenzoic and p-hydroxybenzoic acids are 6.3×10^{-5}, 40×10^{-5} and 2.9×10^{-5}, respectively. Explain. ◄

Factors dispersing the charge increase the acidity. NO_2 resonance produces a + charge at the *para* C, bonded to COO⁻, thereby causing dispersal of − charge from the —CO_2^- group.

OH is electron-releasing by resonance and concentrates the − on the *para* C, bonded to COO⁻. This increases basicity of the anion and decreases acidity of its conjugate acid.

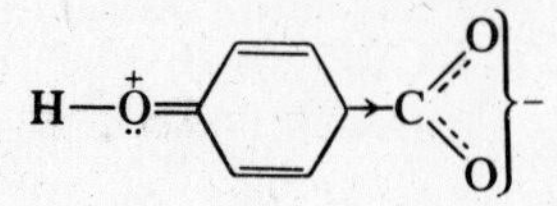

Problem 16.10 Explain why highly branched carboxylic acids such as

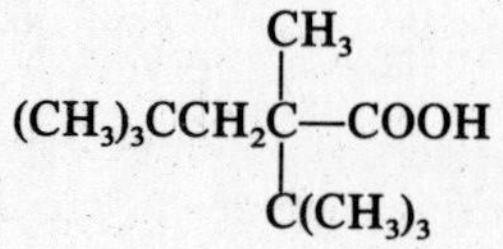

$$(CH_3)_3CCH_2\underset{\underset{C(CH_3)_3}{|}}{\overset{\overset{CH_3}{|}}{C}}-COOH$$

are less acidic than unbranched acids.   ◄

The —CO_2^- group of the branched acid is shielded from solvent molecules and cannot be stabilized by solvation as effectively as can acetate anion.

Problem 16.11 Although *p*-hydroxybenzoic acid is less acidic than benzoic acid, salicylic (*o*-hydroxybenzoic) acid ($K_a = 105 \times 10^{-5}$) is 15 times more acidic than benzoic acid. Explain.   ◄

The enhanced acidity is partly due to very effective H-bonding in the conjugate base, which decreases its basic strength.

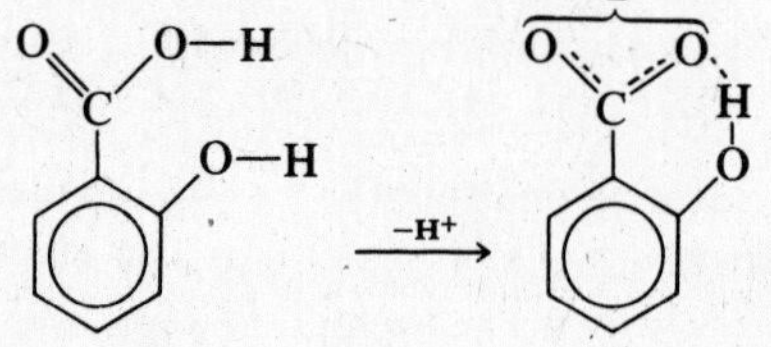

Problem 16.12 The K_2 for fumaric acid (*trans*-butenedioic acid) is greater than for maleic acid, the *cis* isomer. Explain by H-bonding.   ◄

Both dicarboxylic acids have two ionizable H's. The concern is with the 2nd ionization step.

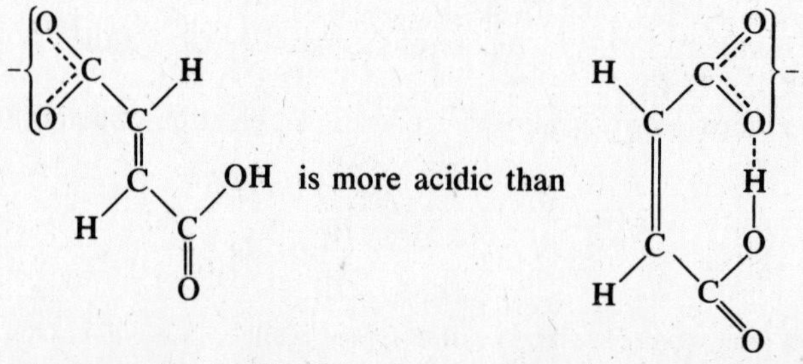

Fumarate monoanion (no H-bond)   Maleate monoanion (H-bond)

Since the 2nd ionizable H of maleate participates in H-bonding, more energy is needed to remove this H because the H-bond must be broken. The maleate monoanion is therefore the weaker acid.

In general, *H-bonding involving the acidic H has an acid-weakening effect*; *H-bonding in the conjugate base has an acid-strengthening effect*.

OH *OF* COOH *MAY BE REPLACED*

Replacement by another group, G, forms a **carboxylic acid derivative**

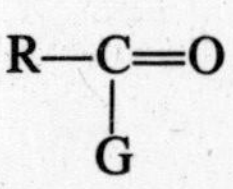

$$R-\underset{\underset{G}{|}}{C}=O$$

1. Acyl Chloride (RCOCl) **Formation**, OH $\longrightarrow$ Cl

(For the **acyl group**, see Problem 16.22.)

$$3\,R\text{---}COOH + PCl_3 \longrightarrow 3\,R\text{---}COCl + H_3PO_3 \quad \text{(OH replaced)}$$

$$R\text{---}COOH + PCl_5 \longrightarrow R\text{---}COCl + HCl(g) + POCl_3$$

$$R\text{---}COOH + SOCl_2 \longrightarrow R\text{---}COCl + HCl(g) + SO_2$$
$$\text{Thionyl chloride}$$

Reaction with $SOCl_2$ is particularly useful because the 2 gaseous products SO_2 and HCl are readily separated from RCOCl.

2. Ester (RCOOR′) **Formation**, OH $\longrightarrow$ OR′ (Problem 3.13)

$$R\text{---}COOH + R'OH \xrightarrow[\text{reflux}]{H_2SO_4} R\text{---}COOR' + H_2O$$

3. Amide (RCONH$_2$) **Formation**, OH $\longrightarrow$ NH$_2$

$$R\text{---}COOH + NH_3 \longrightarrow \underset{\text{ammonium salt}}{R\text{---}COO^-NH_4^+} \xrightarrow{\text{heat}} R\text{---}CONH_2 + H_2O$$
$$\underset{\text{acid}}{\phantom{R\text{---}COOH}}$$

Problem 16.13   Use ethanol to prepare $CH_3COOC_2H_5$, an important commercial solvent.   ◄

$CH_3COOC_2H_5$, ethyl acetate, is the ester of CH_3CH_2OH and CH_3COOH. CH_3CH_2OH is oxidized to CH_3COOH.

$$CH_3CH_2OH \xrightarrow{MnO_4^-,\ H^+} CH_3COOH$$

Ethanol and acetic acid are then refluxed with concentrated H_2SO_4.

$$C_2H_5OH + CH_3COOH \underset{\text{heat}}{\overset{H_2SO_4}{\rightleftharpoons}} CH_3COOC_2H_5 + H_2O$$

With added benzene and a trace of acid, the reversible reaction is driven to completion by distilling off H_2O as an azeotrope (Problem 13.48).

Problem 16.14   Use $CH_3CH_2CH_2OH$ and $H_2C(COOC_2H_5)_2$ as the only organic reagents to synthesize valeramide, $CH_3CH_2CH_2CH_2CONH_2$.   ◄

Since valeric acid is *n*-propylacetic acid,

$$\boxed{CH_3CH_2CH_2}\text{---}CH_2COOH$$

$CH_3CH_2CH_2Br$ is used to alkylate malonic ester.

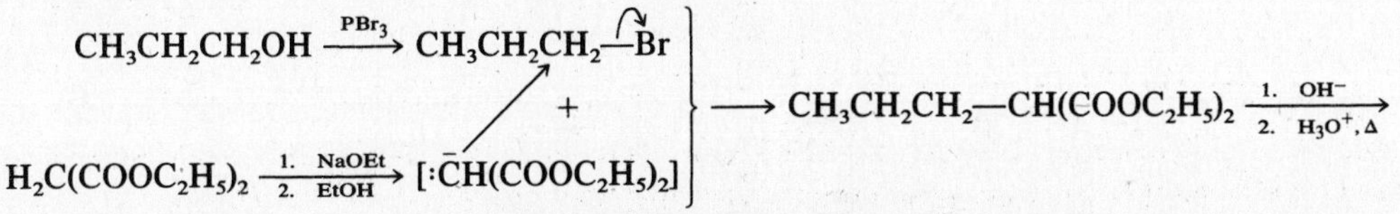

$$CH_3CH_2CH_2OH \xrightarrow{PBr_3} CH_3CH_2CH_2\text{---}Br$$

$$+ \quad \Bigg\} \longrightarrow CH_3CH_2CH_2\text{---}CH(COOC_2H_5)_2 \xrightarrow[\text{2. }H_3O^+,\ \Delta]{\text{1. }OH^-}$$

$$H_2C(COOC_2H_5)_2 \xrightarrow[\text{2. EtOH}]{\text{1. NaOEt}} [:\overset{..}{C}H(COOC_2H_5)_2]$$

$$CH_3CH_2CH_2CH_2COOH \xrightarrow{NH_3} CH_3(CH_2)_3COO^-NH_4^+ \xrightarrow{\Delta} CH_3(CH_2)_3CONH_2$$

REDUCTION OF C$=$O **OF** COOH (RCOOH $\rightarrow$ RCH$_2$OH)

Acids are best reduced to alcohols by $LiAlH_4$.

$$RCOOH \xrightarrow[\text{2. }H_2O]{\text{1. }LiAlH_4\ \text{(ether)}} RCH_2OH$$

Problem 16.15 Prepare *n*-hexyl chloride from *n*-butylmalonic ester. ◄

n-Butylmalonic ester is hydrolyzed with base and decarboxylated to hexanoic acid.

$$n\text{-}C_4H_9CH(COOC_2H_5)_2 \xrightarrow[\text{2. H}^+, \Delta]{\text{1. OH}^-} n\text{-}C_4H_9CH_2COOH$$

This acid and *n*-hexyl chloride have the same number of C's.

$$n\text{-}C_4H_9CH_2COOH \xrightarrow{\text{LiAlH}_4} n\text{-}C_5H_{11}CH_2OH \xrightarrow{\text{SOCl}_2} n\text{-}C_6H_{13}Cl$$

HALOGENATION OF α H's. HELL-VOLHARD-ZELINSKY (HVZ) REACTION

One or more α H's are replaced by Cl or Br by treating the acid with Cl_2 or Br_2, using phosphorus as catalyst.

$$RCH_2COOH \xrightarrow{X_2/P} \underset{\underset{X}{|}}{R}CHCOOH \xrightarrow{X_2/P} RCX_2COOH \quad (X = Cl, Br)$$

α-Halogenated acids react like active alkyl halides and are convenient starting materials for preparing other α-substituted acids by nucleophilic displacement of halide anion.

$$X^- + \underset{\underset{OH}{|}}{R}\text{—CH—COOH} \xleftarrow[\text{2. H}^+]{\text{1. NaOH}} \boxed{\underset{\underset{X}{|}}{R}\text{—CH—COOH}} \xrightarrow[-H^+, X^-]{NH_3} \underset{\underset{NH_3^+}{|}}{R}\text{—CH—COO}^-$$

$$\quad\quad\quad\quad\quad\text{α-hydroxy acid} \quad\quad\quad\quad\quad\quad\quad\quad\quad\quad\quad\quad\quad\quad\quad\quad \text{α-amino acid}$$

Heating an α-halogenated acid with alcoholic KOH leads to α,β-unsaturated acids when there is a β H in the molecule.

$$R\text{—CH}_2\underset{\underset{X}{|}}{C}\text{H—COOH} \xrightarrow{\text{alc. KOH}} R\text{—CH=CH—COO}^-K^+ \xrightarrow{H^+} R\text{—CH=CH—COOH}$$

$$\quad\quad\quad\quad\quad\quad\quad\quad\quad\quad\quad\quad\quad\quad\quad\quad \text{α,β-unsaturated acid}$$
$$\quad\quad\quad\quad\quad\quad\quad\quad\quad\quad\quad\quad\quad\quad\quad\quad\quad\quad \text{salt}$$

Problem 16.16 Prepare malonic acid (propanedioic acid, HOOC—CH₂—COOH) from CH₃COOH. ◄

CH₃COOH is first converted to ClCH₂COOH. The acid is changed to its salt to prevent formation of the very poisonous HCN when replacing the Cl by CN. The C≡N group is then carefully hydrolyzed with acid to prevent decarboxylation.

$$CH_3COOH \xrightarrow{Cl_2/P} \underset{\text{Chloroacetic acid}}{Cl\text{—}CH_2COOH} \xrightarrow{NaOH} \underset{\text{Sodium chloroacetate}}{Cl\text{—}CH_2COO^-\overset{+}{N}a} \xrightarrow{CN^-}$$

$$\underset{\text{Sodium cyanoacetate}}{N{\equiv}C\text{—}CH_2COO^-\overset{+}{N}a} \xrightarrow{H_3O^+} HOOC\text{—}CH_2\text{—}COOH + NH_4^+$$

REACTION OF COOH. DECARBOXYLATION (ArCOOH ⟶ ArH)

$$ArCOOH \xrightarrow{\text{soda lime}} ArH \quad (\textit{poor yield})$$

RING REACTIONS OF AROMATIC CARBOXYLIC ACIDS

The electron-attracting COOH is *meta*-directing and deactivating during electrophilic substitution.

16.4 SUMMARY OF CARBOXYLIC ACID CHEMISTRY

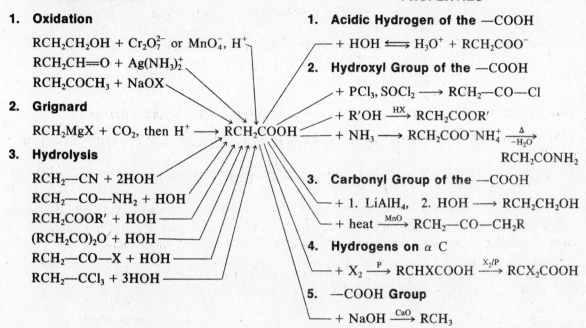

PREPARATION

1. Oxidation

$RCH_2CH_2OH + Cr_2O_7^{2-}$ or MnO_4^-, H^+

$RCH_2CH=O + Ag(NH_3)_2^+$

$RCH_2COCH_3 + NaOX$

2. Grignard

$RCH_2MgX + CO_2$, then $H^+ \longrightarrow RCH_2COOH$

3. Hydrolysis

$RCH_2-CN + 2HOH$

$RCH_2-CO-NH_2 + HOH$

$RCH_2COOR' + HOH$

$(RCH_2CO)_2O + HOH$

$RCH_2-CO-X + HOH$

$RCH_2-CCl_3 + 3HOH$

PROPERTIES

1. Acidic Hydrogen of the —COOH

$+ HOH \rightleftharpoons H_3O^+ + RCH_2COO^-$

2. Hydroxyl Group of the —COOH

$+ PCl_3, SOCl_2 \longrightarrow RCH_2-CO-Cl$

$+ R'OH \xrightarrow{HX} RCH_2COOR'$

$+ NH_3 \longrightarrow RCH_2COO^-NH_4^+ \xrightarrow[-H_2O]{\Delta}$

RCH_2CONH_2

3. Carbonyl Group of the —COOH

$+ 1.\ LiAlH_4,\ \ 2.\ HOH \longrightarrow RCH_2CH_2OH$

$+ heat \xrightarrow{MnO} RCH_2-CO-CH_2R$

4. Hydrogens on α C

$+ X_2 \xrightarrow{P} RCHXCOOH \xrightarrow{X_2/P} RCX_2COOH$

5. —COOH Group

$+ NaOH \xrightarrow{CaO} RCH_3$

16.5 ANALYTICAL DETECTION OF CARBOXYLIC ACIDS

CHEMICAL REACTIONS

Carboxylic acids dissolve in Na_2CO_3, thereby evolving CO_2. The **neutralization equivalent** or **equivalent weight** of a carboxylic acid is determined by titration with standard base.   It is the number of grams of acid neutralized by 1 equivalent of base.   If 40.00 ml of a $0.100N$ base is needed to neutralize 0.500 g of an unknown acid, the number of equivalents of base is

$$\frac{40.00\ \text{ml}}{1000\ \text{ml/liter}} \times 0.100\ \frac{\text{eq}}{\text{liter}} = 0.00400\ \text{eq}$$

The neutralization equivalent is found by dividing the weight of acid by the number of equivalents of base, i.e.

$$\frac{0.500\ \text{g}}{0.00400\ \text{eq}} = 125\ \frac{\text{g}}{\text{eq}}$$

SPECTROSCOPIC METHODS

1. Infrared

COOH has a strong O—H stretching at 2500–3000 cm^{-1} for dimeric acids with hydrogen bonding, and a C=O absorption at 1700–1725 cm^{-1} for aliphatic and 1670–1700 cm^{-1} for aromatic acids.

2. Nmr

The H of COOH is strongly deshielded and absorbs downfield at $\delta = 10.5$–12.0 ppm.

3. Mass Spectra

Carboxylic acids and their derivatives are cleaved into acylium cations and free radicals.

$$R-\overset{\overset{\ddot{O}:}{\|}}{C}-OH \xrightarrow{-e^-} R-\overset{\overset{\cdot\ddot{O}^+}{\|}}{C}\!\!\mid\!\!OH \longrightarrow R-\overset{\overset{\ddot{O}^+}{\|}}{C} + \cdot OH$$

Like other carbonyl compounds, carboxylic acids undergo β cleavage and γ H transfer.

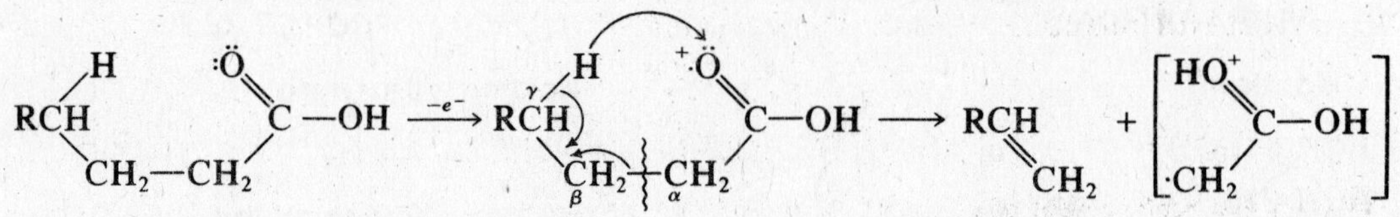

Supplementary Problems

Problem 16.17 Describe the electronic effect of C_6H_5 on acidity if the acid strengths of C_6H_5COOH and HCOOH are 6.3×10^{-5} and 1.7×10^{-4}, respectively. ◄

The weaker acidity of C_6H_5COOH shows that the electron-releasing resonance effect of C_6H_5 outweighs its electron-attracting inductive effect.

Problem 16.18 Outline the preparation of glycine, H_2NCH_2COOH ($\overset{+}{H_3}NCH_2COO^-$), from CH_3CH_2OH. ◄

$$CH_3CH_2OH \xrightarrow{MnO_4^-,\ H^+} CH_3COOH \xrightarrow{Cl_2/P} ClCH_2COOH \xrightarrow{NH_3} H_2NCH_2COOH(\overset{+}{H_3}NCH_2COO^-)$$

Problem 16.19 Prepare 3-phenylpropenoic acid from malonic ester and $C_6H_5CH_2Br$. ◄

$$CH_2(COOC_2H_5)_2 \xrightarrow[\substack{2.\ C_6H_5CH_2Br}]{1.\ NaOEt} \boxed{C_6H_5CH_2}CH(COOC_2H_5)_2 \xrightarrow[\substack{2.\ H_3O^+,\ \Delta}]{1.\ OH^-}$$

$$C_6H_5CH_2CH_2COOH \xrightarrow{Br_2/P} C_6H_5CH_2\underset{\underset{Br}{|}}{CH}-COOH \xrightarrow{alc.\ KOH} C_6H_5CH=CH-COOH$$
Cinnamic acid

Problem 16.20 Use ethanol as the only organic compound to prepare (a) $HOCH_2COOH$, (b) $CH_3CHOHCOOH$. ◄

(a) The acid and alcohol have the same number of C's.

$$CH_3CH_2OH \xrightarrow[H^+]{KMnO_4} CH_3COOH \xrightarrow{Cl_2/P} CH_2ClCOOH \xrightarrow[\substack{2.\ H^+}]{1.\ OH^-} HOCH_2COOH$$

(b) Now the acid has one more C. A one-carbon "step-up" is needed before introducing the OH.

$$C_2H_5OH \xrightarrow{PCl_3} C_2H_5Cl \xrightarrow{KCN} C_2H_5CN \xrightarrow{H_3O^+} CH_3CH_2COOH \xrightarrow[part\ (a)]{as\ in} CH_3CHOHCOOH$$

Problem 16.21 Name the following compounds:

$$(a)\quad CH_3CH_2-\underset{\underset{CH_3}{|}}{\overset{\overset{CH_3}{|}}{C}}-COOH \qquad\qquad (b)\quad CH_3CH=CH-CH=CH-COOH$$

$$(c)\quad CH_3-\underset{\underset{H}{|}}{\overset{\overset{CH_3}{|}}{C}}-\underset{\underset{CH_3}{|}}{\overset{\overset{H}{|}}{C}}-CH_2-COOH \qquad (d)\quad CH_3CH_2CH_2CH_2\underset{\underset{CH_2CH_3}{|}}{CH}CH_2\overset{\overset{CH_3}{|}}{CH}-COOH \quad ◄$$

(a) dimethylethylacetic acid, 2,2-dimethylbutanoic acid; (b) 2,4-hexadienoic acid (sorbic acid); (c) 3,4-dimethylpentanoic acid, β,γ-dimethylvaleric acid; (d) 2-methyl-4-ethyloctanoic acid.

Problem 16.22 What is meant by the term **acylation**? ◄

Removal of OH from RCOOH leaves the **acyl group**,

$$R-\overset{\displaystyle O}{\overset{\|}{C}}-$$

An acylation is a replacement of H by RCO (or ArCO).

Problem 16.23 The ^{18}O of R^{18}OH appears in the ester, not in the water, when the alcohol reacts with a carboxylic acid. Offer a mechanism consistent with this finding. ◄

The O from ROH must bond to the C of —COOH and the OH of COOH ends up in H_2O. The acid catalyst protonates the O of C=O to enhance nucleophilic attack by ROH.

$$R-\overset{O}{\overset{\|}{C}}-OH \xrightarrow{H^+} R-\overset{+OH}{\overset{\|}{C}}-OH \xrightarrow{R'\,^{18}OH} R-\overset{OH}{\overset{|}{\underset{^{18}OH}{C}}}-OH \longrightarrow R-\overset{OH}{\overset{|}{\underset{^{18}O}{C}}}-\overset{+}{O}H_2 \longrightarrow R-\overset{O}{\overset{\|}{C}}-^{18}O-R' + H_3O^+$$

Problem 16.24 When reactive alcohols such as CH_2=$CHCH_2\,^{18}OH$ are esterified in acid, some $H_2{}^{18}O$ is found. Explain. ◄

Protonated allyl alcohol undergoes an S_N1-type substitution reaction.

$$CH_2=CHCH_2\,^{18}OH \xrightarrow{H^+} CH_2=CHCH_2-\overset{+}{\underset{H}{^{18}O}}H \xrightarrow{S_N1} H_2{}^{18}O + CH_2=\overset{+}{C}HCH_2$$

$$CH_2=\overset{+}{C}HCH_2 \xrightarrow[-H^+]{CH_3\overset{O}{\overset{\|}{C}}-OH} CH_3\overset{}{\underset{O}{C}}-OCH_2CH=CH_2$$

Problem 16.25 Use simple, rapid, test tube reactions to distinguish among hexane, hexanol and hexanoic acid. ◄

Only hexanoic acid liberates CO_2 from aqueous Na_2CO_3. Na reacts with hexanol to liberate H_2. Hexane is inert.

Problem 16.26 Prepare 2-methylbutanoic acid from 2-butanol. ◄

The precursor of the acid has a halogen on the site of the COOH. 2-Butanol is converted to this halogenated butane.

$$CH_3CH_2-\overset{H}{\underset{OH}{C}}-CH_3 \xrightarrow{SOCl_2} CH_3CH_2-\overset{H}{\underset{Cl}{C}}-CH_3 \xrightarrow[ether]{Mg} CH_3CH_2-\overset{H}{\underset{MgCl}{C}}-CH_3 \xrightarrow[2.\ H_3O^+]{1.\ CO_2} CH_3CH_2-\overset{H}{\underset{COOH}{C}}-CH_3$$

Do not use the nitrile route because 2-chlorobutane, a 2° halide, may undergo extensive dehydrohalogenation.

Problem 16.27 Compare the products formed on heating the following dicarboxylic acids: (a) oxalic acid, (b) malonic acid, (c) succinic acid, (d) glutaric acid (1,5-pentanedioic acid), (e) longer-chain $HOOC(CH_2)_n COOH$. Acids (c), (d) and (e) undergo dehydration. ◄

(a) Decarboxylation: $HOOC-COOH \xrightarrow{heat} CO_2 + H-COOH \longrightarrow CO + H_2O$

(b) Decarboxylation: $HOOC-CH_2-COOH \xrightarrow{heat} CO_2 + CH_3COOH$

(c) Intramolecular dehydration and ring formation:

$$\begin{array}{c} CH_2-COOH \\ | \\ CH_2-COOH \end{array} \xrightarrow[-H_2O]{heat} \begin{array}{c} CH_2-C=O \\ | \qquad\qquad O \\ CH_2-C=O \end{array} \quad \text{Succinic anhydride}$$

(d) Intramolecular dehydration and ring formation:

$$\begin{array}{c} CH_2-COOH \\ / \\ CH_2 \\ \backslash \\ CH_2-COOH \end{array} \xrightarrow[-H_2O]{heat} \begin{array}{c} CH_2-C=O \\ / \qquad\qquad\quad \\ CH_2 \qquad\qquad O \\ \backslash \qquad\qquad\quad \\ CH_2-C=O \end{array} \quad \text{Glutaric anhydride}$$

(e) Longer-chain α,ω-dicarboxylic acids usually undergo *inter*molecular dehydration on heating to form long-chain polymeric anhydrides. In the following equation, $n > 3$.

$$HOOC-(CH_2)_n-COOH \xrightarrow{heat} HOOC-(CH_2)_n-\underset{O}{\overset{}{C}}-O-\underset{O}{\overset{}{C}}-(CH_2)_n-\underset{O}{\overset{}{C}}-\cdots$$

Problem 16.28 Convert 2-chlorobutanoic acid into 3-chlorobutanoic acid. ◄

2-Chlorobutanoic acid is dehydrohalogenated to 2-butenoic acid and HCl is added. H^+ adds to give a β-carbocation which bonds to Cl^- to form the β-chloroacid.

$$CH_3CH_2\underset{\underset{Cl}{|}}{CH}-COOH \xrightarrow{\text{alc. KOH}} CH_3CH=CH-COOH \xrightarrow{\boxed{H^+}} CH_3\overset{+}{CH}\overset{\boxed{H}}{\underset{}{|}}CHCOOH \xrightarrow{Cl^-} CH_3\underset{\underset{Cl}{|}}{CH}-CH_2-COOH$$
$$\text{a } \beta\text{-R}^+$$

The α-carbocation is not formed because its $+$ charge would be next to the positive C of COOH.

$$CH_3\overset{\boxed{H}}{\underset{}{|}}\overset{+}{CH}CH\overset{O^{\delta-}}{\overset{\|}{C^{\delta+}}}-OH$$

Compare Problem 6.45(c), (d).

Problem 16.29 What is the product when each of the following compounds is reacted with aqueous base (NaOH)? (a) 2-bromobutanoic acid, (b) 3-bromobutanoic acid, (c) 4-bromobutanoic acid, (d) 5-bromopentanoic acid. (In (a) and (b) the initially formed salt is acidified.) ◄

(a) α-Hydroxyacid by S_N2 substitution.

$$CH_3CH_2\underset{\underset{Br}{|}}{CH}-COOH \xrightarrow[2.\ H^+]{1.\ OH^-} CH_3CH_2\underset{\underset{OH}{|}}{CH}-COOH$$
$$\text{2-Bromobutanoic acid} \qquad\qquad \text{2-Hydroxybutanoic acid}$$

(b) Dehydrohalogenation to an α,β-unsaturated acid. The driving force for this easy reaction is the formation of a conjugated system.

$$CH_3\underset{\underset{Br}{|}}{CH}CH_2-\underset{\underset{OH}{|}}{C}=O \xrightarrow[2.\ H^+]{1.\ OH^-} CH_3CH=CH-\overset{\overset{OH}{|}}{C}=O$$
$$\text{2-Butenoic acid}$$

Elimination is a typical reaction of β-substituted carboxylic acid.

$$R\underset{\underset{Y}{|}}{CH}CH_2COOH \xrightarrow{-HY} RCH=CHCOOH \quad (Y = Cl, Br, I, OH, NH_2)$$

(c) γ-Haloacids undergo intramolecular S_N2-type displacement of X^- initiated by the nucleophilic carboxylate anion, to yield inner cyclic esters known as **γ-lactones**.

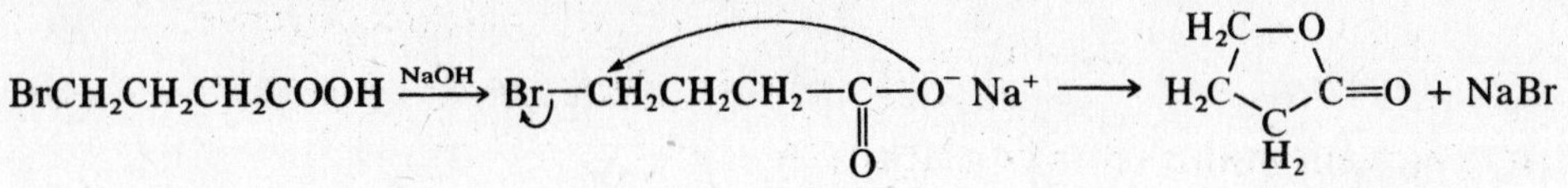

$$BrCH_2CH_2CH_2COOH \xrightarrow{\text{NaOH}} Br-CH_2CH_2CH_2-\overset{\displaystyle O}{\underset{\displaystyle \parallel}{C}}-O^- Na^+ \longrightarrow$$ γ-Butyrolactone + NaBr

(d) Similar to part (c) to give a **δ-lactone**.

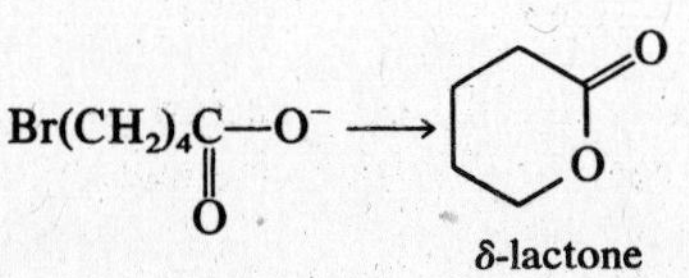

$$Br(CH_2)_4\overset{\displaystyle \parallel}{\underset{\displaystyle O}{C}}-O^- \longrightarrow$$ δ-lactone

Problem 16.30 Write structures for the compounds (A) through (D).

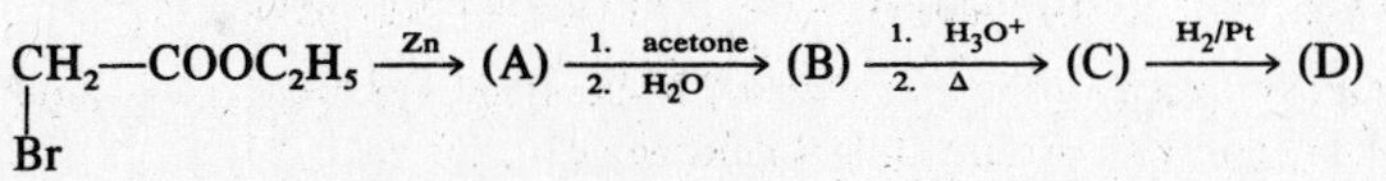

$$\underset{\displaystyle Br}{CH_2}-COOC_2H_5 \xrightarrow{\text{Zn}} (A) \xrightarrow[\text{2. } H_2O]{\text{1. acetone}} (B) \xrightarrow[\text{2. } \Delta]{\text{1. } H_3O^+} (C) \xrightarrow{H_2/Pt} (D)$$

This is a Reformatsky reaction (see Problem 15.51).

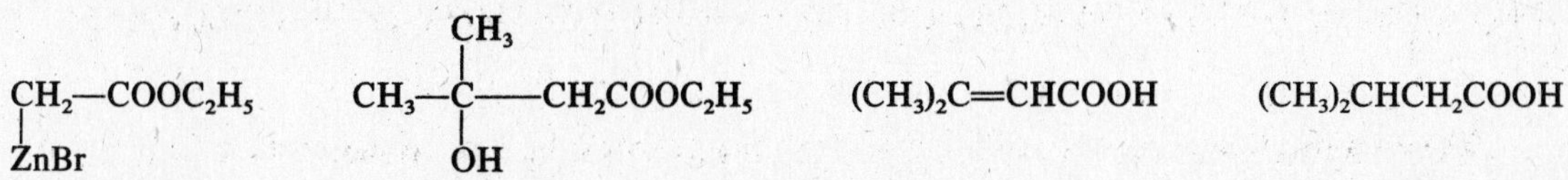

$$\underset{\displaystyle ZnBr}{CH_2}-COOC_2H_5 \qquad CH_3-\overset{\displaystyle CH_3}{\underset{\displaystyle OH}{C}}-CH_2COOC_2H_5 \qquad (CH_3)_2C{=}CHCOOH \qquad (CH_3)_2CHCH_2COOH$$

(A)   (B)   (C)   (D)

Problem 16.31 Use isopropyl alcohol as the only organic compound to synthesize (a) α-hydroxyisobutyric acid, (b) β-hydroxybutyric acid.

(a) 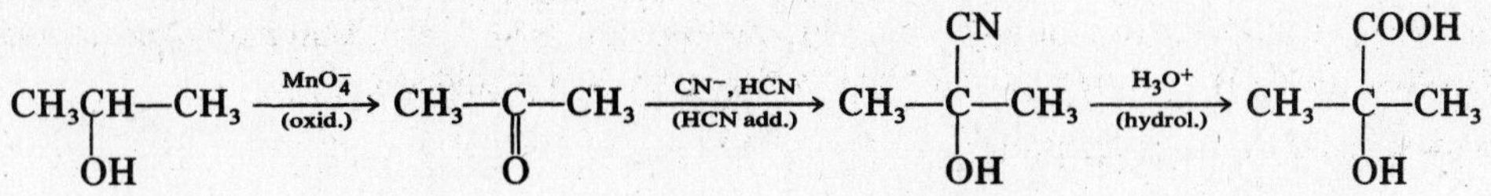
$$CH_3\underset{\displaystyle OH}{CH}-CH_3 \xrightarrow[\text{(oxid.)}]{MnO_4^-} CH_3-\overset{\displaystyle \parallel}{\underset{\displaystyle O}{C}}-CH_3 \xrightarrow[\text{(HCN add.)}]{CN^-, HCN} CH_3-\overset{\displaystyle CN}{\underset{\displaystyle OH}{C}}-CH_3 \xrightarrow[\text{(hydrol.)}]{H_3O^+} CH_3-\overset{\displaystyle COOH}{\underset{\displaystyle OH}{C}}-CH_3$$
                                                                            Cyanohydrin

(b)
$$CH_3-\underset{\displaystyle OH}{CH}-CH_3 \xrightarrow{H_2SO_4} CH_3-CH{=}CH_2 \xrightarrow{HOCl} CH_3-\underset{\displaystyle OH}{CH}-CH_2Cl \xrightarrow{CN^-, H^+}$$

$$CH_3\underset{\displaystyle OH}{CH}-CH_2-CN \xrightarrow{H_3O^+} CH_3\underset{\displaystyle OH}{CH}CH_2-COOH + NH_4^+$$

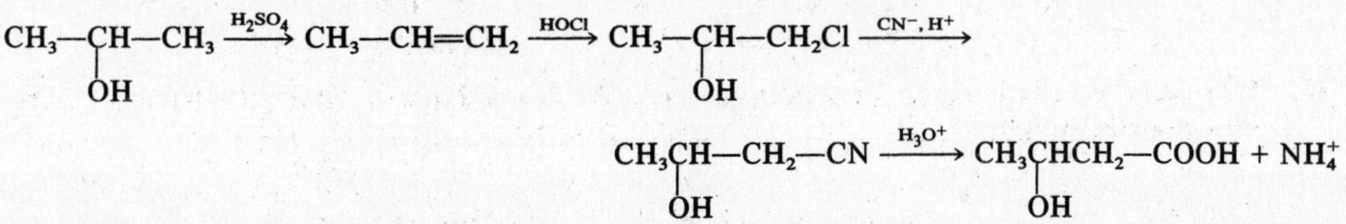

Problem 16.32 How many equivalents of base are needed to neutralize one mole of phthalic acid, $C_6H_4(COOH)_2$, and of mellitic acid, $C_6(COOH)_6$?

These dibasic and hexabasic acids require 2 and 6 equivalents of base, respectively, one for each ionizable H. The neutralization equivalents are the molecular weights divided by the number of equivalents, i.e. $166/2 = 83$ and $342/6 = 57$.

Problem 16.33 Assign numbers from 1 for LEAST to 3 for MOST to show the relative ease of acid-catalyzed esterification of:

	I	II	III

(a) $CH_3CH_2CH_2OH$ by:

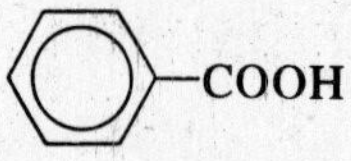

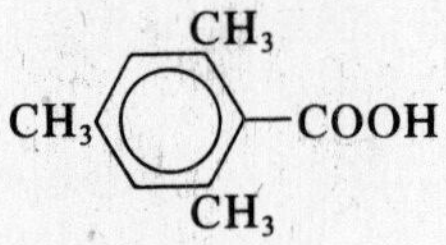

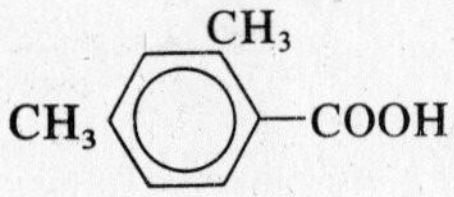

I — C_6H_5—COOH

II — 2,6-dimethyl C_6H_3—COOH (with CH_3 groups) —COOH

III — CH_3—C_6H_4—COOH (para CH_3)

(b) CH_3CH_2OH by:

I CH_3—CH(CH_3)—COOH

II CH_3CH_2COOH

III $(CH_3)_3CCOOH$

(c) C_6H_5COOH by:

I CH_3—C(CH_3)(OH)—CH_2CH_3

II $CH_3CHOHCH_2CH_3$

III $CH_3CH_2CH_2OH$ ◄

Steric factors are chiefly responsible for relative reactivities.

	I	II	III
(a)	3	1	2
(b)	2	3	1
(c)	1	2	3

Chapter 17

Carboxylic Acid Derivatives

17.1 INTRODUCTION

Acid derivatives, RCOY, contain the **acyl group**,

$$R-\overset{|}{C}=O$$

bonded to a functional group other than OH. Some important ones are **acid chloride** (Y = Cl), **amides** (Y = NH$_2$), **acid anhydrides**

$$Y = O-\overset{\overset{\displaystyle O}{\|}}{C}R'$$

and **esters** (Y = OR').

Acid derivatives may be hydrolyzed to the parent acid using dilute OH$^-$ or H$_3$O$^+$. Hydrolysis and many other reactions involve nucleophilic attack on the carbonyl C followed by loss of Y.

Problem 17.1 Compare and explain the difference in reactivity with H$_2$O of RCl and RCOCl. ◄

Alkyl halides are much less reactive than acyl halides in nucleophilic substitution because nucleophilic attack on the tetrahedral C of RX involves a badly crowded transition state. Also, a σ bond must be partly broken to permit the attachment of the nucleophile. Nucleophilic attack on C=O of RCOCl involves a relatively unhindered transition state leading to a tetrahedral intermediate. Nucleophilic substitution of acid derivatives occurs in two steps: the first step resembles addition to carbonyl compounds (page 257); the second step is loss of Y, in this case Cl.

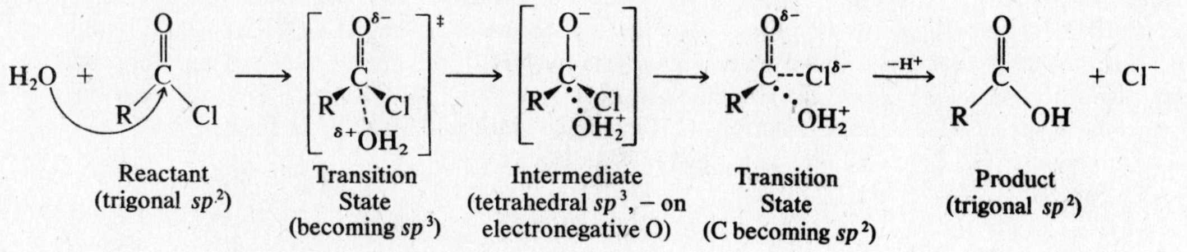

| Reactant (trigonal sp^2) | Transition State (becoming sp^3) | Intermediate (tetrahedral sp^3, − on electronegative O) | Transition State (C becoming sp^2) | Product (trigonal sp^2) |

Problem 17.2 Account for relative reactivity with nucleophiles: ◄

$$RCOX > (RCO_2)_2O > RCOOR' > RCONH_2$$

The relative order of "leavability" is

$$Y^- > R\overset{\overset{\displaystyle O}{\|}}{C}O^- > R'O^- > H_2N^-$$

This is the reverse order of basicities.

301

Problem 17.3 Outline a mechanism for hydrolysis of acid derivatives with   (a) H_3O^+,   (b) NaOH.   ◄

(a) Protonation of carbonyl O makes C more electrophilic and hence more reactive toward weakly nucleophilic H_2O.

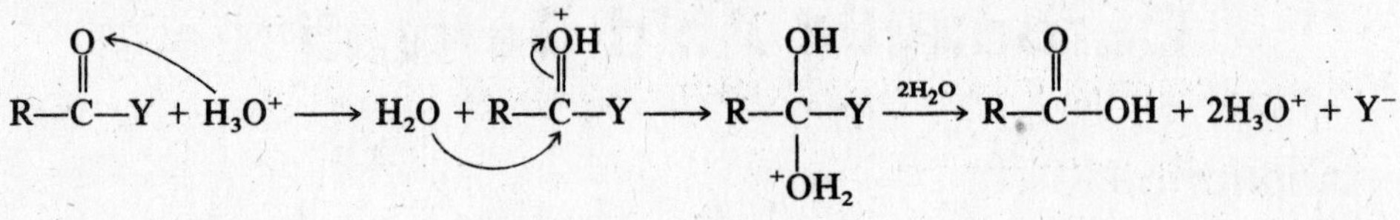

(If Y^- is basic, we get HY.)

(b) Strongly basic OH^- readily attacks the carbonyl C. Unlike acid hydrolysis, this reaction is irreversible, because OH^- removes H^+ from —COOH to form resonance-stabilized $RCOO^-$ (page 32).

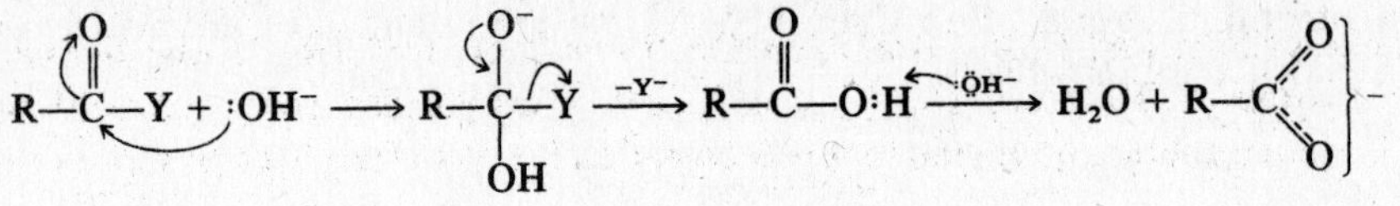

Problem 17.4 Does each of the following reactions take place easily? Explain.

(a) $CH_3COCl + H_2O \longrightarrow CH_3COOH + HCl$

(b) $CH_3COOH + NH_3 \longrightarrow CH_3CONH_2 + H_2O$

(c) $(CH_3CO)_2O + NaOH \longrightarrow CH_3COOH + CH_3COO^-Na^+$

(d) $CH_3COBr + C_2H_5OH \longrightarrow CH_3COOC_2H_5 + HBr$

(e) $CH_3CONH_2 + NaOH \longrightarrow CH_3COO^-Na^+ + NH_3$

(f) $CH_3COOCH_3 + Br^- \longrightarrow CH_3COBr + {}^-OCH_3$ ◄

     Nucleophilic substitution of acyl compounds takes place readily if the incoming group (Nu:) is a stronger base than the leaving group (Y:) or if the final product is a resonance-stabilized $RCOO^-$. (a) Yes. :OH^- is a stronger base than is Cl^-; even the weak base H_2O reacts vigorously. (b) No. NH_3 reacts with RCOOH to form $RCOO^-NH_4^+$, which *does not react*. Amides are prepared from RCOOH by strongly heating dry $RCOO^-NH_4^+$, because reaction is aided by acid catalysis by NH_4^+. (c) Yes. The leaving group $RCOO^-$ is a weaker base than OH^-. (d) Yes. Br^- is a much weaker base than C_2H_5OH. (e) Yes. Even though NH_2^- is a stronger base than OH^-, in basic solution the resonance-stabilized $RCOO^-$ is formed, and this shifts the reaction to completion. (f) No. Br^- is a weaker base than OCH_3^-.

17.2   CHEMISTRY OF ACYL DERIVATIVES

ACYL CHLORIDES   (see page 293 for the preparation of RCOCl)

Acyl chlorides are readily converted into the corresponding acid, amide, or ester by reaction with water, ammonia, or alcohol, respectively.

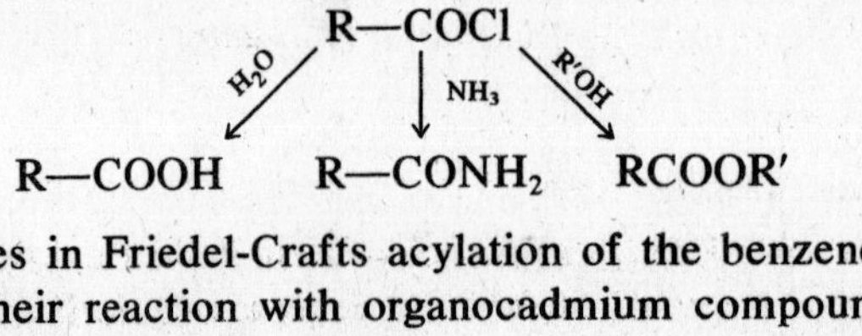

     The use of acyl chlorides in Friedel-Crafts acylation of the benzene ring has been discussed [Problem 15.8(d)], as has their reaction with organocadmium compounds [Problem 15.8(f)] and reduction to aldehydes [Problem 15.8(c)].

ACID ANHYDRIDES

All carboxylic acids have anhydrides,

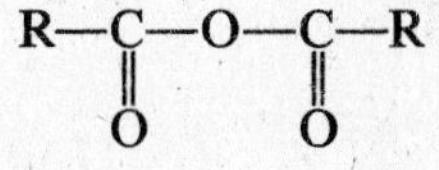

$$R-\underset{\underset{O}{\|}}{C}-O-\underset{\underset{O}{\|}}{C}-R$$

but the one most generally used is acetic anhydride, prepared as follows:

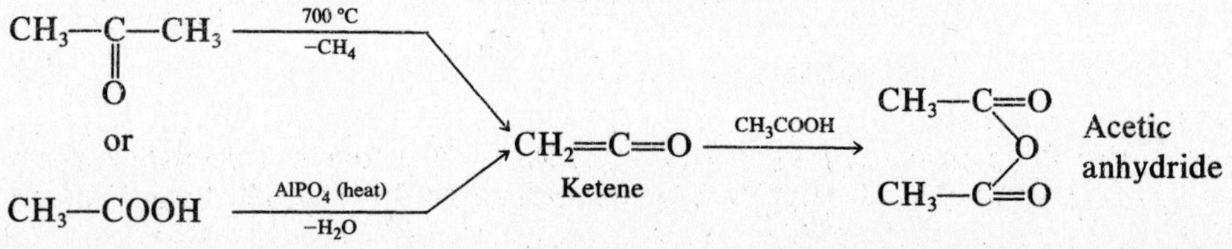

$$CH_3-\underset{\underset{O}{\|}}{C}-CH_3 \xrightarrow[-CH_4]{700\,°C}$$

or

$$CH_3-COOH \xrightarrow[-H_2O]{AlPO_4\ (heat)}$$

$$\searrow CH_2=C=O \xrightarrow{CH_3COOH} \quad \underset{\text{Ketene}}{}$$

$$\underset{CH_3-C=O}{\overset{CH_3-C=O}{\diagdown \diagup}} O \quad \text{Acetic anhydride}$$

Heating dicarboxylic acids, $HOOC(CH_2)_n COOH$ ($n = 2$ or 3), forms cyclic anhydrides by intramolecular dehydration [Problem 16.27(c), (d), (e)]. Anhydrides resemble acid halides in their reactions. Because acetic anhydride reacts less violently, it is often used in place of acetyl chloride.

Problem 17.5 Give the products formed when acetic anhydride reacts with   (a) H_2O,   (b) NH_3,   (c) C_2H_5OH,   (d) C_6H_6 with $AlCl_3$.   ◄

(a) $2CH_3COOH$.   (b) $CH_3CONH_2 + CH_3CO_2^-NH_4^+$.   ($c$) $CH_3COOC_2H_5 + CH_3COOH$. This is a good way to form acetates. (d) $C_6H_5COCH_3$. Friedel-Crafts acetylation.

AMIDES

Unsubstituted amides, $RCONH_2$, are generally prepared in the laboratory as shown:

1. $R-COCl + 2NH_3 \longrightarrow R-CONH_2 + NH_4Cl$

or $(RCO)_2O + 2NH_3 \longrightarrow RCONH_2 + RCO_2^-NH_4^+$

2.   See page 293 for preparation from ammonium carboxylates.

3.   By careful partial hydrolysis of nitriles:

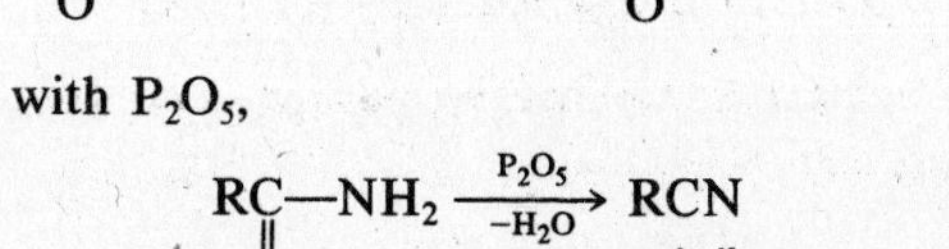

$$H_2O + RC\equiv N \xrightarrow[\substack{2.\ \ \text{cold water}}]{\substack{1.\ \ \text{cold } H_2SO_4}} R-\underset{\underset{O}{\|}}{C}-NH_2$$

Amides are slowly hydrolyzed under either acidic or basic conditions. The mechanisms are those shown in Problem 17.1. Unsubstituted amides are converted to $RCOOH$ with HNO_2,

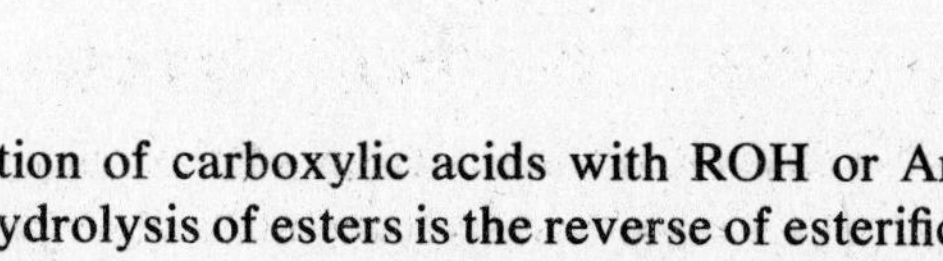

$$R\underset{\underset{O}{\|}}{C}NH_2 \xrightarrow[\substack{(NaNO_2 + aq.\ HCl)}]{HONO} R\underset{\underset{O}{\|}}{C}-OH + N_2$$

and are dehydrated to RCN with P_2O_5,

$$R\underset{\underset{O}{\|}}{C}-NH_2 \xrightarrow[-H_2O]{P_2O_5} \underset{\substack{\textit{a nitrile}}}{RCN}$$

ESTERS

The mechanism of esterification of carboxylic acids with ROH or ArOH is discussed in Problem 17.6. Since acid-catalyzed hydrolysis of esters is the reverse of esterification, it proceeds through the same intermediates and transition states (Principle of Microscopic Reversibility).

$$RCOOH + \begin{Bmatrix} HOR' & & RCOOR' \\ \text{or} & \underset{\text{hydrol.}}{\overset{\text{ester.}}{\rightleftharpoons}} & \text{or} \\ HOAr & & RCOOAr \end{Bmatrix} + H_2O$$

For the role of steric hindrance, see Problem 16.33.

Esters react with the Grignard reagent:

$$R-\underset{\underset{OR'}{|}}{C}=O \xrightarrow{\boxed{R''}MgX} Mg(OR')X + \left[R-\underset{\underset{O}{\|}}{C}-\boxed{R''} \right] \xrightarrow{R''MgX} R-\underset{\underset{OMgX}{|}}{\overset{\boxed{R''}}{C}}-\boxed{R''} \xrightarrow{H_2O} R-\underset{\underset{OH}{|}}{\overset{\boxed{R''}}{C}}-\boxed{R''}$$

not isolated   a 3° alcohol with
at least 2 like R'''s

Reduction of esters gives alcohols:

$$RCOOR' \xrightarrow[\text{2. } H_2O \text{ in ether}]{\text{1. } LiAlH_4} RCH_2OH + R'OH$$

On pyrolysis, esters give alkenes:

$$RCH_2CH_2O\underset{\underset{O}{\|}}{C}-R' \xrightarrow{400 \text{ °C}} RCH=CH_2 + R'COOH$$

Problem 17.6 Use the mechanism of esterification to explain the lower rates of both esterification and hydrolysis of esters when the alcohol, the acid, or both have branched substituent groups. ◄

The carbonyl C of RCOOH and RCOOR' is trigonal sp^2 hybridized, but that of the intermediate is tetrahedral sp^3 hybridized. If R' in R'OH or R in RCOOH is extensively branched, formation of the unavoidably crowded transition state has to occur with greater difficulty and more slowly.

Problem 17.7 Acid-catalyzed hydrolysis with $H_2^{18}O$ of an ester of an optically active 3° alcohol, RCOOC*R'R''R''', yields the partially racemic alcohol containing ^{18}O, R'R''R'''C^{18}OH. Similar hydrolyses of esters of 2° chiral alcohols, RCOOC*HR'R'', produce no change in the optical activity of the alcohol, and ^{18}O is found in RC$^{18}O_2$H. Explain these observations. ◄

Hydrolyses of esters of most 2° (and 1°) alcohols occur by cleavage of the O—acyl bond,

$$R-\underset{\underset{}{\|}}{\overset{O}{C}}\!+\!O-\overset{*}{C}HR'R''$$

Since no bond to C* is broken, no racemization occurs. However, with 3° alcohols, there is an S_N1 O—alkyl cleavage,

$$R-\underset{\underset{}{\|}}{\overset{O}{C}}-O\!+\!CR'R''R'''$$

producing RCOOH and a 3° carbocation, $^+CR'R''R'''$, which reacts with the solvent ($H_2^{18}O$) to form R'R''R'''C^{18}OH. This alcohol is partially racemized because $^+CR'R''R'''$ is partially racemized.

$$R-\underset{\underset{}{\|}}{\overset{O}{C}}-O-CR'R''R''' + H\!:\!A \underset{-A^-}{\rightleftharpoons} R-\underset{\underset{}{\|}}{\overset{O}{C}}-\underset{\underset{+}{}}{\overset{H}{O}}-CR'R''R''' \overset{S_N1}{\rightleftharpoons}$$

$$RCOOH + R'R''R'''C^+ \overset{H_2^{18}O}{\rightleftharpoons} R'R''R'''C^{18}OH + H^+$$

Problem 17.8 Write mechanisms for the reactions of RCOOR' with (a) aqueous OH⁻ to form RCOO⁻ and R'OH, (b) NH_3 to form $RCONH_2$, (c) R''OH in acid, HA, to form a new ester RCOOR'' (**transesterification**). ◄

(a) $R—\overset{\ddot{O}:}{\overset{\|}{C}}—\ddot{O}R' + OH^- \rightleftharpoons R—\overset{:\ddot{O}:^-}{\overset{|}{\underset{OH}{C}}}—\ddot{O}R' \longrightarrow R—\overset{O}{\overset{\|}{C}}—OH + :\ddot{O}R' \longrightarrow R—\overset{O}{\overset{\|}{C}}\Big\} ^- + HOR'$

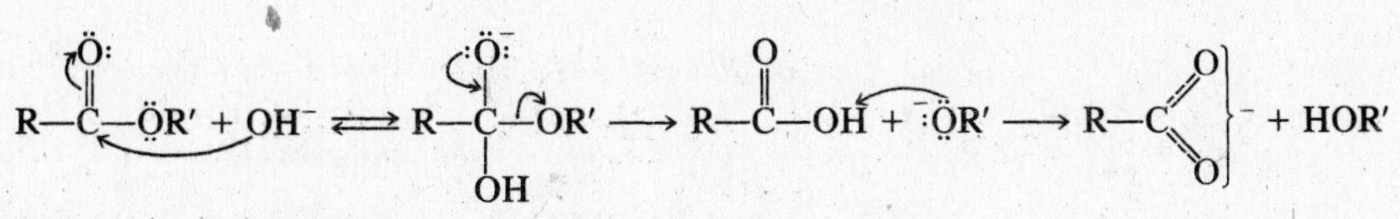

The last step is irreversible and drives the reaction to completion.

(b) $R—\overset{O}{\overset{\|}{C}}—OR' + :NH_3 \rightleftharpoons R—\overset{O^-}{\overset{|}{\underset{\overset{\overset{+}{H}NH}{H}}{C}}}—OR' \overset{\sim H^+}{\longrightarrow} R—\overset{O^-}{\overset{|}{\underset{NH_2\ H}{C}}}—\overset{+}{O}R' \longrightarrow R—\overset{O}{\overset{\|}{C}}—NH_2 + HOR'$

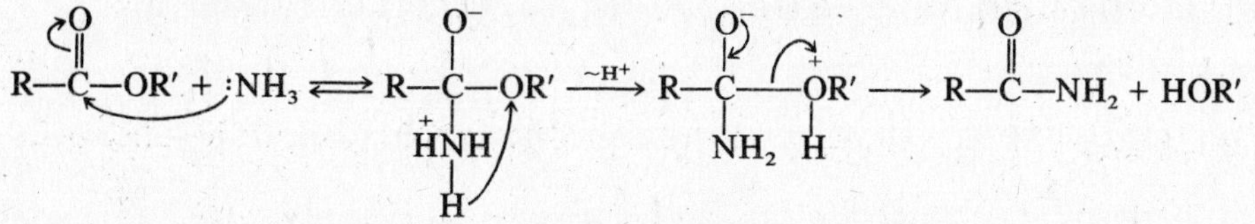

(c) $R—\overset{O}{\overset{\|}{C}}—OR' \underset{-A^-}{\overset{H·A}{\rightleftharpoons}} R—\overset{\overset{+}{O}H}{\overset{\|}{C}}—OR' \underset{HOR''}{\rightleftharpoons} R—\overset{O—H}{\overset{|}{\underset{\overset{+}{HOR''}}{C}}}—OR' \rightleftharpoons R—\overset{O}{\overset{\|}{C}}—OR'' + R'OH + HA$

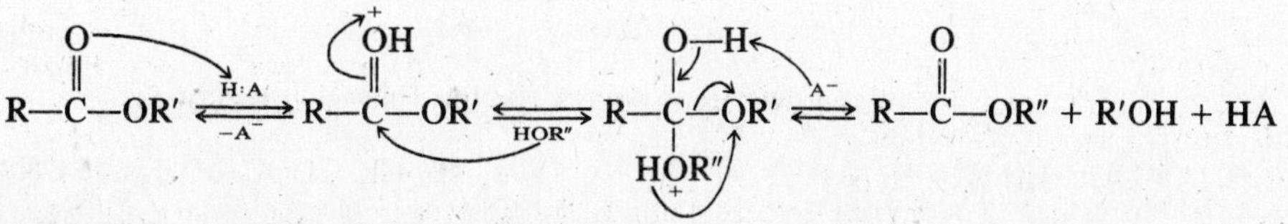

To drive the reaction to completion, a large excess of R″OH is used, and when R′OH is lower-boiling than R″OH, R′OH is removed by distillation.

Problem 17.9 Assign numbers from 1 for LEAST to 4 for MOST to show relative rates of alkaline hydrolysis of compounds I through IV and point out the factors determining the rates.

	I	**II**	**III**	**IV**
(a)	$CH_3COOCH(CH_3)_2$	CH_3COOCH_3	$CH_3COOC(CH_3)_3$	$CH_3COOC_2H_5$
(b)	$HCOOCH_3$	$(CH_3)_2CHCOOCH_3$	CH_3COOCH_3	$(CH_3)_3CCOOCH_3$
(c)	$O_2N—\bigcirc—COOCH_3$	$CH_3O—\bigcirc—COOCH_3$	$\bigcirc—COOCH_3$	$Cl—\bigcirc—COOCH_3$

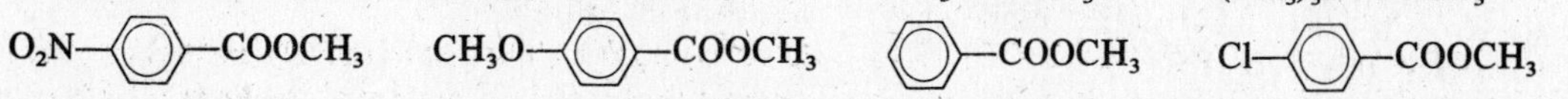

See Table 17-1.

Table 17-1

	\multicolumn{4}{c}{Ranks}	Rate-Determining Factors
	I	II	III	IV	
(a)	2	4	1	3	Steric effects (branching on alcohol portion)
(b)	4	2	3	1	Steric factor (branching on acid portion)
(c)	4	1	2	3	Electron-attracting groups disperse, developing – charge in transition state and increasing reactivity

Problem 17.10 Name the following acid derivatives:

(a) C_6H_5COCl (b) $CH_3CH_2\overset{O}{\overset{\|}{C}}O\overset{O}{\overset{\|}{C}}CH_2CH_3$ (c) $CH_3CH_2CH_2\overset{}{C}—OCH_2CH_3$
$\overset{}{\underset{O}{}}$

(d) $CH_3CH_2\overset{}{\underset{O}{\overset{\|}{C}}}NH_2$ (e) $C_6H_5\overset{}{\underset{O}{\overset{\|}{C}}}—OC_6H_5$ (f) $C_6H_5CH_2COO^-K^+$

(a) Benzoyl chloride (change **-ic acid** to **-yl chloride**). (b) Propionic or propanoic anhydride (change **acid** to **anhydride**). (c) Ethyl butyrate (butanoate) (change **-ic acid** to **-ate** preceded by name of alcohol or phenol). (d) Propanamide or propionamide (change **-oic acid** to **-amide**). (e) Phenyl benzoate. (f) Potassium phenylacetate.

Problem 17.11 Give the structure and name of the principal product formed when propionyl chloride reacts with: (a) H_2O, (b) C_2H_5OH, (c) NH_3, (d) $C_6H_6(AlCl_3)$, (e) $(n\text{-}C_3H_7)_2Cd$, (f) aq. NaOH, (g) $LiAlH(O\text{-}t\text{-}C_4H_9)_3$, (h) H_2NOH, (i) CH_3NH_2, (j) Na_2O_2 (sodium peroxide). ◄

(a) CH_3CH_2COOH, propionic acid; (b) $CH_3CH_2COOC_2H_5$, ethyl propionate; (c) $CH_3CH_2CONH_2$, propanamide; (d) $C_6H_5COCH_2CH_3$, propiophenone or ethyl phenyl ketone; (e) $CH_3CH_2COC_3H_7$, 3-hexanone; (f) $CH_3CH_2COO^-Na^+$, sodium propionate; (g) CH_3CH_2CHO, propanal;

(h) $CH_3CH_2\overset{\displaystyle O}{\overset{\displaystyle \|}{C}}-NHOH$ (i) $CH_3CH_2CONHCH_3$ (j) $CH_3CH_2\overset{\displaystyle O}{\overset{\displaystyle \|}{C}}-O-O-\overset{\displaystyle O}{\overset{\displaystyle \|}{C}}-CH_2CH_3$

Propanehydroxamic   N-Methylpropanamide   Propionyl peroxide
acid   (a substituted amide)

Problem 17.12 Write structures of the organic products for the following reactions:

(a) (R)-$CH_3COOCH(CH_3)CH_2CH_3 + H_2O \xrightarrow{\text{NaOH}}$ (b) (R)-$CH_3COOCH(CH_3)CH_2CH_3 + H_3O^+ \longrightarrow$

(c) $C_6H_5COOC_2H_5 + NH_3 \longrightarrow$ (d) $C_6H_5COOC_2H_5 + n\text{-}C_4H_9OH \xrightarrow{\text{H}^+}$

(e) $C_6H_5COOC_2H_5 + LiAlH_4 \longrightarrow$ (f) $C_6H_5\overset{\displaystyle O}{\overset{\displaystyle \|}{C}}-O-O-\overset{\displaystyle O}{\overset{\displaystyle \|}{C}}-C_6H_5 + CH_3O^-Na^+ \longrightarrow$ ◄

(a) $CH_3COO^-Na^+$ + (R)-$HOCH(CH_3)CH_2CH_3$. The alcohol is 2°; we find O—acyl cleavage and no change in configuration of alcohol. $RCOO^-Na^+$ forms in basic solution.
(b) CH_3COOH + (R)-$HOCH(CH_3)CH_2CH_3$. Again O—acyl cleavage occurs.
(c) $C_6H_5CONH_2 + C_2H_5OH$.
(d) $C_6H_5COO\text{-}n\text{-}C_4H_9 + C_2H_5OH$. Acid-catalyzed transesterification.
(e) $C_6H_5CH_2OH + C_2H_5OH$.

(f) $C_6H_5\overset{\displaystyle O}{\overset{\displaystyle \|}{C}}-O-O^-Na^+ + C_6H_5COOCH_3$
Sodium peroxybenzoate

Problem 17.13 Supply a structural formula for the alcohol formed from

(a) $CH_3CH_2COOCH_3 + 2C_3H_7MgBr$ (b) $C_6H_5CH_2COOCH_3 + 2C_6H_5MgBr$ ◄

Two R or Ar groups bonded to carbinol C come from the Grignard reagent, while the carbonyl C becomes the carbinol C.

(a) $(\overline{CH_3CH_2})COOCH_3 + 2\boxed{C_3H_7}MgBr \longrightarrow (\overline{CH_3CH_2})-\overset{\boxed{C_3H_7}}{\underset{OH}{\overset{|}{\underset{|}{C}}}}-\boxed{C_3H_7}$

(b) $(\overline{C_6H_5CH_2})COOCH_3 + 2\boxed{C_6H_5}MgBr \longrightarrow (\overline{C_6H_5CH_2})-\overset{\boxed{C_6H_5}}{\underset{OH}{\overset{|}{\underset{|}{C}}}}-\boxed{C_6H_5}$

Problem 17.14 Methyl esters can be prepared on a small scale from RCOOH and CH_2N_2. Suggest a mechanism involving S_N2 displacement of N_2. ◄

Diazomethane is a resonance hybrid,

$$H\overset{H}{\overset{|}{C}}-\overset{+}{N}\equiv N\!: \longleftrightarrow H\overset{H}{\overset{|}{C}}=\overset{+}{N}=\ddot{N}\!:$$

and we have

$$\text{RCOOH} + \overset{\text{H}}{\underset{\ddots}{\text{HC}}}-\text{N}\equiv\text{N}: \longrightarrow \text{RCOO}^- + \text{CH}_3\overset{+}{:}\text{N}\equiv\text{N}: \longrightarrow \text{RCOOCH}_3 + :\text{N}\equiv\text{N}:$$

Acid$_1$ Base$_2$ Base$_1$ Acid$_2$

Problem 17.15   Prepare $CH_3CH_2CH_2COOCH_2CH_2CH_2CH_3$ from n-C_4H_9OH.   ◄

The acid portion of the ester has the same number of C's (4) as does the starting material, the alcohol.   Therefore, oxidize the alcohol.   The ester is best formed from the acid chloride:

$$n\text{-}C_4H_9OH \xrightarrow{KMnO_4} CH_3(CH_2)_2COOH \xrightarrow{SOCl_2} CH_3(CH_2)_2COCl \xrightarrow{n\text{-}C_4H_9OH} \text{product}$$

Problem 17.16   **Fats** and **oils (glycerides)** are esters of carboxylic acids and glycerol, $HOCH_2CHOHCH_2OH$. (a) Write a formula for a fat (found in butter) of butanoic acid.   (b) Alkaline hydrolysis of a fat of a high-molecular-weight acid (**saponification**) gives a carboxylate salt (**soap**).   Write the equation for the reaction of the fat of palmitic acid, n-$C_{15}H_{31}COOH$, with aqueous NaOH.   ◄

(a)

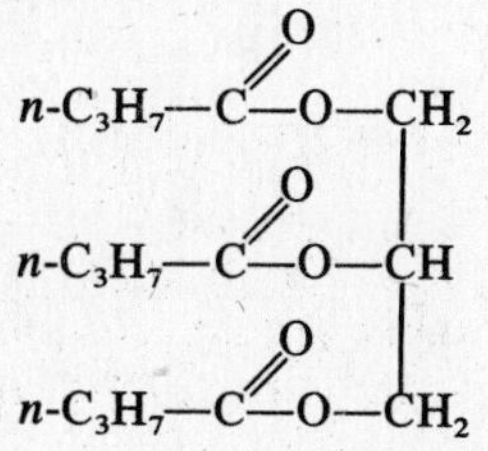

$$n\text{-}C_3H_7-\overset{\overset{\text{O}}{\|}}{C}-O-CH_2$$
$$n\text{-}C_3H_7-\overset{\overset{\text{O}}{\|}}{C}-O-CH$$
$$n\text{-}C_3H_7-\overset{\overset{\text{O}}{\|}}{C}-O-CH_2$$

(b)

$$\begin{array}{l} n\text{-}C_{15}H_{31}COO-CH_2 \\ n\text{-}C_{15}H_{31}COO-CH + 3NaOH \longrightarrow 3\ n\text{-}C_{15}H_{31}COO^-Na^+ + HOCH_2CHOHCH_2OH \\ n\text{-}C_{15}H_{31}COO-CH_2 \qquad\qquad \textit{a soap} \end{array}$$

17.3   DICARBOXYLIC ACID DERIVATIVES

Problem 17.17   Show steps in the following syntheses, using any needed inorganic reagents:

(a)   o-Xylene ⟶

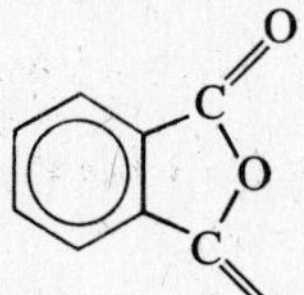

Phthalic anhydride

(b) ⟶ $ClCO(CH_2)_4COCl$

Tetrahydrofuran   Adipoyl dichloride   ◄

(a)

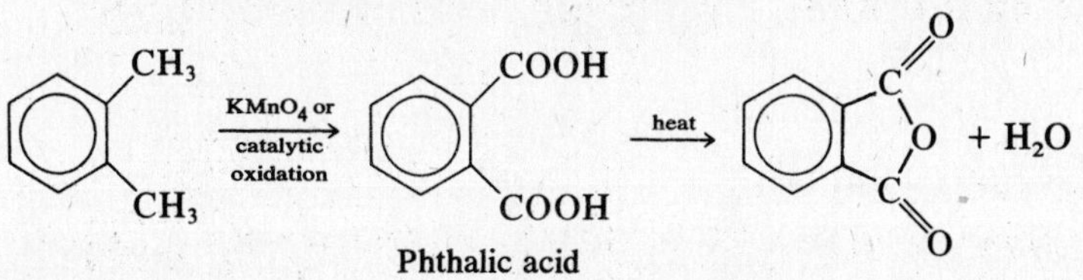

Phthalic acid

(b)   The chain is increased from 4 to 6 C's by forming a dinitrile.

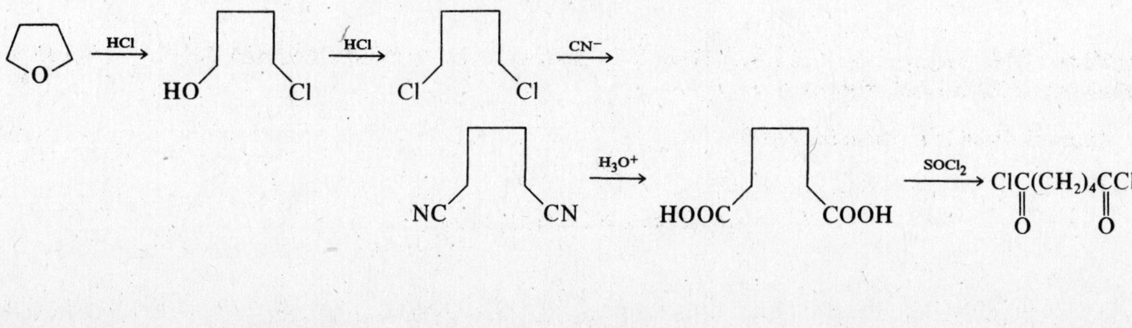

Problem 17.18 Give the structural formula and name for the product formed when 1 mole of succinic anhydride, the cyclic anhydride of succinic acid (butanedioic acid), reacts with (*a*) 1 mole of CH_3OH; (*b*) 2 moles of NH_3 and (*c*) 1 mole of C_6H_6 with $AlCl_3$. ◀

Products are formed in which half of the anhydride forms the appropriate derivative and the other half becomes a COOH.

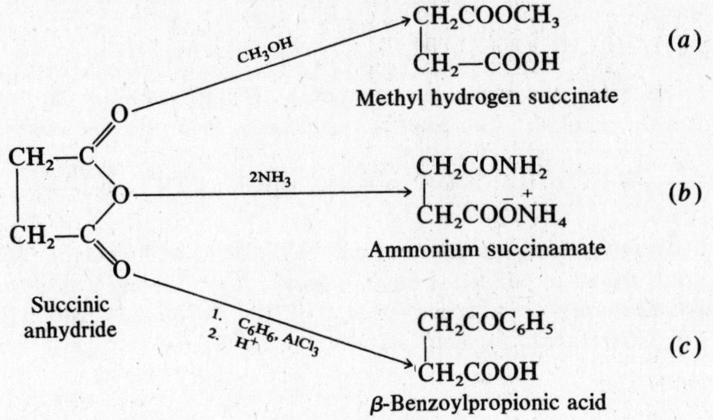
CH_3OH
CH_2COOCH_3
$CH_2—COOH$
Methyl hydrogen succinate
(a)
$CH_2—C$
O
$CH_2—C$
$2NH_3$
CH_2CONH_2
$CH_2CO\bar{O}\overset{+}{N}H_4$
Ammonium succinamate
(b)
Succinic anhydride
1. C_6H_6, $AlCl_3$
2. H^+
$CH_2COC_6H_5$
CH_2COOH
β-Benzoylpropionic acid
(c)

The monoamides of dicarboxylic acids, $HOOC(CH_2)_nCONH_2$ (n = 2, 3), form cyclic imides on heating.

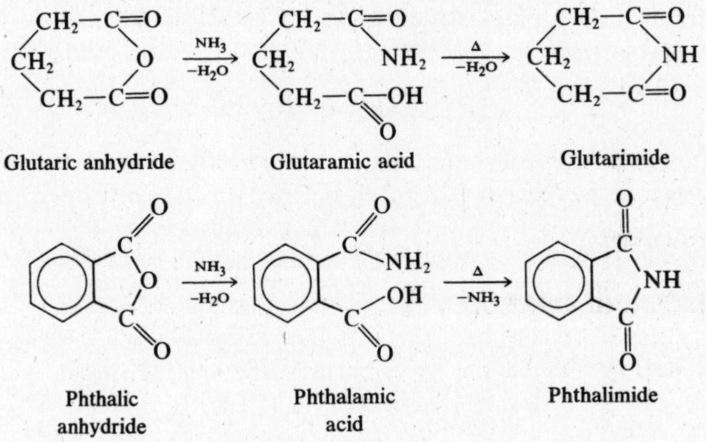
NH_3
$-H_2O$
Δ
$-H_2O$
Glutaric anhydride
Glutaramic acid
Glutarimide
NH_3
$-H_2O$
Δ
$-NH_3$
Phthalic anhydride
Phthalamic acid
Phthalimide

Problem 17.19 Use the concepts of charge delocalization and resonance to account for the acidity of imides. (They dissolve in NaOH.) ◀

The H on N of the imides is acidic because the negative charge on N of the conjugate base is delocalized to each O of the two C=O groups, thereby stabilizing the anion.

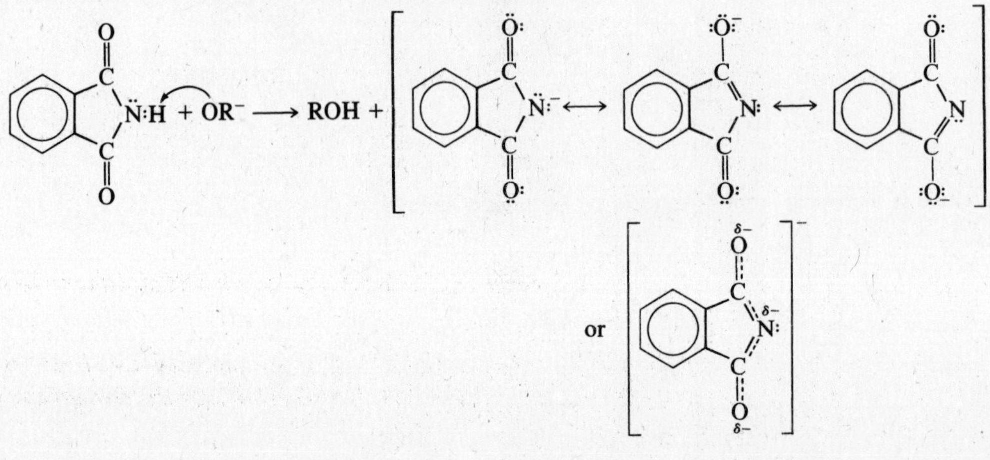
N:H + OR⁻ ⟶ ROH +
or
δ−

Phthalic anhydride reacts with 2 moles of phenol, C_6H_5OH, to eliminate 1 mole of H_2O when heated with anhydrous $ZnCl_2$.

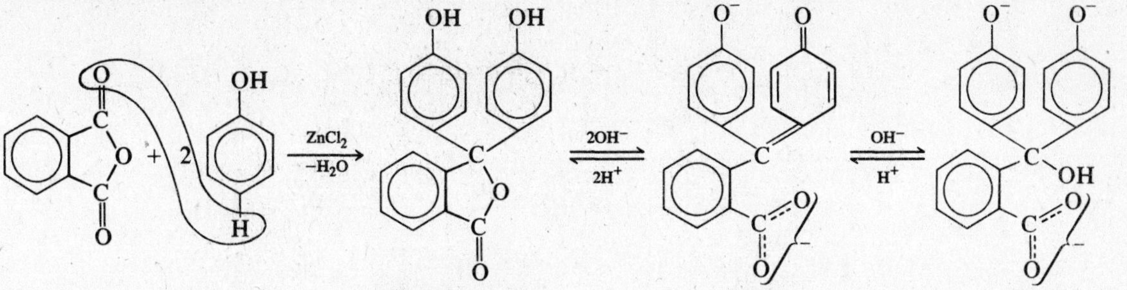

Phenolphthalein
(colorless)　　　　　　　　(red)　　　　　　　　(colorless)

The product, phenolphthalein, is an acid-base indicator.

17.4 CLAISEN CONDENSATION; REACTIONS OF β-KETOESTERS

Esters with α H's condense in strong bases to give β-ketoesters (**Claisen condensation**).

Step 1　Formation of stabilized α-carbanion.

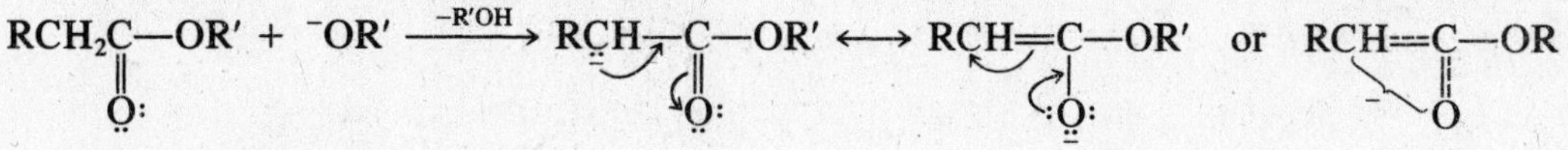

Step 2　Nucleophilic attack by α-carbanion on C=O of ester and displacement of $^-$OR'.

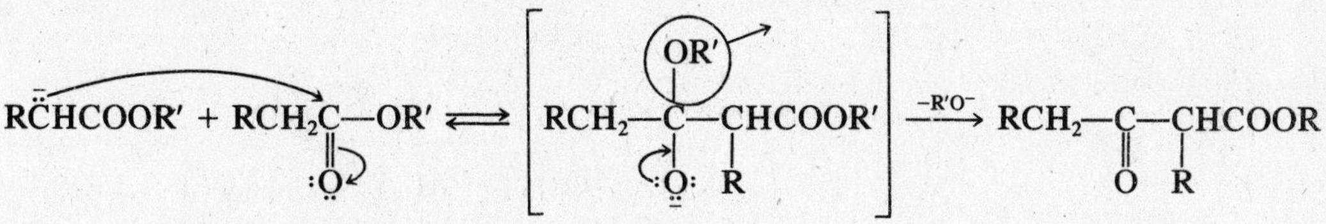

Step 3　The only irreversible step completes the reaction by forming a stable carbanion where – is delocalized to both O's.

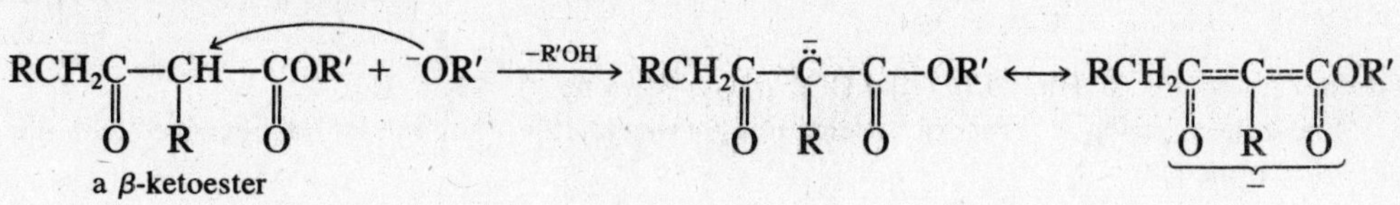

a β-ketoester

Acid is then added to neutralize the carbanion salt.

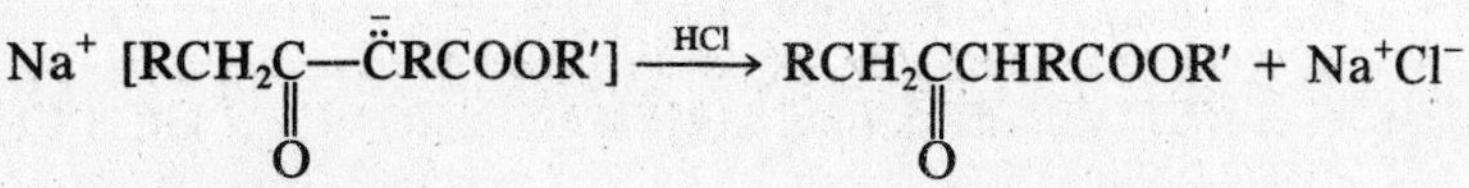

Na$^+$ *salt if*
NaOR' *is used as base*

Problem 17.20　Write structural formulas for the products from the reaction of $C_2H_5O^-Na^+$ with the following esters:

(a) $CH_3CH_2COOC_2H_5$　(b) $C_6H_5COOC_2H_5 + CH_3COOC_2H_5$　(c) $C_6H_5CH_2COOC_2H_5 + O{=}C(OC_2H_5)_2$

In these Claisen condensations $^-OC_2H_5$ is displaced from the $COOC_2H_5$ group by the α-carbanion formed from another ester molecule. Mixed Claisen condensations are feasible only if one of the esters has no α H.

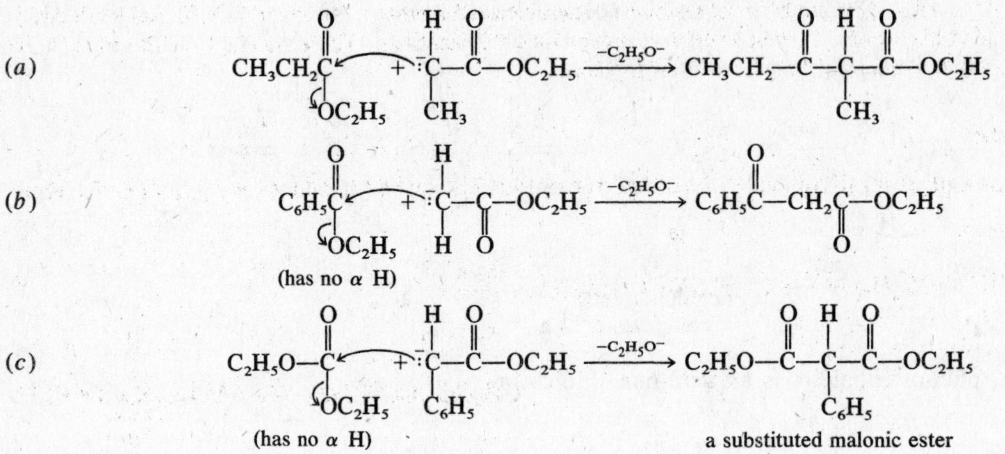

(a)

$$CH_3CH_2\overset{\displaystyle O}{\underset{\displaystyle OC_2H_5}{C}} \; + \; :\overset{\displaystyle H \; O}{\underset{\displaystyle CH_3}{C}}{-}C{-}OC_2H_5 \xrightarrow{-C_2H_5O^-} CH_3CH_2{-}\overset{\displaystyle O}{C}{-}\overset{\displaystyle H}{\underset{\displaystyle CH_3}{C}}{-}\overset{\displaystyle O}{C}{-}OC_2H_5$$

(b)

$$C_6H_5\overset{\displaystyle O}{\underset{\displaystyle OC_2H_5}{C}} \; + \; :\overset{\displaystyle H}{\underset{\displaystyle H \; O}{C}}{-}C{-}OC_2H_5 \xrightarrow{-C_2H_5O^-} C_6H_5{-}\overset{\displaystyle O}{C}{-}CH_2\overset{\displaystyle O}{C}{-}OC_2H_5$$

(has no α H)

(c)

$$C_2H_5O{-}\overset{\displaystyle O}{\underset{\displaystyle OC_2H_5}{C}} \; + \; :\overset{\displaystyle H \; O}{\underset{\displaystyle C_6H_5}{C}}{-}C{-}OC_2H_5 \xrightarrow{-C_2H_5O^-} C_2H_5O{-}\overset{\displaystyle O}{C}{-}\overset{\displaystyle H}{\underset{\displaystyle C_6H_5}{C}}{-}\overset{\displaystyle O}{C}{-}OC_2H_5$$

(has no α H) \hfill a substituted malonic ester

Problem 17.21 Ethyl pimelate, $C_2H_5OOC(CH_2)_5COOC_2H_5$, reacts with $C_2H_5O^-Na^+$ **(Dieckmann condensation)** to form a cyclic keto ester, $C_9H_{14}O_3$. Supply a mechanism for its formation and compare the yields in ethanol and ether as solvents. ◄

We have:

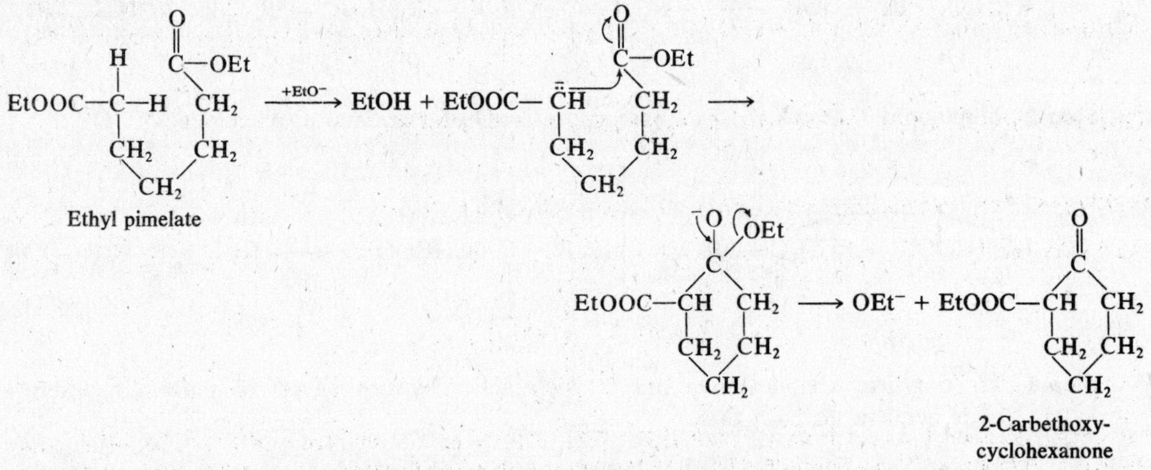

Ethyl pimelate

2-Carbethoxy-cyclohexanone

The net reaction is $C_2H_5OOC(CH_2)_5COOC_2H_5 \longrightarrow$ product $+ HOC_2H_5$.

Since the reaction is reversible, yields are greater in ether than in alcohol because alcohol is a product (Le Chatelier principle).

Intramolecular Claisen cyclizations occur with ethyl adipate and pimelate because 5- and 6-membered rings are formed (Problem 9.11).

PROPERTIES OF β-KETOESTERS (e.g. acetoacetic ester, $CH_3COCH_2COOC_2H_5$)

1. Acidity; Carbanion Formation

Acetoacetic ester is acidic ($pK_a = 10.2$) and forms a resonance-stabilized carbanion whose negative charge is delocalized over one C and two O's.

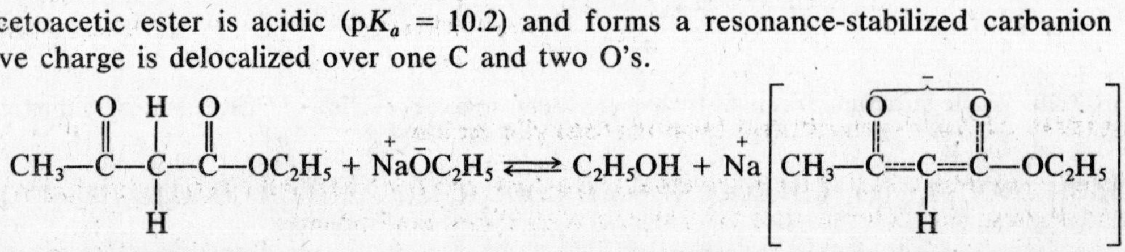

$$CH_3{-}\overset{\displaystyle O}{C}{-}\overset{\displaystyle H}{\underset{\displaystyle H}{C}}{-}\overset{\displaystyle O}{C}{-}OC_2H_5 + Na\overset{+}{O}C_2H_5 \rightleftharpoons C_2H_5OH + \overset{+}{Na}\left[CH_3{-}\overset{\displaystyle O}{C}{=}\overset{\displaystyle H}{C}{-}\overset{\displaystyle O}{C}{-}OC_2H_5\right]^-$$

Problem 17.22 Compare relative acid strengths of $CH_3COCH_2COCH_3$ (I), $CH_3COCH_2COOC_2H_5$ (II) and $C_2H_5OOCCH_2COOC_2H_5$ (III). Explain your ranking. ◄

I > II > III. All 3 compounds afford resonance-stabilized carbanions. However, COOEt has an electron-releasing O bonded to carbonyl C, which decreases resonance stabilization. There are two COOEt groups in III and one in II, while I has only ketonic carbonyl groups.

2. Alkylation

As with malonic ester (Problem 16.5), either one or two R's can be introduced in acetoacetic ester.

$$Na(CH_3CO\overset{..}{C}HCOOC_2H_5) \xrightarrow{RX} CH_3CO\overset{\overset{\displaystyle R}{|}}{C}HCOOC_2H_5 \xrightarrow{OC_2H_5^-}$$

$$CH_3CO\overset{\overset{\displaystyle R}{|}}{\underset{..}{C}}COOC_2H_5 \xrightarrow{R'X} CH_3CO\overset{\overset{\displaystyle R}{|}}{\underset{\underset{\displaystyle R'}{|}}{C}}COOC_2H_5$$

3. Hydrolysis and Decarboxylation

Dilute acid or *base* hydrolyzes the $COOC_2H_5$ group and forms acetoacetic acids, which decarboxylate to methyl ketones.

$$CH_3-\overset{\overset{\displaystyle O}{\|}}{C}-\overset{\overset{\displaystyle H}{|}}{\underset{\underset{\displaystyle R}{|}}{C}}-\overset{\overset{\displaystyle O}{\|}}{C}-OC_2H_5 \xrightarrow{H_3O^+} C_2H_5OH + CH_3-\overset{\overset{\displaystyle O}{\|}}{C}-\overset{\overset{\displaystyle H}{|}}{\underset{\underset{\displaystyle R}{|}}{C}}-\overset{\overset{\displaystyle O}{\|}}{C}-OH \longrightarrow CH_3-\overset{\overset{\displaystyle O}{\|}}{C}-\overset{\overset{\displaystyle H}{|}}{\underset{\underset{\displaystyle R}{|}}{C}}-H + C=O$$

This sequence of steps can be used to synthesize methyl ketones.

Problem 17.23 Prepare 3-methyl-2-pentanone from acetoacetic ester. ◄

In the product

$$CH_3-\overset{}{C}-\overset{\overset{\displaystyle H}{|}}{\underset{\underset{\displaystyle O}{\|}\ \boxed{CH_3}}{\overset{*}{C}}}-\boxed{C_2H_5}$$

the $CH_3C{=}O$ and *C come from CH_3COCH_2COOEt. The H on *C replaced COOEt. The $-CH_3$ and $-CH_2CH_3$ attached to *C are introduced by alkylation of the carbanion of acetoacetic ester with appropriate alkyl halides; in this case use CH_3I and then C_2H_5Br.

$$CH_3CO-\overset{\overset{\displaystyle H}{|}}{\underset{\underset{\displaystyle H}{|}}{C}}-COOC_2H_5 \xrightarrow{Na\bar{O}Et} CH_3CO\overset{..}{C}HCOOC_2H_5 \xrightarrow{CH_3I} CH_3CO\overset{\overset{\displaystyle CH_3}{|}}{C}HCOOC_2H_5 \xrightarrow[2.\ C_2H_5Br]{1.\ Na\bar{O}Et}$$

$$CH_3CO\overset{\overset{\displaystyle CH_3}{|}}{\underset{\underset{\displaystyle C_2H_5}{|}}{C}}COOC_2H_5 \xrightarrow[2.\ -CO_2]{1.\ \text{hydrolysis}} CH_3CO\overset{\overset{\displaystyle CH_3}{|}}{\underset{\underset{\displaystyle C_2H_5}{|}}{C}}H$$

Methylethylacetoacetic   3-Methyl-2-pentanone
ester

4. Synthesis of Alkyl-substituted Monocarboxylic Acids

Hydrolysis of monoalkyl or dialkylacetoacetic esters with *concentrated* NaOH yields carboxylate anions rather than the ketones that are formed with dilute acid or base.

$$CH_3C\!\!\underset{\substack{\| \\ O}}{\overset{\substack{(H) \\ \\}}{-}}\!\!C\!\!\underset{(R)}{-}(COOC_2H_5) \xrightarrow[\text{2. acid}]{\text{1. conc. Na}^+\text{OH}^-} CH_3COOH + H\!\!\underset{(R)}{\overset{(H)}{-}}\!\!C\!\!-(COOH) + C_2H_5OH$$

Monoalkylacetic acid

$$CH_3CO\!\!\underset{(R)}{\overset{(R')}{C}}\!\!(COOC_2H_5) \xrightarrow[\text{2. acid}]{\text{1. conc. NaOH}} CH_3COOH + H\!\!\underset{(R)}{\overset{(R')}{-}}\!\!C\!\!-(COOH) + C_2H_5OH$$

Dialkylacetic acid

Malonic ester (Problem 16.5) is a superior substrate for synthesizing carboxylic acids.

Problem 17.24 What product is formed if methylethylacetoacetic ester (Problem 17.23) is treated with concentrated NaOH and the products acidified?   ◄

Cleavages leading to the products are indicated by the zigzag lines:

$$\underset{\substack{HO \\ H}}{\overset{\substack{I \quad\quad II \quad\quad III \\ O \;\; C_2H_5 \; O}}{CH_3-C-C-C-OC_2H_5}} \xrightarrow[\text{2. } H_3O^+]{\substack{\text{1. conc.} \\ \text{NaOH}}} \underset{\substack{\text{Acetic acid} \\ \\}}{\overset{\substack{I \\ O}}{CH_3-C-OH}} + \underset{\substack{\alpha\text{-Methylbutyric} \\ \text{acid}}}{\overset{\substack{II \\ C_2H_5 \; O}}{H-C-C-OH}} + \underset{\text{Ethanol}}{\overset{III}{HOC_2H_5}}$$

Problem 17.25 (*a*) Write the structural formulas for the stable keto and enol tautomers of ethyl acetoacetate. (*b*) Why is this enol much more stable than that of a simple ketone? (*c*) How can the enol be chemically detected?   ◄

(*a*)

$$\underset{\substack{\quad\quad\;\; \| \;\; \| \\ \quad\quad\;\; O \;\; O \\ \textit{keto form}}}{CH_3CCH_2C-OC_2H_5} \rightleftharpoons \underset{\substack{\quad\quad\quad\;\; \| \\ \quad\quad O-H- O \\ \textit{enol form}}}{CH_3C\!=\!CHC-OC_2H_5} \;\; \text{H-bond}$$

(*b*) There is a stable conjugated C=C—C=O linkage; moreover, intramolecular H-bonding (chelation) adds stability to the enol.

(*c*) The enol decolorizes a solution of Br_2 in CCl_4.

17.5 LACTONES, LACTAMS

A **lactone** is a cyclic ester; a **lactam** is a cyclic amide.   Those with 5- or 6-membered rings are readily formed.

Problem 17.26 Write the structure for the lactone formed on heating in the presence of acid (*a*) γ-hydroxybutyric acid, (*b*) δ-hydroxyvaleric acid.   ◄

Since an OH and a COOH are present in each compound, intramolecular dehydration gives lactones with 5- and 6-membered rings, respectively.

(*a*)

$$\underset{\gamma\text{-Hydroxybutyric acid}}{\overset{\substack{CH_2-CH_2 \\ H_2C \quad\quad\quad C=O \\ | \quad\quad\quad\quad | \\ OH \quad\quad\quad OH}}{}} \xrightarrow[-H_2O]{H^+} \underset{\gamma\text{-Butyrolactone}}{\overset{\substack{CH_2-CH_2 \\ H_2C \quad\quad\quad C=O \\ \quad\quad O}}{}}$$

(b)

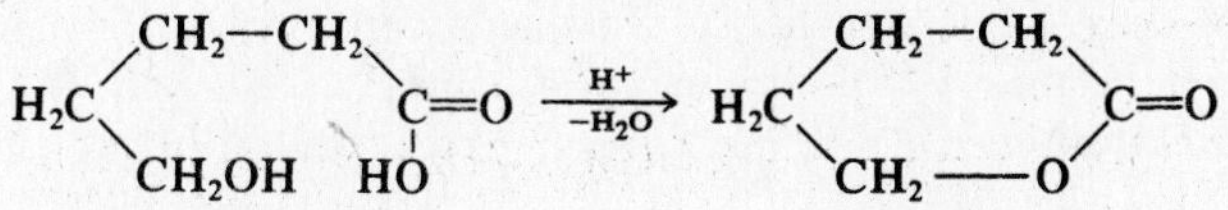

δ-Hydroxyvaleric acid   δ-Valerolactone

Problem 17.27 β-Hydroxyacids readily undergo dehydration but do not yield a lactone.   Give the structure of the product and account for its formation.   ◄

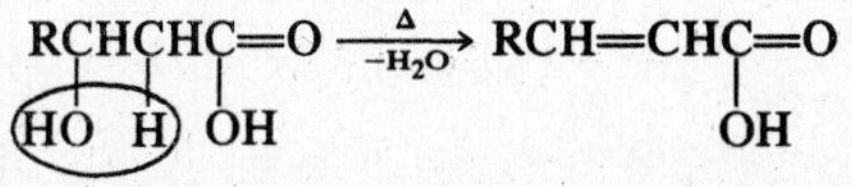

$$RCHCHC{=}O \xrightarrow[-H_2O]{\Delta} RCH{=}CHC{=}O$$

A double bond forms because the product is a stable, conjugated α,β-unsaturated acid [Problem 16.29(b)].

Problem 17.28 (a) When heated, 2 moles of an α-hydroxyacid loses 2 moles of H_2O to give a cyclic diester (a lactide).   Give the structural formulas for two diastereomers obtained from lactic acid, $CH_3CHOHCOOH$, and select the diastereomer which is not resolvable.   (b) Synthesize lactic acid from CH_3CHO.   ◄

(a)

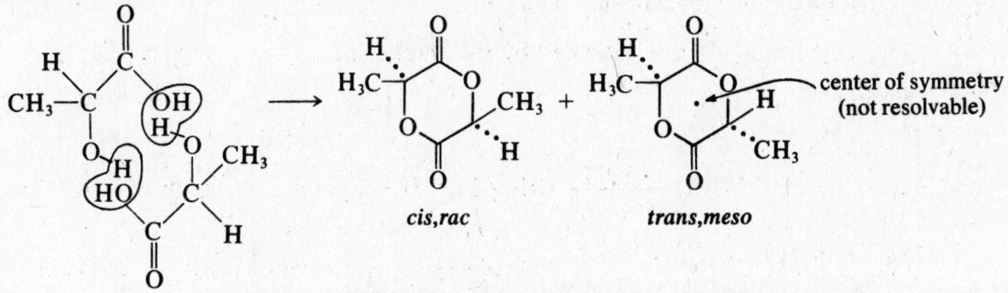

center of symmetry (not resolvable)

cis,rac   trans,meso

(b)

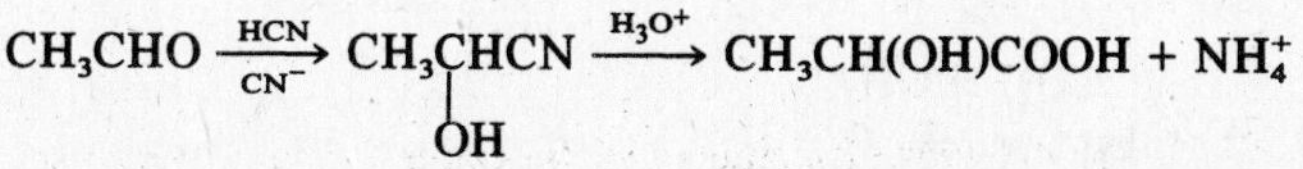

$$CH_3CHO \xrightarrow[CN^-]{HCN} CH_3CHCN \xrightarrow{H_3O^+} CH_3CH(OH)COOH + NH_4^+$$
                                    |
                                    OH

Problem 17.29 Write structural formulas for the products formed from reaction of δ-valerolactone with: (a) $LiAlH_4$, then H_2O; (b) NH_3; (c) CH_3OH and H_2SO_4 catalyst.   ◄

   (a) $HOCH_2CH_2CH_2CH_2CH_2OH$, (b) $HOCH_2CH_2CH_2CH_2CONH_2$, (c) $HOCH_2CH_2CH_2CH_2COOCH_3$.

Problem 17.30 Amino acids react similarly to hydroxyacids.   Predict the products formed on heating the following: (a) $H_2NCHRCOOH$,   (b) $RCH(NH_2)CH_2COOH$,   (c) $RCH(NH_2)CH_2CH_2COOH$, (d) $RCH(NH_2)CH_2CH_2CH_2COOH$.   ◄

(a)

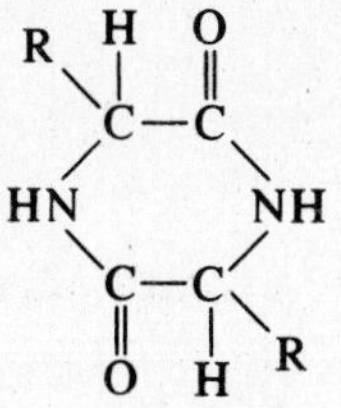

a diketopiperazine

(b) $RCH{=}CHCOOH$

an α,β-unsaturated acid

(c)

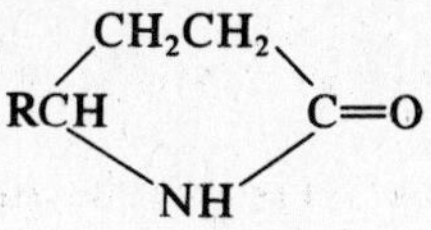

a γ-lactam

(d)

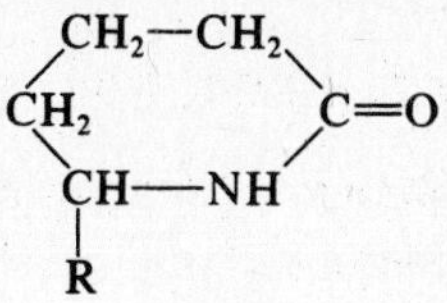

a δ-lactam

17.6 CARBONIC ACID DERIVATIVES

Problem 17.31 The following carboxylic acids are unstable and their decomposition products are shown in parentheses: carbonic acid, $(HO)_2C{=}O$ $(CO_2 + H_2O)$; carbamic acid, H_2NCOOH $(CO_2 + NH_3)$; and chloro-carbonic acid, $ClCOOH$ $(CO_2 + HCl)$.   Indicate how the *stable* compounds below are derived from one or more of these unstable acids.   Name those for which a common name is not given.

(a) $Cl_2C{=}O$ (b) $(H_2N)_2C{=}O$ (c) $ClCOCH_3$ (d) $(CH_3O)_2C{=}O$ (e) $H_2NC{-}OCH_3$

   Phosgene   Urea   ‖   ‖
                                                   O   O   ◄

(*a*) The acid chloride of chlorocarbonic acid; (*b*) amide of carbamic acid; (*c*) ester of chlorocarbonic acid, methyl chlorocarbonate; (*d*) diester of carbonic acid, methyl carbonate; (*e*) ester of carbamic acid, methyl carbamate (called a **urethane**).

Problem 17.32 Give the name and formula for the organic product formed from 1 mole of $COCl_2$ and 2 moles of (*a*) $(C_2H_5)_2NH$, (*b*) C_2H_5OH. ◄

$COCl_2$ behaves like a diacid chloride.

$$(a) \quad \underset{\underset{\displaystyle \|}{\displaystyle O}}{C_2H_5NHCNHC_2H_5} \quad N,N'\text{-Diethylurea} \qquad (b) \quad \underset{\underset{\displaystyle \|}{\displaystyle O}}{C_2H_5OCOC_2H_5} \quad \text{Ethyl carbonate}$$

Problem 17.33 Synthesize the tranquilizer meprobamate (Equanil, Miltown),

$$\underset{\underset{\displaystyle CH_2CH_2CH_3}{\displaystyle |}}{\overset{\overset{\displaystyle O \qquad CH_3 \quad O}{\displaystyle \| \qquad | \quad \|}}{H_2NCOCH_2CCH_2OC-NH_2}}$$

which is a diurethane. ◄

$$\underset{\underset{\displaystyle H}{\displaystyle |}}{\overset{\overset{\displaystyle H}{\displaystyle |}}{EtOOC-C-COOEt}} \xrightarrow[\text{3. EtO}^-, \text{ 4. PrBr}]{\text{1. EtO}^-, \text{ 2. MeBr}} \underset{\underset{\displaystyle Pr}{\displaystyle |}}{\overset{\overset{\displaystyle Me}{\displaystyle |}}{EtOOC-C-COOEt}} \xrightarrow[\text{2. H}_2\text{O}]{\text{1. LiAlH}_4} \underset{\underset{\displaystyle Pr}{\displaystyle |}}{\overset{\overset{\displaystyle Me}{\displaystyle |}}{HOCH_2-C-CH_2OH}}$$

$$\underset{\underset{\displaystyle O}{\displaystyle \|}}{Cl-C-Cl} + \underset{\underset{\displaystyle CH_2CH_2CH_3}{\displaystyle |}}{\overset{\overset{\displaystyle CH_3}{\displaystyle |}}{HOCH_2CCH_2OH}} + \underset{\underset{\displaystyle O}{\displaystyle \|}}{Cl-C-Cl} \longrightarrow \underset{\underset{\displaystyle CH_2CH_2CH_3}{\displaystyle |}}{\overset{\overset{\displaystyle O \quad CH_3 \quad O}{\displaystyle \| \quad | \quad \|}}{Cl-COCH_2CCH_2OC-Cl}} \xrightarrow[-2HCl]{+2NH_3} \text{product}$$

2-Methyl-2-(*n*-propyl)-
1,3-propanediol

Problem 17.34 What products are formed when 1 mole of urea reacts with (*a*) 1 mole, (*b*) a second mole, of methyl acetate (or acetyl chloride)? ◄

$$CH_3COOCH_3 + \underset{\underset{\displaystyle O}{\displaystyle \|}}{H_2N-C-NH_2} \xrightarrow{-CH_3OH} \underset{\underset{\displaystyle \text{Acetylurea}}{}}{\underset{\underset{\displaystyle O}{\displaystyle \|}}{CH_3CO-NH-C-NH_2}} \xrightarrow[-CH_3OH]{CH_3COOCH_3}$$

$$\underset{\underset{\displaystyle O}{\displaystyle \|}}{CH_3CO-NH-C-NH-COCH_3} + \underset{\underset{\displaystyle O}{\displaystyle \|}}{(CH_3CO)_2N-C-NH_2}$$

N,N'-Diacetylurea   N,N-Diacetylurea
(a ureide)

Problem 17.35 Give the structure of the cyclic ureide formed on heating urea and ethyl malonate in the presence of $NaOC_2H_5$. ◄

$$\underset{\underset{\displaystyle NH_2}{}}{\overset{\overset{\displaystyle NH_2}{}}{O=C}} + \underset{\underset{\displaystyle C_2H_5O}{}}{\overset{\overset{\displaystyle C_2H_5O-C=O}{}}{CH_2}} \xrightarrow[\text{NaOC}_2\text{H}_5]{\text{heat}} \underset{\underset{\displaystyle NH}{}}{\overset{\overset{\displaystyle NH-C=O}{}}{O=C}} \underset{\underset{\displaystyle C=O}{}}{\overset{\overset{\displaystyle }{}}{CH_2}} + 2C_2H_5OH$$

Barbituric acid
(a pyrimidine)

Problem 17.36 Synthesize barbital (the diethyl derivative of barbituric acid) from malonic ester and urea. ◀

Malonic ester is alkylated twice with C_2H_5Br to give the needed diethyl malonic ester.

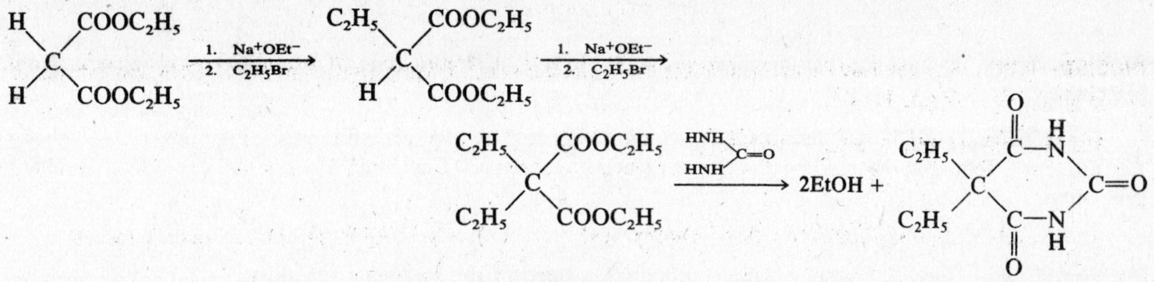

Ethyl diethyl malonate

Barbital

Supplementary Problems

Problem 17.37 How may an acyl chloride such as CH_3COCl be (*a*) reduced to CH_3CHO, acetaldehyde; (*b*) reduced to CH_3CH_2OH; (*c*) esterified to $CH_3COOC_2H_5$; (*d*) converted into N-methylacetamide, $CH_3CONHCH_3$?

(*a*) Hydrogenate in the presence of Pd deactivated with $BaSO_4$. (*b*) Reduce with $LiAlH_4$, followed by reaction with dilute H_3O^+. (*c*) Add C_2H_5OH. (*d*) Add CH_3NH_2.

Problem 17.38 Prepare the **mixed anhydride** of acetic and propionic acids. ◀

Mixed anhydrides,

$$R-\overset{\overset{\displaystyle O}{\|}}{C}-O-\overset{\overset{\displaystyle O}{\|}}{C}-R'$$

are made by reacting the acid chloride of one of the acid portions with the carboxylate salt of the other. Use CH_3COCl and CH_3CH_2COONa, or CH_3COONa and CH_3CH_2COCl. For example:

$$CH_3-\overset{\overset{\displaystyle O}{\|}}{C}-Cl + Na^{+-}O-\overset{\overset{\displaystyle O}{\|}}{C}-CH_2CH_3 \longrightarrow CH_3-\overset{\overset{\displaystyle O}{\|}}{C}-O-\overset{\overset{\displaystyle O}{\|}}{C}-CH_2CH_3 + Na^+Cl^-$$

Problem 17.39 Use $^{14}CH_3CH_2OH$ to synthesize $^{14}CH_3CONH_2$. ◀

$$^{14}CH_3CH_2OH \xrightarrow[KMnO_4]{oxid.} {}^{14}CH_3COOH \xrightarrow{PCl_5} {}^{14}CH_3COCl \xrightarrow{NH_3} {}^{14}CH_3CONH_2$$

Problem 17.40 Name the following compounds:

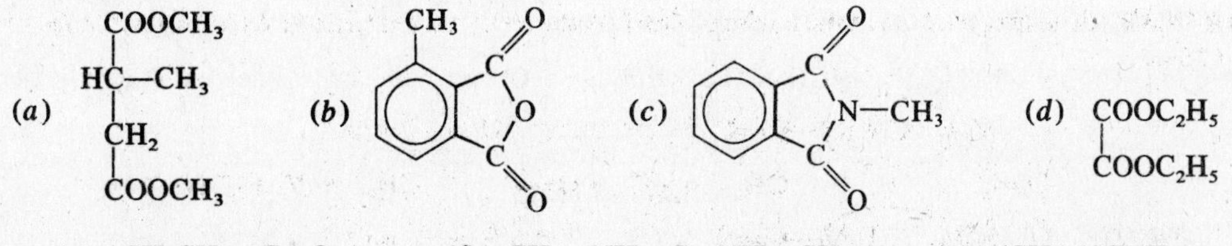

(*e*) $CH_3CH_2-\underset{\underset{\displaystyle N(CH_3)_2}{|}}{C}=O$ (*f*) $CH_3-NH-\underset{\underset{\displaystyle O}{\|}}{C}-NH-CH_3$ (*g*) $(CH_3)_2N\underset{\underset{\displaystyle O}{\|}}{C}NH_2$ ◀

(a) methyl α-methylsuccinate (dimethyl 2-methylbutanedioate), (b) 3-methylphthalic anhydride, (c) N-methylphthalimide, (d) diethyl oxalate (ethyl ethanedioate), (e) N,N-dimethylpropanamide, (f) N,N'-dimethylurea, (g) N,N-dimethylurea.

Problem 17.41 Distinguish by chemical tests (a) CH_3COCl from $(CH_3CO)_2O$, (b) nitrobenzene from benzamide. ◄

(a) With H_2O, CH_3COCl liberates HCl, which is detected by giving a white precipitate of AgCl on adding $AgNO_3$. (b) Refluxing the amide with aqueous NaOH releases NH_3, detected by odor and with moist litmus or pH paper.

Problem 17.42 Name the main organic product(s) formed in the following reactions:

(a) $C_6H_5COOCH_3$ + excess $CH_3CH_2CH_2CH_2MgBr$, then H_3O^+

(b) $(CH_3)_2CHCH_2CH_2OH + C_6H_5COCl$

(c) $HCOOCH_2(CH_2)_4CH_3 + NH_3$

(d) $CH_3CH_2CH_2COCl + CH_3CH_2COONa$

(e) $C_6H_5COBr + 2C_6H_5MgBr$, then H_3O^+ ◄

(a) $C_6H_5C(OH)(CH_2CH_2CH_2CH_3)_2$ (b) $C_6H_5COOCH_2CH_2CH(CH_3)_2$ (c) $HCONH_2 + HO(CH_2)_5CH_3$
   5-Phenyl-5-nonanol   Isoamyl benzoate   Formamide   1-Hexanol

(d) $CH_3CH_2CH_2-\overset{\overset{\displaystyle O}{\|}}{C}-O-\overset{\overset{\displaystyle O}{\|}}{C}-CH_2CH_3$ (e) $(C_6H_5)_3COH$
       Butyric propionic anhydride   Triphenylcarbinol

Problem 17.43 Name and write the structures of the main organic products: (a) $H_2C=CHCH_2I$ heated with NaCN; (b) $CH_3CH_2CONH_2$ heated with P_2O_5; (c) p-iodobenzyl bromide reacted with CH_3COOAg; (d) nitration of benzamide. ◄

(a) allyl cyanide or 3-butenenitrile, $CH_2=CH-CH_2-CN$; (b) propionitrile or ethyl cyanide, CH_3CH_2CN (an intramolecular dehydration); (c) p-iodobenzyl acetate, p-$IC_6H_4CH_2OCOCH_3$; (d) m-nitrobenzamide, m-$NO_2C_6H_4CONH_2$ (carboxylic acid derivatives orient meta during electrophilic substitution).

Problem 17.44 Give steps for the following preparations: (a) 1-phenylpropane from β-phenylpropionic acid, (b) β-benzoylpropionic acid from benzene and succinic acid. ◄

(a) The net change is COOH $\longrightarrow$ CH_3.

$PhCH_2CH_2COOH \xrightarrow{LiAlH_4} PhCH_2CH_2CH_2OH \xrightarrow[-H_2O]{H_2SO_4} PhCH_2CH=CH_2 \xrightarrow{H_2/Pt} PhCH_2CH_2CH_3$
β-Phenylpropionic acid   1-Phenylpropane

(b) $\begin{matrix} CH_2COOH \\ | \\ CH_2COOH \end{matrix} \xrightarrow[-H_2O]{heat} \begin{matrix} CH_2-CO \\ \qquad\quad\diagdown \\ \qquad\qquad O \\ \qquad\quad\diagup \\ CH_2-CO \end{matrix} \xrightarrow{C_6H_6 \ (AlCl_3)} \text{Ph}-COCH_2CH_2COOH$

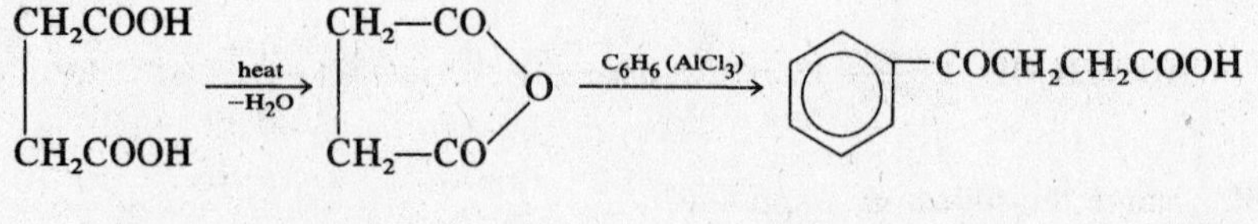

Problem 17.45 Prepare α-methylbutyric acid from ethanol. ◄

Introduce COOH of $CH_3CH_2CH(CH_3)COOH$ through a Cl and build up the needed 4-carbon skeleton

$$C-C-\underset{\underset{\displaystyle Cl}{|}}{C}-C$$

by a Grignard reaction.

(1) $C_2H_5OH \xrightarrow{PBr_3} C_2H_5Br \xrightarrow[ether]{Mg} C_2H_5MgBr$ (use in Step 3)

(2) $C_2H_5OH \xrightarrow[\Delta, Cu]{oxid.} CH_3CH=O$ (use in Step 3)

(3) $CH_3CH=O + C_2H_5MgBr \longrightarrow CH_3CHCH_2CH_3 \xrightarrow{H_3O^+} CH_3CHCH_2CH_3$
$\qquad\qquad\qquad\qquad\qquad\qquad\quad |$
$\qquad\qquad\qquad\qquad\qquad\quad {}^-O(MgBr)^+ \qquad\qquad OH$

$\qquad\qquad\qquad\qquad\qquad\qquad\qquad\qquad\qquad\qquad\qquad \downarrow PCl_3$

$CH_3CHCH_2CH_3 \xleftarrow[\text{2. }H_3O^+]{\text{1. }CO_2} CH_3CHCH_2CH_3 \xleftarrow{Mg} CH_3CHCH_2CH_3$
$\quad |\qquad\qquad\qquad\qquad\qquad\qquad\quad |\qquad\qquad\qquad\qquad\qquad |$
$\quad COOH\qquad\qquad\qquad\qquad\quad MgCl\qquad\qquad\qquad\qquad Cl$

Problem 17.46 What compounds are formed on heating (a) 1,1,2-cyclohexanetricarboxylic acid, and (b) 1,1,2-cyclobutanetricarboxylic acid? ◀

(a)

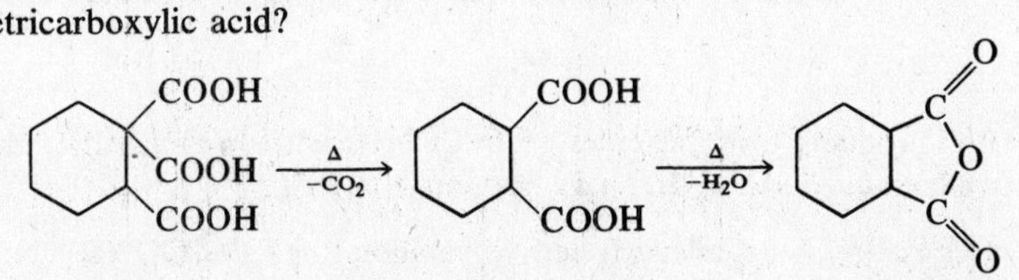

cis- and trans-
Cyclohexane-1,2-dicarboxylic acid

cis- and trans-
anhydride

(b)

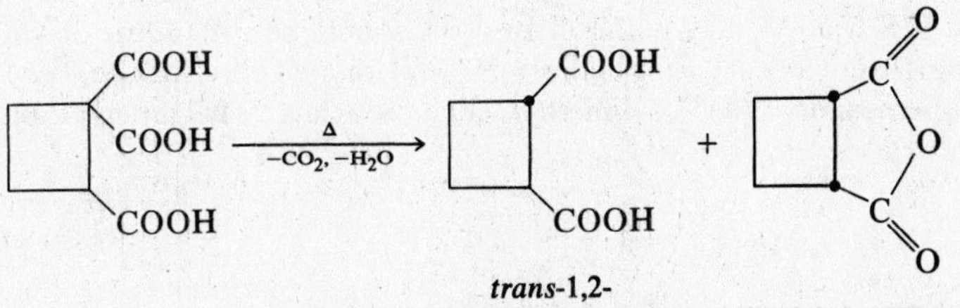

trans-1,2-
Cyclobutanedicarboxylic acid   cis-anhydride

The *trans*-dicarboxylic acid cannot form the anhydride because a 5- and a 4-membered ring cannot be fused *trans*.

Problem 17.47 γ-Butyrolactone is reduced with LiAlH$_4$ and the product acidified. What is the expected final product? ◀

$H_2C-CH_2-CH_2-C=O \xrightarrow[\text{2. }H_3O^+]{\text{1. LiAlH}_4} H_2C-CH_2-CH_2-CH_2-OH$
$\quad\underset{\textstyle O}{\underline{\qquad\qquad\qquad\qquad}}\qquad\qquad\qquad\quad |$
$\qquad\qquad\qquad\qquad\qquad\qquad\qquad\qquad\quad OH$

1,4-Butanediol

Problem 17.48 Indicate the halide(s) required in the malonic ester synthesis of (a) 2-methylbutanoic acid, (b) 2-benzyl-3-phenylpropanoic acid, (c) butyric acid, (d) 1,6-heptadien-4-carboxylic acid. ◀

(a) $\boxed{CH_3CH_2}CHCOOH$ from CH_3Br, C_2H_5Br
$\qquad\boxed{CH_3}$

(b) $\boxed{PhCH_2}CHCOOH$ from $2PhCH_2Br$
$\qquad\boxed{CH_2Ph}$

(c) $\boxed{CH_3CH_2}CH_2COOH$ from C_2H_5Br

(d) $\boxed{H_2C=CHCH_2}CHCOOH$ from $2CH_2=CHCH_2Br$
$\qquad\boxed{CH_2CH=CH_2}$

Problem 17.49 Show how phosgene,

$$Cl-\underset{\underset{\textstyle O}{\|}}{C}-Cl$$

is used to prepare (a) urea, (b) methyl carbonate, (c) ethyl chlorocarbonate, (d) ethyl N-ethylcarbamate (a urethane), (e) ethyl isocyanate ($C_2H_5-N=C=O$).

(a) $\qquad\qquad\qquad\qquad COCl_2 \xrightarrow{NH_3} NH_2CONH_2$

(b) $\qquad\qquad\qquad\qquad COCl_2 \xrightarrow{2CH_3OH} CH_3-O-\underset{\underset{\textstyle O}{\|}}{C}-O-CH_3$

(c)

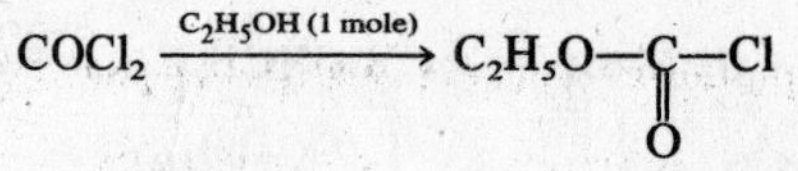

$$COCl_2 \xrightarrow{C_2H_5OH \ (1 \ mole)} C_2H_5O-\overset{\displaystyle O}{\underset{\displaystyle ||}{C}}-Cl$$

(d)

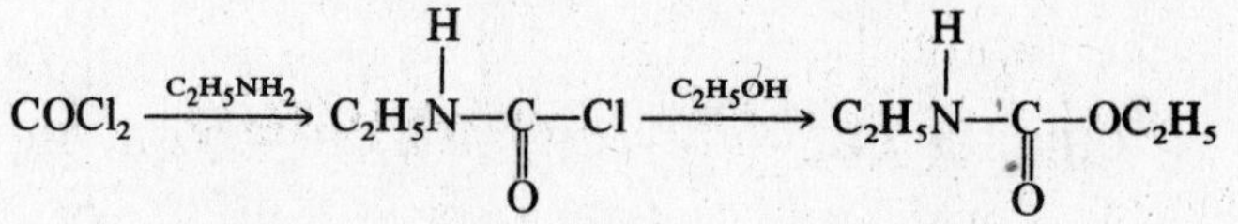

$$COCl_2 \xrightarrow{C_2H_5NH_2} C_2H_5\overset{\displaystyle H}{\underset{\displaystyle |}{N}}-\overset{\displaystyle O}{\underset{\displaystyle ||}{C}}-Cl \xrightarrow{C_2H_5OH} C_2H_5\overset{\displaystyle H}{\underset{\displaystyle |}{N}}-\overset{\displaystyle O}{\underset{\displaystyle ||}{C}}-OC_2H_5$$

(e)

$$COCl_2 \xrightarrow{C_2H_5NH_2} C_2H_5\overset{\displaystyle H}{\underset{\displaystyle |}{N}}-\overset{\displaystyle O}{\underset{\displaystyle ||}{C}}-Cl \xrightarrow{heat} C_2H_5-N=C=O$$

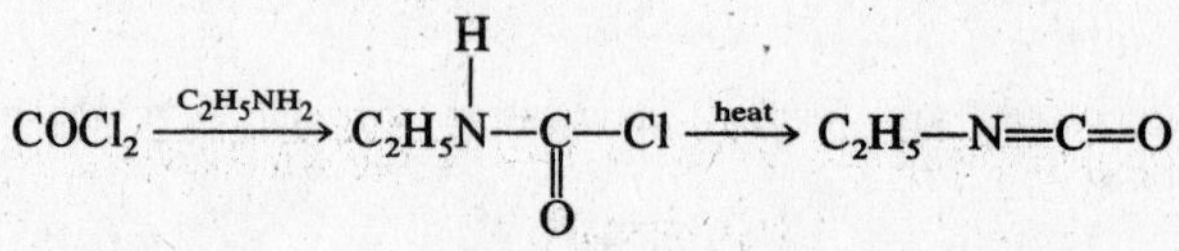

Problem 17.50 Give the products from reaction of $PhCONH_2$ with (a) $LiAlH_4$, then H_3O^+; (b) P_2O_5 [Problem 17.43(b)]; (c) hot aqueous NaOH; (d) hot aqueous HCl. ◀

(a) $PhCH_2NH_2$; (b) PhCN, benzonitrile (phenyl cyanide); (c) $PhCOONa + NH_3$; (d) $PhCOOH + NH_4Cl$.

Problem 17.51 Draw up a table for acetoacetic ester synthesis of the compounds below, showing (1) structure of product, (2) RX and R′X needed, (3) whether hydrolysis should be with dilute or with concentrated NaOH, and (4) whether decarboxylation occurs after hydrolysis. (a) methyl ethyl ketone, (b) 2-methylbutanoic acid, (c) 3-ethyl-4-methyl-2-pentanone, (d) 2,4-dimethylpentanoic acid, (e) 3-methyl-2-butanone, (f) β-phenyl-propanoic acid. ◀

See Table 17-2.

Table 17-2

Structure	RX	R′X	NaOH	Decarboxylation
(a) $CH_3\overset{\displaystyle O}{\overset{\displaystyle ||}{C}}CH_2\boxed{CH_3}$	CH_3Br	none	dil.	yes
(b) $\boxed{CH_3CH_2}\underset{\underset{\displaystyle \boxed{CH_3}}{\displaystyle |}}{CH}\overset{\displaystyle O}{\overset{\displaystyle |}{C}}OH$	CH_3Br	C_2H_5Br	conc.	no
(c) $CH_3\overset{\displaystyle H}{\underset{\displaystyle O \boxed{CH_2} \atop \boxed{CH_3}}{C}}C-\boxed{CH(CH_3)_2}$	C_2H_5Br	$i\text{-}C_3H_7Br$	dil.	yes
(d) $\boxed{(CH_3)_2CHCH_2}\underset{\underset{\displaystyle \boxed{CH_3}}{\displaystyle |}}{CH}\overset{\displaystyle O}{\overset{\displaystyle ||}{C}}OH$	CH_3Br	$i\text{-}C_4H_9Br$	conc.	no
(e) $CH_3\overset{\displaystyle O}{\overset{\displaystyle ||}{C}}\underset{\underset{\displaystyle \boxed{CH_3}}{\displaystyle |}}{CH}\boxed{CH_3}$	CH_3Br	CH_3Br	dil.	yes
(f) $\boxed{C_6H_5CH_2}CH_2COOH$	$C_6H_5CH_2Br$	none	conc.	no

Problem 17.52 Synthesize the tranquilizer Seconal, 4-allyl-4-(2-pentyl)barbituric acid,

$$O=C \underset{NH-CO}{\overset{NH-CO}{\diagdown}} C \underset{CH(CH_3)CH_2CH_2CH_3}{\overset{CH_2CH=CH_2}{\diagup}}$$

from malonic ester and urea.   ◄

First:

$$CH_2(COOC_2H_5)_2 \xrightarrow[\text{2. } CH_2=CHCH_2Br]{\text{1. NaOEt}} \underset{CH_2CH=CH_2}{\overset{|}{CH(COOC_2H_5)_2}} \xrightarrow[\text{2. } CH_3CHBrCH_2CH_2CH_3]{\text{1. NaOEt}} \underset{\underset{CH_2CH=CH_2}{\overset{|}{C(COOC_2H_5)_2}}}{\overset{CH_3CHCH_2CH_2CH_3}{\overset{|}{}}}$$

Now, condense this product with urea:

$$O=C \overset{NH_2}{\underset{NH_2}{\diagup}} + \overset{C_2H_5O-C}{\underset{C_2H_5O-C}{}} \overset{O}{\overset{\parallel}{}} C \overset{CH_2CH=CH_2}{\underset{CH(CH_3)CH_2CH_2CH_3}{}} \xrightarrow[\text{heat}]{NaOC_2H_5} \text{Seconal} + 2C_2H_5OH$$

Urea

Problem 17.53 Can 3,3-dimethyl-2-butanone (pinacolone), $CH_3COC(CH_3)_3$, be synthesized from acetoacetic ester?   ◄

No. Pinacolone has a trimethylated α C. Acetoacetic ester is used for preparing mono- or dialkylated methyl ketones. The compound is made by a pinacol-pinacolone rearrangement of $(CH_3)_2COHCOH(CH_3)_2$.

Problem 17.54 Identify the substances (A) through (E) in the sequence

$$C_6H_5COOH \xrightarrow{PCl_5} (A) \xrightarrow{(B)} C_6H_5CONH_2 \xrightarrow{P_2O_5} (C) \xrightarrow{H_2/Ni} (D) \xrightarrow{(E)} C_6H_5CH_2NH-\underset{\underset{O}{\parallel}}{C}-NHCH_2C_6H_5 \quad ◄$$

(A) C_6H_5COCl, (B) NH_3, (C) C_6H_5CN, (D) $C_6H_5CH_2NH_2$, (E) $COCl_2$.

Problem 17.55 Use acetoacetic ester (AAE) and any needed alkyl halide or dihalide to prepare:
(a) $CH_3COCH_2CH_2COCH_3$, (b) cyclobutyl methyl ketone, (c) $CH_3COCH_2CH_2CH_2COCH_3$, and (d) 1,3-diacetylcyclopentane.   ◄

The portion of all these compounds that comes from AAE is

$$CH_3-\underset{\underset{O}{\parallel}}{C}-\overset{H}{\underset{|}{C}}-$$

Encircling this portion, obtain the rest of the molecule from the alkyl halide(s).

(a)

$$\boxed{CH_3CCH_2} - \boxed{CH_2CCH_3}$$
$$\quad\; \underset{O}{\parallel} \qquad\qquad \underset{O}{\parallel}$$

Bond 2 molecules of AAE at the acidic CH_2 group of each with NaOEt and I_2.

$$2CH_3COCH_2COOEt \xrightarrow[\text{2. } I_2]{\text{1. 2NaOEt}} \underset{CH_3\overset{\parallel}{\underset{O}{C}}CHCOOEt}{\overset{O}{\overset{\parallel}{CH_3CCHCOOEt}}} \xrightarrow[\text{2. } H_3O^+]{\text{1. dil. OH}^-} CH_3\underset{O}{\overset{\parallel}{C}}CH_2CH_2\underset{O}{\overset{\parallel}{C}}CH_3 + CO_2$$

(*b*)

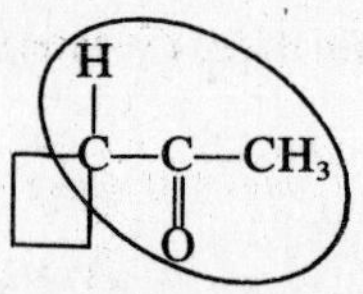

We need 3 C's in the halogen compound, and the 2 terminal C's are bonded to the acidic CH_2 group of AAE. The halide is $BrCH_2CH_2CH_2Br$.

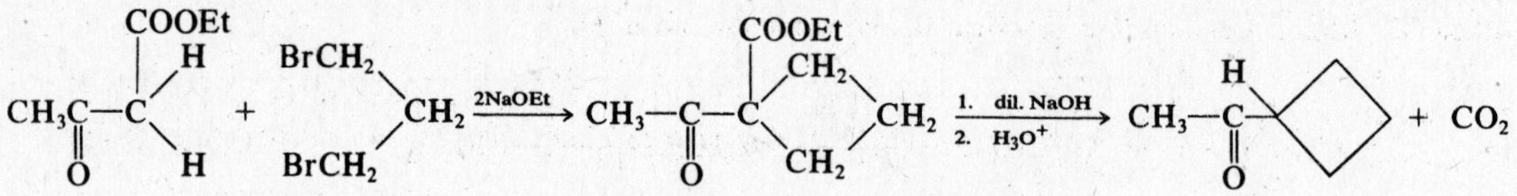

(*c*)

CH_3COCH_2—CH_2—CH_2COCH_3

Join 2 moles of AAE through the acidic CH_2 of each with 2 moles of NaOEt and 1 mole of $BrCH_2Br$.

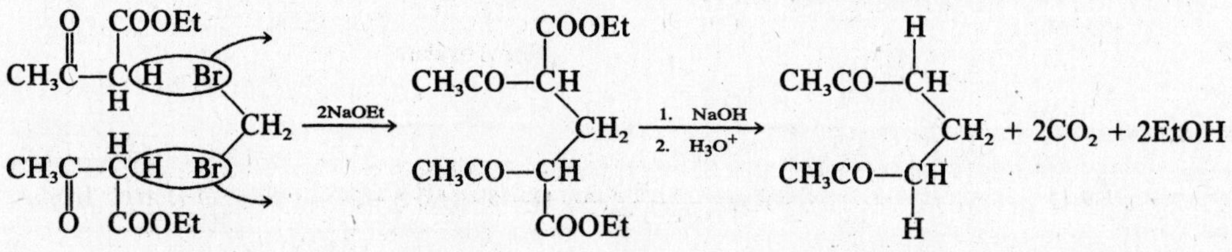

(*d*)

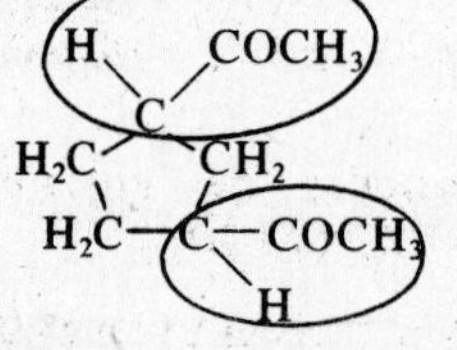

Two molecules of AAE are first bonded together with 1 mole of $BrCH_2CH_2Br$ and then the ring is closed with $BrCH_2Br$.

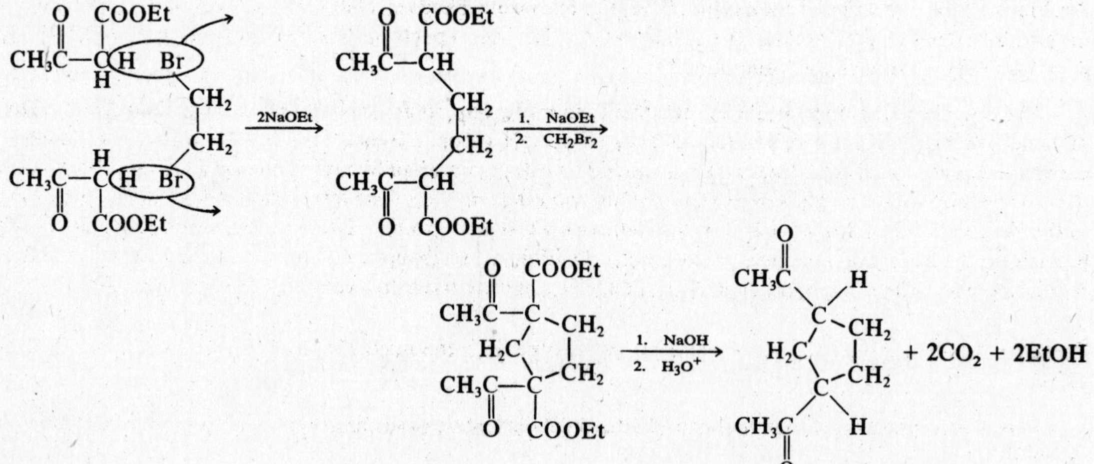

Problem 17.56 Suggest a synthesis of $(CH_3)_3CCH{=}CH_2$ from $(CH_3)_3CCHOHCH_3$. ◀

An attempt to dehydrate the alcohol directly would lead to rearrangement of the R^+ intermediate. The major product would be $(CH_3)_2C{=}C(CH_3)_2$. To avoid this, pyrolyze the acetate ester of this alcohol.

$$CH_3COCl + (CH_3)_3CCHOHCH_3 \longrightarrow (CH_3)_3C\underset{\displaystyle OCOCH_3}{\underset{|}{C}}HCH_3 \xrightarrow{\Delta} \text{product}$$

Problem 17.57 Give the ester or combination of esters needed to prepare the following by a Claisen condensation.

(a) $C_6H_5CH_2CH_2C$—CHCOOEt (b) $EtOC$—C—CHCOOEt (c) H—C—CHCOOEt ◄
           ‖ | ‖ ‖ | ‖ |
           O $CH_2C_6H_5$ O O CH_3 O C_6H_5

In the Claisen condensation the bond formed is between the carbonyl C and the C that is α to COOR. Work backwards by breaking this C—C bond and adding OR to the carbonyl C and adding H to the other C. Mixed Claisens are practical if one ester has no α H.

(a) $C_6H_5CH_2CH_2$—C⫽CHCOOEt $\xleftarrow[\text{2. }-\text{EtOH}]{\text{1. NaOEt}}$ $C_6H_5(CH_2)_2$—C—(OC$_2$H$_5$ + H)CHCOOEt
                                                                                       $CH_2C_6H_5$

Ethyl 3-phenylpropanoate

(b) EtO—C—C⫽CHCOOEt $\xleftarrow[\text{2. }-\text{EtOH}]{\text{1. NaOEt}}$ EtO—C—C—(OEt + H)—CH—COOEt
        ‖ ‖ | ‖ ‖ |
        O O CH_3 O O CH_3

Ethyl oxalate   Ethyl propanoate

(c) H—C⫽CHCOOEt $\xleftarrow[\text{2. }-\text{EtOH}]{\text{1. NaOEt}}$ H—C—(OEt + H)—CHCOOEt
       ‖ | ‖ |
       O C_6H_5 O C_6H_5

Ethyl formate   Ethyl phenylacetate

Problem 17.58 Can the following ketones be prepared by the acetoacetic ester synthesis? Explain. (a) $CH_3COCH_2C_6H_5$, (b) $CH_3COCH_2C(CH_3)_3$. ◄

(a) No. C_6H_5Br is aromatic and does not react in an S_N2 displacement. (b) No. $BrC(CH_3)_3$ is a 3° bromide which undergoes elimination rather than substitution.

Problem 17.59 An acyclic compound, $C_6H_{12}O_2$, has strong bands at 1740 cm^{-1}, 1250 cm^{-1} and 1060 cm^{-1}, with no bands at frequencies greater than 2950 cm^{-1}. The nmr spectrum has two singlets at $\delta = 3.4$ (1 H) and $\delta = 1.0$ (3 H). What is the compound? ◄

The one degree of unsaturation is due to a carbonyl group, indicated by the ir band at 1740 cm^{-1}. The lack of bands above 2950 cm^{-1} shows the absence of an OH group. Hence the compound is not an alcohol or a carboxylic acid. That the compound is probably an ester is revealed by the ir bands at 1250 cm^{-1} and 1060 cm^{-1} (C—O stretch). Two singlets in the nmr means two kinds of H's. The integration of 1:3 means the 12 H's are in the ratio of 3:9. The signal at $\delta = 3.4$ indicates a CH_3—O group. The 9 equivalent protons at $\delta = 1.0$ are present in 3 CH_3's not attached to an electron-withdrawing group. A t-butyl group, $(CH_3)_3C$—, fits these requirements. The compound is $(CH_3)_3CCOOCH_3$, methyl trimethylacetate or methyl pivalate.

Problem 17.60 Predict the base (most prominent) peak in the mass spectrum of the compound in Problem 17.59. ◄

Parent ions of esters resemble those of other acid derivatives and carboxylic acids in that they cleave into an acylium ion.

$$\left[(CH_3)_3CC\text{—}OCH_3\right]^{+\cdot} \longrightarrow (CH_3)_3CC\equiv O^+ + \cdot OCH_3 \text{ or } CH_3C\cdot + \overset{+}{O}\equiv COCH_3$$
$$\quad\quad\quad \underset{O}{\|}$$

The base peak should be $m/e = 85$ or 59.

Problem 17.61 Explain the fact that (R)-CH$_3$CHBrCOO$^-$Na$^+$ reacts with NaOH to give (R)-CH$_3$CHOHCOO$^-$Na$^+$. ◄

On displacing Br by OH there is no change in order of priorities (see page 56). Therefore, since both configurations are R, there was *retention* of configuration rather than the inversion expected from S_N2 reactions. It is believed that the —COO$^-$ group participates in first displacing the Br$^-$ with inversion to give an unstable α-lactone. Then OH$^-$ attacks the α-lactone with a second inversion to give the product. Two inversions add up to retention. Intervention by an adjacent group, called **neighboring group participation**, always leads to retention.

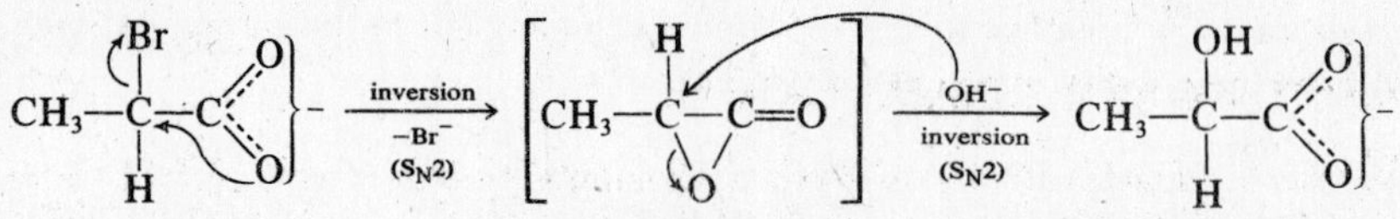

unstable α-lactone

(R)   (S)   (R)

Problem 17.62 Inorganic acids such as H_2SO_4, H_3PO_4 and HOCl (hypochlorous acid) form esters. Write structural formulas for (a) dimethyl sulfate, (b) tribenzyl phosphate, (c) diphenyl hydrogen phosphate, (d) t-butyl hypochlorite, (e) lauryl hydrogen sulfate (lauryl alcohol is n-$C_{11}H_{23}CH_2OH$), (f) sodium lauryl sulfate. ◄

Replacing the H of the OH of an acid gives an ester.

(a) CH_3O—$\underset{\underset{O}{\overset{O}{\|}}}{S}$—$OCH_3$ (b) $(PhCH_2O)_3P$—O   (c) $(PhO)_2\overset{O}{\overset{\|}{P}}$—OH   (d) $(CH_3)_3COCl$

(e) n-$C_{11}H_{23}CH_2O\underset{\underset{O}{\overset{O}{\|}}}{S}$—OH   (f) n-$C_{11}H_{23}CH_2OSO_2O^-Na^+$ (a detergent)

Problem 17.63 Dialkyl sulfates are good alkylating agents. (a) Write an equation for the reaction of dimethyl sulfate with C_6H_5OH in NaOH. (b) Explain why dimethyl sulfate is a good methylating agent. ◄

(a) $C_6H_5OH \xrightarrow{\text{OH}^-} C_6H_5O^- \xrightarrow{CH_3-OSO_2OCH_3} C_6H_5OCH_3 + {}^-OSO_2OCH_3$

Anisole   Methyl
sulfate anion

(b) Methyl sulfate anion ($CH_3OSO_3^-$) is the conjugate base of a strong acid, methyl sulfuric acid (CH_3OSO_3H). It is a very weak base and a good leaving group.

Problem 17.64 Explain why trialkyl phosphates are readily hydrolyzed with OH$^-$ to dialkyl phosphate salts, whereas dialkyl hydrogen phosphates and alkyl dihydrogen phosphates resist alkaline hydrolysis. ◄

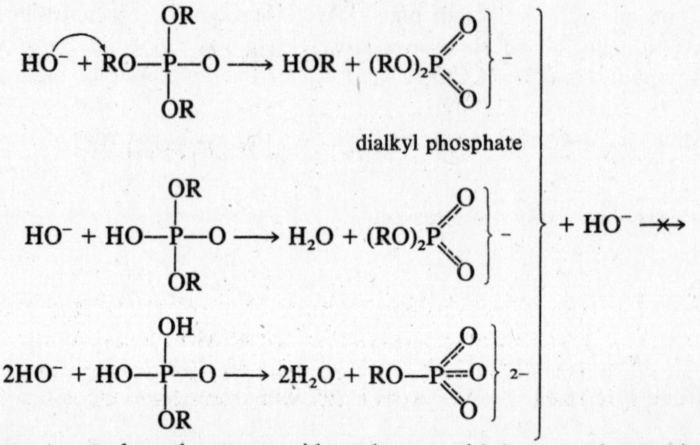

dialkyl phosphate

The hydrogen phosphates are moderately strong acids and react with bases to form anions (conjugate bases). Repulsion between negatively charged species prevents further reaction between these anions and OH$^-$.

Chapter 18

Amines

18.1 INTRODUCTION AND NOMENCLATURE

Amines are alkyl or aryl derivatives of NH_3. Replacing 1, 2 or 3 H's of NH_3 gives **primary** (1°), **secondary** (2°) and **tertiary** (3°) amines, respectively.

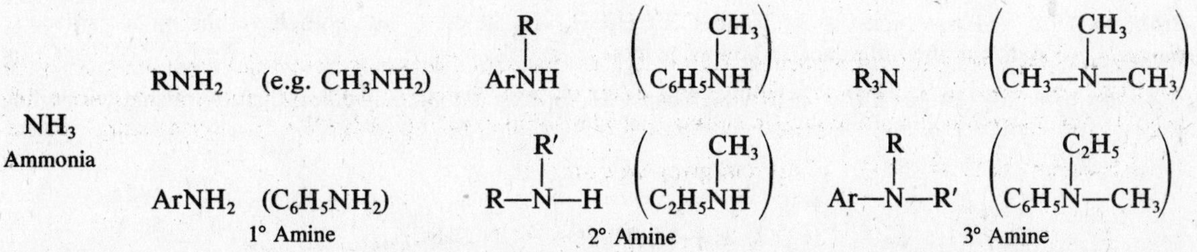

NH_3
Ammonia

RNH_2 (e.g. CH_3NH_2) $ArNH_2$ ($C_6H_5NH_2$)

1° Amine

$ArNH\begin{pmatrix} R \\ | \\ C_6H_5NH \\ | \\ CH_3 \end{pmatrix}$ $R-N-H \begin{pmatrix} R' \\ | \\ C_2H_5NH \\ | \\ CH_3 \end{pmatrix}$

2° Amine

$R_3N \begin{pmatrix} CH_3 \\ | \\ CH_3-N-CH_3 \end{pmatrix}$ $Ar-N-R' \begin{pmatrix} R \\ | \\ C_6H_5N-CH_3 \\ | \\ C_2H_5 \end{pmatrix}$

3° Amine

Amines are named by combining in one word the name of each group on the N with the suffix **-amine**; $(C_6H_5)_2NCH_3$ is **methyldiphenylamine**. Amines are also named by placing the prefix **amino-**, **N-alkylamino-** or **N,N-dialkylamino-** before the name of the parent chain.

Aromatic and cyclic amines often have common names such as **aniline**, $C_6H_5NH_2$; *p*-**toluidine**, *p*-$CH_3C_6H_4NH_2$; and **piperidine** [Problem 18.1(*g*)].

Like the **oxa** method for naming ethers (Problem 14.35), the **aza** method is used for amines. Di-*n*-propylamine, $CH_3CH_2CH_2NHCH_2CH_2CH_3$, is **4-azaheptane** and piperidine is **azacyclohexane**.

The 4 H's of NH_4^+ can be replaced to give a **quaternary** (4°) tetraalkyl (tetraaryl) ammonium ion.

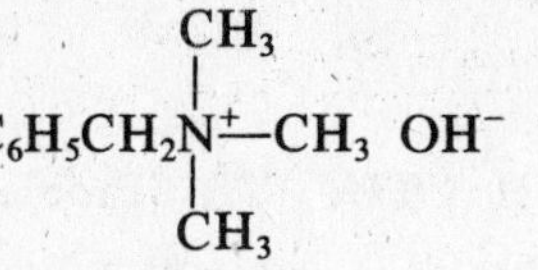

$$C_6H_5CH_2\overset{\overset{\displaystyle CH_3}{|}}{\underset{\underset{\displaystyle CH_3}{|}}{N^+}}-CH_3 \ OH^- \qquad \text{(Triton-B)}$$

is **benzyltrimethylammonium hydroxide**.

Problem 18.1 Name and classify the following amines:

(*a*) $(CH_3)_2NH$

(*b*) $(CH_3)_2NCH(CH_3)_2$

(*c*) $C_6H_5N(CH_3)_2$

(*d*) $H_2NCH_2CH_2CH_2NH_2$

(*e*) $CH_3NHCH(CH_3)CH_2CH_3$

(*f*) $CH_3NHCH_2CH_2NHCH_3$

(*g*) [piperidine ring with NH$_2$ at position 3 and CH$_3$ on N]

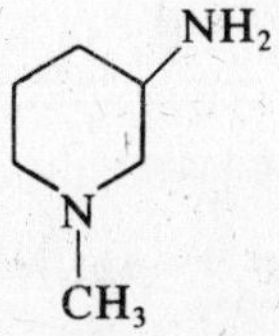

(*h*) $CH_3CH_2CH_2N(CH_3)_3^+Cl^-$ ◄

(*a*) dimethylamine, 2°; (*b*) dimethylisopropylamine, 3°; (*c*) N,N-dimethylaniline, 3° (N indicates substitution on nitrogen); (*d*) 1,3-diaminopropane (or trimethylenediamine), both 1°; (*e*) 2-(N-methylamino)butane, 2°; (*f*) 2,5-diazahexane, both 2°; (*g*) 3-amino-N-methylpiperidine or 1-methyl-3-amino-1-azacyclohexane, 1° (N of NH_2) and 3°; (*h*) *n*-propyltrimethylammonium chloride, 4°.

Problem 18.2 Give names for (a) CH_3NHCH_3, (b) $CH_3NHCH(CH_3)_2$, (c) $CH_3CH_2CHNH_2COOH$,

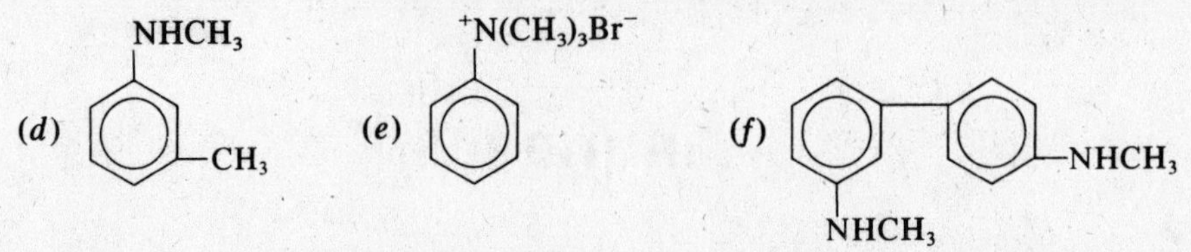

(d) [benzene ring with $NHCH_3$ and CH_3 substituents]   (e) [benzene ring with $^+N(CH_3)_3Br^-$]   (f) [biphenyl with $NHCH_3$ and $NHCH_3$ substituents]   ◀

(a) dimethylamine,   (b) methylisopropylamine or 2-(N-methylamino)propane,   (c) 2-aminobutanoic acid,   (d) N-methyl-m-toluidine or 3-(N-methylamino)toluene,   (e) trimethylanilinium bromide,   (f) 3,4'-N,N'-methylaminobiphenyl (note the use of N and N' to designate the different N's on the separate rings).

Problem 18.3 $(CH_3)_3N$ boils at 3 °C and $CH_3CH_2CH_2NH_2$ at 49 °C but both have the same molecular weight. Account for the difference in boiling points.   ◀

n-Propylamine (1°) has 2 H's on N, which cause H-bonding among molecules, thereby raising the boiling point.

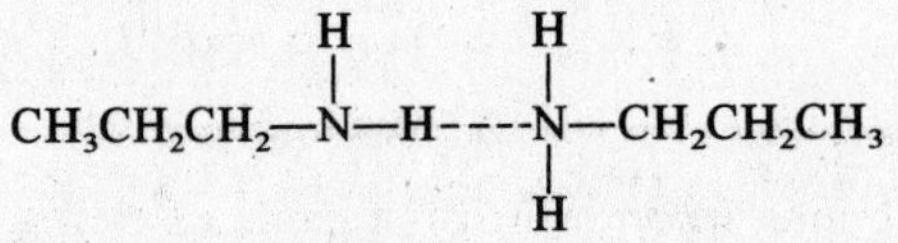

$$CH_3CH_2CH_2\!-\!\underset{\displaystyle H}{\overset{\displaystyle H}{N}}\!-\!H\text{-}\text{-}\text{-}\underset{\displaystyle H}{\overset{\displaystyle H}{N}}\!-\!CH_2CH_2CH_3$$

Such association also occurs in 2° amines. $(CH_3)_3N$ (3°), with no H's on N, cannot form intermolecular H-bonds and does not associate. (See also Problem 4.28.)

Problem 18.4 Does n-propylamine or n-propanol have the higher boiling point?   ◀

H-bonding and intermolecular association are more effective with the more electronegative O (3.5) than with N (3.1). The alcohol boils at 97 °C and n-propylamine at 49 °C.

18.2 PREPARATION OF AMINES

ALKYLATION OF NH_3, RNH_2 AND R_2NH WITH RX [Problem 7.7(f)]

Step 1 $RX + NH_3 \longrightarrow RNH_3^+X^-$ (an S_N2 reaction)
                                                an ammonium
                                                salt

Step 2 $RNH_3^+X^- + NH_3 \longrightarrow RNH_2 + NH_4^+X^-$

Di-, tri- and tetraalkylation:

$$RNH_2 \xrightarrow[-HX]{RX} R_2NH \xrightarrow[-HX]{RX} R_3N \xrightarrow[-HX]{RX} R_4N^+X^-$$
$$\ \ 1° \qquad\qquad 2° \qquad\quad 3° \qquad\quad 4°$$
$$\text{ammonium salt}$$

When RX = MeI, the sequence is called **exhaustive methylation**.

ALKYLATION OF IMIDES; GABRIEL SYNTHESIS OF 1° AMINES

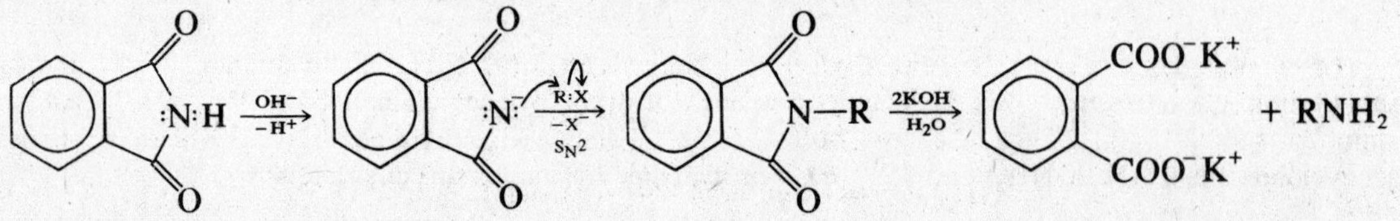

[reaction scheme: Phthalimide $\xrightarrow[-H^+]{OH^-}$ salt $\xrightarrow[-X^-]{R:X}{S_N2}$ N-R $\xrightarrow[H_2O]{2KOH}$ benzene ring with COO^-K^+ and COO^-K^+ + RNH_2]

Phthalimide   salt

REDUCTION OF N-CONTAINING COMPOUNDS

1. Nitro Compounds

$$\underset{\text{Nitrobenzene}}{C_6H_5NO_2} \xrightarrow[\text{or } H_2/Pt]{\text{1. Zn, HCl; 2. OH}^-} \underset{\text{Aniline}}{C_6H_5NH_2}$$

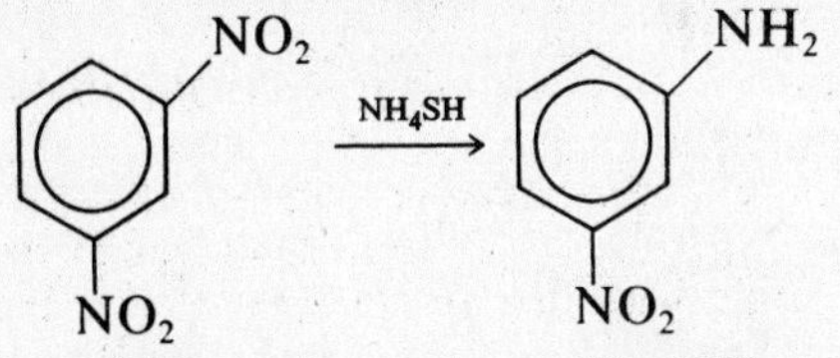

(only one NO_2 is reduced)

m-Dinitrobenzene → *m*-Nitroaniline

2. Nitriles

$$RCN \xrightarrow{LiAlH_4} RCH_2NH_2$$

3. Amides

$$RC(=O){-}NR_2 \xrightarrow{LiAlH_4} RCH_2NHR_2 \quad (R = H, \text{alkyl, aryl})$$

4. Oximes

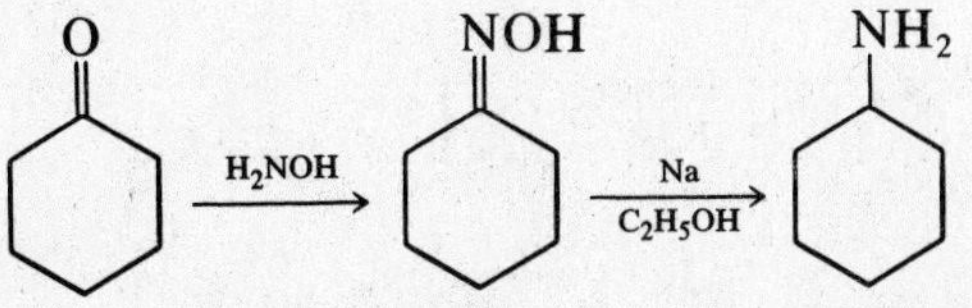

Oxime Cyclohexylamine

5. Carbonyl Compounds

$$CH_3CH{=}O \xrightarrow{NH_3,\, H_2/Ni} CH_3CH_2NH_2 \quad \textbf{(Reductive amination)}$$

$$CH_3CH_2CHO + CH_3CH_2NH_2 \xrightarrow{H_2/Ni} CH_3CH_2CH_2NHCH_2CH_3 \quad (1^\circ \rightarrow 2^\circ \text{ amine})$$

$$RNH_2 + 2\,H_2C{=}O + 2HCOOH \longrightarrow RN(CH_3)_2 + 2H_2O + 2CO_2 \quad (\text{Dimethylation of } 1^\circ \text{ amine})$$

HOFMANN DEGRADATION OF AMIDES

$$RCONH_2 + Br_2 + 4KOH \longrightarrow RNH_2 + K_2CO_3 + 2KBr + 2H_2O$$ (The amine has one less C than the amide.)

Step 1

$$R{-}C(=O){-}\ddot{N}H_2 \xrightarrow[(Br_2,\, OH^-)]{OBr^-} \underset{\text{N-bromoamide}}{R{-}C(=O){-}\ddot{N}(H){-}Br} + OH^-$$

Step 2

$$R{-}C(=O){-}\ddot{N}(H){-}Br + OH^- \longrightarrow \left[R{-}C(=\ddot{O}){-}\ddot{\underset{..}{N}}{-}Br \longleftrightarrow R{-}C(-\ddot{O}{:}^-){=}\ddot{N}{-}Br\right] + H_2O$$

N-bromoamide anion

Step 3

$$R{-}C(=O){-}\ddot{\underset{..}{N}}{-}\ddot{Br}{:} \longrightarrow R{-}C(=O){-}\ddot{N} + {:}\ddot{Br}{:}^-$$

electron-deficient N

Step 4

$$(R{-})C(=O){-}\ddot{N}{:} \xrightarrow{\sim R:} \underset{\text{alkyl isocyanate}}{O{=}C{=}\ddot{N}{:}R}$$

Step 5

$$2OH^- + R{-}N{=}C{=}O \xrightarrow{H_2O} \underset{\text{amine}}{R{-}NH_2} + CO_3^{2-}$$

Problem 18.5 Prepare ethylamine by (a) Gabriel synthesis,   (b) alkyl halide amination,   (c) nitrile reduction,   (d) reductive amination,   (e) Hofmann degradation.   ◄

(a)

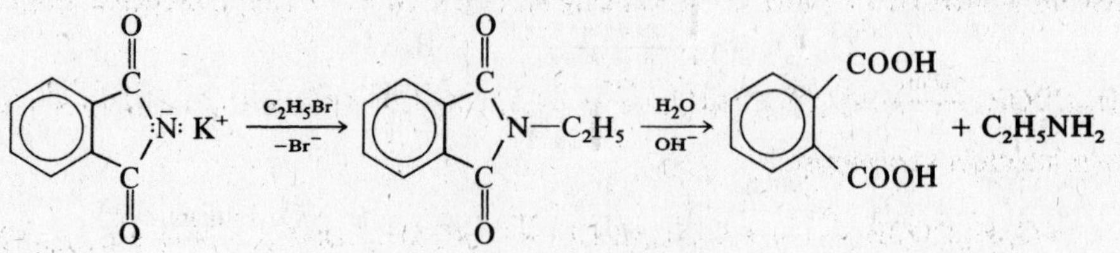

$$\underset{\text{(phthalimide K salt)}}{} \xrightarrow[-Br^-]{C_2H_5Br} \underset{}{} \xrightarrow[OH^-]{H_2O} \underset{}{\overset{COOH}{\underset{COOH}{}}} + C_2H_5NH_2$$

(b) $\qquad\qquad\qquad\qquad C_2H_5Br \xrightarrow[NH_3]{\text{excess}} C_2H_5NH_2$

(c) $\qquad\qquad\qquad\qquad CH_3CN \xrightarrow[2.\ H_2O]{1.\ LiAlH_4} CH_3CH_2NH_2$

(d) $\qquad\qquad\qquad\qquad CH_3CH{=}O \xrightarrow[H_2/Ni]{NH_3} CH_3CH_2NH_2$

(e) $\qquad\qquad\qquad\qquad CH_3CH_2CONH_2 \xrightarrow[NH_3]{Br_2,\ KOH} CH_3CH_2NH_2$

Problem 18.6 Prepare *p*-toluidine from toluene.   ◄

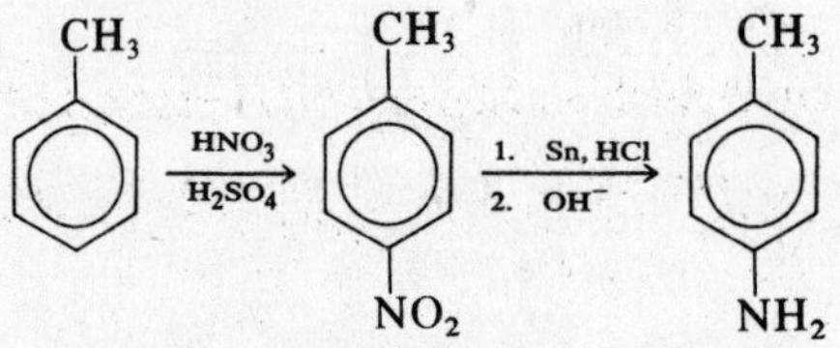

$$\underset{}{\overset{CH_3}{}} \xrightarrow[H_2SO_4]{HNO_3} \underset{NO_2}{\overset{CH_3}{}} \xrightarrow[2.\ OH^-]{1.\ Sn,\ HCl} \underset{NH_2}{\overset{CH_3}{}}$$

This is the best way of substituting an NH_2 on a phenyl ring.

Problem 18.7 Prepare $CH_3NHCH_2CH_3$ from compounds with 1 or 2 C's.   ◄

$$CH_3CHO + CH_3NH_2 \xrightarrow{H_2/Ni} CH_3CH_2NHCH_3$$

Problem 18.8 In preparing 1° amines by alkylation of NH_3, how does one prevent formation of 2° and 3° amines?   ◄

Excess NH_3 reduces the chances of the product, RNH_2, reacting with RX to give R_2NH and R_3N.

Problem 18.9 Prepare $C_6H_5N(CH_3)_2$ from $C_6H_5NH_2$.   ◄

$$C_6H_5NH_2 \xrightarrow[H_2SO_4,\ 230\ °C]{2CH_3OH} C_6H_5N(CH_3)_2 \quad \text{(Industrial method)}$$

With CH_3I, $C_6H_5N(CH_3)_3^+I^-$ is a by-product. H_2CO and $HCOOH$ cannot be used because the intermediate electrophile, $H_2\overset{+}{C}OH$, can attack the activated ring to give p-$H_2NC_6H_4CH_2OH$.

Problem 18.10 Synthesize the following compounds from n-$C_{12}H_{25}COOH$ and inorganic reagents:

$$(a)\ \ C_{14}H_{29}NH_2 \qquad (b)\ \ C_{13}H_{27}NH_2 \qquad (c)\ \ C_{12}H_{25}NH_2$$

$$(d)\ \ C_{12}H_{25}{-}\overset{NH_2}{\underset{}{CH}}{-}C_{13}H_{27} \qquad (e)\ \ C_{12}H_{25}NHC_{13}H_{27}$$

Do not repeat preparation of any needed compound.   ◄

First, note the change, if any, in carbon content.

(*a*) Chain length is increased by one C by reducing RCH_2CN ($R = n\text{-}C_{12}H_{25}$), prepared from RCH_2Br and CN^-.

$$n\text{-}C_{12}H_{25}COOH \xrightarrow{LiAlH_4} n\text{-}C_{12}H_{25}CH_2OH \xrightarrow{PBr_3} n\text{-}C_{12}H_{25}CH_2Br \xrightarrow{KCN} n\text{-}C_{13}H_{27}CN \xrightarrow{LiAlH_4} n\text{-}C_{14}H_{29}NH_2$$

(*b*) The chain length is unchanged.

$$n\text{-}C_{12}H_{25}COOH \xrightarrow{SOCl_2} n\text{-}C_{12}H_{25}COCl \xrightarrow{NH_3} n\text{-}C_{12}H_{25}CONH_2 \xrightarrow{LiAlH_4} n\text{-}C_{13}H_{27}NH_2$$

(*c*) Chain length is decreased by one C; use the Hofmann degradation.

$$n\text{-}C_{12}H_{25}CONH_2 \xrightarrow{NaOBr} n\text{-}C_{12}H_{25}NH_2$$

(*d*) The C content is doubled. The 1° amine is made from the corresponding ketone.

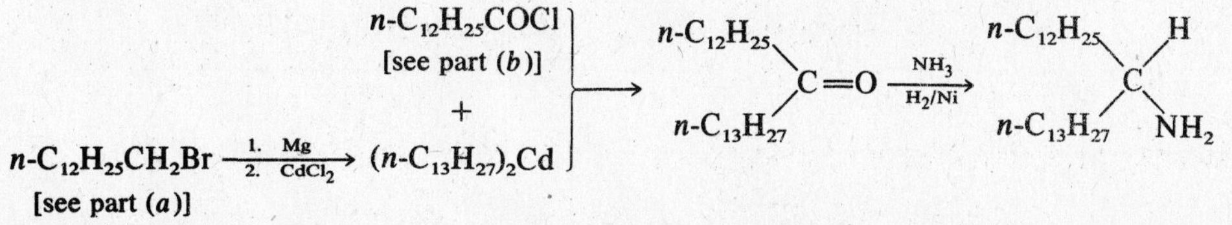

(*e*) 2° Amines may be prepared by reductive amination of an aldehyde (from RCOCl reduction), using a 1° amine.

$$n\text{-}C_{12}H_{25}COCl \xrightarrow[S,\ \Delta]{H_2,\ Pd/BaSO_4} n\text{-}C_{12}H_{25}CHO \xrightarrow[H_2/Ni]{n\text{-}C_{12}H_{25}NH_2} n\text{-}C_{13}H_{27}NHC_{12}H_{25}$$

Problem 18.11 There is no change in the configuration of the chiral C in *sec*-butylamine formed from the Hofmann degradation of (S)-2-methylbutanamide. Explain. ◀

:R migrates with its electron pair to the electron-deficient :N̈, and configuration is retained because C—C is being broken at the same time that C—N is being formed in the transition state.

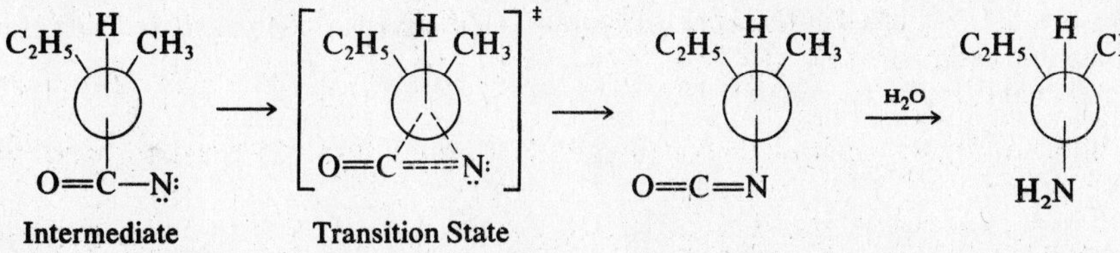

Problem 18.12 A rearrangement of R, from C to an electron-deficient N, occurs in the following reactions. The substrates and conditions are given. Indicate how the intermediate is formed and give the structure of each product.

(*a*) **Curtius,** $RC(=O)—N_3$ (an acylazide) with heat or **Schmidt,** $RCOOH + NH_3$ with H_2SO_4

(*b*) **Lossen,** $RC(=O)—NHOH$ (a hydroxamic acid) with base

(*c*) **Beckmann,** $R—C(=NOH)—R'$ with strong acid ◀

(*a*)

$$R—C(=O)—\ddot{\underset{..}{N}}{}^{-}{:}\overset{+}{N}{\equiv}N{:} \xrightarrow{\Delta} {:}N{\equiv}N{:} + \left[R—C(=O)—\ddot{N}\right] \longrightarrow O{=}C{=}N—R \xrightarrow[H_2O]{OH^-} RNH_2 + CO_3^{2-}$$

Intermediate

(b)

$$R-\underset{\underset{O}{\|}}{C}-\underset{\underset{H}{|}}{\ddot{N}}-OH \xrightarrow[\substack{1. \ -H^+ \\ 2. \ -OH^-}]{OH^-} H_2O + \left[R-\underset{\underset{O}{\|}}{C}-\ddot{N} \right] \longrightarrow O=C=N-R \xrightarrow[H_2O]{OH^-} RNH_2 + CO_3^{2-}$$
Intermediate

(c)

$$R-\underset{\underset{NOH}{\|}}{C}-R' \xrightarrow{H^+} R-\underset{\underset{\overset{+}{N}OH_2}{\|}}{C}-R' \longrightarrow H_2O + \left[R-\underset{\overset{\curvearrowright}{:N^+}}{C}-R' \right] \longrightarrow \underset{R-N:}{\overset{+}{C}}-R' \xrightarrow{H_2O} \underset{R-N-H}{O=C-R'}$$
Intermediate

The group *trans* to the OH (R) migrates as H_2O leaves.

Problem 18.13 Which of the following compounds are (i) chiral, (ii) resolvable? Give reasons in each case.

(a) $[C_6H_5\overset{+}{N}(C_2H_5)_2(CH_3)]Br^-$ (b) $C_6H_5N(CH_3)(C_2H_5)$

(c) $C_6H_5-\underset{\underset{C_2H_5}{|}}{\overset{+}{\underset{|}{N}}}-O^-$ (d) $CH_3CH_2\overset{\overset{\displaystyle CH_3}{|}}{CH}-N(CH_3)(C_2H_5)$ ◄

(a) Not chiral and not resolvable, because N has 2 identical groups (C_2H_5).

(b) Chiral, but the low energy barrier (6 kcal/mole) to inversion of configuration prevents resolution of its enantiomers,

(c) Chiral and resolvable. N has 4 different substituents. The absence of an unshared pair of electrons on N prevents inversion as in part (b).

(d) Chiral and resolvable. An asymmetric C is present.

$$CH_3CH_2\overset{*}{C}H\overset{\overset{\displaystyle CH_3}{|}}{N}(CH_3)(C_2H_5)$$

Problem 18.14 (a) Propose a mechanism for synthesis of $(CH_3)_3CNH_2$ by reaction of $(CH_3)_3COH$ or $(CH_3)_2C=CH_2$ with CH_3CN in conc. H_2SO_4, followed by hydrolysis of the product. (b) Why can this compound not be made by: (1) reductive amination of carbonyl compounds, (2) reduction of nitriles, (3) reduction of oximes, (4) Gabriel synthesis or RX amination? (c) What other synthesis is applicable? ◄

(a)

$$(CH_3)_3COH \xrightarrow[-H_2O]{H^+} \left[(CH_3)_3\overset{+}{C}\right] \xleftarrow{H^+} (CH_3)_2C=CH_2$$
t-Butyl alcohol   Isobutylene

$$(CH_3)_3C^+ + :N\equiv CCH_3 \longrightarrow (CH_3)_3C\overset{+}{N}\equiv CCH_3 \xrightarrow[-H^+]{H_2O} (CH_3)_3C\underset{\underset{H}{|}}{N}-\underset{\underset{O}{\|}}{C}CH_3 \xrightarrow{H_2O} (CH_3)_3CNH_2$$
t-Butylamine

(b) (1) Products can have only CH_3, a 1° C, or 2° C's bonded to N as in CH_3NH_2, RCH_2NH_2 or R_2CHNH_2 respectively. (2) Products can have only a 1° C bonded to N. (3) Products can have only 1° or 2° C's bonded to N. (4) The needed $(CH_3)_3CCl$ would undergo elimination (E2) rather than substitution (S_N2).

(c) Hofmann degradation of $(CH_3)_3CCONH_2$.

18.3 CHEMICAL PROPERTIES OF AMINES

BASICITY AND SALT FORMATION [see Problem 3.24(b)]

Problem 18.15 (a) Why does aqueous CH_3NH_2 turn litmus blue? (b) Why does $C_6H_5NH_2$ dissolve in aqueous HCl? ◄

(a) Methylamine (pK_b = 3.36) is a weak base:

$$CH_3NH_2 + H_2O \rightleftharpoons CH_3NH_3^+ + OH^-$$
$$\text{Base}_1 \quad \text{Acid}_2 \qquad \text{Acid}_1 \qquad \text{Base}_2$$

(b) A water soluble salt forms.

$$C_6H_5NH_2 + H_3O^+ + Cl^- \longrightarrow C_6H_5NH_3^+Cl^- + H_2O$$
$$\text{Anilinium chloride}$$

Problem 18.16 (a) Assign numbers from 1 for LEAST to 5 for MOST to indicate relative base strengths of (i) $C_6H_5NH_2$, (ii) $(C_6H_5)_2NH$, (iii) $(C_6H_5)_3N$, (iv) CH_3NH_2, and (v) NH_3. (b) Explain. ◄

(a) (i) 3, (ii) 2, (iii) 1, (iv) 5, (v) 4. (b) CH_3NH_2, a typical *aliphatic* amine is *more* basic than NH_3 because of the electron-donating inductive effect of CH_3. $C_6H_5NH_2$, an *aromatic* amine, is much *less* basic than NH_3 because the electron density on N is delocalized to the ring mainly to the *ortho* and *para* positions [Fig. 18-1(a)]. With more phenyls bonded to N there is more delocalization and weaker basicity.

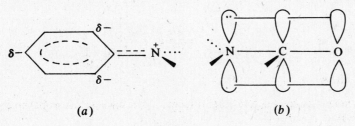

$$(a) \qquad\qquad\qquad\qquad (b)$$

Fig. 18-1

Problem 18.17 Compare the reactions of $(CH_3)_3N$ and $(C_6H_5)_3N$ with BF_3. ◄

We have:

$$(CH_3)_3N\text{:} + \underset{\underset{F}{|}}{\overset{\overset{F}{|}}{B}}\text{:}F \longrightarrow (CH_3)_3\overset{+}{N}\text{:}\underset{\underset{F}{|}}{\overset{\overset{F}{|}}{B}}\text{:}F$$

$$\text{Lewis} \quad \text{Lewis}$$
$$\text{base} \quad\;\; \text{acid}$$

$(C_6H_5)_3N$ does not react, because the electron pair on N needed to bond to B is delocalized to 3 benzene rings.

Problem 18.18 Assign a number from 1 for LEAST to 4 for MOST to indicate the relative base strength of the following:

	I	II	III	IV
(a)	$CH_3\ddot{N}H^-Na^+$	$C_2H_5NH_2$	$(i\text{-}C_3H_7)_3N$	CH_3CONH_2
(b)	$C_6H_5NH_2$	$p\text{-}NO_2C_6H_4NH_2$	$m\text{-}NO_2C_6H_4NH_2$	$p\text{-}H_3COC_6H_4NH_2$

◄

(*a*)

	I	II	III	IV
	4	3	2	1

CH_3NH^- is the conjugate base of the very weak acid CH_3NH_2 and therefore is the strongest base. The 3 bulky *i*-propyl groups on N cause steric strain, but with an unshared electron pair on N, this strain is partially relieved by increasing the normal C—N—C bond angle (109°) to about 112°. If the unshared electron pair forms a bond to H, as in R_3NH^+, relief of strain by angle expansion is prevented. 3° Amines therefore resist forming a 4th bond and suffer a decrease in basicity. Acyl R—C=O groups are strongly electron-withdrawing and base-weakening because electron density from N can be delocalized to O of the carbonyl group by extended π bonding [Fig. 18-1(*b*)].

(*b*)

	I	II	III	IV
	3	1	2	4

The strongly electron-attracting NO_2 group *decreases* electron density on N. It thus also decreases *base strength* by an inductive effect in the *meta* position, and to a greater extent, by both extended π bonding and inductive effects, in *ortho* and *para* positions.

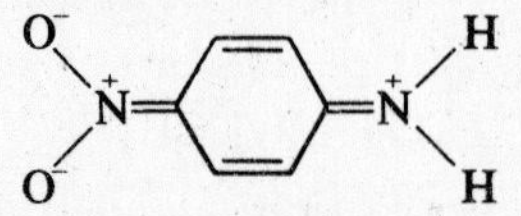

Since OCH_3 is electron-donating through extended π bonding, it *increases* electron density on N and the base strength of the amine because the ring accepts less electron density from N.

REACTION WITH NITROUS ACID, HONO

1. Primary Amines

(*a*) Aromatic (ArNH₂).

$$C_6H_5NH_2 \xrightarrow[\text{HCl, 5 °C}]{\text{HONO}} (C_6H_5\overset{+}{-}N\equiv N\colon)Cl^- \quad \text{Benzenediazonium chloride}$$

(*b*) Aliphatic (RNH₂).

$$CH_3CH_2CH_2NH_2 \xrightarrow[\text{HCl}]{\text{HONO}} [CH_3CH_2CH_2N_2]^+Cl^- \longrightarrow$$
(unstable)

| 1-Propanol | 1-Chloropropane | Propene | 2-Propanol | 2-Chloropropane |

This reaction of RNH_2 has no synthetic utility, but the appearance of N_2 gas signals the presence of NH_2.

2. Secondary Amines

$$Ar(R)NH \text{ (or } R_2NH) + HONO \longrightarrow Ar(R)NNO \text{ (or } R_2N-NO) + H_2O$$
an N-nitrosoamine
(insoluble in acid)

N-Nitrosoamines are cancer-causing agents (**carcinogens**).

3. Tertiary Amines

No reaction except for N,N-dialkyl arylamines.

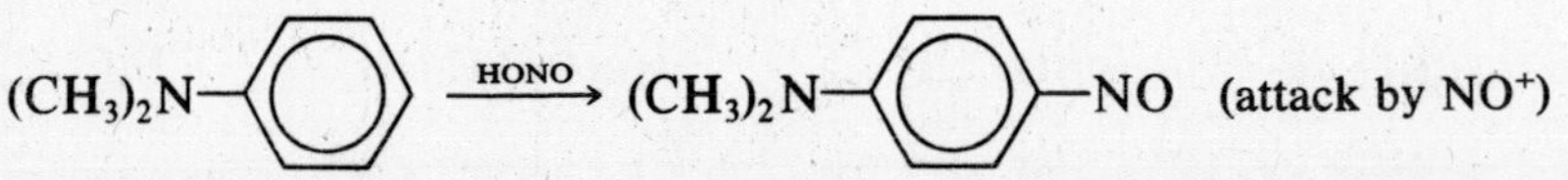

$(CH_3)_2N- \xrightarrow{\text{HONO}} (CH_3)_2N--NO$ (attack by NO⁺)

Problem 18.19 Prepare $p\text{-}H_2NC_6H_4N(CH_3)_2$ from $C_6H_5N(CH_3)_2$.

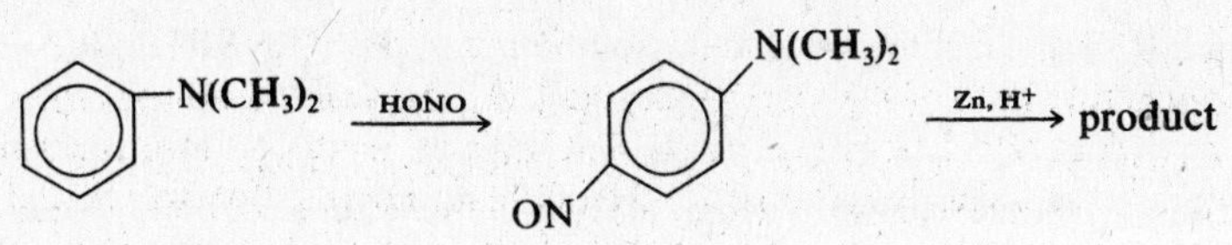

$$\text{[structure]} \xrightarrow{\text{HONO}} \text{[structure]} \xrightarrow{\text{Zn, H}^+} \text{product}$$

With HNO_3, polynitration would occur since NMe_2 is very activating.

REACTIONS WITH CARBOXYLIC ACID DERIVATIVES

$$RNH_2 \begin{cases} \xrightarrow{R'COCl} R'CONHR + HCl \\ \xrightarrow{(R'CO)_2O} R'CONHR + R'COOH \\ \xrightarrow{R'COOR''} R'CONHR + R''OH \end{cases}$$

REACTIONS WITH OTHER ELECTROPHILIC REAGENTS

$$R'CH{=}O + RNH_2 \xrightarrow{-H_2O} R'CH{=}NR \quad \textbf{(Schiff base} \text{ or azomethine)}$$

$$Cl{-}\overset{\overset{\textstyle O}{\|}}{C}{-}Cl + RNH_2 \longrightarrow 2HCl + RNH{-}\overset{\overset{\textstyle O}{\|}}{C}{-}NHR \quad \text{(symmetrical disubstituted urea)}$$

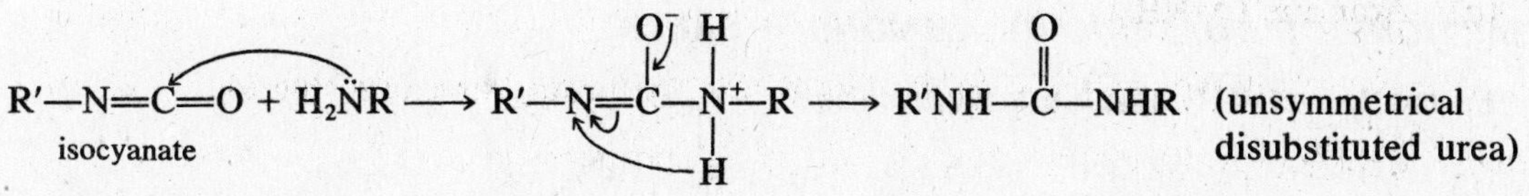

$$R'{-}N{=}C{=}O + H_2\overset{..}{N}R \longrightarrow R'{-}\overset{\overset{\textstyle O^-}{|}}{N}{=}\overset{\overset{\textstyle H}{|}}{\underset{\overset{\textstyle |}{H}}{C}{-}N^+{-}R} \longrightarrow R'NH{-}\overset{\overset{\textstyle O}{\|}}{C}{-}NHR \quad \begin{array}{l}\text{(unsymmetrical}\\ \text{disubstituted urea)}\end{array}$$

isocyanate

$$R'{-}N{=}C{=}S + H_2NR \longrightarrow R'NH{-}\overset{\overset{\textstyle S}{\|}}{C}{-}NHR \quad \text{(a thiourea)}$$

isothiocyanate

NUCLEOPHILIC DISPLACEMENTS

1. Carbylamine Reactions of 1° Amines

$$RNH_2 + CHCl_3 + 3KOH \longrightarrow R{-}\overset{+}{N}{\equiv}\overset{..}{C}: + 3KCl + 3H_2O$$

an isocyanide
(*foul smelling*)

Nucleophilic RNH_2 attacks electrophilic intermediate [$:CCl_2$] [Problem 7.32(c)].

2. Hinsberg Reaction

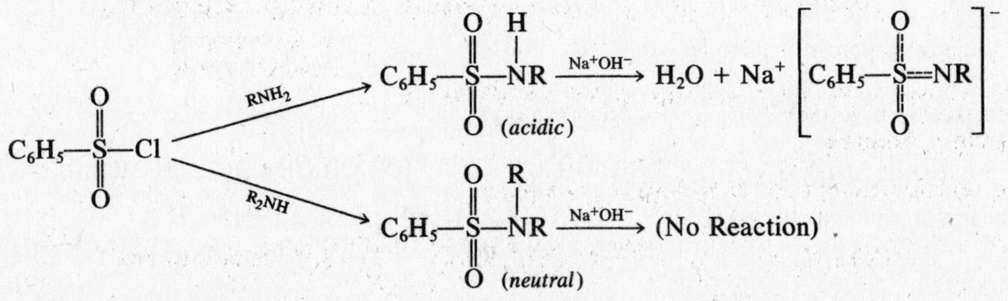

$$C_6H_5{-}\overset{\overset{\textstyle O}{\|}}{\underset{\overset{\textstyle \|}{O}}{S}}{-}Cl \begin{cases} \xrightarrow{RNH_2} C_6H_5{-}\overset{\overset{\textstyle O}{\|}}{\underset{\overset{\textstyle \|}{O}}{S}}{-}\overset{\overset{\textstyle H}{|}}{N}R \xrightarrow{Na^+OH^-} H_2O + Na^+ \left[C_6H_5{-}\overset{\overset{\textstyle O}{\|}}{\underset{\overset{\textstyle \|}{O}}{S}}{=}NR \right]^- \\ \qquad\qquad (acidic) \\ \xrightarrow{R_2NH} C_6H_5{-}\overset{\overset{\textstyle O}{\|}}{\underset{\overset{\textstyle \|}{O}}{S}}{-}\overset{\overset{\textstyle R}{|}}{N}R \xrightarrow{Na^+OH^-} \text{(No Reaction)} \\ \qquad\qquad (neutral) \end{cases}$$

Problem 18.20 How can the Hinsberg test be used to distinguish among liquid RNH_2, R_2NH and R_3N? ◄

R_3N does not react; RNH_2 reacts to give a water solution of $[C_6H_5SO_2\bar{N}R]Na^+$; R_2NH reacts to give a solid precipitate, $C_6H_5SO_2NR_2$.

Problem 18.21 Outline 2 laboratory preparations of *sym*-diphenylurea.   ◄

$$C_6H_5NH_2 + Cl\overset{\overset{O}{\|}}{-C}-Cl + H_2NC_6H_5 \longrightarrow C_6H_5NH\overset{\overset{O}{\|}}{-C}-NHC_6H_5 + 2HCl$$

$$C_6H_5NH_2 + O{=}C{=}NC_6H_5 \longrightarrow C_6H_5NH\overset{\overset{O}{\|}}{C}NHC_6H_5$$

Problem 18.22 Condensation of $C_6H_5NH_2$ with $C_6H_5CH{=}O$ yields compound (A), which is hydrogenated to compound (B). What are compounds (A) and (B)?   ◄

$$C_6H_5NH_2 + O{=}CHC_6H_5 \xrightarrow{-H_2O} C_6H_5N{=}CHC_6H_5 \xrightarrow{H_2/Ni} C_6H_5NHCH_2C_6H_5$$

$$\text{Benzalaniline (A)} \qquad\qquad \text{N-Benzylaniline (B)}$$

Problem 18.23 Account for the following order of decreasing basicity:

$$R\ddot{N}H_2 > R\ddot{N}{=}CHR' > RC{\equiv}N\colon$$ ◄

The hybrid atomic orbitals used by N to accommodate the lone pair of electrons in the above compounds are $R\ddot{N}H_2$ (sp^3), $R\ddot{N}{=}CHR'$ (sp^2), $RC{\equiv}N\colon$ (sp). The nitrile (RCN) N has the most *s* character and is the least basic. The 1° amine has the least *s* character and is the most basic [Problems 8.3 and 8.5(*b*)].

REACTIONS OF QUATERNARY AMMONIUM SALTS

1. Formation of 4° Ammonium Hydroxides

$$2R_4N^+X^- + Ag_2O + H_2O \longrightarrow 2R_4N^+OH^- + 2AgX$$

$$\underset{\substack{\textit{very strong}\\ \textit{bases}\,;\,\textit{like NaOH}}}{}$$

2. Hofmann Elimination of Quaternary Hydroxides

$$[(CH_3)_3NCH(CH_3)CH_2CH_3]^+OH^- \xrightarrow{\Delta} (CH_3)_3N + H_2C{=}CHCH_2CH_3 + H_2O$$

$$\underset{\textit{s}\text{-Butyltrimethylammonium hydroxide}}{} \qquad\qquad\qquad \underset{\text{1-Butene}}{}$$

This E2 elimination (Table 7-3) gives the less substituted alkene (**Hofmann product**) rather than the more substituted alkene (Saytzeff product; page 75).

Problem 18.24 Compare and account for the products obtained from thermal decomposition of (*a*) $[(CH_3)_3N^+(C_2H_5)]OH^-$,   (*b*) $(CH_3)_4N^+OH^-$.   ◄

(*a*) $H_2C{=}CH_2, (CH_3)_3N, H_2O$. Alkenes are formed from C_2H_5 and larger R groups having an H on the β C.

(*b*)
$$H\colon\!\ddot{O}\colon^- + CH_3\!\left(\colon\!\overset{+}{N}(CH_3)_3\right) \longrightarrow HOCH_3 + \colon N(CH_3)_3 \quad (S_N2 \text{ reaction})$$

Alkene formation is impossible with 4 CH_3's on N.

Problem 18.25 Give the alkene formed from heating

$$(a)\ \ [(C_2H_5)(CH_3)_2\overset{+}{N}CH_2CH_2CH_3]OH^- \qquad\qquad (b)\ \ [CH_3CH_2CH_2CH(CH_3)N^+Me_3]OH^-$$ ◄

The less substituted alkene is formed: (*a*) $H_2C{=}CH_2$ rather than $H_2C{=}CHCH_3$,
(*b*) $CH_3CH_2CH_2CH{=}CH_2$ rather than $CH_3CH_2CH{=}CHCH_3$.

Problem 18.26 Deduce the structures of the following amines from the products obtained from exhaustive methylation and Hofmann elimination. (*a*) $C_5H_{13}N$ (A) reacts with 1 mole of CH_3I and eventually yields propene. (*b*) $C_5H_{13}N$ (B) reacts with 2 moles of CH_3I and gives ethene and a 3° amine. The latter reacts with 1 mole of CH_3I and gives propene. ◄

(*a*) (A) is a 3° amine because it reacts with only 1 mole of CH_3I. Since propene is eliminated, C_3H_7 can be *n*- or *iso*-; hence (A) is $(C_3H_7)N(CH_3)_2$.

(*b*) (B) reacts with 2 moles of CH_3I; it is a 2° amine. Separate formation of C_2H_4 and C_3H_6 shows that the alkyl groups are C_3H_7 and C_2H_5. (B) is $(C_3H_7)NHC_2H_5$, where C_3H_7 is *n*-propyl or isopropyl.

Problem 18.27 Outline the reactions and reagents used to establish the structure of 4-methylpyridine by exhaustive methylation and Hofmann elimination. ◄

3-Methyl-1,4-pentadiene

3-Methyl-1,3-pentadiene
more stable
(conjugated diene)

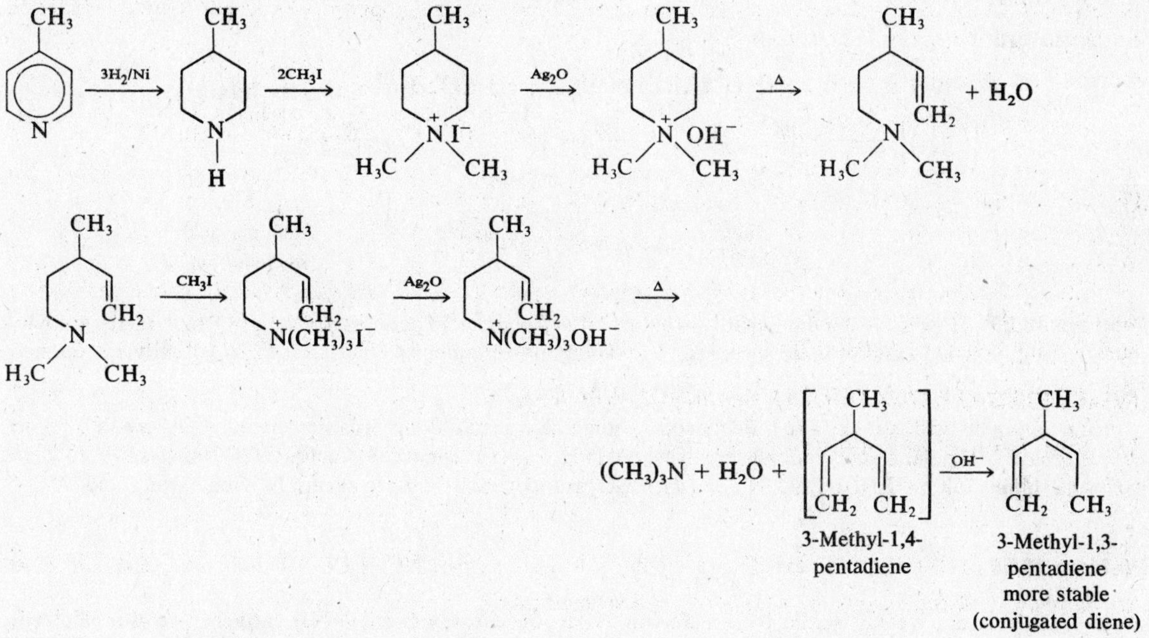

3. Cope Elimination of 3° Amine Oxides

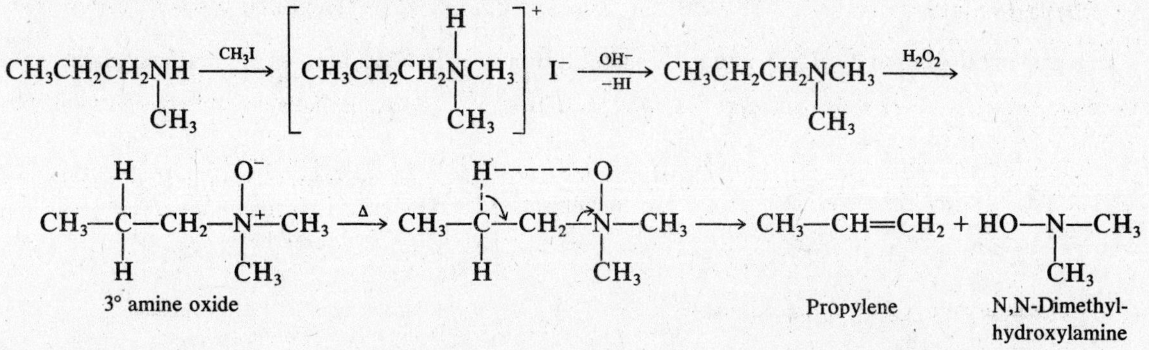

3° amine oxide

Propylene

N,N-Dimethyl-hydroxylamine

Elimination is *cis* and requires lower temperatures than pyrolysis of $[R_4N]^+OH^-$.

RING REACTIONS OF AROMATIC AMINES

—NH_2, —NHR and —NR_2 strongly activate the benzene ring toward electrophilic substitution.

1. Halogenation

For monohalogenation, —NH$_2$ is first acetylated, because

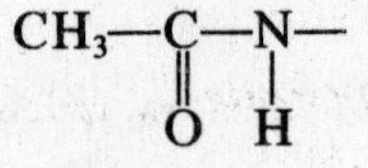

$$CH_3—\underset{\underset{H}{|}}{\overset{\overset{O}{||}}{C}}—N—$$

is only moderately activating [Problem 11.8(b)].

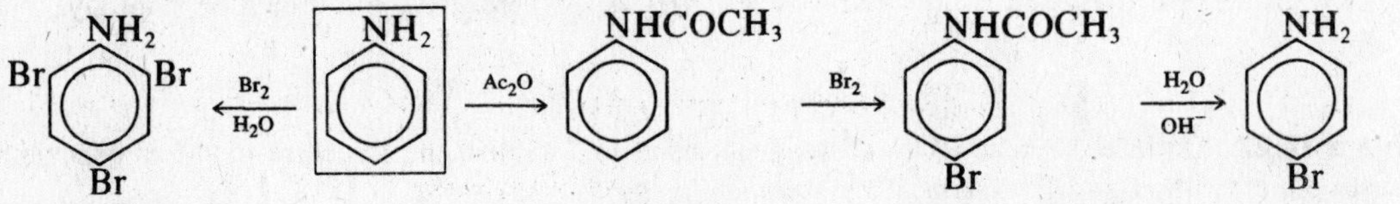

2,4,6-Tribromoaniline   Acetanilide   p-Bromoacetanilide p-Bromoaniline

2. Sulfonation

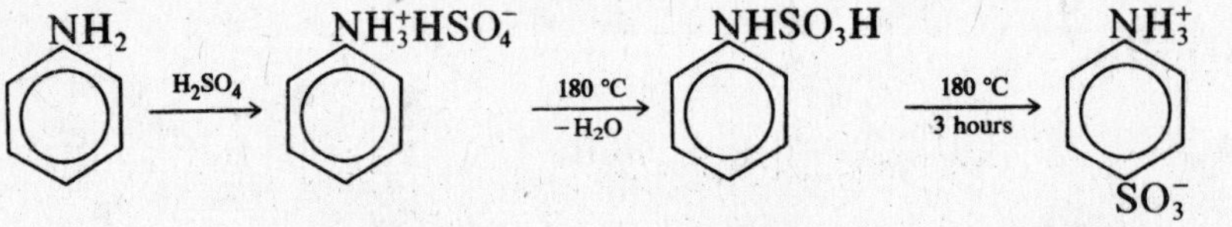

Anilinium   Sulfamic   Sulfanilic acid
sulfate   acid   *a dipolar ion*

Problem 18.28 How does the dipolar ion structure of sulfanilic acid account for its (a) high melting point, (b) insolubility in H$_2$O and organic solvents, (c) solubility in aqueous NaOH, (d) insolubility in aqueous HCl? ◄

(a) Sulfanilic acid is ionic. (b) Because it is ionic, it is insoluble in organic solvents. Its insolubility in H$_2$O is typical of dipolar salts. Not all salts dissolve in H$_2$O. (c) The weakly acidic NH$_3^+$ transfers H$^+$ to OH$^-$ to form a soluble salt, p-H$_2$NC$_6$H$_4$SO$_3^-$Na$^+$. (d) —SO$_3^-$ is too weakly basic to accept H$^+$ from strong acids.

Problem 18.29 H$_3\overset{+}{N}$CH$_2$COO$^-$ exists as a dipolar ion whereas p-H$_2$NC$_6$H$_4$COOH does not. Explain. ◄

—COOH is too weakly acidic to transfer an H$^+$ to the weakly basic —NH$_2$ attached to the electron-withdrawing benzene ring. When attached to an aliphatic C, the NH$_2$ is sufficiently basic to accept H$^+$ from COOH.

3. Nitration

To prevent oxidation by HNO$_3$ and *meta* substitution of C$_6$H$_5$NH$_3^+$, amines are first acetylated.

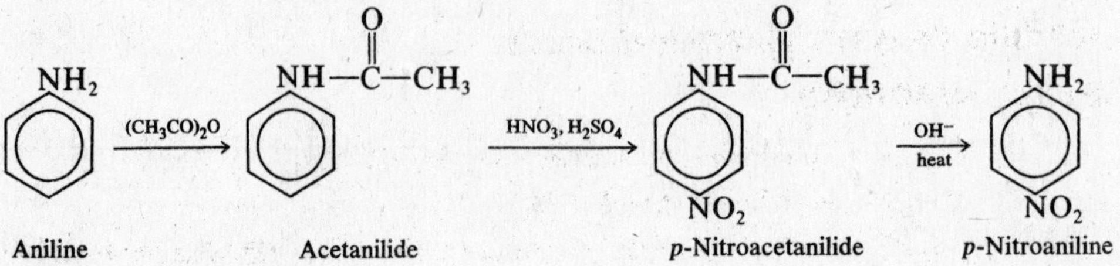

Aniline   Acetanilide   p-Nitroacetanilide   p-Nitroaniline

ANILINE-X REARRANGEMENTS

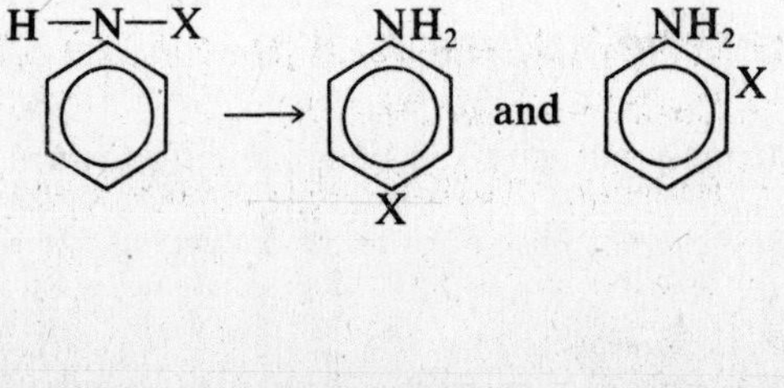

1. Fischer-Hepp

$$
\underset{\substack{\text{N-Nitroso-}\\\text{N-alkylaniline}}}{\text{Ph}-\underset{\underset{\text{NO}}{|}}{\text{N}}-\text{R}} \xrightarrow[\substack{2.\ \text{OH}^-}]{1.\ \text{H}^+}
\underset{\substack{\textit{(neutralize}\\\textit{salt)}}}{} \underset{\substack{p\text{-Nitroso-N-alkylaniline}}}{\text{O}=\text{N}\!\!-\!\!\bigcirc\!\!-\!\!\underset{\underset{}{}}{\overset{\text{H}}{\text{N}}}-\text{R}}
$$

2. Phenylhydroxylamines

$$
\text{Ph}-\underset{\underset{}{\overset{\text{H}}{\text{N}}}}{}-\text{OH} \xrightarrow[\substack{2.\ \text{OH}^-}]{1.\ \text{H}^+} p\text{-HO}\!\!-\!\!\bigcirc\!\!-\!\!\text{NH}_2
$$

$$
\substack{p\text{-Hydroxyaniline}\\(p\text{-Aminophenol})}
$$

$$
(\text{PhNO}_2 \xrightarrow{\text{Zn, NH}_4^+} \text{PhNHOH})
$$

3. Sulfamic Acid

$$
\text{Ph}-\text{NH}-\text{SO}_3\text{H} \xrightarrow{\text{heat}} {}^-\text{O}_3\text{S}\!\!-\!\!\bigcirc\!\!-\!\!\text{NH}_3^+
$$

$$
\text{Sulfanilic acid}
$$

18.4 SPECTRAL PROPERTIES

The N—H stretching and NH_2 bending frequencies occur in the ir spectrum at 3050–3550 cm^{-1} and 1600–1640 cm^{-1}, respectively.   In the N—H stretching region, 1° amines and unsubstituted amides show a pair of peaks for a symmetric and an antisymmetric vibration.   In nmr, N—H proton signals of amines fall in a wide range ($\delta = 1$–5) and are often very broad.   The signals of N—H protons of amides are even broader, appearing at $\delta = 5$–8.   Mass spectra of amines show α,β-cleavage, like alcohols.

$$
-\overset{|}{\underset{|}{\overset{\beta}{\text{C}}}}-\overset{|}{\underset{|}{\overset{\alpha}{\text{C}}}}-\overset{\cdot\ +}{\underset{|}{\text{N}}}- \longrightarrow -\overset{|}{\underset{|}{\overset{\beta}{\text{C}}}}\cdot + \overset{}{\underset{|}{\text{C}}}\!\!=\!\!\overset{+}{\underset{|}{\text{N}}}-
$$

Problem 18.30   Distinguish 1°, 2° and 3° amines by ir spectroscopy.　　　　　　　◄

1° amine, two N—H stretching bands; 2° amine, one N—H stretching band; 3° amine, no N—H stretching band.

18.5 REACTIONS OF ARYL DIAZONIUM SALTS

DISPLACEMENT REACTIONS

$$
\text{ArN}_2^+\text{X}^-
\begin{cases}
+\ \text{HPH}_2\text{O}_2 \text{ or NaBH}_4 \longrightarrow \text{Ar}-\text{H} + \text{N}_2 \\
+\ \text{KI} \longrightarrow \text{Ar}-\text{I} + \text{N}_2 \\
+\ \text{CuCl (CuBr)} \longrightarrow \text{ArCl (ArBr)} + \text{N}_2 \quad \textbf{(Sandmeyer reaction)} \\
+\ \text{HBF}_4 \longrightarrow \text{ArN}_2^+\text{BF}_4^- \xrightarrow{\Delta} \text{ArF} + \text{N}_2 + \text{BF}_3 \\
+\ \text{HOH} \longrightarrow \text{ArOH} + \text{N}_2 \\
+\ \text{HO}-\text{C}_2\text{H}_5 \longrightarrow \text{ArOC}_2\text{H}_5 + \text{ArH} + \text{CH}_3\text{CHO} + \text{N}_2 \\
+\ \text{CuCN} \longrightarrow \text{ArCN} + \text{N}_2 \\
+\ \text{NaNO}_2 + \text{NaHCO}_3 \xrightarrow[\text{or Cu}_2\text{O}]{\text{Cu}^{2+}} \text{ArNO}_2 + \text{N}_2
\end{cases}
$$

COUPLING (G in ArG is an electron-releasing group)

$$ArN_2^+ + C_6H_5G \longrightarrow p\text{-}G\text{-}C_6H_4\text{-}N=N\text{-}Ar \quad (G = OH, NR_2, NHR, NH_2)$$

a weak   a strongly   an azo compound;
electro-   activated   mainly *para*
phile   ring

Azo compounds undergo reduction as shown:

$$ArN=NAr \xrightarrow[NaOH]{Sn} Ar\overset{H}{\underset{}{N}}\text{-}\overset{H}{\underset{}{N}}Ar \quad \text{(mild reduction)}$$

a hydrazo compound

$$ArN=NAr' \xrightarrow[\text{2. OH}^-]{\text{1. } SnCl_2, H_3O^+} ArNH_2 + H_2NAr' \quad \text{(vigorous reduction)}$$

Hydrazo compounds are also made as follows:

$$PhNO_2 \xrightarrow[NaOH]{Sn} PhNHNHPh$$

Diaryl hydrazo compounds undergo the **benzidine rearrangement.**

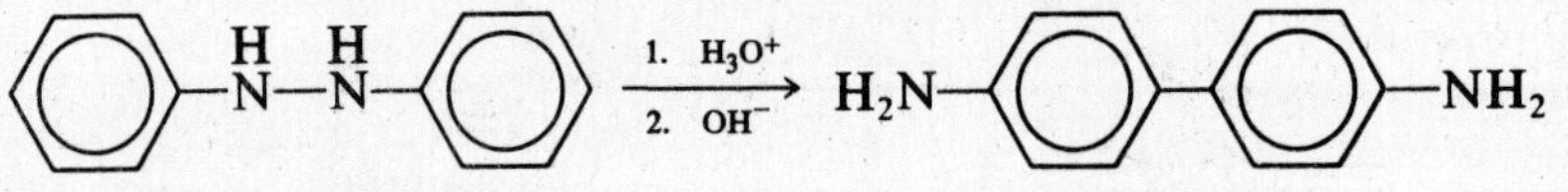

Hydrazobenzene   Benzidine

REDUCTION OF ArN_2^+

$$ArN_2^+Cl^- \xrightarrow[\text{2. NaOH}]{\text{1. } Na_2SO_3, H_3O^+, 100\ °C} ArNHNH_2$$

arylhydrazine

Problem 18.31 Using C_6H_6, $C_6H_5CH_3$, via diazonium salts and any other needed reagents, prepare
(*a*) *o*-chlorotoluene,   (*b*) *m*-chlorotoluene,   (*c*) 1,3,5-tribromobenzene,   (*d*) *m*-bromochlorobenzene,
(*e*) *p*-iodotoluene,   (*f*) *p*-dinitrobenzene   and (*g*) *p*-cyanobenzoic acid.   Do not repeat the synthesis of
intermediate products.   ◀

(*a*) $C_6H_5CH_3$

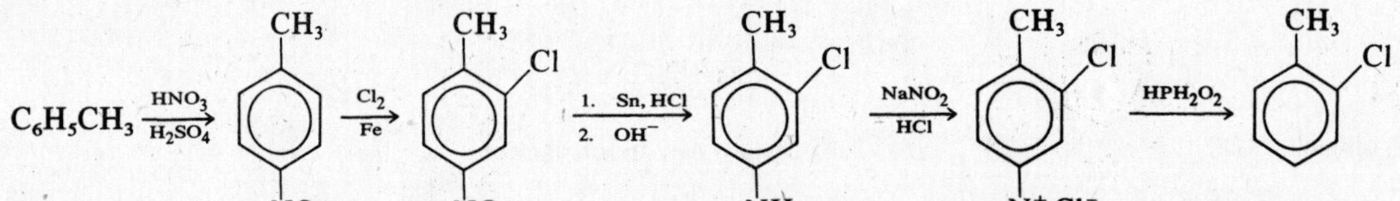

The —NO_2 is used to block the *para* position and it also directs *meta*, so that chlorination will occur only
ortho to CH_3.

(*b*)

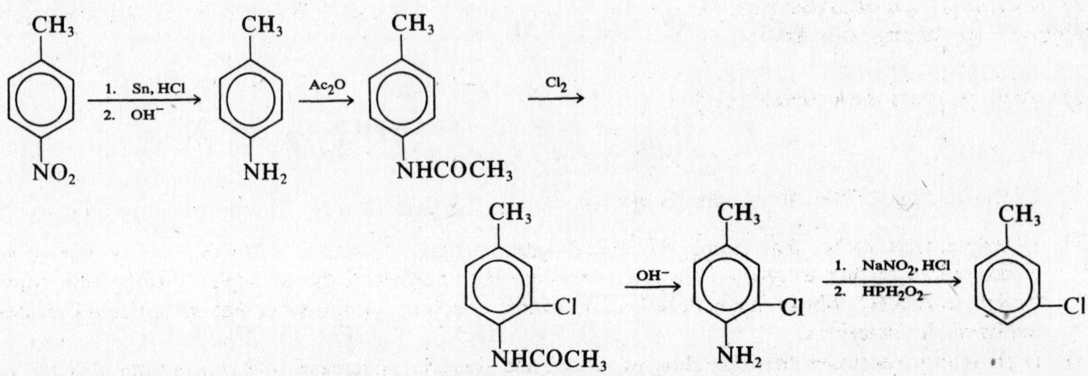

The acetylated —NH_2 is used to direct Cl into its *ortho* position, which is *meta* to CH_3; it is then removed.

(c) Aniline is rapidly and directly tribrominated and the NH_2 removed.

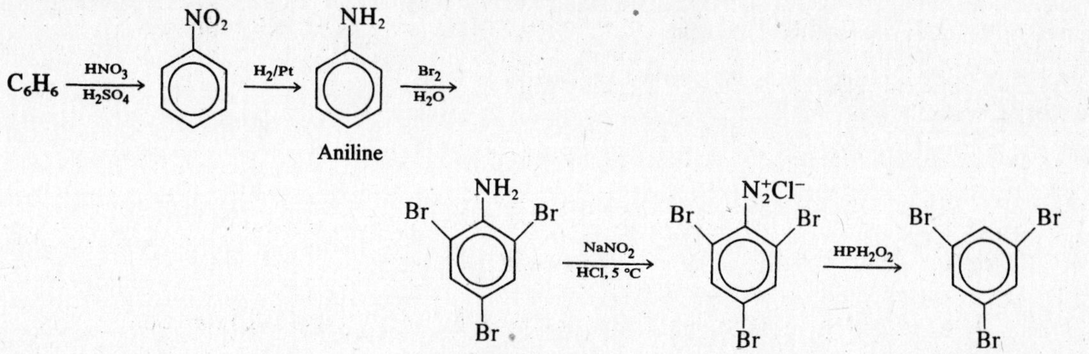

Aniline

Tribromoaniline

1,3,5-Tribromobenzene

(d)

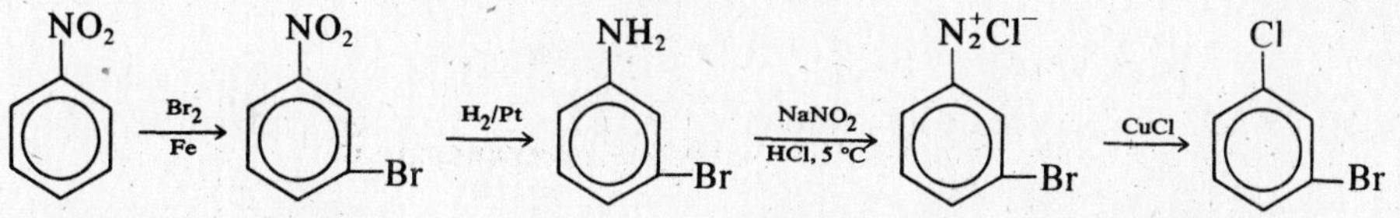

(e)

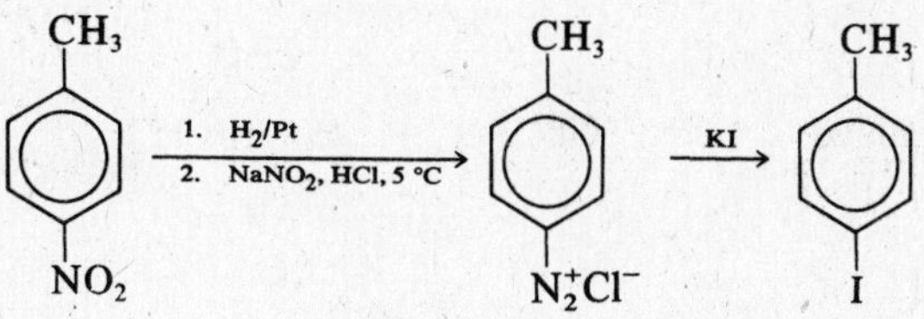

(f)

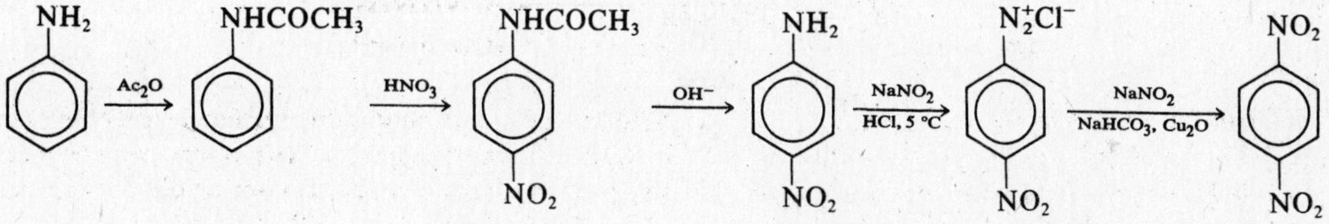

(g)

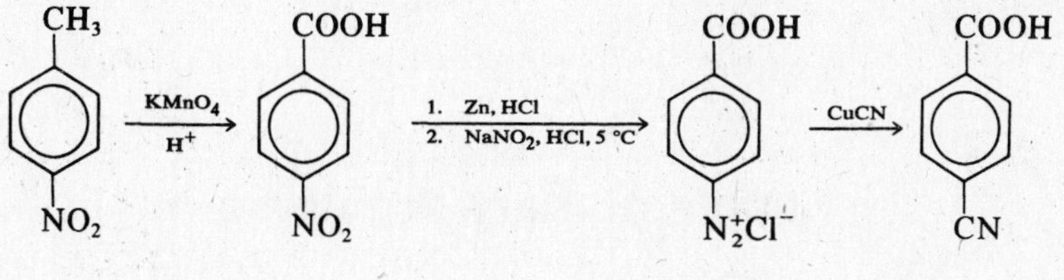

Problem 18.32 Explain the following conditions used in coupling reactions: (a) excess of mineral acid during diazotization of arylamines, (b) weakly acidic medium for coupling with $ArNH_2$, (c) weakly basic solution for coupling with ArOH. ◄

(a) Acid prevents the coupling reaction

$$ArN\equiv N\colon + H_2\overset{..}{N}Ar' \longrightarrow ArN\!=\!N\!-\!NHAr'$$

by converting $Ar'NH_2$ to its salt, $Ar'NH_3^+X^-$.

(b) In strong base, rather than coupling, $ArN\equiv N$ reacts with OH^- to form $ArN\!=\!N\!-\!OH$ (a diazoic acid), which reacts further to give a diazotate, $ArN\!=\!N\!-\!O^-$; neither of these couple. Strong acid converts $ArNH_2$ to $ArNH_3^+$, whose ring is deactivated towards coupling. It turns out that amines couple fastest in mildly acidic solutions.

(c) High acidity represses the ionization of ArOH and therefore decreases the concentration of the more reactive ArO^-. In weak base, ArO^- is formed and $ArN\!=\!N\!-\!OH$ is not.

Problem 18.33 Deduce the structures of the azo compounds that yield the indicated aromatic amines on reduction with $SnCl_2$: (a) p-toluidine and p-aminodimethylaniline, (b) 1 mole of 4,4'-diaminobiphenyl and 2 moles of 2-hydroxy-5-aminobenzoic acid. ◄

NH_2's originate from N's of the cleaved azo bond.

(a)

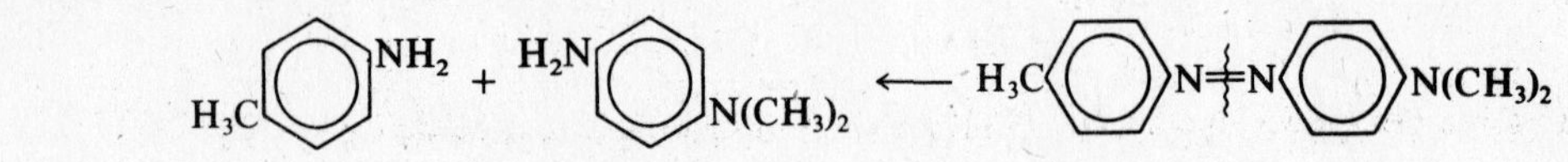

(b)

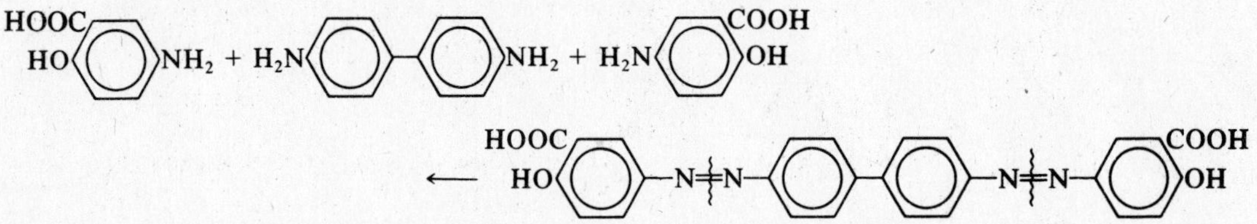

Problem 18.34 Which reactants are coupled to give the azo compounds in Problem 18.33? ◄

(a)

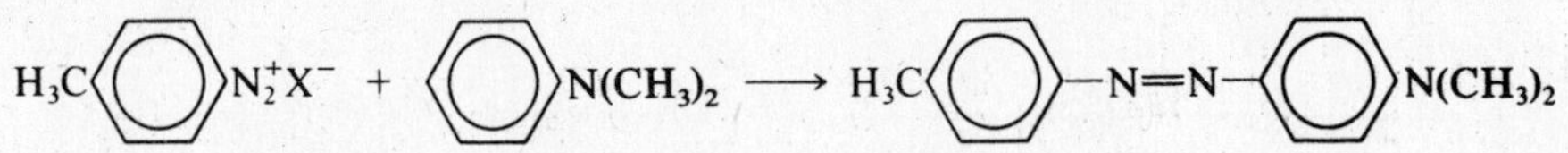

$CH_3C_6H_5$ is not active enough to couple with p-N_2^+—C_6H_4—$N(CH_3)_2$.

(b)

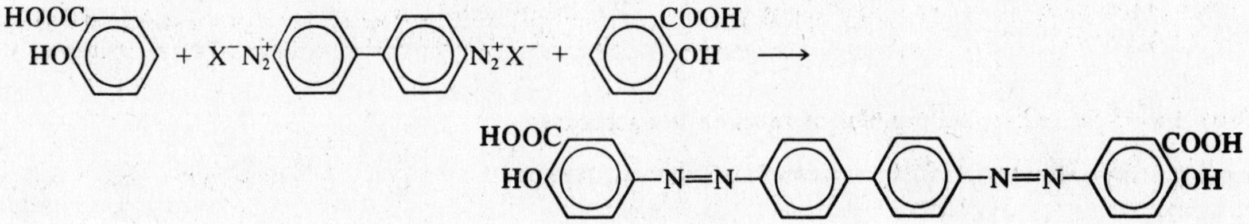

Supplementary Problems

Problem 18.35 Write a structural formula for (a) N,N'-di-p-tolylthiourea, (b) 2,4-xylidene, (c) N-methyl-p-nitrosoaniline, (d) 4-ethyl-3'-methylazobenzene. ◄

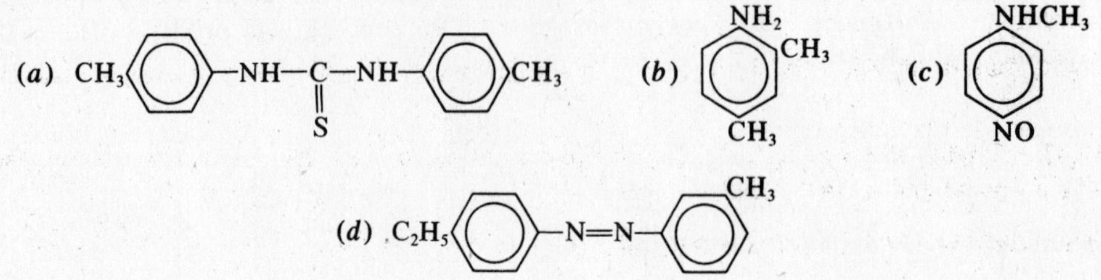

(a)

(b)

(c)

(d)

Problem 18.36 Name: (a) $C_6H_5CH_2CH_2NH_2$, (b) $CH_3NHCH_2CH_3$, (c) $NH_2CH_2CH_2CH(NH_2)CH_3$, (d) $NH_2CH_2CH_2OH$, (e) $(CH_3)_3CNHC(CH_3)_3$. ◄

(a) 2-phenyl-1-aminoethane or β-phenylethylamine, (b) methylethylamine, (c) 1,3-diaminobutane, (d) 2-aminoethanol, (e) di-*tert*-butylamine.

Problem 18.37 Give the structure of   (a) 3-(N-methylamino)-1-propanol,   (b) ethyl 3-(N-methylamino)-2-butenoate,   (c) 2-(N,N-dimethylamino)butane,   (d) allylamine.   ◄

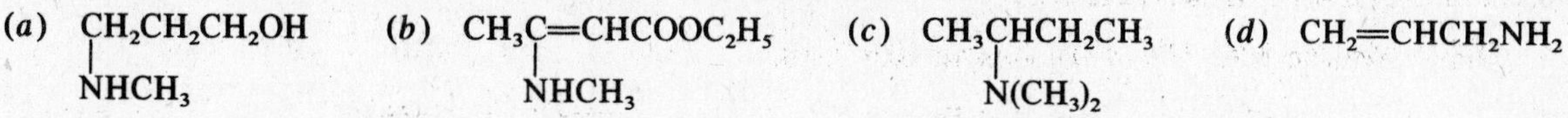

(a) $CH_2CH_2CH_2OH$ (b) $CH_3C=CHCOOC_2H_5$ (c) $CH_3CHCH_2CH_3$ (d) $CH_2=CHCH_2NH_2$
     | | |
     $NHCH_3$ $NHCH_3$ $N(CH_3)_2$

Problem 18.38 Give structures and names of five C_7H_9N amines having one benzene ring.   ◄

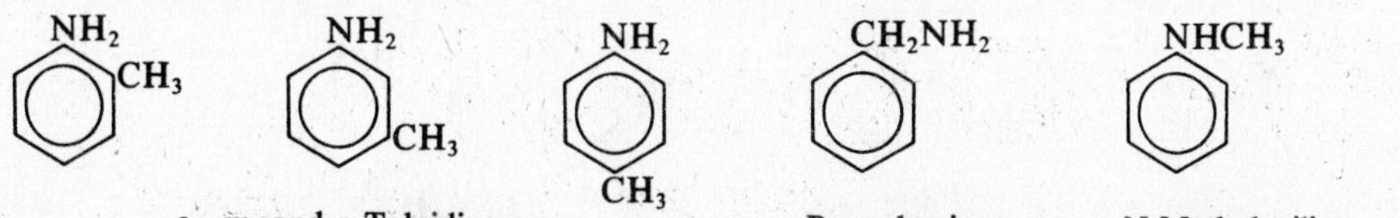

o-, m- and p-Toluidine   Benzylamine   N-Methylaniline

Problem 18.39 Name and give the structural formulas of 5 cyclic compounds having the molecular formula C_4H_9N.   ◄

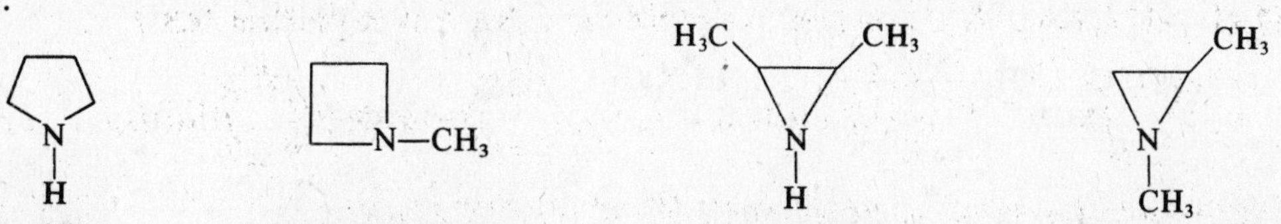

Pyrrolidine or   N-Methylazacyclobutane   cis- or trans-   N-Methyl-2-methylaziridine
Azacyclopentane   or N-Methylazetidine   2,3-Dimethylethylenimine   (no cis-trans isomers
or Azolidine   or 2,3-Dimethylaziridine   at room temperature)

Problem 18.40 Give the product of reaction in each case:

(a) C_2H_5Br + excess NH_3 (b) $CH_2=CHCN$ + H_2/Pt   (c) n-Butyramide + Br_2 + KOH
                                               Acrylonitrile

(d) Dimethylamine + HONO   (e) Ethylamine + $CHCl_3$ + KOH   ◄

(a) $CH_3CH_2NH_2$, (b) $CH_3CH_2CH_2NH_2$, (c) $CH_3CH_2CH_2NH_2$, (d) $Me_2NN=O$, (e) $CH_3CH_2\overset{+}{N}\equiv\overset{-}{C}$:

Problem 18.41 What is the organic product when n-propylamine is treated with   (a) $PhSO_2Cl$;   (b) excess $CH_3CH_2CH_2Cl$;   (c) chlorobenzene;   (d) excess CH_3I, then Ag_2O and heat?   ◄

(a) N-(n-propyl)benzenesulfonamide, $PhSO_2NHCH_2CH_2CH_3$;   (b) tetra-n-propylammonium chloride; (c) no reaction;   (d) propene and trimethylamine.

Problem 18.42 Compare reactions of HNO_2 with: (a) aniline (at 0 °C),   (b) N-methylaniline,   (c) N,N-dimethylaniline.   ◄

(a) Soluble diazonium salt.   (b) N-Methyl-N-nitrosoaniline.   (c) p-Nitroso-N,N-dimethylaniline.

Problem 18.43 What is the product of catalytic hydrogenation of   (a) acetone oxime,   (b) propane-1,3-dinitrile,   (c) propanal and methylamine?   ◄

(a) isopropylamine,   (b) 1,5-diaminopentane,   (c) $CH_3CH_2CH=NCH_3$.

Problem 18.44 Show the steps in the following syntheses:

(a)   o-Nitrotoluene $\longrightarrow$ 2-Methyl-4-hydroxyaniline

(b)   Ethylamine $\longrightarrow$ Methylethylamine

(c)   Ethylamine $\longrightarrow$ Dimethylethylamine

(d)   n-Propyl chloride $\longrightarrow$ Isopropylamine

(e)   Aniline $\longrightarrow$ p-Aminobenzenesulfonamide   (Sulfanilamide)   ◄

(a)

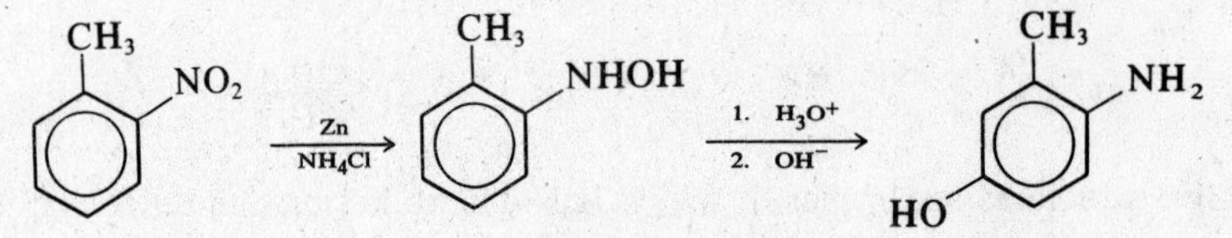

$$\text{(o-CH}_3\text{-C}_6\text{H}_4\text{-NO}_2) \xrightarrow[\text{NH}_4\text{Cl}]{\text{Zn}} (\text{o-CH}_3\text{-C}_6\text{H}_4\text{-NHOH}) \xrightarrow[\text{2. OH}^-]{\text{1. H}_3\text{O}^+} (\text{CH}_3\text{-C}_6\text{H}_3(\text{OH})\text{-NH}_2)$$

(b) $CH_3CH_2NH_2 \xrightarrow[\text{KOH}]{\text{CHCl}_3} CH_3CH_2\overset{+}{N}\overset{-}{C} \xrightarrow{H_2/Pt} CH_3CH_2\overset{\overset{\displaystyle H}{|}}{N}CH_3$

(c) $CH_3CH_2NH_2 \xrightarrow[\text{HCOOH}]{H_2C=O} CH_3CH_2N(CH_3)_2$

(d) $CH_3CH_2CH_2Cl \xrightarrow[\text{KOH}]{\text{alc.}} CH_3CH=CH_2 \xrightarrow{HBr} CH_3CHBrCH_3 \xrightarrow{NH_3} CH_3CH(NH_2)CH_3$

(e)

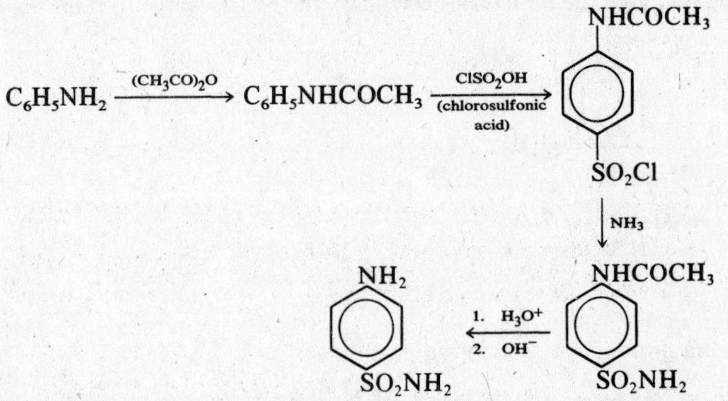

$$C_6H_5NH_2 \xrightarrow{(CH_3CO)_2O} C_6H_5NHCOCH_3 \xrightarrow[\substack{\text{(chlorosulfonic} \\ \text{acid)}}]{\text{ClSO}_2\text{OH}} (\text{p-CH}_3\text{CONH-C}_6\text{H}_4\text{-SO}_2\text{Cl})$$

$$\downarrow NH_3$$

$$(\text{p-H}_2\text{N-C}_6\text{H}_4\text{-SO}_2\text{NH}_2) \xleftarrow[\text{2. OH}^-]{\text{1. H}_3\text{O}^+} (\text{p-CH}_3\text{CONH-C}_6\text{H}_4\text{-SO}_2\text{NH}_2)$$

Problem 18.45 The sulfonamide group of $p\text{-H}_2\text{NO}_2\text{S—C}_6\text{H}_4\text{—CONH}_2$ is much more resistant to base hydrolysis than the carboxamide group.   Explain.   ◄

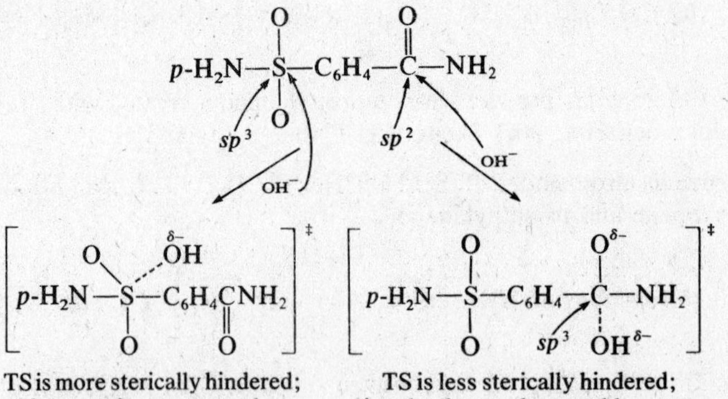

TS is more sterically hindered;   TS is less sterically hindered;
S is becoming pentacovalent   involved atoms have stable octets
(less stable than octet)

Carboxamide hydrolysis is slower than that of esters and requires more vigorous conditions.

Problem 18.46 Prepare aminoacetic acid (glycine) by the Gabriel synthesis.   ◄

Since potassium phthalimide is protonated by α-chloroacetic acid, the ester, ethyl chloroacetate, is used.

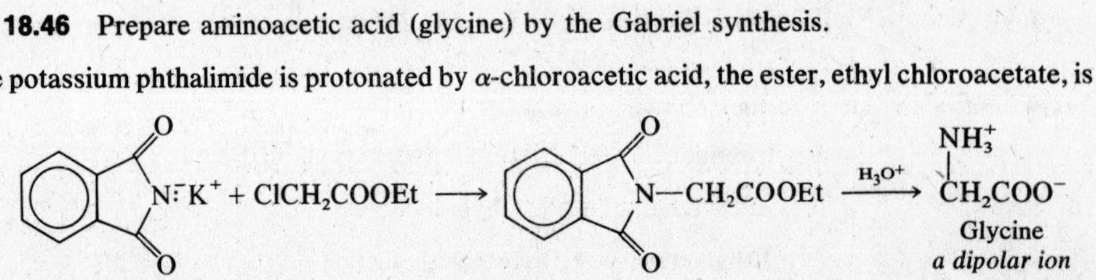

Problem 18.47 Outline the steps in the syntheses of the following compounds from C_6H_6, $C_6H_5CH_3$ and any readily available aliphatic compound: (a) p-aminobenzoic acid, (b) m-nitroacetanilide, (c) 1-amino-1-phenylpropane, (d) 4-amino-2-chlorotoluene. ◄

(a)

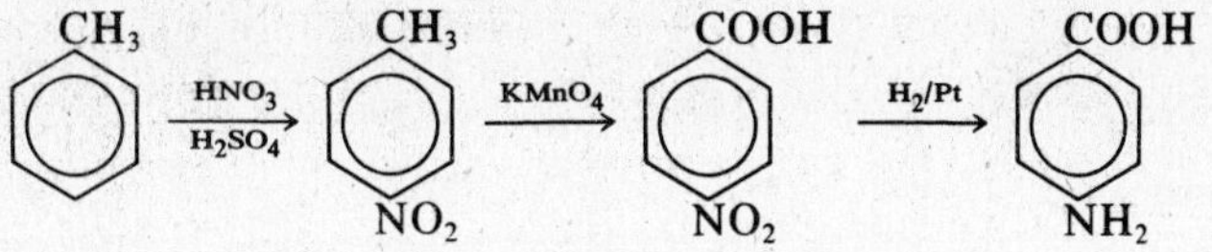

Side chain is oxidized when a deactivating group (NO_2) rather than an activating group (NH_2) is attached to ring.

(b)

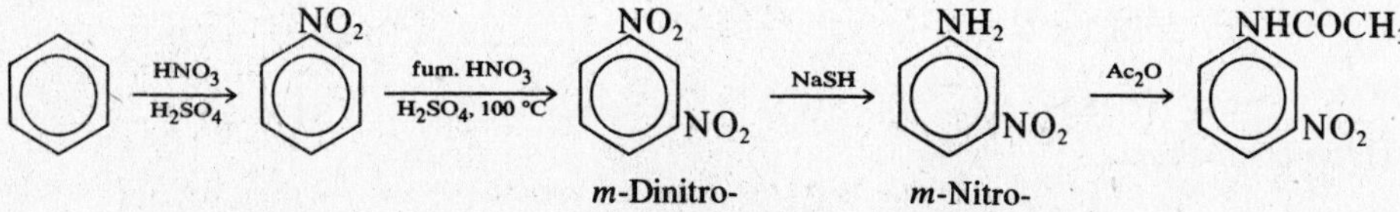

m-Dinitro-benzene   m-Nitro-aniline

(c) $C_6H_6 + CH_3CH_2COCl \xrightarrow{AlCl_3} C_6H_5COCH_2CH_3 \xrightarrow[H_2/Pt]{NH_3} C_6H_5CH(NH_2)CH_2CH_3$

(d)

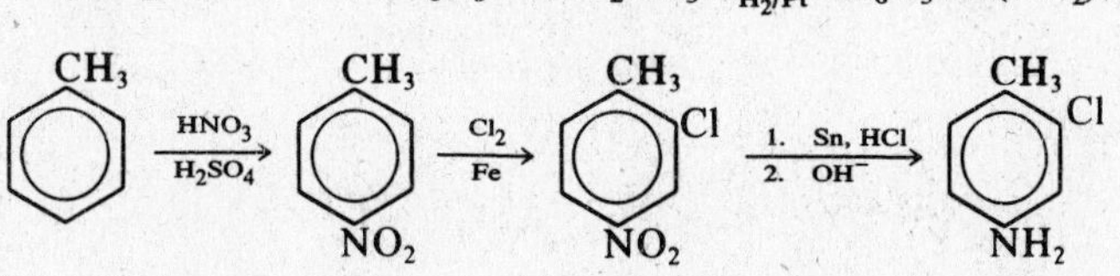

Problem 18.48 Use the benzidine rearrangement to synthesize 2,2'-dichlorobiphenyl from benzene and inorganic reagents. ◄

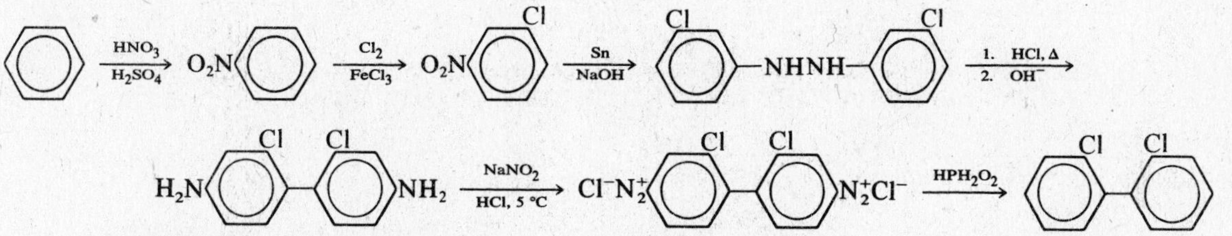

2,2'-Dichloro-4,4'-diaminobiphenyl   bis-Diazonium salt   2,2'-Dichlorobiphenyl

Problem 18.49 Use o- or p-nitroethylbenzene and any inorganic reagents to synthesize the 6 isomeric dichloroethylbenzenes. Do not repeat preparations of intermediate products. ◄

The —NO_2 is used as a blocking group by the sequence —$NO_2 \rightarrow$ —$NH_2 \rightarrow$ —$N_2^+ \rightarrow$ —H or as a source of Cl by —$N_2^+ \rightarrow$ —Cl or is converted to —NHAc, whose directive effect supersedes that of C_2H_5.

(a) 2,3-Dichloroethylbenzene

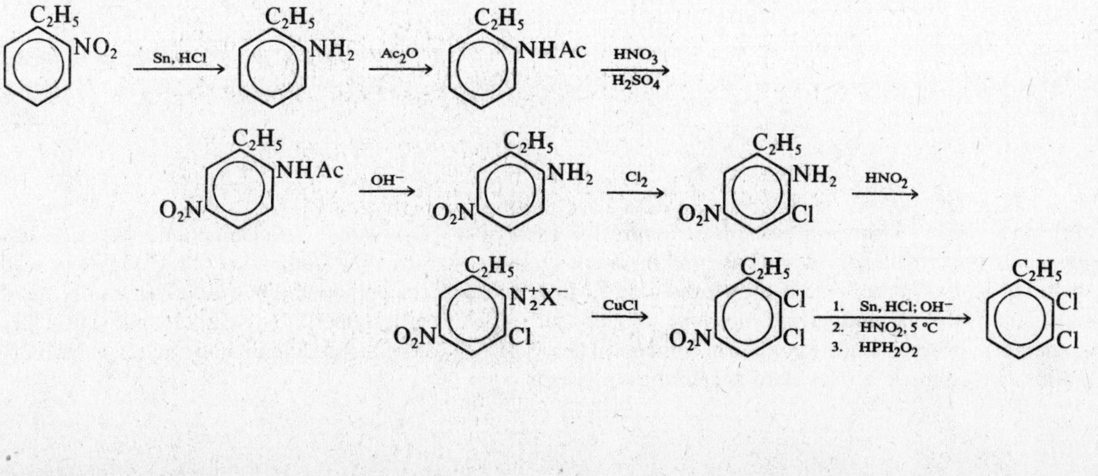

(b) 2,4-Dichloroethylbenzene

C₂H₅ / NO₂ → C₂H₅ / Cl / NO₂ —1. Sn, HCl; OH⁻ 2. HNO₂, 5 °C 3. CuCl→ C₂H₅ / Cl / Cl

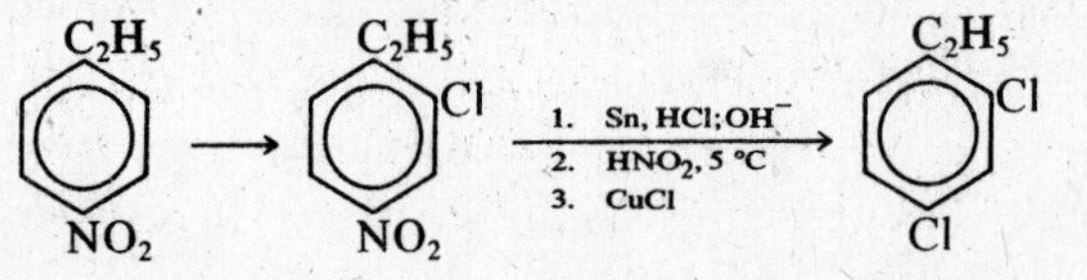

(c) 2,5-Dichloroethylbenzene

C₂H₅ / NHAc —Cl₂→ C₂H₅ / NHAc / Cl —OH⁻→ C₂H₅ / NH₂ / Cl —HNO₂ / 5 °C→ C₂H₅ / N₂⁺X⁻ / Cl —CuCl→ C₂H₅ / Cl / Cl

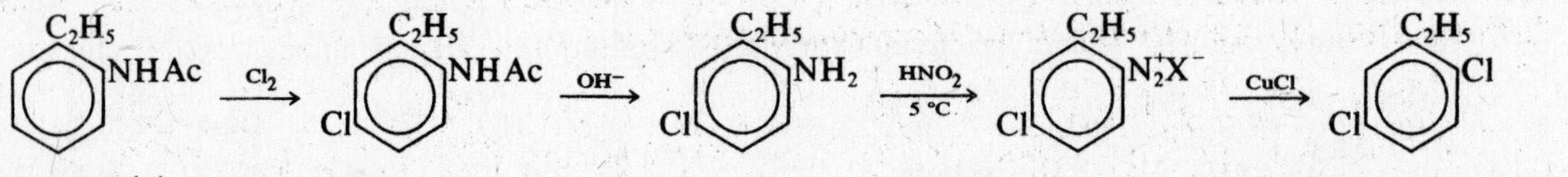

see part (a)

(d) 2,6-Dichloroethylbenzene

C₂H₅ / NO₂ —Cl₂ / Fe→ C₂H₅ / Cl Cl / NO₂ —1. Sn, HCl; OH⁻ 2. HNO₂, 5 °C 3. HPH₂O₂→ C₂H₅ / Cl Cl

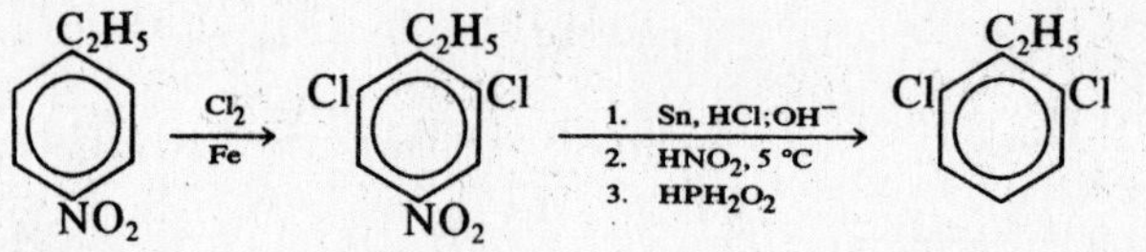

(e) 3,4-Dichloroethylbenzene

C₂H₅ / NO₂ —Sn / HCl→ C₂H₅ / NH₂ —Ac₂O→ C₂H₅ / NHAc —Cl₂→ C₂H₅ / NHAc / Cl —1. OH⁻ 2. HNO₂, 5 °C 3. CuCl→ C₂H₅ / Cl / Cl

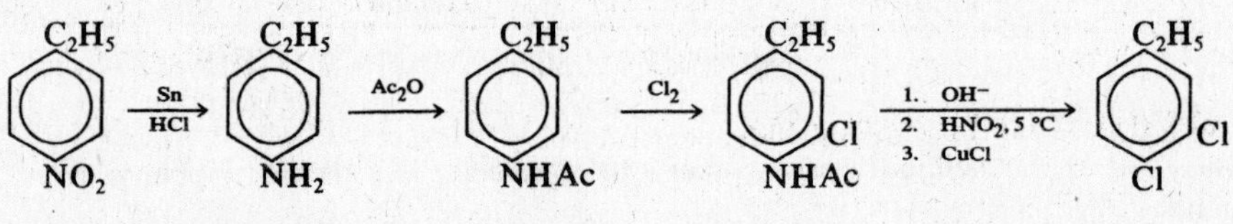

(f) 3,5-Dichloroethylbenzene

C₂H₅ / NH₂ —Cl₂→ C₂H₅ / Cl Cl / NH₂ —1. HNO₂, 5 °C 2. HPH₂O₂→ C₂H₅ / Cl Cl

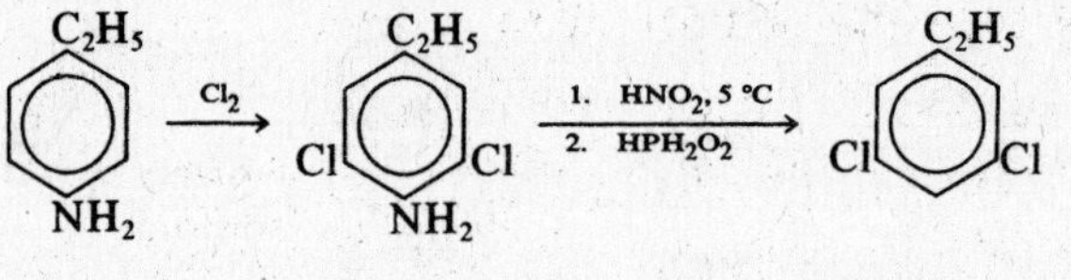

Problem 18.50 Deduce the structural formula of compound (A), $C_8H_9NO_2$, which is reduced by Sn in OH⁻ to product (B). Strong mineral acid rearranges (B) to an aromatic amine (C), which is treated with HNO_2 and then HPH_2O_2 to form 3,3'-diethylbiphenyl (D). ◄

Since (D) is a diphenyl compound, (C) is a substituted benzidine formed from a hydrazobenzene (B) in which the C_2H_5's are *ortho* to the two NH_2's. Compound (A) is *o*-nitroethylbenzene.

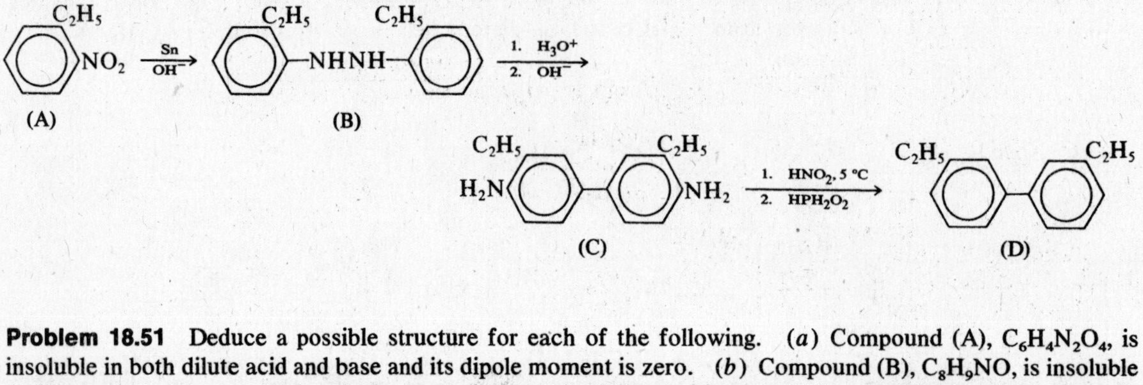

C₂H₅ / NO₂ —Sn / OH⁻→ C₂H₅ — NHNH — C₂H₅ —1. H₃O⁺ 2. OH⁻→

(A)   (B)

C₂H₅ H₂N — — C₂H₅ NH₂ —1. HNO₂, 5 °C 2. HPH₂O₂→ C₂H₅ — — C₂H₅

(C)   (D)

Problem 18.51 Deduce a possible structure for each of the following. (a) Compound (A), $C_6H_4N_2O_4$, is insoluble in both dilute acid and base and its dipole moment is zero. (b) Compound (B), C_8H_9NO, is insoluble in dilute acid and base. (B) is transformed by $KMnO_4$ in H_2SO_4 to compound (C), which is free of N, soluble in aqueous $NaHCO_3$ and gives only one mononitro substitution product. (c) Compound (D), $C_7H_7NO_2$, undergoes vigorous oxidation to form compound (E), $C_7H_5NO_4$, which is soluble in dilute aqueous $NaHCO_3$ and forms two isomeric monochloro substitution products. ◄

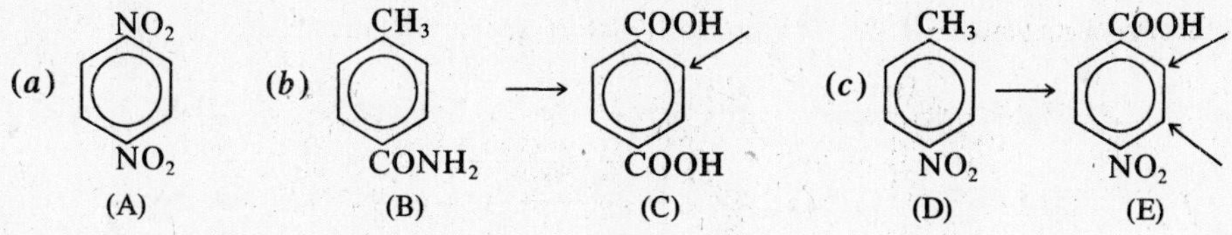

(a) (A) (b) (B) → (C) (c) (D) → (E)

Problem 18.52 Use any toluidine and necessary aliphatic or inorganic reagents to synthesize (a) 3-nitro-4-fluorotoluene, (b) 2,2'-dimethyl-4-nitro-4'-aminoazobenzene. ◄

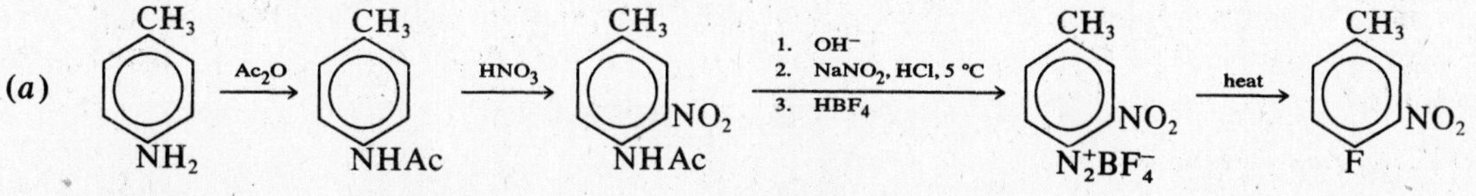

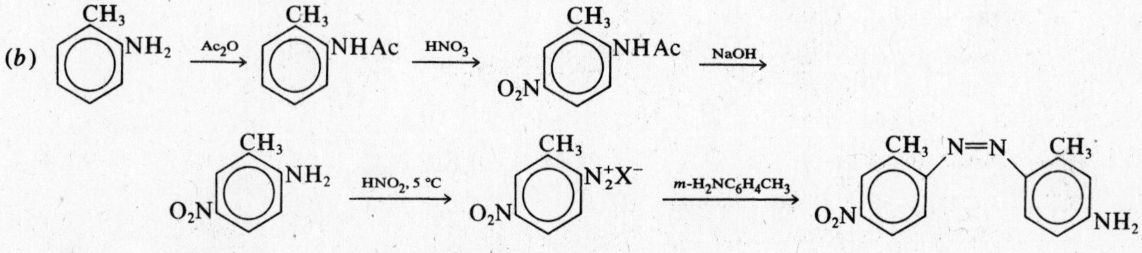

For m-H₂NC₆H₄CH₃ see Problem 18.53(a).

Problem 18.53 Using any aliphatic and inorganic reagents, outline the syntheses of (a) m-HOC₆H₄CH₃ from toluene, (b) 4-bromo-4'-aminoazobenzene from aniline. ◄

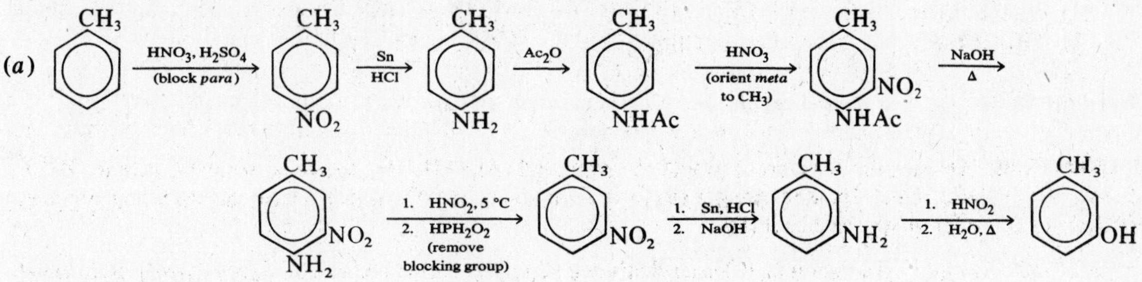

or else

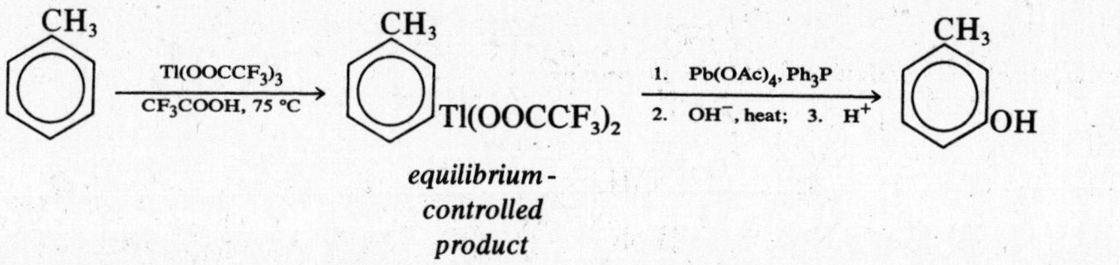

equilibrium-controlled product

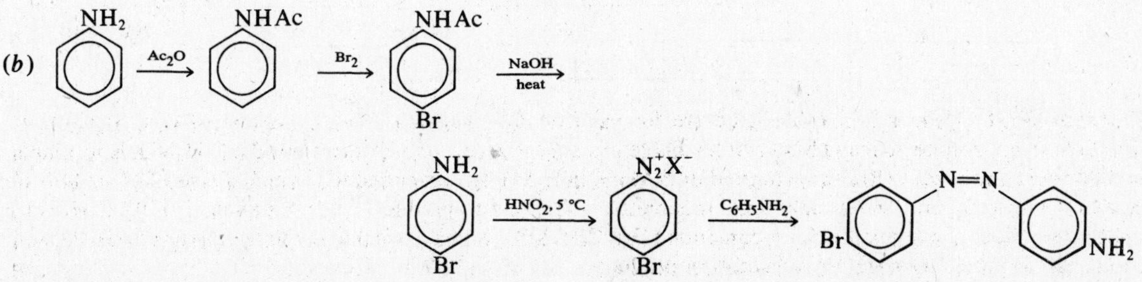

Problem 18.54 Prepare (a) C_6H_5D, (b) optically active *sec*-butylbenzene. ◄

(a) $C_6H_5NH_2 \xrightarrow[\text{2. } DPH_2O_2]{\text{1. } NaNO_2, HCl, 5\,°C} C_6H_5D$ or $C_6H_5MgBr \xrightarrow{D_2O} C_6H_5D$

(b) $C_6H_6 \xrightarrow[AlCl_3]{sec\text{-}BuCl} C_6H_5\text{-}sec\text{-}Bu \xrightarrow[\substack{\text{2. } Sn, HCl \\ \text{3. } OH^-}]{\text{1. } HNO_3, H_2SO_4} p\text{-}H_2N—C_6H_4\text{-}sec\text{-}Bu$
                                    racemate   racemate

which is resolved with an optically active carboxylic acid, such as tartaric acid (Problem 5.7). Then deaminate via diazonium salt and HPH_2O_2.

Problem 18.55 Synthesize novocaine, $p\text{-}H_2NC_6H_4COOCH_2CH_2NEt_2$, from toluene and any aliphatic compound of 4 C's or fewer. ◄

(1) $Et_2NH + H_2C\overset{O}{\underset{}{\frown}}CH_2 \longrightarrow HOCH_2CH_2NEt_2$

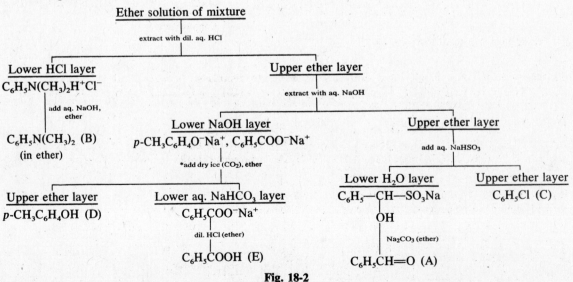

Problem 18.56 An optically active amine is subjected to exhaustive methylation and Hofmann elimination. The alkene obtained is ozonized and hydrolyzed to give an equimolar mixture of formaldehyde and butanal. What is the amine?

The alkene is 1-pentene, $CH_2{=}CHCH_2CH_2CH_3$ ($H_2C{=}\textcircled{O}\ \textcircled{O}{=}CHCH_2CH_2CH_3$). The amine is chiral; $CH_3CH(NH_2)CH_2CH_2CH_3$. The other possibility, $H_2NCH_2CH_2CH_2CH_2CH_3$, is not chiral.

Problem 18.57 Draw a flow sheet to show the separation and recovery in almost quantitative yield of a mixture of the water-insoluble compounds benzaldehyde (A), N,N-dimethylaniline (B), chlorobenzene (C), *p*-cresol (D), benzoic acid (E). ◄

See Fig. 18-2.

Ether solution of mixture
|
extract with dil. aq. HCl

Lower HCl layer   **Upper ether layer**
$C_6H_5N(CH_3)_2H^+Cl^-$ |
| extract with aq. NaOH
add aq. NaOH,
ether

$C_6H_5N(CH_3)_2$ (B)   **Lower NaOH layer**   **Upper ether layer**
(in ether) $p\text{-}CH_3C_6H_4O^-Na^+$, $C_6H_5COO^-Na^+$ |
                                                              add aq. NaHSO_3
                          *add dry ice (CO_2), ether

Upper ether layer   **Lower aq. NaHCO₃ layer**   **Lower H₂O layer**   **Upper ether layer**
$p\text{-}CH_3C_6H_4OH$ (D) $C_6H_5COO^-Na^+$ $C_6H_5—CH—SO_3Na$ C_6H_5Cl (C)
                                                            OH
                          dil. HCl (ether)   Na_2CO_3 (ether)

                          C_6H_5COOH (E) $C_6H_5CH{=}O$ (A)

Fig. 18-2

*$NaOH + CO_2$ gives $NaHCO_3$, in which carboxylic acids dissolve but phenols do not.

Problem 18.58 Synthesize the following compounds from alcohols of four C's or fewer, cyclohexanol and any needed solvents and inorganic reagents. (a) n-hexylamine, (b) triethylamine N-oxide, (c) 4-(N-methyl-amino)heptane, (d) cyclohexyldimethylamine, (e) cyclopentylamine, (f) 6-aminohexanoic acid. ◄

(a) (1) $C_2H_5OH \xrightarrow{H_2SO_4} H_2C=CH_2 \xrightarrow[\text{2. OH}^-]{\text{1. Br}_2, H_2O} H_2C\overset{\displaystyle\frown}{\underset{O}{-}}CH_2$

(2) $n\text{-}C_4H_9OH \xrightarrow{SOCl_2} n\text{-}C_4H_9Cl \xrightarrow{Mg} n\text{-}C_4H_9MgCl \xrightarrow[\text{2. } H_3O^+]{\text{1. } H_2C-CH_2 \text{ from (1)}}$

$n\text{-}C_6H_{13}OH \xrightarrow[\text{heat}]{Cu} n\text{-}C_5H_{11}CHO \xrightarrow[H_2/Pt]{NH_3} n\text{-}C_6H_{13}NH_2$

(b) $C_2H_5OH \xrightarrow{HBr} C_2H_5Br \xrightarrow{NH_3} (C_2H_5)_3N \xrightarrow{H_2O_2} (C_2H_5)_3NO$

(c) (1) $CH_3OH \xrightarrow{PBr_3} CH_3Br \xrightarrow[NH_3]{\text{excess}} CH_3NH_2$

(2) $n\text{-}BuOH \xrightarrow[\text{heat}]{Cu} n\text{-}PrCHO$

$n\text{-}PrOH \xrightarrow{SO_2Cl} n\text{-}PrCl \xrightarrow{Mg} n\text{-}PrMgCl$ $\longrightarrow (n\text{-}Pr)_2\overset{H}{\underset{}{C}}O^-(MgCl)^+ \xrightarrow{H_3O^+}$

$(n\text{-}Pr)_2CHOH \xrightarrow[H_2SO_4]{Na_2Cr_2O_7} (n\text{-}Pr)_2C=O \xrightarrow[CH_3NH_2 \text{ from (1)}]{H_2/Pt} (n\text{-}Pr)_2CHNHCH_3$

(d) (1) $CH_3OH \xrightarrow[\text{heat}]{Cu} H_2CO \xrightarrow[\text{2. } H^+]{\text{1. } Ag(NH_3)_2^+} HCOOH$

(2) Cyclohexanol $\xrightarrow[H_2SO_4]{Na_2Cr_2O_7}$ Cyclohexanone $\xrightarrow[H_2/Pt]{NH_3}$

Cyclohexylamine $\xrightarrow[\text{HCOOH from (1)}]{H_2CO \text{ from (1)}}$ Cyclohexyldimethylamine

(e) Cyclohexanol $\xrightarrow{H_2SO_4}$ Cyclohexene $\xrightarrow[H^+, \text{heat}]{KMnO_4}$

$HOOC(CH_2)_4COOH \xrightarrow[\text{heat}]{BaO}$ Cyclopentanone $\xrightarrow[H_2/Pt]{NH_3}$ Cyclopentylamine

(f) Cyclohexanol $\xrightarrow[H_2SO_4]{Na_2Cr_2O_7}$ Cyclohexanone $\xrightarrow{H_2NOH}$

Cyclohexanone oxime $\xrightarrow{H_2SO_4}$ $\xrightarrow{H_3O^+}$ 6-Aminohexanoic acid

Caprolactam

[See page 312; also Problem 18.12(c).]

Problem 18.59 Use simple, rapid, test tube reactions to distinguish between (a) $CH_3CONHC_6H_5$ and $C_6H_5CONH_2$, (b) $C_6H_5NH_3^+Cl^-$ and $p\text{-}ClC_6H_4NH_2$, (c) $(CH_3)_4N^+OH^-$ and $(CH_3)_2NCH_2OH$, (d) $p\text{-}CH_3COC_6H_4NH_2$ and $CH_3CONHC_6H_5$. ◄

(a) With hot aqueous NaOH, only $C_6H_5CONH_2$ liberates NH_3. (b) Aqueous $AgNO_3$ precipitates AgCl from $C_6H_5NH_3^+Cl^-$. (c) CrO_3 is reduced to green Cr^{3+} by $(CH_3)_2NCH_2OH$. $(CH_3)_4N^+OH^-$ is strongly basic to litmus. (d) Cold dilute HCl dissolves $p\text{-}CH_3COC_6H_4NH_2$, which also gives a positive iodoform test with NaOI.

Problem 18.60 Synthesize from benzene, toluene and any aliphatic or inorganic compounds (a) α-(p-nitrophenyl)ethylamine, (b) β-(p-bromophenyl)ethylamine. ◄

(a)

(b)

Problem 18.61 Suggest a structural formula for a compound, $C_8H_{11}N$ (A), that is optically active, dissolves in dilute aqueous HCl, and releases N_2 with HONO. ◄

The compound has four degrees of unsaturation (if it were a substituted alkane, its formula would be $C_8H_{19}N$), which indicates the presence of a phenyl ring. (A) dissolves in HCl and therefore is an amine. Since (A) releases N_2 with HONO, it is a 1° amine. Since (A) is optically active, it has a chiral C. The $—NH_2$ cannot be on the ring since the two remaining C's cannot be positioned to give a chiral C. The $—NH_2$ must be on the side chain. The compound is $C_6H_5CH(NH_2)CH_3$.

Problem 18.62 How can N-methylaniline and o-toluidine be distinguished by ir spectroscopy? ◄

o-Toluidine is a 1° amine and has a pair of peaks (symmetric and antisymmetric stretches) in the N—H stretch region. N-Methylaniline is a 2° amine and has only one peak.

Problem 18.63 Amines A, B, D and E each have their parent-ion peaks at $m/e = 59$. The most prominent peaks for each are at m/e values of 44 for A and B, 30 for D and 58 for E. Give the structure for each amine and for the ion giving rise to the most prominent peak for each. ◄

Since the parent peak is $m/e = 59$, the formula is C_3H_9N. The major fragmentation of amines is a bond to the α carbons,

$$-N-\overset{\alpha}{C}+$$

A C—C bond is weaker and breaks more easily and more often than a C—H bond. Amines A and B both lose a CH_3 ($m = 15$): $59 - 15 = 44$.
The two isomers are:

$$CH_3NHCH_2CH_3 \xrightarrow{-e^-} [CH_3\overset{\cdot}{N}HCH_2 \vdots CH_3]^+ \longrightarrow CH_3\overset{H}{\underset{|}{\overset{|+}{N}}}=CH_2 + \cdot CH_3$$
$$(A)$$

$$H_2N-\underset{\underset{CH_3}{|}}{CH}-CH_3 \xrightarrow{-e^-} \left[H_2\overset{\cdot}{N}CH \vdots CH_3\atop \underset{CH_3}{|}\right]^+ \longrightarrow H_2\overset{+}{N}=CHCH_3 + \cdot CH_3$$
$$(B)$$

Amine D loses CH_2CH_3 ($59 - 29 = 30$).

$$H_2NCH_2CH_2CH_3 \xrightarrow{-e^-} [H_2\overset{\cdot}{N}CH_2 \vdots CH_2CH_3]^+ \longrightarrow H_2\overset{+}{N}=CH_2 + \cdot CH_2CH_3$$

Amine E loses H ($59 - 1 = 58$).

$$(CH_3)_3N \xrightarrow{-e^-} \left[(CH_3)_2\overset{\cdot}{N}-\overset{H}{\underset{H}{\overset{|}{C}}} \vdots H\right]^+ \longrightarrow (CH_3)_2\overset{+}{N}=CH_2 + H\cdot$$

Problem 18.64 What compound, C_3H_7NO, has the following nmr spectrum: $\delta = 6.5$, broad singlet (2 H's); $\delta = 2.2$, quartet (2 H's); and $\delta = 1.2$, triplet (3 H's)? ◄

The integration ratio 2:2:3 accounts for the 7 H's. The peaks at $\delta = 2.2$ and $\delta = 1.2$ are from a $—CH_2CH_3$, as indicated by the splitting, and the group is attached to a C=O group, as shown by the $\delta = 2.2$ value for the H's on the CH_2. The broad singlet at $\delta = 6.5$ are H's of an amide. The compound is $CH_3CH_2CONH_2$, propanamide.

Chapter 19

Aryl Halides

19.1 INTRODUCTION

Aryl halides have a halogen atom (X) bonded directly to a C of the benzene ring, e.g. C_6H_5Cl, chlorobenzene. Compounds with X in the side chain, such as benzyl chloride ($C_6H_5CH_2Cl$), are arylalkyl halides which have properties of alkyl halides.

19.2 REACTIONS OF ARYL HALIDES

Like vinyl halides, $R_2C{=}CRX$, aryl halides do *not* undergo nucleophilic substitution under the conditions used for alkyl halides. In electrophilic substitution, X is *deactivating* and *ortho-para* directing [Problem 11.7(*c*)].

Problem 19.1 Explain why *p*-dihalobenzenes have higher melting points and lower solubilities than do their *ortho* or *meta* isomers. ◄

The *para* isomer has a higher melting point and lower solubility than its *ortho* or *meta* isomers because it is more symmetrical and therefore fits better into a crystal lattice. Consequently, the intermolecular forces are greater.

Problem 19.2 The reaction of alcoholic $AgNO_3$ with organic halides is used to determine reactivity and structure. For the following halogen compounds, indicate whether they (i) react rapidly at room temperature, (ii) react readily only when heated or (iii) are inert even to hot alcoholic silver nitrate: (*a*) bromobenzene, (*b*) *tert*-butyl bromide, (*c*) *n*-hexyl chloride, (*d*) benzyl bromide, (*e*) carbon tetrachloride, (*f*) *sec*-butyl chloride, (*g*) *n*-hexyl iodide. ◄

The more stable the R^+, the more reactive is RX. (*a*) An aryl halide; Ar^+ is very unstable (iii). (*b*) A 3° RX; stable R^+ (i). (*c*) A 1° RX; unstable R^+; can go by slower S_N2 pathway (ii). (*d*) A benzyl halide; stable R^+ (i). (*e*) $^+CCl_3$; unstable because of electron-withdrawing effect of Cl's. There is too much steric hindrance towards S_N2 attack (iii). (*f*) A 2° RX; not too stable R^+ (ii). (*g*) Although 1° like RX in part (*c*), RI is more reactive than RCl because AgI is more insoluble than AgCl (i).

Problem 19.3 How does resonance theory account for the low reactivity of aryl and vinyl halides? ◄

Their lower reactivity in nucleophilic displacements is attributed to: (1) the partial double bond character of the C—X bond, which consequently is shorter and stronger; (2) the + charge on X.

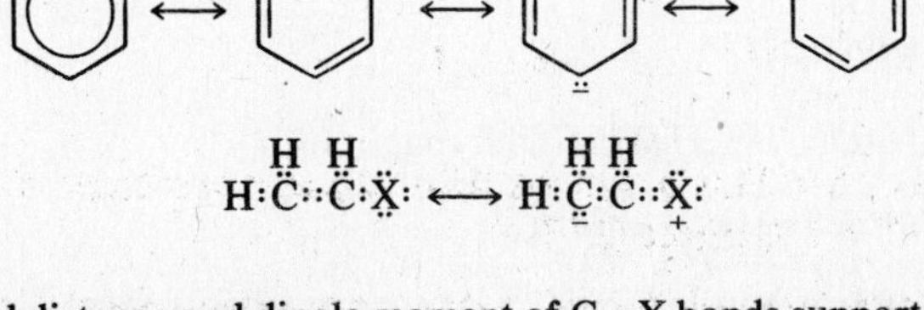

Problem 19.4 How do bond distance and dipole moment of C—X bonds support the resonance explanation of low reactivity of aryl and vinyl halides? ◄

   Double bonds are shorter and stronger than single bonds between the same atoms. The C—Cl bond distance (1.69 Å) of aryl and vinyl chlorides is less than that of most alkyl chlorides (1.8 Å).

   The C—Cl bond is polar and the difference in electronegativity between C (2.5) and Cl (3.5) causes this bond to have a dipole moment of 2.02–2.15 D in alkyl chlorides. Aryl and vinyl chlorides have smaller dipole moments of 1.7 D and 1.4 D, respectively, because of the contributing structures in which Cl has a + and C a − charge. These charges oppose and therefore diminish the polarity of the bond.

Problem 19.5 Which halogen (X) best delocalizes electron density to the benzene ring? ◄

   The smaller the X, the shorter the bond to C. Shorter bonds have more effective overlap of p orbitals (extended π bonding) and therefore more extensive delocalization. The order of effectiveness of delocalization by p-p extended π bonding is

$$F > Cl > Br > I$$

19.3 METHODS OF PREPARATION OF ARYL HALIDES

1. Direct Halogenation by Electrophilic Substitution [Problem 11.2(a)]

2. The Sandmeyer Reaction of Diazonium Salts (page 335)

3. Thallation [Problem 11.2(f)]

$$ArH + Tl(OOCCF_3)_3 \longrightarrow ArTl(OOCCF_3)_2 \xrightarrow{KX} [ArTlX_2] \xrightarrow{\Delta} ArX \quad (X = F, Br, I)$$

   To prepare ArF, $ArTlF_2$ is heated with BF_3.

Problem 19.6 Show the steps in the preparation from benzene of   (a) m-chloronitrobenzene,   (b) p-chloronitrobenzene,   (c) p-bromoaniline. ◄

(a)   benzene $\xrightarrow[H_2SO_4]{HNO_3}$ nitrobenzene $\xrightarrow[FeCl_3]{Cl_2}$ m-chloronitrobenzene

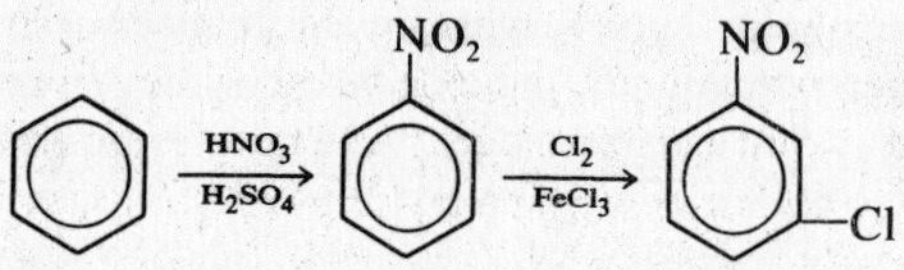

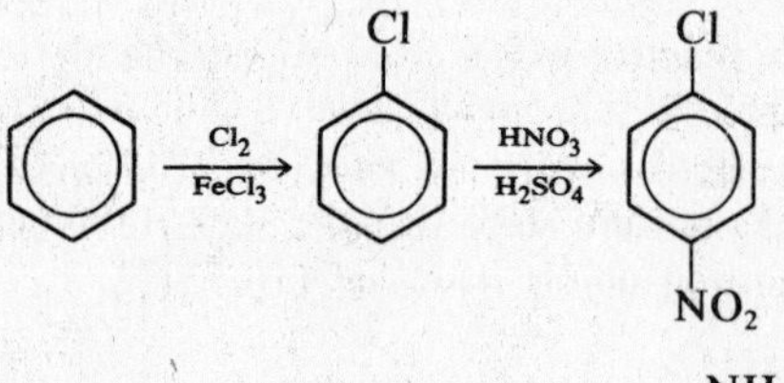

(b)   benzene $\xrightarrow[FeCl_3]{Cl_2}$ chlorobenzene $\xrightarrow[H_2SO_4]{HNO_3}$ p-chloronitrobenzene

(c) $C_6H_5NO_2 \xrightarrow[\text{2. NaOH}]{\text{1. Sn, HCl}} C_6H_5NH_2 \xrightarrow{Ac_2O} C_6H_5NHAc \xrightarrow{Br_2}$ p-bromo(NHAc) $\xrightarrow[\text{2. OH}^-]{\text{1. } H_3O^+}$ p-bromoaniline + AcOH

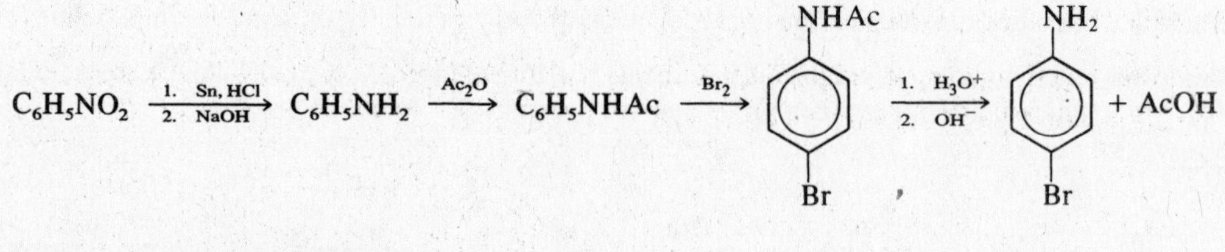

19.4 CHEMICAL PROPERTIES

AROMATIC NUCLEOPHILIC SUBSTITUTION (page 183)

1. Bimolecular Displacement Mechanism

   A good leaving group, such as halide ion (X^-), is more easily displaced than H^- from a benzene ring by nucleophiles. Electron-attracting substituents, such as NO_2 and CN, in *ortho* and *para* positions facilitate the nucleophilic displacement of X of aryl halides. The greater the number of such *ortho* and *para* substituents, the more rapid the reaction and the less vigorous the conditions needed.

$$C_6H_5Cl \xrightarrow{\text{NaOH, 300 °C}} C_6H_5OH$$

$$p\text{-}O_2NC_6H_4Cl \xrightarrow[\text{160 °C}]{\text{15% NaOH}} p\text{-}O_2NC_6H_4OH$$

$$2,4\text{-}(O_2N)_2C_6H_3Cl \xrightarrow[\text{130 °C}]{\text{Na}_2\text{CO}_3} 2,4\text{-}(O_2N)_2C_6H_3OH$$

$$2,4,6\text{-}(O_2N)_3C_6H_2Cl \xrightarrow[\text{warm}]{\text{H}_2\text{O}} 2,4,6\text{-}(O_2N)_3C_6H_2OH$$

Problem 19.7 Write resonance structures to account for activation in bimolecular aromatic nucleophilic substitution from delocalization of the charge of the intermediate carbanion by the following *para* substituent groups: (*a*) —NO₂, (*b*) —CN, (*c*) —N=O, (*d*) CH=O.   ◄

Only the resonance structures with the negative charge on the *para* C are written to show delocalization of charge from ring C to the *para* substituent.

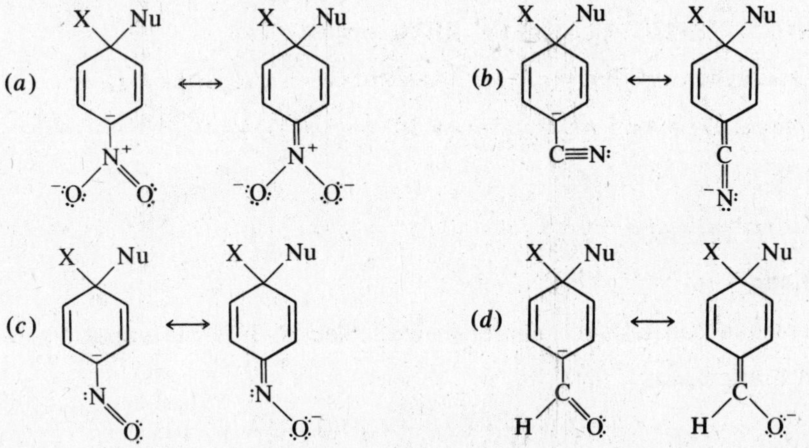

Problem 19.8 Compare bimolecular aromatic nucleophilic and electrophilic substitution reactions with aliphatic S_N2 reactions in terms of (*a*) number of steps and transition states, (*b*) character of intermediates.   ◄

(*a*) Nucleophilic and electrophilic bimolecular aromatic substitutions are *two-step* reactions, having a first slow and rate-determining step followed by a rapid second step. Aliphatic S_N2 reactions have only *one* step. There are 2 transition states for the *aromatic* and 1 for the *aliphatic* substitution. (*b*) S_N2 reactions have no intermediate. In *electrophilic* aromatic substitution the intermediate is a *carbonium ion* (carbocation), while that in *nucleophilic* substitution is an *anion*.

2. Elimination-Addition Reactions

With very strong bases, such as amide ion, NH_2^-, unactivated aryl halides undergo substitution by an elimination-addition (**benzyne**) mechanism.

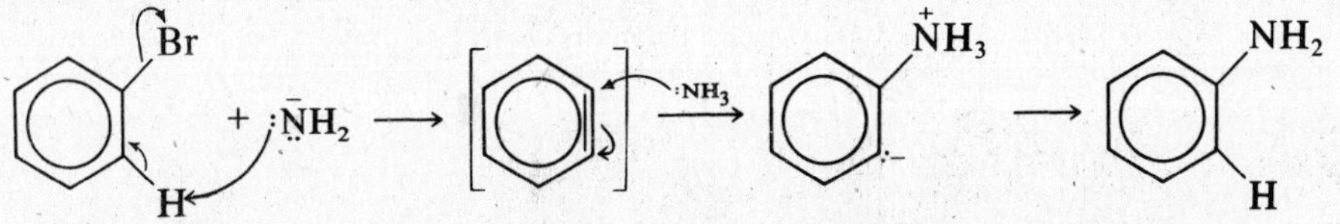

Problem 19.9 How do the following observations support the benzyne mechanism? (*a*) Compounds lacking *ortho* H's, such as 2,6-dimethylchlorobenzene, do not react. (*b*) 2,6-Dideuterobromobenzene reacts more slowly than bromobenzene. (*c*) *o*-Bromoanisole, *o*-CH₃OC₆H₄Br, reacts with NaNH₂/NH₃ to form *m*-anisidine. (*d*) Chlorobenzene with Cl bonded to ¹⁴C gives almost 50% aniline having NH₂ bonded to ¹⁴C and 50% aniline with NH₂ bonded to an *ortho* C.   ◄

(a) With no H *ortho* to Cl, vicinal elimination cannot occur.

(b) This primary isotope effect [Problem 7.29(c)] indicates that a bond to H is broken in the rate-determining step, which is consistent with the first step in the benzyne mechanism being rate-determining.

(c) NH_2^- need not attack the C^2 from which the Br^- left; it can add at C^3.

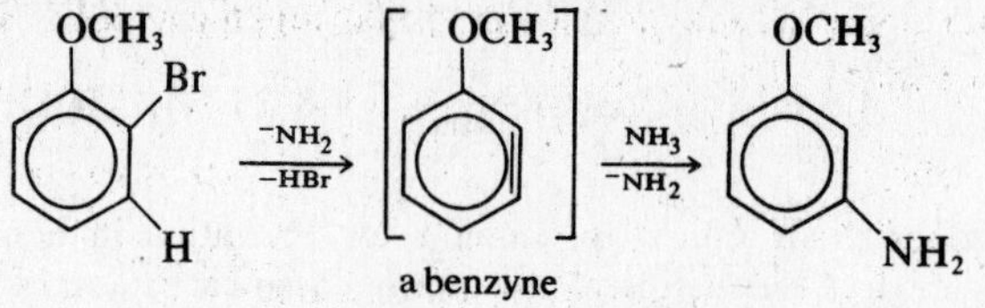

a benzyne

(d)

(50%)

(50%)

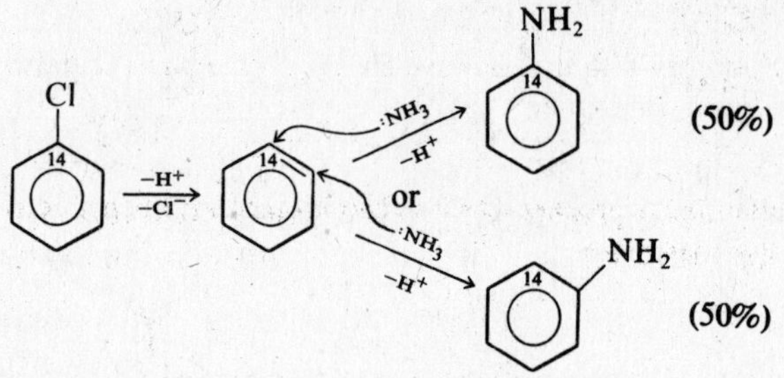

REACTIONS WITH METALS

1. Grignard Reagents

ArX is less reactive than RX in forming Grignards. Tetrahydrofuran is used as a solvent for ArCl.

2. Aryllithium Compounds

$$C_6H_5Cl + 2Li \longrightarrow LiCl + \underbrace{C_6H_5{:}Li \longleftrightarrow C_6H_5{:}^- Li^+}_{\text{Phenyllithium}}$$

ArLi is more reactive than ArMgX because of the greater percentage of ionic character of the C—metal bond.

3. The Ullmann Reaction with Copper

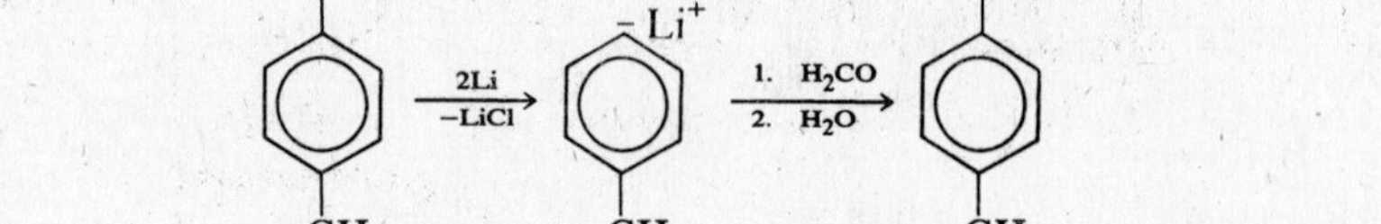

4,4'-Dimethylbiphenyl

Problem 19.10 Use inorganic reagents and any organic compound with one C to prepare (a) p-methylbenzyl alcohol from p-chlorotoluene, (b) p-dichlorobenzene from p-chloronitrobenzene. ◄

(a)

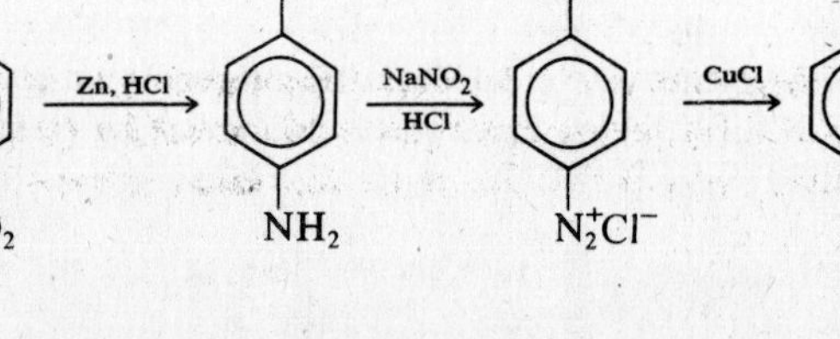

(b)

Supplementary Problems

Problem 19.11 Which Cl in 1,2,4-trichlorobenzene reacts with $^-OCH_2COO^-$ to form the herbicide "2,4-D"? Give the structure of "2,4-D." ◄

Cl's are electron-withdrawing and activate the ring to nucleophilic attack. The Cl at C^1 is displaced because it is *ortho* and *para* to the other Cl's.

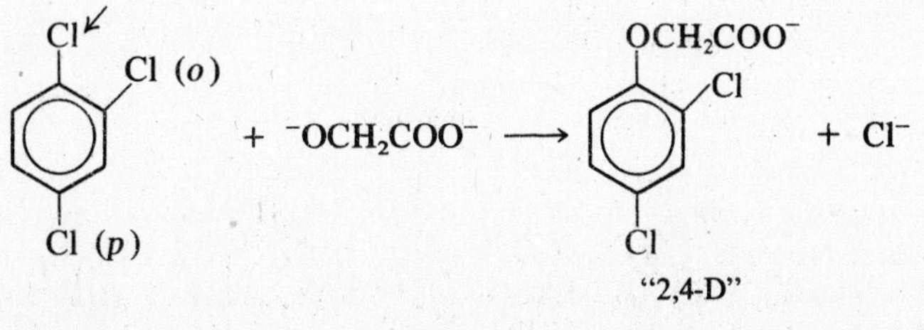

Problem 19.12 Use thallation to prepare (a) *m*-bromoisopropyl benzene from $C_6H_5CHMe_2$, (b) methyl *o*-iodobenzoate from methyl benzoate. ◄

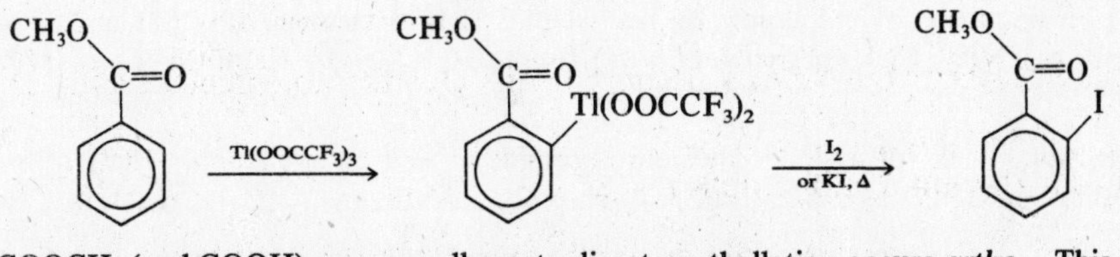

Although $COOCH_3$ (and COOH) are normally *meta*-directors, thallation occurs *ortho*. This is attributed to prior complexation of $^+Tl(OOCCF_3)_2$ with the O of the C=O group.

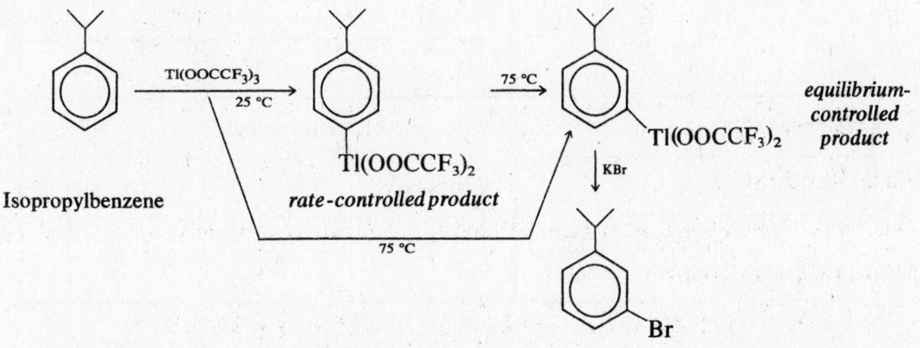

Problem 19.13 Dialkylamines are prepared from *p*-nitroso-N,N-dialkylamines and aqueous KOH. Outline the preparation of $(n$-$C_4H_9)_2NH$ from n-C_4H_9Br and $C_6H_5NH_2$ and classify the nitroso compound reaction with KOH. ◄

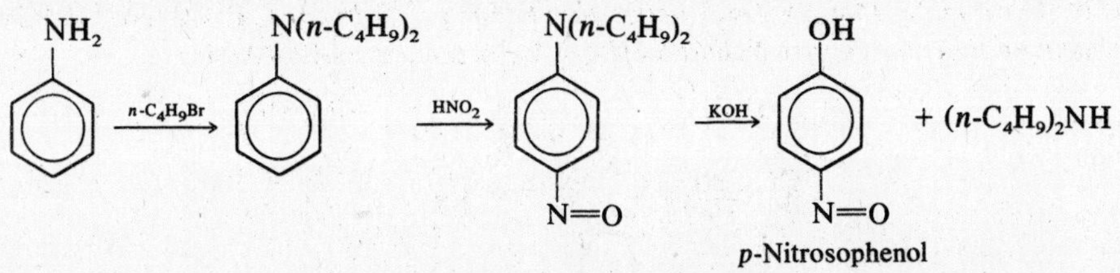

This is an aromatic nucleophilic bimolecular substitution in which NO is an electron-attracting substituent that stabilizes the carbanion by delocalizing the negative charge [see Problem 19.7(c)].

Problem 19.14 Explain these observations: (a) *p*-Nitrobenzenesulfonic acid is formed from the reaction of *p*-nitrochlorobenzene with $NaHSO_3$, but benzenesulfonic acid cannot be formed from chlorobenzene by this reaction. (b) 2,4,6-Trinitroanisole with $NaOC_2H_5$ gives the same product as 2,4,6-trinitrophenetole with $NaOCH_3$. ◄

(a) Nucleophilic aromatic substitution occurs with p-nitrochlorobenzene, but not with chlorobenzene, because NO_2 stabilizes the carbanion [Problem 19.7(a)].

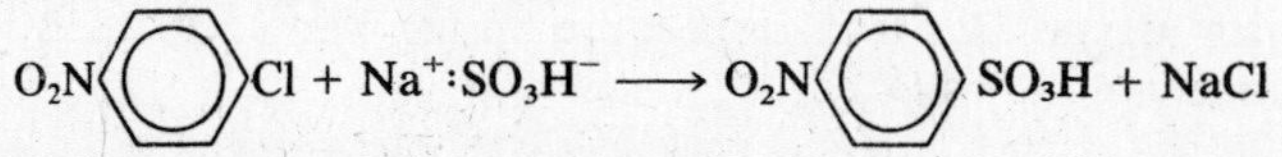

$$O_2N\text{—}\bigcirc\text{—}Cl + Na^+{:}SO_3H^- \longrightarrow O_2N\text{—}\bigcirc\text{—}SO_3H + NaCl$$

(b) The product is a sodium salt formed by addition of alkoxide.

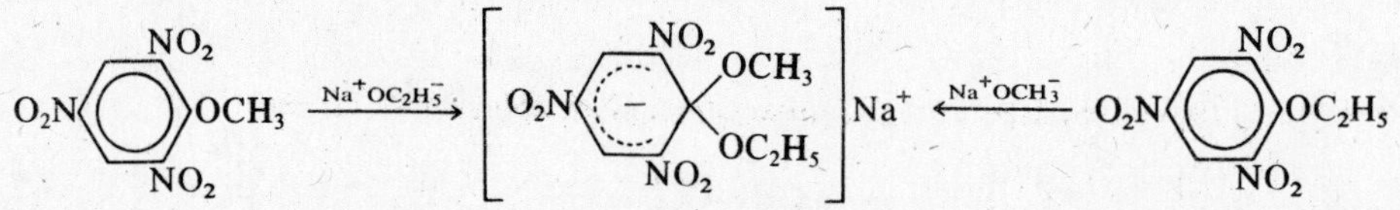

Problem 19.15 Assign numbers from 1 for LEAST to 3 for MOST to show the relative reactivities of the compounds with the indicated reagents: (a) $C_6H_5CH_2CH_2Br$ (I), $C_6H_5CHBrCH_3$ (II) and $C_6H_5CH{=}CHBr$ (III) with alcoholic $AgNO_3$; (b) CH_3CH_2Cl (I), $C_6H_5CH_2Cl$ (II) and C_6H_5Cl (III) with KCN; (c) m-nitro-chlorobenzene (I), 2,4-dinitrochlorobenzene (II) and p-nitrochlorobenzene (III) with sodium methoxide. ◄

See Table 19-1.

Table 19-1

	I	II	III
(a) An S_N1 reaction	2 (primary)	3 (benzylic)	1 (vinylic)
(b) An S_N2 reaction	2 (primary)	3 (benzylic)	1 (aryl)
(c) An aromatic nucleophilic displacement	1 (*meta*)	3 (*o* and *p*)	2 (*para*)

Problem 19.16 Name the products, if any, formed when C_6H_5Br is treated with (a) hot 5% aqueous NaOH, (b) Mg in ether, (c) NH_3 and copper powder, (d) fuming H_2SO_4, (e) Cl_2 and $FeCl_3$, (f) C_6H_6 and $AlCl_3$, (g) Cu at 200 °C. ◄

(a) no reaction, (b) phenyl magnesium bromide, (c) no reaction, (d) *ortho*- and *para*-bromobenzenesulfonic acid, (e) *ortho*- and *para*-chlorobromobenzene, (f) no reaction, (g) biphenyl.

Problem 19.17 Predict the reaction product from C_6D_5Br and $NaNH_2/NH_3$. ◄

2,3,4,5-Tetradeuteroaniline. A D *ortho* to the Br is replaced by an H from NH_3.

Problem 19.18 After o-DC_6H_4F is reacted for a short time with $NaNH_2/NH_3$, the recovered starting material contains some C_6H_5F. Under the same conditions o-DC_6H_4Br leaves no C_6H_5Br. Explain. ◄

F^- is a much poorer leaving group than is Br^-.

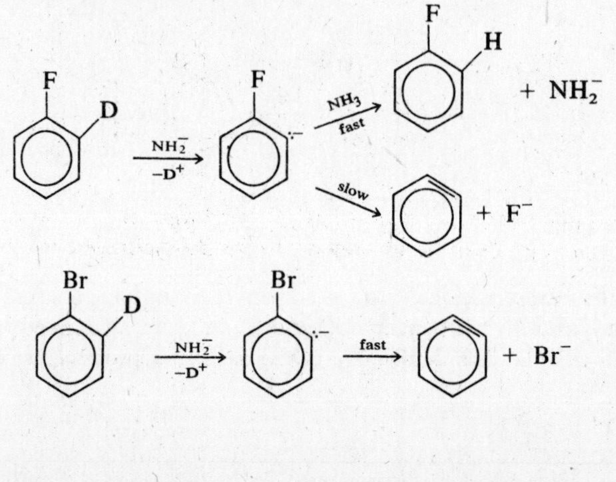

Problem 19.19 Identify the compounds (A) through (D):

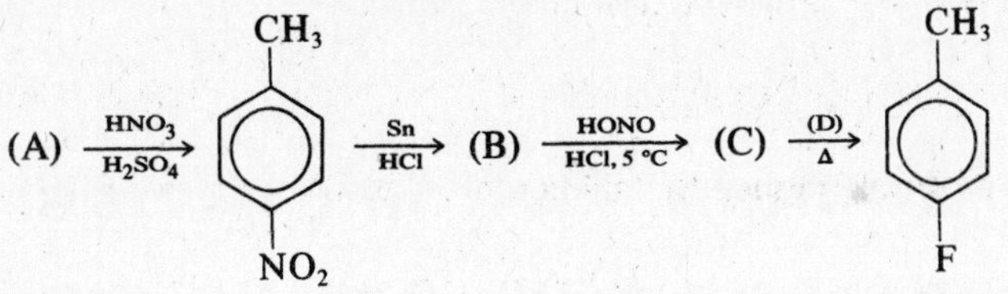

(A) toluene, (B) *p*-toluidine, (C) *p*-toluenediazonium chloride, (D) HBF_4.

Problem 19.20 Deduce the structures of aromatic compounds C_7H_7Cl whose oxidation produces (*a*) an aromatic compound with no halogen, (*b*) an aryl halide.

(*a*) Cl^- must be on the side chain; benzyl chloride. (*b*) Cl must be on the ring; a chlorotoluene.

Problem 19.21 Outline practical laboratory syntheses from benzene or toluene and any needed inorganic reagents of: (*a*) *p*-chlorobenzal chloride, (*b*) 2,4-dinitroaniline, (*c*) *m*-chlorobenzotrichloride, (*d*) 2,5-dibromonitrobenzene.

(*a*)

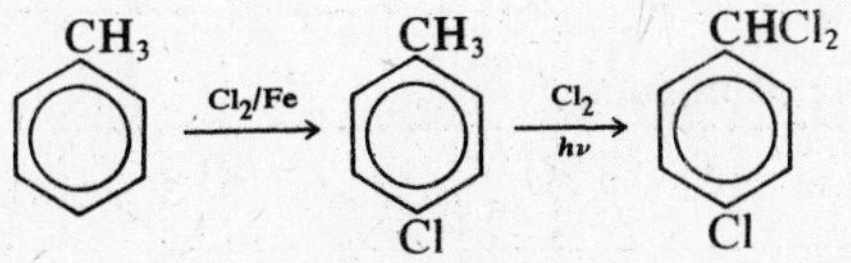

(*b*)

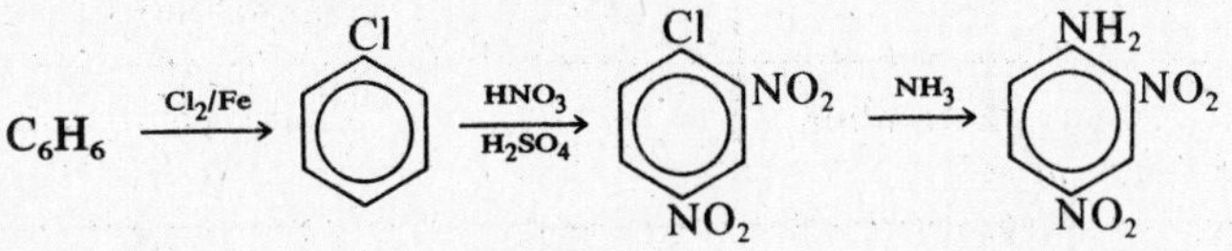

(*c*)

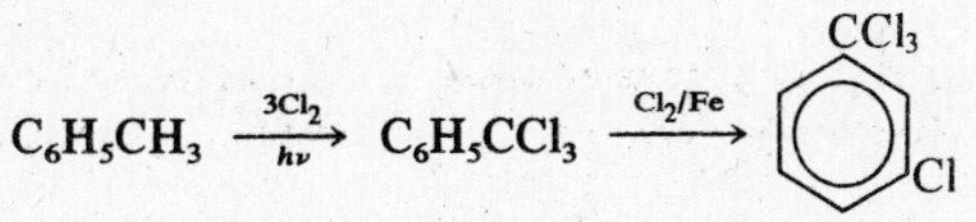

(*d*)

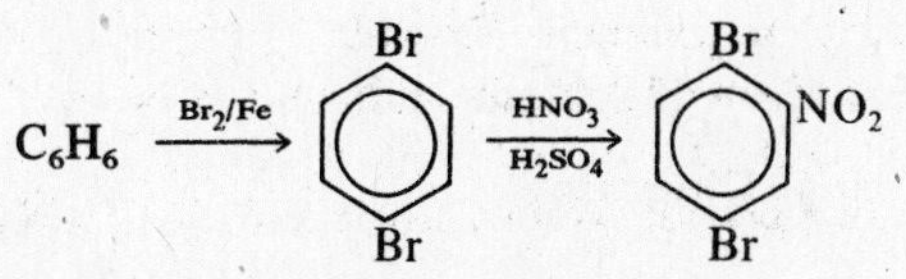

Problem 19.22 Explain the following observations. (*a*) Reaction of *p*-bromotoluene with NaOH at high temperature gives equal amounts of *para* and *meta* cresols. (*b*) A hydrocarbon, $C_{25}H_{20}$, is formed when one mole each of C_6H_5Cl and $(C_6H_5)_3C^-K^+$ are reacted with $K^+NH_2^-$ in NH_3. (*c*) 2,4-Dinitrochlorobenzene is prepared by nitration of C_6H_5Cl but is not isolated if the reaction product is washed with aqueous $NaHCO_3$ to remove the acid.

(*a*)

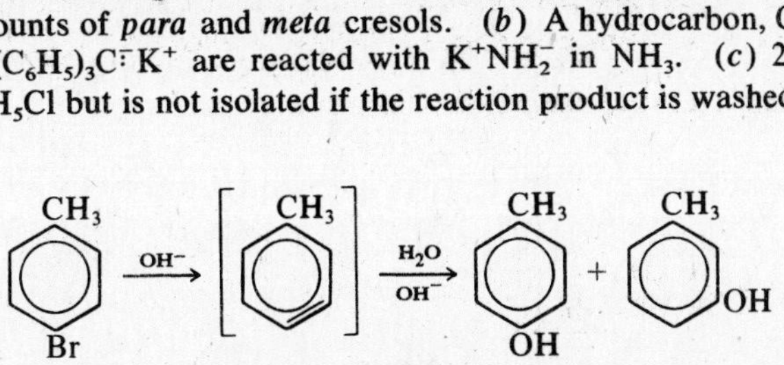

(*b*) Benzyne formed from chlorobenzene adds $(C_6H_5)_3C^-$ and then H^+ to form $C(C_6H_5)_4$.
(*c*) Nucleophilic displacement of Cl in basic solution forms 2,4-dinitrophenol.

Chapter 20

Aromatic Sulfonic Acids; Organosulfur Compounds

20.1 INTRODUCTION

Aromatic sulfonic acids have the —SO_2OH group bonded to an aromatic ring (Ar—SO_2OH or Ar—SO_3H). They differ from organic esters of sulfuric acid, like phenyl hydrogen sulfate (C_6H_5—OSO_3H), in which a ring C is bonded to an O.

To name ArSO_3H, the suffix **-sulfonic acid** is added to the name of the rest of the compound, along with numbers or the letters o, p and m as prefixes to locate ring substituents. SO_3H is also called the **sulfo** group.

Ar—SO_3H is soluble in H_2O.

Problem 20.1 Name the following compounds:

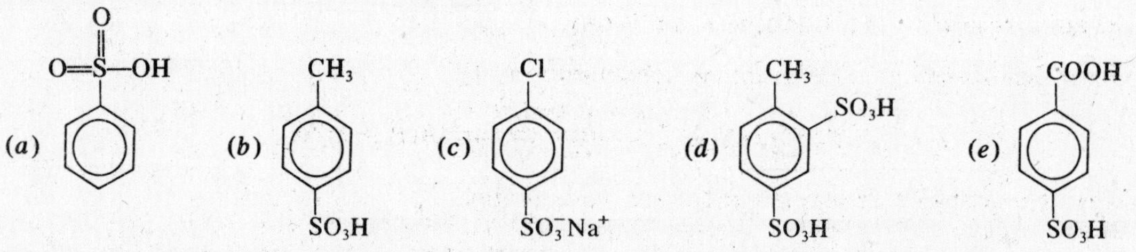

(a) benzenesulfonic acid, *(b)* p-toluenesulfonic acid, *(c)* sodium p-chlorobenzenesulfonate, *(d)* 1-methyl-2,4-benzenedisulfonic acid, *(e)* p-sulfobenzoic acid (COOH has priority over SO_3H in naming).

20.2 PREPARATION

1. Direct Electrophilic Sulfonation [Problem 11.2(c)]

2. Direct Formation of Sulfonyl Chlorides

$$ArH + ClSO_3H \longrightarrow ArSO_2Cl + H_2O$$

Chlorosulfonic   sulfonyl
acid   chloride

Problem 20.2 Sulfonic acids are recovered from the sulfonation mixture (H_2SO_4) as their sodium salts, by addition of $CaCO_3$ and then Na_2CO_3. Use a flow sheet to illustrate steps. ◄

$$\begin{matrix} H_2SO_4 \\ ArSO_3H \end{matrix} \Bigg\} \xrightarrow{CaCO_3} \begin{matrix} CaSO_4\downarrow & (filter) \\ (ArSO_3)_2Ca & (solution) \end{matrix} \xrightarrow{Na_2CO_3} \begin{matrix} CaCO_3\downarrow & (filter) \\ ArSO_3Na & (evaporate) \end{matrix}$$

3. Oxidation of Sulfur Compounds

$$C_6H_5SH \xrightarrow{KMnO_4} C_6H_5SO_3H$$
Thiophenol

$$2\,p\text{-}ClC_6H_4NO_2 \xrightarrow{Na_2S_2} p\text{-}O_2NC_6H_4SSC_6H_4NO_2\text{-}p \xrightarrow{HNO_3} 2\,p\text{-}O_2NC_6H_4SO_3H$$
a disulfide

4. Aromatic Nucleophilic Substitution

$$p\text{-}O_2NC_6H_4Cl \xrightarrow{\text{NaHSO}_3} p\text{-}O_2NC_6H_4SO_3H \quad [\text{see Problem } 19.14(a)]$$

20.3 CHEMICAL PROPERTIES

REACTIONS OF THE BENZENE RING H's

On electrophilic substitution, the strongly electron-attracting —SO$_3$H is *deactivating* and *meta-directing*.

REACTIONS OF SO$_3$H GROUP

1. Strong Acid (used as organic acid catalysts)

2. Reactions of the Hydroxyl Group (ArSO$_2$—OH)

Esters and amides of sulfonic acids (page 357) are difficult to prepare directly from sulfonic acids and are formed from the sulfonyl chlorides.

$$\left.\begin{array}{ll}\text{Acid} & \text{ArSO}_2\text{OH} \\ \text{Salt} & \text{ArSO}_2\text{ONa}\end{array}\right\} + \text{PCl}_5 \longrightarrow \text{ArSO}_2\text{Cl} + \text{POCl}_3 + \left\{\begin{array}{l}\text{HCl} \\ \text{NaCl}\end{array}\right.$$

3. Displacement of the Sulfonic Acid Group

(*a*) **Desulfonation.** Electrophilic substitution by H$^+$.

$$\text{ArSO}_3\text{H} + \text{H}_2\text{O} \underset{\text{fum. H}_2\text{SO}_4}{\overset{\text{50\% aq. H}_2\text{SO}_4, 150\,°C}{\rightleftharpoons}} \text{ArH} + \text{H}_2\text{SO}_4$$

(*b*) Nucleophilic displacements of sulfonate group.

$$\text{ArSO}_3\text{Na} + \text{Na}^+\text{Nu}^- \xrightarrow{\text{heat}} \text{ArNu} + \text{Na}_2\text{SO}_3 \quad (\text{Nu}^- = \text{OH}^-, \text{CN}^-)$$

Problem 20.3 Use the Principle of Microscopic Reversibility to outline a mechanism for desulfonation. ◄

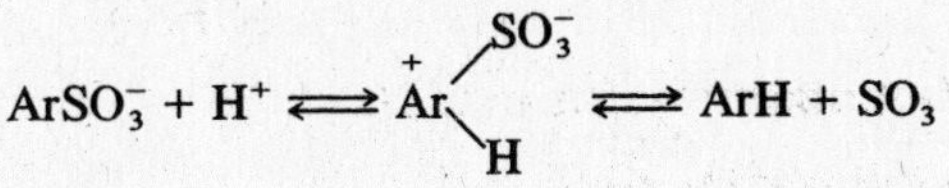

$$\text{ArSO}_3^- + \text{H}^+ \rightleftharpoons \overset{+}{\text{Ar}}\underset{\text{H}}{\overset{\text{SO}_3^-}{\diagdown}} \rightleftharpoons \text{ArH} + \text{SO}_3$$

Problem 20.4 Explain the following reactions:

(*a*) 1,4-Dimethyl-2-benzenesulfonic acid + 2Br$_2$ + H$_2$O ⟶ 1,4-Dimethyl-2,5-dibromobenzene

(*b*) Sulfanilic acid (*p*-Aminobenzenesulfonic acid) + 3Br$_2$ ⟶ 2,4,6-Tribromoaniline

(*c*) 4-Hydroxy-1,3-benzenedisulfonic acid + HNO$_3$ ⟶ 2,4,6-Trinitrophenol ◄

All are electrophilic displacements of SO$_3$H as SO$_3$ and H$^+$ from compounds having an activating group.

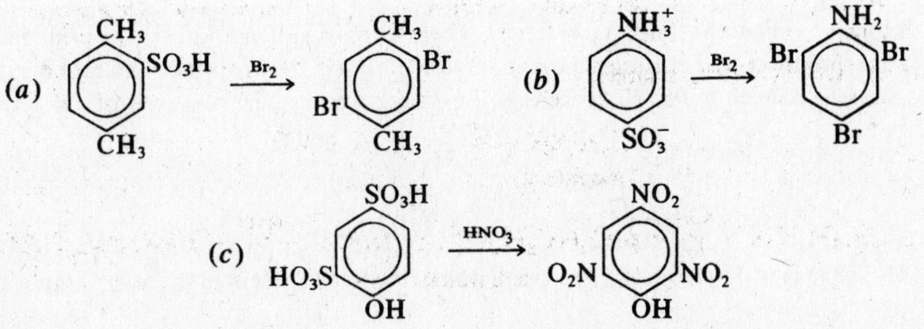

Problem 20.5   Sulfonation resembles nitration and halogenation in being an electrophilic substitution, but differs in being *reversible* and in having a *moderate primary kinetic* isotope effect.   Illustrate with diagrams of enthalpy (H) versus reaction coordinate.   ◄

In nitration (and other irreversible electrophilic substitutions) the transition state (TS) for the reaction wherein

$$\left[Ar \underset{NO_2}{\overset{H}{<}} \right]^+$$

loses H^+ has a considerably smaller $\Delta H^{\ddagger}$ than does the TS for the reaction in which NO_2^+ is lost.   In sulfonation the $\Delta H^{\ddagger}$ for loss of SO_3 from

$$\left[\overset{+}{Ar} \underset{SO_3^-}{\overset{H}{<}} \right]$$

is only slightly more than that for loss of H^+.

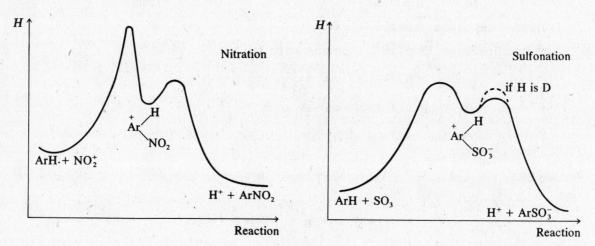

Fig. 20-1

In terms of the specific rate constants

$$ArH + SO_3 \underset{k_{-1}}{\overset{k_1}{\rightleftharpoons}} \overset{+}{Ar} \underset{SO_3^-}{\overset{H}{<}} \overset{k_2}{\longrightarrow} ArSO_3^- + H^+$$

Step (1)   Step (2)

k_2 is about equal to k_{-1}.   (For nitration, $k_2 \gg k_{-1}$.)   Therefore, in sulfonation the intermediate can go almost equally well in either direction, and sulfonation is reversible.   Furthermore, since the rate of Step (2) affects the overall rate, the substitution of D for H decreases the rate because $\Delta H^{\ddagger}$ for loss of D^+ from

$$\overset{+}{Ar} \underset{SO_3^-}{\overset{D}{<}}$$

is greater than $\Delta H^{\ddagger}$ for loss of H^+ from the protonated intermediate.   Hence, there is a modest primary isotope effect.

20.4 DERIVATIVES OF AROMATIC SULFONIC ACIDS

Sulfonyl chlorides (pages 331, 354, 355) react like acyl chlorides but are less reactive.   Sulfonate esters ($ArSO_2OR$ or $ArSO_2OAr$) are formed from alcohols and phenols, while NH_3 and 1° and 2° amines yield amides (Hinsberg reaction, Problem 18.20). OH^- or 3° amines are used to neutralize H^+.

Reduction of sulfonyl chlorides with Zn and acid yields first sulfinic acids and then thiophenols.

$$C_6H_5SO_2Cl \xrightarrow{\text{Zn, HCl}} \underset{\substack{\text{Benzenesulfinic} \\ \text{acid}}}{C_6H_5SO_2H} \xrightarrow{\text{Zn, HCl}} C_6H_5SH$$

Like RX, RSO_2OR' and $ArSO_2OR'$ are good substrates in S_N2 and S_N1 reactions.

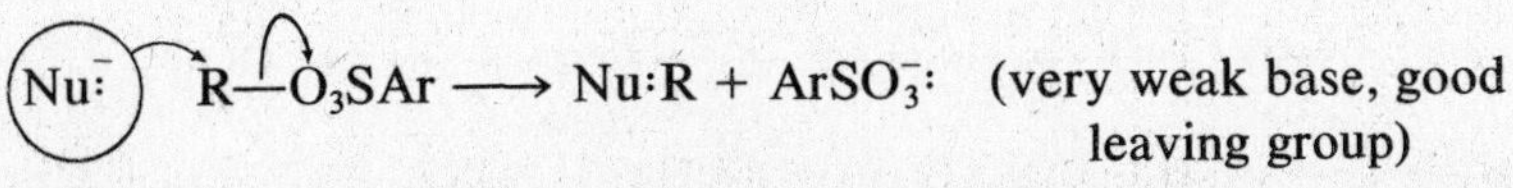

$$\text{Nu:}^- \quad R\!-\!O_3SAr \longrightarrow Nu\!:\!R + ArSO_3^- \quad \text{(very weak base, good leaving group)}$$

Sulfonate esters can be prepared from optically active alcohols without inversion of configuration of the chiral carbinol C.   The reason is that reaction involves cleavage of the H—O bond of the alcohol.

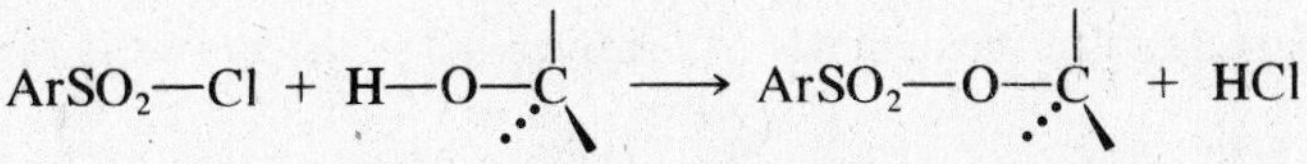

$$\underset{}{ArSO_2\!-\!Cl} + \underset{(R)}{H\!-\!O\!-\!\overset{|}{\underset{}{C}}} \longrightarrow \underset{(R)}{ArSO_2\!-\!O\!-\!\overset{|}{\underset{}{C}}} + HCl$$

20.5 COMPARISON OF SULFONIC AND CARBOXYLIC ACID CHEMISTRY

	SULFONIC	**CARBOXYLIC**
Acids	$ArSO_3H$	RCO_2H
1. Acid strength	Strong	Intermediate
2. Formation of derivatives	Indirect ($ArSO_2Cl$)	Direct
Esters	$ArSO_2OR$	$R'COOR$
1. Preparation	From $ArSO_2Cl$	From $R'COOH$ or $R'COCl$
2. Hydrolysis with $H_2{}^{18}O$	$ArSO_3H + R^{18}OH$ Cleavage of alkyl—oxygen bond	$R'CO^{18}OH + ROH$ Cleavage of acyl— oxygen bond
3. Reaction with nucleophiles	At alkyl C with inversion	At acyl C with retention or occasionally racemization of R. Intermediate is sp^3 and has an octet
Acyl Chlorides	$ArSO_2Cl$	$RCOCl$
Formation of acids, esters and amides	Slow Requires base	Rapid with water ($ArCOCl$ requires base)
Amides	$ArSO_2NH_2$	$RCONH_2$
1. Hydrolysis	Only by acids; slow	By acids or bases; rapid
2. Formation from acyl halides	Slow	Rapid
3. Acidity of H on N	Forms salts with OH^-	No salt formation

Problem 20.6 Give the product formed when $PhSO_2Cl$ is treated with (*a*) phenol, (*b*) aniline, (*c*) water, (*d*) excess Zn and HCl. ◀

(*a*) phenyl benzenesulfonate, $PhSO_2OPh$; (*b*) N-phenylbenzenesulfonamide, $PhSO_2NHPh$; (*c*) benzenesulfonic acid, $PhSO_2OH$, (*d*) thiophenol, C_6H_5SH.

Problem 20.7 Prepare (*a*) tosylamide (Ts = tosyl = *p*-$CH_3C_6H_4SO_2$—) from toluene, (*b*) PhCOOH from $PhSO_2Cl$, (*c*) *o*-methylthiophenol from $PhCH_3$ ◀

(*a*) $C_6H_5CH_3 \xrightarrow{H_2SO_4 \text{ (or } ClSO_3H)} p\text{-}CH_3C_6H_4SO_3H \xrightarrow{PCl_5} p\text{-}CH_3C_6H_4SO_2Cl \xrightarrow{NH_3} p\text{-}CH_3C_6H_4SO_2NH_2$

(*b*) $PhSO_2Cl \xrightarrow{NaOH} PhSO_3Na \xrightarrow[\text{fuse}]{NaCN} PhCN \xrightarrow{H_3O^+} PhCOOH$

(*c*) $PhCH_3 \xrightarrow[0\,^\circ C]{H_2S_2O_7} o\text{-}CH_3C_6H_4SO_3H \xrightarrow{PCl_5} o\text{-}CH_3C_6H_4SO_2Cl \xrightarrow{Zn,\ HCl} o\text{-}CH_3C_6H_4SH$

rate-controlled product

Problem 20.8 Write structures for compounds (A) through (C) in the synthesis of saccharin. ◀

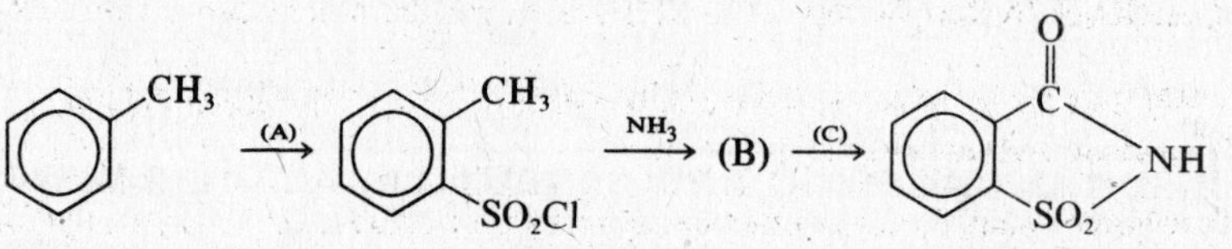

Saccharin

(A) $ClSO_3H$ (B) 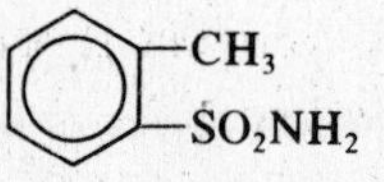 (C) $KMnO_4$

Problem 20.9 Write structural formulas for the organic compounds (I) through (XI) and show the stereochemistry of (X) and (XI). (Write Ts for the tosyl group.)

(*a*) $ArSO_2Cl + CH_3OH \longrightarrow (I) \xrightarrow{H_3{}^{18}O^+} (II) + (III)$

(*b*) 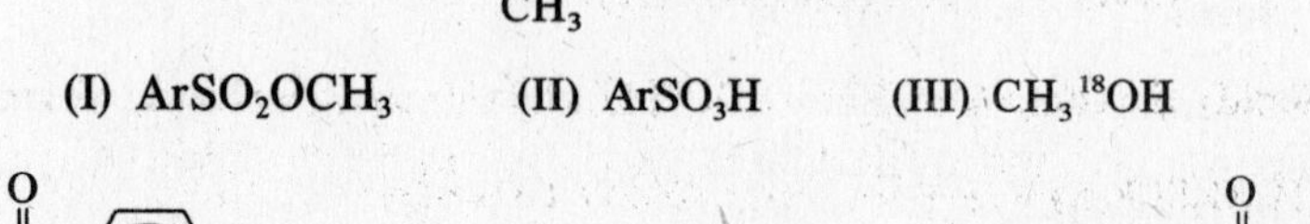

(*c*) $C_6H_5\text{—}SO_2Cl + n\text{-}C_4H_9OH \longrightarrow (VII) \xrightarrow{PhCH_2MgBr} (VIII) + (IX)$

(*d*) TsCl + (S)-HO—C(C_6H_{13})(CH_3)—H ⟶ (X) $\xrightarrow{CH_3COO^-}$ (XI) ◀

(*a*) (I) $ArSO_2OCH_3$ (II) $ArSO_3H$ (III) $CH_3{}^{18}OH$

(*b*) (IV) 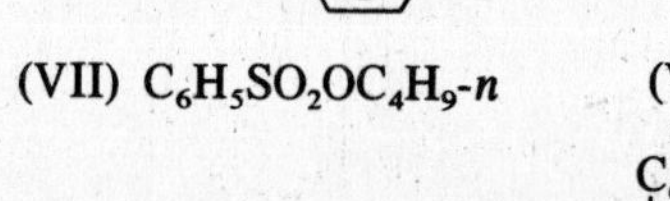 (V) $CH_3\overset{+}{N}H_3X^-$ (VI) 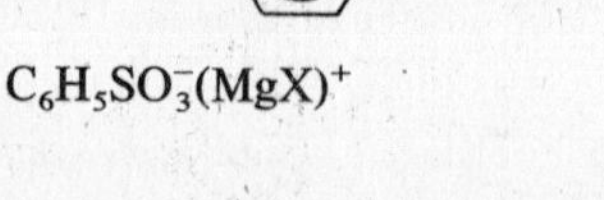

(*c*) (VII) $C_6H_5SO_2OC_4H_9$-*n* (VIII) $PhCH_2\text{—}C_4H_9$-*n* (IX) $C_6H_5SO_3^-(MgX)^+$

(*d*) (X) (S)-TsO—C(C_6H_{13})(CH_3)—H (XI) (R)-H—C(C_6H_{13})(CH_3)—OC(=O)CH_3

Problem 20.10 (a) (S)-sec-C_4H_9OH, which has an optical rotation of $+13.8°$, is reacted with tosyl chloride and the product is saponified. (b) Another sample of this alcohol is treated with benzoyl chloride and the product is also hydrolyzed with base. What is the rotation of the sec-C_4H_9OH from each reaction? ◄

(a)

$$TsCl + HO-\overset{\overset{\displaystyle CH_3}{|}}{\underset{\underset{\displaystyle H}{|}}{C}}-C_2H_5 \longrightarrow TsO-\overset{\overset{\displaystyle CH_3}{|}}{\underset{\underset{\displaystyle H}{|}}{C}}\overset{\longleftarrow}{-}C_2H_5 \xrightarrow[\text{inversion}]{OH^-} C_2H_5-\overset{\overset{\displaystyle CH_3}{|}}{\underset{\underset{\displaystyle H}{|}}{C}}-OH$$

$$\text{(S) (+13.8°)} \qquad\qquad \text{(S) (Tosyl ester)} \qquad\qquad \text{(R) (−13.8°)}$$

(b) $+13.8°$. Reaction with benzoyl chloride causes no change in configuration about the chiral C of the alcohol. Hydrolysis of PhCOOR occurs by attack at the carbonyl group, with retention of alcohol configuration.

Problem 20.11 Reaction of the (R)-tosyl ester of α-phenylethyl alcohol ($C_6H_5CHMeOTs$) with CH_3COOH yields a mixture of (S)- and rac-acetate. Explain. ◄

Inversion and racemization products show operation of an S_N1 reaction. The S_N1 pathway is possible because the alcohol forms the stable $2°$ benzylic $C_6H_5\overset{+}{C}HCH_3$.

Problem 20.12 What reactions prove that $ArSO_3H$ has a C—S rather than a C—O bond? ◄

Reduction of $ArSO_3H$ to thiophenols (Ar—SH) which can be oxidized back to $ArSO_3H$.

Problem 20.13 What is the product of monosulfonation of (a) $PhCH_3$ at 100 °C, (b) $PhNO_2$, (c) p-nitrophenol, (d) m-dimethylbenzene (m-xylene)? ◄

(a) p-toluenesulfonic acid, (b) m-nitrobenzenesulfonic acid, (c) 2-hydroxy-5-nitrobenzenesulfonic acid, (d) 2,4-dimethylbenzenesulfonic acid.

Problem 20.14 Give the reaction product of $PhSO_3H$ and (a) aqueous NaOH, (b) fusion with NaOH, (c) $SOCl_2$, (d) CH_3OH, (e) $HNO_3 + H_2SO_4$. ◄

(a) $PhSO_3^-Na^+$, (b) PhO^-Na^+, (c) $PhSO_2Cl$, (d) no reaction, (e) m-nitrobenzenesulfonic acid.

Problem 20.15 Name (a) $C_6H_5SO_2OCH_3$, (b) $C_6H_5SO_2NH_2$, (c) p-$BrC_6H_4SO_2Cl$. ◄

(a) methyl benzenesulfonate, (b) benzenesulfonamide, (c) p-bromobenzenesulfonyl chloride (brosyl chloride).

Problem 20.16 Prepare o-bromotoluene from toluene without using direct bromination (which gives a large amount of the unwanted p-isomer). ◄

Use —SO_3H as a blocking group and then remove it by desulfonation.

Toluene   p-Toluenesulfonic acid

Problem 20.17 Identify the compound $C_9H_{11}O_2SCl$ (A), and also compounds (B), (C), and (D), in the reactions:

$$AgCl(s) \xleftarrow[\text{rapid}]{AgNO_3} C_9H_{11}O_2SCl \text{ (A)} \xrightarrow[\text{2. } H_3O^+]{\text{1. NaOH}} \text{water soluble (B)} \xrightarrow[\Delta]{H_3O^+}$$

$$C_9H_{12} \text{ (C)} \xrightarrow[\text{Fe}]{Br_2} \text{one monobromo derivative (D)}$$

◄

(D) 1-Bromo-2,4,6-trimethylbenzene   (C) Mesitylene   (B) 2,4,6-Trimethyl-benzenesulfonic acid   (A) 2,4,6-Trimethyl-benzenesulfonyl chloride

Rapid reaction of (A) with $AgNO_3$ indicates an active Cl and possibly a SO_2Cl group. Water soluble compound (B) can be the corresponding SO_3H, which is desulfonated to form (C). The only isomer of (C) with one monobromo derivative is mesitylene (1,3,5-trimethylbenzene).

Problem 20.18 Prepare (a) p-ethylthiophenol from benzene, (b) o-methylanisole from toluene. ◄

(a) $$PhH \xrightarrow{\text{EtCl}}_{\text{AlCl}_3} PhEt \xrightarrow{\text{ClSO}_3\text{H}} p\text{-EtC}_6\text{H}_4\text{SO}_2\text{Cl} \xrightarrow{\text{Zn, HCl}} p\text{-EtC}_6\text{H}_4\text{SH}$$

(b) $$PhMe \xrightarrow[\text{2. NaOH}]{\text{1. } H_2S_2O_7,\, 0\,°C} o\text{-MeC}_6\text{H}_4\text{SO}_3^-\text{Na}^+ \xrightarrow[\text{fuse}]{\text{NaOH}} o\text{-MeC}_6\text{H}_4\text{O}^-\text{Na}^+ \xrightarrow{\text{Me}_2\text{SO}_4} o\text{-CH}_3\text{C}_6\text{H}_4\text{OMe}$$

20.6 SUMMARY OF ALIPHATIC SULFUR COMPOUNDS

Some homologous series of organosulfur compounds, such as **thioalcohols** or **mercaptans** (RSH), **thioethers** (RSR), **thioaldehydes** (RCHS), **thioketones** (R_2CS), **thiocarboxylic acids** (RCSOH, RCSSH), and **disulfides** (RSSR), are sulfur analogs of the oxygen compounds.

Unlike O, S exists in positive oxidation states, and we find other homologous series, such as **sulfoxides** (R_2SO), **sulfones** (R_2SO_2) and **sulfinic** (RSO_2H) **acids**, for which there are no O-analogs.

PREPARATION

1. Mercaptans

(a) Alkyl Halide

  RX + KSH

(b) Olefin

  $R'CH{=}CH_2 + H_2S$

(c) Grignard

  $RMgX + S \longrightarrow RSMgX$

2. Disulfides

(a) Mercaptans

  $RSH + H_2O_2$

(b) $2RX + S_2^{2-}$

3. Thioethers

  Williamson Reaction

  $R{-}X + Na^+SR^- \longrightarrow RSR$

4. Sulfoxides

  Oxidation

  $R_2S + H_2O_2$ (30%)

5. Sulfones

(a) Sulfide Oxidation

  $R{-}S{-}R + KMnO_4$ or H_2O_2

(b) Sulfoxide Oxidation

  $R{-}SO{-}R + KMnO_4$

PROPERTIES

1. Acidic Hydrogen

  $+ OH^- \longrightarrow H_2O + RS^-$

  $+ Pb^{2+} \longrightarrow Pb(SR)_2$

2. Sulfur Atom as Nucleophile

  $+ R'{-}COCl \longrightarrow R'{-}COSR$

  $+ R'{-}CH{=}O \longrightarrow R'{-}CH(SR_2)$ (thioacetal)

3. Oxidation of Sulfur Atom

  $+ KMnO_4 \longrightarrow R{-}SO_3H$

1. Oxidation of Sulfur Atom

  $+ X_2 \longrightarrow 2RSX$ (sulfenyl halide)

2. Reduction of Sulfur Atom

  $+ Zn, H^+ \longrightarrow 2RSH$

  $+ HgX_2 \longrightarrow (R_2SHg)^{2+} + 2X^-$

  $+ RX \longrightarrow (R_3S)^+ + X^-$ (sulfonium salt)

  $\overset{\text{O}}{\underset{}{\overset{\|}{R\!-\!\overset{}{S}\!-\!R}}} + 2HI \longrightarrow R{-}S{-}R + H_2O + I_2$

  $+ OH^- \longrightarrow -C{=}C- + RSO_2^- + H_2O$ (E2)

  Have acidic α H's

PREPARATION **PROPERTIES**

6. Thiocarboxylic Acids and Thioesters

(a) Grignard Reactions

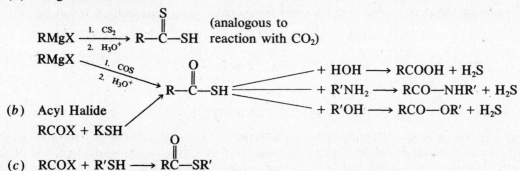

$$RMgX \xrightarrow[\text{2. } H_3O^+]{\text{1. } CS_2} R-\overset{\overset{\displaystyle S}{\|}}{C}-SH \quad \begin{matrix}\text{(analogous to}\\ \text{reaction with } CO_2)\end{matrix}$$

$$RMgX \xrightarrow[\text{2. } H_3O^+]{\text{1. } COS}$$

$$R-\overset{\overset{\displaystyle O}{\|}}{C}-SH \quad \begin{cases} + HOH \longrightarrow RCOOH + H_2S \\ + R'NH_2 \longrightarrow RCO-NHR' + H_2S \\ + R'OH \longrightarrow RCO-OR' + H_2S \end{cases}$$

(b) Acyl Halide

$$RCOX + KSH$$

(c) $RCOX + R'SH \longrightarrow RC\overset{\overset{\displaystyle O}{\|}}{}-SR'$

7. Dithiocarbonates (Xanthates)

$$(RO)^-K^+ \xrightarrow{CS_2} RO-\overset{\overset{\displaystyle S}{\|}}{C}-S^-K^+ \xrightarrow{H_3O^+} ROH + CS_2 + H_2O + K^+$$

8. Alkanesulfonic Acids

(a) Oxidation

$$RSH \text{ or } RSSR + KMnO_4$$

(b) Alkyl Halide

$$RX + K_2SO_3$$

(c) Alkene

$$R'CH=CH_2 + NaHSO_3, O_2$$
(anti-Markovnikov)

$$R-\overset{\overset{\displaystyle O}{\|}}{\underset{\underset{\displaystyle O}{\|}}{S}}-OH$$

1. Acidic Hydrogen

$$+ H_2O \longrightarrow RSO_3^- + H_3O^+$$
$$+ NaOH \longrightarrow RSO_3^- + Na^+ + H_2O$$

2. Hydroxyl

$$+ PCl_5 \longrightarrow RSO_2Cl \xrightarrow{H_2} RSO_2H$$

9. Sulfinic Acids

Grignard
$$RMgX \xrightarrow[\text{2. } H_3O^+]{\text{1. } SO_2} R-\overset{\overset{\displaystyle O}{\|}}{S}-OH$$

$$ArSO_2Cl \xrightarrow{Zn, H_2O} Ar-\overset{\overset{\displaystyle O}{\|}}{S}-OH$$

1. Acidic Hydrogen

$$+ OH^- \longrightarrow RSO_2^- + H_2O$$

2. Hydroxyl

$$+ PCl_3 \longrightarrow RSOCl$$

10. Monoalkyl Sulfates

(a) Alcohol

$$ROH + H_2SO_4 \xrightarrow{0\,°C} RO-SO_2-OH$$

(b) Alkenes

$$R-CH=CH_2 + H_2SO_4 \xrightarrow{0\,°C} RCH(OSO_3H)CH_3$$

11. Dialkyl Sulfates

Alkyl Acid Sulfate

$$RO-SO_2-OH + heat \longrightarrow RO-SO_2-OR \begin{cases} + R'OH \longrightarrow ROR' + ROSO_3H \\ + R'MgBr \longrightarrow R-R' + Mg(ROSO_3)Br \\ + R_2'NH \longrightarrow R_2'N-R + ROSO_3H \end{cases}$$

PREPARATION   **PROPERTIES**

12. Thioureas

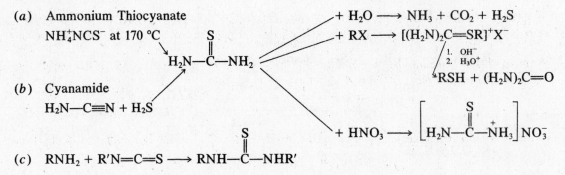

(a)   Ammonium Thiocyanate
$NH_4^+NCS^-$ at 170 °C

(b)   Cyanamide
$H_2N—C≡N + H_2S$

(c) $RNH_2 + R'N=C=S \longrightarrow RNH—\overset{\overset{\displaystyle S}{\|}}{C}—NHR'$

Supplementary Problems

Problem 20.19   Supply a structural formula for   (a) tetronal (3,3-di(ethylsulfonyl)pentane),   (b) methyldiethyl mercaptal,   (c) British Anti-Lewisite (2,3-dimercapto-1-propanol),   (d) ethyl thioacetate,   (e) potassium ethyl sulfonate.   ◄

(a) $(C_2H_5)_2C(SO_2C_2H_5)_2$,   (b) $CH_3CH(SC_2H_5)_2$,   (c) $HOCH_2CHSHCH_2SH$,   (d) $CH_3\overset{\overset{\displaystyle O}{\|}}{C}SC_2H_5$,
(e) $C_2H_5SO_3^-K^+$.

Problem 20.20   Indicate the compounds designated by (?) in

(a) $CH_3—CHBr—C_2H_5 + (?) \longrightarrow CH_3—CH(SCH_3)C_2H_5 + NaBr$

(b) $(?) + C_2H_5X \longrightarrow [(CH_3)_2SC_2H_5]^+X^-$

(c) $(?) + C_4H_9NH_2 \longrightarrow CH_3SO_2NHC_4H_9$

(d) $Me_2S^+—O^- \xrightarrow{KMnO_4} (?)$ ◄

(a) $NaSCH_3$,   (b) $(CH_3)_2S$,   (c) CH_3SO_2Cl,   (d) Me_2SO_2.

Problem 20.21   Why are: (a) mercaptans more acidic $(K_a \approx 10^{-11})$ than alcohols $(K_a \approx 10^{-17})$? (b) mercaptans and thioethers more nucleophilic than alcohols and ethers?   ◄

(a) There are more and stronger H-bonds in alcohols, thus producing an acid-weakening effect. Also, in the conjugate bases RS^- and RO^- the charge is more dispersed over the larger S, thereby making RS^- the weaker base and RSH the stronger acid (Problem 3.29). (b) The larger S is more easily polarized than the smaller O and therefore is more nucleophilic. For example, RSH and RSR participate more rapidly in S_N2 reactions than the corresponding O-analogs. Recall that among the halide anions, nucleophilicity also increases as size increases: $F^- < Cl^- < Br^- < I^-$.

Problem 20.22   Sulfoxides and sulfonium salts having different R or Ar groups and sulfinate esters are resolvable into optical isomers. Explain.   ◄

These molecules are chiral and, unlike chiral amines, are not capable of having their enantiomers rapidly interconverted.

$$R—\overset{*}{\underset{}{S}}—R' \longleftrightarrow R—\overset{*}{\underset{}{S}}^+—R' \qquad R'—\overset{\overset{\displaystyle R''}{|}}{\underset{}{S}}^+—R''' \ \ :\ddot{X}:^-$$
$$\text{sulfoxide} \qquad\qquad\qquad \text{sulfonium salt}$$

Problem 20.23 The ir spectrum of RSH shows a weak —S—H stretching band at about 2600 cm^{-1} that does not shift significantly with concentration or nature of solvent. Explain the difference in behavior of S—H and O—H bonds. ◄

The S—H bond is weaker than the O—H bond and therefore absorbs at a lower frequency. There is little or no H-bonding in S—H and, unlike the O—H bond, there is little shifting of absorption frequency on dilution.

Problem 20.24 Prepare (a) mustard gas, $(ClCH_2CH_2)_2S$, from ethylene; (b) N-(1-propyl)-1-propane-sulfonamide from n-PrBr; (c) di-n-octyl sodium sulfosuccinate,

$$n\text{-octyl}—OOCCH_2CH—COO—n\text{-octyl}$$
$$\underset{SO_3Na}{|}$$

(aerosol detergent) from maleic anhydride and n-octyl alcohol; (d) $C_6H_5SOCH_3$ from C_6H_6 and alkyl halides. ◄

(a)
$$H_2C{=}CH_2 + S_2Cl_2 \longrightarrow (ClCH_2CH_2)_2S + S$$

(b)
$$n\text{-PrBr} \xrightarrow{KSH} n\text{-PrSH} \xrightarrow{H_2O_2} n\text{-PrSO_3H} \xrightarrow{PCl_5} n\text{-PrSO_2Cl} \xrightarrow[\text{NH}_3]{n\text{-PrNH}_2} n\text{-Pr}—SO_2NH—n\text{-Pr}$$

(c)
$$\underset{HC—CO}{\overset{HC—CO}{|}}O + 2\,n\text{-}C_8H_{17}OH \xrightarrow{H^+} \underset{HC—COO—n\text{-}C_8H_{17}}{\overset{HC—COO—n\text{-}C_8H_{17}}{|}} \xrightarrow{H_2S}$$

$$\underset{CH_2—COO—n\text{-}C_8H_{17}}{\overset{HS—CH—COO—n\text{-}C_8H_{17}}{|}} \xrightarrow{KMnO_4} \underset{CH_2—COO—n\text{-}C_8H_{17}}{\overset{HO_3S—CH—COO—n\text{-}C_8H_{17}}{|}} \xrightarrow{NaOH} \text{salt}$$

(d) $C_6H_6 \xrightarrow[\text{2. Mg}]{\text{1. Br}_2/\text{Fe}} C_6H_5MgBr \xrightarrow[\text{2. H}_3O^+]{\text{1. S}} C_6H_5SH \xrightarrow{NaOH} C_6H_5S^-Na^+ \xrightarrow{CH_3I} C_6H_5SCH_3 \xrightarrow{H_2O_2} C_6H_5SOCH_3$

Problem 20.25 Offer mechanisms for

(a) $CH_3CH{=}CH_2$ $\xrightarrow{H_2S, H^+}$ $CH_3CH(SH)CH_3$
$\xrightarrow[h\nu]{H_2S}$ $CH_3CH_2CH_2SH$

(b) $C_6H_5COOC_2H_5 + CH_3SO_2Ph \xrightarrow{NaOEt} C_6H_5COCH_2SO_2Ph + C_2H_5OH$ ◄

(a) Acid-catalyzed addition has an ionic mechanism (Markovnikov):

$$CH_3CH{=}CH_2 + H^+ \longrightarrow CH_3\overset{+}{C}HCH_3 \xrightarrow[-H^+]{H_2S} CH_3\overset{\underset{|}{SH}}{C}HCH_3$$

The peroxide- or light-catalyzed reaction has a free-radical mechanism (anti-Markovnikov):

$$H—S—H \xrightarrow{h\nu} HS\cdot + \cdot H$$

$$CH_3CH{=}CH_2 + \cdot SH \longrightarrow CH_3\dot{C}HCH_2SH$$

$$CH_3\dot{C}HCH_2SH \xrightarrow{H_2S} CH_3CH_2CH_2SH + \cdot SH$$

(b) This is a Claisen-type reaction. Like $\overset{\diagup}{\underset{\diagup}{C}}{=}O$,

$$\underset{O}{\overset{O}{-\overset{\|}{\underset{\|}{S}}-}}$$

enhances the acidity of an α H.

(1) $CH_3SO_2Ph \xrightarrow{OEt^-} \ddot{C}H_2\overset{O}{\underset{O}{\overset{\|}{\underset{\|}{S}}}}Ph$ or $\left[H_2C{=\!=\!=}\overset{O}{\underset{O}{\overset{\|}{\underset{\|}{S}}}}Ph \right]^-$

(2) $PhSO_2\ddot{C}H_2 + C_6H_5\overset{O}{\underset{\|}{C}}{-}OC_2H_5 \longrightarrow PhSO_2CH_2COC_6H_5 + \bar{O}C_2H_5$

Problem 20.26 Alcohols such as Ph_2CHCH_2OH rearrange on treatment with acid; they can be dehydrated by heating their methyl xanthates **(Tschugaev reaction)**. The pyrolysis proceeds by a cyclic transition state. Outline the steps, using Ph_2CHCH_2OH. ◄

$$Ph_2CHCH_2OH \xrightarrow{Na} Ph_2CHCH_2O^-Na^+ \xrightarrow{S=C=S} Ph_2CHCH_2-O-\overset{\overset{\textstyle S}{\|}}{C}-S^-Na^+ \xrightarrow{CH_3I}$$

a xanthate

$$\underset{\text{an }s\text{-methylxanthate}}{Ph_2\overset{\displaystyle C}{\underset{\displaystyle H}{|}}\!\!-\!\!CH_2\!\!-\!\!O\!\!\diagdown_{\displaystyle C-SCH_3}^{}} \longrightarrow Ph_2C=CH_2 + S=C=O + HSCH_3$$

The elimination is *cis*.

Problem 20.27 Since $(CH_3)_2\ddot{S}=\ddot{O}$ is an ambident nucleophile (Problem 7.26), reaction with CH_3I could give $[(CH_3)_3\overset{+}{S}=O]I^-$ or $[(CH_3)_2S=O^+CH_3]I^-$. (*a*) What type of spectroscopy can be used to distinguish between the two products? (*b*) Use the answer to Problem 20.21(*b*) to select the almost exclusive product. ◄

(*a*) Use nmr spectroscopy. In $[(CH_3)_3\overset{+}{S}=O]I^-$ all H's are equivalent and a single peak is observed. Note that $^{32}_{16}S$ has an even number of protons and neutrons and so shows *no* nuclear spin absorption. $[(CH_3^a)_2S=O^+CH_3^b]I^-$ has two different kinds of H's and therefore two peaks would be observed. (*b*) Since S is a better nucleophile than O, $[(CH_3)_3\overset{+}{S}=O]I^-$ is the major product.

Problem 20.28 In living cells, alcohols are converted to acetate esters (acetylated) by the thiol ester $CH_3COS-(CoA)$, acetyl coenzyme A. CoA is an abbreviation for a very complex moiety. Illustrate this reaction using glycerol-1-phosphate. ◄

$$\begin{array}{l} CH_2OH \\ | \\ CHOH \\ | \\ CH_2OPO_3H_2 \end{array} + 2\,CH_3\overset{\overset{\textstyle O}{\|}}{C}-S-(CoA) \longrightarrow \begin{array}{l} CH_2O\overset{\overset{\textstyle O}{\|}}{C}CH_3 \\ | \\ CHOCOCH_3 \\ | \\ CH_2OPO_3H_2 \end{array} + 2\,(CoA)-SH$$

Glycerol-1-phosphate   a phosphatidic   Coenzyme A
                              acid

Problem 20.29 The biosynthesis of fatty acids involves coenzyme A. The suggested steps are (1) carboxylation with CO_2 of acetyl coenzyme A, $CH_3COS-(CoA)$; (2) condensation of the resulting thiomalonyl (CoA) with acetyl coenzyme A; (3) decarboxylation of the resulting acetyl thiomalonyl (CoA); (4) reduction of carbonyl $C=O$ to CH_2. This sequence introduces a unit of 2 C's. The sequence is terminated by hydrolysis of acyl$-S-$(CoA) to the carboxylic acid after the necessary number of repetitions. Formulate the synthesis of hexanoic acid. ◄

$$\overset{2}{C}H_3\overset{3}{C}OS-(CoA) \xrightarrow{CO_2} HO_2\overset{1}{C}\overset{2}{C}H_2\overset{3}{C}OS-(CoA) \xrightarrow[-HS-(CoA)]{\overset{5}{C}H_3\overset{4}{\overset{O}{\|}}{C}S-(CoA)}$$

Thiomalonyl (CoA)   *a Claisen-type*
                      condensation

$$HO_2\overset{1}{C}\overset{2}{C}H\overset{3}{C}OS-(CoA) \xrightarrow{-CO_2} CH_3\boxed{\overset{5}{C}\overset{4}{O}}CH_2\overset{2}{C}OS-(CoA) \xrightarrow{red.} CH_3\boxed{\overset{5}{C}H_2}\overset{4}{C}H_2\overset{2}{C}H_2\overset{3}{C}OS-(CoA) \xrightarrow[-HS-(CoA)]{HO_2\overset{1'}{C}\overset{2'}{C}H_2\overset{3'}{C}OS-(CoA)}$$
$$\underset{\overset{4}{C}\overset{5}{O}CH_3}{|}$$

$$\overset{5}{C}H_3\overset{4}{C}H_2\overset{2}{C}H_2\overset{3}{\overset{\overset{2'}{O}}{\|}}{C}\overset{3'}{C}HCOS-(CoA) \xrightarrow[\text{2. red.}]{\text{1. }-CO_2} \overset{5}{C}H_3\overset{4}{C}H_2\overset{2}{C}H_2\overset{3}{C}H_2\overset{2'}{C}H_2\overset{3'}{C}OS-(CoA) \longrightarrow CH_3(CH_2)_4COOH$$
$$\underset{\underset{1'}{COOH}}{|}$$

Chapter 21

Phenols

21.1 INTRODUCTION

Phenols (ArOH) and alcohols (ROH) are similar in properties but they differ sufficiently so that phenols may be considered as a separate homologous series.

Problem 21.1 Name the following phenols by the IUPAC system:

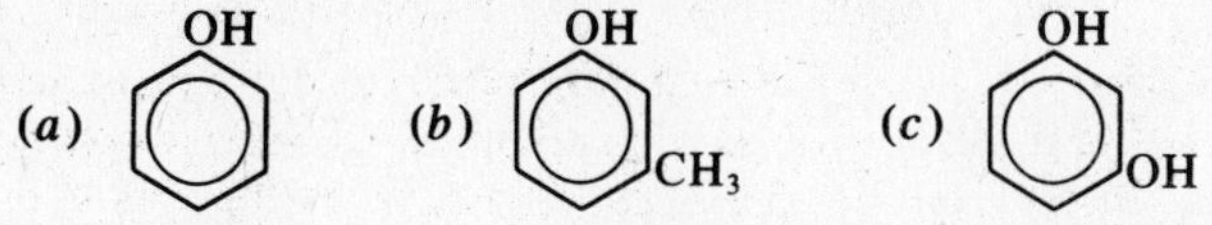

(a) Phenol (b) *m*-Cresol (c) Resorcinol

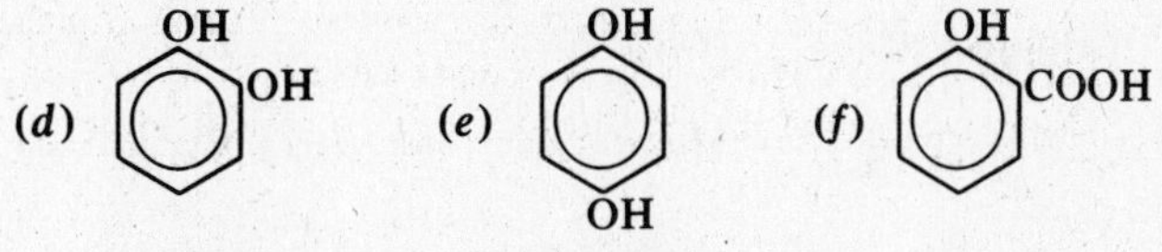

(d) Catechol (e) Hydroquinone (f) Salicylic acid ◀

(a) hydroxybenzene, (b) *m*-hydroxytoluene, (c) 1,3-dihydroxybenzene, (d) 1,2-dihydroxybenzene, (e) 1,4-dihydroxybenzene, (f) *o*-hydroxybenzoic acid.

Problem 21.2 Name the following compounds:

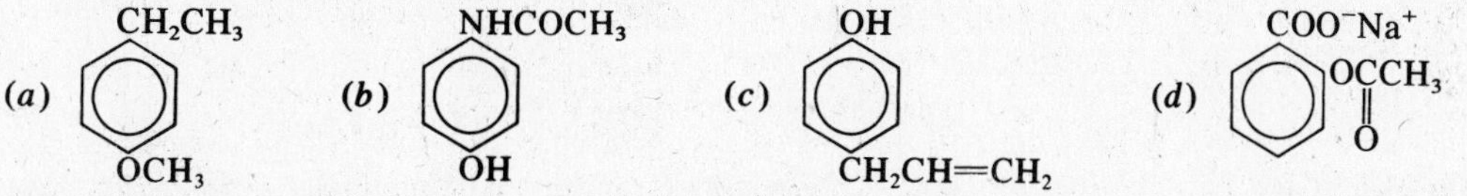

(a) (b) (c) (d) ◀

(a) *p*-methoxyethylbenzene, (b) *p*-hydroxyacetanilide, (c) *p*-allylphenol, (d) sodium acetylsalicylate (sodium salt of aspirin).

Problem 21.3 Write structural formulas for (a) *p*-cresol, (b) 2-nitro-4-bromophenol, (c) 4-*n*-hexylresorcinol, (d) ethyl salicylate. ◀

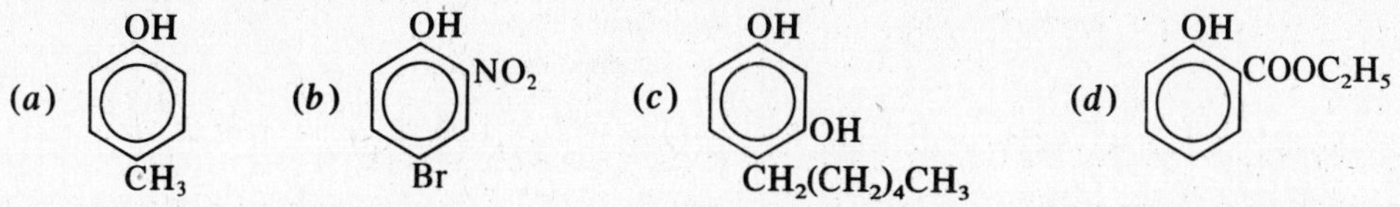

(a) (b) (c) (d)

Problem 21.4 Compared to toluene, phenol (a) has a higher boiling point and (b) is more soluble in H$_2$O. Explain. ◀

Ph—O⋯O—Ph

(a) Intermolecular H-bonding

Ph—O⋯O

(b) H-bonding with H$_2$O

Problem 21.5 Account for the lower b.p. and decreased H_2O solubility of *o*-nitrophenol and *o*-nitrobenzaldehyde as compared with their *m* and *p* isomers. ◀

In some *ortho*-substituted phenols, intramolecular H-bonding (chelation) forms a 6-membered ring. This inhibits H-bonding with water and reduces solubility in H_2O. Since chelation diminishes *inter*molecular H-bonding attraction, the boiling point is decreased.

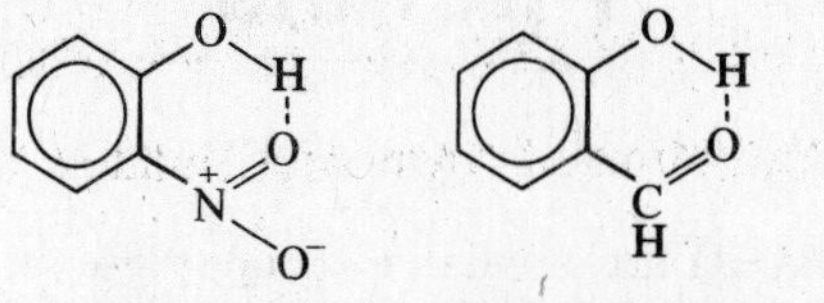

o-Nitrophenol   *o*-Nitrobenzaldehyde
Chelation

Higher boiling points of *para* and *meta* isomers are attributed to their *inter*molecular H-bonding. Their greater H_2O solubility is caused by coassociation with water molecules through H-bonding.

21.2 PREPARATION

INDUSTRIAL METHODS

1. Dow Process (by Benzyne Mechanism)

$$C_6H_5Cl + 2NaOH \xrightarrow[320\ atm]{360\ °C} H_2O + NaCl + C_6H_5O^-Na^+ \xrightarrow{H^+} C_6H_5OH$$

Chlorobenzene   Sodium   Phenol
                                                          phenoxide

2. From Cumene Hydroperoxide

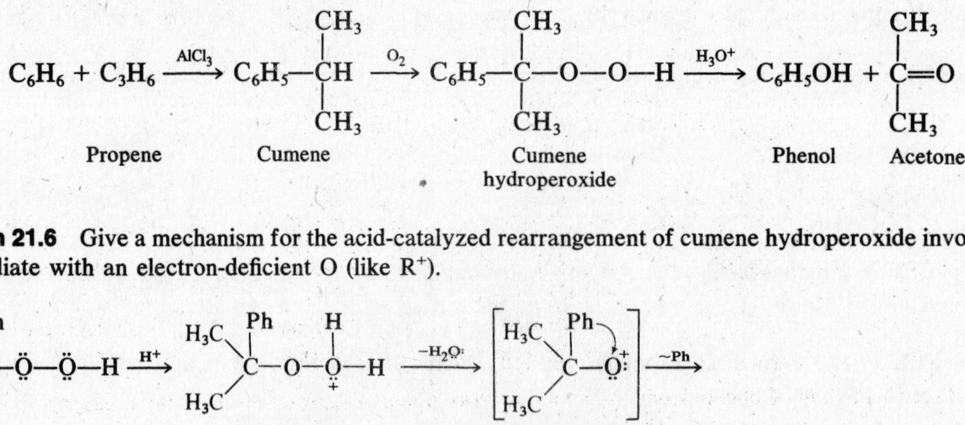

Propene   Cumene   Cumene   Phenol   Acetone
                                 hydroperoxide

Problem 21.6 Give a mechanism for the acid-catalyzed rearrangement of cumene hydroperoxide involving an intermediate with an electron-deficient O (like R^+). ◀

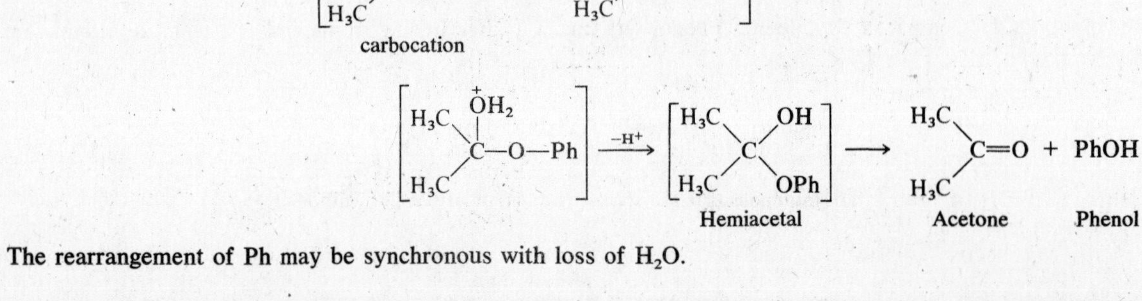

an oxonium ion   electron-deficient
                        intermediate

carbocation

                        Hemiacetal   Acetone   Phenol

The rearrangement of Ph may be synchronous with loss of H_2O.

3. Alkali Fusion of Aryl Sulfonate Salts (page 355)

LABORATORY METHODS

1. Hydrolysis of Diazonium Salts (Section 18.5)

2. Via Thallation

$$ArH + Tl(OOCCF_3)_3 \longrightarrow ArTl(OOCCF_3)_2 \xrightarrow[\text{2. } Ph_3P]{\text{1. } Pb(OAc)_4} ArOOCCF_3 \xrightarrow[\text{2. } H^+]{\text{1. } H_2O, OH^-} ArOH$$

3. Aromatic Nucleophilic Substitution of Nitro Aryl Halides (Problem 19.7)

4. Ring Oxidation

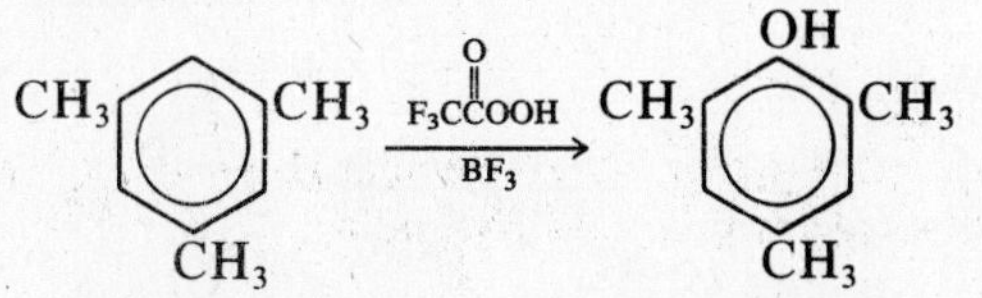

Mesitylene   Mesitol
(an activated ring)

(electrophilic substitution by OH^+)

Problem 21.7 Outline reactions and reagents for industrial synthesis of the following from benzene and inorganic reagents: (a) catechol, (b) resorcinol, (c) picric acid (2,4,6-trinitrophenol). ◄

(a)

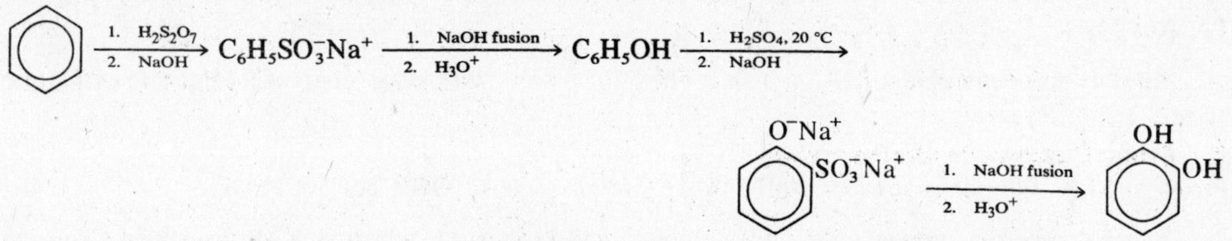

(b)

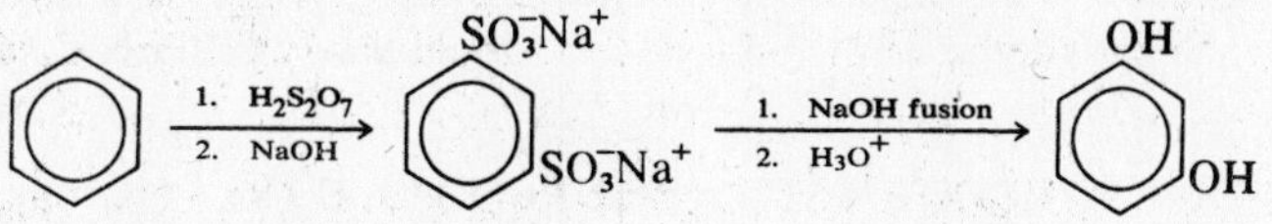

(c)

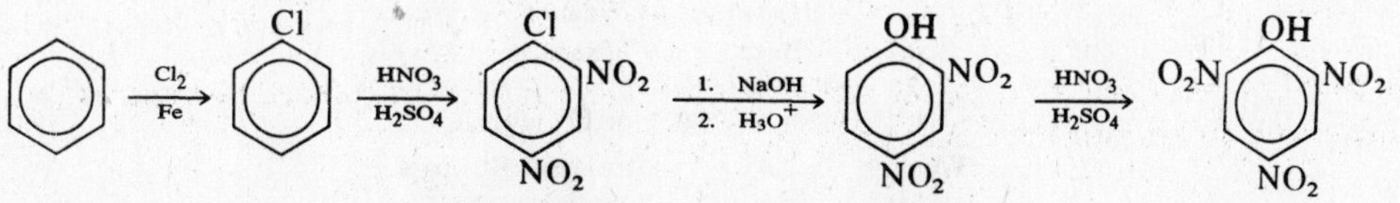

2,4-Dinitro-chlorobenzene   2,4-Dinitro-phenol   Picric acid

Direct nitration of phenol leads to excessive oxidation and destruction of material because HNO_3 is a strong oxidizing agent and the OH activates the ring.

Problem 21.8 Devise practical laboratory syntheses of the following phenols from benzene or toluene and any inorganic or aliphatic compounds: (a) m-iodophenol, (b) 3-chloro-4-methylphenol, (c) 2-bromo-4-methylphenol, (d) p-chlorophenol. ◄

(a)

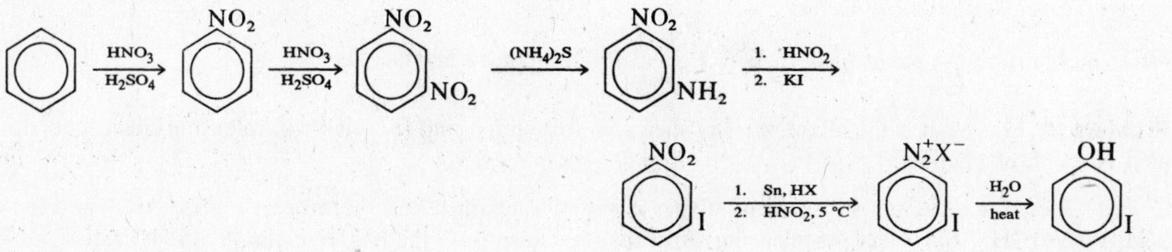

m-Iodophenol

(b) CH_3 (toluene) $\xrightarrow[H_2SO_4]{HNO_3}$ CH_3, NO_2 $\xrightarrow[Fe]{Cl_2}$ CH_3, Cl, NO_2 $\xrightarrow[\text{2. } HNO_2,\ 5\ °C]{\text{1. } Sn,\ HX}$ CH_3, Cl, $N_2^+X^-$ $\xrightarrow{H_2O}$ CH_3, Cl, OH

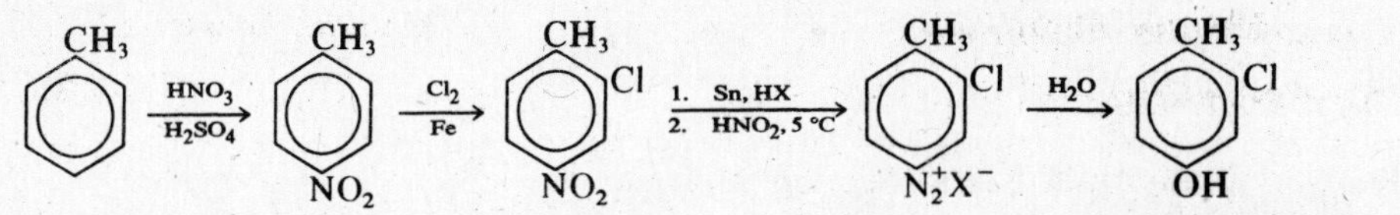

3-Chloro-4-methylphenol

(c) benzene CH_3 $\xrightarrow[H_2SO_4]{HNO_3}$ NO_2, CH_3 $\xrightarrow[\text{2. } Ac_2O]{\text{1. } Sn,\ HCl}$ $NHAc$, CH_3 $\xrightarrow{Br_2}$ $NHAc$, Br, CH_3 $\xrightarrow[\text{3. } H_2O]{\substack{\text{1. } H_3O^+ \\ \text{2. } HNO_2}}$ OH, Br, CH_3

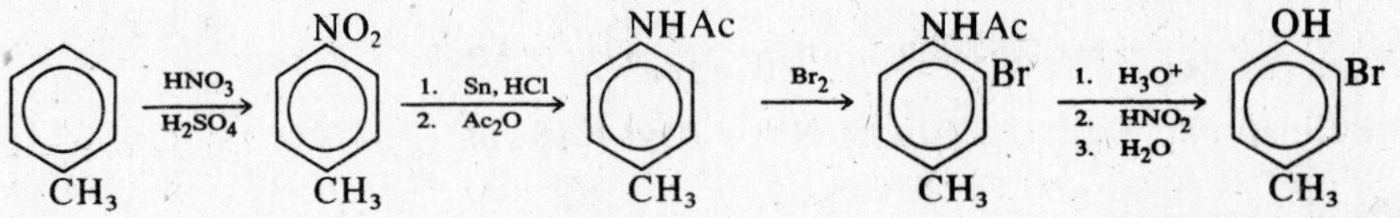

2-Bromo-4-methylphenol

(d) benzene $\xrightarrow[Fe]{Cl_2}$ Cl $\xrightarrow{Tl(OOCCF_3)_3}$ Cl, $Tl(OOCCF_3)_2$ $\xrightarrow[\substack{\text{2. } OH^-,\ H_2O \\ \text{3. } H^+}]{\text{1. } Pb(OAc)_4,\ Ph_3P}$ Cl, OH

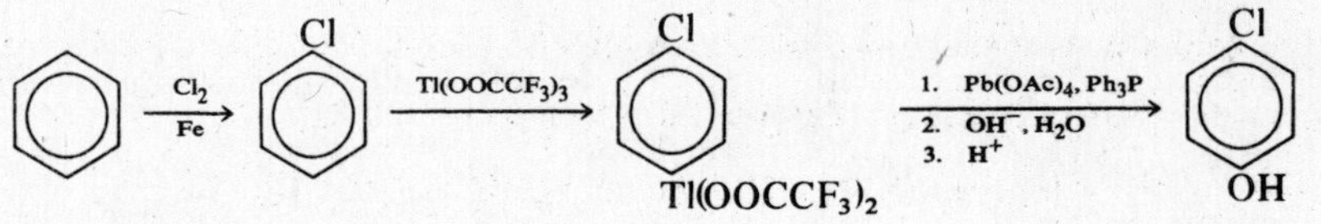

21.3 CHEMICAL PROPERTIES

REACTIONS OF H OF THE OH GROUP

1. Acidity

Phenols are weak acids ($pK_a = 10$). They form salts with aqueous NaOH but not with aqueous $NaHCO_3$.

Problem 21.9 Why does aqueous $NaHCO_3$ solution dissolve RCOOH but not PhOH?   ◄

In both cases the product would be carbonic acid ($pK_a = 6$), which is a stronger acid than phenols ($pK_a = 10$) but weaker than carboxylic acids ($pK_a = 4.5$). Acid-base equilibria lie towards the weaker acid and weaker base.

$$RCOOH + HCO_3^- \rightleftharpoons RCOO^- + H_2CO_3$$

Stronger	Stronger	Weaker	Weaker
Acid$_1$	Base$_2$	Base$_1$	Acid$_2$

$$ArOH + HCO_3^- \leftrightharpoons ArO^- + H_2CO_3$$

Weaker	Weaker	Stronger	Stronger
Acid$_1$	Base$_1$	Base$_2$	Acid$_1$

Problem 21.10 Account for the considerably greater acid strength of PhOH ($pK_a = 10$) than that of ROH ($pK_a = 18$).   ◄

The negative charge on the alkoxide anion, RO^-, cannot be delocalized, but on PhO^- the $-$ charge is delocalized to the *ortho* and *para* ring positions as indicated by the starred sites in the resonance hybrid,

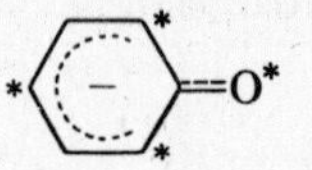

$$*\langle(-)\rangle=O^*$$

PhO^- is therefore a weaker base than RO^-, and PhOH is a stronger acid.

Problem 21.11 What is the effect of (a) electron-attracting and (b) electron-releasing substituents on the acid strength of phenols?   ◄

(a) Electron-attracting substituents disperse negative charges and therefore stabilize ArO^- and increase acidity of ArOH. (b) Electron-releasing substituents concentrate the negative charge on O, destabilize ArO^- and decrease acidity of ArOH.

Electron-Attracting

$(-NO_2, -CN, -CHO, -COOH, -NR_3^+, -X)$

Electron-Releasing

$(-CH_3, -C_2H_5, -OR, -NR_2)$

More
acidic

Charge dispersed;
ion stabilized

Less
acidic

Charge concentrated;
ion destabilized

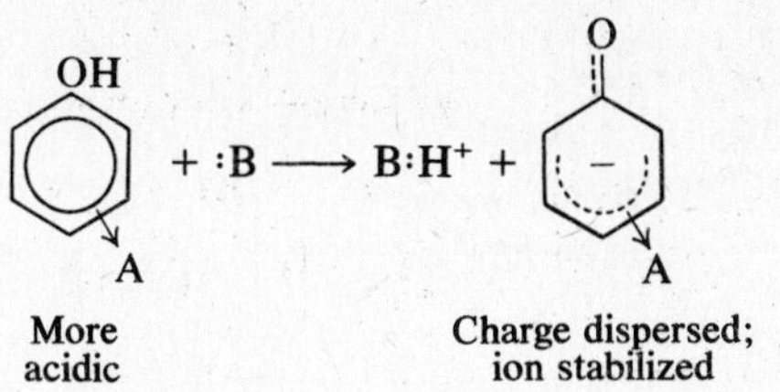

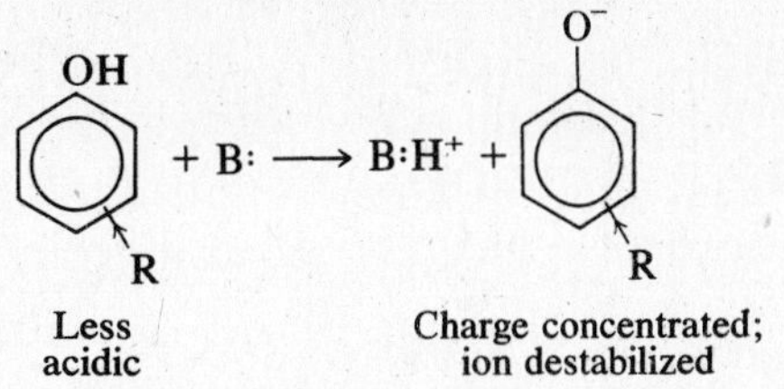

Problem 21.12 In terms of resonance and inductive effects, account for the following relative acidities.

$$(a) \quad p\text{-}O_2NC_6H_4OH > m\text{-}O_2NC_6H_4OH > C_6H_5OH$$

$$(b) \quad m\text{-}ClC_6H_4OH > p\text{-}ClC_6H_4OH > C_6H_5OH$$

(a) The $-NO_2$ is electron-withdrawing and acid-strengthening. Its resonance effect, which occurs only from *para* and *ortho* positions, predominates over its inductive effect, which occurs also from the *meta* position. Other substituents in this category are

$$\text{\Large$>$}C{=}O \qquad -CN \qquad -COOR \qquad -SO_2R$$

(b) Cl is electron-withdrawing by induction. This effect diminishes with increasing distance between Cl and OH. The *meta* is closer than the *para* position and m-Cl is more acid-strengthening than p-Cl. Other substituents in this category are F, Br, I, $\overset{+}{N}R_3$.

Problem 21.13 Assign numbers from 1 for LEAST to 4 for MOST to indicate the relative acid strengths in the following groups: (a) phenol, m-chlorophenol, m-nitrophenol, m-cresol; (b) phenol, benzoic acid, p-nitrophenol, carbonic acid; (c) phenol, p-chlorophenol, p-nitrophenol, p-cresol; (d) phenol, o-nitrophenol, m-nitrophenol, p-nitrophenol; (e) phenol, p-chlorophenol, 2,4,6-trichlorophenol, 2,4-dichlorophenol; (f) phenol, benzyl alcohol, benzenesulfonic acid, benzoic acid.

(a) 2, 3, 4, 1. Because

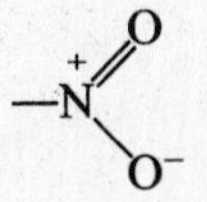

has + on N, it has a greater electron-withdrawing inductive effect than has Cl.

(b) 1, 4, 2, 3.

(c) 2, 3, 4, 1. The resonance effect of p-NO_2 exceeds the inductive effect of p-Cl. p-CH_3 is electron-releasing.

(d) 1, 3, 2, 4. Intramolecular H-bonding makes the o isomer weaker than the p isomer.

(e) 1, 2, 4, 3. Increasing the number of electron-attracting groups increases the acidity.

(f) 2, 1, 4, 3.

2. Formation of Ethers

(a) Williamson synthesis.

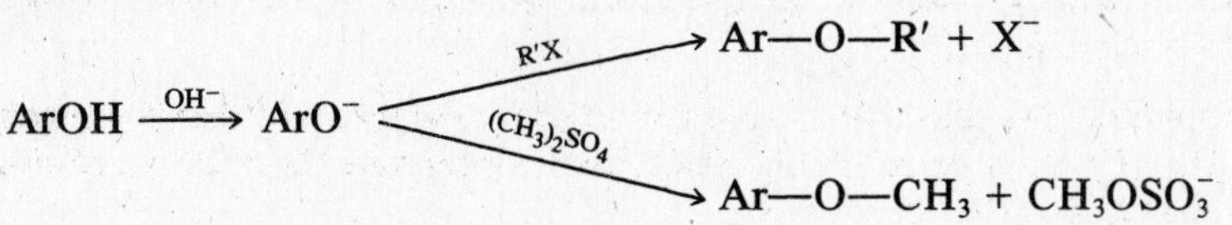

(*b*) Aromatic nucleophilic substitution.

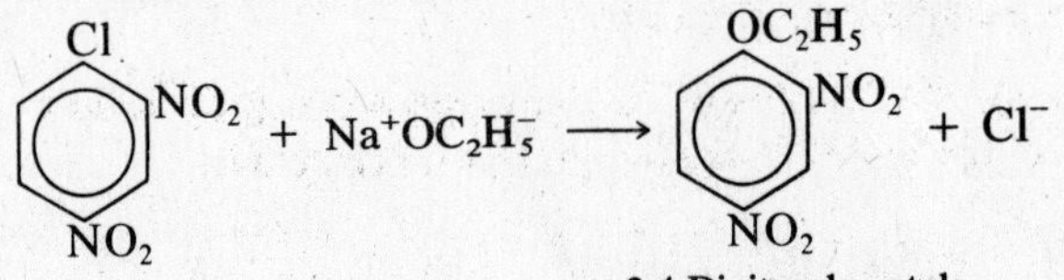

2,4-Dinitrophenetole
(2,4-Dinitrophenyl ethyl ether)

3. Formation of Esters

Phenyl esters (RCOOAr) are not formed directly from RCOOH. Instead, acid chlorides or anhydrides are reacted with ArOH in the presence of strong base.

$$(CH_3CO)_2O + C_6H_5OH + NaOH \longrightarrow CH_3COOC_6H_5 + CH_3COO^-Na^+ + H_2O$$
Phenyl acetate

$$C_6H_5COCl + C_6H_5OH + NaOH \longrightarrow C_6H_5COOC_6H_5 + Na^+Cl^- + H_2O$$
Phenyl benzoate

OH^- converts ArOH to the more nucleophilic ArO^- and also neutralizes the acids formed.

Problem 21.14 Phenyl acetate undergoes the **Fries rearrangement** with $AlCl_3$ to form *ortho*- and *para*-hydroxyacetophenone. The *ortho* isomer is separated from the mixture by its volatility with steam.

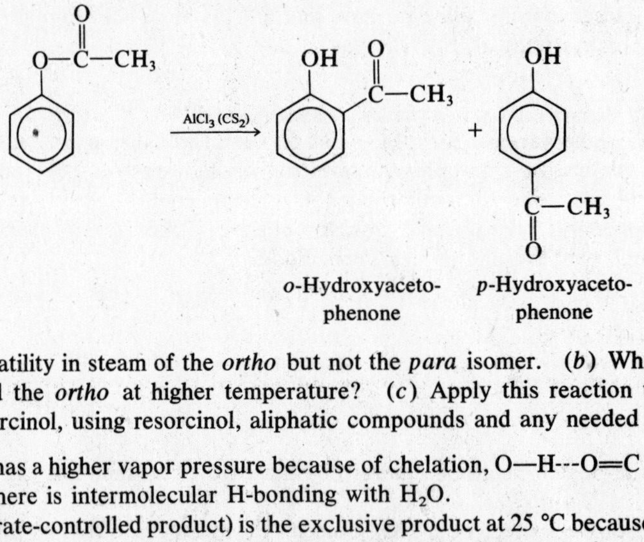

o-Hydroxyaceto-   p-Hydroxyaceto-
phenone   phenone

(*a*) Account for the volatility in steam of the *ortho* but not the *para* isomer. (*b*) Why does the *para* isomer predominate at low and the *ortho* at higher temperature? (*c*) Apply this reaction to the synthesis of the antiseptic 4-*n*-hexylresorcinol, using resorcinol, aliphatic compounds and any needed inorganic reagents. ◄

(*a*) The *ortho* isomer has a higher vapor pressure because of chelation, O—H---O=C (see Problem 21.5). In the *para* isomer there is intermolecular H-bonding with H_2O.

(*b*) The *para* isomer (rate-controlled product) is the exclusive product at 25 °C because it has a lower $\Delta H^{\ddagger}$ and is formed more rapidly. Its formation is reversible, unlike that of the *ortho* isomer which is stabilized by chelation. Although it has a higher $\Delta H^{\ddagger}$, the *ortho* isomer (equilibrium-controlled product) is the chief product at 165 °C because it is more stable.

(*c*) Two activating OH groups in *meta* positions reinforce each other in electrophilic substitution and permit Friedel-Crafts reactions of resorcinol directly with RCOOH and $ZnCl_2$.

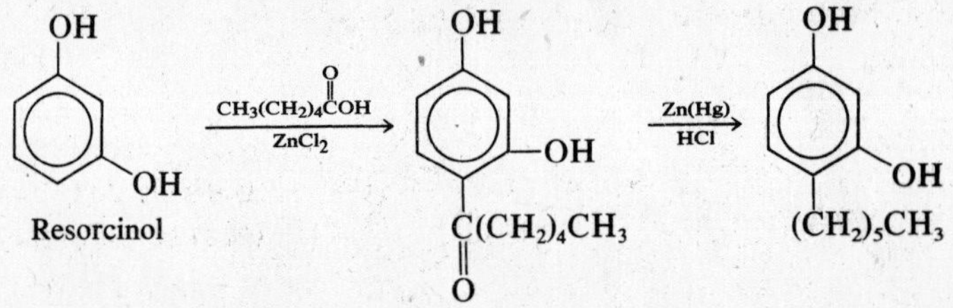

Resorcinol

Problem 21.15 Alkylation of PhO^- with an active alkyl halide such as $CH_2{=}CHCH_2Cl$ gives, in addition to phenyl allyl ether, some *o*-allylphenol. Explain. ◀

PhO^- is an ambident anion with negative charge on both O and the *ortho* C atoms of the ring. Attack by O gives the ether; attack by the *ortho* carbanion gives *o*-allylphenol.

DISPLACEMENT OF OH *GROUP*

Phenols resemble aryl halides in that the functional group resists displacement. Unlike ROH, phenols do not react with HX, $SOCl_2$, or phosphorus halides. Phenols are reduced to hydrocarbons but the reaction is used for structure proof and not for synthesis.

$$ArOH + Zn \xrightarrow{\Delta} ArH + ZnO \quad (\textit{poor yields})$$

REACTIONS OF THE BENZENE RING

1. Hydrogenation

Cyclohexanol

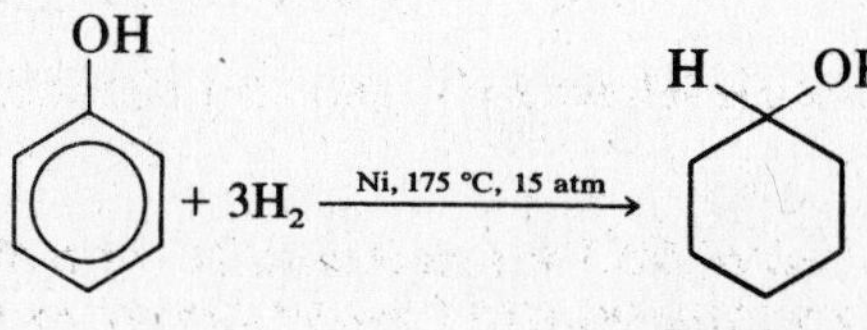

2. Oxidation to Quinones

Phenol Quinone

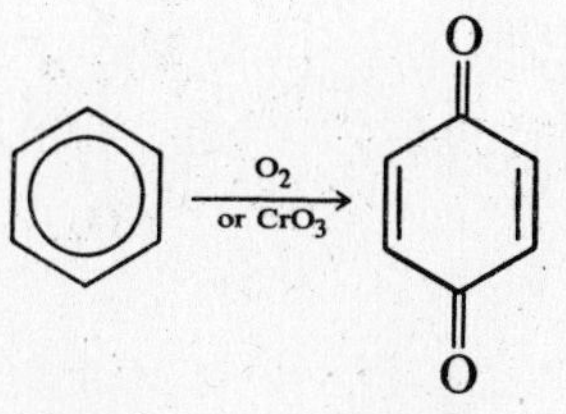

3. Electrophilic Substitution

The —OH and even more so the —O^- (**phenoxide**) are strongly activating and *op*-directing.

Phenol oxonium cations Phenoxide anion Fairly stable unsaturated ketones

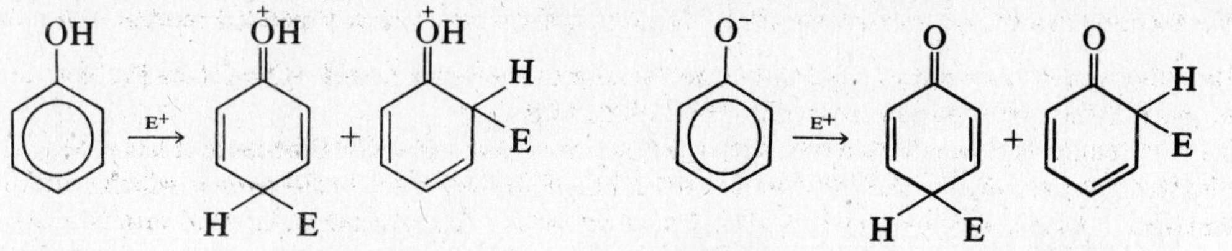

Special mild conditions are needed to achieve electrophilic monosubstitution in phenols because their high reactivity favors both *polysubstitution* and *oxidation.*

(*a*) Halogenation.

(goes via ArO^-; no monobromophenols)

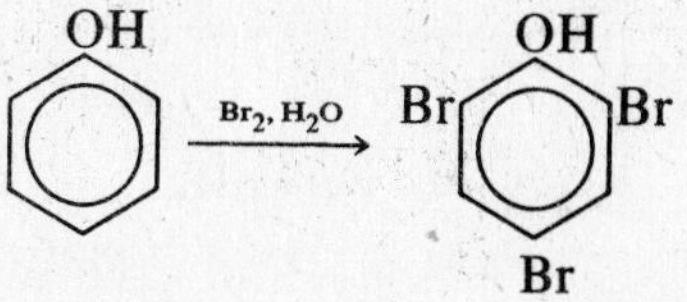

Monobromination is achieved with nonpolar solvents such as CS_2 to decrease the electrophilicity of Br_2 and also to minimize phenol ionization.

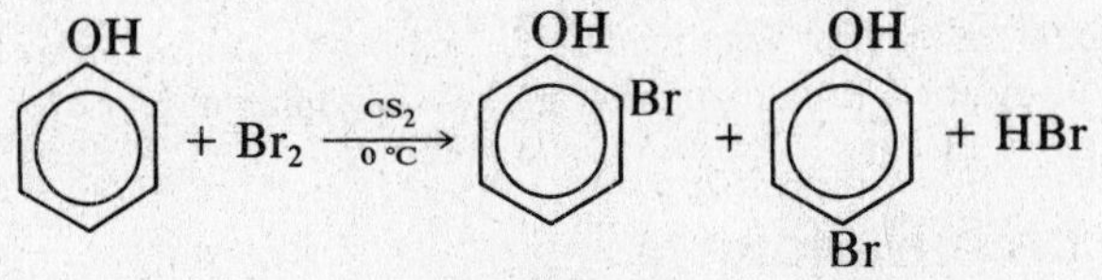

o-Bromophenol p-Bromophenol
      (major)

(b) Nitrosation.

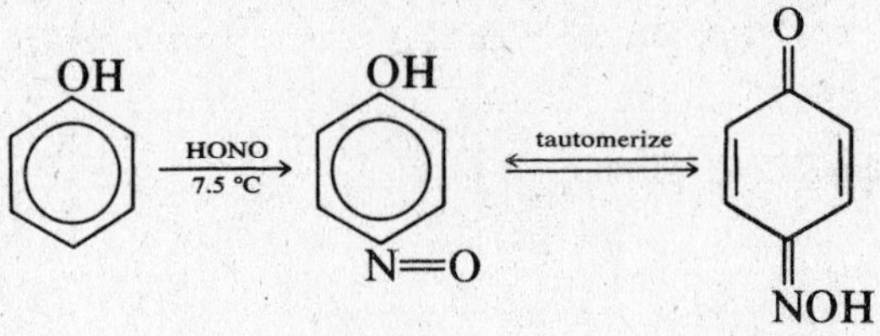

   p-Nitroso-   Quinone monoxime
   phenol

(c) Nitration. See Problem 21.16.

Problem 21.16 Low yields of p-nitrophenol are obtained from direct nitration of PhOH because of ring oxidation. Suggest a better synthetic method. ◄

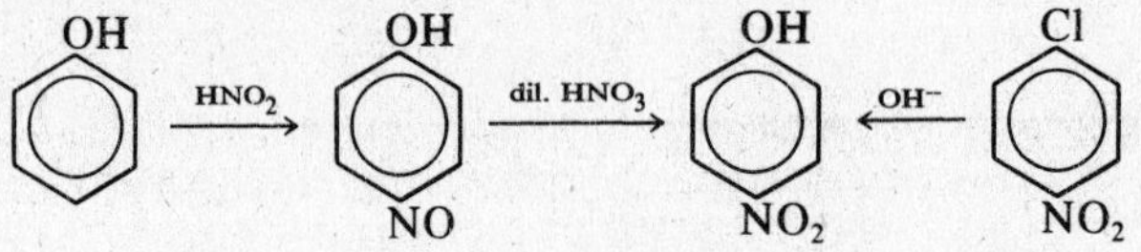

(d) Sulfonation.

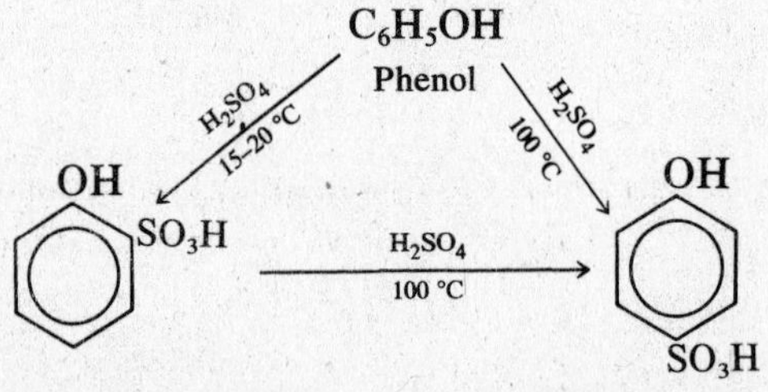

o-Phenolsulfonic acid   p-Phenolsulfonic acid
(rate-controlled)   (equilibrium-controlled)

(e) Diazonium salt coupling to form azophenols (Section 18.5).

(f) Mercuration. Mercuriacetate cation, ^+HgOAc, is a weak electrophile which substitutes in *ortho* and *para* positions of phenols. This reaction is used to introduce an I on the ring.

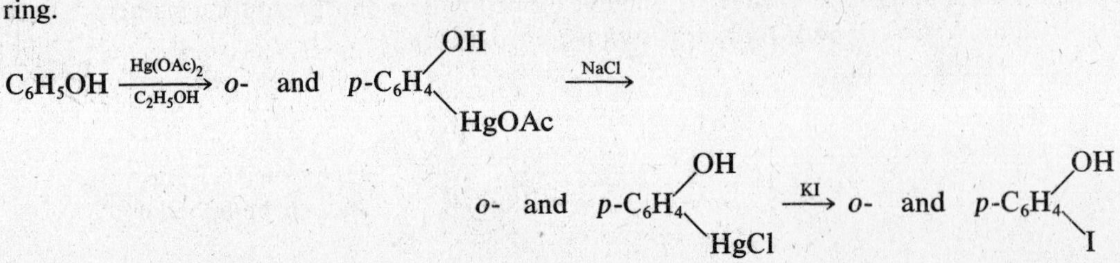

(g) **Ring alkylation.**

$$C_6H_5OH + \begin{cases} CH_3CH{=}CH_2 \\ (CH_3)_2CHOH \end{cases} \xrightarrow[\text{or HF}]{H_2SO_4} \text{o- and } p\text{-}C_6H_4 \overset{OH}{\underset{CH(CH_3)_2}{\big\langle}} + H_2O$$

RX and $AlCl_3$ give poor yields because $AlCl_3$ coordinates with O.

(h) **Ring acylation.** Phenolic ketones are best prepared by the Fries rearrangement (Problem 21.14).

(i) **Kolbe synthesis of phenolic carboxylic acids.**

$$C_6H_5O^-Na^+ + O{=}C{=}O \xrightarrow[\text{6 atm}]{125\ ^\circ C} \text{o-}C_6H_4 \overset{OH}{\underset{COONa}{\big\langle}} \xrightarrow{H_3O^+} \text{o-}C_6H_4 \overset{OH}{\underset{COOH}{\big\langle}}$$

$$\text{Sodium salicylate} \qquad\qquad \text{Salicylic acid}$$

(j) **Reimer-Tiemann synthesis of phenolic aldehydes.**

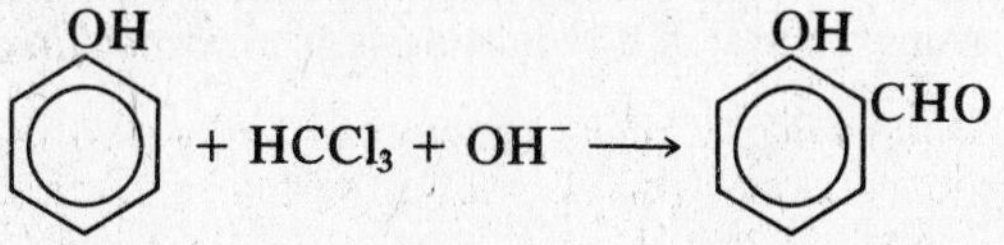

+ $HCCl_3$ + $OH^- \longrightarrow$

Problem 21.17 Outline an acceptable mechanism for the (a) Kolbe reaction, (b) Reimer-Tiemann reaction. ◄

(a) Phenoxide carbanion adds at the electrophilic carbon of CO_2.

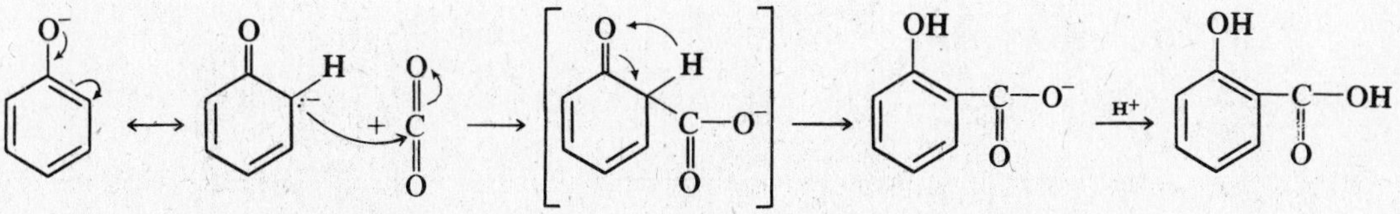

The conjugated ketonic diene tautomerizes to reform the more stable benzenoid ring.

(b) The electrophile is the carbene $:CCl_2$.

$$:\!\overset{..}{O}H^- + H\!:\!\overset{Cl}{\underset{Cl}{\overset{..}{C}}}\!:\!Cl \longrightarrow H\!:\!OH + \overset{..}{:}\!\overset{Cl}{\underset{Cl}{\overset{..}{C}}}\!:\!Cl \longrightarrow :\!Cl^- + \overset{Cl}{:\!\overset{..}{C}\!:\!Cl}$$

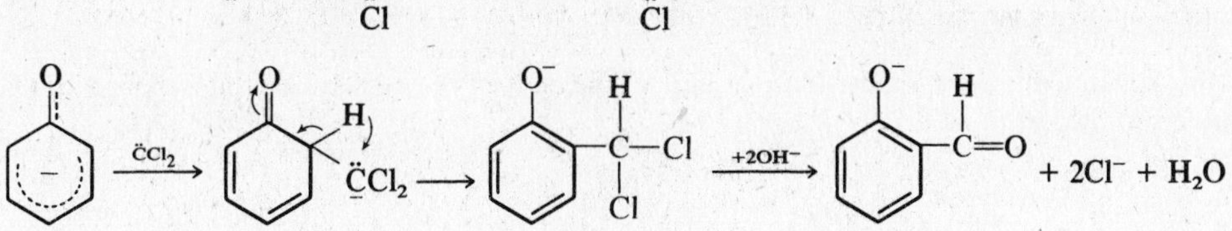

a benzal chloride

Problem 21.18 Use phenol and any inorganic or aliphatic reagents to synthesize (a) aspirin (acetylsalicylic acid), (b) oil of wintergreen (methyl salicylate). Do not repeat the synthesis of any compound. ◄

(a)

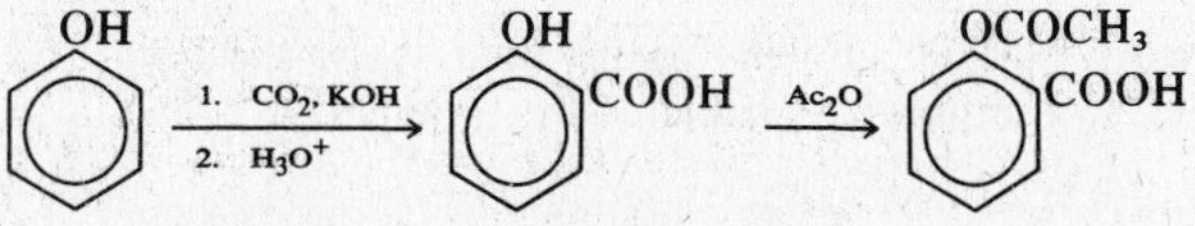

(b)

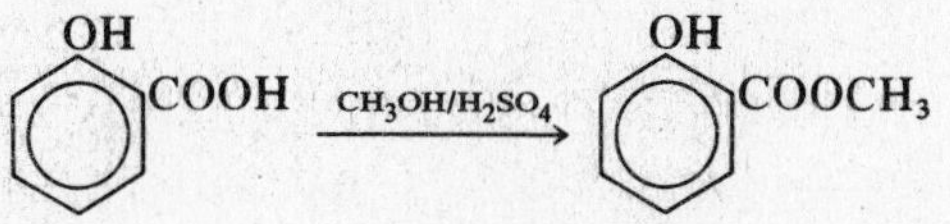

(k) Condensations with carbonyl compounds; phenol-formaldehyde resins. Acid or base catalyzes electrophilic substitution of carbonyl compounds in *ortho* and *para* positions of phenols to form phenol alcohols (**Lederer-Manasse reaction**).

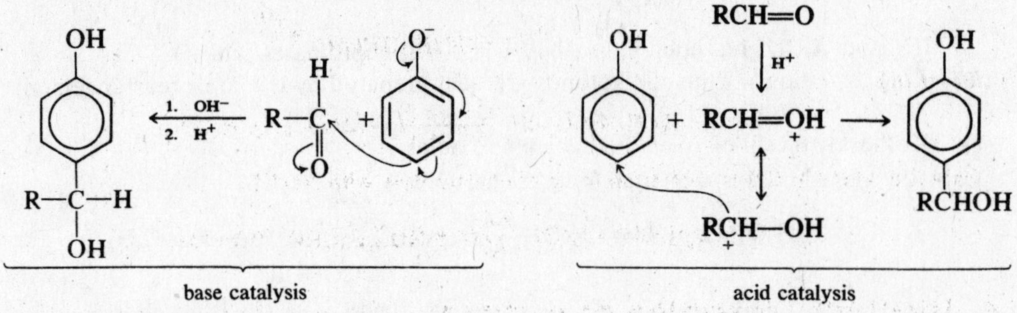

base catalysis   acid catalysis

(l) Rearrangements from O to ring.
  (1) Fries rearrangement of phenolic esters to phenolic ketones (Problem 21.14).
  (2) Claisen rearrangement. The reaction is *intramolecular* and has a cyclic mechanism.

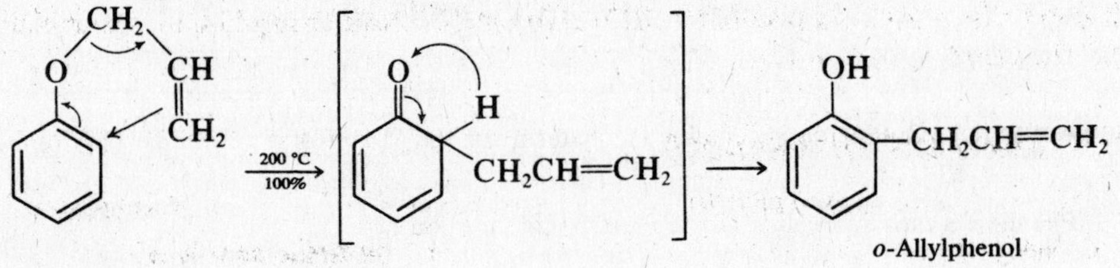

o-Allylphenol

(3) Alkyl phenyl ethers.

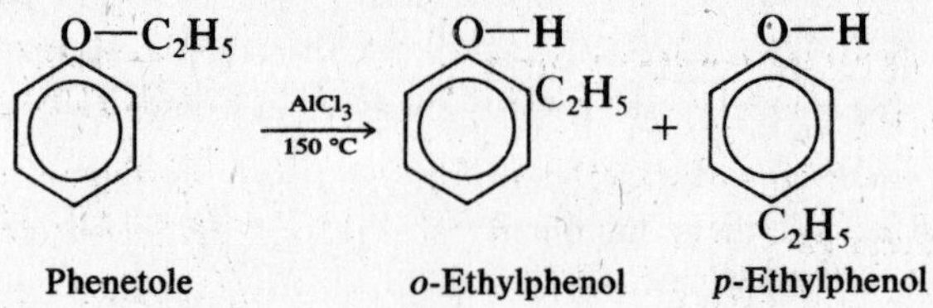

Phenetole   o-Ethylphenol   p-Ethylphenol

Problem 21.19 Predict the product of the Claisen rearrangement of (a) allyl-3-^{14}C phenyl ether, (b) 2,6-dimethylphenyl allyl-3-^{14}C ether.   ◀

(a) In this concerted intramolecular rearrangement, the ends of the allyl system interchange so that the γ C is bonded to the *ortho* C.

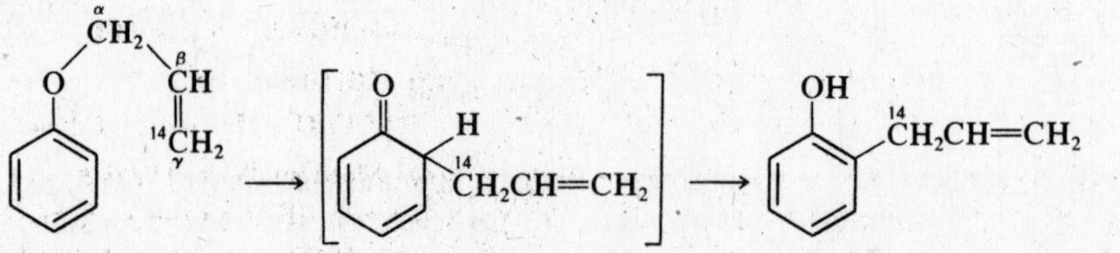

(b) When the *ortho* position is blocked, the allyl group migrates to the *para* position by two consecutive rearrangements and the ^{14}C is in the γ position of the product.

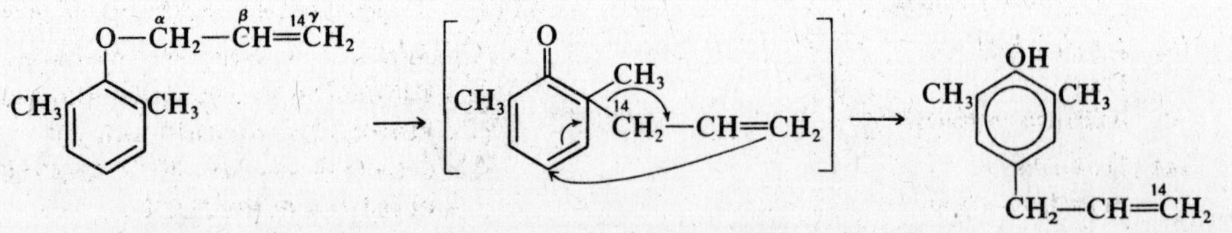

Problem 21.20   Use phenol and any aliphatic and inorganic compounds to prepare: (*a*) *o*-chloromethylphenol, (*b*) *o*-methoxybenzyl alcohol,   (*c*) *p*-hydroxybenzaldehyde.   ◄

(*a*)
$$C_6H_5OH + H_2C{=}O \xrightarrow[\text{aq. OH}^-]{\text{HCl}} o\text{-HOC}_6H_4CH_2Cl$$
$$o\text{-HOC}_6H_4CH_2OH \xrightarrow{\text{SOCl}_2}$$

(*b*)   From (*a*),

$$o\text{-HOC}_6H_4CH_2OH \xrightarrow[(CH_3)_2SO_4]{\text{NaOH}} o\text{-CH}_3OC_6H_4CH_2OH$$

Only the phenolic OH is converted to its conjugate base with NaOH.

(*c*)
$$C_6H_5OH + HCN \xrightarrow[\text{HCl}]{\text{ZnCl}_2} p\text{-HOC}_6H_4CHO \quad (\text{page 251})$$

21.4 ANALYTICAL DETECTION OF PHENOLS

Phenols are soluble in NaOH but not in $NaHCO_3$.   With Fe^{3+} they produce complexes whose characteristic colors are green, red, blue and purple.

Infrared stretching bands of phenols are 3200–3600 cm^{-1} for the O—H (like alcohols), but 1230 cm^{-1} for the C—O (alcohols: 1050–1150 cm^{-1}).   Nmr absorption of OH depends on H-bonding and the range is $\delta = 4\text{--}12$.

21.5 SUMMARY OF PHENOLS

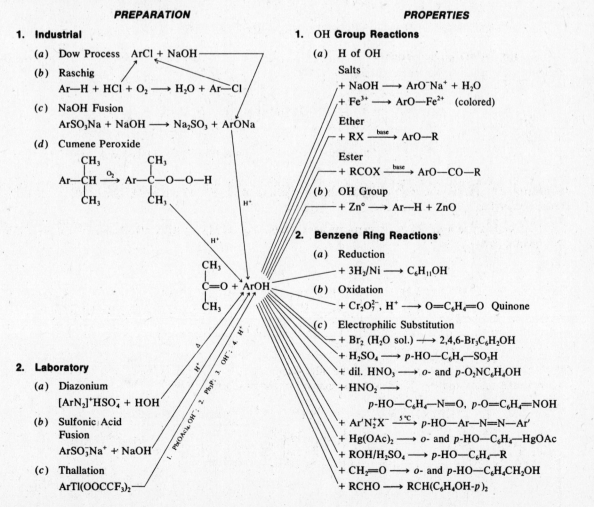

PREPARATION

1. Industrial

(*a*) Dow Process   ArCl + NaOH

(*b*) Raschig

Ar—H + HCl + O_2 ⟶ H_2O + Ar—Cl

(*c*) NaOH Fusion

$ArSO_3Na$ + NaOH ⟶ Na_2SO_3 + ArONa

(*d*) Cumene Peroxide

$$Ar{-}\underset{\underset{CH_3}{|}}{\overset{\overset{CH_3}{|}}{C}}H \xrightarrow{O_2} Ar{-}\underset{\underset{CH_3}{|}}{\overset{\overset{CH_3}{|}}{C}}{-}O{-}O{-}H$$

$$\underset{\underset{CH_3}{|}}{\overset{\overset{CH_3}{|}}{C}}{=}O + ArOH$$

2. Laboratory

(*a*) Diazonium

$[ArN_2]^+HSO_4^-$ + HOH

(*b*) Sulfonic Acid Fusion

$ArSO_3Na^+$ + NaOH

(*c*) Thallation

$ArTl(OOCCF_3)_2$

PROPERTIES

1. OH Group Reactions

(*a*) H of OH

Salts

+ NaOH ⟶ ArO^-Na^+ + H_2O

+ Fe^{3+} ⟶ ArO—Fe^{2+} (colored)

Ether

+ RX $\xrightarrow{\text{base}}$ ArO—R

Ester

+ RCOX $\xrightarrow{\text{base}}$ ArO—CO—R

(*b*) OH Group

+ $Zn°$ ⟶ Ar—H + ZnO

2. Benzene Ring Reactions

(*a*) Reduction

+ $3H_2$/Ni ⟶ $C_6H_{11}OH$

(*b*) Oxidation

+ $Cr_2O_7^{2-}$, H^+ ⟶ O=C_6H_4=O Quinone

(*c*) Electrophilic Substitution

+ Br_2 (H_2O sol.) ⟶ 2,4,6-$Br_3C_6H_2OH$

+ H_2SO_4 ⟶ p-HO—C_6H_4—SO_3H

+ dil. HNO_3 ⟶ o- and p-$O_2NC_6H_4OH$

+ HNO_2 ⟶

p-HO—C_6H_4—N=O, p-O=C_6H_4=NOH

+ $Ar'N_2^+X^- \xrightarrow{5°C} p$-HO—Ar—N=N—Ar'

+ $Hg(OAc)_2$ ⟶ o- and p-HO—C_6H_4—HgOAc

+ ROH/H_2SO_4 ⟶ p-HO—C_6H_4—R

+ $CH_2{=}O$ ⟶ o- and p-HO—$C_6H_4CH_2OH$

+ RCHO ⟶ $RCH(C_6H_4OH\text{-}p)_2$

21.6 SUMMARY OF ETHERS AND ESTERS OF PHENOLS

$$\text{ArO}^-\text{Na}^+ + \text{X—CH}_2\text{CH═CH}_2 \longrightarrow \text{Ar—O—CH}_2\text{CH═CH}_2 \xrightarrow{\text{heat}} o\text{-HO—C}_6\text{H}_4\text{CH}_2\text{CH═CH}_2$$

$$\text{ArO}^-\text{Na}^+ + \text{X—CH}_2\underset{\substack{|\\ \text{CH}_3}}{\text{CH}}\text{—CH}_3 \longrightarrow \text{Ar—O—CH}_2\text{CH(CH}_3)_2 \xrightarrow{\text{AlCl}_3} p\text{-HO—C}_6\text{H}_4\text{C(CH}_3)_3$$

$$\text{ArO}^-\text{Na}^+ + \text{X—}\overset{\overset{\text{O}}{\|}}{\text{C}}\text{—CH}_3 \longrightarrow \text{Ar—O—}\overset{\overset{\text{O}}{\|}}{\text{C}}\text{—R} + \text{AlCl}_3 \longrightarrow p\text{- or } o\text{-HO—C}_6\text{H}_4\overset{\overset{\text{O}}{\|}}{\text{C}}\text{—R}$$

Supplementary Problems

Problem 21.21 Name the following compounds:

(*a*) 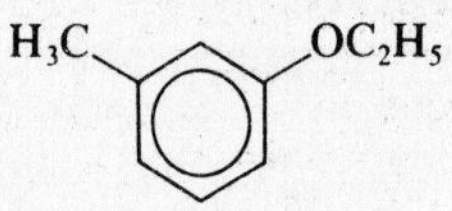
H3C
OC2H5

(*b*) 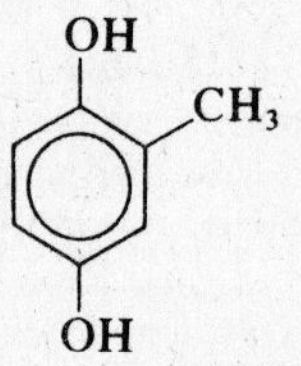
OH
CH3
OH

(*c*) 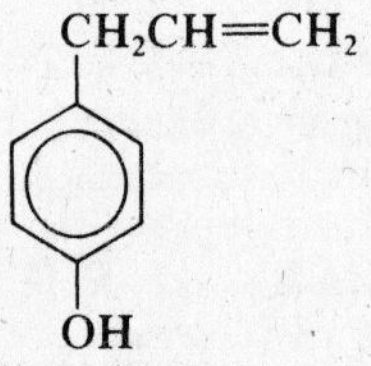
CH2CH=CH2
OH

◀

(*a*) *m*-ethoxytoluene, (*b*) methylhydroquinone, (*c*) *p*-allylphenol.

Problem 21.22 Write the structure for (*a*) phenoxyacetic acid, (*b*) phenyl acetate, (*c*) 2-hydroxy-3-phenylbenzoic acid, (*d*) *p*-phenoxyanisole. ◀

(*a*) 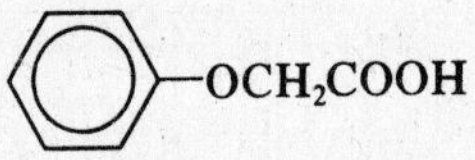
OCH2COOH

(*b*) 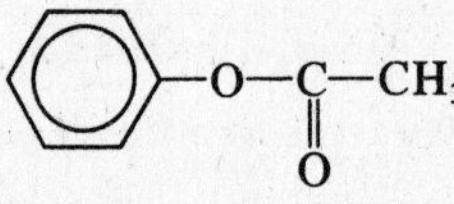
O—C—CH3
O

(*c*) 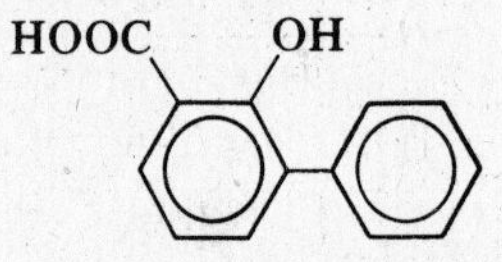
HOOC
OH

(*d*) 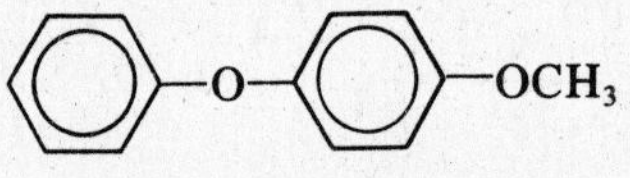
O
OCH3

Problem 21.23 Draw a flow sheet for the separation of a mixture of PhOH, $PhCH_2OH$ and PhCOOH. ◀

See Fig. 21-1.

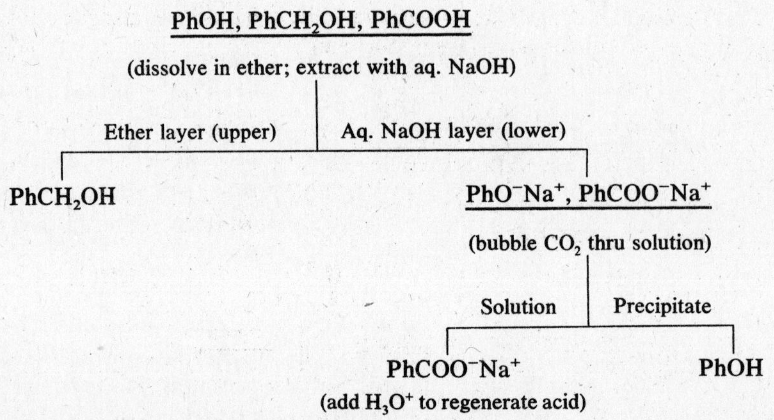
PhOH, PhCH2OH, PhCOOH
(dissolve in ether; extract with aq. NaOH)
Ether layer (upper)
Aq. NaOH layer (lower)
PhCH2OH
PhO⁻Na⁺, PhCOO⁻Na⁺
(bubble CO2 thru solution)
Solution
Precipitate
PhCOO⁻Na⁺
PhOH
(add H3O⁺ to regenerate acid)

Fig. 21-1

Problem 21.24 What product is formed when *p*-cresol is reacted with (*a*) (CH₃CO)₂O, (*b*) PhCH₂Br and base, (*c*) aqueous NaOH, (*d*) aqueous NaHCO₃, (*e*) bromine water? ◄

(*a*) *p*-CH₃C₆H₄OCOCH₃, *p*-cresyl acetate; (*b*) *p*-CH₃C₆H₄OCH₂Ph, *p*-tolyl benzyl ether; (*c*) *p*-CH₃C₆H₄O⁻Na⁺, sodium *p*-cresoxide; (*d*) no reaction; (*e*) 2,6-dibromo-4-methylphenol.

Problem 21.25 Use simple, test tube reactions to distinguish (*a*) *p*-cresol from *p*-xylene, (*b*) salicylic acid from aspirin (acetylsalicylic acid). ◄

(*a*) Aqueous NaOH dissolves the cresol. (*b*) Salicylic acid is a phenol which gives a color (purple in this case) with FeCl₃.

Problem 21.26 Identify compounds (A) through (D).

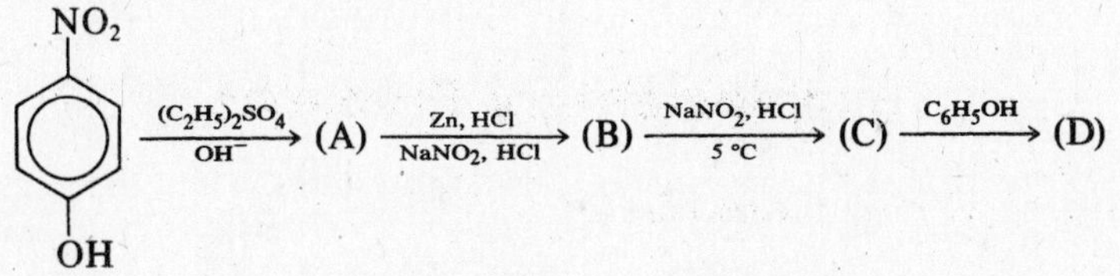

(A) *p*-C₂H₅OC₆H₄NO₂   (B) *p*-C₂H₅OC₆H₄NH₂ (*p*-phenetidine)   (C) *p*-C₂H₅OC₆H₄N₂⁺Cl⁻

(D) 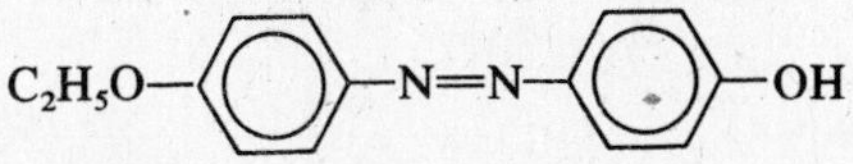 C₂H₅O—⟨ ⟩—N=N—⟨ ⟩—OH

Problem 21.27 Prepare (*a*) *o*-bromo-*p*-hydroxytoluene from toluene, (*b*) 2-hydroxy-5-methylbenz-aldehyde from *p*-toluidine, (*c*) *m*-methoxyaniline from benzenesulfonic acid. ◄

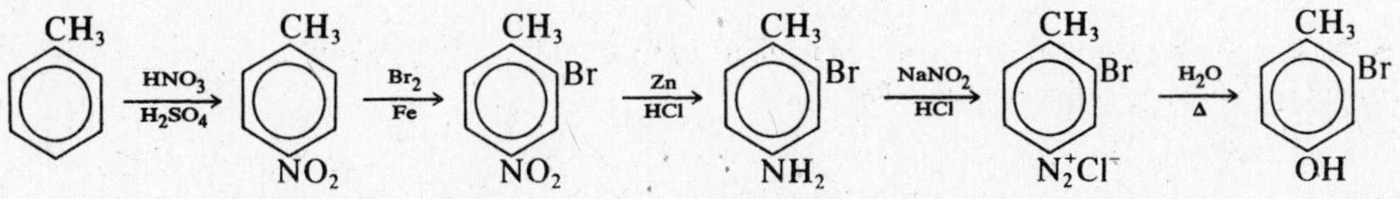

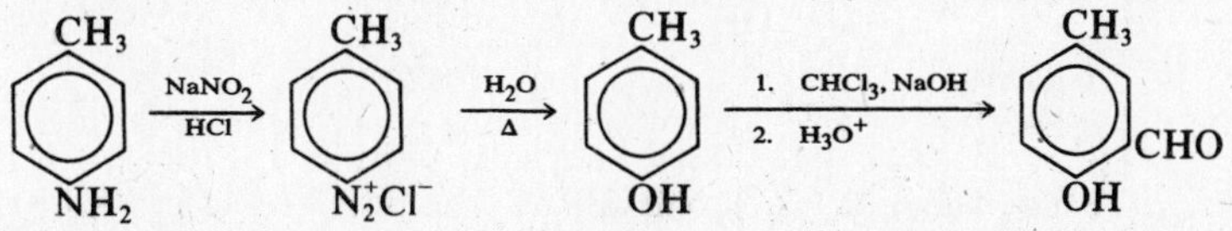

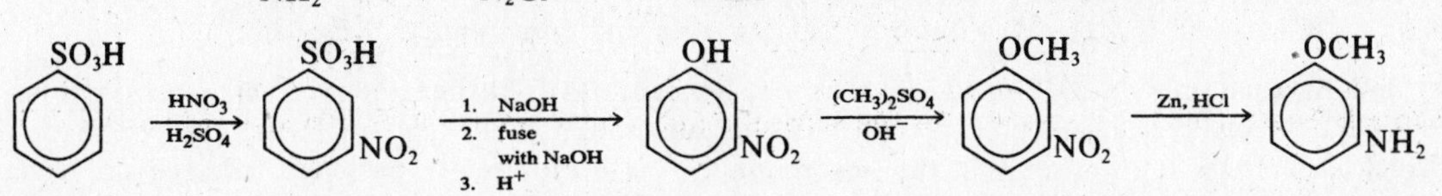

Problem 21.28 From phenol prepare (*a*) *p*-benzoquinone, (*b*) *p*-benzoquinone dioxime, (*c*) quinhy-drone (a 1:1 complex of *p*-benzoquinone and hydroquinone). ◄

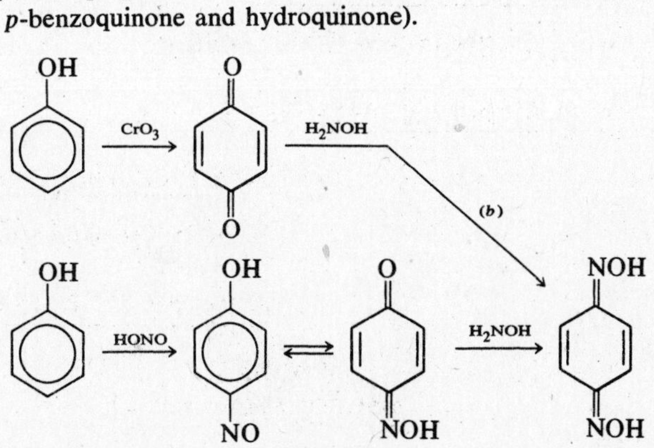

tautomers

(c)

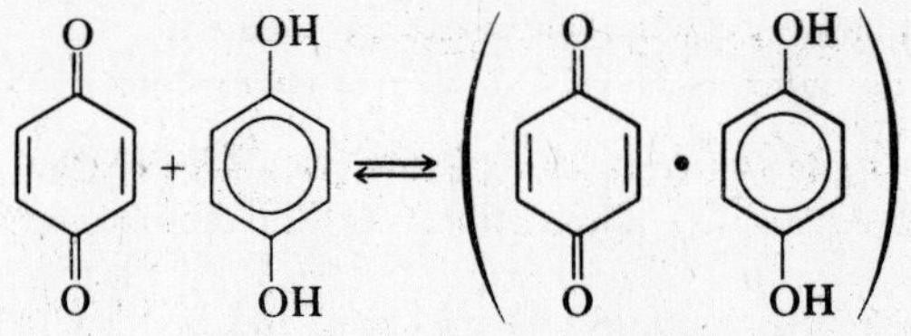

a charge-transfer complex

Problem 21.29 HCl adds to *p*-benzoquinone by a 1,4-addition. Show the steps and predict the structure of the phenolic product. ◄

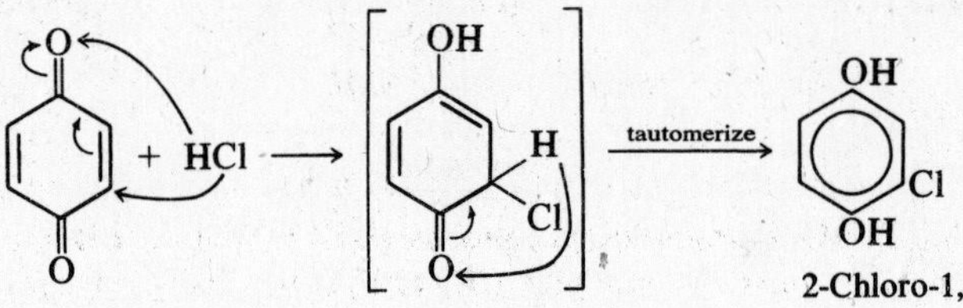

2-Chloro-1,4-dihydroxybenzene

Problem 21.30 Write a structural formula for the product of the Diels-Alder reaction of *p*-benzoquinone with:

(*a*) Butadiene   (*b*)    (*c*)

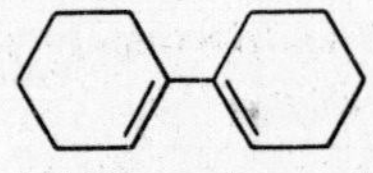

1,3-Cyclohexadiene   1,1′-Bicyclohexenyl   ◄

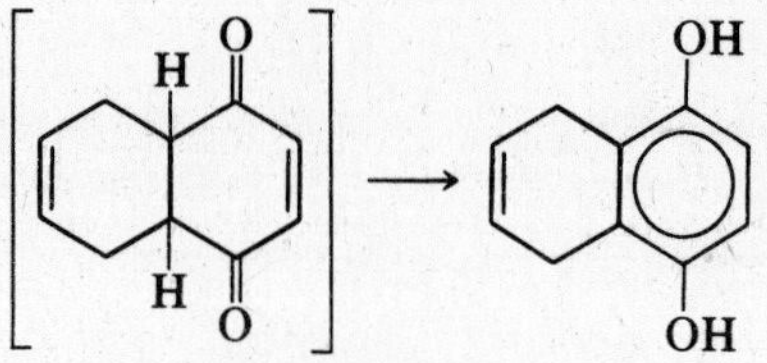

tautomers

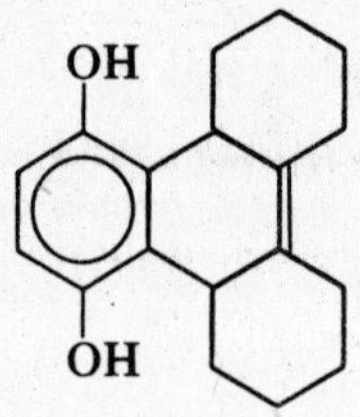

Problem 21.31 The ir OH stretching bands for the 3 isomeric nitrophenols in KBr pellets and dilute CCl₄ solution are identical for the *ortho* but different for *meta* and *para* isomers. Explain. ◄

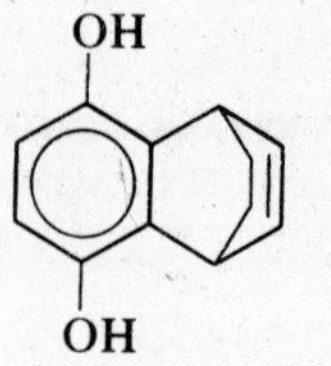

In KBr (solid state) the OH for all 3 isomers is H-bonded. In CCl₄, the H-bonds of *meta* and *para* isomers, which are intermolecular, are broken. Their ir OH absorption bands shift to higher frequencies (3325 to 3520 cm^{-1}). There is no change in the absorption of the *ortho* isomer (3200 cm^{-1}), since intramolecular H-bonds are not broken by dilution by solvent.

Problem 21.32 Show all major products for the following reactions:

(*a*) CH₃—⬡—OH + CHCl₃ + NaOH $\xrightarrow{70\ °C}$ (*b*) C₆H₅CH₂OOH + acid $\longrightarrow$

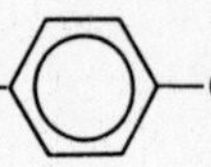

(*c*) Br—⬡—OOCCH₃ + C₂H₅—⬡—OOCC₂H₅ $\xrightarrow{AlCl_3}$ ◄

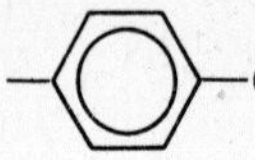

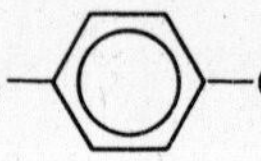

(*a*) Reimer-Tiemann reaction;

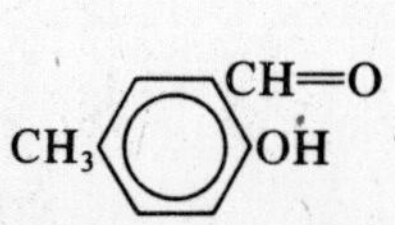

(b) This is a 1,2-shift of phenyl to an electron-deficient O.

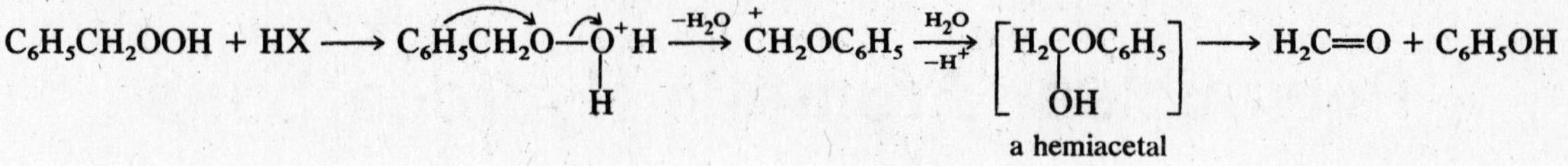

$$C_6H_5CH_2OOH + HX \longrightarrow C_6H_5CH_2O\overset{\frown}{\underset{H}{O^+}}H \xrightarrow[\ ]{-H_2O} \overset{+}{CH_2OC_6H_5} \xrightarrow[-H^+]{H_2O} \left[\underset{OH}{H_2COC_6H_5} \right] \longrightarrow H_2C{=}O + C_6H_5OH$$

a hemiacetal

(c) Intramolecular rearrangement gives two products, (A) and (B), and intermolecular rearrangement gives two more products, (C) and (D). These are Fries rearrangements.

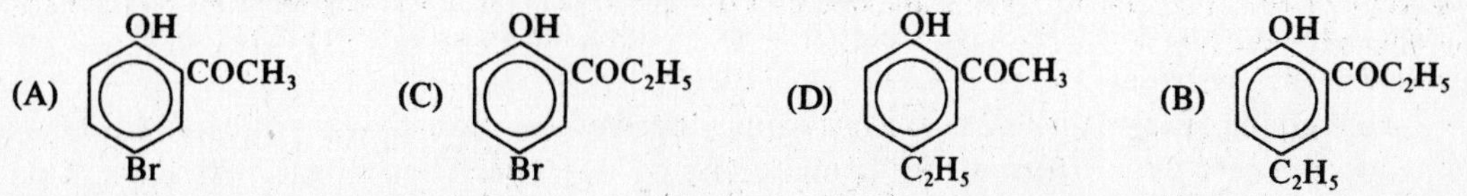

(A) [benzene ring with OH top, COCH$_3$ right, Br bottom]   (C) [benzene ring with OH top, COC$_2$H$_5$ right, Br bottom]   (D) [benzene ring with OH top, COCH$_3$ right, C$_2$H$_5$ bottom]   (B) [benzene ring with OH top, COC$_2$H$_5$ right, C$_2$H$_5$ bottom]

Problem 21.33 (a) Compound (A), C_7H_8O, is insoluble in aqueous $NaHCO_3$ but dissolves in NaOH. When treated with bromine water, (A) rapidly forms compound (B), $C_7H_5OBr_3$. Give structures for (A) and (B). (b) What would (A) have to be if it did not dissolve in NaOH? ◄

(a) With four degrees of unsaturation, (A) has a benzene ring. From its solubility, (A) must be a phenol with a methyl substituent to account for the seventh C. Since a tribromo compound is formed, (A) must be m-cresol and (B) is 2,4,6-tribromo-3-methylphenol. (b) (A) could not be a phenol. It would have to be an ether, $C_6H_5OCH_3$ (anisole).

Chapter 22

Polynuclear Aromatic Hydrocarbons

22.1 INTRODUCTION

Polynuclear aromatic compounds have **more than one** benzene ring. **Biphenyl**, C_6H_5—C_6H_5, and **triphenylmethane**, $(C_6H_5)_3CH$, have **isolated rings**. Benzene rings sharing 2 *ortho* C's are **fused** or **condensed ring systems**.

Although the Hückel $4n + 2$ rule is rigorously derived for monocyclic systems, it is also applied in an approximate way to fused ring compounds. Since two fused rings must share a pair of π electrons, the aromaticity and the conjugation (delocalization) energy per ring is less than that of benzene itself. Decreased aromaticity of polynuclear aromatics is also revealed by the different C—C bond lengths.

22.2 ISOLATED RING SYSTEMS

BIPHENYL AND DERIVATIVES

1. Nomenclature

The numbering system in biphenyl is:

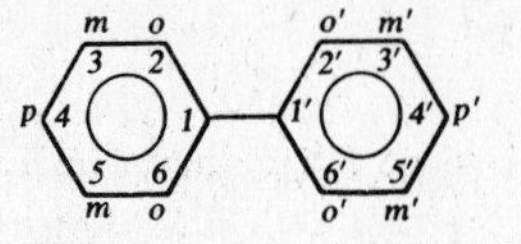

Problem 22.1 Write names for:

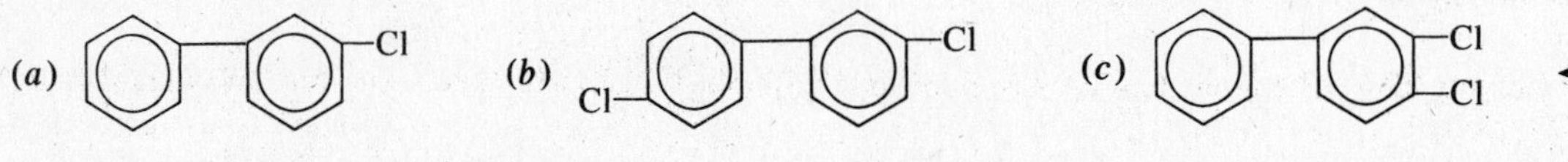

(*a*) 3-chlorobiphenyl, (*b*) 3,4′-dichlorobiphenyl, (*c*) 3,4-dichlorobiphenyl.

Problem 22.2 Write the structures for (*a*) 2-chloro-3-nitrobiphenyl, (*b*) 2,3′-dibromobiphenyl, (*c*) *p,p′*-di(N,N-dimethylamino)biphenyl.

2. Preparation

 (*a*) Ullmann reaction (see page 350).
 (*b*) From diazonium compounds. Free-radical cleavage forms an aryl radical which attacks the aromatic compound (Section 11.2, Problem 11.26).

(1) Gomberg reaction (for unsymmetrical biphenyls)

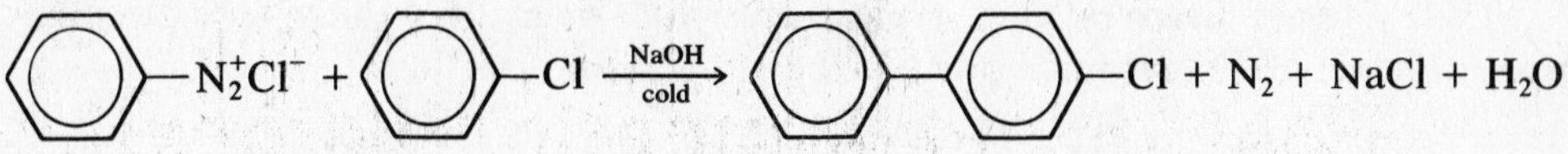

$$C_6H_5{-}N_2^+Cl^- + Cl{-}C_6H_4 \xrightarrow[\text{cold}]{\text{NaOH}} C_6H_5{-}C_6H_4{-}Cl + N_2 + NaCl + H_2O$$

$$ArN_2^+X^- \xrightarrow{OH^-} [Ar{-}N{=}N{-}OH] \longrightarrow \cdot OH + N_2 + \cdot Ar \xrightarrow{C_6H_5Cl} Ar{-}C_6H_4Cl + H_2O$$
a diazoic acid

(2) Gatterman reaction

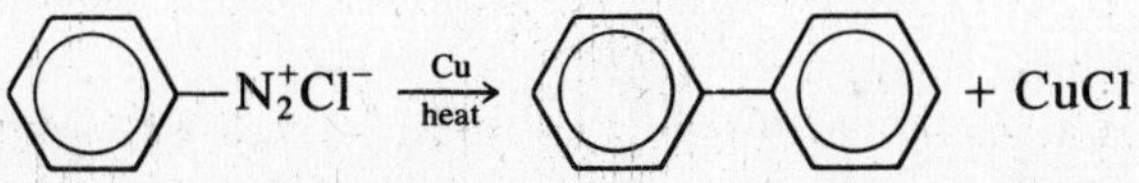

$$C_6H_5{-}N_2^+Cl^- \xrightarrow[\text{heat}]{Cu} C_6H_5{-}C_6H_5 + CuCl$$

(c) **From benzidine** (see page 336).

Problem 22.3 Prepare *p,p'*-dimethylbiphenyl by (a) Ullmann, (b) Gatterman and (c) Gomberg reactions. ◄

(a) $$CH_3{-}C_6H_4{-}I + Cu \xrightarrow{\text{heat}} H_3C{-}C_6H_4{-}C_6H_4{-}CH_3$$

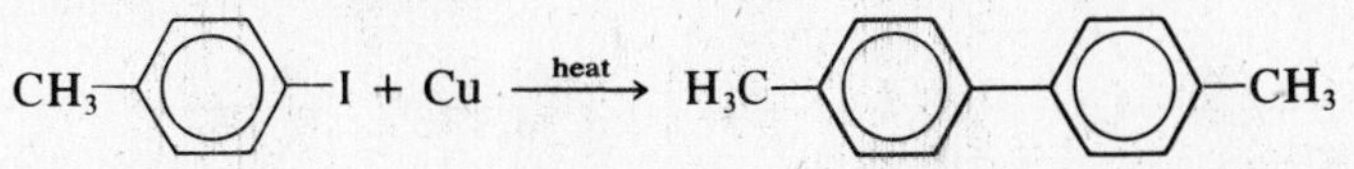

(b) $$CH_3{-}C_6H_4{-}N_2^+Cl^- \xrightarrow[\text{heat}]{Cu} H_3C{-}C_6H_4{-}C_6H_4{-}CH_3$$

(c) $$CH_3{-}C_6H_4{-}N_2^+Cl^- + C_6H_5{-}CH_3 \xrightarrow{\text{NaOH}} H_3C{-}C_6H_4{-}C_6H_4{-}CH_3$$

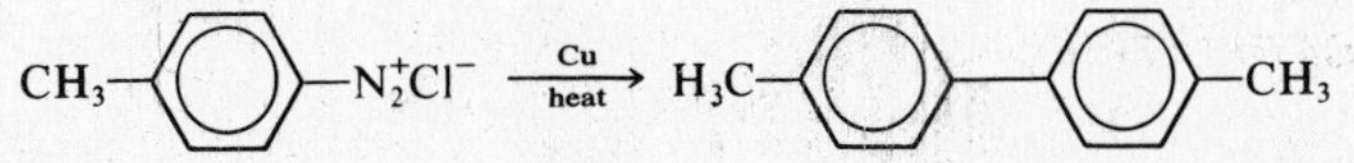

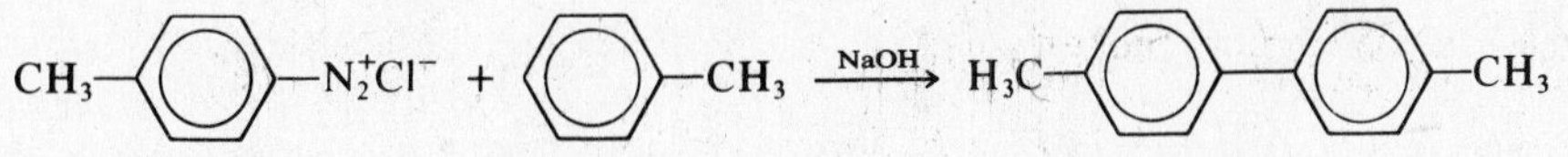

Problem 22.4 Distinguish between $p\text{-}CH_3C_6H_4C_6H_5$ and $(C_6H_5)_2CH_2$ by a rapid chemical reaction. ◄

Mild oxidation with CrO_3 leaves the diphenyl unchanged but converts $(C_6H_5)_2CH_2$ to $(C_6H_5)_2C{=}O$, which forms an oxime and semicarbazone. Vigorous oxidation of diphenylmethane gives C_6H_5COOH, while $p\text{-}CH_3C_6H_4C_6H_5$ forms $p\text{-}HOOC{-}C_6H_4C_6H_5$. Note the stability of the rings toward typical oxidation conditions.

Problem 22.5 Prepare (a) 4,4'-bis(chloromethyl)biphenyl from toluene, (b) *p,p'*-dibromobiphenyl from $C_6H_5NO_2$. ◄

(a) $$CH_3{-}C_6H_5 \xrightarrow[\text{2. } H_2/Pt]{\text{1. } HNO_3, H_2SO_4} CH_3{-}C_6H_4{-}NH_2 \xrightarrow[\text{2. KI}]{\text{1. } NO_2^-, H^+} CH_3{-}C_6H_4{-}I \xrightarrow{Cu}$$

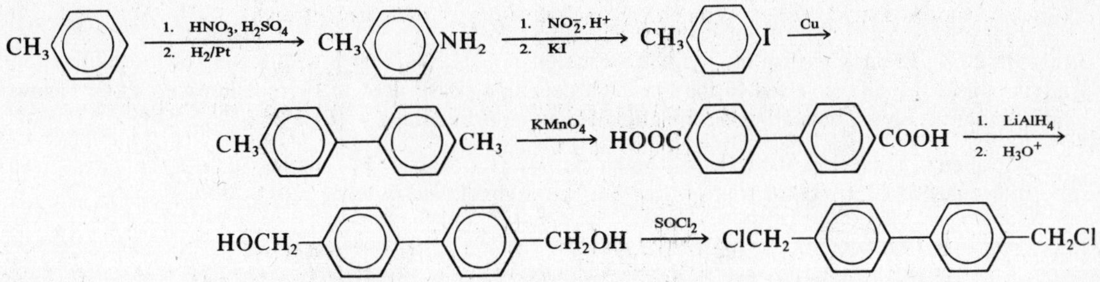

$$CH_3{-}C_6H_4{-}C_6H_4{-}CH_3 \xrightarrow{KMnO_4} HOOC{-}C_6H_4{-}C_6H_4{-}COOH \xrightarrow[\text{2. } H_3O^+]{\text{1. } LiAlH_4}$$

$$HOCH_2{-}C_6H_4{-}C_6H_4{-}CH_2OH \xrightarrow{SOCl_2} ClCH_2{-}C_6H_4{-}C_6H_4{-}CH_2Cl$$

(b) $$C_6H_5{-}NO_2 \xrightarrow{Zn, NaOH} C_6H_5{-}NH{-}NH{-}C_6H_5 \xrightarrow[\Delta]{HCl}$$

$$2Cl^- + H_3N^+{-}C_6H_4{-}C_6H_4{-}NH_3^+ \xrightarrow[\text{2. } 2NaNO_2, H_2SO_4, 5\,°C]{\text{1. } 2NaOH}$$

$$HSO_4^-N_2^+{-}C_6H_4{-}C_6H_4{-}N_2^+HSO_4^- \xrightarrow{CuBr} Br{-}C_6H_4{-}C_6H_4{-}Br$$

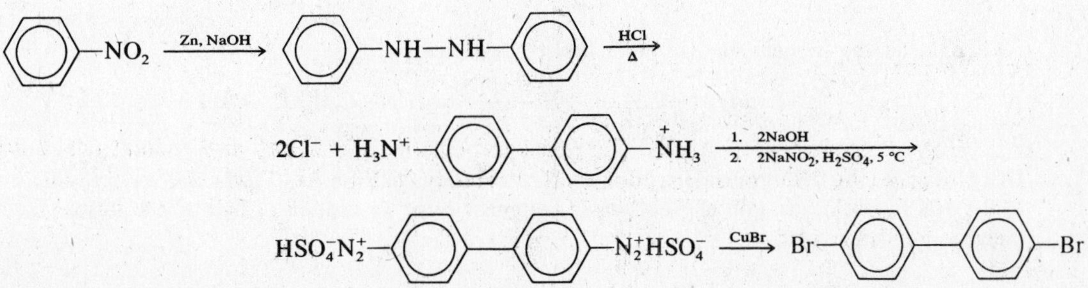

3. Reactions of Biphenyls

Each phenyl behaves as a typical aromatic ring. It weakly activates the other benzene ring. Phenyl and substituted phenyls are *ortho,para*-directing.

Problem 22.6 Explain the products formed when biphenyl is mononitrated and when it is dinitrated. ◄

Mononitration gives mainly the *para* isomer, since the steric effect of the other phenyl inhibits *ortho* substitution. The second —NO$_2$ substitutes in the *para* position of the other ring because the first —NO$_2$ deactivates the ring on which it is substituted. Although NO$_2$ is *meta*-directing, p-O$_2$NC$_6$H$_4$ is *para*-directing.

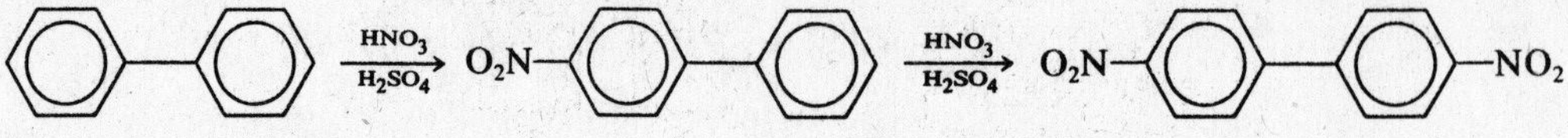

Problem 22.7 Give the main products, (A) through (E), in the following sequence.

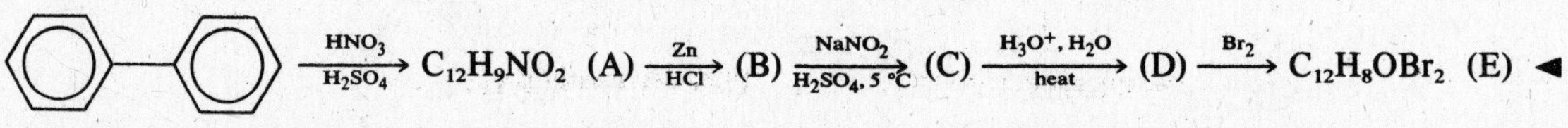

(A) 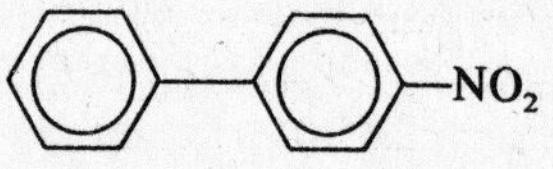 (B) 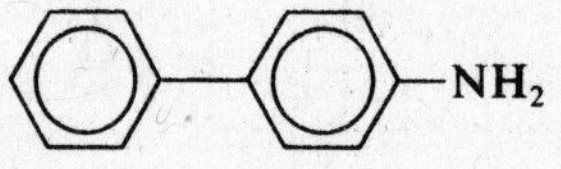 (C)

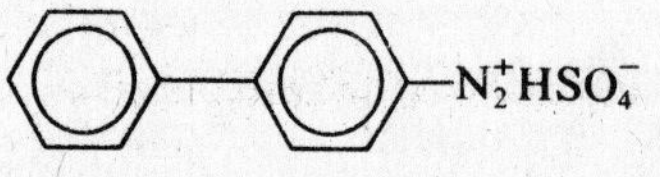

(D) 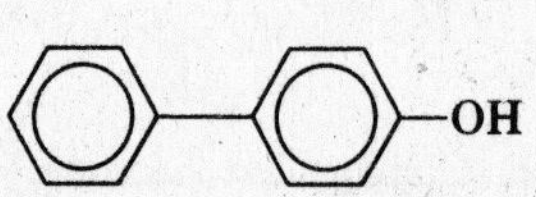 (E)

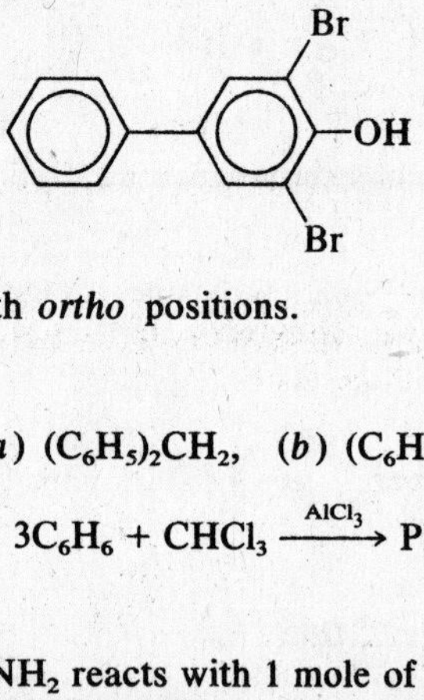

The —OH activates the ring and facilitates substitution in both *ortho* positions.

Problem 22.8 Use the Friedel-Crafts reaction to prepare (a) (C$_6$H$_5$)$_2$CH$_2$, (b) (C$_6$H$_5$)$_3$CH. ◄

(a) $2C_6H_6 + CH_2Cl_2 \xrightarrow{AlCl_3} Ph_2CH_2$ (b) $3C_6H_6 + CHCl_3 \xrightarrow{AlCl_3} Ph_3CH$

Problem 22.9 In the presence of C$_6$H$_5$NO$_2$, 2 moles of C$_6$H$_5$NH$_2$ reacts with 1 mole of *p*-toluidine to give a triarylmethane that is converted to the dye pararosaniline (Basic Red 9) by reaction with PbO$_2$ followed by acid. Show the steps, indicating the function of (a) nitrobenzene, (b) PbO$_2$, (c) HCl. ◄

(a) Nitrobenzene oxidizes the CH$_3$ of *p*-toluidine to CHO, whose O is eliminated with the *para* H's from 2 molecules of C$_6$H$_5$NH$_2$ to form *p*-triaminotriphenylmethane (a leuco base).

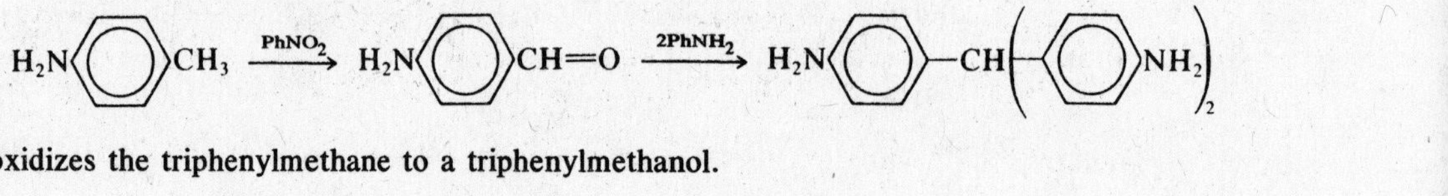

(b) PbO$_2$ oxidizes the triphenylmethane to a triphenylmethanol.

$$(p\text{-}H_2N\text{--}C_6H_4)_3C\text{--}H \xrightarrow{PbO_2} (p\text{-}H_2N\text{--}C_6H_4)_3C\text{--}OH$$
$$\text{leuco base} \qquad\qquad\qquad \text{color base}$$

(c) HCl protonates the OH, thus making possible the loss of H$_2$O to form Ar$_3$C$^+$, whose + charge is delocalized to the 3 N's. Delocalization of electrons is responsible for absorption of light in the visible spectrum, thereby producing color.

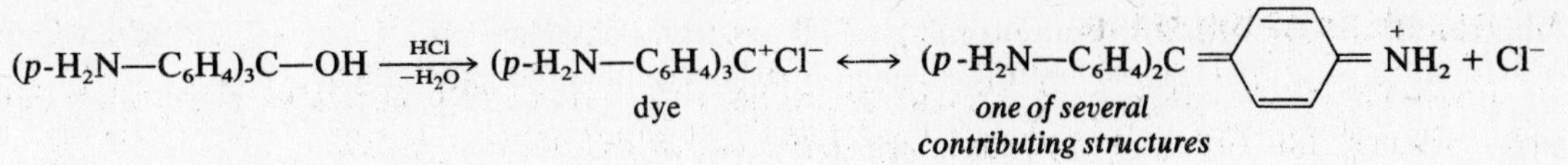

(p-H₂N—C₆H₄)₃C—OH
HCl
−H₂O
(p-H₂N—C₆H₄)₃C⁺Cl⁻
dye
(p-H₂N—C₆H₄)₂C
NH₂ + Cl⁻
one of several
contributing structures

Problem 22.10 The indicator phenolsulfonphthalein is made by using concentrated H_2SO_4 to condense 2 moles of phenol and one mole of sulfobenzoic anhydride by eliminating 1 mole of H_2O. What is its formula? ◄

OH
OH
H O
H
C
O
S
O₂
H₂SO₄
heat
+ H₂O

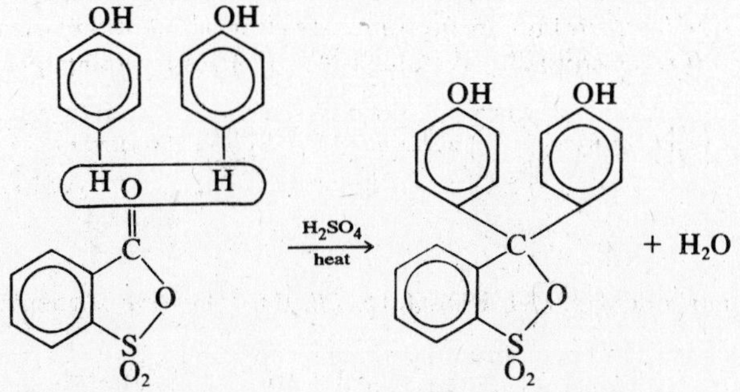

Sulfobenzoic anhydride Phenolsulfonphthalein

22.3 NAPHTHALENE

α
α
β
β
β
β
α
α
1
2
3
4
5
6
7
8

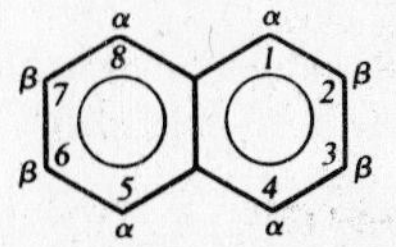

Position designation for naphthalene

Problem 22.11 Write structures for (*a*) β-bromonaphthalene, (*b*) α-naphthol, (*c*) 1,5-dinitronaphthalene, (*d*) β-naphthoic acid. ◄

(a)
Br
(b)
OH
(c)
NO₂
NO₂
(d)
COOH

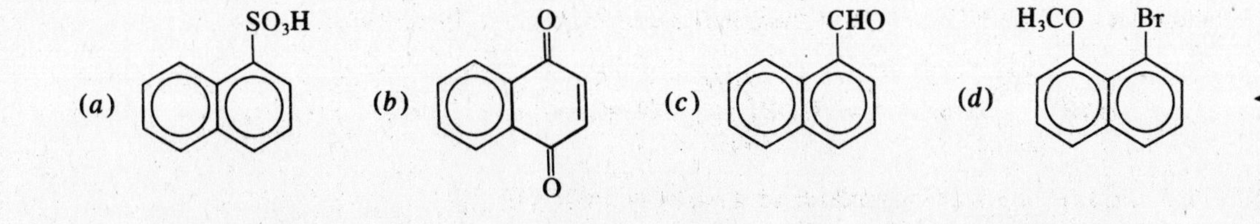

Problem 22.12 Name the following compounds:

(a)
SO₃H
(b)
O
O
(c)
CHO
(d)
H₃CO
Br

◄

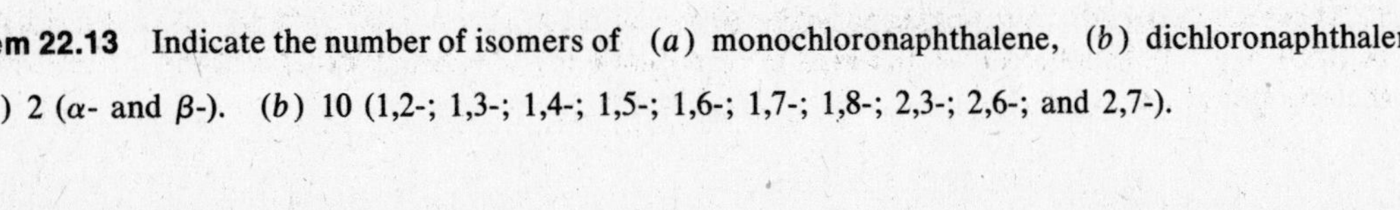

(*a*) 1-naphthalenesulfonic acid or α-naphthalenesulfonic acid, (*b*) 1,4-naphthoquinone, (*c*) 1-naphthaldehyde or α-naphthaldehyde, (*d*) 1-bromo-8-methoxynaphthalene.

Problem 22.13 Indicate the number of isomers of (*a*) monochloronaphthalene, (*b*) dichloronaphthalene.◄

(*a*) 2 (α- and β-). (*b*) 10 (1,2-; 1,3-; 1,4-; 1,5-; 1,6-; 1,7-; 1,8-; 2,3-; 2,6-; and 2,7-).

Problem 22.14   Oxidation of 1-nitronaphthalene yields 3-nitrophthalic acid.   However, if 1-nitronaphthalene is reduced to α-naphthylamine and if this amine is oxidized, the product is phthalic acid.

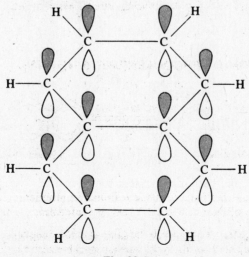

How do these reactions establish the gross structure of naphthalene?   ◄

The electron-attracting —NO_2 stabilizes ring A of 1-nitronaphthalene to oxidation, and ring B is oxidized to form 3-nitrophthalic acid.   The electron-releasing —NH_2 makes ring A more susceptible to oxidation, and α-naphthylamine is oxidized to phthalic acid.   The NO_2 labels one ring and establishes the presence of 2 fused benzene rings in naphthalene.

Problem 22.15   Draw a conclusion about the stability and aromaticity of naphthalene from the fact that the experimentally determined heat of combustion is 61 kcal/mole less than that calculated from the structural formula.   ◄

The difference of 61 kcal/mole is naphthalene's resonance energy.   Naphthalene is less aromatic than benzene because a resonance energy of 30.5 kcal/mole per ring is less than the 36 kcal/mole resonance energy of benzene.

Problem 22.16   Deduce an orbital picture (like Fig. 10-2) for naphthalene, a planar molecule with bond angles of 120°.   ◄

See Fig. 22-1.   The C's use sp^2 hybrid atomic orbitals to form σ bonds with each other and with the H's.   The remaining p orbitals at right angles to the plane of the C's overlap laterally to form a π electron cloud.

Fig. 22-1

CHEMICAL PROPERTIES

1. Oxidation

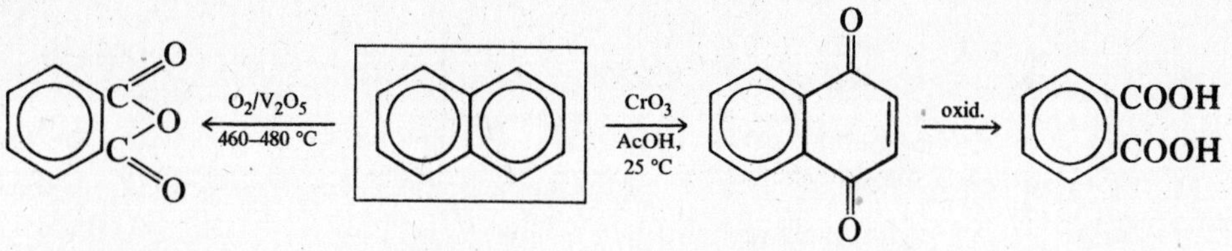

Phthalic anhydride ← O_2/V_2O_5, 460–480 °C — Naphthalene — CrO_3, AcOH, 25 °C → 1,4-Naphthoquinone — oxid. → Phthalic acid

2. Reduction

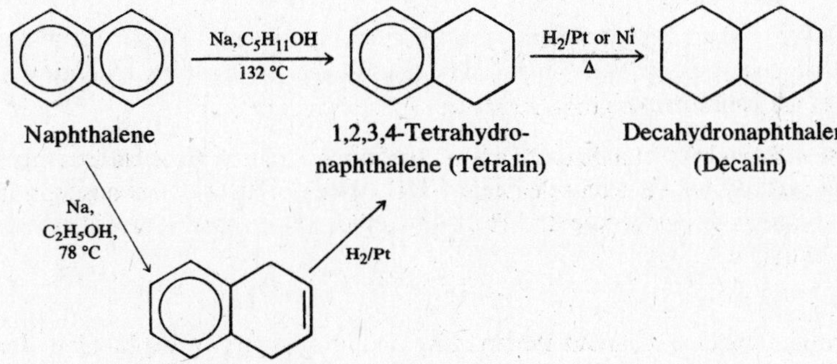

Naphthalene — Na, $C_5H_{11}OH$, 132 °C → 1,2,3,4-Tetrahydro-naphthalene (Tetralin) — H_2/Pt or Ni, Δ → Decahydronaphthalene (Decalin)

Na, C_2H_5OH, 78 °C ↓

1,4-Dihydronaphthalene — H_2/Pt →

ELECTROPHILIC SUBSTITUTION

1. Occurs preferentially at α position.

2. Examples of β-substitution are: (a) sulfonation at high temperatures (at low temperatures α-substitution occurs); (b) acylation with RCOCl and $AlCl_3$, in $C_6H_5NO_2$ as solvent (in CS_2 or CH_2ClCH_2Cl α-substitution occurs).

3. Substitution occurs in a ring holding an activating (electron-releasing) group: (a) *para* to an α-substituent; (b) *ortho* to an α-substituent if the *para* position is blocked; (c) to α *ortho* position if the activating group is a β-substituent.

4. A deactivating group (electron-withdrawing) directs electrophiles into the other ring, usually at α positions.

BUCHERER REACTION FOR INTERCONVERTING β-NH$_2$ AND β-OH

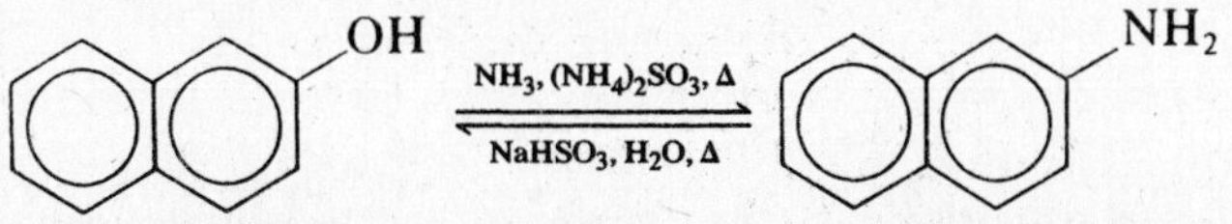

β-Naphthol ⇌ [NH_3, $(NH_4)_2SO_3$, Δ / $NaHSO_3$, H_2O, Δ] ⇌ β-Naphthylamine

Problem 22.17 Account for (a) formation of the α-isomer in nitration and halogenation of naphthalene, (b) formation of α-naphthalenesulfonic acid at 80 °C and β-naphthalenesulfonic acid at 160 °C. ◄

(a) The mechanism of electrophilic substitution is the same as that for benzene. Attack at the α position has a lower $\Delta H^{\ddagger}$ because intermediate I, an allylic R^+ with an intact benzene ring, is more stable than intermediate II from β-attack.

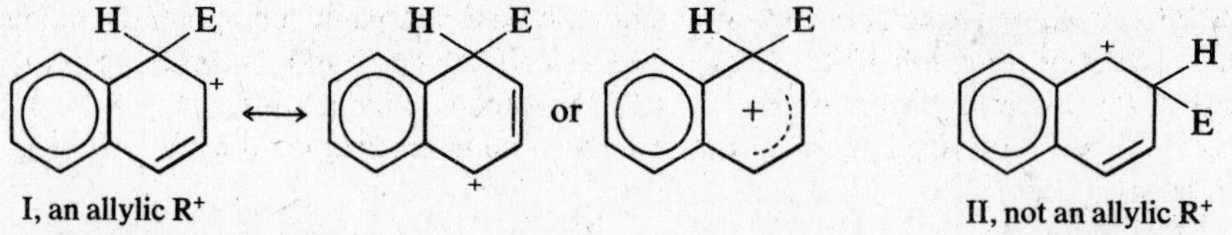

I, an allylic R$^+$
α-substitution

II, not an allylic R$^+$
β-substitution

In II the + charge is isolated from the remaining double bond and hence there is no direct delocalization of charge to the double bond without involvement of the stable benzene ring. In both I and II the remaining aromatic ring has the same effect on stabilizing the + charge. Since I is more stable than II, α-substitution predominates.

(b) α-Naphthalenesulfonic acid is the kinetic-controlled product [see part (a)]. However, sulfonation is a reversible reaction and at 160 °C the *thermodynamic*-controlled product, β-naphthalenesulfonic acid, is formed.

Problem 22.18 (a) Why cannot 2-naphthalenecarboxylic acid be formed by oxidation of 2-methylnaphthalene? (b) How is 2-naphthalenecarboxylic acid prepared? (c) Why does vigorous oxidation of naphthalene stop with the formation of phthalic anhydride? ◄

(a) Oxidation is an electrophilic attack and CH$_3$ is activating. The CH$_3$-substituted benzene ring is more susceptible to oxidation than is the side chain CH$_3$. The product is a quinone:

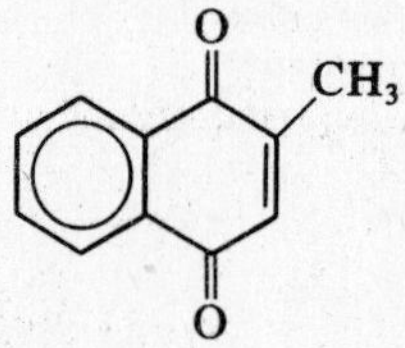

(b)

$$C_{10}H_8 \xrightarrow[\text{AlCl}_3,\ \text{PhNO}_2]{\text{CH}_3\text{COCl}} \beta\text{-}C_{10}H_7CCH_3 \xrightarrow[\text{2. } H_3O^+]{\text{1. NaOH, I}_2} \beta\text{-}C_{10}H_7COOH$$

The second step (NaOH + I$_2$) is an iodoform reaction.
(c) The benzene ring is more stable to oxidation than are side chains, and electron-withdrawing C=O groups make the ring even more resistant to oxidation.

Problem 22.19 Give the product(s) from (a) mononitration (i) and dinitration (ii) of naphthalene; (b) monobromination (i) and dibromination (ii) of α-naphthol. ◄

(a) (i) α-Substitution; α-C$_{10}$H$_7$NO$_2$. (ii) Ring without NO$_2$ is deactivated less and undergoes α-substitution:

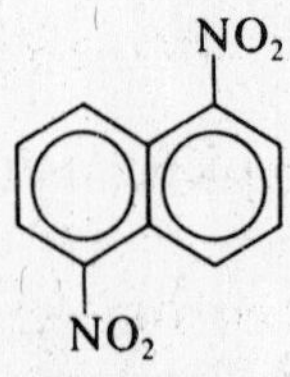

and

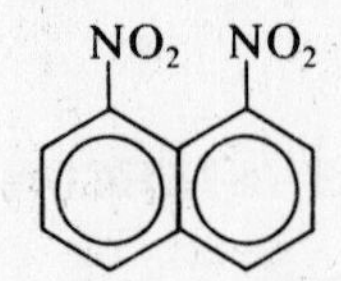

1,5-Dinitronaphthalene   1,8-Dinitronaphthalene

(b) (i) Ring with activating OH is substituted in the remaining α position. (ii) Second Br enters *ortho* to OH.

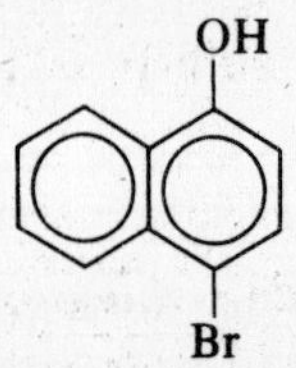

(i)   4-Bromo-1-naphthol 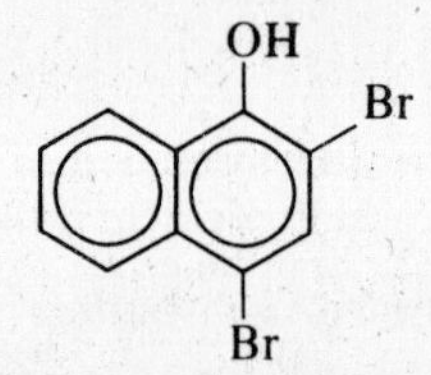   (ii)   2,4-Dibromo-1-naphthol

Problem 22.20 From naphthalene and any aliphatic and inorganic compounds prepare: (*a*) 1-bromo-naphthalene,   (*b*) 1-nitronaphthalene,   (*c*) 1-aminonaphthalene,   (*d*) 1-naphthoic acid,   (*e*) 1-*n*-propyl-naphthalene,   (*f*) 1-naphthylacetic acid,   (*g*) 1-(*α*-aminomethyl)naphthalene,   (*h*) 1-naphthaldehyde, (*i*) 1-naphthol.   Once synthesized, a compound may be used in ensuing syntheses.   Use Np as the symbol for the naphthyl group.   ◀

(*a*)
$$\text{NpH} \xrightarrow[\text{50\% HOAc}]{\text{Br}_2} \text{1-NpBr}$$

(*b*)
$$\text{NpH} \xrightarrow[\text{H}_2\text{SO}_4]{\text{HNO}_3} \text{1-NpNO}_2$$

(*c*)
$$\text{1-NpNO}_2 \xrightarrow[\text{2. OH}^-]{\text{1. Sn, HCl}} \text{1-NpNH}_2$$

(*d*)
$$\text{1-NpBr} \xrightarrow{\text{Mg}} \text{1-NpMgBr} \xrightarrow[\text{2. H}^+]{\text{1. CO}_2} \text{1-NpCOOH}$$

(*e*)
$$\text{NpH} \xrightarrow[\text{CS}_2]{\text{CH}_3\text{CH}_2\text{COCl}} \text{1-NpCOCH}_2\text{CH}_3 \xrightarrow[\text{HCl}]{\text{Zn-Hg}} \text{1-NpCH}_2\text{CH}_2\text{CH}_3$$

(*f*) $\text{1-NpCOOH} \xrightarrow[\text{2. H}_3\text{O}^+]{\text{1. LiAlH}_4} \text{1-NpCH}_2\text{OH} \xrightarrow{\text{SOCl}_2} \text{1-NpCH}_2\text{Cl} \xrightarrow{\text{KCN}} \text{1-NpCH}_2\text{CN} \xrightarrow{\text{H}_3\text{O}^+} \text{1-NpCH}_2\text{COOH}$

(*g*) $\text{1-NpNH}_2 \xrightarrow[\text{HCl, 5 °C}]{\text{NaNO}_2} \text{1-NpN}_2^+\text{Cl}^- \xrightarrow{\text{CuCN}} \text{1-NpCN} \xrightarrow[\text{2. H}_3\text{O}^+]{\text{1. LiAlH}_4} \text{1-NpCH}_2\text{NH}_2$

(*h*)
$$\text{1-NpCOOH} \xrightarrow{\text{SOCl}_2} \text{1-NpCOCl} \xrightarrow{\text{H}_2/\text{Pd(BaSO}_4)} \text{1-NpCHO}$$

(*i*) $\text{NpH} \xrightarrow[\text{80 °C}]{\text{H}_2\text{SO}_4} \text{1-NpSO}_3\text{H} \xrightarrow[\substack{\text{2. NaOH, fuse} \\ \text{3. H}_3\text{O}^+}]{\text{1. NaOH}} \text{1-NpOH}$ or $\text{1-NpN}_2^+\text{Cl}^- \xrightarrow[\text{heat}]{\text{H}_2\text{O}} \text{1-NpOH}$

Problem 22.21 Use instructions in Problem 22.20 and prepare (*a*) 2-naphthalenesulfonic acid, (*b*) 2-ethylnaphthalene,   (*c*) 2-naphthol,   (*d*) 2-naphthylamine,   (*e*) 2-bromonaphthalene,   (*f*) 2-cyanonaph-thalene, (*g*) 2-nitronaphthalene, (*h*) *α*-naphthylcarbinol.   ◀

(*a*)
$$\text{NpH} \xrightarrow[\text{180 °C}]{\text{H}_2\text{SO}_4} \text{2-NpSO}_3\text{H}$$

(*b*)
$$\text{NpH} \xrightarrow[\text{PhNO}_2, \text{AlCl}_3]{\text{CH}_3\text{COCl}} \text{2-NpCOCH}_3 \xrightarrow[\text{HCl}]{\text{Zn/Hg}} \text{2-NpCH}_2\text{CH}_3$$

(*c*) $\text{2-NpSO}_3\text{H} \xrightarrow[\substack{\text{2. NaOH, fuse} \\ \text{3. H}_3\text{O}^+}]{\text{1. NaOH}} \text{2-NpOH}$

(*d*)
$$\text{2-NpOH} \xrightarrow[\text{heat}]{\text{NH}_3, (\text{NH}_4)_2\text{SO}_3} \text{2-NpNH}_2 \quad (\text{Bucherer reaction})$$

(*e*)
$$\text{2-NpNH}_2 \xrightarrow[\text{HCl}]{\text{NaNO}_2} \text{2-NpN}_2^+\text{Cl}^- \xrightarrow{\text{CuBr}} \text{2-NpBr}$$

(*f*)
$$\text{2-NpN}_2^+\text{Cl}^- \xrightarrow{\text{CuCN}} \text{2-NpCN}$$

(*g*) $\text{2-NpN}_2^+\text{Cl}^- \xrightarrow{\text{NaNO}_2} \text{2-NpNO}_2$ or $\text{2-NpNH}_2 \xrightarrow[\text{90\% H}_2\text{O}_2]{\text{CF}_3\text{COOH}} \text{2-NpNO}_2$

(*h*)
$$\text{NpH} \xrightarrow[\text{Fe}]{\text{Br}_2} \alpha\text{-BrNp} \xrightarrow{\text{Mg}} \alpha\text{-BrMgNp} \xrightarrow[\text{2. H}^+]{\text{1. H}_2\text{CO}} \alpha\text{-HOCH}_2\text{Np}$$

Problem 22.22 Use the Bucherer reaction to prepare 2-(N-methyl)- and 2-(N-phenyl)naphthylamines.   ◀

CH$_3$NH$_2$ and C$_6$H$_5$NH$_2$ replace NH$_3$.

Problem 22.23 (*a*) Give products, where formed, for reaction of $PhN_2^+Cl^-$ with (i) α-naphthol, (ii) β-naphthol, (iii) 4-methyl-1-naphthol, (iv) 1-methyl-2-naphthol. (*b*) How can these products be used to make the corresponding aminonaphthols? ◀

(*a*) Structural formulas are given below. Coupling occurs (i) *para* (α) to OH; (ii) *ortho* (α, not β) to OH; (iii) *ortho* (β) to OH, since *para* (α) position is blocked. (iv) No reaction. An activating β-substituent cannot activate other β positions.

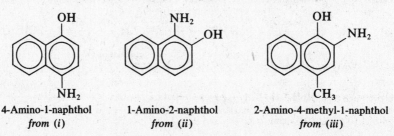

| (i) 4-Phenylazo-1-naphthol | (ii) 1-Phenylazo-2-naphthol | (iii) 4-Methyl-2-phenylazo-1-naphthol | (iv) no reaction |

(*b*) Reduction of the azo compounds with $LiAlH_4$ or $Na_2S_2O_4$ or Sn, HCl yields the amines.

| 4-Amino-1-naphthol *from* (*i*) | 1-Amino-2-naphthol *from* (*ii*) | 2-Amino-4-methyl-1-naphthol *from* (*iii*) |

Problem 22.24 Name the product and account for the orientation in the following electrophilic substitution reactions:

(*a*) 1-Methylnaphthalene + Br_2, Fe   (*b*) 2-Ethylnaphthalene + Cl_2, Fe

(*c*) 2-Ethylnaphthalene + C_2H_5COCl + $AlCl_3$ (*d*) 1-Methylnaphthalene + CH_3COCl, $AlCl_3(CS_2)$

(*e*) 2-Methoxynaphthalene + HNO_3 + H_2SO_4 (*f*) 2-Nitronaphthalene + Br_2, Fe   ◀

(*a*) 1-Methyl-4-bromonaphthalene. Br substitutes in the more reactive α position of the activated ring. (*b*) 1-Chloro-2-ethylnaphthalene. C^1 (*ortho* and α) is activated by C_2H_5 since C^4, which is also α, is *meta* to C_2H_5. (*c*) 1-(2-Ethylnaphthyl) ethyl ketone. Same reason as in (*b*). (*d*) 4-(1-Methylnaphthyl) methyl ketone. Same reason as in (*a*). (*e*) 1-Nitro-2-methoxynaphthalene. Same reason as in (*b*). (*f*) 1-Bromo-6-nitronaphthalene and 1-bromo-7-nitronaphthalene. NO_2 deactivates its ring and bromination occurs at the α positions of the other ring.

Problem 22.25 Name and account for the products from the reaction

2-Ethylnaphthalene + 96% H_2SO_4

at (*a*) 40 °C, (*b*) 140 °C. ◀

(*a*) 2-Ethyl-1-naphthalenesulfonic acid. The activating group directs substitution to the reactive *ortho* position. (*b*) 2-Ethyl-6-naphthalenesulfonic acid. Higher-temperature β-sulfonation in the less hindered ring gives the thermodynamic product.

Problem 22.26 Synthesize from naphthalene and any other reagents: (*a*) naphthionic acid (4-amino-1-naphthalenesulfonic acid), (*b*) 4-amino-1-naphthol, (*c*) 1,3-dinitronaphthalene, (*d*) 1,4-diaminonaphthalene, (*e*) 1,2-dinitronaphthalene. Do not repeat the synthesis of any compound. ◀

(*a*)

(b)

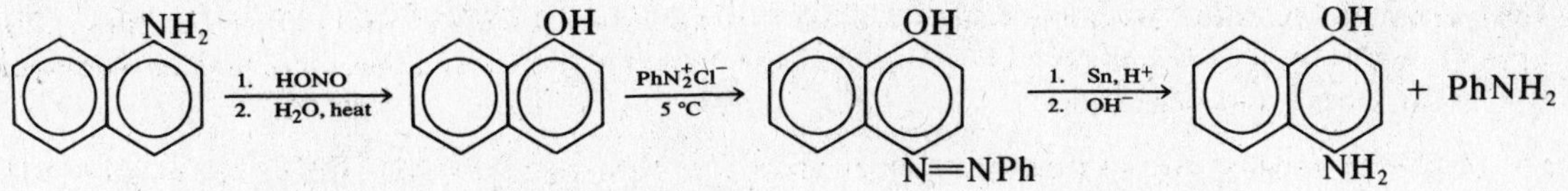

(c)

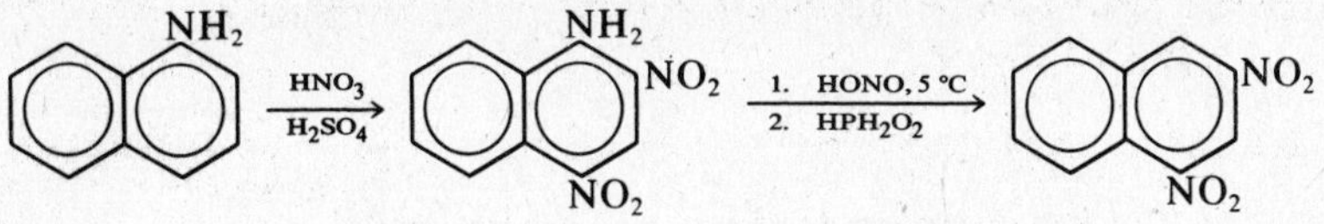

(d)

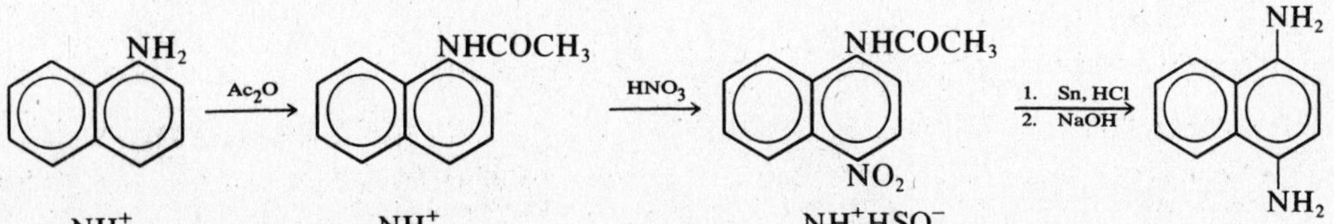

(e)

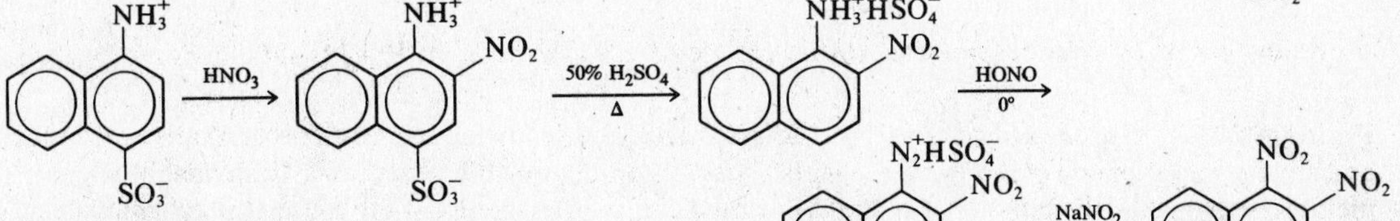

—SO_3^- blocks C^4 position

1. Haworth Synthesis of Naphthalenes by Ring Closure

(a) β-Substituted naphthalenes.

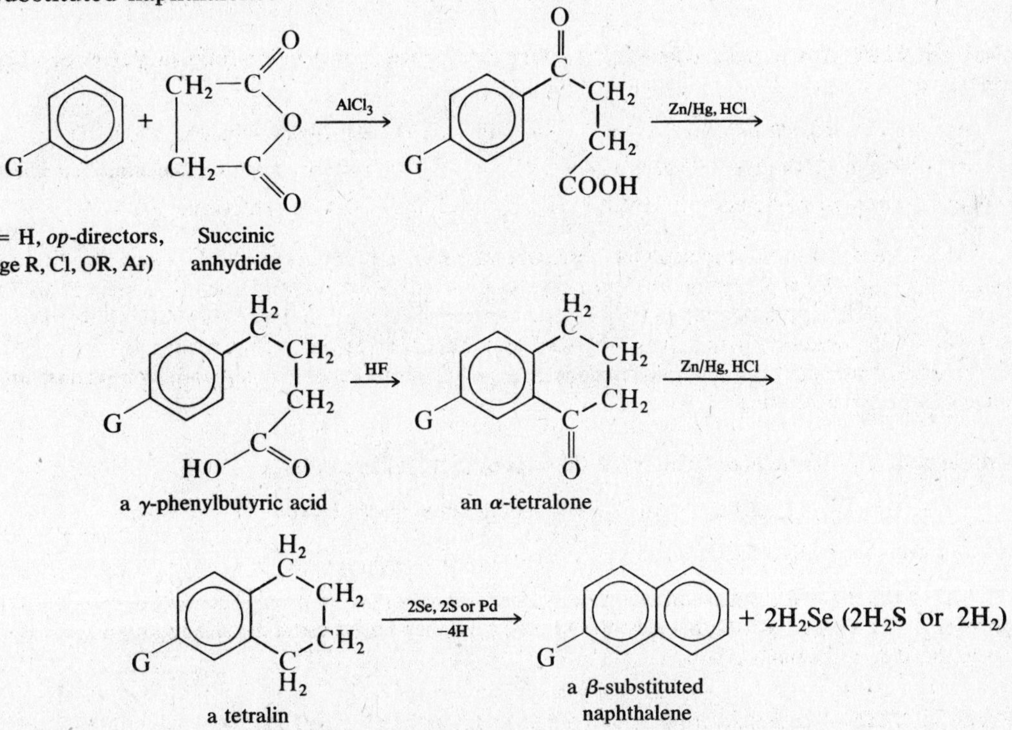

(G = H, op-directors,   Succinic
large R, Cl, OR, Ar)   anhydride

a γ-phenylbutyric acid   an α-tetralone

a tetralin   a β-substituted
                                naphthalene

(b) α-Alkyl (aryl) naphthalenes (G=H); otherwise get 1,7-disubstituted products.

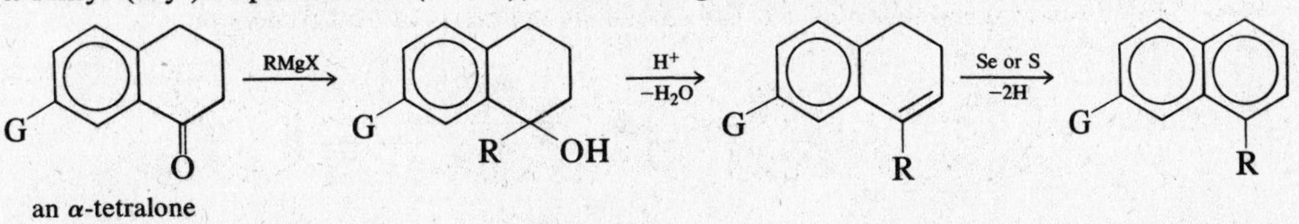

an α-tetralone

(c) Grignard reaction with ester of the keto acid (get 1,6-disubstituted naphthalenes).

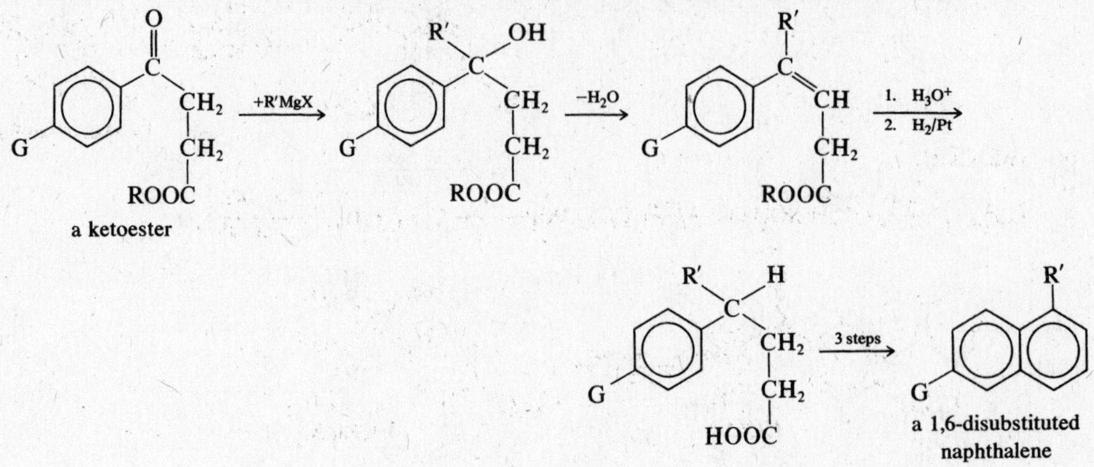

a ketoester

a 1,6-disubstituted
naphthalene

Problem 22.27 Use the Haworth method to synthesize the following naphthalene derivatives from benzene and any needed aliphatic or inorganic reagents: (a) 2-isopropylnaphthalene, (b) 1-ethylnaphthalene, (c) 1-methyl-7-ethylnaphthalene, (d) 1,6-diethylnaphthalene, (e) 1-methyl-4-ethyl-6-methoxynaphthalene. (SA is succinic anhydride). ◄

Only reactions in which substituents are introduced are shown.

(a) G is isopropyl (i-Pr).

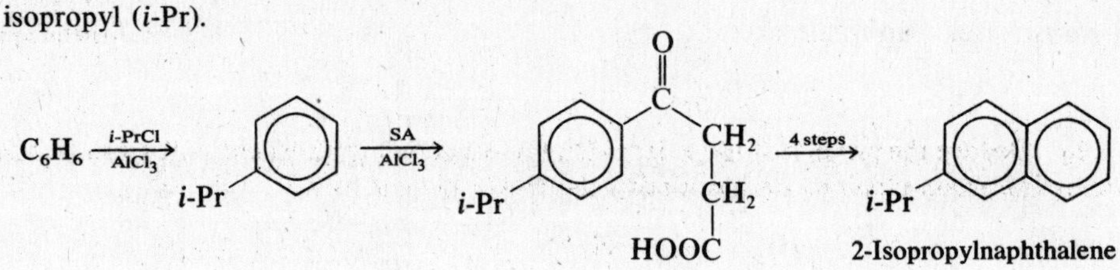

2-Isopropylnaphthalene

(b) G is H.

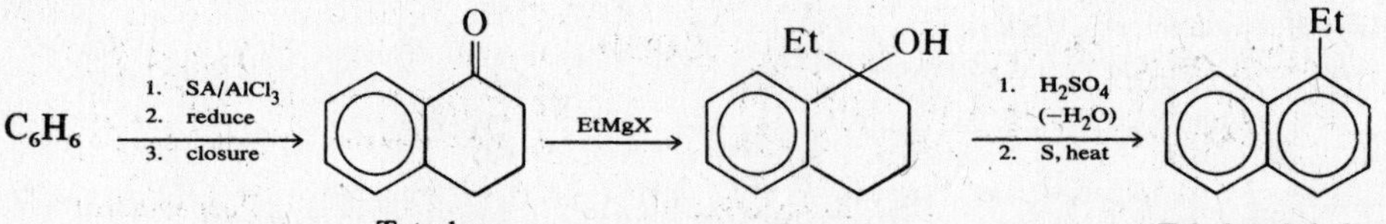

α-Tetralone   1-Ethylnaphthalene

(c) G is Et.

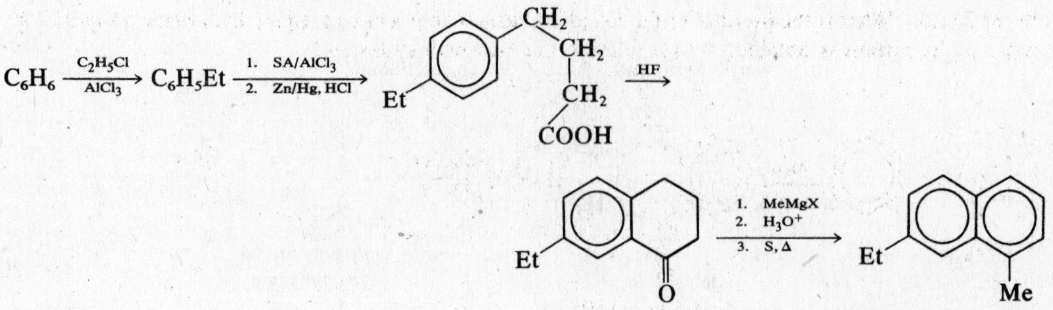

1-Methyl-7-ethylnaphthalene

(d) G is Et.

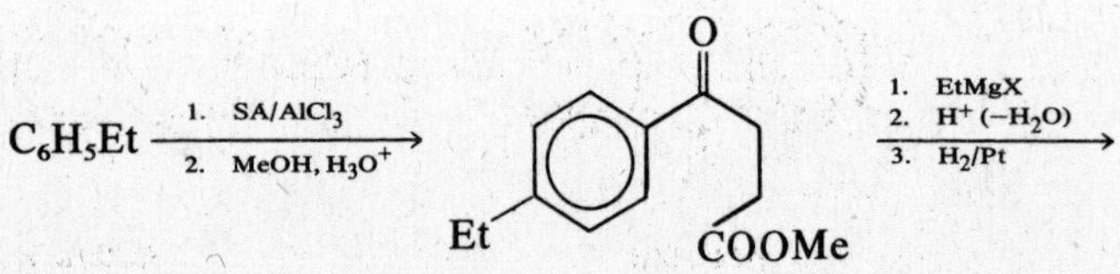

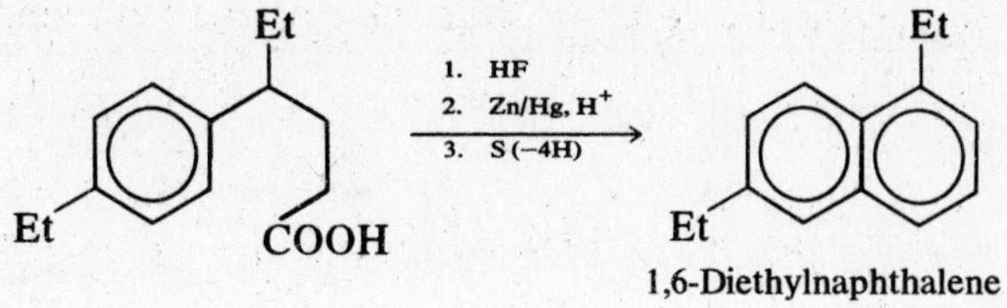

1,6-Diethylnaphthalene

(e) G is OCH$_3$.

$$C_6H_6 \xrightarrow[\text{2. NaOH}]{\text{1. H}_2\text{SO}_4} C_6H_5SO_3^-Na^+ \xrightarrow[\text{fuse}]{\text{NaOH}} C_6H_5\bar{O}\overset{+}{N}a \xrightarrow{\text{CH}_3\text{I}} C_6H_5OCH_3 \xrightarrow[\text{2. ROH, H}^+]{\text{1. SA/AlCl}_3}$$

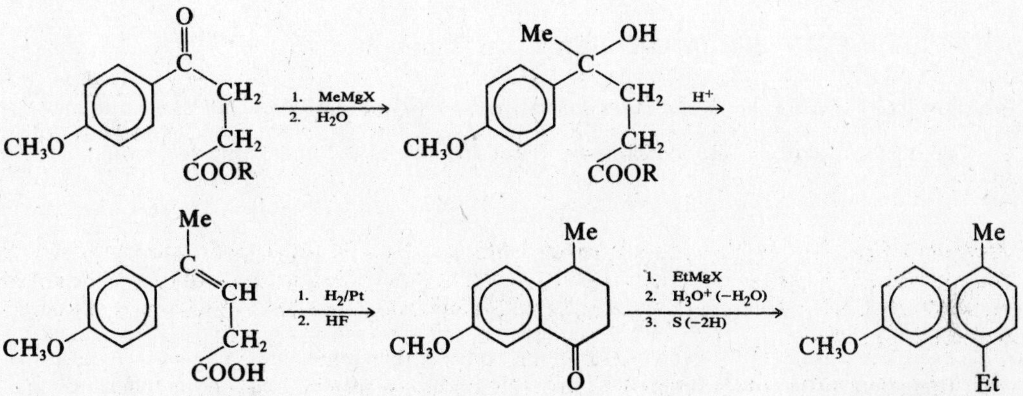

1-Methyl-4-ethyl-
6-methoxynaphthalene

Problem 22.28 Deduce the product of the Friedel-Crafts reaction of naphthalene with succinic anhydride and the aromatic hydrocarbon obtained by completing the Haworth synthesis. ◄

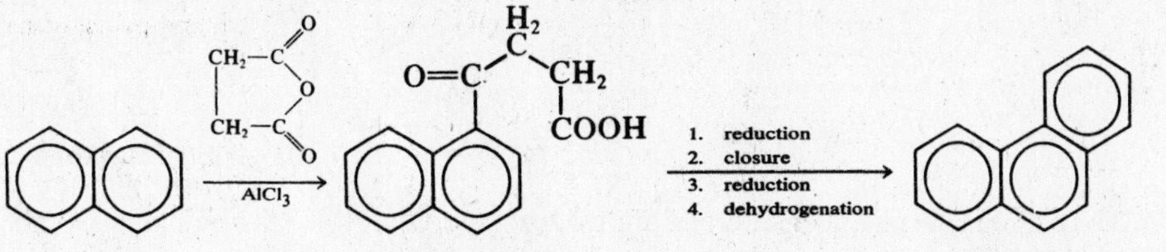

α-acylation   Phenanthrene

Problem 22.29 What is the product of the Friedel-Crafts reaction of benzene with phthalic anhydride? What aromatic hydrocarbon is obtained by completing the Haworth synthesis? ◄

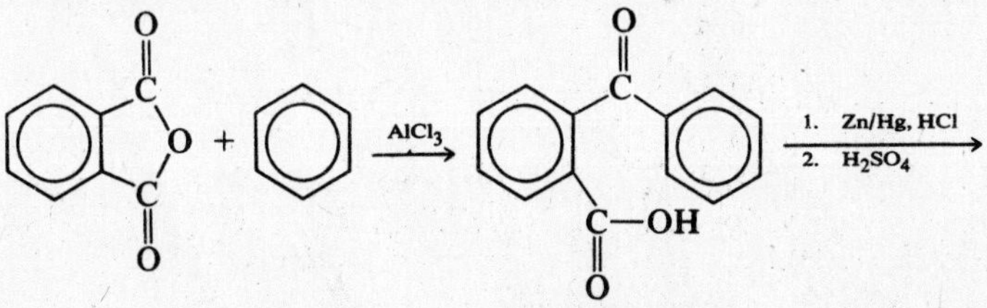

o-Benzoylbenzoic
acid

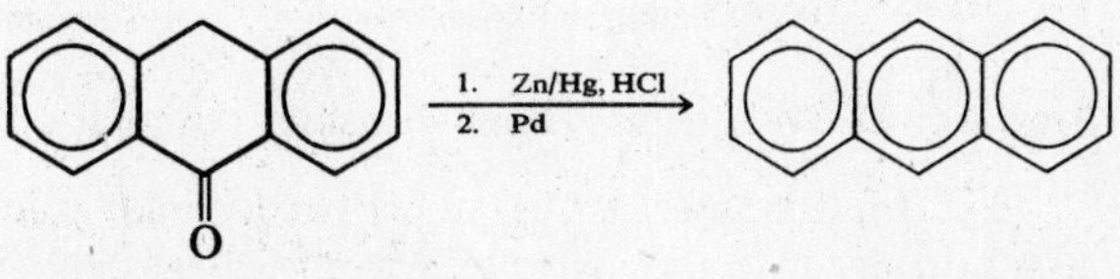

Anthrone   Anthracene

Problem 22.30   In Problem 22.27(d) and (e), why is the unsaturated acid hydrogenated before ring closure? ◄

If the ring is closed first, the resulting compound is the keto tautomer of 4-R-1-naphthol.

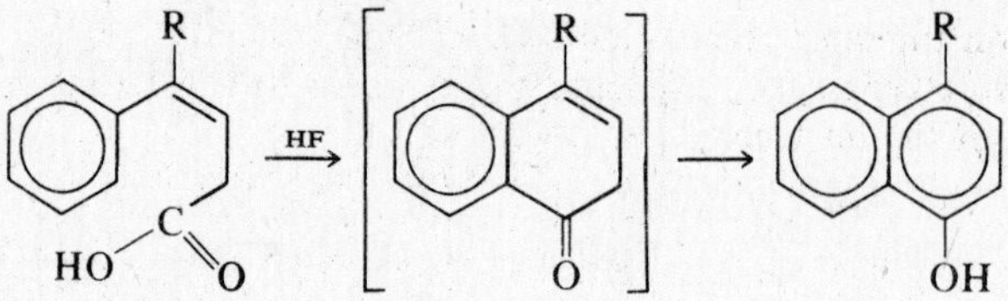

ANTHRACENE AND PHENANTHRENE

Problem 22.31   Name the monobromo derivatives of   (a) anthracene,   (b) phenanthrene.   ◄

The different positions are indicated by Greek letters and the numbering is shown.

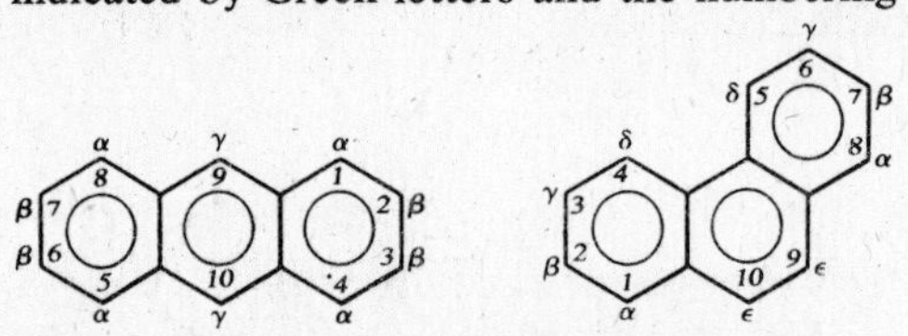

(a)   There are 3 isomers: 1-bromo-, 2-bromo-, and 9-bromoanthracene.
(b)   There are 5 isomers: 1-bromo-, 2-bromo-, 3-bromo-, 4-bromo-, and 9-bromophenanthrene.

Problem 22.32   Which anthracene derivative is prepared by a Diels-Alder reaction of 1 mole of p-benzoquinone (Problem 21.30) with 2 moles of 1,3-butadiene?   Show steps in the synthesis.   ◄

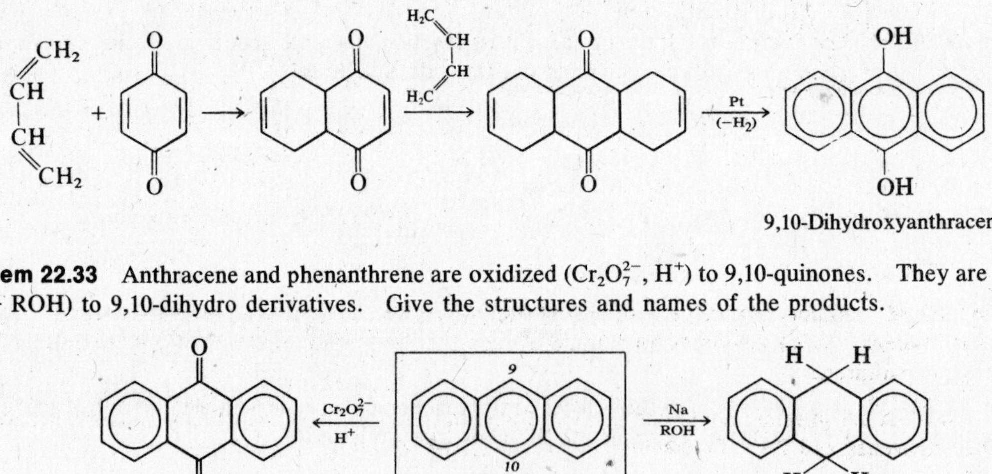

9,10-Dihydroxyanthracene

Problem 22.33   Anthracene and phenanthrene are oxidized ($Cr_2O_7^{2-}$, H^+) to 9,10-quinones.   They are reduced (Na + ROH) to 9,10-dihydro derivatives.   Give the structures and names of the products.   ◄

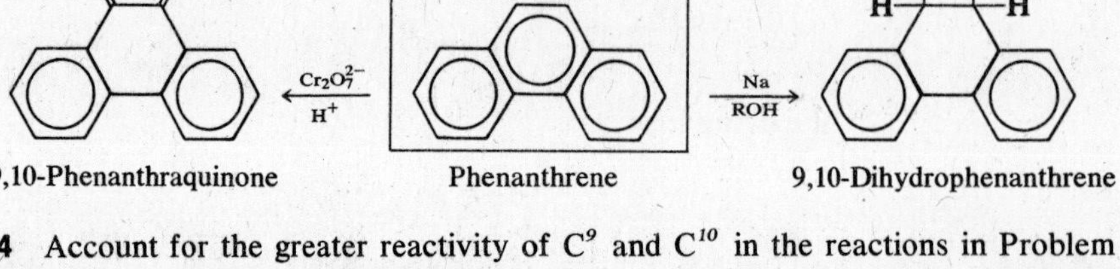

9,10-Anthraquinone   9,10-Dihydroanthracene

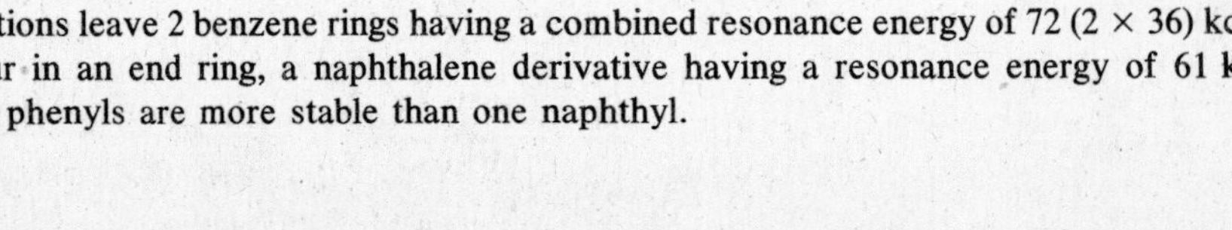

9,10-Phenanthraquinone   Phenanthrene   9,10-Dihydrophenanthrene

Problem 22.34   Account for the greater reactivity of C^9 and C^{10} in the reactions in Problem 22.33.   ◄

Such reactions leave 2 benzene rings having a combined resonance energy of 72 (2 × 36) kcal/mole.   Were attack to occur in an end ring, a naphthalene derivative having a resonance energy of 61 kcal/mole would remain.   Two phenyls are more stable than one naphthyl.

Problem 22.35 At low temperatures, anthracene and phenanthrene undergo C^9—C^{10} addition with Br_2. At high temperatures, Br substitution at C^9 occurs. Outline the mechanism of addition and substitution to anthracene and phenanthrene.   ◀

Br^+ attacks C^9 to form the most stable intermediate, a carbocation at C^{10}. With this R^+, the aromatic sextet is retained in the two other rings. Elimination of a proton from C^9 yields a substitution product, while addition occurs by attack of the nucleophile Br^- at the C^{10} R^+.

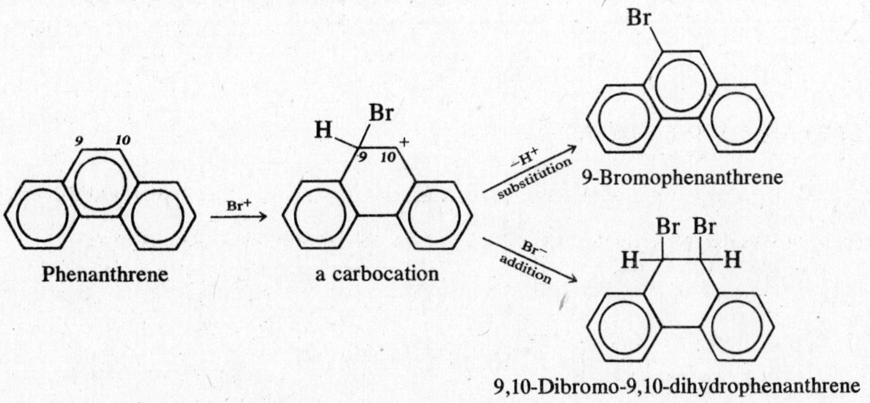

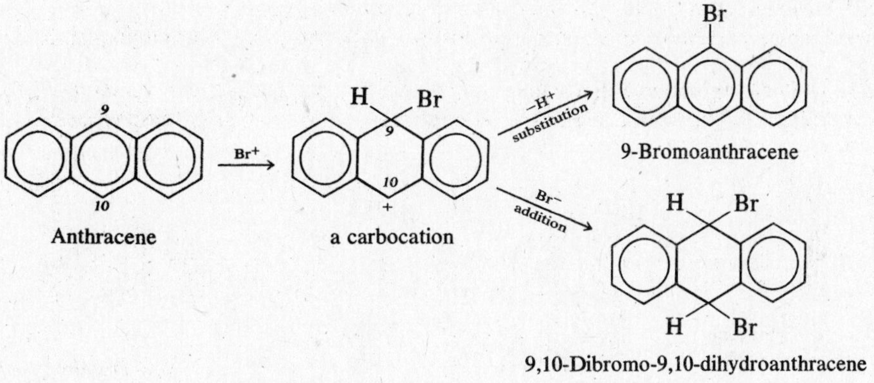

Problem 22.36 Outline syntheses of   (a) 9-methylanthracene,   (b) 2-ethylanthracene,   (c) 1,4-dimethylanthracene,   (d) 2,9-dimethylanthracene,   (e) 1-ethylphenanthrene,   (f) 4-ethylphenanthrene, (g) 9-ethylphenanthrene.   ◀

We write SA for succinic anhydride and PA for phthalic anhydride. Anthracenes are best made from substituted benzenes and PA. Phenanthrenes are made from naphthalenes and SA.

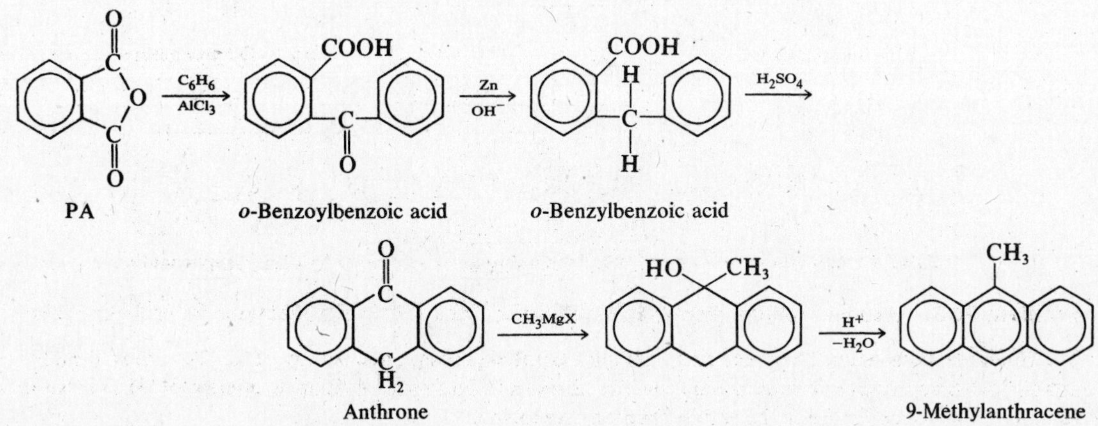

(b) PA + [Et-substituted benzene] $\xrightarrow{\text{AlCl}_3}$ [diaryl ketone with COOH and Et] $\xrightarrow{\begin{array}{l}1.\ \text{Zn/Hg, HCl}\\2.\ \text{HF}\end{array}}$

[anthracenone with Et] $\xrightarrow{\begin{array}{l}1.\ \text{Zn/Hg, HCl}\\2.\ \text{Pd(C)}\end{array}}$ [ethylanthracene with Et]

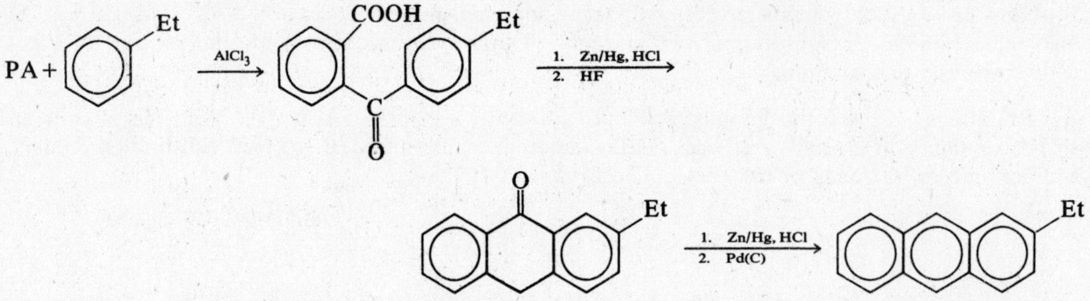

2-Ethylanthracene

(c) PA + [p-xylene, CH₃/CH₃] $\xrightarrow{\text{AlCl}_3}$ [ketone with COOH, CH₃, CH₃] $\xrightarrow{\begin{array}{l}1.\ \text{Zn/Hg, HCl}\\2.\ \text{HF}\\3.\ \text{Zn/Hg, HCl}\\4.\ \text{Pd}\end{array}}$ [1,4-dimethylanthracene]

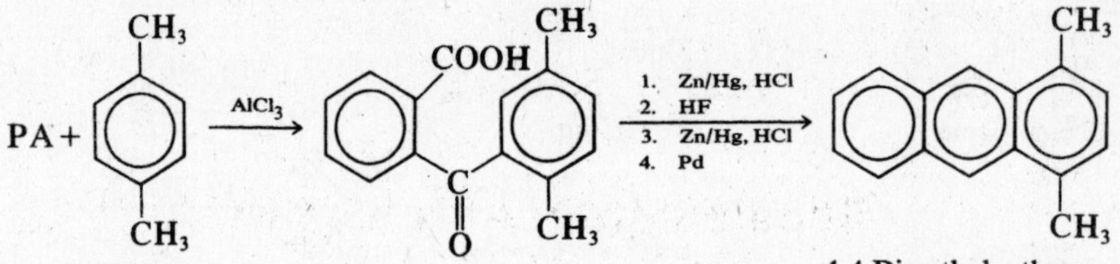

p-Xylene   1,4-Dimethylanthracene

(d) PA + [toluene, CH₃] $\xrightarrow{\begin{array}{l}1.\ \text{AlCl}_3\\2.\ \text{Zn, OH}^-\\3.\ \text{HF}\end{array}}$ [anthracenone with CH₃] $\xrightarrow{\begin{array}{l}1.\ \text{CH}_3\text{MgX}\\2.\ \text{H}^+\ (-\text{H}_2\text{O})\end{array}}$ [2,9-dimethylanthracene]

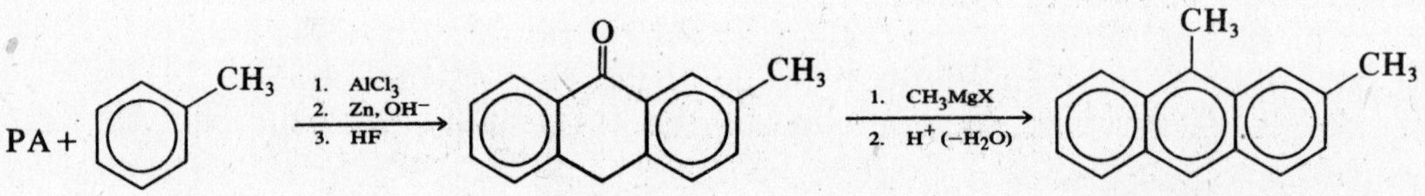

2,9-Dimethylanthracene

(e) [naphthalene] $\xrightarrow[\substack{\text{AlCl}_3\ (\text{CS}_2)\\(\alpha\text{-acylation})}]{\text{SA}}$ [naphthyl ketone with COOH] $\xrightarrow{\begin{array}{l}1.\ \text{Zn/Hg, HCl}\\2.\ \text{HF}\end{array}}$ [phenanthrenone ketone] $\xrightarrow{\begin{array}{l}1.\ \text{EtMgX}\\2.\ \text{H}_2\text{O}\end{array}}$

[tetrahydro with Et, OH] $\xrightarrow[-\text{H}_2\text{O}]{\text{H}^+}$ [with Et] $\xrightarrow[-\text{H}_2]{\text{Pd}}$ [1-ethylphenanthrene with Et]

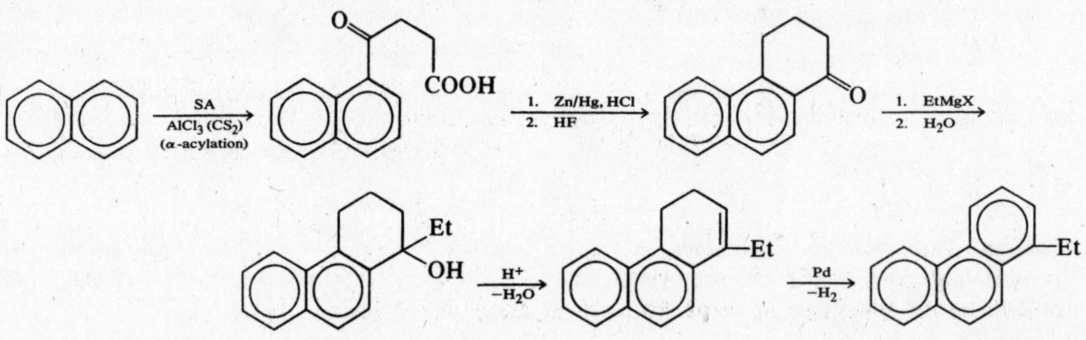

1-Ethylphenanthrene

(f) [naphthalene] $\xrightarrow[\substack{(\text{PhNO}_2)\\(\beta\text{-acylation})}]{\text{SA/AlCl}_3}$ [HOOC, ketone substituted naphthalene] $\xrightarrow{\begin{array}{l}1.\ \text{Zn/Hg, HCl}\\2.\ \text{HF}\end{array}}$ [ketone ring system] $\xrightarrow{\text{EtMgX}}$

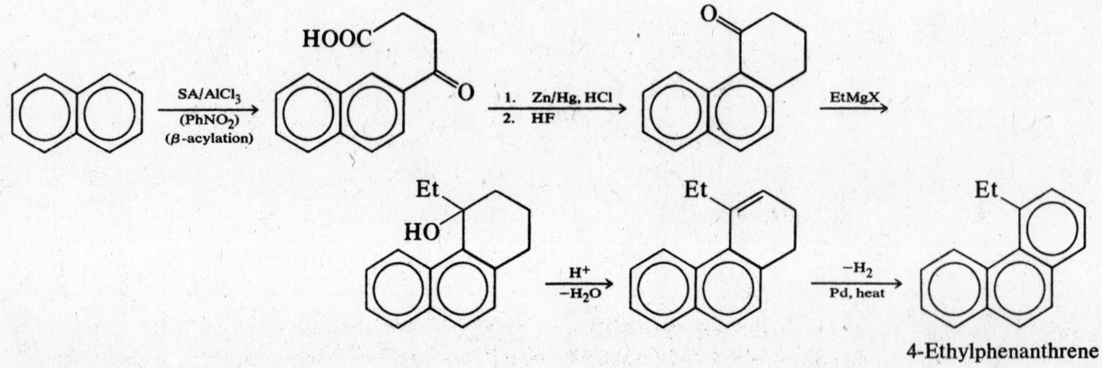

[Et, HO substituted ring] $\xrightarrow[-\text{H}_2\text{O}]{\text{H}^+}$ [Et substituted] $\xrightarrow[\text{Pd, heat}]{-\text{H}_2}$ [4-ethylphenanthrene with Et]

4-Ethylphenanthrene

(g)

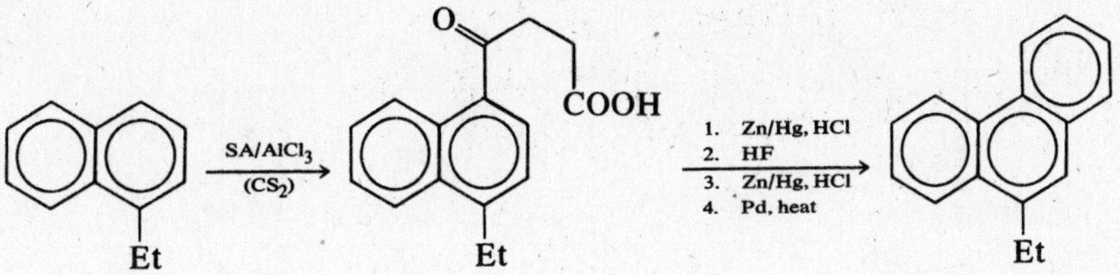

9-Ethylphenanthrene

22.4 SUMMARY OF NAPHTHALENE REACTIONS

See Fig. 22-2, page 396.

Supplementary Problems

Problem 22.37 Write structures for   (a) α,α'-binaphthyl,   (b) 2,6-naphthoquinone,   (c) 2-cyclopropyl-naphthalene,   (d) 1,2,3,4-dibenzanthracene.   ◄

(a)

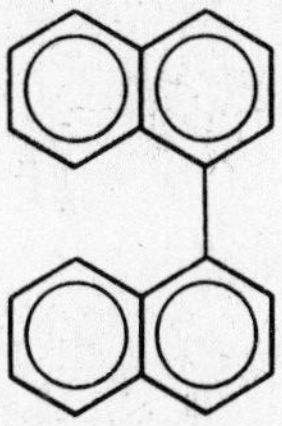

(b)

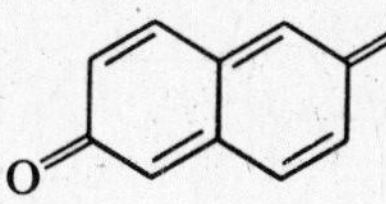

(c)

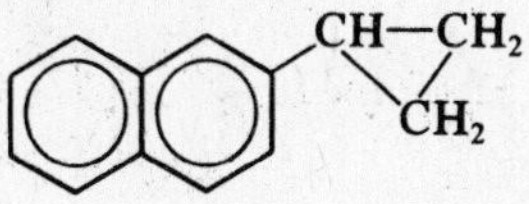

(d)

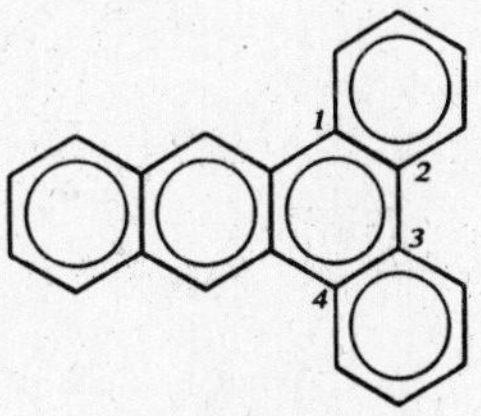

a carcinogen

Problem 22.38 What kind of isomerism exists in decalin (decahydronaphthalene)?   ◄

Geometric or *cis-trans* isomerism due to the possible diaxial and axial-equatorial configurations of the H's on C^9 and C^{10} shared by the 2 rings.

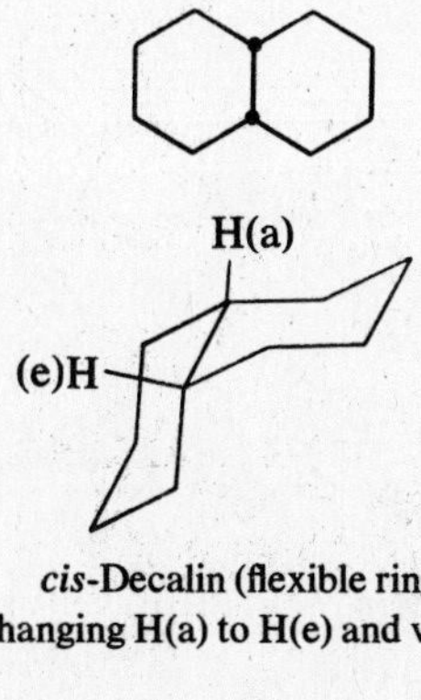

cis-Decalin (flexible rings
flip, changing H(a) to H(e) and vice versa)

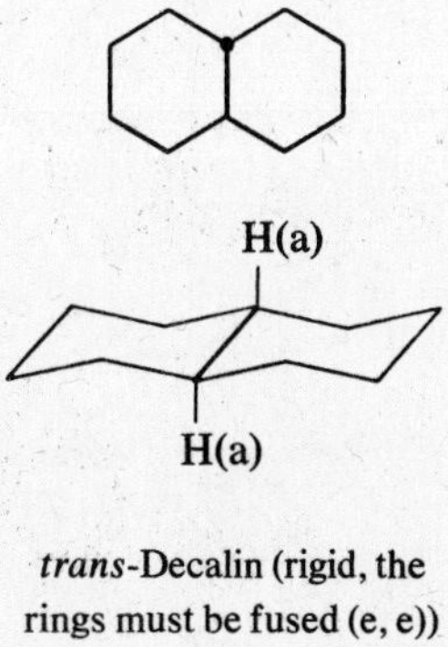

trans-Decalin (rigid, the
rings must be fused (e, e))

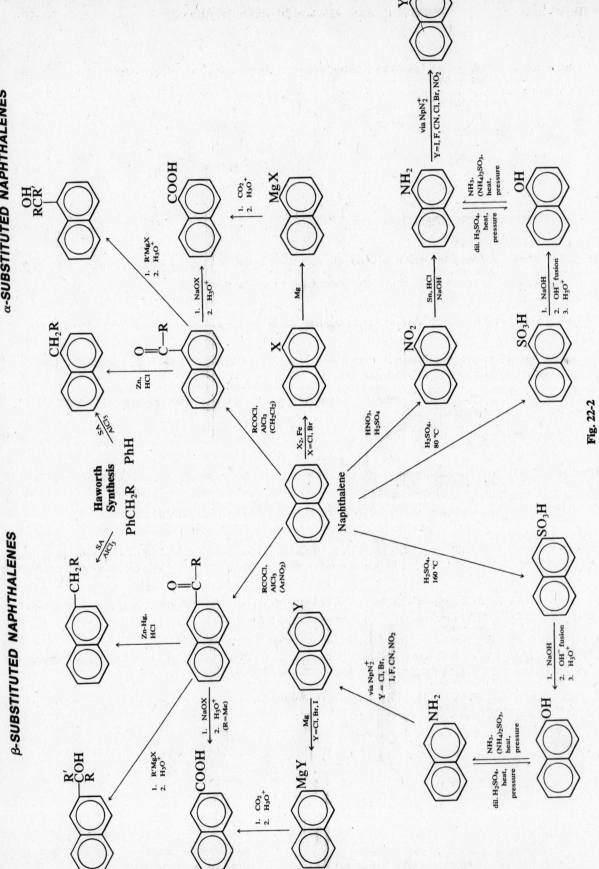

Fig. 22-2

Problem 22.39   What is the Diels-Alder addition product of anthracene and maleic anhydride?   ◄

Addition occurs at the most reactive C^9 and C^{10} positions (see Problem 22.32).

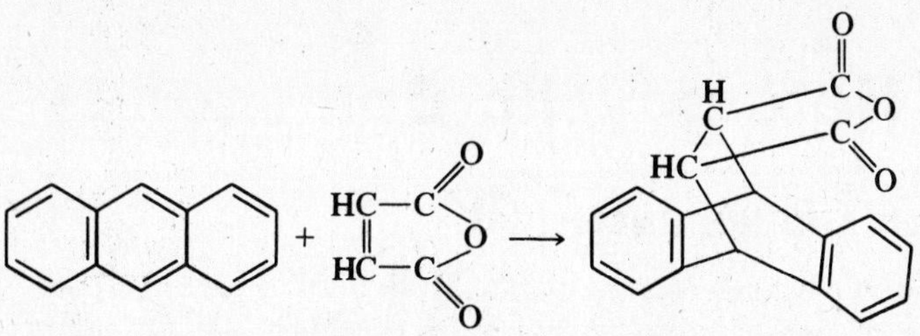

Problem 22.40   What are the oxidation products of   (a) α-naphthoic acid,   (b) α-naphthol?   ◄

(a) 1,2,3-benzenetricarboxylic acid,   (b) phthalic acid (see Problem 22.14).

Problem 22.41   What compound results from treating the diazonium salt of p-toluidine with copper bronze powder?   ◄

4,4'-dimethylbiphenyl (Gatterman reaction).

Problem 22.42   What starting compound would you use in the Ullmann reaction to prepare 4,4'-dinitrobiphenyl-2,2'-dicarboxylic acid?   ◄

2-iodo-5-nitrobenzoic acid.

Problem 22.43   Give the main product of reaction of   (a) heating α-naphthylammonium sulfate,   (b) α-naphthoic acid + HNO_3.   ◄

(a) naphthionic acid (4-amino-1-naphthalenesulfonic acid),   (b) 5-nitro-1-naphthoic acid.

Problem 22.44   Compound (A), $C_{14}H_{14}$, is mildly oxidized to compound (B), $C_{14}H_{10}O_4$, which on vigorous oxidation forms compound (C), $C_9H_6O_6$, having a neutralization equivalent of 70.   Suggest structures for (A), (B), and (C).   ◄

(A) is a dimethylated biphenyl with both —CH₃'s on the same ring.   (B) is a biphenyldicarboxylic acid with both —COOH's on the same ring.   (C) is a benzene tricarboxylic acid.

Problem 22.45   Give the steps in the preparation of DDT, $(p\text{-}ClC_6H_4)_2CHCCl_3$, from chloral (trichloroacetaldehyde) and chlorobenzene in the presence of H_2SO_4.   ◄

$$Cl_3CCHO \xrightarrow{H_2SO_4} Cl_3C\overset{+}{C}H(OH) \xrightarrow{PhCl} p\text{-}Cl\text{—}C_6H_4CHOHCCl_3 \xrightarrow{H_2SO_4} p\text{-}Cl\text{—}C_6H_4\overset{+}{C}HCCl_3 \xrightarrow{PhCl} DDT$$

Chapter 23

Heterocyclic Compounds

23.1 INTRODUCTION AND NOMENCLATURE

Heterocyclic compounds have a **heteroatom** such as O, S, or N as part of a carbon ring. When there is more than one kind of ring heteroatom, the atom of *higher* atomic number receives the *lower* number in naming the compound.

The **ring index system** combines (1) the prefix **oxa-** for O, **aza-** for N or **thia-** for S; and (2) a stem for ring size and saturation or unsaturation. These are summarized in Table 23-1.

Table 23-1. Ring Index Heterocyclic Nomenclature

Ring Size	Stem	
	Saturated	Unsaturated
3	irane	irine
4	etane	ete
5	olane	ole
6	ixane	ixine
7	epane	epine
8	ocane	ocine

Problem 23.1 Name the following compounds systematically:

(a) (b) (c) (d) (e) (f) ◄

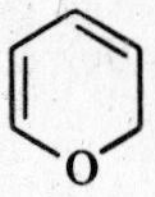

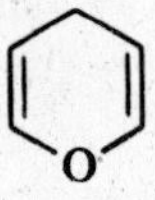

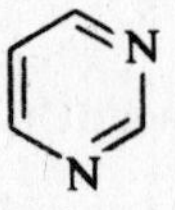

(a) azole (pyrrole), (b) 1,3-thiazole, (c) 2H-oxirine, (d) 4H-oxirine (pyran), (e) 1,4-dithixine, (f) 1,3-dithiazixine (pyrimidine).

2H- and 4H- are used in (c) and (d) to differentiate the position of the saturated sp^3 atom. Common names are given in parentheses.

Problem 23.2 Write structures for (a) oxirane, (b) 1,2-oxazole, (c) 1,4-diazixine (pyrazine), (d) 1-thia-4-oxa-6-azocine, (e) 3H-1,2,4-triazole, (f) azepane. ◄

(a) H_2C-CH_2 (c) (e)

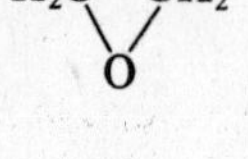

(b) 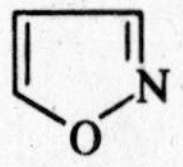 (d) (f)

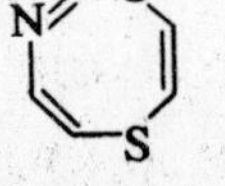

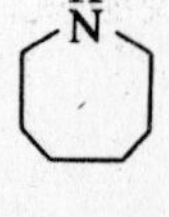

Problem 23.3 Supply structural formulas and trivial names for (a) azetane-2-one, (b) oxolane-2-one, (c) azixine-4-carboxylic acid, (d) oxole-2-carboxaldehyde (2-formyloxole).   ◀

(a)

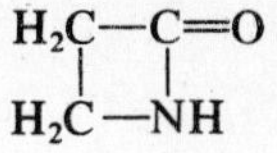

H$_2$C—C=O
| |
H$_2$C—NH

β-Propiolactam

(b)

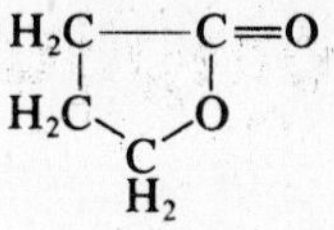

H$_2$C——C=O
| |
H$_2$C   O
\   C   /
    H$_2$

γ-Butyrolactone

(c)

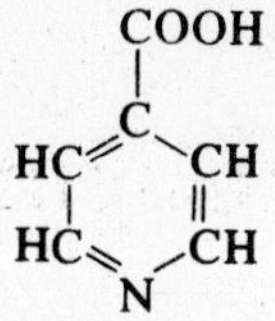

COOH
|
C
HC   CH
|| ||
HC   CH
  \ N /

Isonicotinic acid

(d)

HC——CH
|| ||
HC   C—CH=O
   \ O /

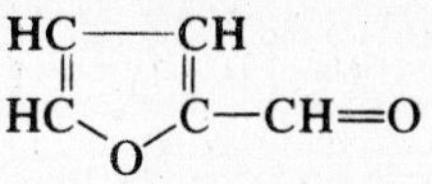

Furfural

23.2 FIVE-MEMBERED AROMATIC HETEROCYCLIC COMPOUNDS. FURAN (WITH O), THIOPHENE (WITH S), PYRROLE (WITH N)

Problem 23.4 Name the following compounds, using (1) numbers and (2) Greek letters.

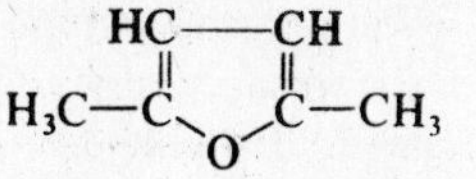

(a)

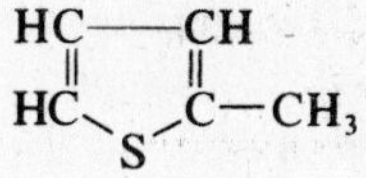

HC——CH
|| ||
HC   C—CH$_3$
   \ S /

(b)

HC——CH
|| ||
H$_3$C—C   C—CH$_3$
     \ O /

(c)

HC——C—CH$_3$
|| ||
H$_3$C—C   CH
     \ O /

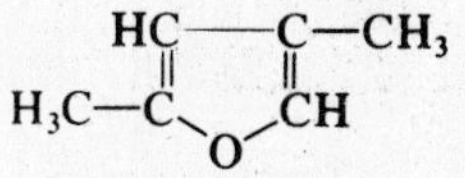

(d)

HC——CH
|| ||
Br—C   C—COOH
     \ N /
        |
     C$_2$H$_5$

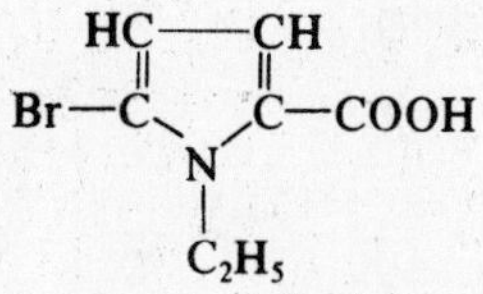

(a) 2-methylthiophene (2-methylthiole) or α-methylthiophene, (b) 2,5-dimethylfuran (2,5-dimethyloxole) or α,α'-dimethylfuran, (c) 2,4-dimethylfuran or α,β'-dimethylfuran (2,4-dimethyloxole), (d) 1-ethyl-5-bromo-2-pyrrolecarboxylic acid or N-ethyl-α-bromo-α'-pyrrolecarboxylic acid (N-ethyl-2-bromazole-5-carboxylic acid).

Problem 23.5 Write structures for (a) 2-benzoylthiophene, (b) 3-furansulfonic acid, (c) α,β'-dichloropyrrole, (d) 4-propionyl-1-oxa-2,3,5-triazole.   ◀

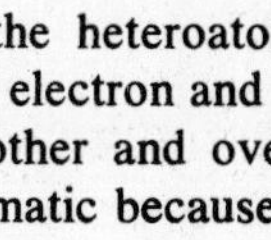

(a)

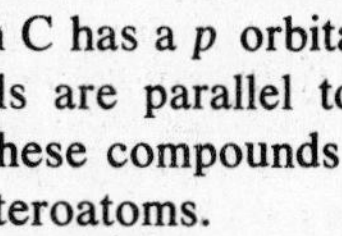

(thiophene ring)
S
  \
   C—Ph
   ||
   O

(b)

(furan ring)
SO$_3$H
O

(c)

Cl
(pyrrole ring)
Cl
N
|
H

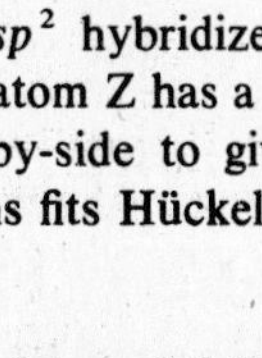

(d)

O
||
C—CC$_2$H$_5$
N
||
N   O
 \ N /

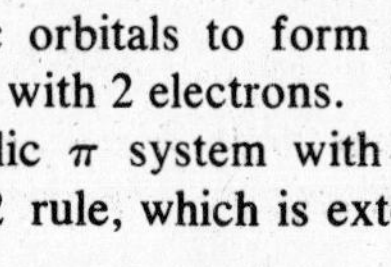

Problem 23.6 Account for the aromaticity of furan, pyrrole and thiophene, which have planar molecules with bond angles of 120°.   ◀

See Fig. 23-1. The 4 C's and the heteroatom Z use sp^2 hybridized atomic orbitals to form the σ bonds. Each C has a p orbital with 1 electron and the heteroatom Z has a p orbital with 2 electrons. These five p orbitals are parallel to each other and overlap side-by-side to give a cyclic π system with six p electrons. These compounds are aromatic because 6 electrons fits Hückel's $4n + 2$ rule, which is extended to include heteroatoms.

Problem 23.7 Account for the following dipole moments: furan, 0.7 D (away from O); tetrahydrofuran, 1.7 D (toward O).   ◀

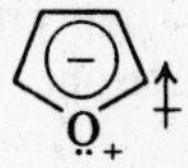

In tetrahydrofuran the greater electronegativity of O directs the moment of the C—O bond toward O. In furan, delocalization of an electron pair from O makes the ring C's − and O +; the moment is away from O.

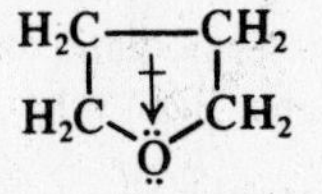

H$_2$C——CH$_2$
| ↑ |
H$_2$C ↓ CH$_2$
   \ O /
     ..

Tetrahydrofuran

(furan ring)
O
−
+
↑

Furan

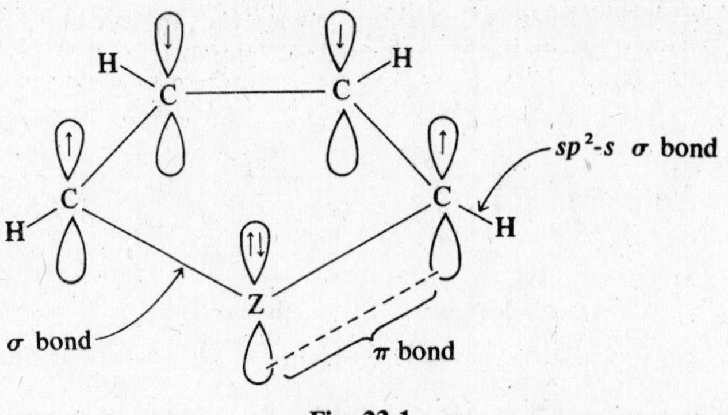

Fig. 23-1

PREPARATION

Problem 23.8 Pyrroles, furans and thiophenes are made by heating 1,4-dicarbonyl compounds with $(NH_4)_2CO_3$, P_2O_5 and P_2S_5, respectively. Which dicarbonyl compound is used to prepare   (a) 3,4-dimethylfuran,   (b) 2,5-dimethylthiophene,   (c) 2,3-dimethylpyrrole? ◄

The carbonyl C's become the α C's in the heterocyclic compound.

(a)

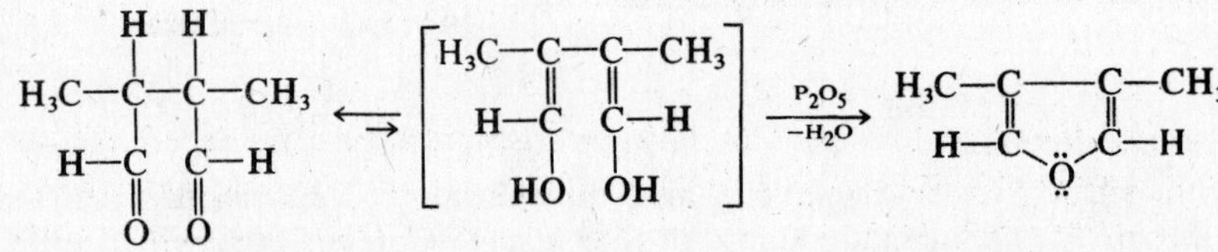

2,3-Dimethylbutanedial   a dienediol   3,4-Dimethylfuran

(b)

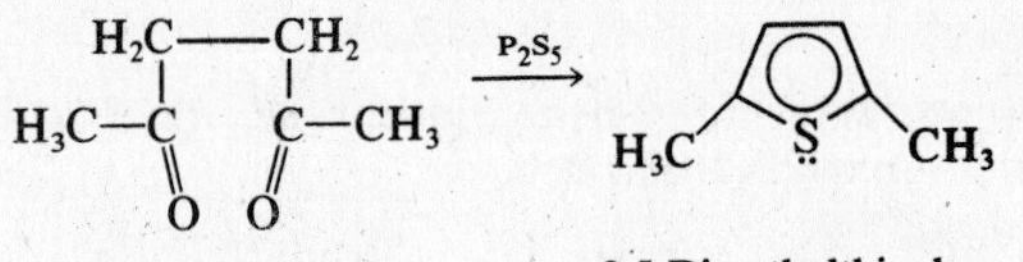

Acetonylacetone   2,5-Dimethylthiophene

(c)

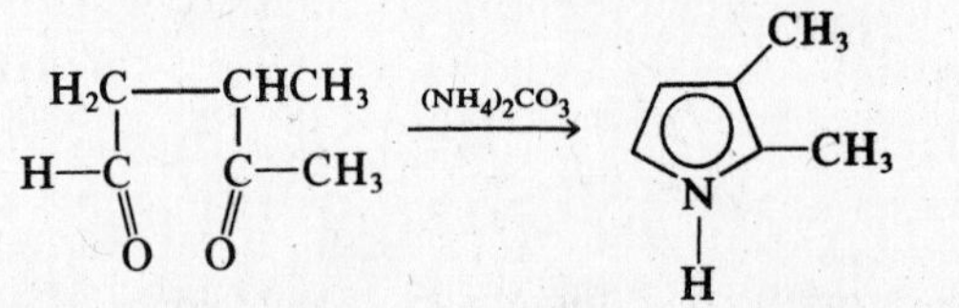

3-Methyl-4-oxopentanal   2,3-Dimethylpyrrole

Problem 23.9 Prepare pyrrole from succinic anhydride. ◄

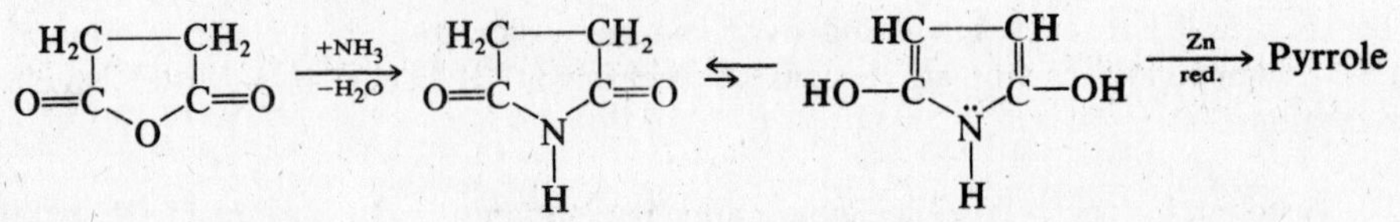

Succinic anhydride   Succinimide   2,5-Dihydroxypyrrole

CHEMICAL PROPERTIES

Problem 23.10   (a) In terms of relative stability of the intermediate, explain why an electrophile (E^+) attacks the α rather than the β position of pyrrole, furan and thiophene.   (b) Why are these heterocyclics more reactive than C_6H_6 to E^+-attack? ◄

(a) The transition state and the intermediate R^+ formed by α-attack is a hybrid of 3 resonance structures which possess less energy; the intermediate from β-attack is less stable and has more energy because it is a hybrid of only 2 resonance structures. I and II are also more stable allylic carbocations; V is not allylic.

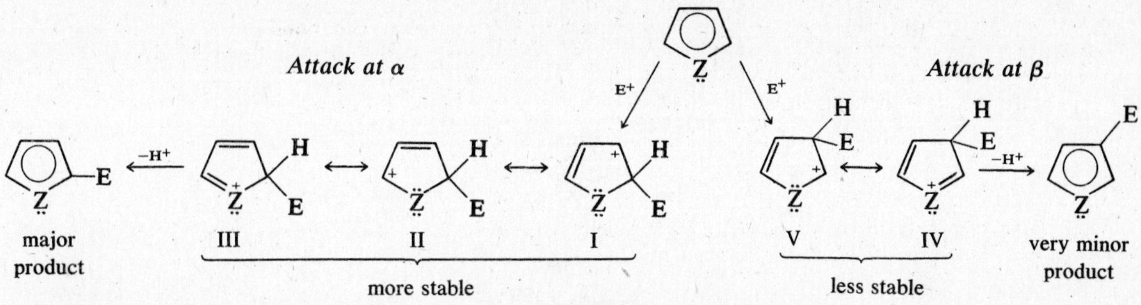

Attack at α   *Attack at β*

| major product | III | II | I | V | IV | very minor product |

              more stable   less stable

(b) This is ascribed to resonance structure III, in which Z has $+$ charge and in which all ring atoms have an octet of electrons. These heterocyclics are as reactive as PhOH and $PhNH_2$.

Problem 23.11 Explain why pyrrole is not basic.   ◄

The unshared pair of electrons on N is delocalized in an "aromatic sextet." Adding an acid to N could prevent delocalization and could destroy the aromaticity.

Problem 23.12 Give the type of reaction and the structure and name of the products obtained from: (a) furfural and concentrated aq. KOH; (b) furan with (i) $CH_3CO—ONO_2$ (acetyl nitrate), (ii) $(CH_3CO)_2O$ and BF_3 and then H_2O, (iii) $HgCl_2$, CH_3COONa and then I_2; (c) pyrrole with (i) SO_3 and pyridine,   (ii) $CHCl_3$ and KOH,   (iii) $PhN_2^+Cl^-$,   (iv) Br_2 and C_2H_5OH;   (d) thiophene   and (i) H_2SO_4,   (ii) $(CH_3CO)_2O\cdot CH_3COONO_2$,   (iii) Br_2 in benzene.   ◄

(a) Cannizzaro reaction;

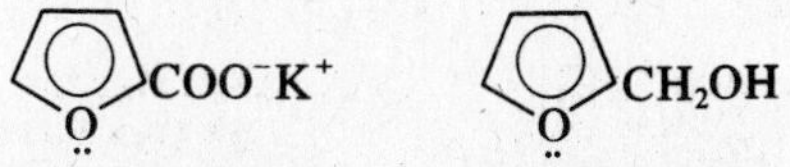

Potassium furoate   Furfuryl alcohol

(b) (i) Nitration; 2-nitrofuran,

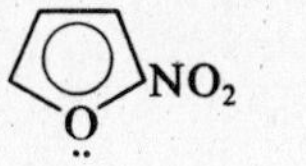

(ii) Acetylation; 2-acetylfuran,

(iii) Mercuration; 2-furanmercuric chloride and displacement of HgCl to form 2-iodofuran.

(c) (i) Sulfonation; 2-pyrrolesulfonic acid,

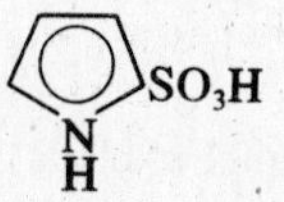

(H_2SO_4 alone destroys ring). (ii) Reimer-Tiemann formylation; 2-pyrrolecarboxaldehyde (2-formyl-pyrrole),

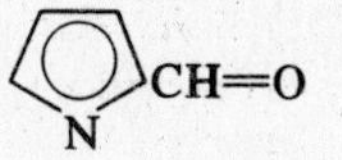

(iii) Coupling; 2-phenylazopyrrole,

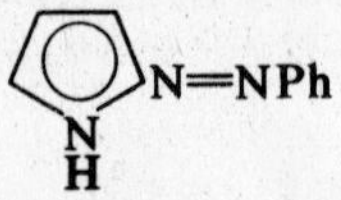

(iv) Bromination; 2,3,4,5-tetrabromopyrrole.
(d) (i) Sulfonation; thiophene-2-sulfonic acid,

(ii) Nitration; 2-nitrothiophene,

(iii) Bromination; 2,5-dibromothiophene. (Thiophene is less reactive than pyrrole and furan.)

Problem 23.13 Write structures for the mononitration products of the following compounds and explain their formation: (a) 3-nitropyrrole, (b) 3-methoxythiophene, (c) 2-acetylthiophene, (d) 5-methyl-2-methoxythiophene, (e) 5-methylfuran-2-carboxylic acid. ◄

(a) Nitration at C^5 to form 2,4-dinitropyrrole. After nitration C^5 becomes C^2, and C^3 becomes C^4. Nitration at C^2 would form an intermediate with a + on C^3, which has the electron-attracting —NO_2 group.

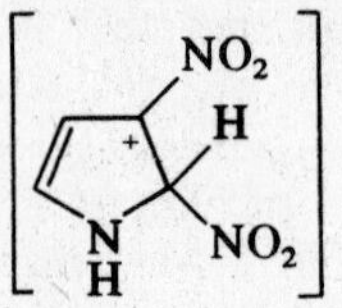

(b)

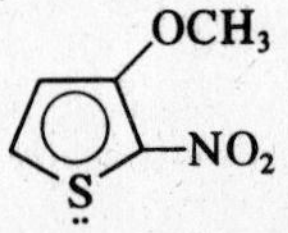

2-Nitro-3-methoxy-
thiophene
(α and *ortho* to OCH₃)

(c)

2-Acetyl-5-nitro-
thiophene
(α-attack)

(d)

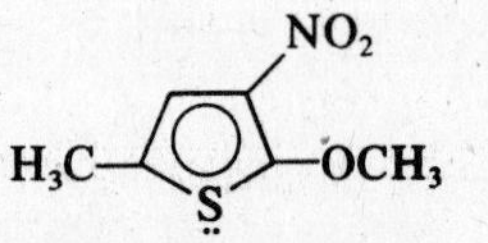

2-Ethoxy-3-nitro-5-methyl-
thiophene
(*ortho* to OCH₃, a stronger
activating group than CH₃)

(e)

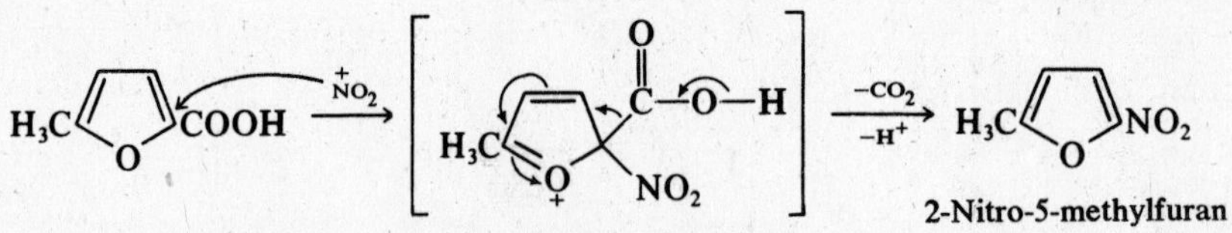

2-Nitro-5-methylfuran

Attack of NO_2^+ at C^2, followed by elimination of CO_2 and H^+.

Problem 23.14 Name the products formed when (a) furan, (b) pyrrole are catalytically hydrogenated. ◄

(a)   (tetrahydrofuran, oxacyclopentane, oxolane)

(b) 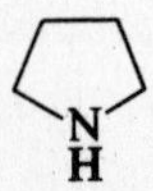   (pyrrolidine, azacyclopentane, azolidine)

Problem 23.15 Give the Diels-Alder product for the reaction of furan and maleic anhydride.   ◄

Furan is the least aromatic of the five-membered ring heterocyclics and acts as a diene towards *strong* dienophiles.

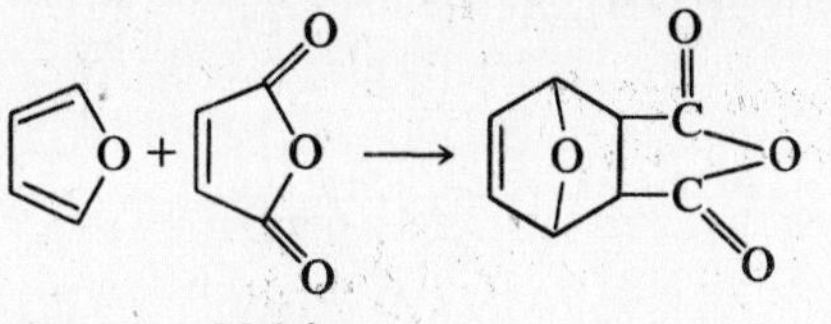

Furan   Maleic
anhydride

Problem 23.16 Identify compounds (A) through (D).

$$CH_3COOC_2H_5 + (A) \xrightarrow{NaOEt} C_6H_5COCH_2COOC_2H_5 \xrightarrow[I_2]{NaOEt} (B) \xrightarrow[2.\ H_3O^+]{1.\ dil.\ NaOH} (C) \xrightarrow{P_2O_5} (D)$$ ◄

(A) PhCOOEt

(B) $C_6H_5COCHCOOC_2H_5$ (via $C_6H_5COCH(I)COOC_2H_5$)
      $C_6H_5COCHCOOC_2H_5$

(C)

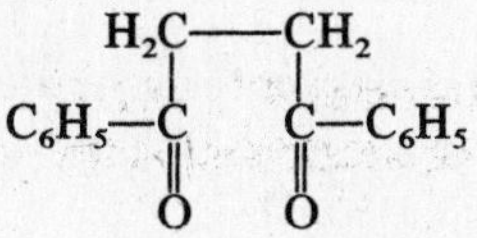

(D)

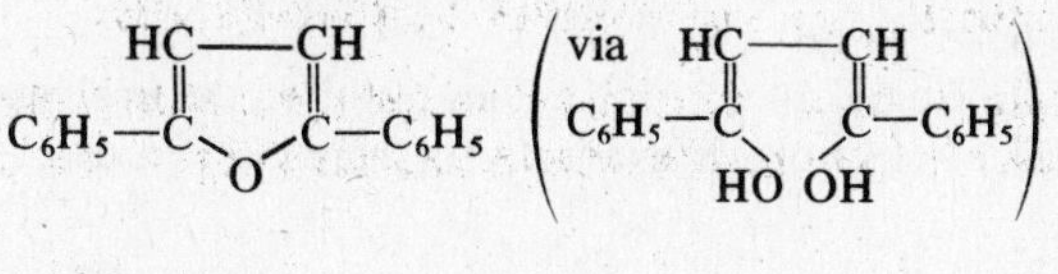

Problem 23.17 Use any necessary inorganic reagents to prepare (a) α,α'-dimethylthiophene from $CH_3COOC_2H_5$, (b) phenyl 2-thienyl ketone from PhCOOH and $OHCCH_2CH_2CHO$.   ◄

(a) $2CH_3COOC_2H_5 \xrightarrow{NaOEt} CH_3COCH_2COOC_2H_5 \xrightarrow[I_2]{NaOEt} CH_3CO-CH-CHCOCH_3 \xrightarrow[2.\ H_3O^+]{1.\ dil.\ OH^-}$
                                                                                                    $C_2H_5OOC \quad COOC_2H_5$

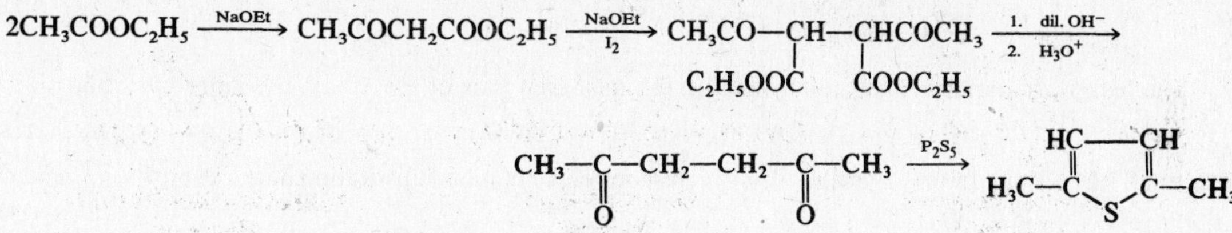

(b)

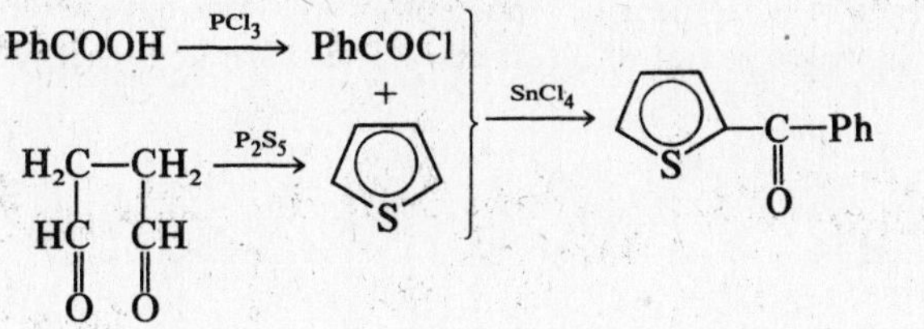

Problem 23.18 Give the products of reaction of pyrrole with (a) I_2 in aqueous KI; (b) $CH_3CN + HCl$, followed by hydrolysis; (c) CH_3MgI.   ◄

(a) 2,3,4,5-Tetraiodopyrrole   (b) α-Acetylpyrrole [see Problem 21.20(c)]

(c)

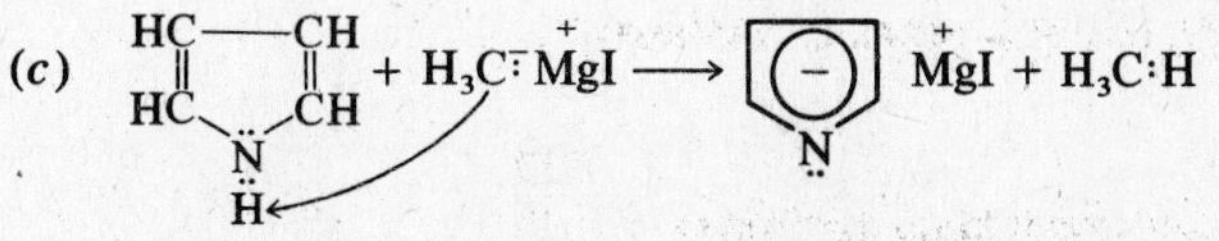

A_1 (stronger B_2 (stronger B_1 (weaker A_2 (weaker acid)
   acid)   base)   base)

23.3 SIX-MEMBERED HETEROCYCLIC COMPOUNDS

COMPOUNDS WITH ONE HETEROATOM. PYRIDINE

Problem 23.19 Write the structural formulas and give the names of the isomeric methylpyridines. ◄

There are 3 isomers.

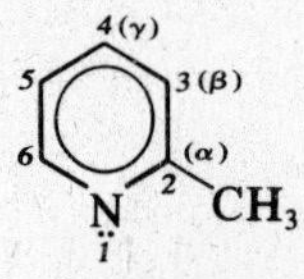

2- or α-Methylpyridine   3- or β-Methylpyridine   4- or γ-Methylpyridine
(α-Picoline)   (β-Picoline)   (γ-Picoline)

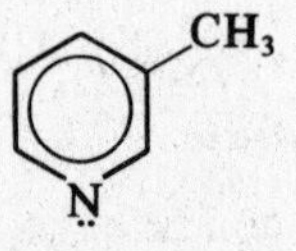

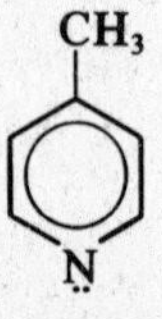

Problem 23.20 (*a*) Account for the aromaticity of pyridine, a planar structure with 120° bond angles. (*b*) Is pyridine basic? Explain. (*c*) Explain why piperidine (azacyclohexane) is more basic than pyridine. ◄

(*a*) Pyridine (azabenzene) is the nitrogen analog of benzene, and both have the same *orbital* picture (Figs. 10-1 and 10-2). The three double bonds furnish six *p* electrons for the delocalized π system, in accordance with Hückel's rule.

(*b*) Yes. Unlike pyrrole, pyridine does not need the unshared pair of electrons on N for its aromatic sextet. The pair of electrons is available for bonding to acids.

(*c*)

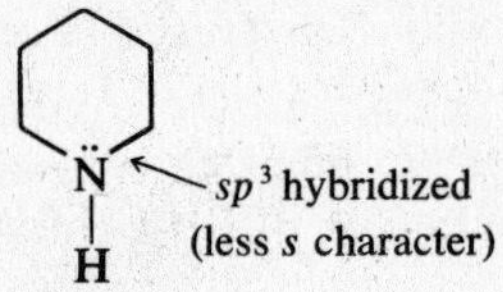

sp^3 hybridized sp^2 hybridized
(less *s* character)   (more *s* character)

Piperidine 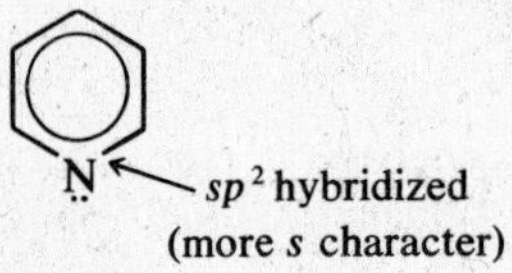   Pyridine

The less *s* character in the orbital holding the unshared pair of electrons, the more basic the site.

Problem 23.21 Explain why pyridine (*a*) undergoes electrophilic substitution at the β position, and (*b*) is less reactive than benzene. ◄

(*a*) The R⁺'s formed by attack of E⁺ at the α or γ positions of pyridine have resonance structures (I, IV) with a positive charge on N having a sextet of electrons. These are high-energy structures.

α-Attack   *γ-Attack*

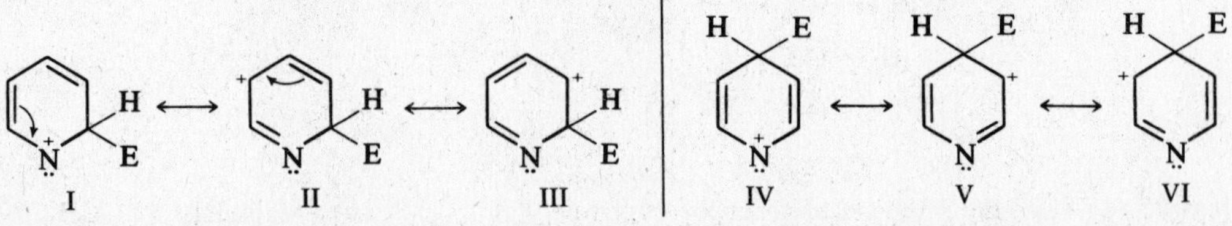

I   II   III   IV   V   VI

With β-attack, the + charge in the intermediate is distributed only to C's. A + on C with 6 electrons is not as unstable as a + on N with 6 electrons, since N is more electronegative than C. β-Electrophilic substitution gives the more stable intermediate.

β-Attack

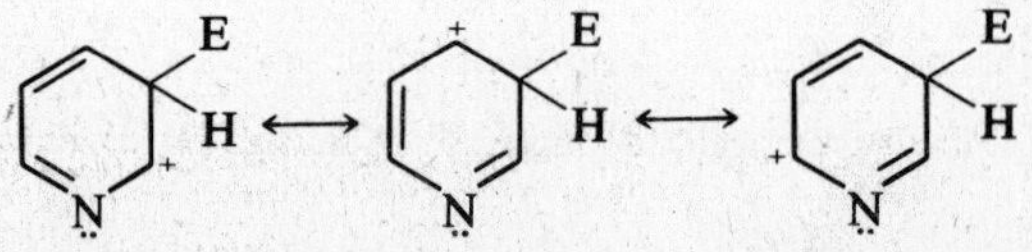

(*b*) N withdraws electrons by induction and destabilizes the R⁺ intermediates formed from pyridine.

Problem 23.22 Give structural formulas and the names of the products formed when pyridine reacts with (a) Br$_2$ at 300 °C; (b) KNO$_3$, H$_2$SO$_4$ at 300 °C and then KOH; (c) H$_2$SO$_4$ at 350 °C; (d) CH$_3$COCl/AlCl$_3$.

(a)

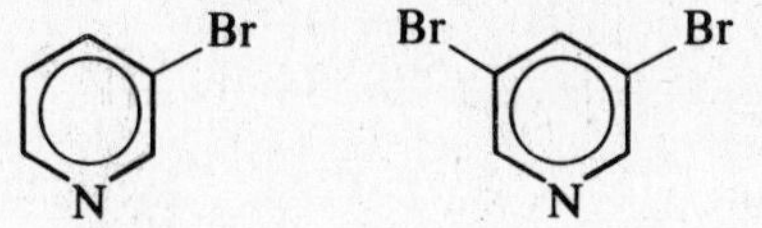

3-Bromo- and 3,5-Dibromopyridine

(b)   (KOH neutralizes the formed salt)

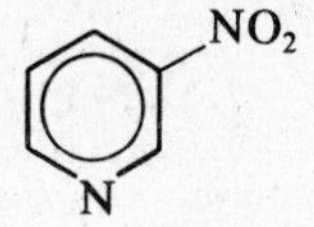

3-Nitropyridine

(c)

(d) No reaction. Pyridine is too unreactive to undergo Friedel-Crafts acylations.

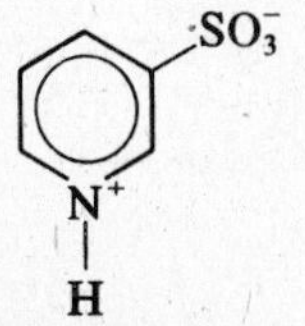

3-Pyridinesulfonic acid (a dipolar ion)

Problem 23.23 Compare and explain the difference between pyridine and pyrrole with respect to reactivity toward electrophilic substitution.

Pyrrole is more reactive than pyridine because its intermediate is more stable. For both compounds the intermediate has a + on N. However, the pyrrole intermediate is *relatively stable* because every atom has a *complete octet*, while the pyridine intermediate is *very unstable* because N has only *six* electrons.

Problem 23.24 If oxidation of an aromatic ring is an electrophilic attack, predict the product formed when α-phenylpyridine is oxidized.

The pyridine ring is less reactive, and the benzene ring is oxidized to α-picolinic acid (α-NC$_5$H$_4$COOH).

Problem 23.25 Predict and account for the product obtained and conditions used in nitration of 2-aminopyridine.

The product is 2-amino-5-nitropyridine because substitution occurs preferentially at the sterically less hindered β position *para* to NH$_2$. The conditions are milder than those for pyridine because NH$_2$ is activating.

Problem 23.26 Explain why (a) pyridine and NaNH$_2$ give α-aminopyridine, (b) 4-chloropyridine and NaOMe give 4-methoxypyridine, (c) 3-chloropyridine and NaOMe give no reaction.

Electron-attracting N facilitates attack by strong nucleophiles in α and γ positions. The intermediate is a carbanion stabilized by delocalization of − to the electronegative N. The intermediate carbanion readily reverts to a stable aromatic ring by ejecting an H$^-$ in (a) or a :Cl$^-$ in (b).

(a)

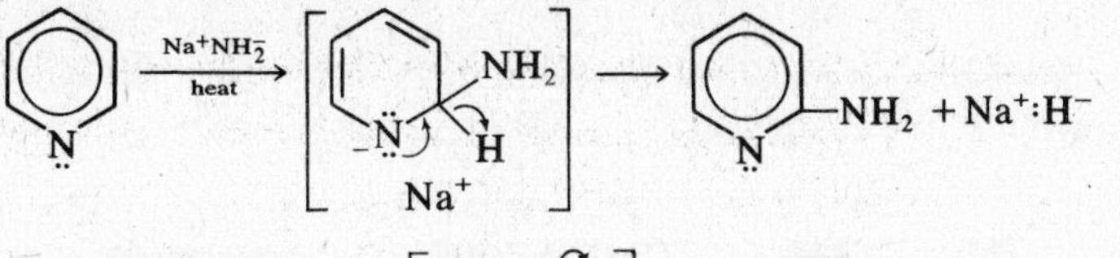

(b)

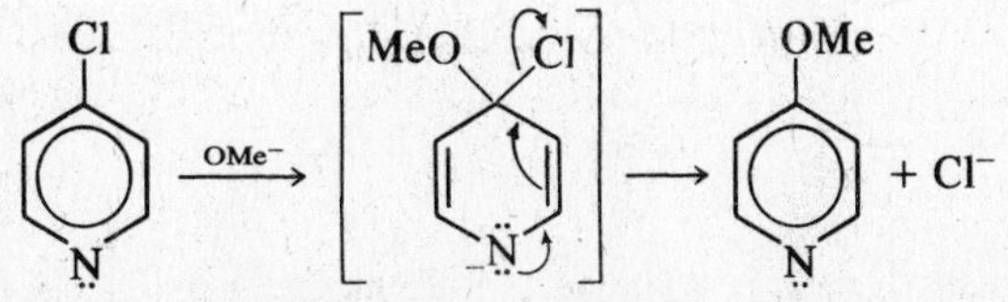

(c) β-Nucleophilic attack does not give an intermediate with − on N.

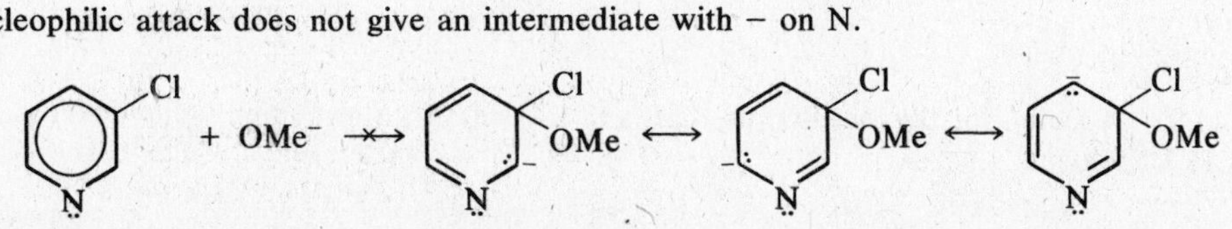

Problem 23.27 Give the product(s) formed when pyridine reacts with (a) BMe_3, (b) H_2SO_4, (c) EtI, (d) t-BuBr,

$$\text{(e) } Ph\overset{\overset{\displaystyle O}{\|}}{C}-OOH \quad \text{Peroxybenzoic acid} \qquad \blacktriangleleft$$

Pyridine is a typical 3° amine. (a) $C_5H_5\overset{+}{N}-\overset{-}{B}Me_3$; (b) $(C_5H_5NH)_2^+SO_4^{2-}$, pyridinium sulfate; (c) $[C_5H_5\overset{+}{N}-Et]I^-$, N-ethylpyridinium iodide; (d) $C_5H_5NH^+Br^- + Me_2C{=}CHMe$ (the 3° halide undergoes elimination rather than S_N2 displacement);

(e) Pyridine-N-oxide

Problem 23.28 Account for the following orders of reactivity:

(a) Toward H_3O^+: 2,6-dimethylpyridine (2,6-lutidine) > pyridine

(b) Toward the Lewis acid BMe_3: pyridine > 2,6-lutidine ◀

(a) Alkyl groups are electron-donating by induction and are base-strengthening. (b) BMe_3 is bulkier than an H_3O^+. The Me's at C^2 and C^6 flanking the N sterically inhibit the approach of BMe_3, causing 2,6-lutidine to be less reactive than pyridine. This is an example of **F-strain** (Front strain).

Problem 23.29 Compare and account for the products formed from nitration of pyridine and pyridine N-oxide. ◀

Pyridine is nitrated in the β position only under vigorous conditions [see Problem 23.22(b)]. However, its N-oxide is nitrated readily in the γ position; the intermediate formed by such attack is very stable because all atoms have an octet of electrons. The N-oxide behaves like phenoxide ion, $C_6H_5O^-$ (page 371).

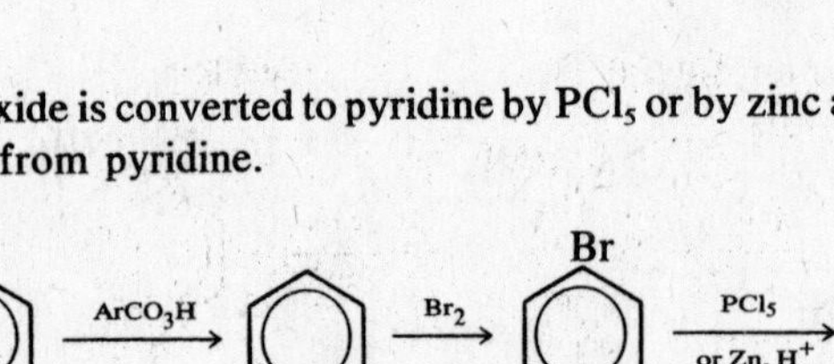

Intermediate   4-Nitropyridine N-oxide

Problem 23.30 Pyridine N-oxide is converted to pyridine by PCl_5 or by zinc and acid. Use this reaction for the synthesis of 4-bromopyridine from pyridine. ◀

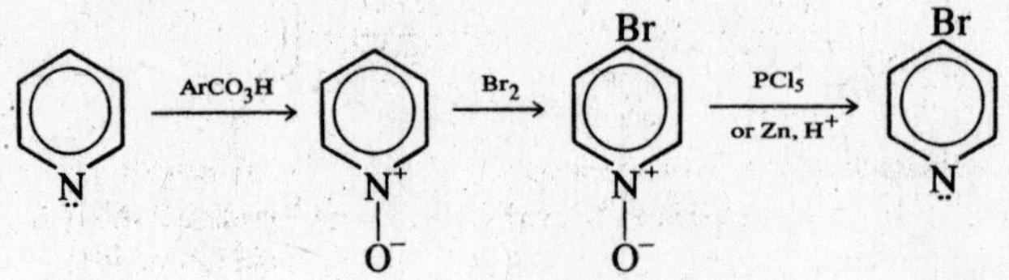

Pyridine   Pyridine N-oxide   4-Bromopyridine N-oxide   4-Bromopyridine

Problem 23.31 Account for the fact that the CH_3's of α- and γ-picolines (methylpyridines) are more acidic than the CH_3 of toluene. ◀

They react with strong bases to form resonance-stabilized anions with − on N.

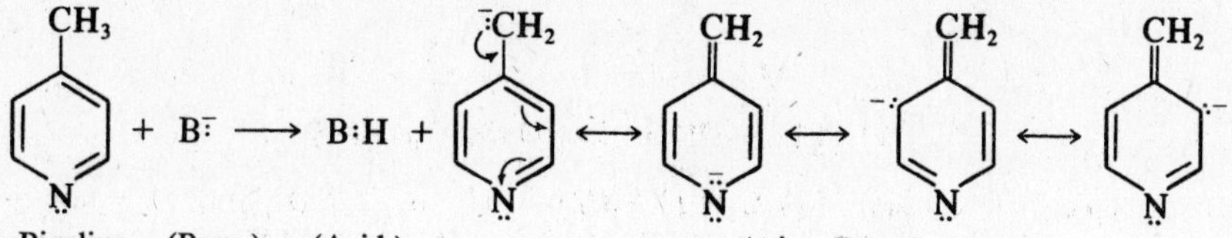

γ-Picoline　(Base₂)　(Acid₂)　　　　　　　　　　　　　　Anion (Base₁)
(Acid₁)

α-Picoline　(Base₂)　(Acid₂)　　　　　　　　　　　　Anion (Base₁)
(Acid₁)

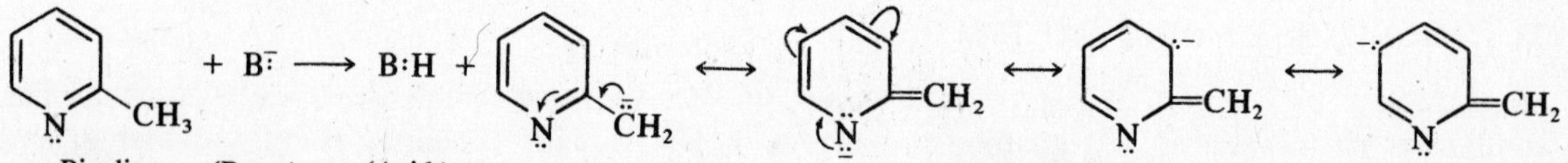

Problem 23.32　Give the products formed when γ-picoline reacts with C_6H_5Li and then with　(a) CO_2 then H_3O^+;　(b) C_6H_5CHO then H_3O^+.　　◄

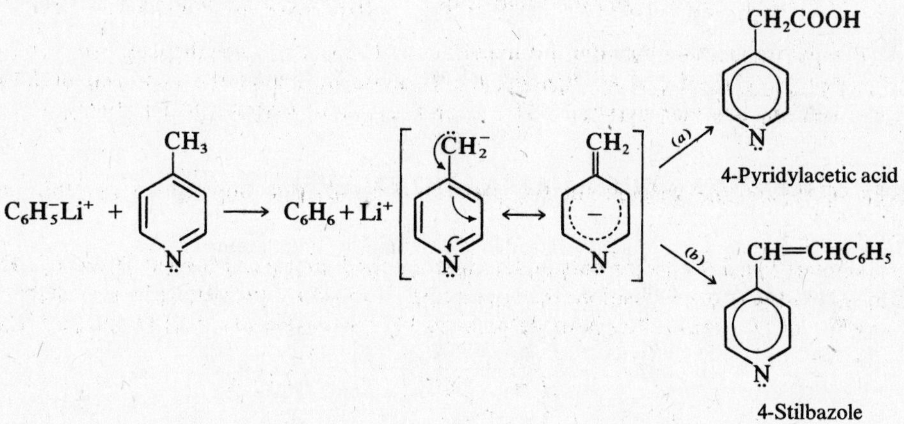

4-Pyridylacetic acid

4-Stilbazole

Problem 23.33　From picolines prepare　(a) the vitamin niacin (3-pyridinecarboxylic acid),　(b) the anti-tuberculosis drug isoniazide (4-pyridinecarboxylic acid hydrazide).　　◄

(a)

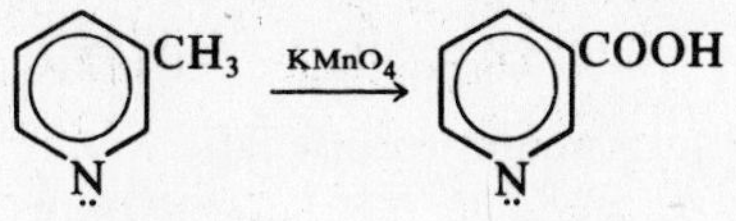

3-Picoline　　　　　　Niacin

(b)

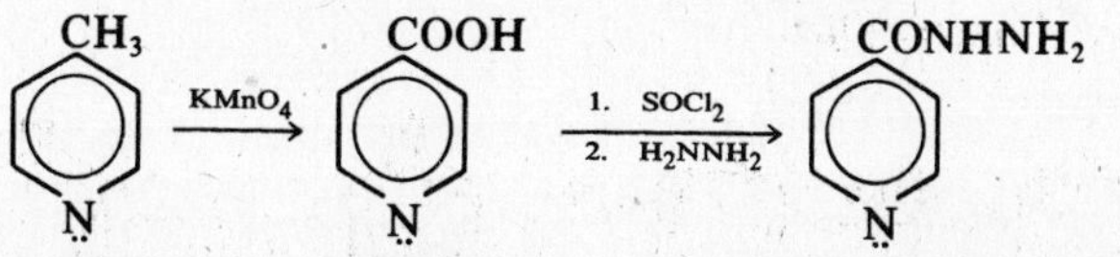

4-Picoline　　　4-Pyridine-　　　　　Isoniazide
　　　　　　carboxylic acid　　(4-Pyridinecarboxylic
　　　　　　　　　　　　　　　acid hydrazide)

COMPOUNDS WITH TWO HETEROATOMS.　PYRIMIDINE [Problem 23.1(f)]

Three pyrimidines are among the constituents of nucleic acids. In the more stable keto forms, they are cytosine, uracil and thymine:

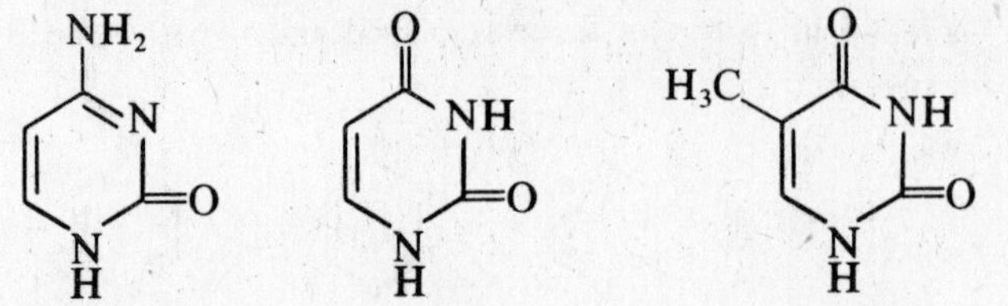

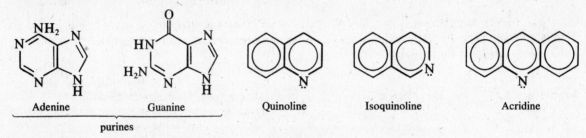

Cytosine   Uracil   Thymine

23.4 CONDENSED RING SYSTEMS

Many biologically important heterocyclic compounds have **fused (condensed)** ring systems. The condensed rings adenine and guanine are found in DNA (with cytosine, 5-methylcytosine, and thymine) and also in RNA (with cytosine and uracil).

Adenine   Guanine   Quinoline   Isoquinoline   Acridine

purines

QUINOLINE (2,3-BENZOPYRIDINE, 1-AZANAPHTHALENE)

Problem 23.34 Which dicarboxylic acid is formed on oxidation of quinoline?   ◄

The pyridine ring is more stable.

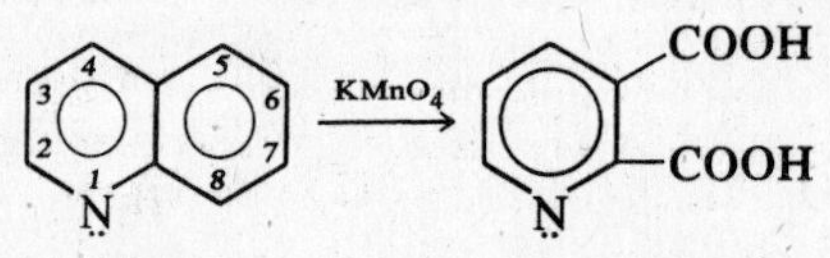

$$\text{Quinoline} \xrightarrow{\text{KMnO}_4} \text{Quinolinic acid}$$

Quinoline   Quinolinic acid

1. Skraup Synthesis

The steps in the reaction of aniline, glycerol and $C_6H_5NO_2$ are:

(*a*) Dehydration of glycerol to acrolein (propenal).

$$\text{H}_2\text{COHCHOHCH}_2\text{OH} \xrightarrow{\text{H}_2\text{SO}_4} \text{H}_2\text{C}=\text{CHCHO} + 2\text{H}_2\text{O}$$

Glycerol   Acrolein

(*b*) Michael-type addition (page 277).

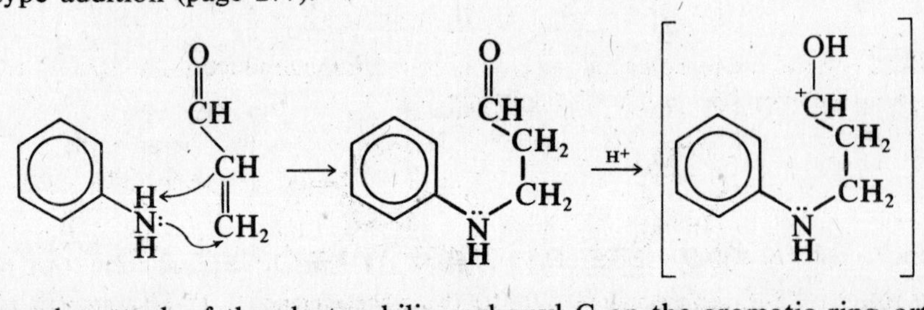

(*c*) Ring closure by attack of the electrophilic carbonyl C on the aromatic ring *ortho* to the electron-releasing —NH. The 2° alcohol formed is dehydrated to 1,2-dihydroquinoline by the strong acid.

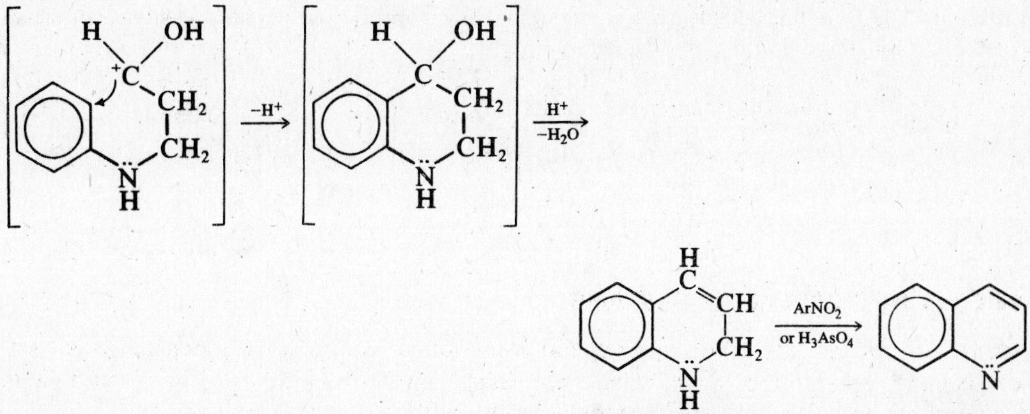

1,2-Dihydroquinoline   Quinoline

(d) PhNO$_2$ oxidizes the dihydroquinoline to the aromatic compound quinoline. PhNO$_2$ is reduced to PhNH$_2$, which then reacts with more acrolein. This often violent reaction is moderated by added FeSO$_4$.

Problem 23.35 Give the Skraup product from (a) p-nitroaniline and crotonaldehyde (trans-CH$_3$CH=CHCHO), (b) 3-bromo-4-aminotoluene and glycerol, (c) 1-amino-naphthalene and glycerol. ◄

(a)

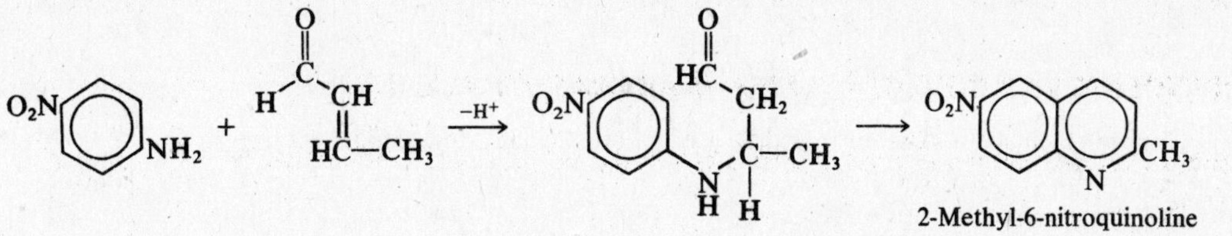

2-Methyl-6-nitroquinoline

(b)

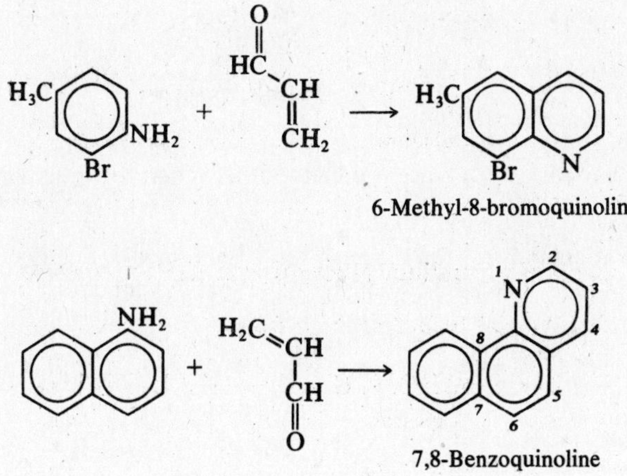

6-Methyl-8-bromoquinoline

(c)

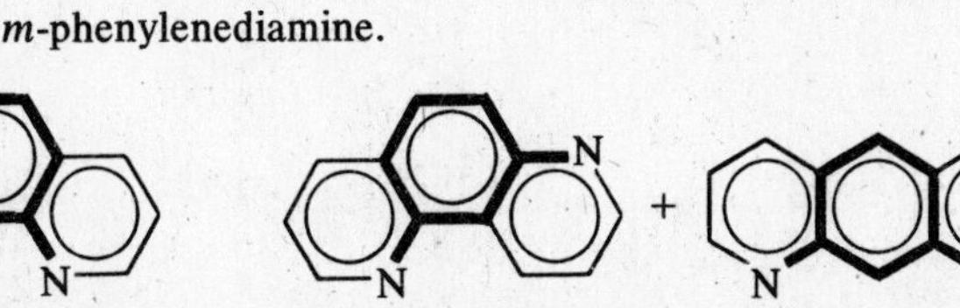

7,8-Benzoquinoline

Problem 23.36 Write the name and structural formula for the product of the Skraup reaction of glycerol with (a) o-phenylenediamine, (b) m-phenylenediamine. ◄

4,5-Diazaphenanthrene   1,5-Diazaphenanthrene   1,8-Diazanthracene
(a)   (b)

(The heavy-lined ring comes from the diaminobenzenes.)

Problem 23.37 Starting with benzene or aniline and any inorganic and aliphatic reagents, outline a synthesis of (a) 8-ethylquinoline, (b) 6-bromoquinoline. ◄

(a)

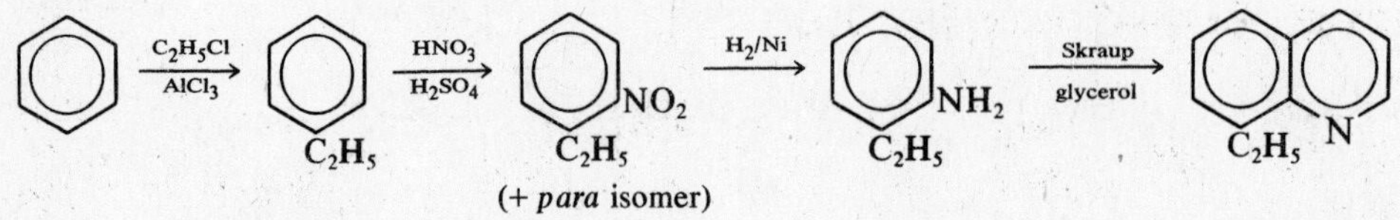

$$\text{benzene} \xrightarrow[\text{AlCl}_3]{\text{C}_2\text{H}_5\text{Cl}} \text{C}_2\text{H}_5 \xrightarrow[\text{H}_2\text{SO}_4]{\text{HNO}_3} \underset{\text{C}_2\text{H}_5}{\text{NO}_2} \xrightarrow{\text{H}_2/\text{Ni}} \underset{\text{C}_2\text{H}_5}{\text{NH}_2} \xrightarrow[\text{glycerol}]{\text{Skraup}} \underset{\text{C}_2\text{H}_5}{\text{N}}$$

(+ *para* isomer)

(b)

$$\underset{\text{NH}_2}{} \xrightarrow{\text{Ac}_2\text{O}} \underset{\text{NHAc}}{} \xrightarrow{\text{Br}_2} \text{Br} \underset{\text{NHAc}}{} \xrightarrow{\text{H}_3\text{O}^+} \text{Br} \underset{\text{NH}_2}{} \xrightarrow[\text{glycerol}]{\text{Skraup}} \text{Br} \underset{\text{N}}{}$$

2. Properties

Problem 23.38 Give the expected products for (a) monobromination of quinoline, (b) catalytic reduction of quinoline with 2 moles of H_2. ◄

(a) The more reactive phenyl ring undergoes E^+-attack in the α positions, giving

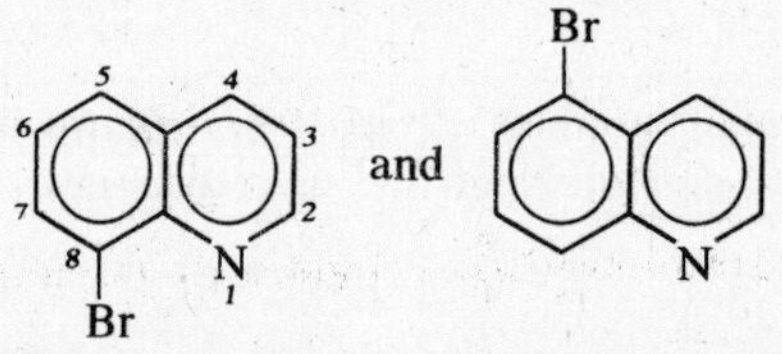

8-Bromoquinoline   5-Bromoquinoline

(b) The more electron-deficient pyridine ring is more easily reduced.

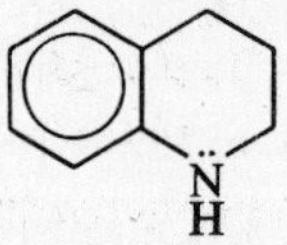

1,2,3,4-Tetrahydroquinoline

Problem 23.39 Give structures for the products of reaction of quinoline with (a) HNO_3, H_2SO_4; (b) $NaNH_2$; (c) PhLi. ◄

(a) 5-nitro- and 8-nitroquinoline; (b) 2-amino- and 4-aminoquinoline (like pyridine, quinoline undergoes nucleophilic substitution in the 2 and 4 positions); (c) 2-phenylquinoline.

Supplementary Problems

Problem 23.40 Supply systematic names for:

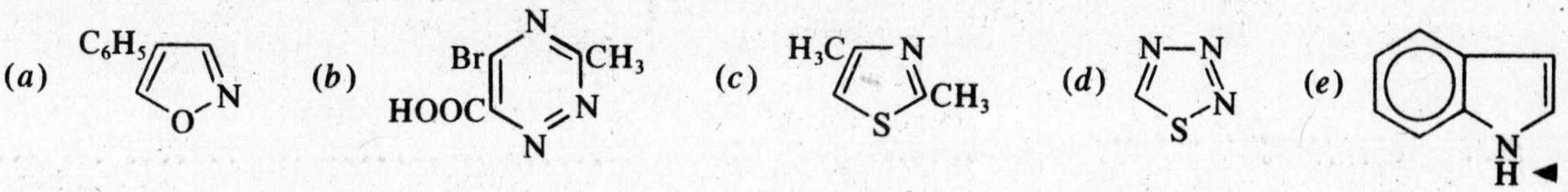

(a) [structure: C₆H₅-substituted isoxazole ring]   (b) [structure: triazine with Br, CH₃, HOOC]   (c) [structure: H₃C-N thiazole with CH₃]   (d) [structure: N-N thiatriazole]   (e) [structure: indole bicyclic with NH]

(a) 4-phenyl-1,2-oxazole,   (b) 3-methyl-5-bromo-1,2,4-triazine-6-carboxylic acid,   (c) 2,4-dimethyl-1,3-thiazole,   (d) 1,2,3,4-thiatriazole,   (e) 2,3-benzazole (indole).

Problem 23.41 Give the structural formulas of   (a) quinoline-5,8-quinone,   (b) 1-methyl-2-(3-pyridyl)pyrrolidine (nicotine),   (c) N-methyl-4-phenyl-4-carbethoxypiperidine (Demerol).   ◄

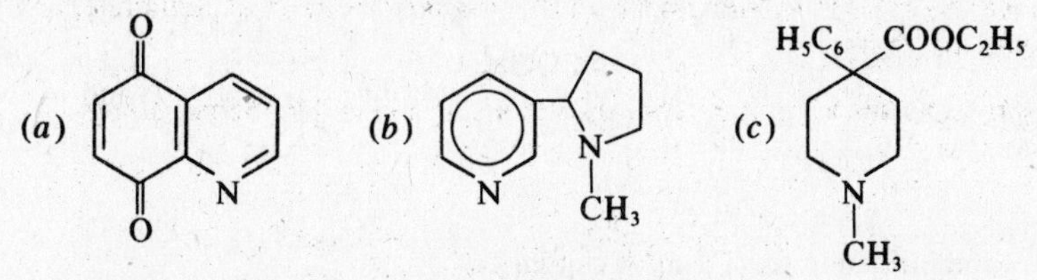

(a) [structure: quinoline-5,8-quinone]   (b) [structure: nicotine]   (c) [structure: Demerol with H₅C₆, COOC₂H₅, CH₃]

Problem 23.42 How many thiophenyl-thiophenes (**bithienyls**) are possible?   ◄

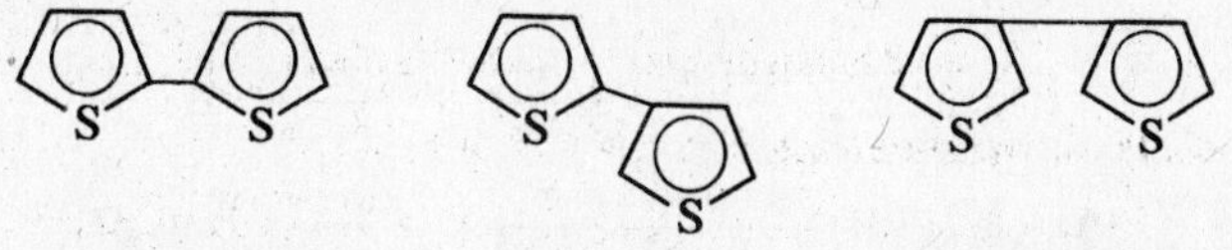

[structures of bithienyls]

2,2′-Bithienyl   2,3′-Bithienyl   3,3′-Bithienyl

Problem 23.43 Outline the reactions in the preparation from tetrahydrofuran of both hexamethylenediamine and adipic acid (for use in making Nylon 66).   ◄

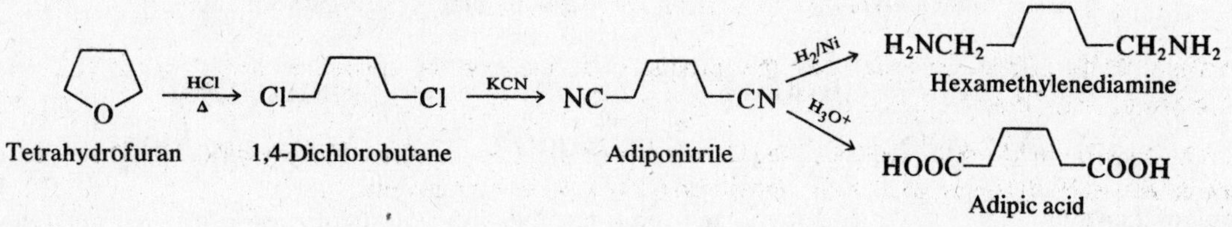

[reaction scheme]

Tetrahydrofuran   1,4-Dichlorobutane   Adiponitrile

H_2NCH_2——————CH_2NH_2
Hexamethylenediamine

$HOOC$——————$COOH$
Adipic acid

Problem 23.44 Which heterocycle is formed when δ-bromocaproic acid is treated with a mild base such as NaHCO₃?   ◄

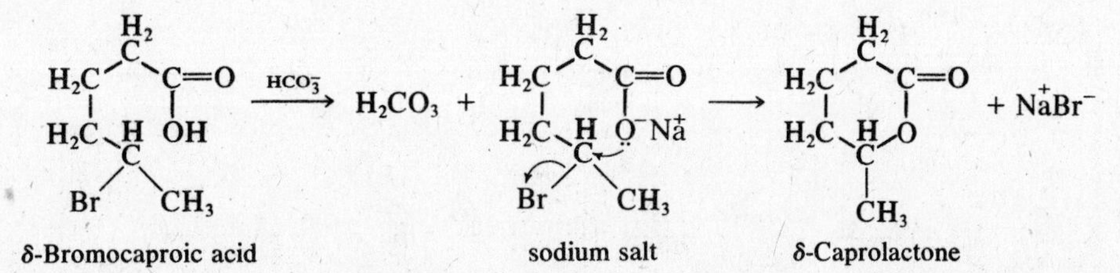

[reaction scheme]

δ-Bromocaproic acid   sodium salt   δ-Caprolactone   + Na⁺Br⁻

Problem 23.45 Which methylquinoline is vigorously oxidized to a tricarboxylic acid that is then dehydrated to a mixture of two carboxylic acid anhydrides?   ◄

If the CH₃ is on the benzene ring, oxidation gives *only* 2,3-pyridinedicarboxylic acid, regardless of the position of the CH₃. If the CH₃ is on the pyridine ring, there are three possible tricarboxylic acids:

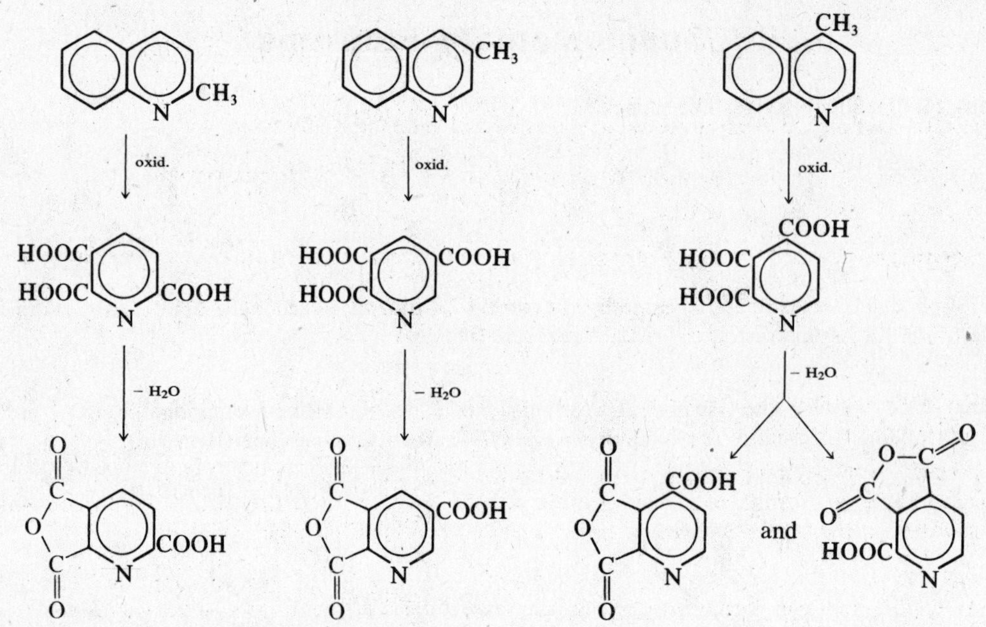

Two anhydrides are obtained only from 4-methylquinoline.

Problem 23.46 Furnish the missing compounds.

(a) $\quad$ Quinoline $\xrightarrow{\text{C}_6\text{H}_5\text{CO}_2\text{H}}$ I $\xrightarrow{\text{HNO}_3}$ II $\xrightarrow{\text{PCl}_3}$ III

(b) $\quad$ Pyrrole (PyH) $\xrightarrow{p\text{-HO}_3\text{SC}_6\text{H}_4\text{N}_2\text{X}^-}$ I $\xrightarrow{\text{Sn, HCl}}$ II + III

(c) $\quad$ Furan (FuH) $\xrightarrow[(\text{C}_2\text{H}_5)_2\text{O, BF}_3]{(\text{CH}_3\text{CO})_2\text{O}}$ I $\xrightarrow{\text{NaOI}}$ II $\xrightarrow{\text{fum. H}_2\text{SO}_4}$ III

	I	II	III
(a)	Quinoline N-oxide	4-Nitroquinoline N-oxide	4-Nitroquinoline
(b)	2-PyN=NC$_6$H$_4$SO$_3$H-p	2-PyNH$_2$	p-$\overset{+}{\text{N}}$H$_3$C$_6$H$_4$SO$_3^-$
(c)	2-FuCOCH$_3$	2-FuCOO$^-$Na$^+$	5-HO$_3$S-2-FuCOOH

Problem 23.47 Prepare (a) 3-aminopyridine from β-picoline, (b) 4-aminopyridine from pyridine, (c) 8-hydroxyquinoline from quinoline, (d) 5-nitro-2-furoic acid from furfural, (e) 2-pyridylacetic acid from pyridine.

(a) $\xrightarrow[\text{KMnO}_4]{\text{oxid.}}$ $\xrightarrow{\text{SOCl}_2}$ $\xrightarrow{\text{NH}_3}$ $\xrightarrow[\text{OH}^-]{\text{NaOBr}}$

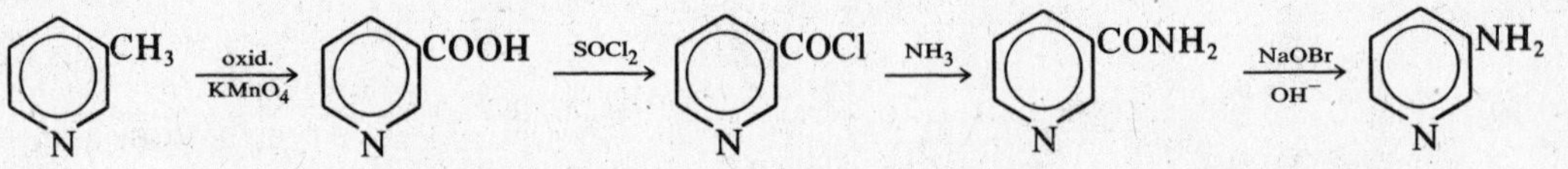

(b) $\xrightarrow{\text{PhCOOOH}}$ $\xrightarrow[\text{H}_2\text{SO}_4]{\text{HNO}_3}$ $\xrightarrow{\text{PCl}_3}$ $\xrightarrow[\text{2. OH}^-]{\text{1. Sn, H}^+}$

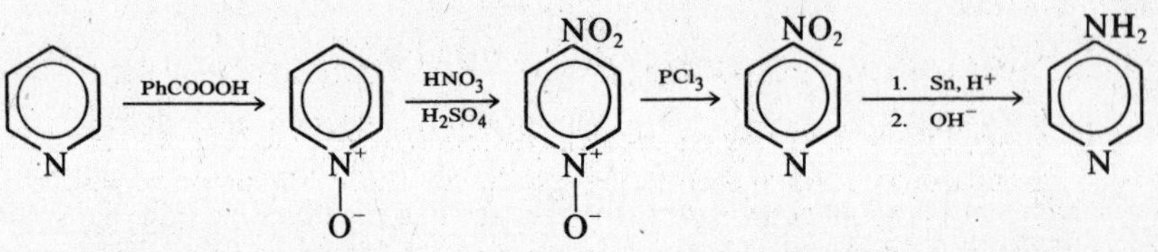

(c) $\quad$ Quinoline $\xrightarrow[220\,°\text{C}]{\text{H}_2\text{SO}_4}$ Quinoline-8-sulfonic acid $\xrightarrow[\substack{\text{2. NaOH, fuse} \\ \text{3. H}_3\text{O}^+}]{\text{1. OH}^-}$ product

(d)

$$\text{(furan)}\text{CHO} \xrightarrow{\text{Ag(NH}_3)_2^+} \text{(furan)}\text{COOH} \xrightarrow{\text{HNO}_3} \text{O}_2\text{N}\text{(furan)}\text{COOH}$$

(COOH stabilizes the ring towards acid cleavage of the ether bond.)

(e)

$$\text{Pyridine} \xrightarrow{\text{NH}_2^-} \text{2-Aminopyridine} \xrightarrow[\text{2. CuBr}]{\text{1. NaNO}_2, \text{H}^+, 5\,°\text{C}} \text{2-Bromopyridine}$$

2-Bromopyridine + [CH(COOC$_2$H$_5$)$_2$]$^-$ Na$^+$ $\longrightarrow$ (pyridine ring)—CH(COOC$_2$H$_5$)$_2$ $\xrightarrow[\text{3. }\Delta, -\text{CO}_2]{\substack{\text{1. NaOH} \\ \text{2. H}_3\text{O}^+}}$ (pyridine ring)—CH$_2$COOH

nucleophilic displacement

Problem 23.48 (a) Explain why pyran [Problem 23.1(d)] is not aromatic. (b) What structural change would theoretically make it aromatic. ◄

(a) There are 6 electrons available: 4 from the 2 π bonds and 2 from the O atom. However, C^4 is sp^3 hybridized and has no p orbital available for cyclic p orbital overlap. (b) Convert C^4 to a carbonium ion. C^4 would now be sp^2 hybridized and would have an empty p orbital for cyclic overlap.

Problem 23.49 How can pyridine and piperidine be distinguished by infrared spectroscopy? ◄

Piperidine has an N—H bond absorbing at 3500 cm^{-1} and H—C(sp^3) stretch below 3000 cm^{-1}. Pyridine has no N—H; has H—C(sp^2) stretch above 3000 cm^{-1}; C=C and C=N stretches near 1600 cm^{-1} and 1500 cm^{-1}, respectively; aromatic ring vibrations near 1200 cm^{-1} and 1050 cm^{-1}; and C—H deformations at 750 cm^{-1}. The peak at 750 varies with substitution in the pyridine ring.

Problem 23.50 How can nmr spectroscopy distinguish among aniline, pyridine and piperidine? ◄

The NH$_2$ of aniline is electron-donating and shields the aromatic H's; their chemical shift is $\delta = 6.5$–7.0 (for benzene the chemical shift is $\delta = 7.1$). The N of pyridine is electron-withdrawing and deshields the aromatic H's ($\delta = 7.5$–8.0). Piperidine is not aromatic and has no signals in these regions.

Problem 23.51 From pyridine (PyH), 2-picoline (2-PyMe) and any reagent without the pyridine ring prepare (a) 2-acetylpyridine, (b) 2-vinylpyridine, (c) 2-cyclopropylpyridine, (d) 2-PyCH$_2$CH$_2$CH$_2$COOH, (e) 2-PyC(Me)=CHCH$_3$, (f) 2-pyridinecarboxaldehyde. Any synthesized compound can be used in ensuing steps. ◄

(a) By a crossed-Claisen condensation:

$$\text{2-PyMe} \xrightarrow{\text{KMnO}_4} \text{2-PyCOOH} \xrightarrow[\text{H}^+]{\text{EtOH}} \text{2-PyCOOEt} \xrightarrow[\text{NaOEt}]{\text{CH}_3\text{COOEt}}$$

$$\text{2-PyCOCH}_2\text{COOEt} \xrightarrow[\text{heat}]{\text{OH}^-} \text{2-PyCOCH}_2\text{COO}^- \xrightarrow{\text{H}^+} \text{2-PyCOCH}_2\text{COOH} \xrightarrow{-\text{CO}_2} \text{2-PyCOCH}_3$$

(b)

$$\text{2-PyCOCH}_3 \xrightarrow{\text{NaBH}_4} \text{2-PyCHOHCH}_3 \xrightarrow{\text{P}_2\text{O}_5} \text{2-PyCH=CH}_2$$

(c)

$$\text{2-PyCH=CH}_2 + \text{CH}_2\text{N}_2 \xrightarrow{\text{uv}} \text{product}$$

(d) By a Michael addition:

$$\text{2-PyCH=CH}_2 + (\text{EtOOC})_2\text{CH}^-\text{Na}^+ \longrightarrow \text{2-Py}\overset{-}{\text{C}}\text{HCH}_2\text{CH(COOEt)}_2\text{Na}^+$$

This α-pyridylcarbanion is stabilized by charge delocalization to the ring N (Problem 23.31). Refluxing the salt in HCl causes decarboxylation and gives the pyridinium salt of the product, which is then neutralized with OH$^-$.

(e) Use the Wittig synthesis: 2-PyCOMe + Ph$_3$P=CHCH$_3$ $\longrightarrow$ products (*trans* and *cis*).

(f)

$$\text{2-PyCH=CH}_2 \xrightarrow[\text{2. Zn, HOAc}]{\text{1. O}_3} \text{2-PyCHO}$$

Chapter 24

Amino Acids and Proteins

24.1 INTRODUCTION

Amino acids are dipolar ions (**Zwitterions**), $RCH(\overset{+}{N}H_3)COO^-$, as indicated by their crystallinity, high melting point and solubility in water rather than in nonpolar organic solvents. The amino acids displayed in Table 24-1 all have the L or S configuration at the α C. An asterisk (∗) denotes an **essential** amino acid, i.e. one that cannot be synthesized in the body and so must be in the diet.

Table 24-1. Natural α-Amino Acids

Name	Symbol	Formula
Monoaminomonocarboxylic		
Glycine	Gly	$\overset{+}{H_3}NCH_2COO^-$
Alanine	Ala	$\overset{+}{H_3}NCH(CH_3)COO^-$
Valine*	Val	$\overset{+}{H_3}NCH(i\text{-}Pr)COO^-$
Leucine*	Leu	$\overset{+}{H_3}NCH(i\text{-}Bu)COO^-$
Isoleucine*	Ileu	$\overset{+}{H_3}NCH(s\text{-}Bu)COO^-$
Serine	Ser	$\overset{+}{H_3}NCH(CH_2OH)COO^-$
Threonine*	Thr	$\overset{+}{H_3}NCH(CHOHCH_3)COO^-$
Monoaminodicarboxylic and Amide Derivatives		
Aspartic acid	Asp	$HOOCCH_2CH(\overset{+}{N}H_3)COO^-$
Asparagine	Asp(NH$_2$)	$H_2NCOCH_2CH(\overset{+}{N}H_3)COO^-$
Glutamic acid	Glu	$HOOC(CH_2)_2CH(\overset{+}{N}H_3)COO^-$
Glutamine	Glu(NH$_2$)	$H_2NCOCH_2CH_2CH(\overset{+}{N}H_3)COO^-$
Diaminomonocarboxylic		
Lysine*	Lys	$\overset{+}{H_3}N(CH_2)_4CH(NH_2)COO^-$
Hydroxylysine	Hylys	$\overset{+}{H_3}NCH_2\underset{\overset{\mid}{OH}}{CH}CH_2CH_2CH(NH_2)COO^-$
Arginine*	Arg	$\overset{H_2N}{\underset{H_2N}{>}}\overset{+}{C}\!=\!NH(CH_2)_3CH(NH_2)COO^-$
Sulfur-Containing		
Cysteine	CySH	$\overset{+}{H_3}NCH(CH_2SH)COO^-$
Cystine	CySSCy	$^-OOCCH(\overset{+}{N}H_3)CH_2S\!-\!SCH_2CH(\overset{+}{N}H_3)COO^-$
Methionine*	Met	$CH_3SCH_2CH_2CH(\overset{+}{N}H_3)COO^-$

414

Table 24-1. Natural α-Amino Acids (Continued)

Name	Symbol	Formula
Aromatic		
Phenylalanine*	Phe	$PhCH_2CH(\overset{+}{NH_3})COO^-$
Tyrosine	Tyr	$p\text{-}HOC_6H_4CH_2CH(\overset{+}{NH_3})COO^-$
Heterocyclic		
Histidine*	His	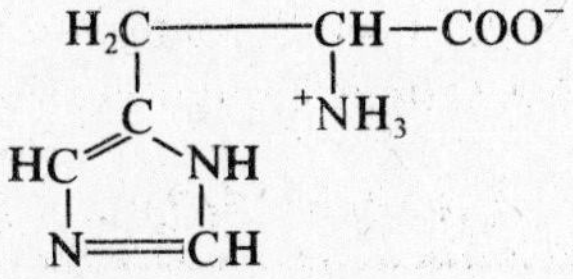
Proline	Pro	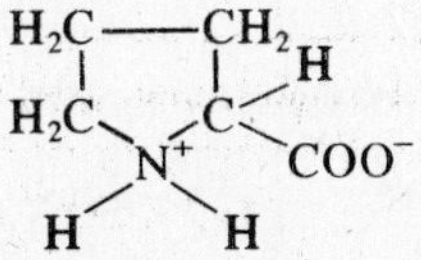
Hydroxyproline	Hypro	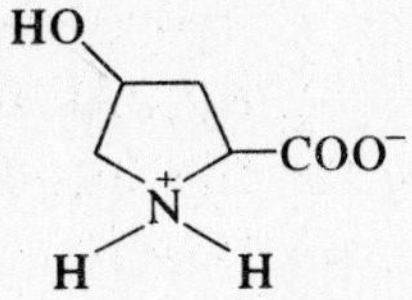
Tryptophane*	Try	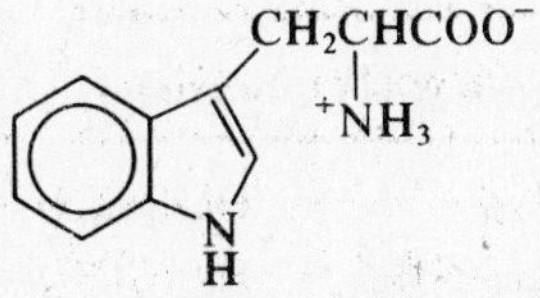

24.2 PREPARATION OF α-AMINO ACIDS

Problem 24.1 Synthesize leucine, $(CH_3)_2CHCH_2CH(\overset{+}{NH_3})COO^-$, by (a) Hell-Volhard-Zelinsky reaction followed by ammonolysis, (b) Gabriel synthesis, (c) phthalimidomalonic ester synthesis, (d) reductive amination of a keto acid, (e) Strecker synthesis (addition of NH_3, HCN to RCH=O), (f) acetyl-aminomalonic ester. ◄

(a) $(CH_3)_2CH-CH_2-CH_2-COOH \xrightarrow{Br_2/P} (CH_3)_2CH-CH_2CHBrCOOH \xrightarrow{xs\,NH_3} Leu$

(b)

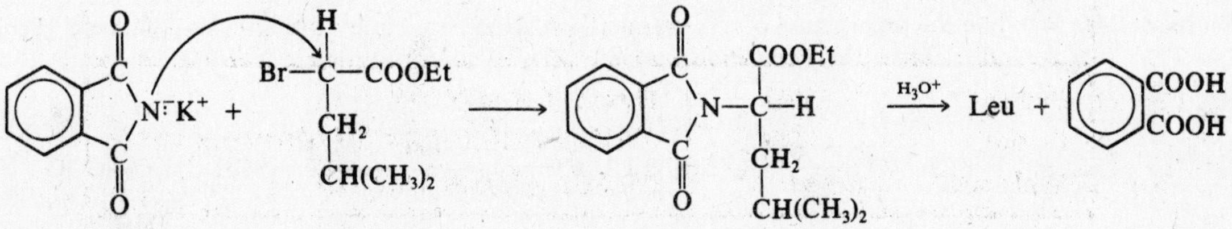

Use ester rather than acid to prevent conversion of anion to phthalimide.

(c) $CH_2(COOEt)_2 \xrightarrow{Br_2} BrCH(COOEt)_2$ (use below)

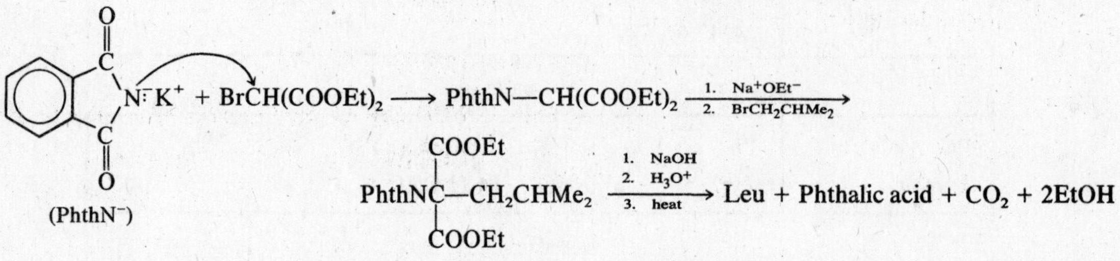

$(PhthN^-)$ $+ BrCH(COOEt)_2 \longrightarrow PhthN-CH(COOEt)_2 \xrightarrow[\text{2. } BrCH_2CHMe_2]{\text{1. } Na^+OEt^-}$

$$PhthN\underset{\underset{COOEt}{|}}{\overset{\overset{COOEt}{|}}{C}}-CH_2CHMe_2 \xrightarrow[\text{3. heat}]{\substack{\text{1. NaOH} \\ \text{2. } H_3O^+}} Leu + Phthalic\ acid + CO_2 + 2EtOH$$

(d) $Me_2CH-CH_2-\overset{\overset{O}{\|}}{C}COOH \xrightarrow[H_2/Pt]{NH_3} Leu$

(e) $Me_2CHCH_2\overset{\overset{H}{|}}{C}=O \xrightarrow[NH_3]{HCN} Me_2CHCH_2\underset{\underset{NH_2}{|}}{C}HCN \xrightarrow{H_3O^+} Leu + NH_4^+$

In the presence of NH_3 an aminonitrile is formed instead of a cyanohydrin.

(f) $H_2C(COOEt)_2 \xrightarrow{HONO} HON=C(COOEt)_2 \xrightarrow[\text{2. } Ac_2O]{\text{1. } H_2/Pt} AcNHCH(COOEt)_2 \xrightarrow[BrCH_2CHMe_2]{NaOEt}$

             an oximino   Acetylaminomalonic
               ester   ester

$$AcNH\underset{\underset{COOEt}{|}}{\overset{\overset{COOEt}{|}}{C}}CH_2CHMe_2 \xrightarrow[\text{3. heat}]{\substack{\text{1. NaOH} \\ \text{2. } H_3O^+}} Leu + CO_2 + 2EtOH$$

Problem 24.2 Write structural formulas for compounds (A) through (C) and point out the reaction type.

$$\overset{1}{C}H_3\overset{2}{C}OOC_2H_5 + \underset{\overset{4}{C}OOC_2H_5}{\overset{\overset{3}{C}OOC_2H_5}{|}} \xrightarrow{NaOEt} (A) \xrightarrow[\text{dil.}]{H_2SO_4} (B) \xrightarrow[H_2/Pt]{NH_3} (C) \quad \blacktriangleleft$$

(A) $\overset{1}{C}H_2\overset{2}{C}OOC_2H_5$
      $\overset{3}{C}=O$ (crossed-Claisen condensation)
      $\overset{4}{C}OOC_2H_5$

(B) $\overset{1}{C}H_3-\overset{3}{\underset{\underset{O}{\|}}{C}}-\overset{4}{C}OOH$ (ester hydrolysis and decarboxylation
                          of a β-keto acid; see page 311)

(C) $CH_3-\underset{\underset{{}^+NH_3}{|}}{C}H-COO^-$ (reductive amination)

Problem 24.3 Outline the preparation of (a) methionine from acrolein by the Strecker synthesis [Problem 24.1(e)], (b) glutamic acid by the phthalimidomalonic ester synthesis [Problem 24.1(c)]. $\blacktriangleleft$

(a) $H_2C=CH-CH=O \xrightarrow[\substack{Michael \\ addition}]{CH_3SH} CH_3SCH_2CH_2CH=O \xrightarrow[\text{2. } H_3O^+]{\text{1. } NH_3, HCN} CH_2SCH_2CH_2\underset{\underset{NH_3^+}{|}}{\overset{\overset{H}{|}}{C}}-COO^-$

                                                                         Methionine

(b)   PhthN—CH(COOEt)$_2$ $\xrightarrow{\text{OEt}^-}$ PhthN$\overset{\cdot\cdot}{\text{C}}$(COOEt)$_2$ $\xrightarrow[\text{2. }H_3O^+;\ Michael\atop addition]{\text{1. }H_2C=CHCOOEt}$

Phthalimidomalonic ester

$$\underset{\text{Glutamic acid}}{PhthN—\overset{\displaystyle COOEt}{\underset{\displaystyle COOEt}{C}}CH_2CHCOOEt} \xrightarrow{H^+} H_3\overset{+}{N}—\overset{\displaystyle}{\underset{\displaystyle COO^-}{C}}HCH_2CH_2COOH$$

24.3 ACID-BASE (AMPHOTERIC) PROPERTIES (see Problem 3.45)

Problem 24.4 Write equilibria showing the amphoteric behavior of an H_2O solution of glycine. All monoaminomonocarboxylic acids are slightly acidic.   ◄

$$H_3O^+ + H_2NCH_2COO^- \underset{\xrightarrow{\hspace{1cm}}}{\overset{K_{a2}=1.6\times10^{-10}}{\longleftarrow}} \boxed{H_3\overset{+}{N}CH_2COO^-} + H_2O \underset{K_b=2.5\times10^{-12}}{\overset{K_{a1}=10^{-2.4}}{\rightleftharpoons}} H_3\overset{+}{N}CH_2COOH + OH^-$$

| Anion | Ampholyte | Cation |
| (acts as RNH$_2$) | | (acts as RCOOH) |

Since the solution is acidic, the reaction to the left is favored and the anion predominates.

Problem 24.5 (a) How does one repress the ionization of an α-amino acid?   (b) The pH at which [anion] = [cation] is called the **isoelectric point**.   In an electrolysis experiment, do α-amino acids migrate at the isoelectric point?   ◄

(a) Since the α-amino acid solution is slightly acidic, the anion predominates, and acid is added to repress the ionization.   (b) No.

Problem 24.6 Why is H$_3\overset{+}{N}$CH$_2$COOH (pK_{a1} = 2.4) more acidic than RCH$_2$COOH (pK_a = 4–5)?   ◄

α-H$_3$N$^+$— increases acidity because of its electron-withdrawing inductive effect.

Problem 24.7 (a) Why does sulfanilic (p-aminobenzenesulfonic) acid exist as a dipolar ion, while p-aminobenzoic acid does not?   (b) Why is sulfanilic acid soluble in bases but not in acids?   ◄

(a) —SO$_3$H is strongly acidic and donates H$^+$ to the weakly basic arylamino group.   ArCOOH is not acidic enough to transfer H$^+$ to the arylamino group.   (b) In dipolar sulfanilic acid, p-H$_3\overset{+}{N}$C$_6$H$_4$SO$_3^-$, H$_3\overset{+}{N}$— is acidic enough to transfer H$^+$ to bases to give the soluble anion, p-H$_2$NC$_6$H$_4$SO$_3^-$.   —SO$_3^-$ is too feebly basic and cannot accept H$^+$ from acids.

Problem 24.8 Predict whether the isoelectric points for the following α-amino acids are considerably acidic, slightly acidic, or basic: (a) alanine, (b) lysine, (c) aspartic acid, (d) cystine, (e) tyrosine. (See Table 24-1 and Problem 24.4.)   ◄

(a) Alanine is a monoaminomonocarboxylic acid.   Its isoelectric point is slightly acidic (pH = 6.0). (b) Lysine is a diaminomonocarboxylic acid.

$$H_3\overset{+}{N}(CH_2)_4CH(NH_2)COO^- + H_2O \rightleftharpoons H_3\overset{+}{N}(CH_2)_4CH(\overset{+}{N}H_3)COO^- + OH^-$$

Additional base is needed to repress this ionization.   The isoelectric point of lysine is basic (pH = 9.6). (c) Aspartic acid is a monoaminodicarboxylic acid.

$$HOOCCH_2CH(\overset{+}{N}H_3)COO^- + H_2O \rightleftharpoons {}^-OOCCH_2CH(\overset{+}{N}H_3)COO^- + H_3O^+$$

A more strongly acidic solution is needed to repress this ionization.   Aspartic acid's isoelectric point is strongly acidic (pH = 2.7). (*d*) Cystine is a diaminodicarboxylic acid and behaves like a monoaminomonocarboxylic acid. The isoelectric point is slightly acid (pH = 4.6). (*e*) Tyrosine is a monoaminomonocarboxylic acid containing a phenolic OH, which however is too weakly acidic to ionize to any significant extent. The isoelectric point is slightly acidic (pH = 5.6).

Problem 24.9 Why is the isoelectric point of arginine (pH = 10.7) greater than that of lysine (pH = 9.6)? ◄

The second basic site in arginine is an R-substituted guanidino group,

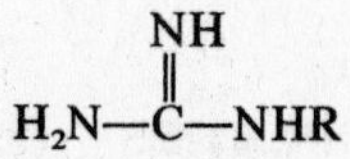

$$
\underset{\text{H}_2\text{N}-\overset{\displaystyle\|}{\underset{}{\text{C}}}-\text{NHR}}{\overset{\text{NH}}{}}
$$

This group is more basic than the ε NH$_2$ of lysine because, on accepting H$^+$, the guanidino group gives a charge-delocalized cation:

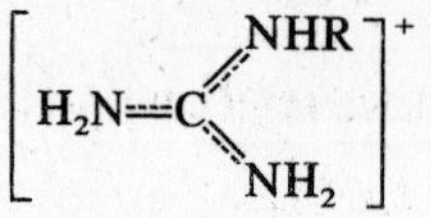

$$
\left[\text{H}_2\text{N}=\!\!\!=\!\!\!\text{C}\underset{\text{NH}_2}{\overset{\text{NHR}}{<}} \right]^+
$$

Problem 24.10 How can lysine be separated from glycine? ◄

The isoelectric points are different: pH = 9.6 for lysine and pH = 5.97 for glycine. An aqueous solution of the mixture is placed between two electrodes, the pH adjusted to either 5.97 or 9.6, and an electric current applied. At pH 5.97, glycine doesn't migrate, while cationic lysine migrates to the cathode, where it is collected. At pH 9.6, lysine doesn't migrate, while anionic glycine migrates to the anode.

Problem 24.11 The ir spectra of amino acids have bands near 1400 cm^{-1} and 1600 cm^{-1}, which correspond to the COO$^-$ group. Why does a strong peak appear at 1720 cm^{-1} in highly acidic solution? ◄

The 1400 cm^{-1} and 1600 cm^{-1} peaks are caused by the symmetrical and antisymmetrical C—O stretching in COO$^-$. Acid converts COO$^-$ to COOH, whose C=O produces the 1720 cm^{-1} peak.

Problem 24.12 On being heated, amino acids behave like hydroxyacids (Problems 17.26 and 17.27), giving N-analog products. Write the structural formulas for the products formed on heating (*a*) α-RCHNH$_3^+$COO$^-$, (*b*) β-RCHNH$_3^+$CH$_2$COO$^-$, (*c*) γ-H$_3$N$^+$(CH$_2$)$_3$COO$^-$, (*d*) δ-H$_3$N$^+$(CH$_2$)$_4$COO$^-$, (*e*) ε-H$_3$N$^+$(CH$_2$)$_5$COO$^-$. ◄

(*a*)

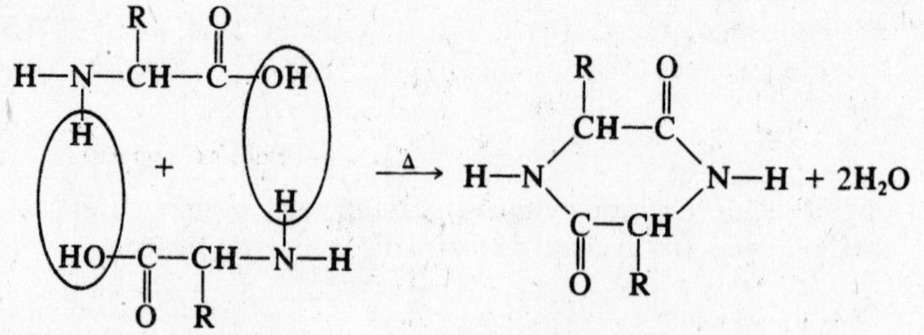

an α-amino acid　　　　　　　a diketopiperazine

(*b*) RCH=CHCOO$^-$NH$_4^+$

(*c*)

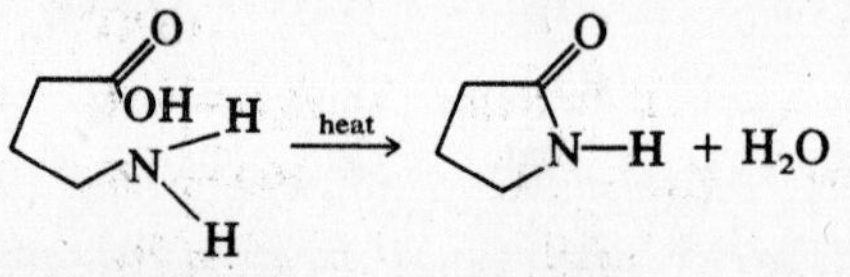

a γ-amino acid　　　　　a γ-lactam
(a cyclic amide)

(d)

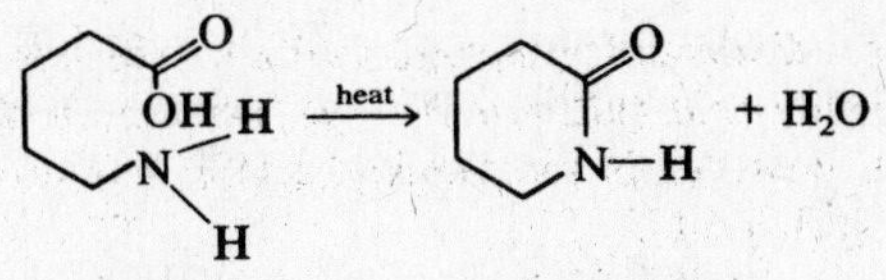

a δ-amino acid   a δ-lactam

(e) $\overset{+}{N}H_3(CH_2)_5COO^-$ ⟶ $-N(CH_2)_5C-NH(CH_2)_5C-NH(CH_2)_5C-$

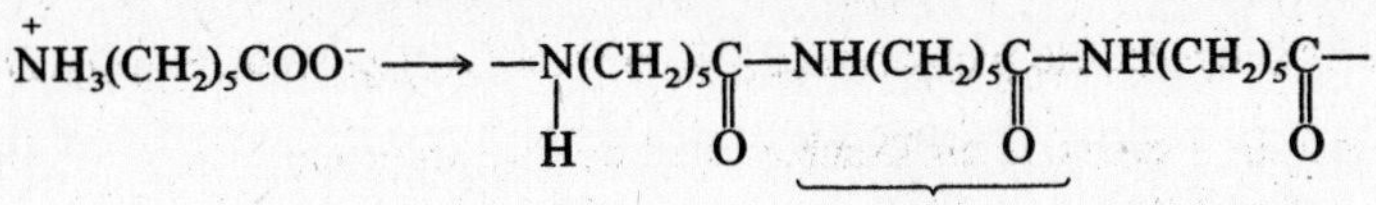

$\underbrace{\qquad\qquad\qquad\qquad}$
mer of
Nylon 6

Intramolecular cyclization would give a 7-membered ring, which is formed with difficulty. Hence the more facile intermolecular reaction occurs (see Problems 9.11 and 9.12, page 141). Since the substrate is bifunctional, polymerization occurs.

Problem 24.13 Give the products of the reaction of glycine with (a) aqueous KOH, (b) aqueous HCl. ◄

(a) $H_2NCH_2COO^-K^+$, (b) $Cl^-\overset{+}{H}_3NCH_2COOH$.

Problem 24.14 Show that amino acids have properties typical both of —COOH and —NH₂ by giving the products formed when glycine reacts with (a) CH₃COCl; (b) NaNO₂ + HCl (HONO); (c) EtOH + HCl; (d) PhCOCl + NaOH; (e) Ba(OH)₂, heat. ◄

(a) $CH_3CONHCH_2COOH$ (b) $HOCH_2COOH + N_2$ (c) $Cl^-\overset{+}{H}_3NCH_2COOEt$

(d) $PhCONHCH_2COO^-Na^+$ (e) $CH_3NH_2 + BaCO_3 + H_2O$ (decarboxylation)

Problem 24.15 Many α-amino acids react with HONO to give a quantitative volume of N₂ (**Van Slyke method** of analysis). Which amino acids in Table 24-1 cannot be analyzed by this method? ◄

The amino acid must have an —NH₂. Proline and hydroxyproline are 2° amines and do not evolve N₂ with HONO.

Problem 24.16 When the methyl ester of *rac*-alanine is heated, two diastereomeric dimethyldiketopiperazines [Problem 24.12(a)] are obtained. One of them cannot be resolved. Give structures for the two products, and account for their stereochemistry. ◄

The *cis* is resolvable.

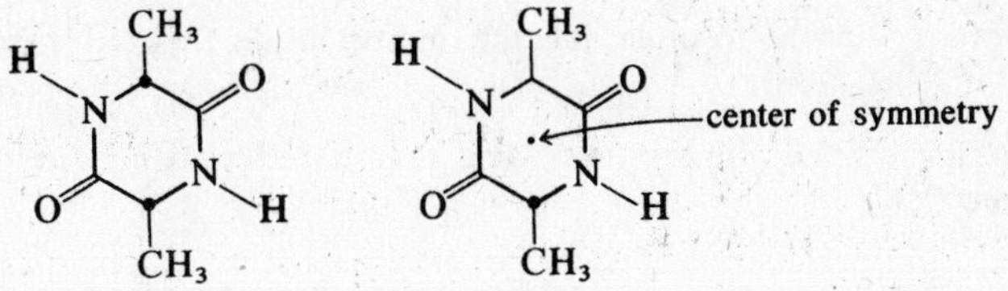

cis ; racemic   trans ; meso

Problem 24.17 Threonine has 2 chiral C's. Write Fischer projections for its stereoisomers. ◄

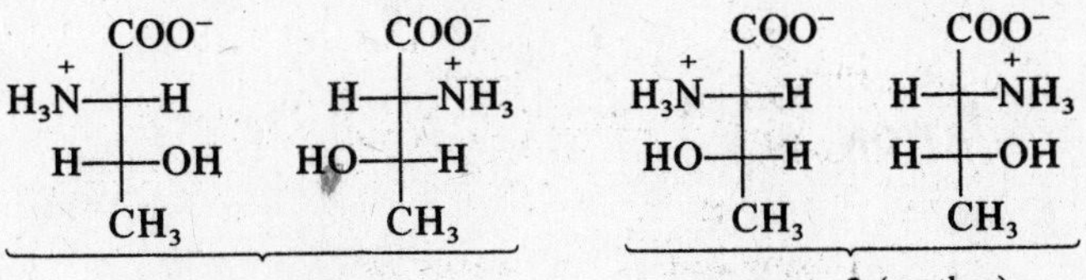

racemate 1 (*threo*)   racemate 2 (*erythro*)

Problem 24.18   Write a structural formula for the product of the reaction of: (a) alanine + carbobenzoxy chloride ($C_6H_5CH_2OCOCl$);   (b) tyrosine + aq. Br_2, and the product reacted with $(CH_3)_2SO_4$ + NaOH.   ◄

(a)

$$C_6H_5CH_2O\overset{\overset{\displaystyle O}{\|}}{C}-\overset{\overset{\displaystyle H}{|}}{N}-\overset{\overset{\displaystyle H}{|}}{\underset{\underset{\displaystyle CH_3}{|}}{C}}-COOH$$

(b)

$$CH_3O\underset{Br}{\overset{Br}{\bigcirc}}CH_2-\underset{\underset{\displaystyle NH_2}{|}}{CH}-\overset{\overset{\displaystyle O}{\|}}{C}\underset{O^-Na^+}{}$$

Problem 24.19   Can the acid chloride of an α-amino acid be made by adding $SOCl_2$?   ◄

No.   Two or more molecules of amino acid chloride react to give peptides.

$$H_2NCHRCOCl + H_2NCHRCOCl \longrightarrow H_2NCHRCONHCHRCOCl$$

24.4   PEPTIDES

INTRODUCTION

Problem 24.20   **Peptides** are polyamides, as in Problem 24.12(e).   Draw the structural formula for   (a) a dipeptide,   (b) a polypeptide unit of alanine (indicate the mer).   ◄

(a)

$$\overset{+}{H_3}N-CH-\underset{\underset{\displaystyle Me}{|}}{C}-\underset{\underset{\displaystyle O}{}}{}N-\underset{\underset{\displaystyle H}{}}{}CH-COO^-$$
$$\qquad\quad Me\quad\ O\ \ H\ \ Me$$

(b)

$$-\overset{\overset{\displaystyle H}{|}}{N}-\underset{\underset{\displaystyle Me}{|}}{CH}-\overset{\overset{\displaystyle}{}}{\underset{\underset{\displaystyle O}{}}{C}}-\overset{\overset{\displaystyle H}{|}}{N}-\underset{\underset{\displaystyle Me}{|}}{CH}-\underset{\underset{\displaystyle O}{}}{C}-\overset{\overset{\displaystyle H}{|}}{N}-\underset{\underset{\displaystyle Me}{|}}{CH}-\underset{\underset{\displaystyle O}{}}{C}-\overset{\overset{\displaystyle H}{|}}{N}-\underset{\underset{\displaystyle Me}{|}}{CH}-\underset{\underset{\displaystyle O}{}}{C}-$$

mer

a polypeptide

Problem 24.21   Draw the structural formula for the pentapeptide glycylglycylalanylphenylalanylleucine (Gly.Gly.Ala.Phe.Leu).   ◄

The sequence starts at the left end with the amino acid having the free α $\overset{+}{N}H_3$ and ends on the right with the amino acid having the free COO^-.

$$\overset{+}{H_3}NCH_2\underset{\underset{\displaystyle O}{}}{C}-\underset{\underset{\displaystyle H}{}}{N}-CH_2\underset{\underset{\displaystyle O}{}}{C}-\underset{\underset{\displaystyle H}{}}{N}-\underset{\underset{\displaystyle CH_3}{}}{CH}\overset{\overset{\displaystyle O}{\|}}{C}-\underset{\underset{\displaystyle H}{}}{N}-\underset{\underset{\displaystyle CH_2}{}}{CH}\overset{\overset{\displaystyle O}{\|}}{C}-\underset{\underset{\displaystyle H}{}}{N}-\underset{\underset{\displaystyle CH_2CH(CH_3)_2}{}}{CH}COO^-$$

Ph

STRUCTURE DETERMINATION

1.   Terminal Amino Acids

(a)   Dinitrofluorobenzene for H_2N terminal amino acid (Sanger).

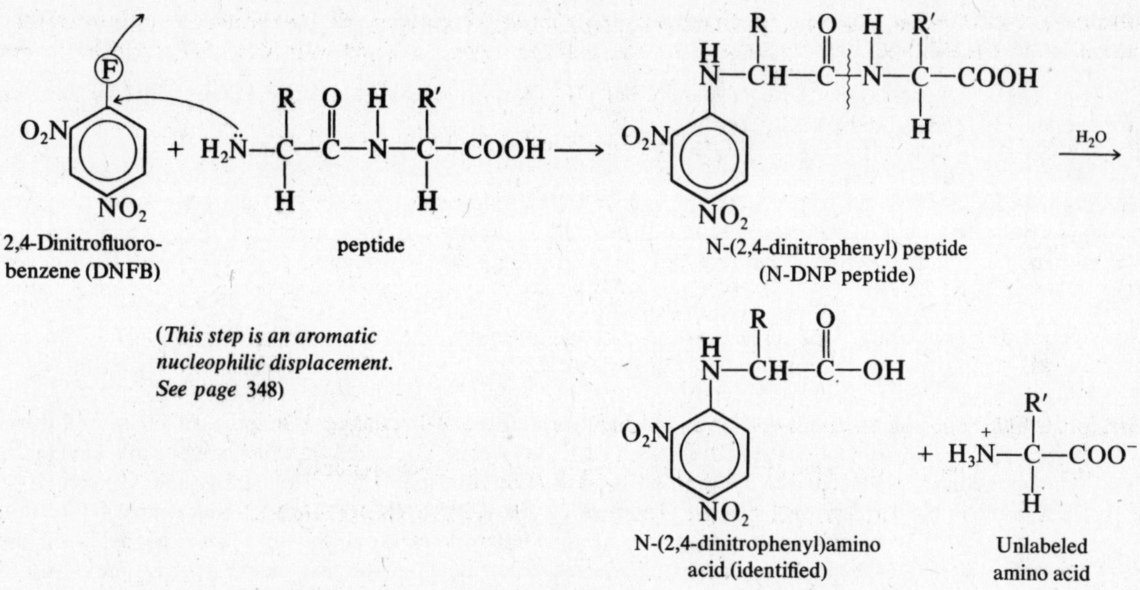

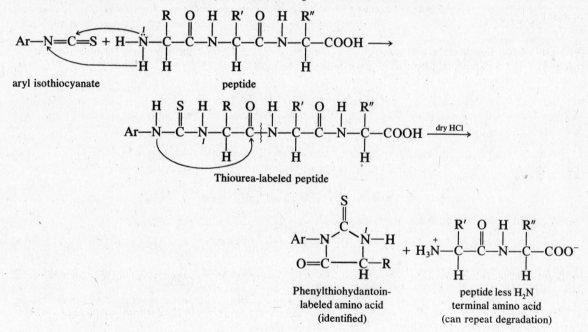

(*This step is an aromatic nucleophilic displacement. See page* 348)

2,4-Dinitrofluoro-benzene (DNFB)   peptide   N-(2,4-dinitrophenyl) peptide (N-DNP peptide)

N-(2,4-dinitrophenyl)amino acid (identified)   Unlabeled amino acid

(*b*) **Phenylthiourea and thiohydantoin for H₂N terminal amino acid.**

aryl isothiocyanate   peptide

Thiourea-labeled peptide

Phenylthiohydantoin-labeled amino acid (identified)   peptide less H₂N terminal amino acid (can repeat degradation)

(*c*) **Carboxypeptidase enzyme for COOH terminal amino acid.**

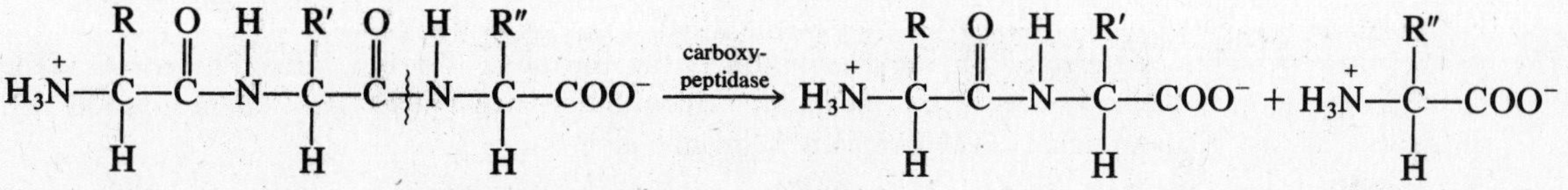

2. Partial Hydrolysis to Smaller Polypeptide Chains

Large peptides are partially hydrolyzed with enzymes or acid to mixtures of di- and tripeptides. From the composition of these small units we establish the sequence of amino acids in the large polypeptide.

Problem 24.22 What products are expected from (a) complete and (b) partial hydrolysis of the tetrapeptide Ala.Glu.Gly.Leu? ◄

(a) The amino acids Ala, Glu, Gly and Leu. (b) The dipeptides Ala.Glu, Glu.Gly, Gly.Leu and the tripeptides Ala.Glu.Gly and Glu.Gly.Leu.

Problem 24.23 What is the sequence of amino acids in the following peptides? (a) a tripeptide Gly,Leu,Asp (commas indicate an unknown sequence) that is partially hydrolyzed to the dipeptides Gly.Leu and Asp.Gly; (b) a heptapeptide Gly,Ser,His$_2$,Ala$_2$,Asp that is hydrolyzed to the tripeptides Gly.Ser.Asp, His.Ala.Gly and Asp.His.Ala. ◄

(a) Since Gly is bonded to Leu in one dipeptide and to Asp in the other, Gly must be in the middle. The free —$\overset{+}{N}H_3$ is in aspartic acid in Asp.Gly and the free —COO$^-$ is in leucine in Gly.Leu. The sequence is Asp.Gly.Leu.

(b) The first amino acid (free —$\overset{+}{N}H_3$) in the heptapeptide must be the first amino acid in one of the 3 tripeptides. The possibilities are Gly, His, or Asp. However, Gly and Asp, which are one of a kind in the heptapeptide, are also found in the third position of different tripeptides. Hence, Gly and Asp cannot be first in the original compound. In the heptapeptide, there are 2 His units, one of which must be the first amino acid. His is first in His.Ala.Gly and this unit comes from the first 3 amino acids of the heptapeptide. The terminal amino acid (free —COO$^-$) must be the last amino acid in one of the 3 tripeptides. The possibilities are Asp, Gly, or Ala. Since the one-of-a-kind Gly and Asp are multi-positioned in the tripeptides, they cannot be terminal in the heptapeptide. Ala must be the end amino acid in the heptapeptide and the tripeptide Asp.His.Ala comes from the last 3 amino acids. The seventh and middle amino acid is Ser. The tripeptides are arranged in order, and cross lines remove duplicating amino acids.

$$\text{His.Ala.Gly} + \cancel{\text{Gly}}.\text{Ser}.\cancel{\text{Asp}} + \text{Asp.His.Ala} = \text{His.Ala.Gly.Ser.Asp.His.Ala}$$

Problem 24.24 Determine the amino acid sequence of a peptide which (a) has the composition Leu$_2$,Ala$_2$,Tyr$_2$,Gly, reacts with DNFB to produce N-DNP tyrosine on hydrolysis, yields alanine with carboxypeptidase, and undergoes partial hydrolysis to the peptides

$$\text{Leu.Ala} + \text{Tyr.Ala} + \text{Ala.Tyr} + \text{Gly.Leu.Leu} + \text{Tyr.Gly}$$

(b) is shown by complete hydrolysis to have amino acids in the ratio

$$\text{Leu}_2,\text{Arg,CySH,Glu,Ileu,Val}_2,\text{Tyr,Phe}$$

while partial hydrolysis produces the tripeptides

$$\text{Leu.Val.Val} + \text{Leu.Arg.CySH} + \text{Tyr.Ileu.Phe} + \text{Phe.Glu.Leu} + \text{Arg.CySH.Leu} \quad ◄$$

(a) The first amino acid (free —$\overset{+}{N}H_3$) in the heptapeptide is Tyr and the last one (free —COO$^-$) is Ala. The dipeptides show that 2 heptapeptides are possible: one has Tyr.Ala at the end and Tyr.Gly at the beginning, and the other has Leu.Ala at the end and Tyr.Ala at the beginning. The sequences are, respectively,

$$\text{Tyr.Gly} + \cancel{\text{Gly}}.\text{Leu}.\cancel{\text{Leu}} + \text{Leu.Ala} + \cancel{\text{Ala}}.\text{Tyr} + \cancel{\text{Tyr}}.\text{Ala} = \text{Tyr.Gly.Leu.Leu.Ala.Tyr.Ala}$$

$$\text{Tyr}.\cancel{\text{Ala}} + \text{Ala}.\cancel{\text{Tyr}} + \text{Tyr.Gly} + \cancel{\text{Gly}}.\text{Leu.Leu} + \cancel{\text{Leu}}.\text{Ala} = \text{Tyr.Ala.Tyr.Gly.Leu.Leu.Ala}$$

The polypeptide cannot have Tyr.Gly at the beginning and Leu.Ala at the end.

(b) In the isolated tripeptides, the sole amino acid having only a first position is Tyr, in Tyr.Ileu.Phe; these are the first 3 amino acids in the sequence. The sole amino acid with only a last position is Val, in Leu.Val.Val; these are the last 3 amino acids. The complete sequence is:

$$\text{Tyr.Ileu}.\cancel{\text{Phe}} + \text{Phe.Glu}.\cancel{\text{Leu}} + \text{Leu.Arg.CySH} + \cancel{\text{Arg.CySH}}.\text{Leu} + \cancel{\text{Leu}}.\text{Val.Val}$$

$$= \text{Tyr.Ileu.Phe.Glu.Leu.Arg.CySH.Leu.Val.Val}$$

3. Synthesis of Peptides

The problem in combining different amino acids in a specific sequence to synthesize polypeptides is to prevent reaction between the COOH (or COX) and NH$_2$ of the same amino acid (Problem

24.19).   This reaction is blocked by attaching a group (B) to N of NH_2.   After the desired peptide is constructed, the blocking group is removed without destroying the peptide linkages.   The general scheme is:

$$\overset{+}{H_3}NCHRCOO^- \xrightarrow[+(B)]{block\ NH_2} (B)\!-\!NHCHRCOOH \xrightarrow[COOH]{activate} (B)\!-\!NHCHRCOX \xrightarrow[form\ peptide]{\overset{+}{H_3}NCHR'COO^-}$$

$$(B)\!-\!NHCHR\overset{O}{\overset{\|}{C}}NHCHR'COOH \xrightarrow[-(B)]{unblock} \overset{+}{H_3}NCHR\overset{O}{\overset{\|}{C}}NHCHR'COO^-$$

Table 24-2

	BLOCKING			UNBLOCKING	
(B) Group	Reagent	Name	Reagent	Products	
$PhCH_2O\overset{\|}{C}{=}O$	$PhCH_2OCOCl$	Carbobenzoxy chloride	H_2/Pt or Cold HBr, HOAc	$PhCH_3 + CO_2 +$ peptide $PhCH_2Br + CO_2 +$ peptide	
$t\text{-}BuO\overset{\|}{C}{=}O$	$t\text{-}BuO\overset{\|}{\underset{O}{C}}N_3$	t-Butyl azidoformate	HCl, HOAc or Et_2O	$(CH_3)_2C{=}CH_2 + CO_2$ + peptide	

In synthesizing polypeptides, the X groups used to activate COOH while minimizing racemization of the chiral α C are $-N_3$ (to form an acid azide), $-OC_6H_4NO_2\text{-}p$ (to form a p-nitrophenyl ester) and

$$-O\!-\!\overset{O}{\overset{\|}{C}}\!-\!OR$$

(to form a mixed anhydride of alkylcarbonic acid).

In automated solid-phase syntheses, the growing peptide chain is bound to a resin and finally removed with dilute HBr.

Problem 24.25   Prepare Gly.Val by the carbobenzoxy method using the   (a) azido group,   (b) p-nitrophenyl ester,   (c) t-butyl carbonate group to activate COOH.   ◄

(a) $\underbrace{PhCH_2OCOCl}_{(B)} + H_2NCH_2COOMe \longrightarrow PhCH_2OCONHCH_2COOMe \xrightarrow[\substack{1.\ H_2NNH_2 \\ 2.\ HONO}]{}$

$$\underset{(acid\ azide)}{PhCH_2OCONHCH_2CON_3} \xrightarrow[(valine)]{\overset{+}{H_3}NCH(i\text{-}Pr)COO^-} \underset{(I)}{PhCH_2OCONHCH_2CONHCH(i\text{-}Pr)COOH} \xrightarrow[-PhCH_3,\ -CO_2]{H_2/Pt}$$

$$\overset{+}{H_3}NCH_2CONHCH(i\text{-}Pr)COO^- \quad (product)$$

(b) $\underbrace{PhCH_2OCOCl}_{(B)} + \overset{+}{H_3}NCH_2COO^- \xrightarrow{-HCl} (B)\!-\!NHCH_2COOH \xrightarrow[(DCC)]{p\text{-}O_2NC_6H_4OH}$

$$(B)\!-\!NHCH_2COOC_6H_4NO_2\text{-}p \xrightarrow[-(HOC_6H_4NO_2\text{-}p)]{\overset{+}{H_3}NCH(i\text{-}Pr)COO^-} (I) \xrightarrow{H_2/Pt} product$$

$$\substack{see \\ part\ (a)}$$

In the above esterification, (DCC) is cyclohexyl$-N{=}C{=}N-$cyclohexyl (N,N-dicyclohexylcarbodimide), a powerful dehydrating agent that forms the urea derivative

$$cyclohexyl\!-\!NH\!-\!\overset{O}{\overset{\|}{C}}\!-\!NH\!-\!cyclohexyl$$

(c)   (B)—NHCHCOOH + $\boxed{t\text{-BuO}-\overset{\overset{\displaystyle O}{\|}}{C}-Cl}$ ⟶

   see part (b)   *t*-Butyl
                           chloroformate

(B)—NHCH$_2$$\overset{\overset{\displaystyle O}{\|}}{C}$—O—$\boxed{\overset{\overset{\displaystyle O}{\|}}{C}\text{—O}(t\text{-Bu})}$ $\xrightarrow[-CO_2,\ t\text{-BuOH}]{H_3\overset{+}{N}CH(i\text{-Pr})COO^-}$ (I) $\xrightarrow{H_2/Pt}$ product

        a mixed anhydride   *see*
                                                                       part (a)

24.5   PROTEINS

Proteins with molecular weights in the millions are the major constituents of all living cells.

CLASSIFICATION

1.   By Hydrolysis Products

   (a)   **Simple proteins** are hydrolyzed only to amino acids.
   (b)   **Conjugated proteins** are hydrolyzed to amino acids and nonpeptide substances known as **prosthetic groups**.   These prosthetic groups include nucleic acids of **nucleoproteins**, carbohydrates of **glycoproteins**, pigments (such as **hemin** and **chlorophyll**) of **chromoproteins**, and fats or lipids of **lipoproteins**.

2.   By Structure

   (a)   **Fibrous proteins** are threadlike and are insoluble in water.   They include **fibroin** (found in silk), **keratin** (in hair, skin, feathers, etc.) and **myosin** (in muscle tissue).
   (b)   **Globular proteins (globulins)** are folded in spherelike shapes.   Globulins are soluble in H_2O and dissolve in dilute salt solutions.   Globulins include all enzymes, antibodies, albumin of eggs, hemoglobin and many hormones such as insulin.

PROPERTIES OF PROTEINS

1.   Amphoteric Properties.   Isoelectric Points and Electrophoresis

Proteins have different isoelectric points, and in an electrochemical cell they migrate to one of the electrodes (depending on their charge, size and shape) at different speeds.   This difference in behavior is used in **electrophoresis** for the separation and analysis of protein mixtures.

2.   Hydrolysis

Proteins are hydrolyzed to α-amino acids by heating with aqueous strong acids or, at room temperature, by digestive enzymes such as trypsin and pepsin.

3.   Denaturation

Heat, strong acids or bases, ethanol, or heavy metal ions cause irreversible precipitation of proteins.   This process, known as **denaturation**, is exemplified by the heat-induced coagulation and hardening of egg white (albumin).   Denaturation destroys the physiological activity of proteins.

STRUCTURE OF PROTEINS

The sequence and number of amino acids constitute the **primary** structure of proteins.   **Secondary** structures arise from different conformations of the protein chains.   A common secondary structure, the α-**helix**, is a coiled arrangement maintained by H-bonds between an amide H and a carbonyl O that are four peptide bonds apart.   The helices can be right- or left-handed.   The secondary structures of proteins are best determined by X-ray analysis.

Problem 24.26   Which α-amino acid can act to cross-link peptide chains?

Cysteine.   Cysteine linkages also occur between distant parts of the same chain.

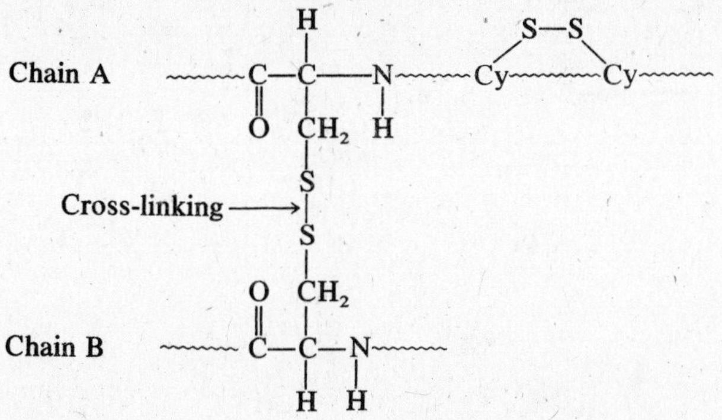

Supplementary Problems

Problem 24.27   How many dipeptides can be synthesized from glycine and alanine?

Four: Gly.Gly, Ala.Ala, Ala.Gly and Gly.Ala.

Problem 24.28   Write all the possible amino acid sequences for a tripeptide having the empirical formula Gly.Ala$_2$.

Gly.Ala.Ala, Ala.Gly.Ala, Ala.Ala.Gly.

Problem 24.29   Write the structural formula for tyrosylglycylalanine.

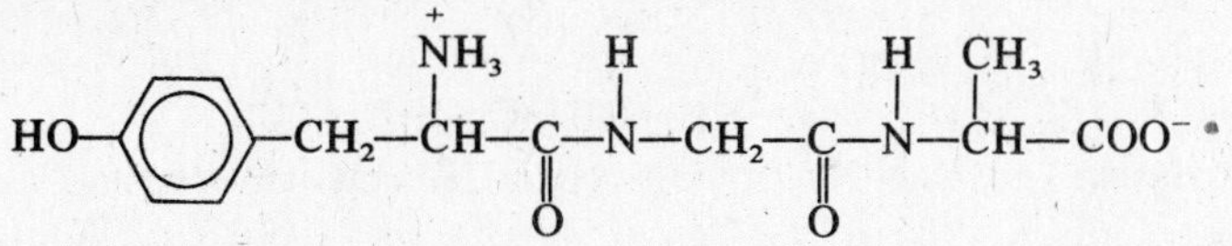

Problem 24.30   Identify the compounds (I), (II), (III).

$$\text{(I)} + \overset{+}{\text{H}_3}\text{NCHMeCOO}^- \xrightarrow{\text{NaOH}} \text{CH}_3\text{CONHCHMeCOO}^-\text{Na}^+$$

$$\text{(II)} + \text{HONO} \longrightarrow \text{N}_2 + \text{HOCH}_2\text{CH}(n\text{-Pr})\text{COOH}$$

$$\text{(III)} + \text{NH}_3 + \text{H}_2 \longrightarrow \text{MeCH}(\overset{+}{\text{NH}_3})\text{COO}^-$$

(I) $(\text{CH}_3\text{CO})_2\text{O}$;   (II) $\overset{+}{\text{H}_3}\text{NCH}_2\text{CH}(n\text{-Pr})\text{COO}^-$;   (III) MeCOCOOH.

Problem 24.31   Identify the compounds (I) through (IV).

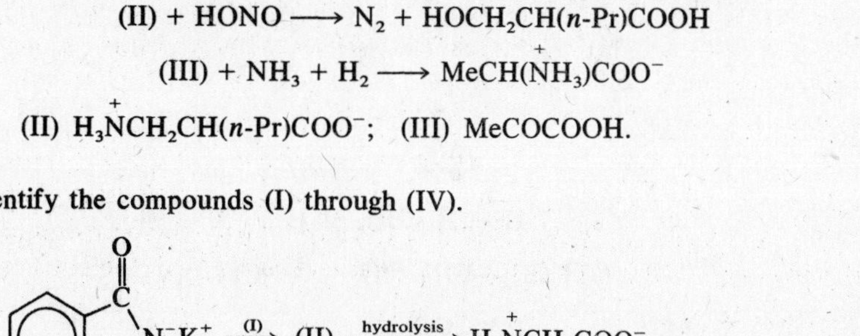

(I) ethyl bromoacetate (or any haloacetate);   (II) PhthN—CH$_2$—COOC$_2$H$_5$ [Problem 24.1(c)];
(III) PhCOCl;   (IV) hippuric acid, PhCONHCH$_2$COOH.

Problem 24.32 Supply structural formulas for principal organic compounds (I) through (XIV).

(a) O_2N—(NO₂)—F $+ H_3\overset{+}{N}$—CH(COO⁻)—CH₂—CH₂—S—CH₃ ⟶ (I)

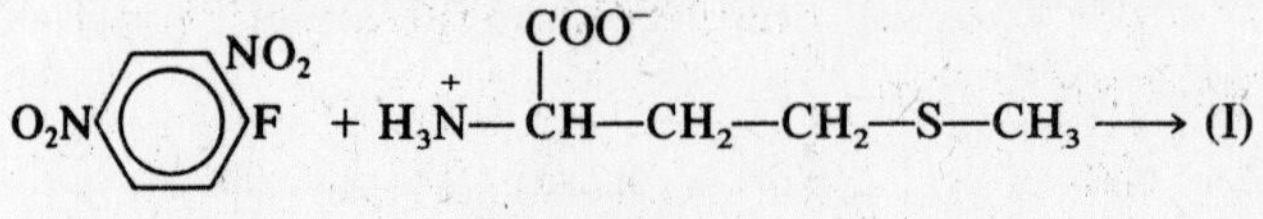

(b) Ph—N=C=S $+ (CH_3)_2CHCH(\overset{+}{N}H_3)(COO^-) \xrightarrow{OH^-}$ (II) $\xrightarrow{H^+}$ (III)

(c) 2 Ph—COCl $+ ^-OOCCH(CH_2)_4NH_2 \xrightarrow{OH^-}$ (IV)
     with $^+NH_3$

(d) $C_6H_5CH_2$—CH($\overset{+}{N}H_3$)—COO⁻ $+ (CH_3CO)_2O \xrightarrow{OH^-}$ (V)

(e) HO—(pyrrolidine ring)—COO⁻ $+ C_2H_5OH \xrightarrow{HX}$ (VI)
     with $\overset{+}{N}H$—H

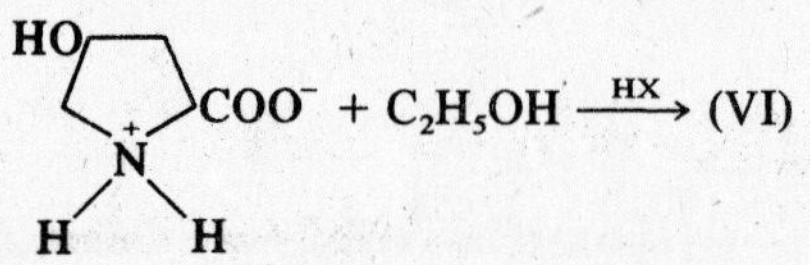

(f) ^-OOC—CH(CH₂)₄NH₂ with NH₂ $\underset{OH^-}{\overset{H^+}{\rightleftharpoons}}$ $\begin{Bmatrix}(VII)\\ \updownarrow \\ (VIII)\end{Bmatrix}$ $\underset{OH^-}{\overset{H^+}{\rightleftharpoons}}$ (IX) $\underset{OH^-}{\overset{H^+}{\rightleftharpoons}}$ (X)

(g) ^-OOC—CHCH₂COO⁻ with NH₂ $\underset{OH^-}{\overset{H^+}{\rightleftharpoons}}$ (XI) $\underset{OH^-}{\overset{H^+}{\rightleftharpoons}}$ $\begin{Bmatrix}(XII)\\ \updownarrow \\ (XIII)\end{Bmatrix}$ $\underset{OH^-}{\overset{H^+}{\rightleftharpoons}}$ (XIV)   ◄

(a) O_2N—(NO₂)—NH—CH(COOH)—CH₂—CH₂SCH₃   (I)

(b) (II): ring with HO—C(=O), Ph—N, NH, CHCH(CH₃)₂, C=S
     (III): ring with O=C, Ph—N, NH, CHCH(CH₃)₂, C=S

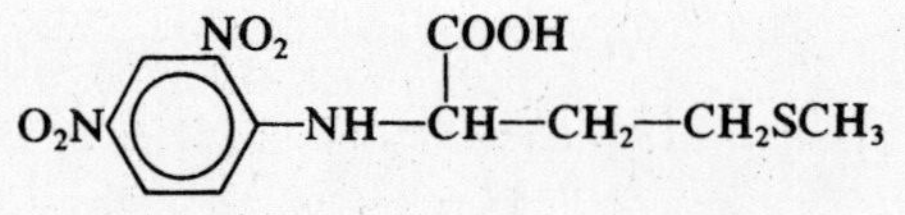

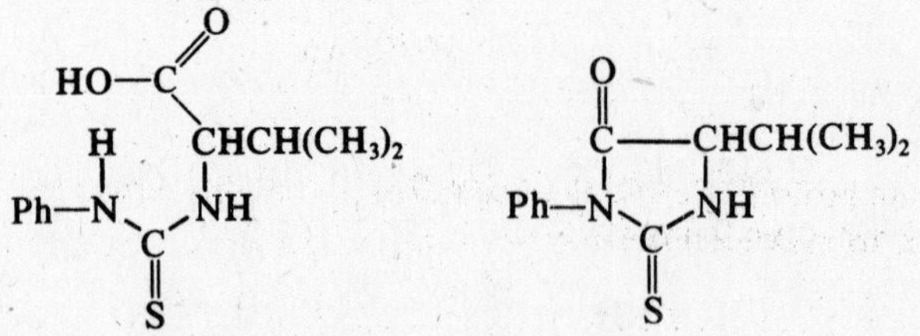

(c) PhCO—NH—CH(COO⁻)(CH₂)₄—NH—COPh   (IV)

(d) $C_6H_5CH_2$—CH(COO⁻)—NH—CO—CH₃   (V)

(e) HO—(pyrrolidine ring)—COOC₂H₅   (VI)
     with $\overset{+}{N}H$—H

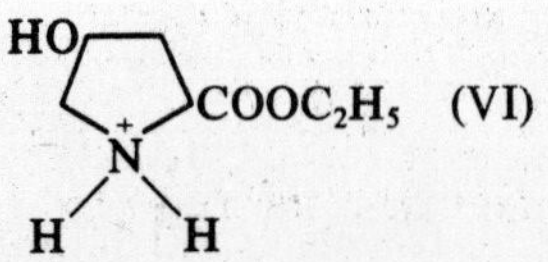

(f)

(VII) $^-OOC-CH(CH_2)_4NH_2$
$\quad\quad\quad\quad | $
$\quad\quad\quad\quad {}^+NH_3$
$\quad\quad\quad\quad \Updownarrow$
(VIII) $^-OOC-CH(CH_2)_4\overset{+}{N}H_3$
$\quad\quad\quad\quad | $
$\quad\quad\quad\quad NH_2$

$\rightleftharpoons \;\; {}^-OOC-CH(CH_2)_4\overset{+}{N}H_3 \;\; \rightleftharpoons \;\; HOOC-CH(CH_2)_4\overset{+}{N}H_3$
$\quad\quad\quad\quad\quad | \quad\quad\quad\quad\quad\quad\quad\quad\quad\quad | $
$\quad\quad\quad\quad\quad {}^+NH_3 \quad\quad\quad\quad\quad\quad\quad\quad\quad {}^+NH_3$
$\quad\quad\quad\quad\quad (IX) \quad\quad\quad\quad\quad\quad\quad\quad\quad\quad (X)$

(g)

$\overset{\overset{+}{N}H_3}{\underset{|}{{}^-OOC-CH-CH_2COO^-}} \rightleftharpoons$

(XII) $\quad \overset{NH_3}{\underset{|}{{}^-OOC-CH-CH_2COOH}}$
$\quad\quad\quad\quad \Updownarrow$
(XIII) $\quad \underset{|}{HOOC-CH-CH_2COO^-}$
$\quad\quad\quad\quad NH_3^+$

$\rightleftharpoons \overset{\overset{+}{N}H_3}{\underset{|}{HOOC-CH-CH_2COOH}}$
$(XI) \quad\quad\quad\quad\quad\quad\quad\quad\quad\quad\quad\quad\quad\quad\quad (XIV)$

Problem 24.33 Starting with acetamidomalonic acid, prepare phenylalanine. ◄

$$\underset{COOEt}{\overset{COOEt}{\underset{|}{\overset{|}{HCNHCOCH_3}}}} \xrightarrow[\text{2. PhCH}_2\text{Cl}]{\text{1. NaOEt}} \underset{COOEt}{\overset{COOEt}{\underset{|}{\overset{|}{PhCH_2-C-NHCOCH_3}}}} \xrightarrow[\text{2. H}_3\text{O}^+]{\text{1. NaOH}} \underset{}{\overset{{}^+NH_3}{\underset{|}{PhCH_2-CH-COO^-}}}$$

Problem 24.34 Use p-methoxytoluene and any other needed reagents to prepare tyrosine. ◄

$$CH_3O\text{—}\langle\bigcirc\rangle\text{—}CH_3 \xrightarrow[\text{uv}]{Br_2} CH_3O\text{—}\langle\bigcirc\rangle\text{—}CH_2Br \xrightarrow{\overset{+}{Na}\cdot\overset{-}{C}H(COOEt)_2} CH_3O\text{—}\langle\bigcirc\rangle\text{—}CH_2-CH(COOEt)_2 \xrightarrow[\substack{\text{2. H}_3\text{O}^+\\ \text{3. heat}}]{\text{1. NaOH}}$$

$$CH_3O\text{—}\langle\bigcirc\rangle\text{—}CH_2CH_2COOH \xrightarrow[\text{2. NH}_3 \text{ (excess)}]{\text{1. Br}_2/\text{P}} CH_3O\text{—}\langle\bigcirc\rangle\text{—}CH_2\overset{\overset{+}{N}H_3}{\underset{|}{CHCOO^-}} \xrightarrow[-\text{CH}_3\text{Br}]{\text{conc. HBr, }\Delta}$$

$$HO\text{—}\langle\bigcirc\rangle\text{—}CH_2\overset{\overset{+}{N}H_3Br^-}{\underset{|}{CHCOOH}} \xrightarrow{OH^-} HO\text{—}\langle\bigcirc\rangle\text{—}CH_2\overset{\overset{+}{N}H_3}{\underset{|}{CHCOO^-}}$$
$$\text{Tyrosine}$$

p-CH$_3$C$_6$H$_4$OH itself cannot be used because the ring would be brominated and the acidic H of OH would convert basic malonate carbanion to malonic ester.

Problem 24.35 Prepare glycylalanyltyrosine from the free amino acids by the carbobenzoxy method. ◄

$$C_6H_5CH_2OCOCl + \overset{+}{H_3N}CH_2COO^- \longrightarrow \underset{\text{Gly}}{\underline{C_6H_5CH_2OCONHCH_2COOH}} \xrightarrow[\text{DCC}]{\text{alanine}}$$

$$\underset{\text{Gly}}{\underline{C_6H_5CH_2OCONHCH_2}}\underset{\text{Ala}}{\underline{CONHCH(CH_3)COOH}} \xrightarrow[\text{DCC}]{\text{tyrosine}}$$

$$\underset{\text{Gly}}{\underline{C_6H_5CH_2OCONHCH_2}}\underset{\text{Ala}}{\underline{CONHCH(CH_3)}}\underset{\text{Tyr}}{\underline{CONH\ldots COOH}} \xrightarrow{\text{H}_2/\text{Pd}} C_6H_5CH_3 + CO_2 + \text{Gly.Ala.Tyr}$$

(DCC, Problem 24.25(b), forms a peptide from the acid itself.)

Problem 24.36 Synthesize valine (a) from isobutyl alcohol by a Strecker reaction, (b) by malonic ester synthesis. ◄

(a)

$$\underset{CH_3CHCH_2OH}{\overset{CH_3}{|}} \xrightarrow[heat]{Cu} \underset{CH_3CHCH=O}{\overset{CH_3}{|}} \xrightarrow[NaCN]{NH_4Cl} \underset{CH_3CH-CHCN}{\overset{CH_3\ NH_2}{|\ \ \ |}} \xrightarrow{H_3O^+} \underset{CH_3CH-CHCOO^-}{\overset{CH_3\ \overset{+}{N}H_3}{|\ \ \ |}}$$

Valine

(b)

$$\underset{CH_3CHBr}{\overset{CH_3}{|}} \xrightarrow{Na^+\overset{-}{:}CH(COOR)_2} \underset{CH_3CHCH(COOR)_2}{\overset{CH_3}{|}} \xrightarrow[\substack{1.\ NaOH,\ heat \\ 2.\ HCl \\ 3.\ Br_2/P \\ 4.\ NH_3\ (xs)}]{} \underset{CH_3CH-C-COO^-}{\overset{CH_3\ H}{|\ \ |}}$$

$$\overset{}{\underset{\overset{+}{N}H_3}{}}$$

Valine

Problem 24.37 What occurs when an electric current is passed through an aqueous solution, buffered at pH = 6.0, of alanine (6.0), glutamic acid (3.2) and arginine (10.7)? The isoelectric points are shown in parentheses. ◄

Glutamic acid is present as an anion and migrates to the anode. Arginine is a cation and migrates to the cathode. Alanine is a dipolar ion and remains uniformly distributed in solution.

Problem 24.38 Use DNFB (page 421) to distinguish between Lys.Gly and Gly.Lys.

$$\text{DNFB} + \underset{\text{Lys.Gly}}{\overset{+}{H_3}N(CH_2)_4CH(NH_2)CONHCH_2COO^-} \longrightarrow \underset{\underset{\text{DNP DNPNH}}{}}{HN(CH_2)_4CHCONHCH_2COOH} \xrightarrow{H_3O^+}$$

$$\underset{\underset{\text{DNP DNPNH}}{}}{HN(CH_2)_4CHCOOH} + \underset{\text{Glycine}}{\overset{+}{H_3}NCH_2COOH}$$

The α and ϵ NH$_2$'s are bonded to DNP.

$$\text{DNFB} + \underset{\text{Gly.Lys}}{\overset{O\ \ H\ \ COO^-}{H_2NCH_2C-N-C(CH_2)_4\overset{+}{N}H_3}} \longrightarrow \underset{\overset{}{\underset{\text{DNP\ \ O\ H\ \ \ DNP}}{}}}{H-NCH_2C-NCH(CH_2)_4-NH} \xrightarrow{H_3O^+}$$

(with COOH above, H below)

$$\underset{\underset{\text{DNP}}{}}{HNCH_2COOH} + \underset{\overset{+}{N}H_3\ \ DNP}{HOOC-CH(CH_2)_4NH}$$

After hydrolysis of the DNP derivative of Gly.Lys, both amino acids have a DNP attached. However, only the amino acid with DNP attached to the α NH$_2$ (in this case, glycine) is the NH$_2$ terminal amino acid. When a diaminomonocarboxylic acid is part of a peptide chain, but not in an NH$_2$ terminal position, it will be derivatized by DNP but *not* at the α NH$_2$.

Problem 24.39 Among the hydrolysis products of the pentapeptide Gly$_3$,Ala,Phe are Ala.Gly and Gly.Ala. Give 2 possible structures if this substance gives no N$_2$ with HONO (Van Slyke method). ◄

Since N$_2$ is not obtained, there is no free NH$_2$. The compound must be a *cyclic* peptide. It has the partial sequence Gly.Ala.Gly. The remaining Gly and Phe units can be placed in two sequences, giving

Gly.Ala.Gly   Gly.Ala.Gly

     or

Gly.Phe   Phe.Gly

Problem 24.40 Which amino acids in Table 24-1 cannot participate in the H-bonding involved in the α-helix structure of proteins? ◄

Proline and hydroxyproline. They are 2° amines, and after forming the peptide bond they have no H attached to N for H-bonding.

Problem 24.41 Account for the stereochemical specificity of enzymes with chiral substrates.   ◄

Since enzymes are proteins made up of optically active amino acids, enzymes are themselves optically active and therefore react with only one enantiomer of a chiral substrate.

Problem 24.42 Explain how the following denature proteins: (a) Pb^{2+} and Ag^+, (b) EtOH, (c) urea, and (d) heat.   ◄

(a) These heavy metal cations form insoluble salts with the free COO^- groups. (b) EtOH interferes with H-bonding by offering its own competing H. (c) Urea is a good H-bond acceptor and interferes with H-bonding in the protein. (d) Heat causes more random conformations in the protein and decreases the conformational population needed for helix formation.

Chapter 25

Carbohydrates

25.1 INTRODUCTION

Carbohydrates (saccharides) are aliphatic polyhydroxyaldehydes (**aldoses**), polyhydroxyketones (**ketoses**), or compounds that can be hydrolyzed to them. The suffix **-ose** denotes this class of compounds.

The **monosaccharide** D-(+)-glucose, an aldohexose, is formed by plants in photosynthesis and is converted to the **polysaccharides** cellulose and starch. Simple saccharides are called **sugars**.

Problem 25.1 Classify the following monosaccharides:

(a) CH_2—C—CH_2—OH (b) HO—CH_2—CH—CH—CH—CH—CH=O

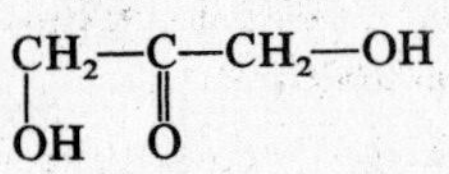

  OH O   OH OH OH OH

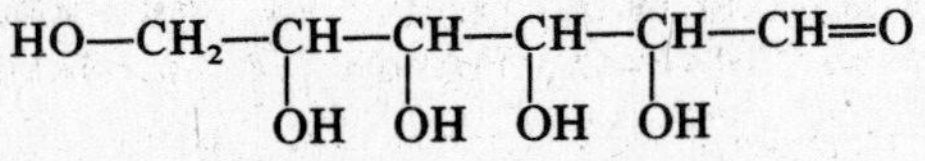

(c) HO—CH_2—CH—C—CH—CH_2—OH ◀

    OH O OH

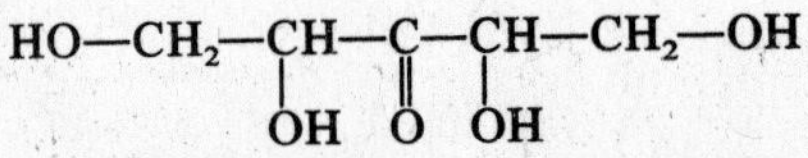

The number of C's in a chain is indicated by the affix **-di-**, **-tri-**, etc. (a) ketotriose, (b) aldohexose, (c) ketopentose.

Problem 25.2 (a) Write a structural formula for (i) an aldotriose, (ii) a ketohexose. (b) Write molecular formulas for (i) a hexose trisaccharide, (ii) a pentose polysaccharide. (c) What is the general formula for most carbohydrates? ◀

(a) (i) 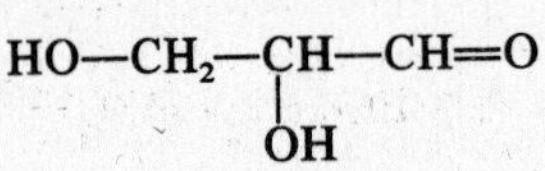 HO—CH_2—CH—CH=O (ii) 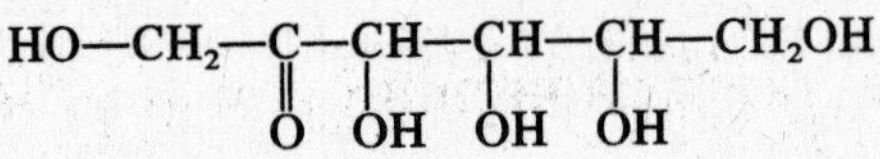 HO—CH_2—C—CH—CH—CH—CH_2OH

      OH   O OH OH OH

(b) Because

$$n \text{ moles monosaccharide} = 1 \text{ mole polysaccharide} + (n - 1) \text{ moles } H_2O$$

we have:

(i) $n = 3$, $C_{18}H_{32}O_{16}$; (ii) $(C_5H_8O_4)_n$

(c) $C_x(H_2O)_y$. This formula accounts for the name **carbohydrate**.

Problem 25.3 Deduce the gross structure of glucose from the following data. (a) The elementary composition is C = 40.0%, H = 6.7%, O = 53.3%. (b) A solution of 18 g in 100 g of H_2O freezes at −1.86 °C. (c) Mild oxidation with Br_2/H_2O, $Ag(NH_3)_2^+$ (Tollens' reagent) or Cu^{2+} (tartrate) complex (Fehling's test) gives a carboxylic acid with the same number of C's. (d) This acid, a **glyconic acid**, reacts with HI and P to give n-hexanoic acid. (e) One mole of glucose reacts with 5 moles of Ac_2O to form an ester. (f) Vigorous oxidation of glucose with HNO_3 gives a dicarboxylic acid (a **glycaric acid**) with the same number of C's. (g) Reduction of glucose with H_2/Ni or Na/Hg gives a product that reacts with 6 moles of Ac_2O. ◀

(a) The molar amounts of C, H and O in 100 g of glucose would be

$$\frac{40.0}{12} = 3.33 \qquad \frac{6.7}{1} = 6.7 \qquad \frac{53.3}{16} = 3.33$$

Thus, $C:H:O = 1:2:1$, and the empirical formula is CH_2O (empirical mol. wt. $= 30$).   (b) 18 g in 100 g of H_2O is equivalent to 180 g in 1 kg of H_2O.   Since $\Delta f.p. = 1.86$, this solution is one-molal ($\Delta f.p. = K_{f.p.}m$), and 180 g is the gram molecular weight.   The molecular formula is $(CH_2O)_n$, where

$$n = \frac{180 \,(\text{molecular weight})}{30 \,(\text{empirical weight})} = 6$$

Thus glucose is $C_6H_{12}O_6$.   (c) Glucose has a $HC{=}O$ group, which undergoes mild oxidation to $COOH$. (d) The 6 C's of glucose are in a chain sequence (there are no branching C's).   (e) Glucose has 5 OH's that are acetylated to a pentaacetate.   (f) One of the OH's is 1°, since it was oxidized to a $COOH$ along with CHO. (g) CHO is reduced to CH_2OH to give a hexahydroxy compound (a **glykitol**).   The gross structure of glucose is $HOCH_2CHOHCHOHCHOHCHOHCHO$.

Problem 25.4   (a) The monosaccharide D-(−)-fructose undergoes the reaction sequence

$$\text{Fructose} \xrightarrow{\text{HCN}} \xrightarrow{\text{H}_3\text{O}^+} \xrightarrow{\text{HI/P}} CH_3CH_2CH_2CH_2CH(CH_3)COOH$$

Classify fructose.   (b) Which carboxylic acid is formed when glucose is submitted to the same sequence?   ◀

(a)   Fructose is a hexose since the final product has 7 C's, one of which came from HCN addition to a $C{=}O$.   The precursor of COOH is CN, which added to $C{=}O$.   COOH is attached to the C that was originally the C of $C{=}O$.   Fructose is a 2-ketohexose.   The sequence is:

CH_2OH	CH_2OH	CH_2OH	CH_3
$C{=}O$	$HO{-}C{-}CN$	$HO{-}C{-}COOH$	$H{-}C{-}COOH$
$(CHOH)_3$	$(CHOH)_3$	$(CHOH)_3$	$(CH_2)_3$
CH_2OH	CH_2OH	CH_2OH	CH_3
2-ketohexose	cyanohydrin	hydroxy acid	2-Methylhexanoic acid

with arrows HCN, H_3O^+, HI/P between them.

(b)   By this sequence, glucose gives n-heptanoic acid.

Problem 25.5   (a) How many chiral C's are there in a typical   (i) aldohexose ($C_6H_{12}O_6$),   (ii) 2-ketohexose? (b) How many stereoisomers should an aldohexose have?   ◀

(a)   (i) 4: $HO\overset{6}{C}H_2\overset{5}{C}HOH\overset{4}{C}HOH\overset{3}{C}HOH\overset{2}{C}HOH\overset{1}{C}HO$ (ii) 3: $HO\overset{6}{C}H_2\overset{5}{C}HOH\overset{4}{C}HOH\overset{3}{C}HOH\overset{2}{C}OCH_2OH$

(b)   There are 4 different chiral C's and $2^4 = 16$ stereoisomers.

25.2   CHEMICAL PROPERTIES OF MONOSACCHARIDES

1.   Oxidation (see Problem 25.3(c) and (f), and page 431)

2.   Reaction in Base

Problem 25.6   (a) Explain how in basic solution an equilibrium is established between an aldose, its C^2 **epimer** (a diastereomer with a different configuration at one chiral C) and a 2-ketose.   (b) Will fructose give a positive Fehling's test, which is done in a basic solution?   ◀

(a)   In basic solutions, aldoses and ketoses tautomerize to a common intermediate enediol, and the following equilibria are established:

$H{-}C{=}O$	$H{-}C{-}OH$	$H{-}C{-}OH$
$H{-}C^2{-}OH$	$C{-}OH$	$C{=}O$
$H{-}C{-}OH$	$H{-}C{-}OH$	$H{-}C{-}OH$
aldose	enediol	ketose

chiral, sp^3 achiral, sp^2

When the aldose is reformed from the enediol, H^+ can attack the now achiral C^2 from either side of the double bond to give C^2 epimers.

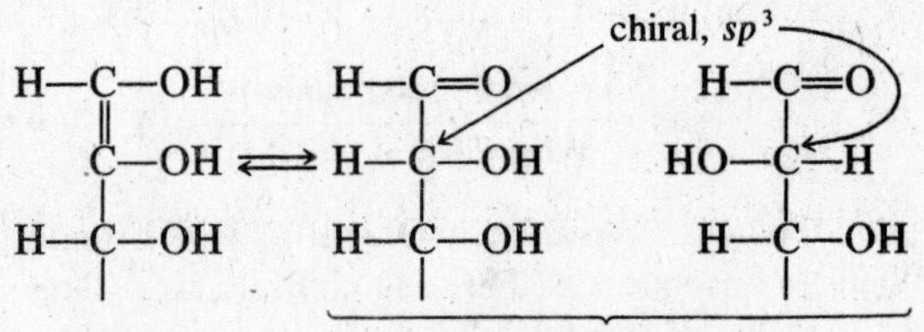

$$\text{enediol} \qquad \text{chiral, } sp^3 \qquad C^2 \text{ epimeric aldoses}$$

(b) Since fructose isomerizes to an aldohexose (glucose) in base, it gives a positive test:

$$\text{Cupric tartrate } (blue\ solution) \xrightarrow{OH^-} Cu_2O\ (red\ solid)$$

3. Reduction to Glykitols [Problem 25.3(g)]

Problem 25.7 What are the products of reaction of $HO\overset{6}{C}H_2(CHOH)_4\overset{1}{C}HO$ (I) and

$$HO\overset{6}{C}H_2(CHOH)_3-\overset{O}{\overset{\|}{\underset{2}{C}}}-\overset{1}{C}H_2OH \quad \text{(II)}$$

with (a) $Br_2 + H_2O$, (b) HNO_3, (c) HIO_4?

	I	II
(a)	$HOCH_2(CHOH)_4COOH$	No reaction (ketones are not oxidized)
(b)	$HOOC(CHOH)_4COOH$	$HOOC(CHOH)_3-\overset{O}{\overset{\|}{C}}-COOH$ (1° OH's react)
(c)	$\overset{6}{C}H_2{=}O + 4HCOOH + \overset{1}{H}COOH$	$\overset{6}{C}H_2{=}O + 3HCOOH + \overset{2}{C}O_2 + \overset{1}{C}H_2{=}O$

Problem 25.8 Glucose is reduced to a single glykitol; fructose is reduced to 2 epimers, one of which is identical to the glykitol from glucose. Explain in terms of conformations. ◄

In glucose no new chiral C is formed on going from CHO to CH_2OH. In fructose the carbonyl C (C^2) becomes chiral; there are 2 configurations and we get epimers.

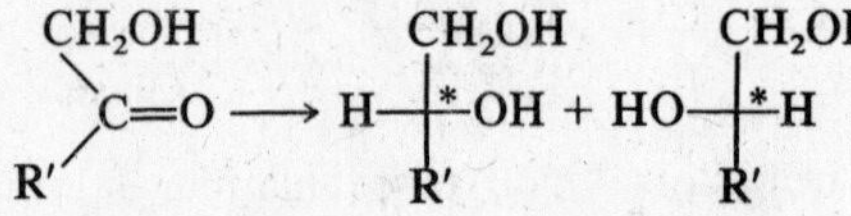

$$\begin{array}{ccc} CH_2OH & CH_2OH & CH_2OH \\ \underset{R'}{\overset{}{C}}{=}O & \longrightarrow & H\underset{R'}{\overset{*}{-}}OH + HO\underset{R'}{\overset{*}{-}}H \end{array}$$

The chiral C^3, C^4 and C^5 [see Problem 25.5(a)] of glucose and fructose have the same configuration. The identical glykitols also have the same configuration at C^2.

4. Ruff Degradation (Loss of One Carbon)

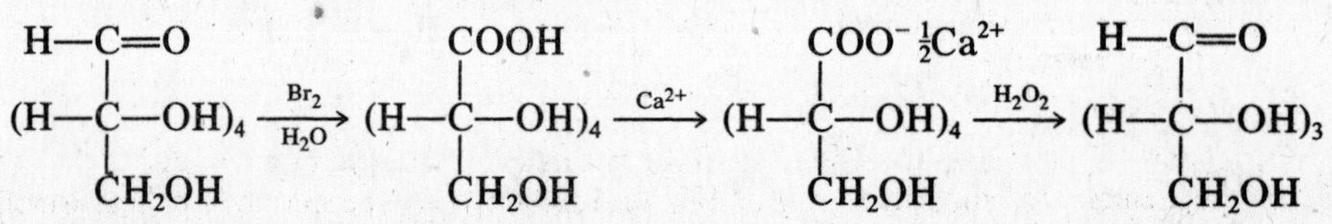

$$\underset{\text{aldohexose}}{\begin{array}{c} H-C{=}O \\ (H-C-OH)_4 \\ CH_2OH \end{array}} \xrightarrow[H_2O]{Br_2} \underset{\text{glyconic acid}}{\begin{array}{c} COOH \\ (H-C-OH)_4 \\ CH_2OH \end{array}} \xrightarrow{Ca^{2+}} \begin{array}{c} COO^-\frac{1}{2}Ca^{2+} \\ (H-C-OH)_4 \\ CH_2OH \end{array} \xrightarrow{H_2O_2} \underset{\text{aldopentose}}{\begin{array}{c} H-C{=}O \\ (H-C-OH)_3 \\ CH_2OH \end{array}}$$

5. Kiliani-Fischer Step-up Method

The aldohexose reacts with HCN, NaCN to give two epimeric cyanohydrins, which are converted by H_3O^+ to glyconic acids, then by $-H_2O$ to glyconolactones, and by Na-Hg, CO_2 to aldoheptoses.

Top pathway:

$$\begin{array}{c} CN \\ H-C-OH \\ (H-C-OH)_3 \\ H-C-OH \\ CH_2OH \end{array} \xrightarrow{H_3O^+} \begin{array}{c} COOH \\ H-C-OH \\ (H-C-OH)_3 \\ H-C-OH \\ CH_2OH \end{array} \xrightarrow{-H_2O} \begin{array}{c} C=O \\ H-C-OH \\ (H-C-OH)_3 \\ H-C \\ CH_2OH \end{array} \xrightarrow[CO_2]{Na-Hg} \begin{array}{c} H-C=O \\ H-C-OH \\ (H-C-OH)_3 \\ H-C-OH \\ CH_2OH \end{array}$$

Aldohexose:

$$\begin{array}{c} H-C=O \\ H-C-OH)_3 \\ H-C-OH \\ CH_2OH \end{array} \xrightarrow{HCN,\ NaCN}$$

Bottom pathway (+):

$$\begin{array}{c} CN \\ HO-C-H \\ (H-C-OH)_3 \\ H-C-OH \\ CH_2OH \end{array} \xrightarrow{H_3O^+} \begin{array}{c} COOH \\ HO-C-H \\ (H-C-OH)_3 \\ H-C-OH \\ CH_2OH \end{array} \xrightarrow{-H_2O} \begin{array}{c} C=O \\ HO-C-H \\ (H-C-OH)_3 \\ H-C \\ CH_2OH \end{array} \xrightarrow[CO_2]{Na-Hg} \begin{array}{c} H-C=O \\ HO-C-H \\ (H-C-OH)_3 \\ H-C-OH \\ CH_2OH \end{array}$$

aldohexose   cyanohydrins   glyconic acids   glyconolactones   aldoheptoses

Diastereomeric and epimeric

6. Reaction with Phenylhydrazine. Osazone Formation

$$\begin{array}{c} CH=O \\ CHOH \\ (CHOH)_3 \\ CH_2OH \end{array} \xrightarrow{PhNHNH_2} \left[\begin{array}{c} H-C=NNHPh \\ H-C-OH \\ (H-C-OH)_3 \\ CH_2OH \end{array}\right] \xrightarrow[-PhNH_2,\ -NH_3]{PhNHNH_2} \left[\begin{array}{c} H-C=NNHPh \\ C=O \\ (H-C-OH)_3 \\ CH_2OH \end{array}\right] \xrightarrow{PhNHNH_2} \begin{array}{c} H-C=NNHPh \\ C=NNHPh \\ (H-C-OH)_3 \\ CH_2OH \end{array}$$

aldohexose   phenylhydrazone   osazone

Problem 25.9 The C^2-epimeric aldohexoses glucose and mannose give the same osazone as fructose. Write equations and explain the configurational significance. ◄

Since mannose and glucose are C^2 epimers, they are identical at the C^3, C^4, C^5 and C^6 portions of the molecule, which are unaltered during osazone formation. The chiral C^2, which differs in each hexose, loses chirality in the osazone and becomes identical for both mannose and glucose. 2-Ketohexoses give osazones in which the C^1 CH_2OH is oxidized. Since C^3, C^4, C^5 and C^6 of fructose and glucose are identical (Problem 25.8), fructose also gives the same osazone. Identical portions are in the boxes below.

$$\begin{array}{c} CHO \\ H-C-OH \\ \boxed{\begin{array}{c}(CHOH)_3 \\ CH_2OH\end{array}} \end{array} \text{ or } \begin{array}{c} CHO \\ HO-C-H \\ \boxed{\begin{array}{c}(CHOH)_3 \\ CH_2OH\end{array}} \end{array} \text{ or } \begin{array}{c} CH_2OH \\ C=O \\ \boxed{\begin{array}{c}(CHOH)_3 \\ CH_2OH\end{array}} \end{array} \xrightarrow{3PhNHNH_2} \begin{array}{c} H-C=NNHPh \\ C=NNHPh \\ \boxed{\begin{array}{c}(CHOH)_3 \\ CH_2OH\end{array}} \end{array}$$

Glucose   Mannose   Fructose   *same osazone*

C^2 epimers

Problem 25.10 Osazones are converted by PhCHO to 1,2-dicarbonyl compounds called **osones**. Use this reaction to change glucose to fructose. ◄

$$
\begin{array}{ccccc}
\begin{array}{c}
\text{H—C=O} \\
\text{H—C—OH} \\
\text{(H—C—OH)}_3 \\
\text{CH}_2\text{OH} \\
\text{Glucose}
\end{array}
& \xrightarrow{\text{3PhNHNH}_2}
&
\begin{array}{c}
\text{H—C=NNHPh} \\
\text{C=NNHPh} \\
\text{(H—C—OH)}_3 \\
\text{CH}_2\text{OH} \\
\text{osazone}
\end{array}
& \xrightarrow[\text{(−2 PhCH=NNHPh)}]{\text{PhCHO}}
&
\begin{array}{c}
\text{H—C=O} \\
\text{C=O} \\
\text{(H—C—OH)}_3 \\
\text{CH}_2\text{OH} \\
\text{osone}
\end{array}
\end{array}
$$

osone $\xrightarrow{\text{Zn, HAc}}$

$$
\begin{array}{c}
\text{CH}_2\text{OH} \\
\text{C=O} \\
\text{(H—C—OH)}_3 \\
\text{CH}_2\text{OH} \\
\text{Fructose}
\end{array}
$$

—CHO rather than C=O of the osone is reduced.

25.3 EVIDENCE FOR HEMIACETAL FORMATION AS EXEMPLIFIED WITH GLUCOSE

1. Some Negative Aldehyde Tests

Glucose does *not* form a bisulfite adduct or give a magenta color with the Schiff reagent.

2. Mutarotation

Naturally occuring (+)-glucose is obtained in 2 forms: m.p. = 146 °C, $[\alpha]_D$ = +112° and m.p. = 150 °C, $[\alpha]_D$ = +19°. The specific rotation of each changes (**mutarotates**) in water and both reach a constant value of +52.7°.

3. Acetal (Glycoside) Formation

Unlike typical aldehydes, glucose reacts with only 1 mole of ROH in dry HCl to give 2 isomeric acetals (alkyl **glycosides**). See Problem 15.25.

To explain the above results, glucose is assumed to exist as a cyclic hemiacetal in equilibrium with a small amount of the open-chain aldehyde. In the hemiacetal, C^1 is chiral and 2 diastereomers (**anomers**) are possible.

$$
\begin{array}{c}
\overset{1}{\text{H—C=O}} \\
\text{H—C—OH} \\
\text{HO—C—H} \\
\text{H—C—OH} \\
\text{H—}\overset{5}{\text{C}}\text{—OH} \\
\text{CH}_2\text{OH} \\
\text{Glucose} \\
\text{(open-chain)}
\end{array}
\rightleftharpoons
\begin{array}{c}
\overset{1}{\text{H—C—OH}} \\
\vdots \quad \text{O} \\
\text{—}\overset{5}{\text{C}}\text{—}
\end{array}
\xrightarrow{\text{CH}_3\text{OH/dry HCl}}
\begin{array}{c}
\overset{1}{\text{H—C—OCH}_3} \\
\vdots \quad \text{O} \\
\text{—}\overset{5}{\text{C}}\text{—}
\end{array}
$$

α-Glucose: $[\alpha]_D$ = +112°,
C^1 OH on right side

Methyl α-glucoside

$$
\begin{array}{c}
\overset{1}{\text{HO—C—H}} \\
\vdots \quad \text{O} \\
\text{—}\overset{5}{\text{C}}\text{—}
\end{array}
\xrightarrow{\text{CH}_3\text{OH/dry HCl}}
\begin{array}{c}
\overset{1}{\text{CH}_3\text{O—C—H}} \\
\vdots \quad \text{O} \\
\text{—}\overset{5}{\text{C}}\text{—}
\end{array}
$$

β-Glucose: $[\alpha]_D$ = +19°,
C^1 OH on left side

Methyl β-glucoside

Hemiacetal Anomers

Problem 25.11 Why do aldoses give positive Fehling and osazone reactions but negative Schiff and bisulfite tests?   ◄

These reactions are typical of the —CHO group. The Schiff and bisulfite reactions are reversible, and the equilibrium favors unreacted hemiacetal. Since osazone and Fehling reactions are irreversible, equilibrium is shifted to restore the low concentration (0.02%) of open-chain aldehyde as soon as some reaction occurs, and eventually all the aldose reacts.

Problem 25.12   Glycosides do not react with either Fehling's or Tollens' reagents and do not mutarotate. Explain. ◄

Glycosides are acetals.   They are stable in the basic Fehling's and Tollens' solutions and in aqueous solutions used in mutarotation.   Glycosides are *nonreducing*.

Problem 25.13   Calculate the percentages of α- and β-glucose present in the equilibrium mixture whose rotation is +52.7°. ◄

Let a and b be the mole fractions of the α- and β-anomers.   Solving the simultaneous equations

$$a + b = 1$$
$$112°a + 19°b = 52.7°$$

gives $a \times 100\% = 36.2\%$ and $b \times 100\% = 63.8\%$.

Problem 25.14   Acetylation of glucose produces 2 isomeric pentaacetates that do not react with either phenylhydrazine or Tollens' solution.   Explain. ◄

The C^1 OH of the hemiacetal is acetylated instead of the OH that forms the ring.   Since equilibrium with the open-chain aldehyde is prevented, these reactions are negative.

25.4 STEREOCHEMISTRY OF GLUCOSE

Problem 25.15   (+)-Glucose, (+)-mannose and (+)-galactose are among the hexoses isolated after 3 Kiliani-Fischer step-ups (page 433) with D-(+)-glyceraldehyde (page 59).   Do these hexoses belong in the D or L families? ◄

During the step-ups the D configuration of the chiral C^2 of glyceraldehyde (glycerose) does not change. This C becomes C^5 of (+)-glucose, (+)-mannose and (+)-galactose.   Configuration of C^5 determines the family, and these monosaccharides are D-sugars.   See Fig. 25-1.

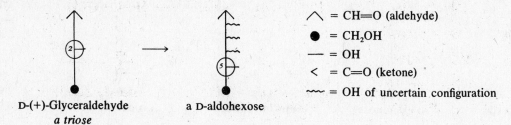

D-(+)-Glyceraldehyde   a D-aldohexose
a triose

= CH=O (aldehyde)
= CH₂OH — rendered: $= CH_2OH$
— = OH
< = C=O (ketone)
~~~ = OH of uncertain configuration

Fig. 25-1

Problem 25.16   Write shorthand formulas for   (*a*) the tetroses prepared from L-glycerose by the Kiliani-Fischer synthesis,   (*b*) their enantiomers. ◄

See Fig. 25-2.

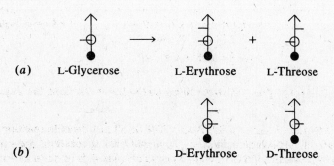

(*a*)   L-Glycerose   L-Erythrose   L-Threose

(*b*)   D-Erythrose   D-Threose

Fig. 25-2

Problem 25.17 How can oxidation with HNO_3 be used to distinguish between D-erythrose and D-threose? ◄

See Fig. 25-3.

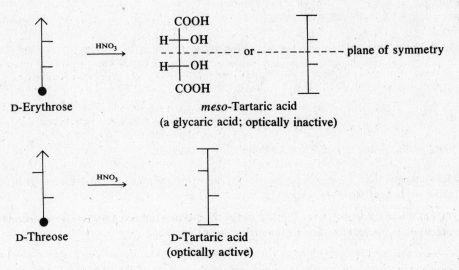

Fig. 25-3

Problem 25.18 Two Ruff degradations on an aldohexose give an aldotetrose that is oxidized by HNO_3 to *meso*-tartaric acid. What can be the family configuration of the aldohexose? ◄

See Fig. 25-4. The C^4 and C^5 OH's must be on the same side. The aldohexose can belong to either the D or L family.

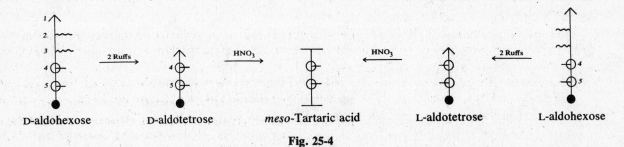

Fig. 25-4

Problem 25.19 Which D-aldohexoses give *meso*-glycaric acids on oxidation with HNO_3? ◄

There is a plane of symmetry (-----) in the dicarboxylic acids formed by HNO_3 oxidation of D-allose and D-galactose. See Fig. 25-5.

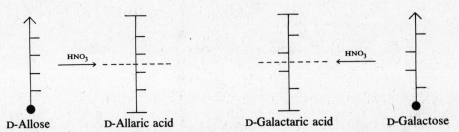

Fig. 25-5

Problem 25.20 The reaction sequence $\xrightarrow{\text{Ruff}}$ $\xrightarrow{\text{HNO}_3}$ on D-allose gives a *meso*—but on D-galactose gives an optically active—pentaglycaric acid.   Assign the structures.   ◄

See Fig. 25-6.

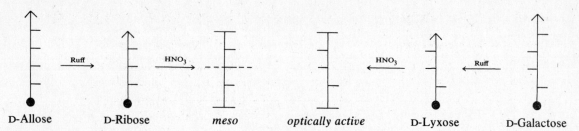

D-Allose   D-Ribose   *meso*   *optically active*   D-Lyxose   D-Galactose

Fig. 25-6

Problem 25.21 D-Allose, D-glucose and D-talose (the C^2 epimer of D-galactose) each give a *meso*-heptaglycaric acid after the sequence $\xrightarrow{\text{Kiliani}}$ $\xrightarrow{\text{HNO}_3}$.   Assign all structures.   ◄

There are 3 *meso*-heptaglycaric acids, as shown in Fig. 25-7.

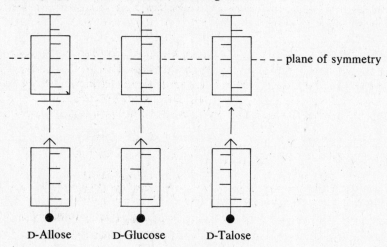

-- -- -- plane of symmetry

D-Allose   D-Glucose   D-Talose

Fig. 25-7

Problem 25.22 What monosaccharide is obtained from a Ruff degradation of D-mannose (Problem 25.9), the C^2 epimer of D-glucose (Problem 25.21)?   ◄

See Fig. 25-8.

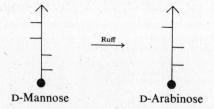

D-Mannose   D-Arabinose

Fig. 25-8

Problem 25.23 What is the structure of the aldopentose D-ribose, a constituent of ribonucleic acids (RNA), if it gives the same osazone as D-arabinose (Problem 25.22) and is reduced to an optically inactive glykitol?   ◄

Ribose and arabinose are C^2 epimers, since they give the same osazone (Fig. 25-9).

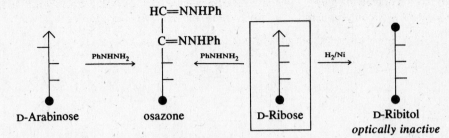

Fig. 25-9

Problem 25.24 The pentoses D-xylose and D-lyxose give the same osazone.   Xylose and lyxose are oxidized to an optically inactive and optically active dicarboxylic acid, respectively.   Give their shorthand structural formulas.   ◄

They are D-sugars; the C^4 OH is on the right side.   They differ from D-arabinose and D-ribose (Problem 25.23), both of which have C^3 OH on the right side; therefore D-xylose and D-lyxose have the C^3 OH on the left side.   Since D-xylose gives the inactive (*meso*) dicarboxylic acid, its C^2 OH must be on the right side, the same side as C^4 OH. C^2 OH is on the left side in D-lyxose.   See Fig. 25-10.

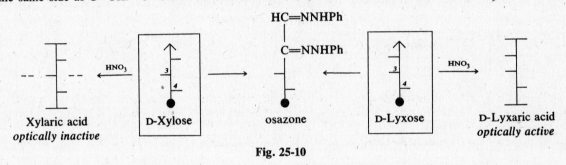

Fig. 25-10

Problem 25.25 Fischer prepared L-gulose by oxidizing D-glucose to 2 separable lactones of glucaric acid.   These were reduced to lactones of gluconic acid, which were further reduced.   Give all the structures in these reactions.   ◄

See Fig. 25-11.

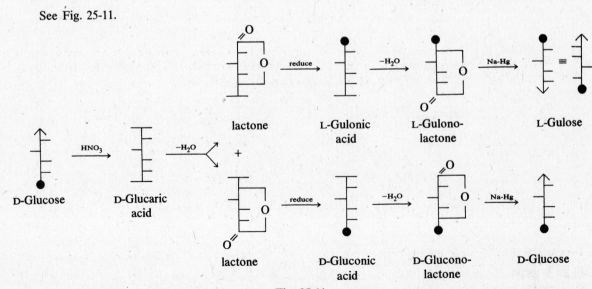

Fig. 25-11

Problem 25.26   An aldohexose (I) is oxidized by HNO_3 to a *meso*-glycaric acid (II).   Ruff degradation of (I) yields (III), which is oxidized to an optically active dicarboxylic acid (IV).   Ruff degradation of (III) yields (V), which is oxidized to L-(+)-tartaric acid (VI).   Represent compounds (I) through (VI).   ◄

See Fig. 25-12.   Since L-(+)-tartaric acid (VI) is isolated, (I) is an L-sugar.

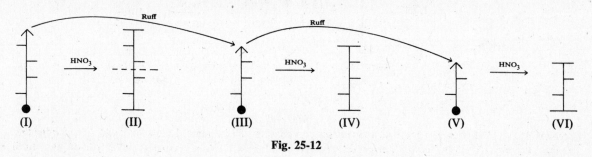

Fig. 25-12

Problem 25.27   Deduce whether the hemiacetal ring of glucose is 5- or 6-membered from the following data and write equations for all steps.   Glucose is first monomethylated at C^1 with MeOH and HCl, then tetramethylated with Me_2SO_4 and NaOH.   The pentamethyl derivative is treated with aq. HCl, then vigorously oxidized with HNO_3 to give 2,3-dimethoxysuccinic acid and 2,3,4-trimethoxyglutaric acid.   ◄

We have:

$$\text{Glucose} \xrightarrow[\text{HCl}]{\text{MeOH}} \text{Methyl glucoside (acetal)} \xrightarrow[\text{NaOH}]{\text{Me}_2\text{SO}_4} \text{Methyl tetra-O-methyl glucoside} \xrightarrow[\text{H}_2\text{O}]{\text{HCl}} \text{(A)}$$
(other 4 OH's
form ethers)

(Only acetal C^1 —OMe is hydrolyzed; the other ether linkages are stable.)   The unmethylated OH's of (A) are those involved in ring formation.   They are C^1 and C^4 if the ring is 5-membered (**furanose**) or C^1 and C^5 for a 6-membered ring (**pyranose**).   See Fig. 25-13.

Vigorous oxidation in the last step causes cleavage on both sides of the 2°

$$\text{H}\overset{|}{\underset{|}{\text{C}}}\text{—OH}$$

the routes being (*a*) and (*b*) for the pyranose, and (*c*) and (*d*) for the furanose.   Only the 6-membered ring gives the observed degradation products.

Problem 25.28   (*a*) Draw the chair conformations for α- and β-D-(+)-glucopyranose.   The bulky CH_2OH is usually assigned an equatorial position.   In flat Fischer formulas, OH's on the right are *trans* to CH_2OH, and those on the left are *cis*.   (*b*) Why are most naturally occurring glycopyranosides β-anomers?   ◄

(*a*)

β-D-(+)-Glucopyranose   α-D-(+)-Glucopyranose

(*b*)   Only in β-glucopyranosides are all substituents equatorial.   In the α-anomer, the C^1 —OR is axial.

Problem 25.29   How can the sequence $\xrightarrow{\text{HIO}_4} \xrightarrow{\text{Br}_2,\text{H}_2\text{O}} \xrightarrow{\text{H}_3\text{O}^+}$ show if a methyl glycoside is a furanoside or a pyranoside?   ◄

See Fig. 25-14.   The acetal formed by HIO_4 is hydrolyzed by H_3O^+.

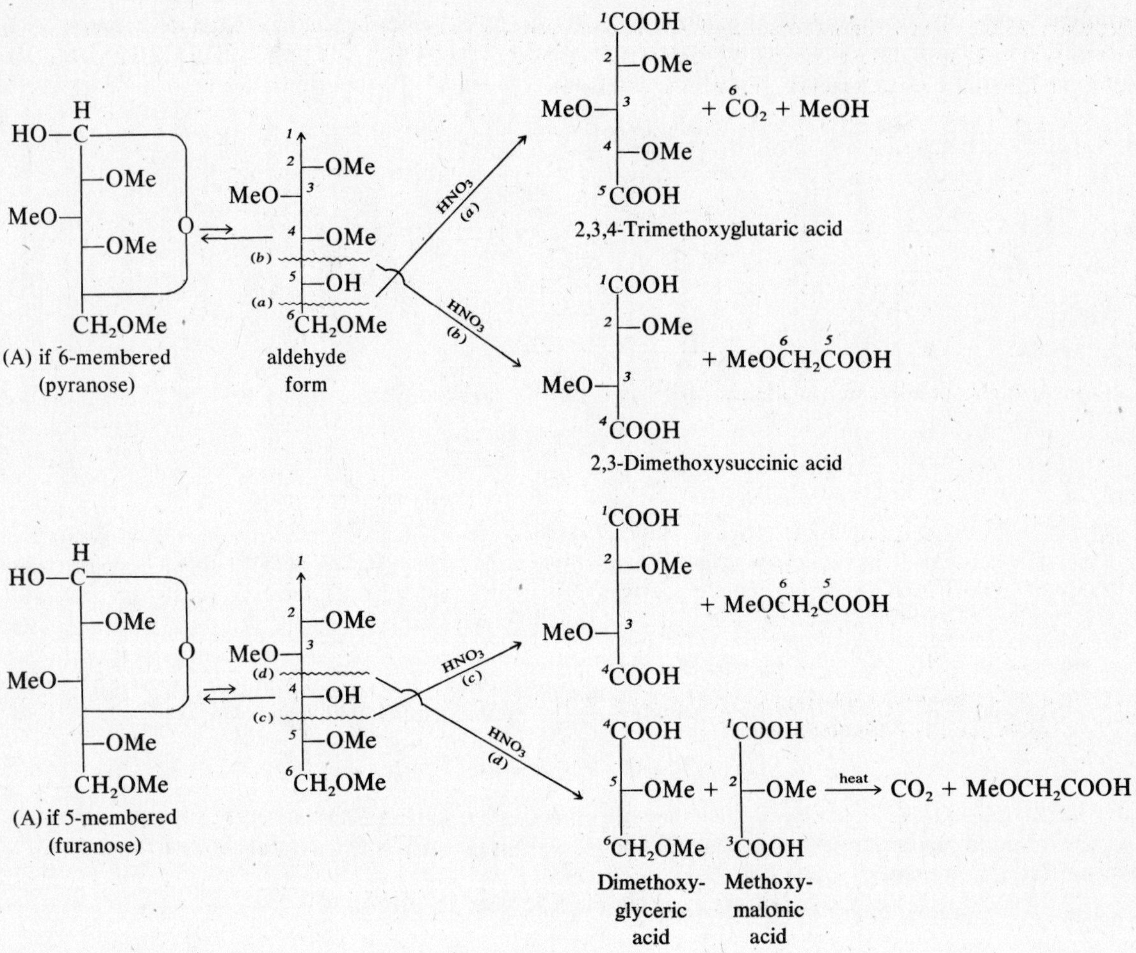

Fig. 25-13

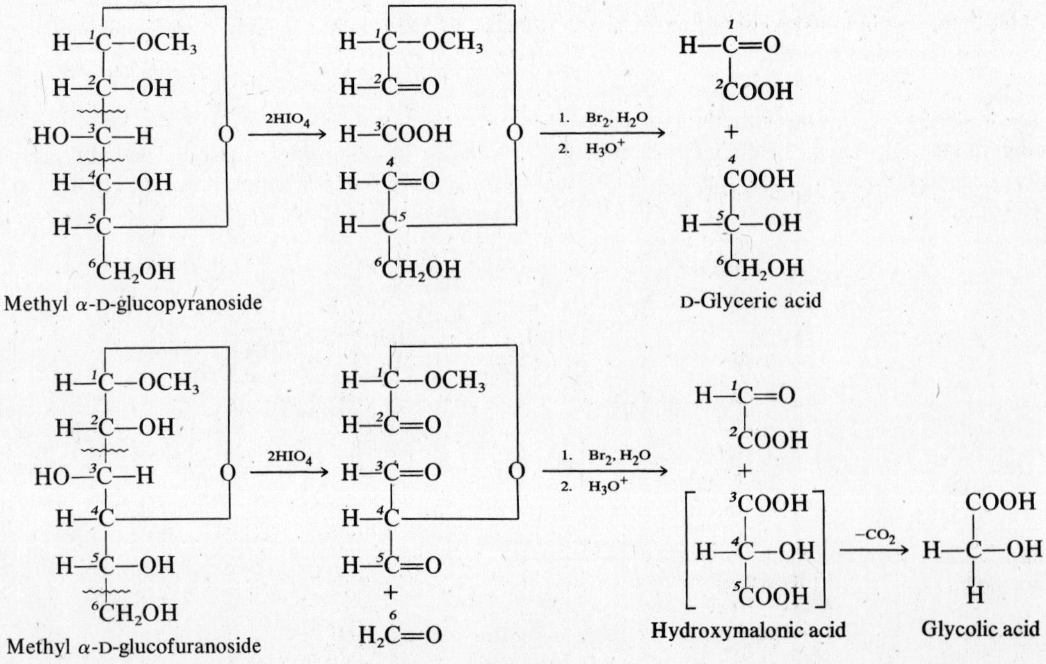

Fig. 25-14

Problem 25.30 Draw the more and less stable conformations of (a) β-D-mannopyranose (Problem 25.9), (b) α-D-idopyranose (idose is the C^2 epimer of gulose; see Problem 25.25), (c) β-L-glucopyranose (β-L- and β-D- are enantiomers). Explain each choice.

(a)

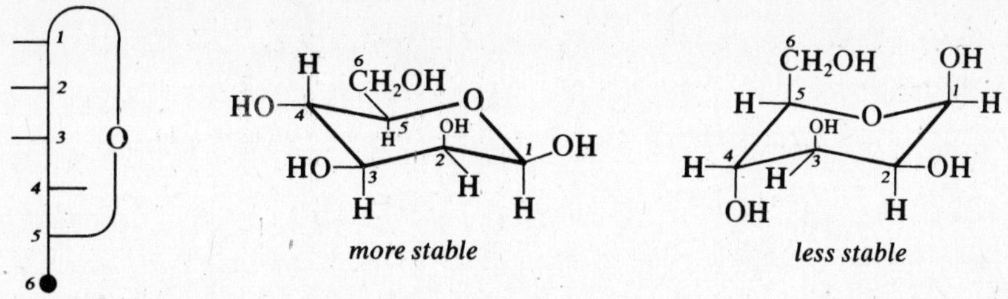

more stable   *less stable*

In the more stable conformation, CH_2OH and 3 OH's are (e).

(b)

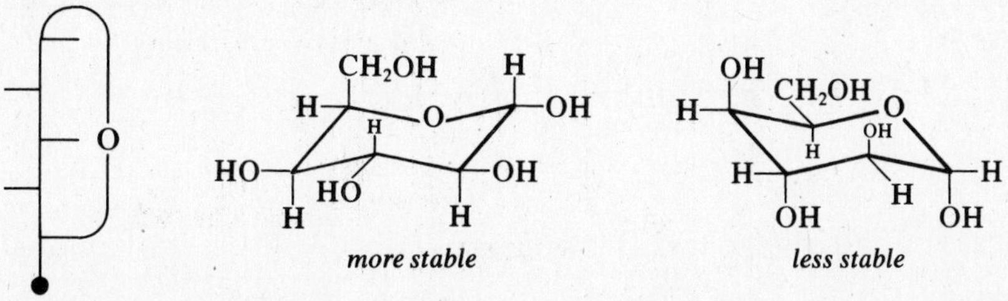

more stable   *less stable*

The less stable conformation has CH_2OH (e) but 4 axial OH's, while the more stable conformation has CH_2OH (a) but 4 equatorial OH's.

(c)

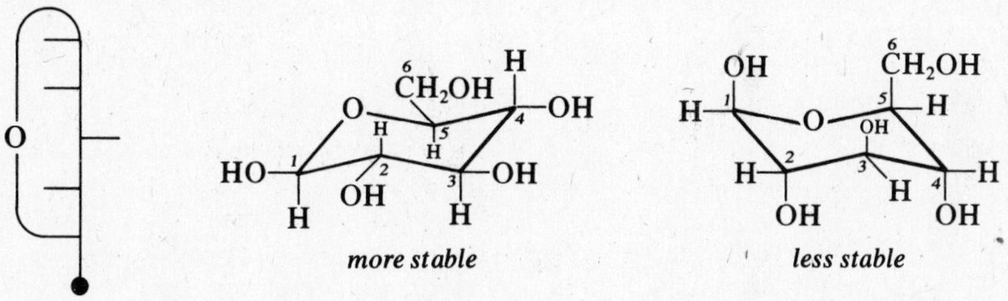

more stable   *less stable*

The more stable conformation has all groups (e); the less stable has all groups (a).

Problem 25.31 Write a structural formula for (a) β- and α-D-(−)-fructofuranose, (b) β-D-(−)-fructopyranose. ◄

In this case the anomeric position is C^2.

(a)

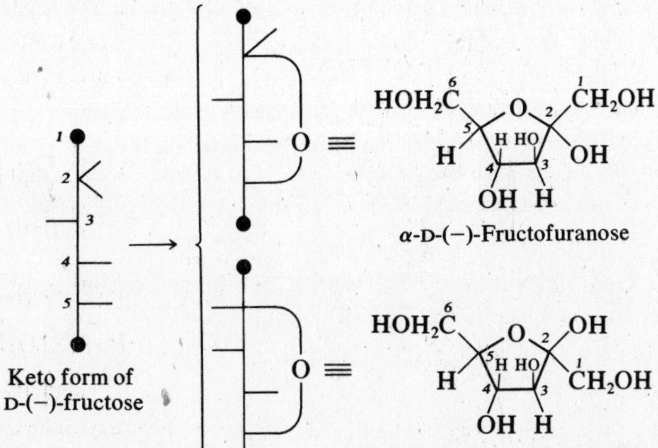

Keto form of
D-(−)-fructose

α-D-(−)-Fructofuranose

β-D-(−)-Fructofuranose

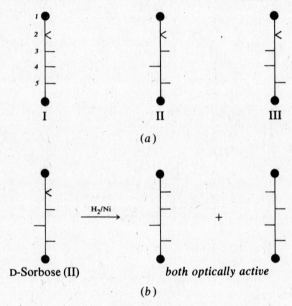

Problem 25.32 (*a*) Write shorthand structures for the three D-2-ketohexoses other than D-fructose. (*b*) Which one does not give a *meso* glykitol on reduction? (*c*) Which one gives the same osazone as D-galactose (Problem 25.20)? ◄

(*a*) See Fig. 25-15(*a*). The D family has the C^5 OH on the right side. (*b*) See Fig. 25-15(*b*). (*c*) Compound III is D-tagatose, since its C^3, C^4 and C^5 OH's have the same configuration as the corresponding OH's in D-galactose.

Fig. 25-15

25.5 DISACCHARIDES

Disaccharides have an OH of one monosaccharide (the **aglycone**, denoted *A*) bonded to the anomeric C of a second monosaccharide, *B*. The disaccharide is a glycoside of *B*.

Problem 25.33 (*a*) Maltose (I) is hydrolyzed to D-glucose by maltase, a specific enzyme for α-glycosides. (*b*) (I) reduces Fehling's solution, undergoes mutarotation and is oxidized by Br_2 in H_2O to D-maltobionic acid, $C_{11}H_{21}O_{10}COOH$ (II). (*c*) (II) reacts with Me_2SO_4 and NaOH to form an octamethyl derivative (III) that is then hydrolyzed to 2,3,4,6-tetra-O-methyl-D-glucose (IV) and 2,3,5,6-tetra-O-methyl-D-gluconic acid (V). Deduce the structure of maltose. ◄

(*a*) Maltose is composed of 2 glucose molecules with an α-glycosidic linkage. (*b*) Maltose is a reducing sugar and therefore has 1 free hemiacetal unit *A*. The anomeric C^1, of A is oxidized to COOH to give (II).   (c) (V) comes from unit A (the aglycone).   Unit A is linked through its C^4 —OH, since this group in (V) is not methylated.   (IV) comes from B.   The C^5 —OH of (IV) is not methylated and must be involved in forming a pyranose ring.   Unit A is also a pyranose; a furanose is ruled out because the C^4 —OH needed for this ring size is part of the glycosidic linkage.   The only uncertainty in the structure is the configuration of the anomeric C of A, which is assumed to have the more stable β-configuration.   See Fig. 25-16.

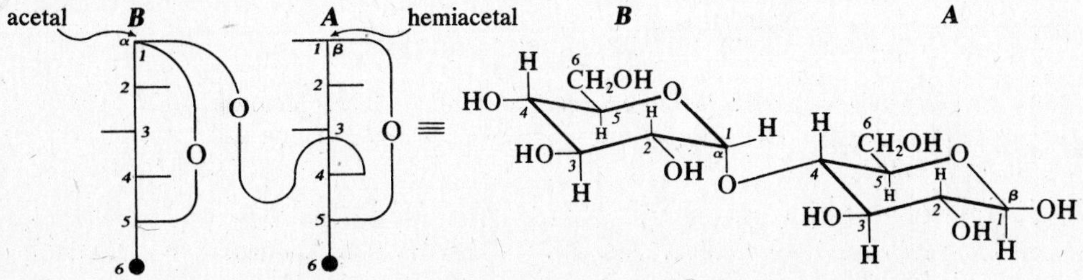

(I) β-Anomer of (+)-maltose, 4-O-(α-D-glucopyranosyl)-D-glucopyranose

Br₂, H₂O

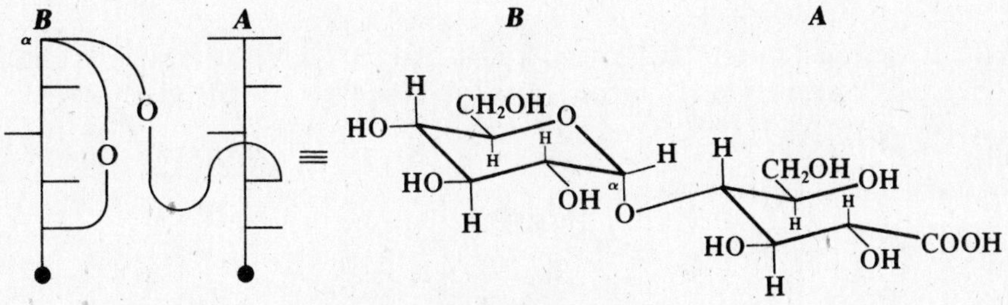

(II)   D-Maltobionic acid

(CH₃)₂SO₄, NaOH

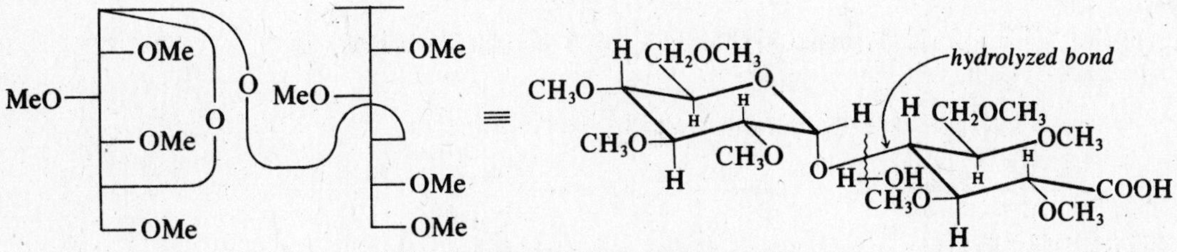

(III)   Octa-O-methyl-D-maltobionic acid

H₃O⁺

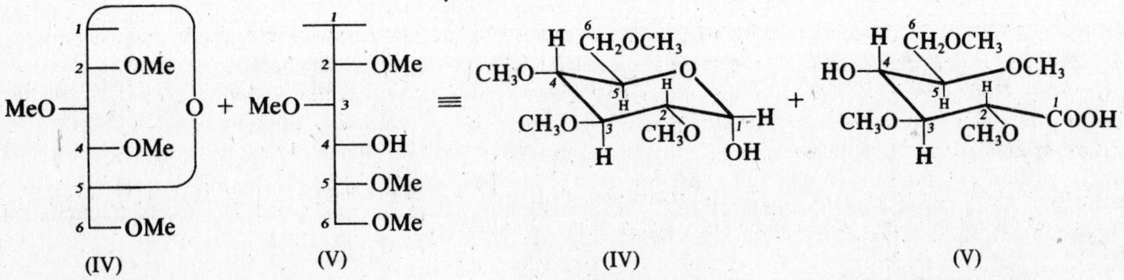

(IV)
(α-anomer)

(V)

(IV)
2,3,4,6-Tetra-O-methyl-D-glucopyranose

(V)
2,3,5,6-Tetra-O-methyl-D-gluconic acid

Fig. 25-16

Problem 25.34 Cellobiose, isolated from the polysaccharide cellulose, has the same chemistry as maltose except that it is hydrolyzed by emulsin. Give the structure of this disaccharide. ◄

Unlike maltose, cellobiose is a β-glycoside. (The aglycone unit is turned 180° to permit a reasonable bond angle for the glycosidic linkage.)

Problem 25.35 Deduce the structure of the disaccharide lactose. (*a*) It is hydrolyzed by emulsin to D-glucose and D-galactose. (*b*) It is a reducing sugar and undergoes mutarotation. (*c*) It forms an osazone that is hydrolyzed to D-galactose and D-glucosazone. (*d*) Mild oxidation, methylation and then hydrolysis gives results analogous to those observed with D-maltose. ◄

(*a*) Lactose is a β-glycoside made up of D-glucose and D-galactose. (*b*) It has a free hemiacetal group. (*c*) Since the glucose unit forms the osazone, it is **A** and must have the hemiacetal linkage. Lactose is a β-galactoside. (*d*) Both units are pyranoses joined through the C^4 —OH of the glucose unit.

Problem 25.36 Use shorthand formulas to show how osazone formation establishes the glucose unit of lactose to have the hemiacetal linkage (Problem 25.35). ◄

Lactosazone is hydrolyzed to D-(+)-galactose and D-glucosazone.

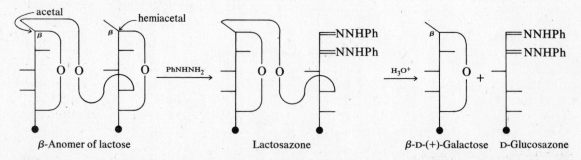

Problem 25.37 (*a*) Give the structure of sucrose (cane and beet sugar) from the following information: (1) It is hydrolyzed by maltase or emulsin to a mixture of D-(+)-glucose and D-(−)-fructose. (2) It does *not* reduce Fehling's solution and does not mutarotate. (3) Methylation and hydrolysis gives 2,3,4,6-tetra-O-methyl-D-glucose and a tetramethyl-D-fructose. (*b*) What structural features are uncertain? ◄

(*a*) (1) Sucrose has a glucose linked to a fructose unit. The linkage is α to one monosaccharide and β to the other. (2) It has no hemiacetal group. The anomeric OH's on C^1 of glucose and C^2 of fructose are joined to form 2 acetals. (3) The glucose unit is a pyranose; the C^5 —OH is unmethylated.

(b) *The glycosidic linkages*:   the linkage to glucose is α and the one to fructose is β.   *The fructose ring size*:   it is a furanose.   (+)-Sucrose is α-D-glucopyranosyl-β-D-fructofuranoside or β-D-fructofuranosyl-α-D-glucopyranoside.

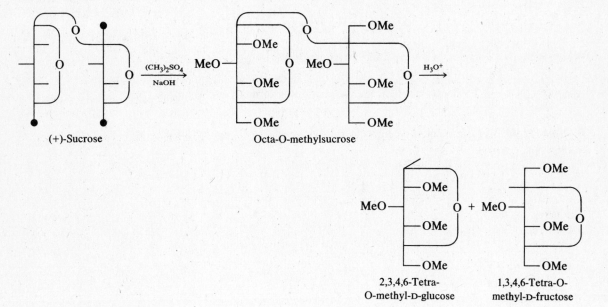

Problem 25.38 Hydrolysis of (+)-sucrose gives a mixture of D-(+)-glucose ($[\alpha]_D = 52.7°$) and D-(−)-fructose ($[\alpha]_D = -92.4°$) called **invert sugar**.   Calculate the specific rotation of invert sugar.   ◄

One mole of sucrose produces 1 mole each of glucose and fructose, and the specific rotation is one-half the sum of those of the two monosaccharides, i.e.

$$\frac{1}{2}[+52.7° + (-92.4°)] = -19.9°$$

Problem 25.39 Write the reactions for methylation and hydrolysis of sucrose.   ◄

25.6 POLYSACCHARIDES

Problem 25.40 The polysaccharide amylose, the water soluble component of starch, is hydrolyzed to (+)-maltose and D-(+)-glucose.   Methylated and hydrolyzed, amylose gives mainly 2,3,6-tri-O-methyl-D-glucopyranose.   Deduce the structure of amylose.   ◄

It has D-glucose units joined by α-glycosidic linkages [Problem 25.33(a)] to the C^4 of the next unit.   This is revealed by isolating maltose, an α-glycoside, and the 2,3,6-tri-O-methyl derivative.   The C^5 —OH forms the pyranoside ring.

Problem 25.41   Methylation and hydrolysis of amylose (Problem 25.40) also gives 0.2–0.4% 2,3,4,6-tetra-O-methyl-D-glucose.   Explain the origin of this compound.   ◄

The end glycosidic glucose unit has a free C^4 —OH and gives this tetra-O-methyl derivative.

Problem 25.42   Amylopectin, the water insoluble fraction of starch, behaves like amylose, except that more 2,3,4,6-tetra-O-methyl-D-glucose (5%), and an equal amount of 2,3-di-O-methyl-D-glucose, is formed.   Deduce the structure of amylopectin.   ◄

Like amylose, amylopectin has chains of D-glucose units formed by an α-glycosidic linkage to C^4 of the next glucose unit.   Occasionally along the chain, the C^6 —OH is not free but is used to branch 2 chains at the aglycone end of one chain.   Since more 2,3,4,6-tetra-O-methyl glucose is formed, the amylopectin chains are shorter than those of amylose.

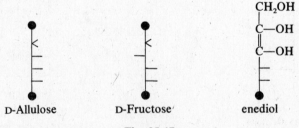

2,3-Di-O-methyl-D-glucose

Supplementary Problems

Problem 25.43   How do epimers and anomers differ?   ◄

Epimers differ in configuration about a single chiral center in molecules with more than 1 chiral center.   Anomers are epimers in which the chiral site was formerly a carbonyl C.

Problem 25.44   What enediol is common to D-fructose and its C^3 epimer, D-allulose?   ◄

See Fig. 25-17.

CH₂OH
C—OH
C—OH

D-Allulose   D-Fructose   enediol

Fig. 25-17

Problem 25.45   A monosaccharide is treated with HCN, the product hydrolyzed and reduced to a carboxylic acid by heating with HI and P.   What carboxylic acid is formed if the monosaccharide is   (a) D-glucose, (b) D-mannose,   (c) D-fructose,   (d) D-arabinose,   (e) D-erythrose?   ◄

(a) and (b) n-heptanoic acid,   (c) 2-methylhexanoic acid,   (d) n-hexanoic acid,   (e) n-pentanoic acid.

Problem 25.46   Prepare D-talose from its C^2 epimer, D-galactose (Problem 25.20).   ◄

See Fig. 25-18.

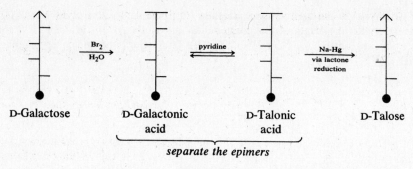

Fig. 25-18

Problem 25.47 Convert D-glucose to hexa-O-acetyl-D-sorbitol (a glykitol polyester). ◀

See Fig. 25-19.

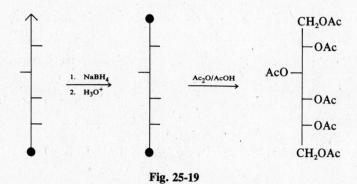

Fig. 25-19

Problem 25.48 Give names and structural formulas for the products of the reaction of α-D-mannose (Problem 25.9) with: (*a*) Ac₂O, pyridine; (*b*) HIO₄; (*c*) CH₃OH, HCl, then HIO₄; (*d*) H₂NOH; (*e*) CH₃OH, HCl, then (CH₃)₂SO₄, NaOH, and then aq. HCl; (*f*) HCN/KCN, then H₃O⁺, and then Na-Hg, CO₂; (*g*) PhNHNH₂, then PhCH=O. ◀

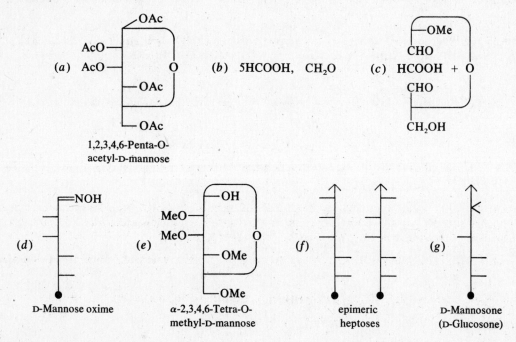

Problem 25.49 (*a*) Write shorthand open structures for a ketohexose (I) and an aldohexose (II) which form the same osazone as D-(−)-altrose (III), the C^3 epimer of D-mannose, and for the aric acid (IV) formed by HNO_3 oxidation of the D-aldohexoses (III) and D-talose (V). (*b*) Write shorthand and conformational structures for β-D-altropyranose (VI). ◄

See Fig. 25-20.

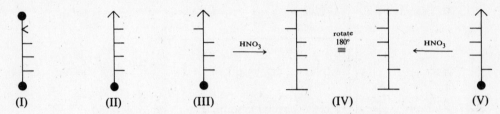

C^3, C^4 and C^5 have the same configuration in (I), (II) and (III).

(*a*)

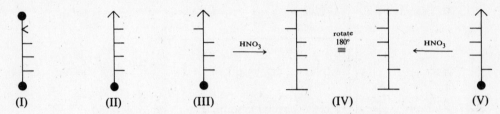

(VI)

(*b*)

Fig. 25-20

Problem 25.50 Distinguish 2-deoxyglucose from 3-deoxyglucose (**deoxy-** means H in place of OH). ◄

3-Deoxyglucose forms an osazone with 3 moles of $PhNHNH_2$, but 2-deoxyglucose has no C^2 OH and forms only a phenylhydrazone.

Problem 25.51 What hexose (I) forms a cyanohydrin (II) that is hydrolyzed and then reduced with HI and P to a carboxylic acid (III), where (III) is synthesized from *n*-PrI and $Na^+[CC_2H_5(COOEt)_2]^-$? ◄

We have:

$$\overset{6}{C}H_3\overset{5}{C}H_2\overset{4}{C}H_2I + Na^+\begin{bmatrix} \overset{3}{C}(COOEt)_2 \\ | \\ \underset{2}{C}H_2\underset{1}{C}H_3 \end{bmatrix}^- \xrightarrow[\text{steps}]{\text{several}} \overset{6}{C}H_3\overset{5}{C}H_2\overset{4}{C}H_2\overset{3}{C}HCOOH$$
$$\underset{|}{\underset{\underset{2}{C}H_2\underset{1}{C}H_3}{}}$$

2-Ethylpentanoic acid (III)

For COOH to be placed on C^3 by the cyanohydrin reaction, C^3 must be a carbonyl C in (I). (I) is therefore a 3-ketohexose,

and (II) is

(A wavy line indicates unknown configuration.)

Problem 25.52 Which D-pentose is oxidized to an optically inactive dibasic acid and undergoes a Ruff degradation to a tetrose whose glycaric acid is *meso*-tartaric acid? ◄

Ribose; see Fig. 25-21.

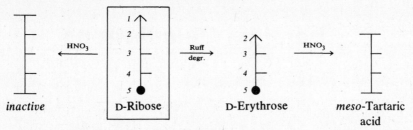

Fig. 25-21

Problem 25.53 Deduce possible structures for the following disaccharides. (*a*) A compound, $C_{10}H_{18}O_9$, undergoes oxidation with aq. Br_2, followed by methylation and hydrolysis by maltase, to yield 2,3,4-tri-O-methyl-D-xylose and 2,3-di-O-methyl-L-arabonic acid. (*b*) (+)-Gentiobiose, $C_{12}H_{22}O_{11}$, reduces Fehling's solution, undergoes mutarotation and is hydrolyzed by emulsin to D-glucose. Methylation followed by hydrolysis yields 2,3,4,6-tetra-O-methyl-D-glucose and 2,3,4-tri-O-methyl-D-glucose. (*c*) Trehalose, $C_{12}H_{22}O_{11}$, is nonreducing and is hydrolyzed only to D-glucose by either maltase or aqueous acid. Methylation followed by hydrolysis yields only 2,3,4,6-tetra-O-methyl-D-glucose. (*d*) Isotrehalose is identical with trehalose except that it is hydrolyzed by either emulsin or maltase. ◄

D-Xylose portion   L-Arabinose portion

(*a*)   4-O-(α-D-Xylopyranosyl)-L-arabinose

(*b*)   6-O-(β-D-Glucopyranosyl)-D-glucopyranose

(*c*)   α-D-Glucopyranosyl-α-D-glucopyranoside

(*d*)   α-D-Glucopyranosyl-β-D-glucopyranoside

Problem 25.54 What is the enantiomer of α-D-(+)-glucose? ◄

α-L-(−)-Glucose. In the D family the more dextrorotatory anomer is named α-D-. In the L family the more levorotatory anomer is named α-L-.

Chapter 26

Chemical Calculations

26.1 FORMULAS AND MOLECULAR WEIGHTS

Problem 26.1 What is the lowest **gram molecular weight** (GMW) possible for a compound with 14.4% carbon? ◄

One **gram atomic weight** (GAW) of an element is the least amount of that element that can be present in a mole of any of its compounds. Therefore 14.4% of the GMW of the present compound must be the GAW of the element C.

$$14.4\% \text{ of GMW} = \text{GAW of C}$$

$$(0.144)(\text{GMW}) = 12.01$$

$$\text{GMW} = \frac{12.01}{0.144} = 83.4 \text{ g per mole}$$

Problem 26.2 A gas has the elementary composition C = 84.41%, H = 15.60%. At standard conditions (SC) one liter (1 ℓ = 10^{-3} m^3) of this gas weighs (has a mass of) 6.34 g. What is its molecular formula? ◄

Since the sum of the percentages of C and H is 100.01%, no other elements are present in the compound. The **empirical formula** of a compound is the simplest integral ratio of the numbers of each kind of atom present. These integers are obtained from the percentages, which are masses of atoms in 100 g of compound, by dividing each percentage by the GAW. Thus,

Element	Mass of element in 100 g compound (g)		GAW		Relative number of gram atoms
C	84.41	÷	12.01	=	7.028
H	15.60	÷	1.008	=	15.48

Divide both quotients by the smaller one to get the lowest ratio:

$$C = \frac{7.028}{7.028} = 1.000 \qquad H = \frac{15.48}{7.028} = 2.203$$

Since this C : H ratio is *still* not integral, both numbers of this ratio are multiplied by integers to give the values:

	×1	×2	×3	×4	×5	×6	×7
C	1.000	2.000	3.000	4.000	5.000	6.000	7.000
H	2.203	4.406	6.609	8.812	11.015	13.218	15.421

The multiplier giving most nearly two integers is 5.   The empirical formula for this compound is C$_5$H$_{11}$, and its **gram empirical formula weight** (GEFW) is

$$(12.01)(5) + (1.01)(11) = 71.16$$

The molecular weight of the compound is calculated from the mass μ of a sample of the gas at SC using the ideal gas law:

$$PV = \frac{\mu}{(\text{GMW})} RT$$

450

Thus,

$$GMW = \frac{\mu RT}{PV} = \frac{(6.34 \text{ g})(0.0821 \text{ } \ell \text{ atm } °K^{-1} \text{ mole}^{-1})(273 °K)}{(1 \text{ atm})(1.00 \text{ } \ell)} = 142 \text{ g/mole}$$

Alternatively, at SC one GMW (mole) of an ideal gas occupies 22.4 ℓ, so that

$$GMW = 6.34 \frac{g}{\ell} \times 22.4 \frac{\ell}{\text{mole}} = 142 \text{ g/mole}$$

We have GMW = GEFW $\times$ k, where k is an integer. In this case

$$k = \frac{142}{71.16} = 2$$

and we double the empirical formula, C_5H_{11}, to get the molecular formula, $C_{10}H_{22}$.

Problem 26.3   A 0.322 g sample of an organic vapor at 100 °C and 740 torr occupies 62.7 mℓ. Two analyses give elementary compositions C = 65.60%, H = 5.65% and C = 65.25%, H = 5.35%. What is its molecular formula?   ◄

The two analyses are close enough to be averaged. Hence,

$$C = \frac{1}{2}(65.60\% + 65.25\%) = 65.43\% \qquad H = \frac{1}{2}(5.65\% + 5.35\%) = 5.50\%$$

The sum of the percentages of C and H is 70.93%. The remaining 29.07% is presumed to be O. The empirical formula is calculated in Table 26-1 by the method of Problem 26.2.

Table 26-1

(1)	(2)	(3)	(4)	(5)
E (Element)	M_E ÷ (E in 100 g ÷ compound)	W_E = (GAW of = element)	n_E (Relative number of gram atoms)	n_E/least n_E
C	65.43	12.01	5.45	3.00
H	5.50	1.01	5.45	3.00
O	29.07	16.00	1.82	1.000

Divide the number of gram atoms (column 4) by the lowest value (1.82) to get the values in column 5. The empirical formula is C_3H_3O (GEFW = 55.06).

The GMW is obtained from the vapor density data.

$$GMW = \frac{\mu RT}{PV} = \frac{(0.322 \text{ g})(0.0821 \text{ } \ell \text{ atm } °K^{-1} \text{ mole}^{-1})(373 °K)}{(740 \text{ torr}/760 \text{ torr atm}^{-1})(0.0627 \text{ } \ell)} = 161 \text{ g/mole}$$

Pressure is converted from torr to atmospheres by dividing torr by 760 torr per atmosphere, and volume is converted from mℓ to liters to make the units consistent. Since the GMW nearly equals 3 × GEFW, the molecular formula must be 3 times the empirical formula, i.e. $C_9H_9O_3$.

Problem 26.4   Complete combustion of 0.858 g of an organic compound (X) gave 2.63 g of CO_2 and 1.28 g of H_2O. (a) Find the percent composition of (X). (b) What is the lowest molecular weight this compound could have?   ◄

(a)   The mass of C in 2.63 g of CO_2 and in the sample is

$$\frac{\text{GAW of C}}{\text{GMW of } CO_2} \times (\text{mass of } CO_2) = \frac{12.01}{44.01}(2.63 \text{ g}) = 0.718 \text{ g}$$

which is $(0.718 \text{ g}/0.858 \text{ g})(100\%) = 83.7\%$ of (X). The mass of H in the sample and in 1.28 g of H_2O is

$$\frac{2.016}{18.015}(1.28 \text{ g}) = 0.143 \text{ g}$$

which is $(0.143 \text{ g}/0.858 \text{ g})(100\%) = 16.7\%$ of (X). Since the sum of C and H percentages is 100.4%, O is absent.

(b) By the method of Problem 26.2:

E	M_E	$\div$	W_E	$=$	n_E	$n_E/6.97$	Multiplier is 3
C	83.7		12.01		6.97	1.00	3.00
H	16.7		1.01		16.5	2.37	7.11

From this, the empirical formula would seem to be C_3H_7. But C_3H_7 doesn't fit any general formula for a hydrocarbon; a hydrocarbon cannot have an odd number of H's. Therefore, double this formula to get C_6H_{14} (an alkane, C_nH_{2n+2}). The corresponding molecular weight is

$$(12 \times 6) + (1 \times 14) = 86$$

Had we used 8 as the multiplier in the above table, we would have obtained the empirical formula C_8H_{19}, in which the number of H's is also odd.

Problem 26.5 Analysis of a compound (X) gives the following data. (a) Combustion of 11.9 mg of (X) produces 27.94 mg of CO_2 and 4.68 mg of H_2O. (b) 23.9 mg of (X) is oxidized with concentrated HNO_3, and the oxidation products are treated with excess $AgNO_3$ solution. 30.3 mg of AgCl is obtained as a precipitate. (c) A solution of 4.48 g of (X) in 50.0 g of benzene freezes at 3.45 °C ($K_{f.p.}$ for benzene is 5.12 °C kg/mole). What is the molecular formula of the compound? ◄

(a)　　$\dfrac{12.01}{44.01} \times 27.94 \text{ mg of } CO_2 = 7.625 \text{ mg of C}$　　　　$\dfrac{7.625 \text{ mg of C}}{11.9 \text{ mg of (X)}} \times 100\% = 64.1\% \text{ C}$

　　　　$\dfrac{2.016}{18.015} \times 4.68 \text{ mg of } H_2O = 0.524 \text{ mg of H}$　　　　$\dfrac{0.524 \text{ mg of H}}{11.9 \text{ mg of (X)}} \times 100\% = 4.40\% \text{ H}$

(b) Here the Cl in (X) is converted to AgCl. The fraction of Cl in AgCl is 35.45/143.32. The weight of the AgCl obtained is 30.3 mg. Therefore, the weight of Cl in (X) is

$$\frac{35.45}{143.32}(30.3 \text{ mg}) = 7.49 \text{ mg}$$

The total weight of the sample in this procedure was 23.9 mg, and so the % Cl in (X) is

$$\frac{7.49 \text{ mg}}{23.9 \text{ mg}} \times 100\% = 31.3\%$$

The percentages of C (64.1), H (4.40) and Cl (31.3) total 99.8%; hence (X) has no O. The empirical formula is obtained as before:

E	M_E	$\div$	W_E	$=$	n_E	$n_E/0.883$	Multiplier is 1
C	64.1		12.01		5.34	6.05	6.05
H	4.40		1.01		4.36	4.94	4.94
Cl	31.3		35.45		0.883	1.00	1.00

The empirical formula is C_6H_5Cl.

(c) The freezing point data are used to find the GMW of (X).   The freezing point of pure benzene is 5.50 °C, and the solution described freezes at 3.45 °C.   The observed freezing point depression,

$$\Delta T_f = 5.50\ °C - 3.45\ °C = 2.05\ °C$$

and the **molality** (m) of the solution are related by **Raoult's law**:

$$\Delta T_f = K_{f.p.}(m)$$
$$2.05\ °C = (5.12\ °C\ kg/mole)(m)$$
$$m = \frac{2.05}{5.12} = 0.400\ \frac{mole}{kg\ solvent}$$

This means that 0.400 mole of the compound is dissolved in 1000 g of benzene.   Since there is 4.48 g in 50 g of benzene, a solution of the same molality in 1000 g of benzene has (4.48)(1000/50.0) = 89.6 g of (X). The solution is 0.400-molal, and so 89.6 g of (X) must be 0.400 mole:

$$GMW = \frac{89.6\ g}{0.400\ mole} = 224\ g/mole$$

The empirical formula weight of C_6H_5Cl is 112.56 g.   Since the molecular weight of our compound is 224 g/mole, there are twice as many atoms in the molecular formula as in the empirical formula.   The molecular formula is $C_{12}H_{10}Cl_2$.

Problem 26.6   An 11.75 g sample of a hydrocarbon is volatilized at 1 atm and 100 °C to a gas that occupies 5.0 liters.   If the compound has only one monobromo substitution product, what is its structure?   ◄

$$GMW = \frac{(11.75\ g)(0.0821\ \ell\ atm\ °K^{-1}\ mole^{-1})(373\ °K)}{(1\ atm)(5.0\ \ell)} = 72\ g/mole$$

For a GMW of 72, the maximum number of C's is 5 (weighing 5 × 12 = 60), leaving 12 as the possible number of H's.   The compound is a pentane (C_5H_{12}).   Only neopentane (2,2-dimethylpropane), $(CH_3)_4C$, gives one monobromo substitution product.

Problem 26.7   The GMW of tyrosine is 181.   A protein is found to contain 0.22% of this amino acid.   What is the lowest molecular weight the protein could have?   ◄

One mole of the protein will contain k moles of tyrosine, weighing $181k$ g.   Its molecular weight is then

$$\frac{181k}{0.0022} = 82,000k\ g/mole$$

and this is a minimum for $k = 1$.

Problem 26.8   Hemoglobin is a chromoprotein having 4 atoms of Fe (GAW = 56) in each molecule.   Analysis discloses 0.35% Fe.   What is the molecular weight of hemoglobin?   ◄

Four moles of Fe weigh 4 × 56 g = 224 g, and this is 0.35% of the GMW of the protein.   Therefore:

$$GMW = \frac{224\ g/mole}{0.0035} = 64,000\ g/mole$$

Problem 26.9   When 1.78 g of an optically active α-amino acid (A) is treated with HONO, 448 mℓ of N_2 (at SC) is evolved.   What is its molecular formula?   ◄

The amount of N_2 evolved is

$$\frac{0.448\ \ell}{22.4\ \ell/mole} = 0.0200\ mole$$

Since 1 mole of (A) gives 1 mole of N_2, 1.78 g of (A) is 0.0200 mole of (A). The GMW is

$$\frac{1.78\ g}{0.0200\ mole} = 89.0\ g/mole$$

The fundamental α-amino acid structure,

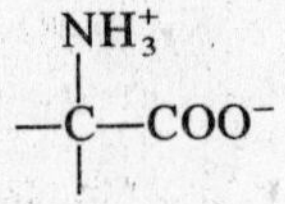

$$\begin{array}{c} NH_3^+ \\ | \\ -C-COO^- \\ | \end{array}$$

weighs 73 g. Since the molecular weight is 89 g, the substituents attached to the α C weigh 16 g. This corresponds to one CH_3 (15 g) and one H (1 g). The compound is alanine (L-α-aminopropionic acid), which is optically active.

26.2 COLLIGATIVE PROPERTIES

Problem 26.10 The molal freezing point depression constant for camphor as a solvent is 40 °C kg/mole. Pure camphor melts at 179 °C. When 0.108 g of compound (A) is dissolved in 0.90 g of camphor, the solution melts at 166 °C. Find the molecular weight of (A). ◄

Raoult's law [Problem 26.5(c)] gives the molality as

$$m = \frac{\Delta T_f}{K_{f.p.}} = \frac{179\ °C - 166\ °C}{40\ °C\ kg/mole} = 0.325\ mole/kg$$

0.325 mole of (A) is dissolved in 1 kg of camphor. Now, 0.108 g of (A) in 0.90 g camphor is the same concentration as 120 g of (A) in 1000 g camphor. Therefore, 120 g of (A) is 0.325 mole, and so the mass of one mole of (A) is

$$\frac{120\ g}{0.325\ mole} = 369\ g/mole$$

Problem 26.11 The molecular weights of macromolecules such as proteins, starches and synthetic polymers are best determined from osmotic pressures. The osmotic pressure of a solution of polypropylene in benzene at 25 °C is 0.155 torr. The solution contains 0.20 g of polypropylene per 100 mℓ of solution. Find the molecular weight of the polymer. ◄

The equation for osmotic pressure is $\Pi = MRT$, where Π is expressed in atmospheres, M is **molarity** (moles solute/liter solution), R is the gas constant (0.0821 ℓ atm °K^{-1} mole^{-1}) and T is the absolute temperature (in °K). From the data,

$$T = 25\ °C + 273\ °C = 298\ °K$$

$$\Pi = \frac{0.155\ torr}{760\ torr/atm} = 2.04 \times 10^{-4}\ atm$$

whence

$$M = \frac{\Pi}{RT} = \frac{2.04 \times 10^{-4}}{(0.0821)(298)} = 8.33 \times 10^{-6}\ mole/\ell$$

The solution has 0.20 g solute per 100 mℓ of solution, which is 2.0 g per liter.

$$GMW = \frac{2.0\ g/\ell}{8.33 \times 10^{-6}\ mole/\ell} = 240,000\ g/mole$$

26.3 YIELDS

Problem 26.12 In a preparation of cyclohexene from cyclohexanol, 50 g of cyclohexanol yields 20 g of cyclohexene. What is the percentage yield of cyclohexene? ◄

The equation is $C_6H_{11}OH \longrightarrow C_6H_{10} + H_2O$. One mole of cyclohexanol (GMW = 100) would give one mole of cyclohexene (GMW = 82) if the yield were 100% (**theoretical yield**). In this reaction, 50 g (0.50 mole) of cyclohexanol is used, so that the theoretical yield of cyclohexene is 0.50 mole (41 g). Since only 20 g of cyclohexene is actually obtained, the **percentage yield** is

$$\frac{\text{actual yield}}{\text{theoretical yield}} \times 100\% = \frac{20\ g}{41\ g} \times 100\% = 49\%$$

Problem 26.13 In a nitration of benzene, 10 g benzene gave a 79.1% yield of nitrobenzene. How much nitrobenzene was prepared?   ◄

The equation

$$C_6H_6 + HNO_3 \xrightarrow{\text{H}_2\text{SO}_4} C_6H_5NO_2 + H_2O$$

shows that one mole of benzene (78 g) gives one mole of nitrobenzene (123 g) if the yield is 100%. Starting with 10 g of benzene (10/78 mole), a 100% yield would be 10/78 mole of nitrobenzene, which is

$$\left(\frac{10}{78}\ \text{mole}\right)\left(123\ \frac{g}{\text{mole}}\right) = 16\ g$$

The actual yield is 79.1% of this, or $0.791 \times 16\ g = 13\ g$.

Problem 26.14 Phenol and P_2S_5 react to form thiophenol:

$$5C_6H_5OH + P_2S_5 \longrightarrow 5C_6H_5SH + P_2O_5$$

If 23.5 g of phenol and 10 g of P_2S_5 are reacted, what is the theoretical weight of PhSH obtained?   ◄

The balanced equation shows that 5 moles of phenol (5 × GMW = 470) reacts with one mole of P_2S_5 (GMW = 222) to give 5 moles of PhSH (5 × GMW = 550). The reaction mixture has 23.5 g (0.25 mole) phenol and 10 g (0.045 mole) of P_2S_5. The ratio of 5 moles C_6H_5OH to 1 mole P_2S_5 is used to show that 0.25 mole of phenol requires (0.25)/5 = 0.050 mole of P_2S_5 for complete reaction. Inasmuch as only 0.045 mole of P_2S_5 is reacted, phenol is present in excess. The theoretical yield is based on P_2S_5, the "limiting" reactant. The amount of PhSH formed from 0.045 mole of P_2S_5 is 5 × 0.045 mole = 0.225 mole, or 24.8 g.

Problem 26.15 In the series of reactions shown below, 180 g of ethane is converted into 2-bromobutane ($CH_3CHBrCH_2CH_3$). How much 2-bromobutane is obtained?

(1) $C_2H_6 + Br_2 \longrightarrow C_2H_5Br$ (50% yield) + HBr

(2) $2C_2H_5Br + 2Na \longrightarrow C_4H_{10}$ (33% yield) + 2NaBr

(3) $C_4H_{10} + Br_2 \longrightarrow CH_3CHBrCH_2CH_3$ (40% yield) + HBr   ◄

The amount of ethane used is

$$\frac{180\ g}{30\ g/\text{mole}} = 6.0\ \text{moles}$$

In 50% yield from reaction (1), 3.0 moles of C_2H_5Br are obtained. From (2), 2 moles of C_2H_5Br yields 1 mole of C_4H_{10} if the yield is 100%. Hence, 3.0 moles of C_2H_5Br theoretically give 1.5 moles of C_4H_{10}. Since the yield in (2) is 33%, only 0.33×1.5 mole = 0.50 mole of C_4H_{10} is obtained. From (3), a 1:1 molar reaction, the 40% yield from 0.50 mole of C_4H_{10} gives 0.20 mole of $CH_3CHBrCH_2CH_3$, which is

$$(0.20\ \text{mole})(137\ g/\text{mole}) = 27\ g$$

26.4 NEUTRALIZATION EQUIVALENT (NE)

Problem 26.16 A carboxylic acid has GMW = 118, and 169.6 mℓ of 1.000 N KOH neutralize 10.0 g of the acid. When heated, 1 mole of this acid loses 1 mole of H_2O without loss of CO_2. What is the acid?   ◄

The volume of base (0.1696 ℓ) multiplied by the normality of the base (1.000 equivalents/liter) gives 0.1696 equivalents, which is the **number of equivalents** of acid titrated.   Since the weight of acid is 10.0 g, one equivalent of acid weighs

$$\frac{10.0 \text{ g}}{0.1696 \text{ eq.}} = 59.0 \text{ g/eq.} = \text{NE}$$

Because the molecular weight of the acid (118) is twice this **equivalent weight**, there must be two equivalents per mole.   The number of equivalents gives the number of ionizable hydrogens.   This carboxylic acid has two COOH groups.

The 2 COOH's weigh 90 g, leaving 118 g − 90 g = 28 g (2 C's and 4 H's) as the weight of the rest of the molecule.   Since no CO_2 is lost on heating, the COOH groups must be on separate C's.   The compound is succinic acid, $HOOC—CH_2—CH_2—COOH$.

Problem 26.17   What carboxylic acid (A) has an NE of 52 and decomposes on heating to yield CO_2 and a carboxylic acid (B) with an NE of 60?   ◄

The NE of a carboxylic acid is its equivalent weight (molecular weight divided by the number of COOH groups).   Since CO_2 is lost on heating, there are at least 2 COOH's in acid (A), and loss of CO_2 can produce a monocarboxylic acid, (B).   Since one COOH weighs 45 g, the rest of (B) weighs 60 g − 45 g = 15 g, which is a CH_3 group.   (B) is CH_3COOH, and (A) is malonic acid ($HOOC—CH_2—COOH$), whose NE is

$$\frac{104 \text{ g}}{2 \text{ eq.}} = 52 \text{ g/eq.}$$

26.5   CHEMICAL EQUILIBRIUM

Problem 26.18   The equilibrium constant for the reaction

$$CH_3COOH + C_2H_5OH \rightleftharpoons CH_3COOC_2H_5 + H_2O$$

is 4.0 at room temperature.   One mole each of C_2H_5OH and CH_3COOH are mixed.   How many moles of ethyl acetate are present at equilibrium?   ◄

The equilibrium constant for the reaction is given by

$$K_e = \frac{[CH_3COOC_2H_5][H_2O]}{[CH_3COOH][C_2H_5OH]}$$

The original amounts of alcohol and acid are 1 mole each.   Assuming that x mole of ester is present at equilibrium, then there will also be x mole of H_2O; $(1 − x)$ mole each of alcohol and acid remain.   Let the volume of solution at equilibrium be V.   Then

$$4 = \frac{(x/V)(x/V)}{\left(\dfrac{1-x}{V}\right)\left(\dfrac{1-x}{V}\right)} = \frac{x^2}{1 - 2x + x^2}$$

$$4 - 8x + 4x^2 = x^2$$

$$3x^2 - 8x + 4 = 0$$

Solving this quadratic equation for x gives:

$$x = \frac{-b \pm \sqrt{b^2 - 4ac}}{2a} = \frac{8 \pm \sqrt{64 - 48}}{6} = 2 \text{ and } 0.67$$

The solution $x = 2$ is impossible, since no more than one mole of ester is present even if the reaction goes to completion.   Therefore, 0.67 mole of ethyl acetate is present at equilibrium.   Not knowing the value of V, we cannot solve for a numerical value for $[CH_3COOC_2H_5]$.   In fact, if in the balanced equation the number of moles of reactants did not equal the number of moles of products, we could not even determine the number of moles of $CH_3COOC_2H_5$ present, for V would not drop out of the expression for K_e.

Problem 26.19 The equilibrium constant for the reaction

$$6CH_2O \rightleftharpoons C_6H_{12}O_6$$
$$\text{Formaldehyde} \quad \text{Glucose}$$

is 6×10^{22}. How many moles of CH_2O are theoretically present at equilibrium in 1 liter of an aqueous $1M$ glucose solution?   ◄

Let us write:

	Original Concentration (M)	Equilibrium Concentration (M)
$C_6H_{12}O_6$	1	$1 - x$
CH_2O	0	$6x$

Then:

$$K_e = \frac{[C_6H_{12}O_6]}{[HCHO]^6}$$

$$6 \times 10^{22} = \frac{1 - x}{(6x)^6}$$

Assuming that x is negligible compared to 1 gives $1 - x \approx 1$ and

$$x^6 \approx \frac{1}{6^7(10^{22})} = \frac{10^8/6^7}{10^{30}} = 357 \times 10^{-30}$$

whence $x = 2.66 \times 10^{-5}$ (which is indeed much less than 1). The equilibrium concentration of CH_2O is then $6x = 1.44 \times 10^{-4}M$.

26.6 DISTILLATION PROBLEMS

Problem 26.20 Two miscible volatile liquids are mixed at 25 °C. At this temperature, the vapor pressure of pure A (p_A°) is 100 torr and that of pure B (p_B°) 240 torr. What is the mole fraction of A in the vapor in equilibrium with an ideal solution of 3 moles of A and 5 moles of B?   ◄

In the liquid, the number of moles of (A + B) is 8. N_A (the mole fraction of A) is $3/8 = 0.375$, and N_B is $5/8 = 0.625$. The composition of the vapor depends both on the mole fractions and on the vapor pressures of A and B in the solution. The vapor pressures of A and B, p_A and p_B, are calculated from Raoult's law: $p_A = p_A^\circ N_A$ and $p_B = p_B^\circ N_B$.

$$p_A = (100 \text{ torr})(0.375) = 37.5 \text{ torr}$$
$$p_B = (240 \text{ torr})(0.625) = \underline{150.0 \text{ torr}}$$

$$\text{Total vapor pressure} = 187.5 \text{ torr}$$

The mole fraction of A in the vapor is the fraction of the total vapor pressure that is due to A:

$$\frac{37.5 \text{ torr}}{187.5 \text{ torr}} = 0.200$$

The mole fraction of the more volatile component, i.e. B, of a mixture is usually greater in the vapor phase than in the liquid phase with which it is in contact.

Problem 26.21 Water (GMW = 18; v.p. at 80 °C = 355 torr) and n-octane (GMW = 108; v.p. at 80 °C = 175 torr) are immiscible. What is the vapor pressure at 80 °C of an equimolar mixture of the two liquids?   ◄

Each immiscible liquid contributes to the vapor pressure of the mixture as though the other were not present. The vapor pressure of the mixture, regardless of the relative quantities of each liquid present, is the sum of the individual vapor pressures,

$$355 \text{ torr} + 175 \text{ torr} = 530 \text{ torr}$$

Problem 26.22 An immiscible mixture of equal masses of an organic liquid, (X), and H_2O distills at 98 °C when the barometric pressure is 732 torr. At this temperature, the vapor pressure of water is 712 torr. The distillate collected after a few minutes is found to contain 5 times as much H_2O, by weight, as of liquid (X). What is the molecular weight of (X)? ◄

Liquid (X) and H_2O are immiscible, and in this process, known as a **steam distillation**, each contributes its own vapor pressure to the total. The amount of each component in the liquid is not important. At the boiling point of the mixture, 98 °C, the sum of the vapor pressures of H_2O and of (X) equals the external pressure, 732 torr. Water has a vapor pressure of 712 torr at 98 °C; therefore, the vapor pressure of (X) is $732 - 712 = 20$ torr.

The number of moles distilled of a component is directly proportional to its vapor pressure:

$$\frac{n_{H_2O}}{n_{(X)}} = \frac{p_{H_2O}}{p_{(X)}}$$

Number of moles (n) is equal to mass (μ) in grams divided by the GMW; therefore,

$$\frac{\mu_{H_2O}/GMW_{H_2O}}{\mu_{(X)}/GMW_{(X)}} = \frac{p_{H_2O}}{p_{(X)}}$$

Solving for $GMW_{(X)}$ gives:

$$GMW_{(X)} = \frac{(\mu_X)(p_{H_2O})(GMW_{H_2O})}{(\mu_{H_2O})(p_{(X)})}$$

and substituting 1/5 for $\mu_{(X)}/\mu_{H2O}$ we obtain:

$$GMW_{(X)} = \frac{(1)(712 \text{ torr})(18 \text{ g/mole})}{(5)(20 \text{ torr})} = 128 \text{ g/mole}$$

Problem 26.23 Figure 26-1 is the boiling point curve of a binary ideal mixture of liquids A and B. (a) Does the upper (broken) or lower (solid) curve represent liquid composition? (b) At what temperature does a liquid having a 0.62 mole fraction B boil? (c) What are the mole fractions of A and B in the vapor in equilibrium with the liquid containing 0.62 mole fraction B? (d) If the vapor from (c) is condensed to liquid, what is its composition? (e) What is the composition of the vapor in equilibrium with the liquid condensed in (d)? ◄

(a) The liquid composition is the lower (solid) line, since it indicates a higher mole fraction of the higher-boiling component B at every temperature. (b) About 60 °C. (c) 0.29 B and 0.71 A (sum must be 1). (d) Same as (c), since condensation of vapor to liquid does not change composition. (e) 0.10 B and 0.90 A.

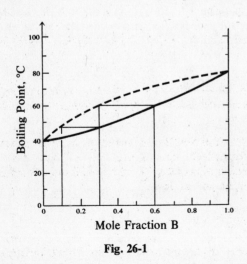

Fig. 26-1

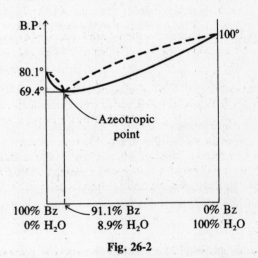

Fig. 26-2

Problem 26.24 Benzene and H_2O boil at 80.1 °C and 100 °C, respectively. An azeotrope of 91.1% benzene (Bz) and 8.9% H_2O distills at 69.4 °C. Draw a diagram of boiling point vs. composition. ◄

See Fig. 26-2.

Problem 26.25   Why does the addition of NaCl increase the efficiency of steam distillation of aniline? ◄

Because the partial vapor pressure of the salt solution is reduced.   The boiling point of the aqueous salt solution is greater than that of H_2O.   At the higher temperature, aniline has a greater vapor pressure and is more efficiently distilled.

26.7 SOLVENT EXTRACTION

Problem 26.26   An organic compound (X) is soluble in H_2O and in ether.   When 10 mℓ of ether is used to extract (X) from 20 mℓ of an aqueous solution containing 6.0 g of (X), it is found that 4.0 g of (X) is extracted by the ether.   What is the distribution constant K_D of (X) between ether and water? K_D of a solute in immiscible solvents is:

$$K_D = \frac{\text{g solute per m}\ell \text{ of solvent A}}{\text{g solute per m}\ell \text{ of solvent B}}$$ ◄

H_2O is taken as solvent B.

$$K_D = \frac{4.0 \text{ g}/10 \text{ m}\ell \text{ ether}}{2.0 \text{ g}/20 \text{ m}\ell \text{ H}_2\text{O}} = 4.0$$

Problem 26.27   How many grams of solute (X) (Problem 26.26) is extracted from 20 mℓ of H_2O containing 6.0 g of (X) by   (a) 20 mℓ of ether,   (b) 2 successive 10 mℓ portions of ether? ◄

(a) Let x be the number of grams of solute in ether.   Therefore $(6.0 - x)$ grams of solute remain in H_2O.

$$K_D = 4.0 = \frac{(x \text{ g})/(20 \text{ m}\ell \text{ ether})}{[(6.0 - x) \text{ g}]/(20 \text{ m}\ell \text{ H}_2\text{O})} = \frac{x}{6.0 - x}$$
$$x = 4.8 \text{ g of solute in ether}$$

(b) From Problem 26.26, 10 mℓ of ether extracts 4.0 g of (X), leaving 2.0 g of (X) in the 20 mℓ of H_2O.   The amount x extracted by the second 10 mℓ portion of ether is calculated as shown:

$$K_D = 4.0 = \frac{(x \text{ g})/(20 \text{ m}\ell \text{ ether})}{[(6.0 - x) \text{ g}]/(60 \text{ m}\ell \text{ H}_2\text{O})} = \frac{x/20}{(2.0 - x)/20}$$
$$x = 1.3 \text{ g of solute in ether}$$

The two successive 10 mℓ ether portions extract a total of 4.0 g + 1.3 g = 5.3 g of solute.   In part (a), a single 20 mℓ portion of ether extracted only 4.8 g.

Problem 26.28   50 mℓ of an aqueous solution of RCOOH is neutralized by 50 mℓ of NaOH.   A 30 mℓ sample of this same solution upon extraction with 25 mℓ of ether leaves an aqueous solution that is neutralized by only 20 mℓ of NaOH.   Find K_D for RCOOH between ether and H_2O. ◄

Since the RCOOH in 50 mℓ of solution is equivalent to 50 mℓ of NaOH, the amounts of RCOOH in the final 30 mℓ aqueous solution and in the ether must be respectively equivalent to 20 mℓ and 10 mℓ of NaOH.   Consequently, since we are dealing with a fixed concentration of RCOOH, the masses of RCOOH in the 30 mℓ aqueous solution and in the ether are in the ratio 20/10, and we have:

$$K_D = \frac{10/(25 \text{ m}\ell \text{ ether})}{20/(30 \text{ m}\ell \text{ H}_2\text{O})} = 0.60$$

(Strictly speaking, 30 mℓ is the volume of the aqueous *solution* and not that of the *solvent*, but the difference is negligible.)

Supplementary Problems

Problem 26.29 (a) Find the empirical formula of a hydrocarbon whose percentage composition is C = 85.63%, H = 14.37%. (b) To what class of hydrocarbons does it belong? ◄

(a) $$C_{85.63/12.01}H_{14.37/1.008} = C_{7.130}H_{14.26} = C_1H_{14.26/7.130} = CH_2$$

(b) Alkenes; CH_2 fits the general formula C_nH_{2n}.

Problem 26.30 Derive the empirical formula of a compound containing 63.1% C, 11.92% H and 24.97% F. ◄

$$C_{63.1/12.01}H_{11.92/1.008}F_{24.97/18.99} = C_{5.25}H_{11.83}F_{1.315} = C_{5.25/1.315}H_{11.83/1.315}F_1 = C_4H_9F$$

Problem 26.31 If 1.500 g of a monohydroxycarboxylic acid is oxidized completely to give 1.738 g of CO_2 and 0.711 g of H_2O, determine the molecular formula and suggest a structural formula. ◄

The 1.500 g of the acid is made up of

$$\frac{12.01}{44.01}(1.738 \text{ g}) = 0.4743 \text{ g C}$$

$$\frac{2.016}{18.016}(0.711 \text{ g}) = 0.0796 \text{ g H}$$

and

$$1.500 \text{ g} - (0.4743 \text{ g} + 0.0796 \text{ g}) = 0.9461 \text{ g O}$$

Hence, the formula is

$$C_{0.4743/12.01}H_{0.0796/1.008}O_{0.9461/16.00} = C_{0.04}H_{0.08}O_{0.06} = C_2H_4O_3$$

or $HOCH_2COOH$.

Problem 26.32 A sample of potato starch (A) has 0.086% phosphorus. Find the average GMW of the material if each molecule has one atom of P. ◄

$$\frac{31.0 \text{ g P/mole A}}{0.00086 \text{ g P/g A}} = 36,000 \frac{\text{g A}}{\text{mole A}}$$

Problem 26.33 The freezing point of a solution made by dissolving 0.512 g of substance (A) in 8.03 g of naphthalene is 75.2 °C. (m.p. of naphthalene is 80.6 °C; $K_{f.p.}$ for naphthalene is 6.8 °C kg/mole). Find the GMW of (A). ◄

$$\frac{\dfrac{0.512 \text{ g (A)}}{8.03 \text{ g Np}} \times \dfrac{1000 \text{ g Np}}{1 \text{ kg Np}}}{(80.6 \text{ °C} - 75.2 \text{ °C})/6.8 \text{ °C kg Np/mole (A)}} = 81 \text{ g (A)/mole (A)}$$

Problem 26.34 At 0 °C the osmotic pressure of an aqueous solution containing 46.0 g of substance (B) per liter of solution is 11.2 atm. Find the GMW of (B). ◄

$$\frac{(46.0)(0.0821)(273)}{(11.2)(1.00)} = 92.0 \text{ g/mole}$$

Problem 26.35 How many grams of phenol (P) can be obtained by hydrolyzing the diazonium salt made from 106 g of aniline (A) if the overall yield is 42%? ◄

The yield from $C_6H_5NH_2 \longrightarrow C_6H_5N_2^+ \longrightarrow C_6H_5OH$ is

$$\frac{(106 \text{ g A})(94 \text{ g P/mole P})(0.42 \text{ mole P/mole A})}{93 \text{ g A/mole A}} = 45 \text{ g P}$$

Problem 26.36 What is the maximum amount of CH_3Cl that can be prepared from 20.0 g of CH_4 and 10.0 g of Cl_2 by the reaction $CH_4 + Cl_2 \longrightarrow CH_3Cl + HCl$? ◄

20.0 g of CH_4 is 1.25 mole and 10.0 g of Cl_2 is 0.141 mole, so Cl_2 is the limiting reactant. The reaction equation gives the molar ratio of $Cl_2:CH_3Cl$ as $1:1$. Thus the theoretical (maximum) yield is 0.141 mole of CH_3Cl, or

$$0.141 \text{ mole} \times 50.48 \text{ g/mole} = 7.12 \text{ g}$$

Problem 26.37 What is the percent yield if 14.8 g of propanoic acid is formed by hydrolysis of 12.5 g of propionitrile? ◄

The equation

$$CH_3CH_2CN + 2H_2O \longrightarrow CH_3CH_2COOH + NH_3$$
$$\text{(GMW = 55.1)} \qquad\qquad \text{(GMW = 74.1)}$$

shows a $1:1$ molar ratio. Thus the theoretical yield of CH_3CH_2COOH is

$$\frac{12.5 \text{ g nitrile}}{55.1 \text{ g nitrile/mole nitrile}} \times \frac{1 \text{ mole acid}}{1 \text{ mole nitrile}} \times 74.1 \frac{\text{g acid}}{\text{mole acid}} = 16.8 \text{ g acid}$$

and

$$\% \text{ yield} = \frac{14.8 \text{ g}}{16.8 \text{ g}} \times 100\% = 88.1\%$$

Problem 26.38 If 50.0 mℓ of 0.0500 N KOH neutralizes 0.225 g of an organic acid, calculate (a) the neutralization equivalent (NE) of the acid and (b) the possible molecular weights. ◄

(a)
$$\frac{(0.225 \text{ g})(1000 \text{ m}\ell/\ell)}{(50.0 \text{ m}\ell)(0.0500 \text{ eq.}/\ell)} = 90.0 \text{ g/eq.}$$

(b) Any integral multiple of 90.0.

Problem 26.39 Neutralization of 0.38 g of a carboxylic acid having the empirical formula $C_9H_{18}O_4$ requires 20 mℓ of 0.10 N base. (a) Calculate the NE. (b) How many COOH's are present in a molecule of this acid? ◄

(a)
$$\frac{(0.38)(1000)}{(20)(0.10)} = 190 \text{ g}$$

(b) One COOH (molecule has only one degree of unsaturation).

Problem 26.40 The ir spectrum of an optically active amine (NE = 75) has a twin peak at about 3400 cm^{-1}. With CH_3COCl, the amine forms a neutral compound, $C_5H_{11}O_2N$. Write a possible structural formula for the amine. ◄

$CH_3OCH(NH_2)CH_3$. The twin peak indicates a 1° amine.

Problem 26.41 Pentane (GMW = 72; v.p. at 20 °C is 420 torr) and heptane (GMW = 100; v.p. at 20 °C is 360 torr) form an ideal solution. If 144 g of pentane and 400 g of heptane are mixed at 20 °C, (a) what is the vapor pressure of the solution? (b) what is the mole fraction of pentane in the liquid and in the vapor? ◄

(a)
$$\frac{\dfrac{144}{72}(420) + \dfrac{400}{100}(360)}{\dfrac{144}{72} + \dfrac{400}{100}} = 377 \text{ torr}$$

(b) In liquid: $\dfrac{\dfrac{144}{72}}{\dfrac{144}{72} + \dfrac{400}{100}} = 0.33$ In vapor: $\dfrac{\dfrac{144}{72}(420)}{\dfrac{144}{72}(420) + \dfrac{400}{100}(360)} = 0.36$

Problem 26.42 An aromatic organic acid (A), NE = 136, is oxidized with $KMnO_4$ to an aromatic acid (B), NE = 83. Supply structural formulas for these acids. ◄

Acid (A) could be phenylacetic or a toluic (methylbenzoic) acid. Oxidation of $PhCH_2COOH$ would give $PhCOOH$ (NE = 122). (A) is a toluic acid because it is oxidized to a benzene dicarboxylic acid (NE = 83), which is acid (B). Acid (A) can be the *o*-, *m*- or *p*-isomer.

Problem 26.43 An ideal solution consisting of 58 mole percent benzene and 42 mole percent toluene boils at 90 °C at 760 torr. At this temperature and composition, the vapor contains 77 mole percent benzene in equilibrium with the solution. Calculate the vapor pressures of pure benzene and toluene at 90 °C. ◄

First find the pressures of benzene and toluene in the vapor phase by multiplying their percentages by the total pressure:

$$p_b = (0.077)(760 \text{ torr}) = 585.2 \text{ torr}$$
$$p_t = (0.23)(760 \text{ torr}) = 174.8 \text{ torr}$$

Now determine the pressures of the pure components using Raoult's law:

$$p_b^\circ = \frac{p_b}{N_b} = \frac{585.2 \text{ torr}}{0.58} = 1008.9 \text{ torr}$$

$$p_t^\circ = \frac{p_t}{N_t} = \frac{174.8 \text{ torr}}{0.42} = 416.2 \text{ torr}$$

Problem 26.44 Calculate the mole fractions in the vapor phase of an ideal mixture of 3 moles of compound (A) and 7 moles of compound (B) if the vapor pressures of pure (A) and (B) are 300 torr and 360 torr, respectively, at this temperature. ◄

$$N_A = \frac{\frac{3}{10}(300)}{\frac{3}{10}(300) + \frac{7}{10}(360)} = 0.26 \qquad N_B = \frac{\frac{7}{10}(360)}{\frac{7}{10}(360) + \frac{3}{10}(300)} = 0.74$$

Problem 26.45 The immiscible mixture of 1-bromobutane and H_2O distills at 95 °C at 760 torr. What mass of 1-bromobutane distills per gram of H_2O if the vapor pressures of 1-bromobutane and H_2O at 95 °C are 125 torr and 633.9 torr respectively? ◄

$$\frac{\text{mass } C_4H_9Br}{\text{mass } H_2O} = \frac{(125)(137)}{(633.9)(18.0)} = 1.5$$

Problem 26.46 The equilibrium constant for hydration of fumaric acid ($H_2C_4H_2O_4$) to malic acid ($H_2C_4H_4O_5$) is 3.5. How many moles of fumaric acid must be dissolved in water to give a liter of solution in which 0.20 mole of malic acid is present at equilibrium? ◄

Let x be the required initial concentration of $H_2C_4H_2O_4$. At equilibrium,

$$[H_2C_4H_4O_5] = 0.20 \qquad [H_2C_4H_2O_4] = x - 0.20$$

and so

$$\frac{0.20}{x - 0.20} = 3.5 \qquad \text{or} \qquad x = 0.26 \text{ mole}/\ell$$

Problem 26.47 Calculate the equilibrium constant for the mutarotation of α-D-glucose to β-D-glucose if equilibrium is achieved after 64% of the alpha form is converted to the beta form. ◄

$$\frac{0.64}{0.36} = 1.8$$

Problem 26.48 20 mℓ of an aqueous solution of an amine is neutralized by 30 mℓ of standard HCl. A fresh 20 mℓ portion of this amine solution is extracted with 40 mℓ of ether, and the distribution constant for this amine between ether and water is 4. How many mℓ of the standard HCl would be required to titrate the base remaining in the aqueous solution after extraction? ◄

Let

$$x = \text{mass of amine in ether} \qquad y = \text{mass of amine in } H_2O$$

Then
$$K_D = 4 = \frac{x/40}{y/20} \quad \text{or} \quad x/y = 8$$

where the volume of H_2O has been approximated by the volume of aqueous solution (see Problem 26.28). Thus, only 1/9 of the amine is present in the 20 mℓ of solution. This leads to the proportion

$$\frac{20 \text{ m}\ell \text{ amine soln}}{30 \text{ m}\ell \text{ standard HCl}} = \frac{(1/9)(20 \text{ m}\ell \text{ amine soln})}{z} \qquad \text{or} \qquad z = 3.3 \text{ m}\ell \text{ standard HCl}$$

Index

An asterisk (*) following a page number indicates a method for the preparation of the listed compound(s).

Catalog

If you are interested in a list of SCHAUM'S
OUTLINE SERIES send your name
and address, requesting your free catalog, to:

SCHAUM'S OUTLINE SERIES, Dept. C
McGRAW-HILL BOOK COMPANY
1221 Avenue of Americas
New York, N.Y. 10020